LABORATORY OUTLINES IN BIOLOGY-IV

LABORATORY OUTLINES IN BIOLOGY-IV

Peter Abramoff / Robert G. Thomson

Marquette University

W. H. FREEMAN AND COMPANY
New York

Library of Congress Cataloging-in-Publication Data

Abramoff, Peter, 1927–
Laboratory outlines in biology—IV.
1. Biology—Laboratory manuals. 2. Biology—Experiments. I. Thomson, Robert G. II. Title.
QH317.A275 1986 574'.028 85-24540
ISBN 0-7167-1758-1 (pbk.)

Printed in the United States of America

1 2 3 4 5 6 7 8 9 0 Web 4 3 2 1 0 8 9 8 7 6

Contents

Preface

Tell me and I will forget.
Show me and I might remember.
Involve me and I will understand.

Chinese proverb

These words convey our basic philosophy that it is experiences in the laboratory that help develop students' critical thinking and creativity and that increase their appreciation of the mechanisms by which biologists obtain and analyze information. The laboratory accomplishes this by having students become intensely and personally involved in the knowledge that they acquire.

In this, our fourth edition of *Laboratory Outlines in Biology,* we have retained many of the popular exercises and features of previous editions. In addition, several features are new. These include new or extensively rewritten laboratory studies on

- Light microscopy
- Cell structure and function
- Subcellular structure and function
- Movement of materials through cell membranes
- Regulation of enzyme activity by light
- Plant anatomy, including fruits
- Experimental studies on amphibian embryos
- Experimental surgery on chick embryos

The text also contains new appendixes on anatomical terminology and aseptic techniques.

In addition, the taxonomic studies, while still following the Whittaker five-kingdom system of classification, have been modified to reflect some of the contemporary schemes in current textbooks. Taxonomic studies new to this edition include those on lichens.

While the basic sequence of topics still reflects our own approach, the exercises are sufficiently comprehensive and independent to be adapted to any laboratory program. In these times of monetary restraint we have made every effort to keep expenditures for supplies and equipment within the budgets of most colleges and universities.

The accompanying Instructor's Handbook has been designed to save time for you and your laboratory assistants. This handbook contains detailed lists of materials and equipment, sources of supplies, and instructions for the preparation of reagents and for the culture and maintenance of the various organisms used in these studies.

With this manual we hope to (1) introduce the complexity and diversity of living organisms and their relationships to one another, (2) provide a solid foundation for further study for those electing a career in science, and (3) convey something of the meaning, scope, and excitement of biology as a significant perspective from which to view the world.

Our special thanks go to the following reviewers, whose comments proved very helpful to us: David J. Cotter, Georgia College; William E. Brabson, Jr., Lord Fairfax Community College; Alice Walrath, Barnard College; F. Scott Orcutt, Jr., University of Akron; George Edick, Rensselaer Polytechnic Institute; Stephen Fuller, Mary Washington College.

We appreciate many of the comments offered to us over the years by both faculty and students. In our desire to continue to improve this fourth edition, we invite constructive comments from those using it.

September 1985

Peter Abramoff
Robert G. Thomson

EXERCISE 1

Light Microscopy

The light microscope is capable of extending our ability to "see detail" by 1000 times, so that objects as small as 0.1 **micrometer** (μm) or 100 **nanometers** (nm) can be seen. The transmission electron microscope extends this viewing capability to objects as small as 0.5 nm in diameter, enabling us to see objects that are $\frac{1}{200,000}$th the size of those that can be seen by the human eye. Without microscopes, our understanding of the structure and function of cells and tissues would be severely limited.

The ability of the microscope to reveal the structure of small objects, however, is not so much a function of its ability to magnify as its ability to distinguish detail. Merely magnifying an object, without increasing the amount of detail seen is of little value to the observer. The ability to see detail is called **resolving power** and depends on the wavelength (λ) of light used and a value called the **numerical aperture** (NA), an important characteristic that determines how much light will enter the lens. In its simplest form, resolving power, or **resolution,** may be expressed by the formula

$$RP = \frac{\lambda}{2 \times NA}$$

λ = wavelength of light used
NA = numerical aperture

Under normal viewing conditions, resolution is *increased* by decreasing the wavelength of the light source. For example, if you use a green filter that permits a wavelength of 500 nm to pass through a microscope lens having a numerical aperture of 1, then the resolving power would be 500 nm/2 × 1 or 250 nm. This means that two objects that are 250 nm or farther apart would be seen as distinct objects; if closer than 250 nm, they would appear very fuzzy or as one object.

If you use blue light, or a blue filter that provides light at a wavelength of 400 nm and a lens having a NA of 1, the resolving power would be equal to 400 nm/2 × 1 or 200 nm. The two objects observed under these conditions could be 50 nm closer together and still be seen as separate objects.

Knowing the significance of the wavelength of light to the ability to distinguish detail, you can appreciate the role of electron microscopes and microscopes utilizing ultraviolet light in elucidating the structure and function relationships of cells and subcellular organelles.

A. PARTS OF A COMPOUND MICROSCOPE

Your microscope may have all or most of the features described below. Referring to Fig. 1-1, locate the following features of the microscope available in your laboratory.

1. Ocular Lens

The **oculars** are the lenses you look through. If there is only one ocular, you are using a **monocular** microscope; if there are two, it is a **binocular** microscope. In many binocular microscopes, the oculars can be adjusted to compensate for differences in distance between your eyes **(interpupillary adjustment).** One of the oculars may have a knurled adjustment mechanism for moving it in and out to compensate for focusing disabilities between each eye. Your instructor will describe how this is done. Oculars on different microscopes may have different magnifications. You may have to remove the ocular from its holder to determine its magnification. What is the magnification stamped on the housing of the oculars on your microscope?

____

The ocular contains a series of several magnifying lenses and may also include an **ocular micrometer** (a scale for measuring objects) and a pointer (to point out objects to your instructor or other students).

2. Objective Lens

Attached to a rotating nosepiece, or turret, at the base of the body tube are a group of three or four **objectives.** Rotate the nosepiece and notice that a "click" is heard as each objective comes into position.

The magnifying lenses of the objectives focus light that comes from the specimen and passes it up the body tube and through the oculars.

Each objective has numbers stamped on it. One of these numbers identifies the magnification of the objective (e.g., 43 ×). What are the magnifications of each of the objectives on your microscope?

____

The **total magnification** is calculated by multiplying the magnification of the ocular and objective lenses on the microscope being used. In Table 1-1 calculate the total magnification for each ocular/objective combination on your microscope.

TABLE 1-1
Calculation of total magnification for various ocular/objective combinations.

Ocular	×	Objective	=	Total magnification
________		________		________
________		________		________
________		________		________
________		________		________

Note: Objective lenses are usually named according to their magnifying power, as follows:

scanning power— 4 ×
low power— 10 ×
high or high dry power— 43 ×
oil immersion— 93 ×

A second set of numbers, usually given as a decimal, represents the numerical aperture for that lens; the abbreviation NA may precede the number. In Table 1-2 list the magnification and numerical aperture for each objective on your microscope.

TABLE 1-2
Numerical aperture and magnification for various objectives.

Magnification of objective	Numerical aperture (NA)
________	________
________	________
________	________
________	________

3. Body Tube

Light travels from the objectives through a series of magnifying lenses in the body tube to the ocular. In some microscopes, the body tube is straight. In

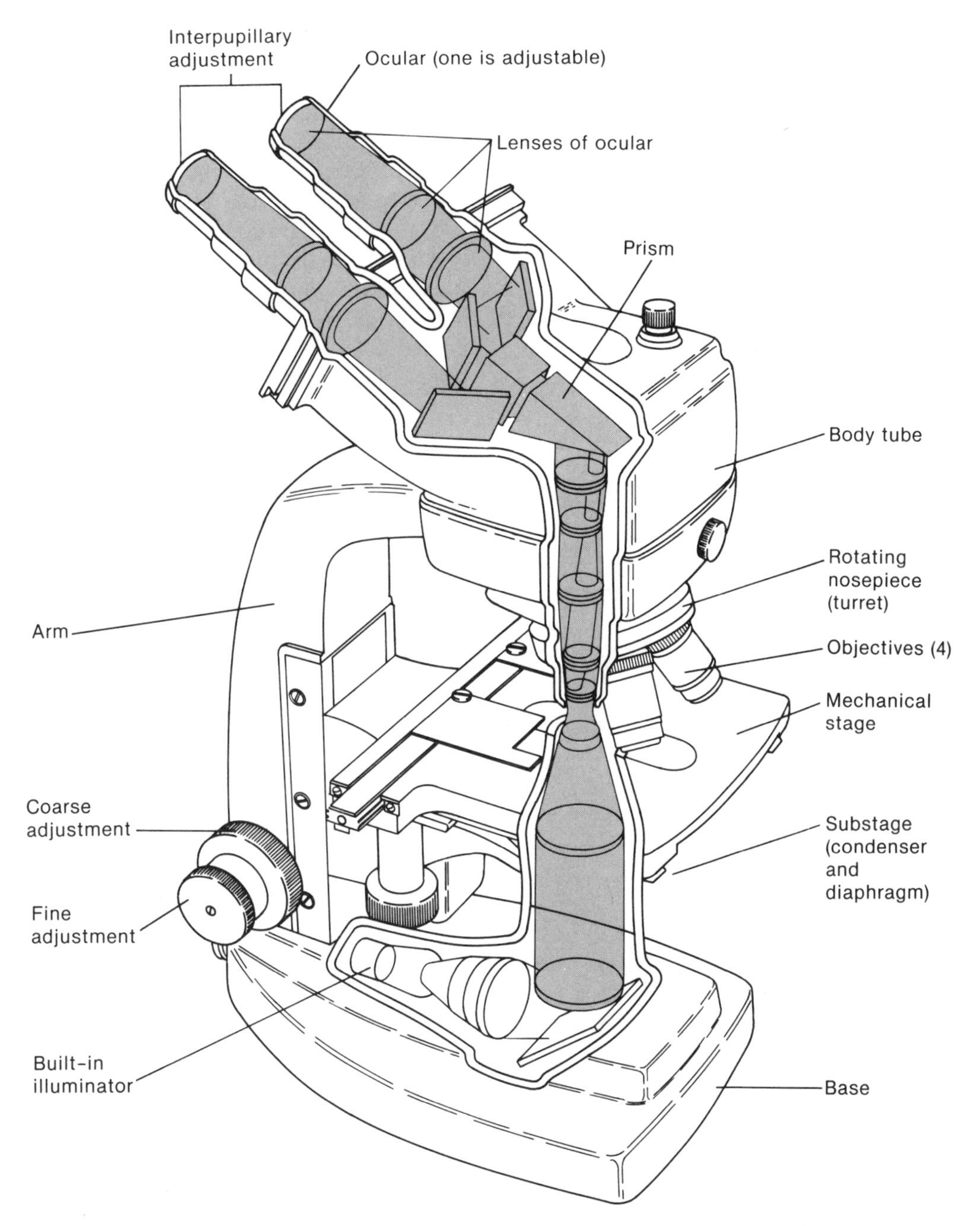

FIG. 1-1
Bausch and Lomb binocular microscope sectioned to show pathway of light from illuminator through various lenses and prisms.

others, the oculars are held at an angle, as in Fig. 1-1, and the body tube contains a prism that bends the light rays coming through the objectives so they will be transmitted through the oculars.

With some compound microscopes, loosening a lock screw allows you to rotate the body tube 180°. What is the advantage of being able to turn the body tube?

____

____

____

4. Stage

The surface or platform on which the microscope slide is placed is the **stage.** Note the opening **(stage aperture)** in the center of the stage. On some microscopes, the stage may be stationary and have clips to hold the slide in place. On other microscopes, the stage can be moved and is therefore called a **mechanical stage.** Movement is controlled by two knobs located on top, on the side, or on the bottom of the stage. Note the **horizontal** and **vertical scales** on the mechanical stage. What is the function of these scales?

How are slides held in position on a mechanical stage?

5. Substage

The area beneath the stage is the substage and may be occupied by one, or both, of the following:

a. Diaphragm

The diaphragm regulates the *amount* of light passing from the light source through the specimen and through the lens system of the microscope. By properly adjusting the diaphragm you provide better contrast with the surrounding medium, thus greatly improving your image of the specimen. The diaphragm may be either

1. An **annular** type of diaphragm consisting of a circular plate with holes of different diameters. This plate is rotated so that the various holes may be positioned in the light path to regulate the amount of light passing from the light source through the object under observation.

2. An **iris** type of diaphragm that consists of a series of overlapping thin metal plates. A lever projecting from the side of the diaphragm opens and closes these plates to regulate the amount of light entering the microscope.

What type of diaphragm does your microscope have?

b. Condenser

The condenser consists of a series of lenses that focus light onto the specimen. Movement of the condenser is regulated by a knob at its side, or a lever projecting from the condenser housing. By properly adjusting the condenser you greatly improve your observation of the specimen.

Attached to the bottom of the condenser may be a filter holder, which normally contains a blue filter. Why would you use a blue filter, instead of a green or red filter, when making microscopic observations? Indeed, why use a filter at all?

6. The Light Source

Your microscope may have an attached mirror or a built-in illuminator. If your microscope uses a mirror, one surface is usually concave and the other is flat. The flat side of the mirror is normally used with the scanning and low power objectives and the concave mirror with higher power objectives. The light source for the mirror is usually a lamp. Natural light may be used, but it is not preferred because the light's intensity will vary greatly, depending on the source of light in your laboratory.

In most compound microscopes, the illuminator is built into the base of the microscope and controlled by an on/off switch. You can control the amount of light entering the specimen by adjusting the diaphragm. You can also control the light intensity by adjusting a transformer attached to the illuminator, whose knob can be turned to regulate the voltage to the light bulb. Use low or medium transformer settings for most microscopic observations. You will need a higher setting when using the oil immersion lens. Why?

7. Focusing

You can focus your microscope by using the coarse and fine adjustment knobs that raise or lower either the body tube or the stage, depending on the type of microscope you are using.

With the low-power objective in position, rotate the coarse adjustment knob one half turn clockwise. Do the same with the fine adjustment knob. Based upon your observations, why should you not use the coarse adjustment knob for focusing when the high-power objective or oil immersion objective is in position?

____

____

____

8. Eyeglasses and Microscope Usage

Should you wear your eyeglasses when using a microscope? The answer to this question is qualified. If you are near- or farsighted, you need not wear your glasses for microscopic observations. The adjustments made in focusing the microscope will compensate for these eye problems. On the other hand, wear your glasses if you have astigmatism (a defect in the eye's refractive surface), since this problem is not corrected by the lenses of the microscope.

In either case, when using a monocular microscope, you should keep both eyes open, despite a tendency to close one eye. Eye strain will develop if you do this for any length of time.

B. PROPER USE OF MICROSCOPES

Before using your microscope, thoroughly clean the oculars and objectives using lens paper in a circular motion to prevent scratching. When using the microscope, keep eyelashes from touching the ocular. Oil from the lashes will adhere to the ocular lenses smearing them. When using salt solutions or other harsh chemicals to prepare wet mounts, thoroughly clean the oculars and objectives, stage, and microscope slides after use to prevent damage to the microscope.

Despite its sturdy appearance a microscope is a delicate, precision instrument. It should be handled carefully and with common sense. The following suggestions will help you avoid some common mishaps that occur when using a microscope.

1. To avoid dropping a microscope, banging it against a laboratory bench, or having the oculars fall out,

   a. Carry the microscope upright using both hands.

   b. Place the microscope away from the edge of the bench, particularly when not in use.

   c. Move power cords out of the way, so that you can't trip on them and pull the microscope or transformer down.

2. To avoid breaking a coverslip and/or microscope slide by an objective,

   a. First locate the specimen using the low-power objective, and then switch to the higher-power objectives.

   b. Never focus the high-power objective with the coarse adjustment knob and never use these lenses when examining thick specimens or whole mounts of specimens.

3. To avoid mechanical difficulties with various parts of a microscope,

   a. Never force microscope parts to work.

   b. When changing the bulb in the built-in illuminator, never force it, since it might shatter in your fingers.

   c. Never try to dismantle the microscope.

C. USING A COMPOUND MICROSCOPE

1. Focusing

1. Cut out a lower case letter *e* from a newspaper or other printed page. Clean a microscope slide and prepare a "wet mount" of the letter, using the procedure described in Fig. 1-2. Put the scanning (4 ×) or low-power (10 ×) objective in position and then place the slide on the stage in its normal viewing position.

2. Clean the oculars and objectives using lens paper.

3. Turn on the illuminator and open the diaphragm fully. If there is a condenser, position it as high as it will go, so that the top lens of the condenser unit is level with the stage aperture.

4. Center the specimen over the stage aperture.

5. Position the scanning objective (4 ×) as close to the slide as possible and then, while looking through the oculars, use the coarse adjustment knob to back off slowly until the specimen comes into focus.

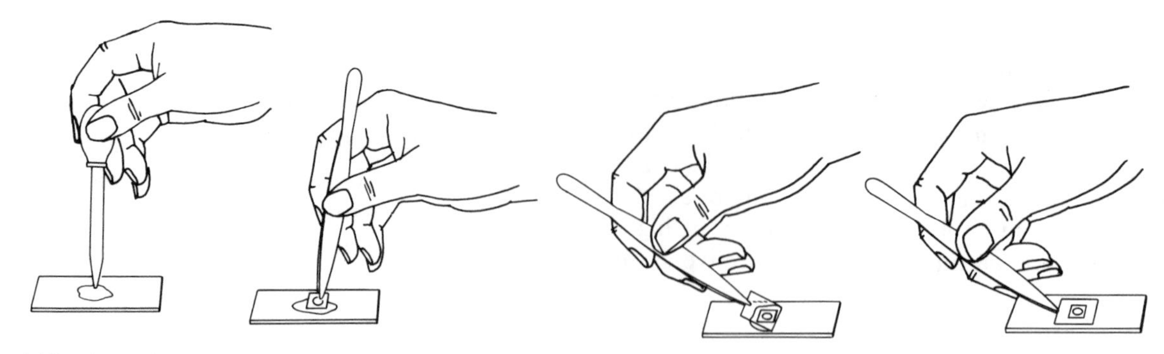

A. Add a drop of water to a slide.

B. Place the specimen in the water.

C. Place the edge of a coverslip on the slide so that it touches the edge of the water.

D. Slowly lower the coverslip to prevent forming and trapping air bubbles.

FIG. 1-2
Preparation of a wet mount slide.

6. Using the diaphragm (and/or adjustment of the transformer voltage), readjust the light intensity as necessary and again center the specimen by moving the slide.

7. Switch from the scanning lens to the low-power objective (10 ×). Make certain the objective "clicks" into position. If the specimen stays in focus, your microscope is **parfocal.** You can sharpen the focus by small adjustments of the fine adjustment knob.

If your microscope is not in focus after changing objectives, you may have to use the coarse adjustment knob followed by the fine adjustment knob. But remember, do not do this with the high-power or oil immersion objectives in position. Ask your instructor for help if you have difficulty focusing your microscope.

Recenter the specimen, adjust the diaphragm, and adjust the position of the condenser to increase the contrast of the specimen.

8. Switch to the high-power objective (43 ×) and adjust the focus using the fine adjustment knob.

These are the procedures usually used when examining a wet mount or a commercially prepared microscope slide. Always make your preparation using clean microscope slides. Always proceed from the lowest-power to the highest-power objectives, making minor corrections in focus and light as necessary. Learn to "fine tune" your microscope.

2. The Microscopic Image

The image you view in the microscope is affected by several factors: the orientation of the image, total magnification, the size and brightness of the field of view, the plane of focus, the depth of focus, and the contrast of the materials being examined.

a. Orientation of the Image

Hold the slide you made of the letter *e* so that the letter is in a normal reading position. Then, place it on the stage in the same position and examine it with the low-power objective. What difference is there, if any, in the way the image is oriented when viewed through the oculars as compared to looking at it directly with your eyes?

____

While looking through the microscope, attempt to make the image move to the right. In which direction did you have to move the slide?

____

Try to move the image up (away from you). Which way did you have to move the slide?

____

In what direction do you have to move the letter to make the image move right then up?

____

There will be times when you will want to show someone something of interest in the **field of view.** One way to do this is to describe its approximate location by referring to the field of view as a clock. Thus, you could tell them to "look at three o'clock," or "look just off center toward nine o'clock," and so forth. Alternately, some microscopes have what appears to be a thin black line cutting across the field. This is a **pointer** that has been added to the ocular of your microscope so you can point out something by moving the object under observation to the end of the pointer.

b. Brightness of the Field of View and Working Distance

Examine your slide starting with the lowest-power objective and progressing to the highest-power objective. Describe any changes in the brightness of the field when you change objectives.

In Fig. 1-3 shade in the appropriate circles to correspond with any change in brightness you observed. When the object on your slide is in focus for each objective, the **working distance** between the slide and the objective lens decreases as the objective magnification increases. Of what value is such information to you?

If you know the diameter of the field for each magnification, you can use this information to estimate the size of object you are examining. To determine the diameter of the field, place a transparent millimeter rule on the stage, focus on the rule, and measure the diameter of the field for the scanning and low-power objectives. It will be very difficult to measure the high-power and oil immersion fields, but you can get a good approximation using the following formula, where

D_L = diameter of the field at a *lower* magnification
D_H = diameter of the field at the *higher* magnification
X_L = magnification of the *lower* power objective lens
X_H = magnification of the *higher* power objective lens

Thus

$$\frac{D_H}{D_L} = \frac{X_L}{X_H} \quad \text{or} \quad D_H = \frac{D_L \times X_L}{X_H}$$

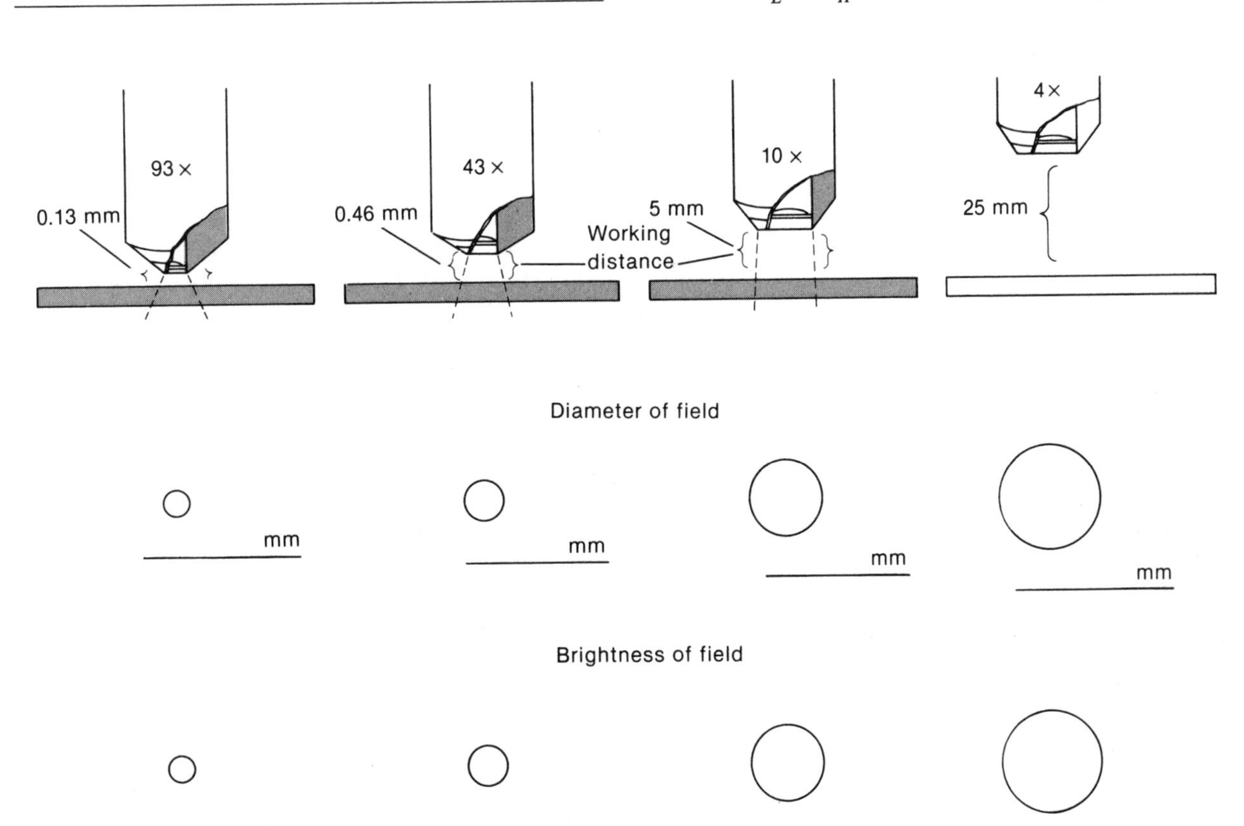

FIG. 1-3
Relationships between working distance, diameter, and brightness of field.

Insert the appropriate values in the formula and determine the diameters of the high power and oil immersion objective fields of your microscope. Record your data in Fig. 1-3.

c. Depth of Focus

Just like the human eye, the lenses of your microscope provide a limited **depth of focus.** This means that only part of the object will be in sharp focus, while areas above and below that part will be slightly out of focus or not in focus at all.

To become familiar with the concept of depth of focus, take your metric rule and, pointing it lengthwise away from you, hold it about 30 cm (12 inches) in front of you and about 7.5 cm (3 inches) below your eyes. Looking down the length of the rule, focus your eyes on the 10-cm mark. When this mark is sharply in focus, what numbers above and below the 10-cm mark are also in focus? What then is the depth of focus for your eyes?

___

In practice, you will find that as magnification *increases,* the depth of focus *decreases.* You will have to learn to constantly use the fine adjustment knob when the higher power objective is in position. This will help you to determine something about the three-dimensional shapes of the objects under observation.

To visualize three-dimensional form and the concept of depth of focus, place a small strand of your hair and a white and a yellow thread across each other on a microscope slide. Add a drop of water and a coverslip. Using the scanning objective (4 ×), focus where the strands of hair intersect and determine the depth of focus at this magnification.

Change to the low-power objective (10 ×). Describe any changes in the depth of focus.

___

___

___

Switch to the high-power objective (43 ×) and describe any changes in the depth of focus.

___

___

___

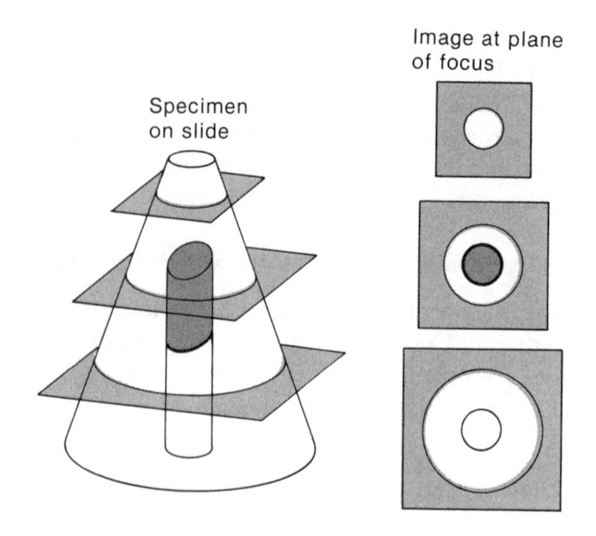

FIG. 1-4
Determining three-dimensional image through "optical" sectioning.

At this higher magnification, it is difficult but not impossible to determine three-dimensional form. You can do this by building a series of **optical sections** in your mind as you focus through the specimen. Fig. 1-4 demonstrates how this is done.

Try to determine the three-dimensional structure of your preparation at high power by making and visualizing a series of optical sections. Begin by focusing on the surface of the top thread and working through to the lower surface of the bottom thread or hair.

d. Contrast

Even with sufficient magnification and resolution, you can only visualize an object under a microscope if there is sufficient contrast between the object and its surroundings or between the various parts of the object.

Cells or subcellular structures may contain naturally occurring pigments (e.g., chlorophyll in chloroplasts, hemoglobin in red blood cells) that provide contrast and make these structures visible. Frequently, however, cells and parts of cells are highly translucent. One way to improve contrast is to use dyes or stains that bind to or are taken up by various subcellular structures and thus absorb enough light to provide the necessary contrast. In addition to staining, or in combination with it, you can improve image contrast by regulating the opening of the diaphragm. This deflects the light rays from edges of the diaphragm and causes them to enter the specimen at an angle. Such scattering of light makes the specimen look darker, since some of the light takes longer to reach the eyes.

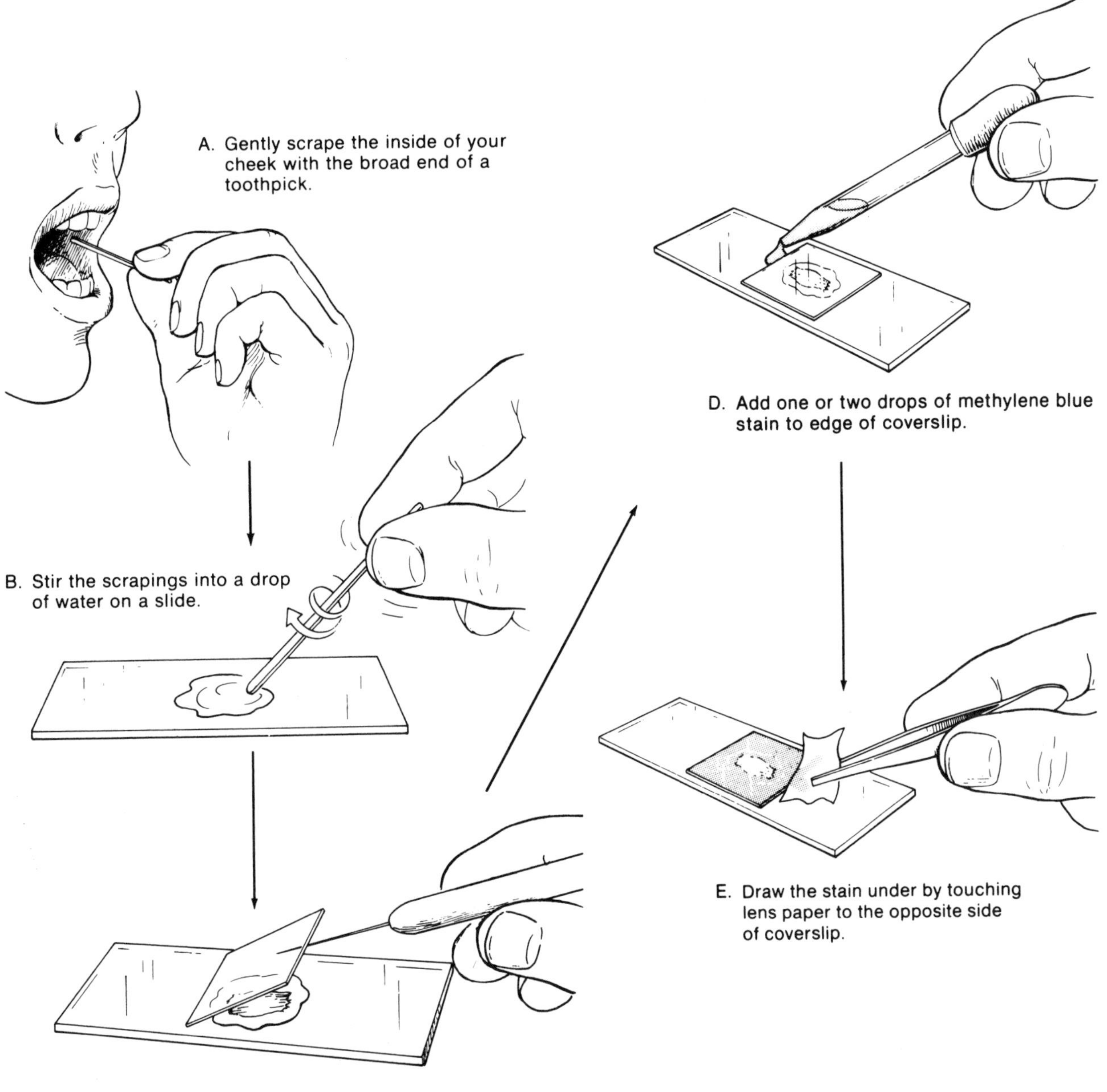

FIG. 1-5
Staining cells to improve image contrast.

Following the instructions in Fig. 1-5A–C, examine cells obtained from the inner epithelial lining of your cheek. Try to determine something of their structure by adjusting the diaphragm and the condenser. Add a drop of methylene blue stain to the edge of the coverslip and draw it under as shown in Figure 1-5D, E. Describe any changes in contrast, or visibility of the structures, in the cell.

____

____

____

____

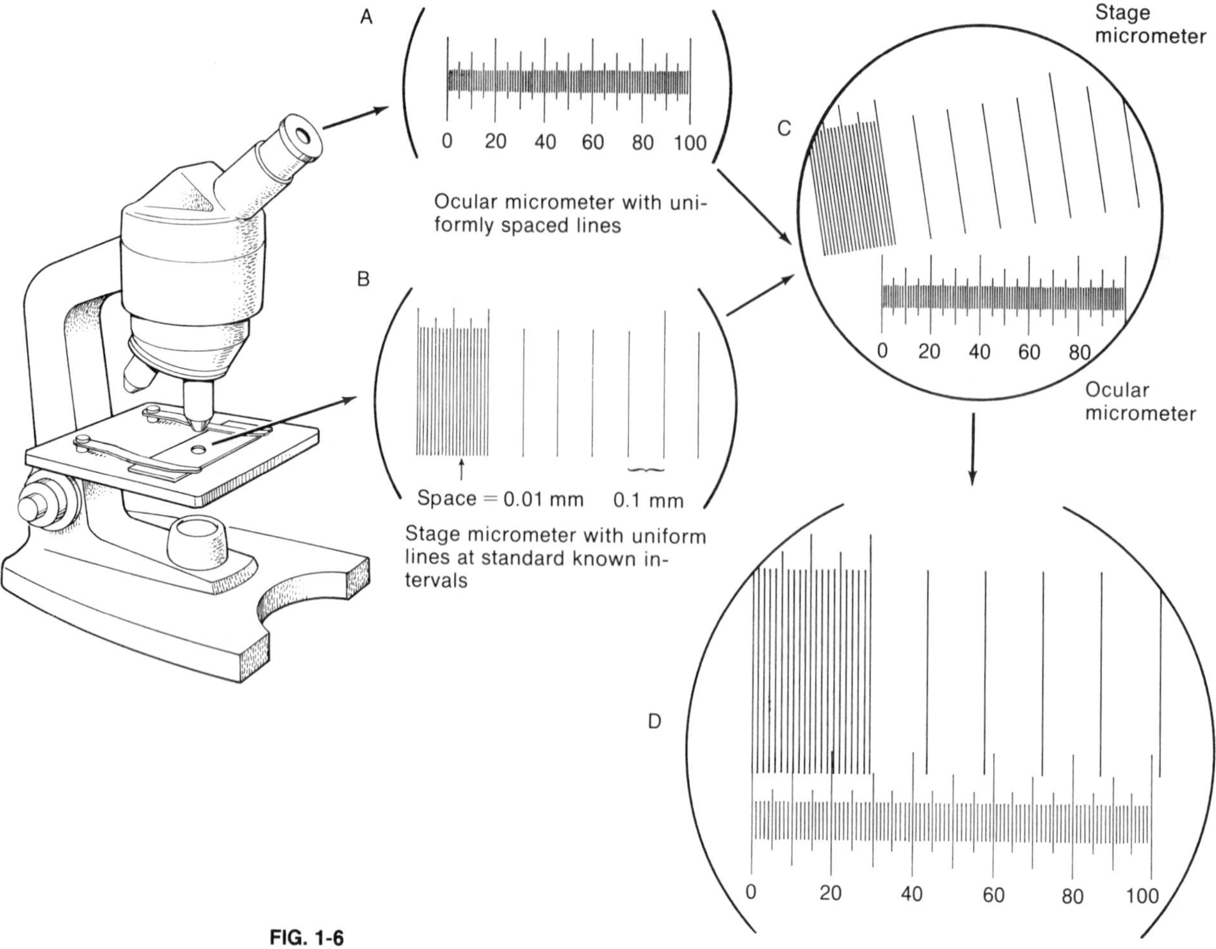

FIG. 1-6
Using an ocular micrometer to determine the size of microscopic objects.

e. Measurement of Microscopic Specimens

Recall that you earlier determined the diameter of the field of view of the various objective lenses on your microscope and, comparing the size of the specimen with the diameter, you obtained a rough estimate of the size of the object.

A more precise method involves using an **ocular micrometer,** a small glass disc on which uniformly spaced lines of *unknown distance* are etched. The ocular micrometer is inserted into the ocular of the microscope and then calibrated against a **stage micrometer,** which has uniformly spaced lines of *known* distances (Fig. 1-6). To calibrate the ocular micrometer use the following procedure:

1. If you were to observe the stage micrometer without the ocular micrometer in place, it would appear as shown in Fig. 1-6B. If you were to observe the stage micrometer with the ocular micrometer in place, it would appear as shown in Fig. 1-6C.

2. Turn the ocular in the body tube until the lines of the ocular micrometer are parallel with those of the stage micrometer. Match the lines at the left edges of the two micrometers by moving the stage micrometer (Fig. 1-6D).

3. Calculate the actual distance in micrometers (μm) between the lines of the ocular micrometer by observing how many spaces of the stage micrometer are included within a given number of spaces on the ocular micrometer. Since the smallest space on the stage micrometer equals 0.01 millimeter (mm), you can calibrate the ocular micrometer using the following:

10 spaces on ocular micrometer
= X spaces on stage micrometer.

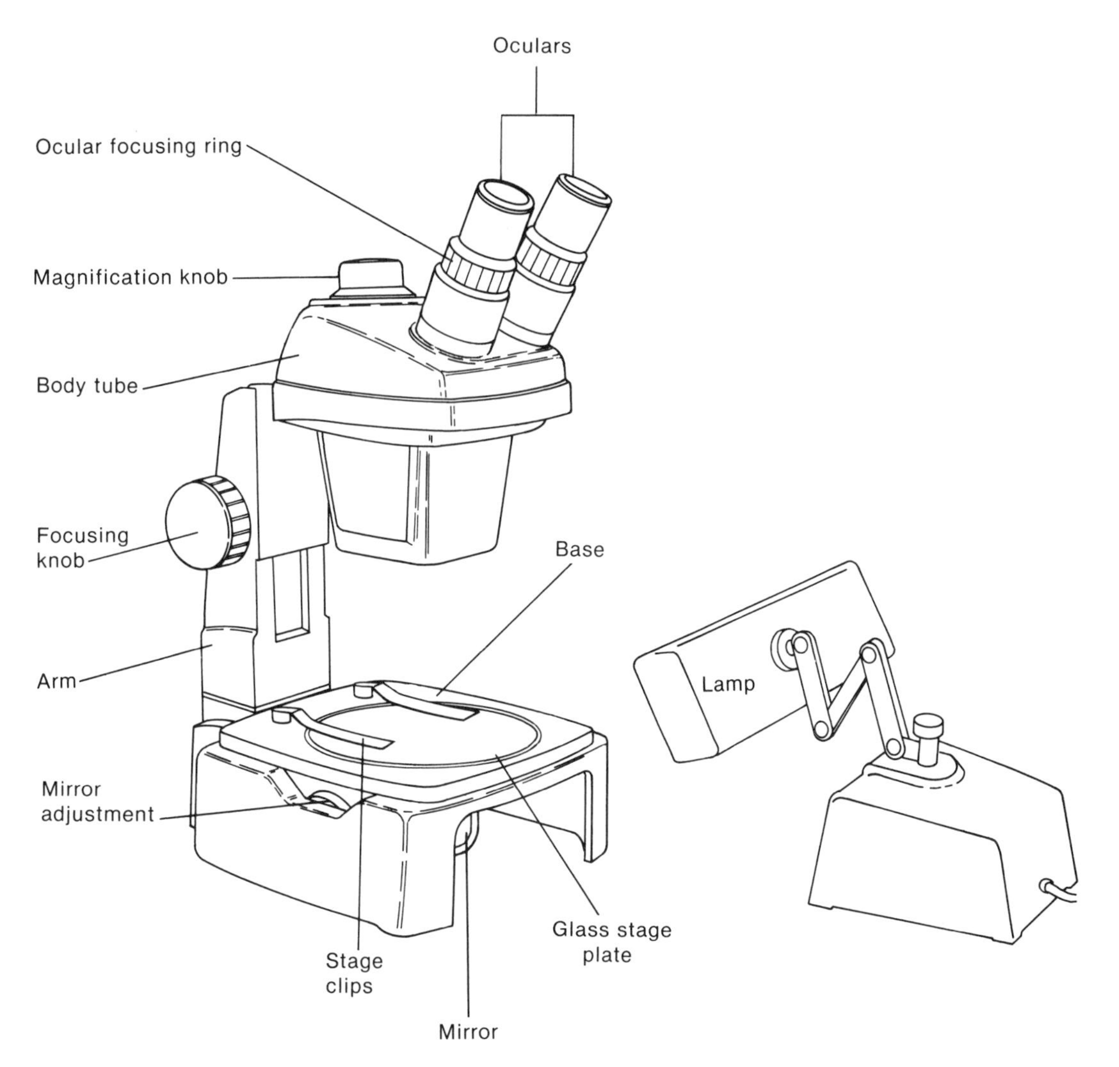

FIG. 1-7
Parts of a stereoscopic (dissecting) microscope.

Since the smallest space on a stage micrometer = 0.01 mm, then

$$10 \text{ spaces on the "ocular"} = X \text{ spaces on the "stage"} \times 0.01 \text{ mm}$$

and

$$1 \text{ space on the "ocular"} = \frac{X \text{ spaces on the "stage"} \times 0.01 \text{ mm}}{10}$$

Example: If 10 spaces on the "ocular" = 6 spaces on the "stage" then

$$1 \text{ ocular space} = \frac{6 \times 0.01 \text{ mm}}{10}$$

Of course, the numerical value obtained holds only for the specific objective–ocular lens combination used. Each time the objective or ocular lens is changed, the ocular micrometer will have to be recalibrated.

D. USE AND CARE OF THE STEREOSCOPIC (DISSECTING) MICROSCOPE

The stereoscopic dissecting microscope shown in Fig. 1-7 has two distinct advantages over the compound microscope: (1) it enables you to examine objects that are too large or too thick to be seen with the higher magnifications of the compound microscope and (2) it gives you a three-dimensional view of the specimen.

The stereoscopic microscope is often used when dissecting specimens. The light source may be reflected from an illuminator above the specimen or, on some microscopes, transmitted through the specimen from a mirror below the stage. The choice of the light source depends upon whether the specimen is transparent or opaque.

Using your dissecting microscope, examine your fingers or some other opaque object. Adjust the oculars for interpupillary distance and focus as previously for the compound microscope (Part A-1). Change the magnification using the magnification knob on the top of the body tube. On other stereoscopic microscopes, the magnification is varied by switching ocular lenses, as with the compound microscope. How does the movement of the image compare to that of the compound microscope (Part C-2 of this exercise)?

How do you adjust the brightness of the field?

Examine the previously prepared slide of the crossed threads or hair. First use reflected light from the mirror, then use transmitted light from a lamp. Describe any advantage of one type of lighting over the other.

E. STUDY OF POND WATER

In your laboratory work, many observations made by the microscope will be on living organisms or on tissues or parts of organisms that you will want to keep alive. To allow them to dry out would greatly distort them, to say nothing of the effect death would have on a study of their movements. To observe living material prepare a wet mount of a drop of pond water as shown in Fig. 1-2.

Excess water under the coverslip can be soaked up by carefully placing a piece of paper toweling to the edge of the coverslip. However, if your preparation begins to dry out while under observation, add one drop of water at the edge of the coverslip.

Under low power and with reduced light, survey the drop of pond water. Identify as many of the organisms as you can. Carefully study their differences in structure and their method of movement. Figs. 1-8, 1-9, 1-10, and 1-11 should help you identify what you see.

Prepare additional wet mounts by taking samples from different parts of the jar of pond water. Do not be too hasty in discarding a slide because you don't find any microorganisms; a systematic survey of the preparation is often necessary to locate the organisms. Why do the organisms often accumulate at the edge of the coverslip?

To identify the smaller organisms, you may have to use the high-power objective. When your work is completed, clean and dry any slides and coverslips used. Wipe the lenses of the microscope with lens paper, clean the stage, and return the microscope to the cabinet.

REFERENCES

Eddy, S., and A. C. Hodson. 1958. *Taxonomic Keys.* Burgess.

Hartley, W. G. 1964. *How to Use a Microscope.* Natural History Press.

John, T. L. 1949. *How to Know the Protozoa.* Wm. C. Brown.

Needham, G. H. 1977. *The Practical Use of the Microscope.* Thomas.

Needham, J., and P. Needham. 1941. *Guide to the Study of Fresh-Water Biology.* Comstock.

Palmer, M. C. 1959. *Algae in Water Supplies.* Public Health Service, Publication No. 657.

Taft, C. E. 1961. A Revised Key for the Field Identification of Some Genera of Algae. *Turtox News* 39(4):98–103.

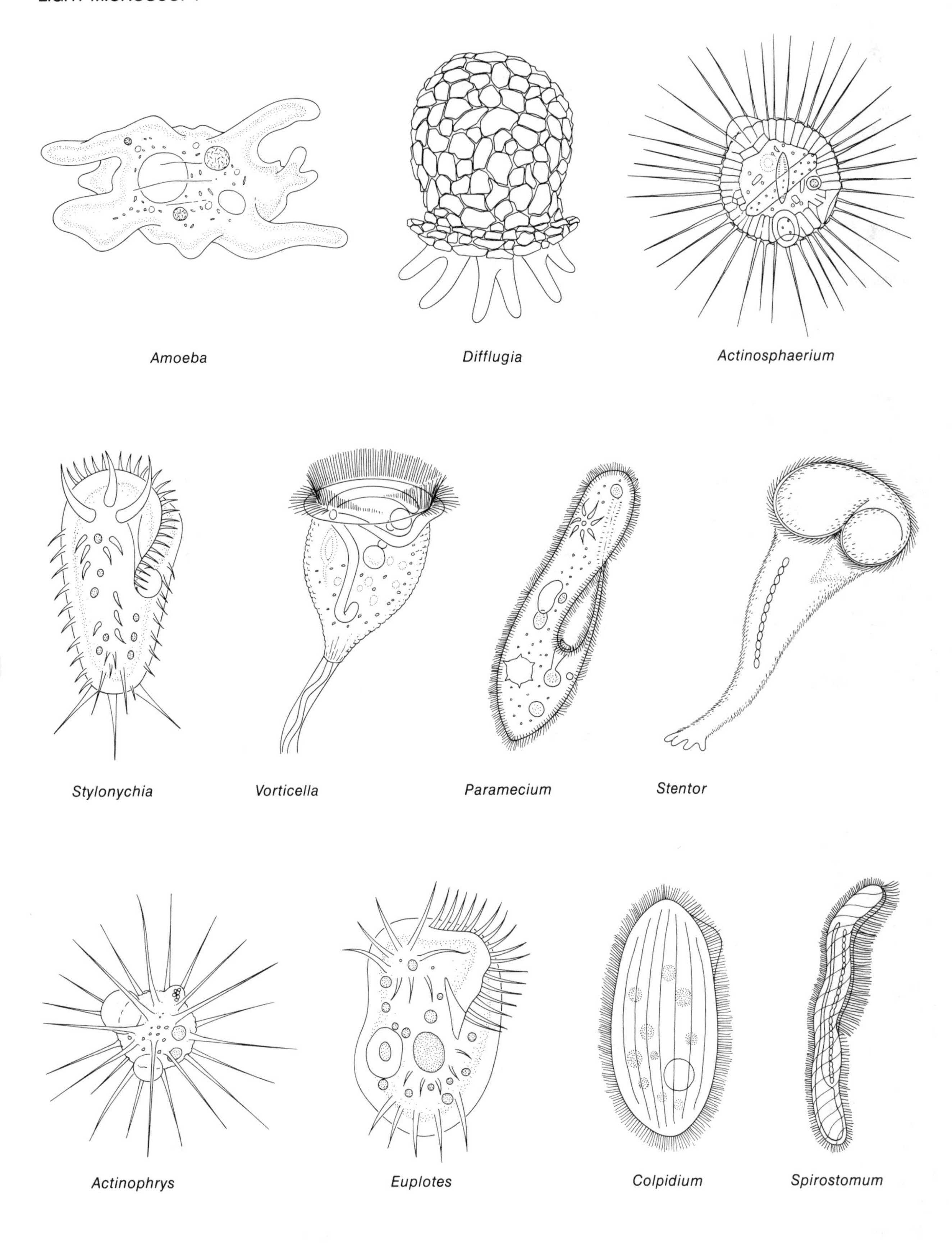

FIG. 1-8
Protozoans commonly found in pond water.

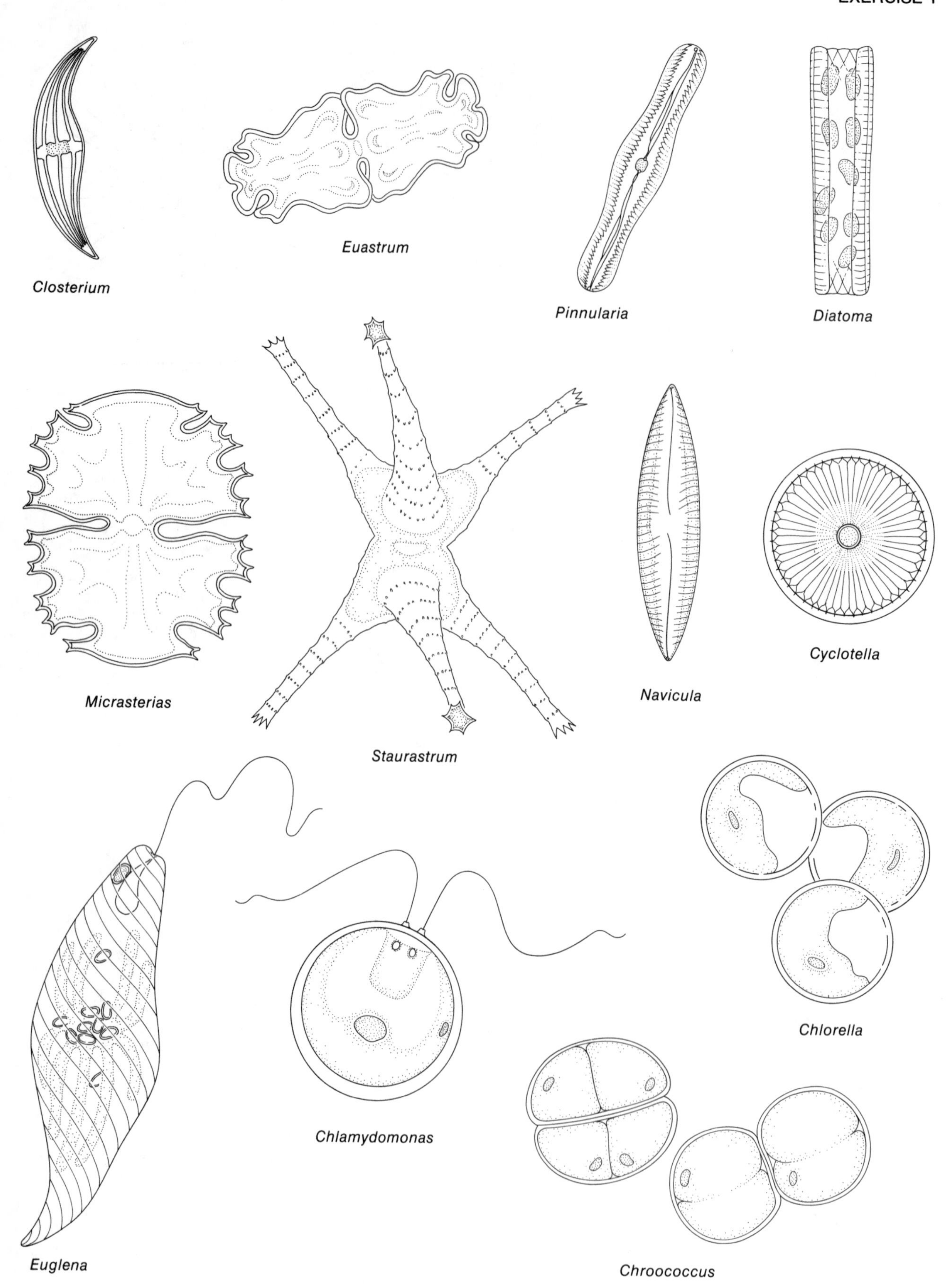

FIG. 1-9
Unicellular algae commonly found in pond water.

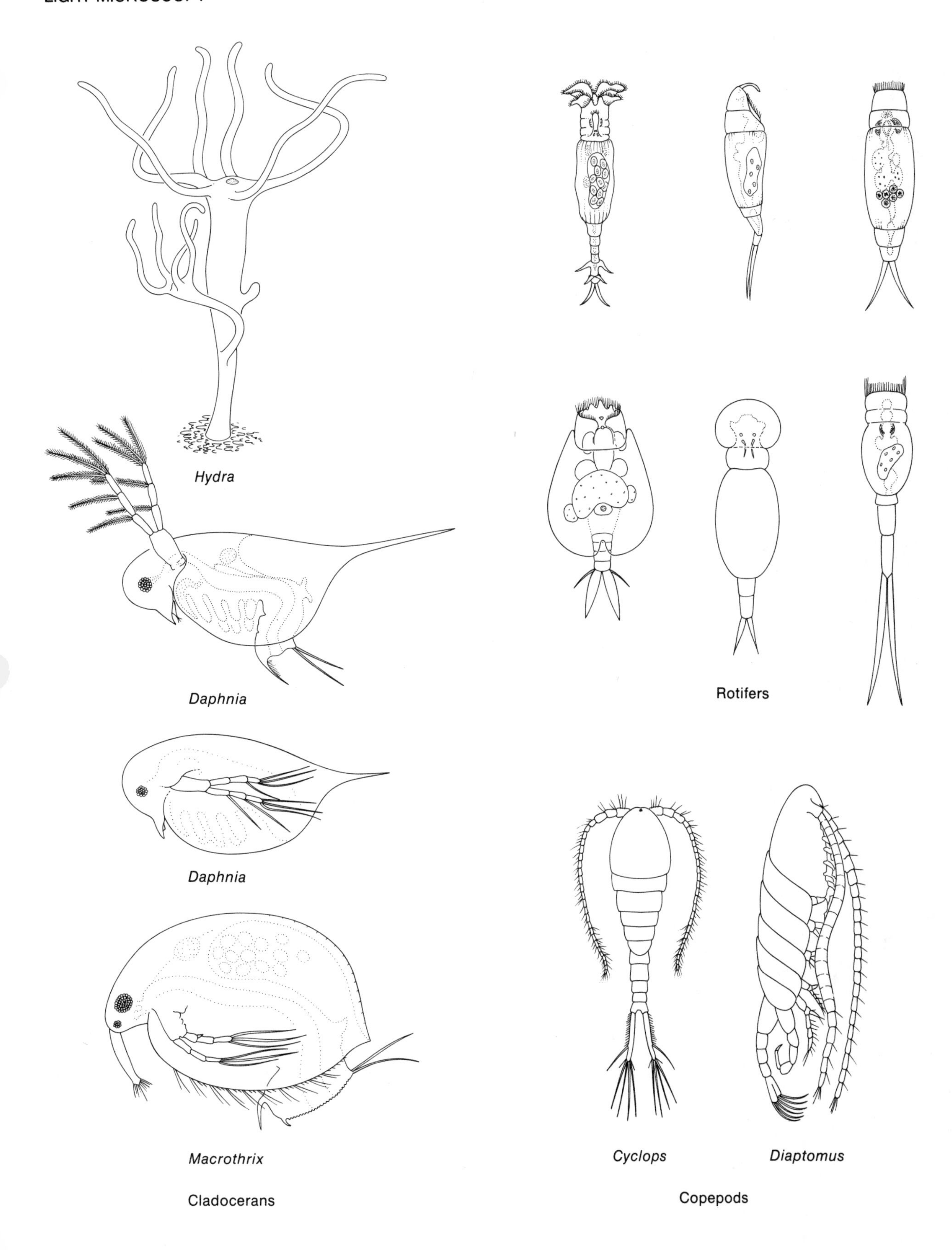

FIG. 1-10
Invertebrates commonly found in pond water.

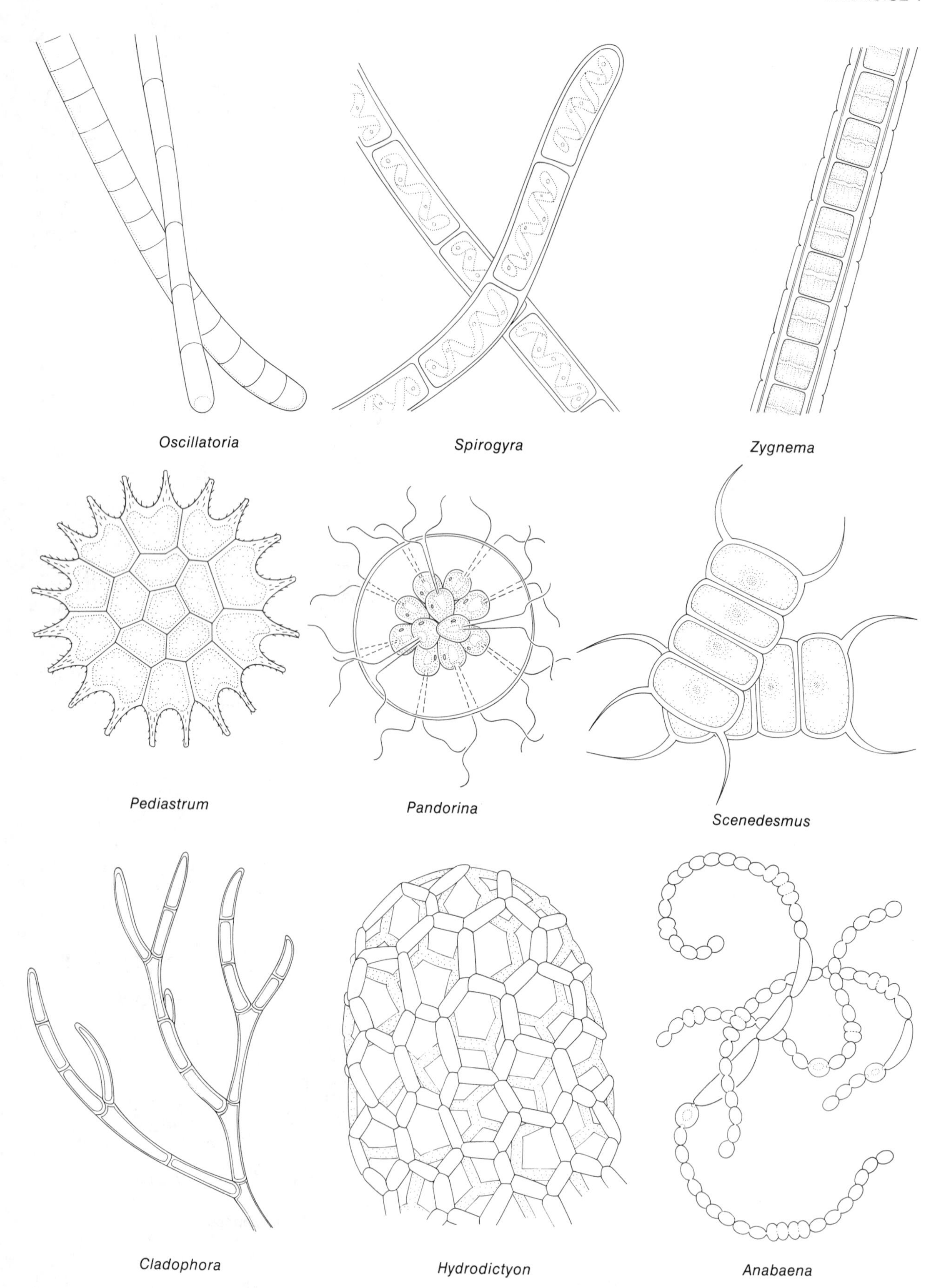

FIG. 1-11
Multicellular algae and cyanobacteria commonly found in pond water.

EXERCISE 2

Cell Structure and Function

Just as there is diversity of form in life, so there is in the form and function of the cells that make up living organisms. Single cells, such as *Amoeba* and *Paramecium,* can be free-living organisms capable of carrying on an independent existence. Some cells live as part of a loosely organized colony of cells that move from place to place. Others are immovably fixed as part of the tissues of higher plants and animals and depend on the closely integrated activities of other cells for their existence.

Cells vary in size. For example, the length of many bacteria is roughly 1 micrometer (μm), which is equal to 10^{-6} meter. On the other hand, the yolk of an ostrich egg, also a single cell, is the size of a small orange. Some cells, such as red blood cells, transport oxygen and carbon dioxide. Other cells have different specialties. Whatever its form or function, the cell is now recognized as the basic unit of living matter, containing all those properties and processes that are collectively called *life.*

Contemporary biologists recognize two basic types of cells. **Prokaryotic cells,** as exemplified by bacteria and cyanobacteria (formerly the blue-green algae), lack a nuclear membrane and membrane-bound cytoplasmic organelles. **Eukaryotic cells** exhibit the structural characteristics lacking in prokaryotic cells: they have a well-defined nucleus surrounded by a membrane and a variety of membrane-bound organelles. Examples of eukaryotic cells include those of protozoa, fungi, plants, and animals.

While the differences between prokaryotic and eukaryotic cells are striking, both types of cells have several common characteristics. They are surrounded by a cell (plasma) membrane that regulates the movement of materials into and out of the cell. Both have similar enzyme systems, DNA as the genetic material, and ribosomes that function in protein synthesis.

While eukaryotic cells are considered to be more advanced (they developed after prokaryotic cells and have a more complex organization), the many similar characteristics of eukaryotic and prokaryotic cells suggest a common ancestor in the evolutionary past.

In this exercise, you will examine examples of prokaryotic and eukaryotic cells and become familiar with the diversity within these organismic groups.

A. PROKARYOTIC CELLS

1. Bacteria

Examine Fig. 2-1A, an electron micrograph of *Azotobacter vinelandii,* a bacterium commonly found in garden soils. In this micrograph, the cell has nearly divided into two; each cell is approximately 1 × 1.5 μm in size. The diagram of this bacterium (Fig. 2-1B) shows the cell wall, cell membrane, ribosomes, and the nucleoid region containing DNA.

Because of their small cell size, you will not be able to see with the light microscope the structural details that you would with the electron microscope. However, you can observe some features if the cells are stained. To do this, add a drop of the stain crystal violet to a clean slide. Then, using an inoculating loop, transfer a loopful of a broth culture of *Bacillus subtilis* to the drop of stain on the slide. Mix the bacteria in the drop, then add a coverslip. Examine the preparation microscopically. What cellular structures and organelles can you identify that are also

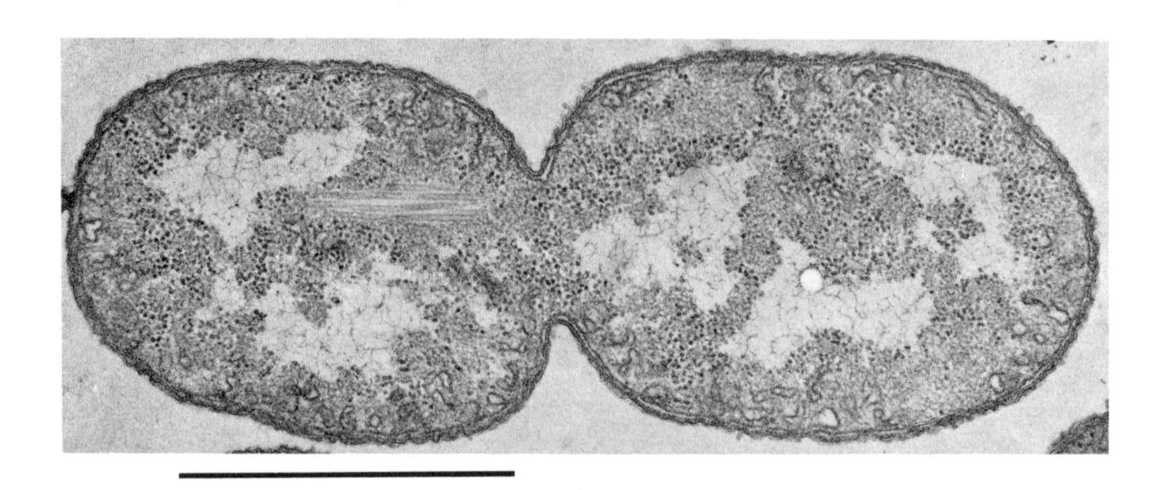

A

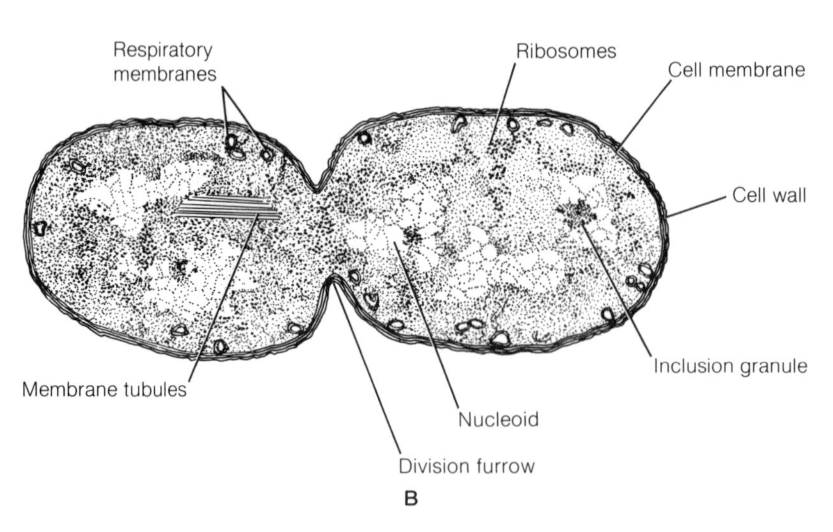

B

FIG. 2-1
(A) *Azotobacter vinelandii,* a bacterium commonly found in garden soils. In this transmission electron micrograph, division into two cells is nearly complete. Scale bar = 1 μm. (Micrograph courtesy of W. J. Brill.) (B) Diagram based upon electron micrograph shows, in greater detail, the subcellular structure of this bacterium. (Diagram by I. Atema—From *Five Kingdoms* by Margulis and Schwartz, p. 47. W. H. Freeman and Company. Copyright © 1982.)

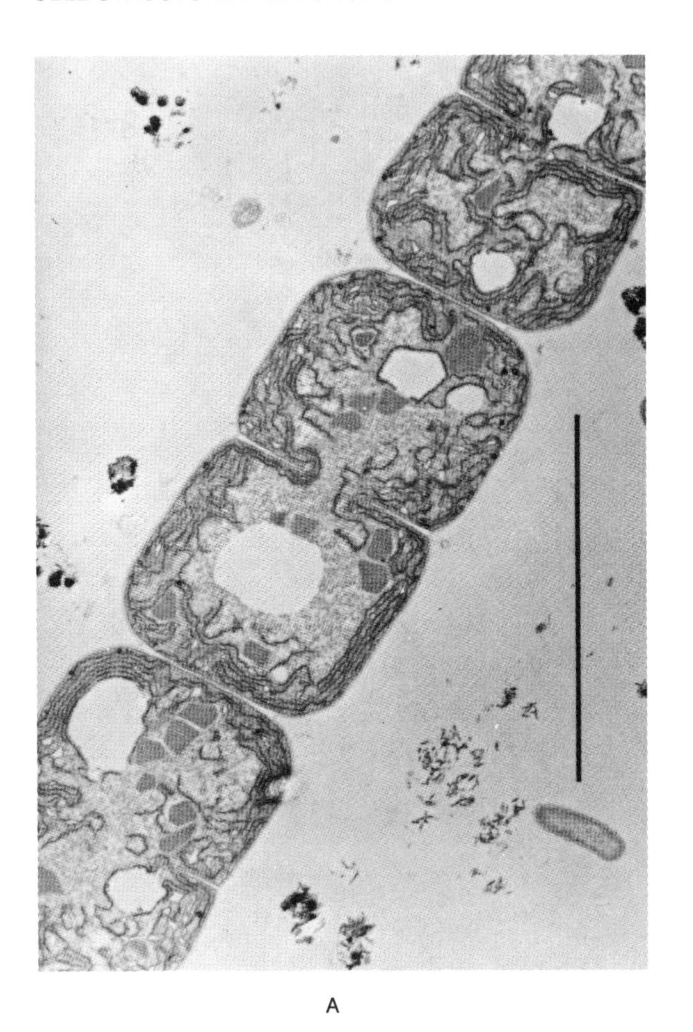

A

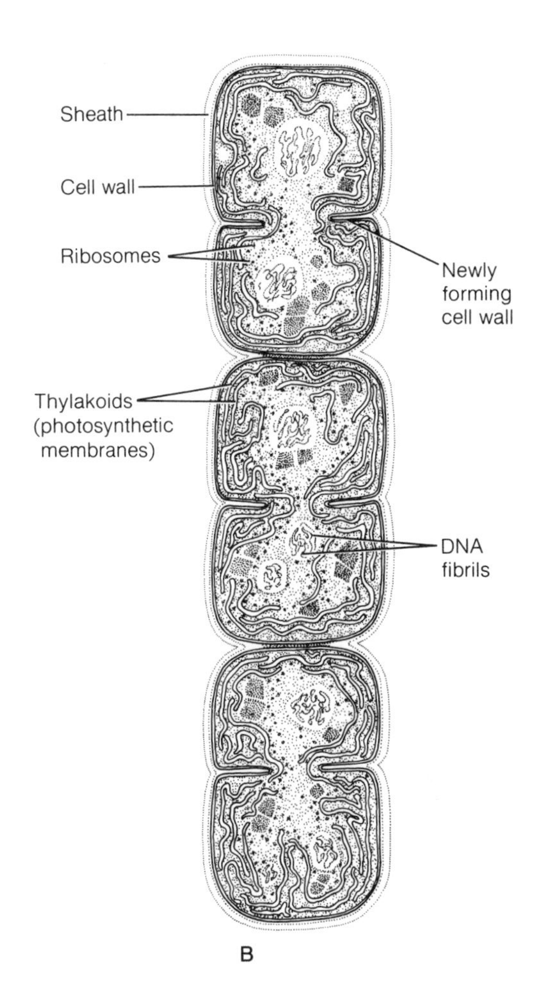

B

FIG. 2-2
(A) Transmission electron micrograph of *Anabaena,* a common filamentous cyanobacterium that grows in freshwater ponds and lakes. Within the gelatinous sheath the cells divide by forming crosswalls. Scale bar = 5 μm. (Micrograph courtesy of N. J. Lang.) (B) Diagram of *Anabaena* showing in greater detail the cellular ultrastructure of this organism. (Diagram courtesy of R. Golder.) (From *Five Kingdoms* by Margulis and Schwartz, p. 43. W. H. Freeman and Company. Copyright © 1982.)

evident in the electron micrograph of *Azotobacter*?

____

____

____

2. Cyanobacteria (Formerly Blue-Green Algae)

The internal organization of cyanobacteria is basically similar to that of bacteria. A major difference is that the cells of cyanobacteria contain membranous structures, called **thylakoids,** in which are embedded the photosynthetic pigments (Fig. 2-2).

Cyanobacteria are among the common nuisance inhabitants of surface waters. As population and industrial demands increase, villages and cities are turning from groundwater to surface waters, such as lakes, streams, and reservoirs, for their source of water. Groundwaters are essentially free from contaminating organisms. Surface waters, on the other hand, contain many organisms that in one way or another contribute to the unpalatability of the water supply. Such organisms, especially some of the cyanobacteria, affect the odor and taste of water; clog filters; grow in pipes, cooling towers, or on reservoir walls; form mats or blooms on the surface of the water; produce toxic materials; and so forth. Furthermore, present methods of waste disposal intensify the effects of these nuisance organisms in water.

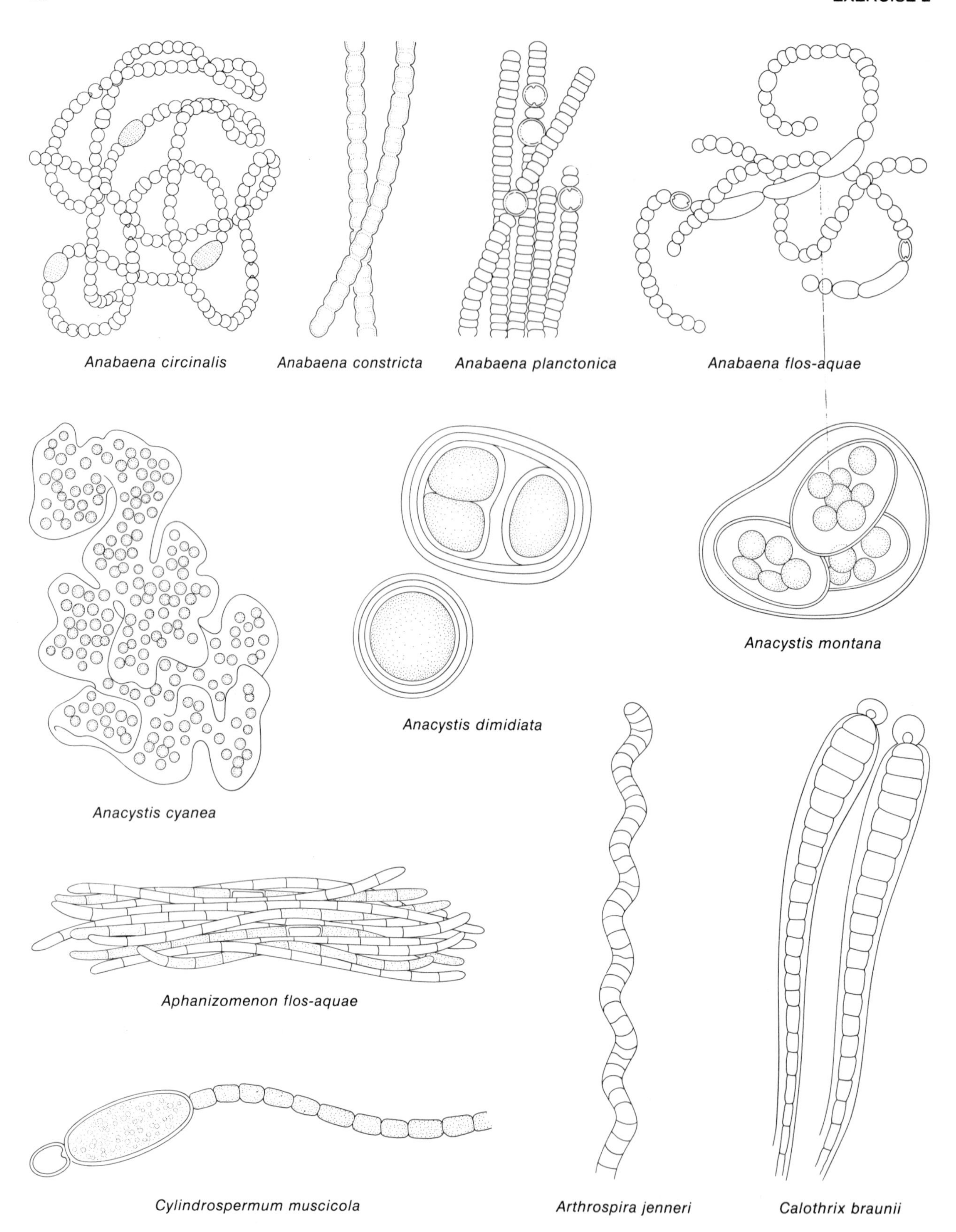

FIG. 2-3
Cyanobacteria that contaminate water supplies.

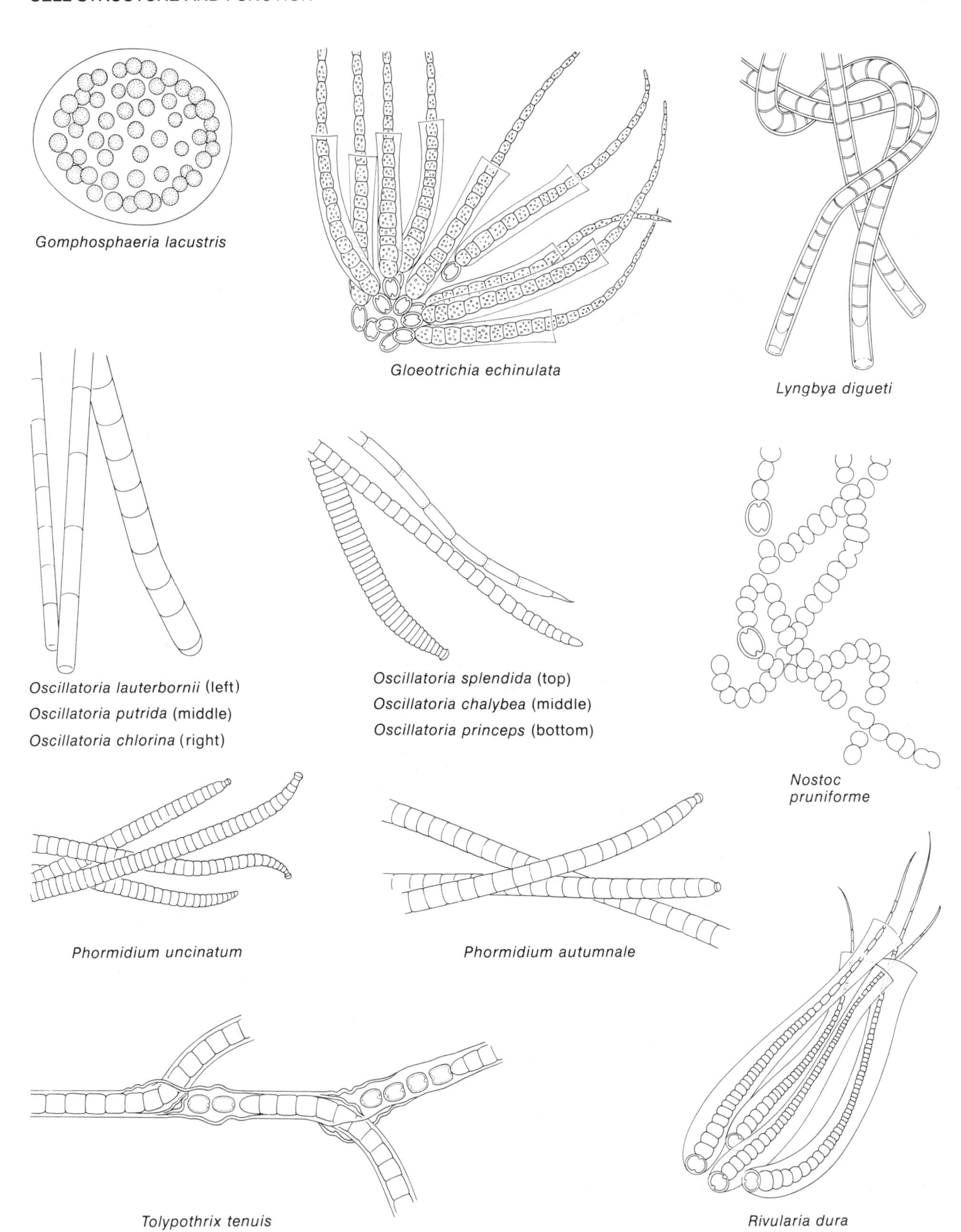
Gomphosphaeria lacustris
Gloeotrichia echinulata
Lyngbya digueti
Oscillatoria lauterbornii (left)
Oscillatoria putrida (middle)
Oscillatoria chlorina (right)
Oscillatoria splendida (top)
Oscillatoria chalybea (middle)
Oscillatoria princeps (bottom)
Nostoc pruniforme
Phormidium uncinatum
Phormidium autumnale
Tolypothrix tenuis
Rivularia dura

Materials such as sewage and organic wastes from paper mills, fish-processing factories, slaughterhouses, and milk plants greatly increase the growth of algae and other organisms. Many of these cause problems when they become abundant.

Examine a commercially prepared slide that has several different species of cyanobacteria on it. Alternately, collect surface water samples from different sources: lakes, streams, ponds, reservoirs (including the walls), swimming pool filters, or water treatment plants, for example. Compare the organisms you observe with Fig. 2-3 to identify some of the cyanobacteria found in these water supplies. Table 2-1 characterizes some of the more common problems associated with an overabundance of these cyanobacteria in the water.

B. EUKARYOTIC CELLS

The primary difference between prokaryotic and eukaryotic cells is that, in the latter, the DNA is associated with proteins and organized into large, complex structures called **chromosomes.** The chromosomes are surrounded by a double membrane, the **nuclear envelope,** that separates them from the cytoplasm in a well-defined nucleus. Hence the name *eu* (true) *karyon* (nucleus) versus *pro* (before) *karyon* (nucleus).

In this exercise, you will become familiar with the diversity in eukaryotic cells. As you examine such cells, compare each with the "generalized" cell shown in Fig. 2-4. Not all the structures shown in this drawing will be evident in the cells you examine, because it takes special staining procedures and high magnification to bring out such subcellular organelles as mitochondria, chloroplasts, the endoplasmic reticulum, and the Golgi complex.

1. Onion Cells

Following the procedure outlined in Fig. 2-5A–E, prepare a wet mount of onion epidermal tissue. Examine this tissue with the low-power objective (10 ×). The "lines" forming the network between the individual cells are nonliving cell walls composed chiefly of cellulose. The **cell wall** immediately surrounds the **plasma membrane,** which encloses the *cytoplasm.* The central part of many plant cells (which is difficult to observe in living cells) is taken up by a fluid-filled **vacuole** containing mostly water and salts.

Next examine the cells under high power. Locate the **nucleus,** which appears as a dense body in the translucent cytoplasm. In some cells, the nucleus seems to be lying in the central part of the cell and looks circular. In other cells, it seems to be compressed and pushed against the cell wall. Explain this apparent discrepancy in the shape and position of the nucleus.

____

____

The cytoplasm is separated from the central vacuole, nucleus, and cell wall by membranes, but the membranes are difficult to observe in this preparation.

In the onion cell, it is possible to observe **mitochondria**—those cell organelles involved in cellular respiration. To do this, select another piece of onion epidermal tissue and cut it with a sharp razor so that it is approximately 1 × 3 mm in size. On a clean slide, mount the tissue in a drop of 5% sucrose solution to which you have added from three to five drops of the stain Janus Green B. Add a coverslip.

If your preparation is truly one cell in thickness, it will have the appearance of a nearly transparent brick wall when viewed under the low power of your microscope. Using the high-power objective, locate mitochondria, which look like very small rods or

TABLE 2-1
Problems caused by cyanobacteria in water supplies.

Problem	Organism
Taste and odor	*Anabaena circinalis* *Anacystis cyanea* *Aphanizomenon flos-aquae* *Cylindrospermum muscicola* *Gomphosphaeria lacustris*
Clogging of filters	*Anabaena flos-aquae* *Anacystis dimidiata* *Gloeotrichia eschinulata* *Oscillatoria princeps* *Oscillatoria chalybea* *Oscillatoria splendida* *Rivularia duro*
Growth on reservoir wall	*Calothrix braunii* *Nostoc pruniforme* *Phormidium uncinatum* *Tolypothrix tenuis*
Polluted water	*Anabaena constricta* *Anacystis montana* *Arthrospira jenneri* *Lyngbya digireti* *Oscillatoria chlorina* *Oscillatoria putrida* *Oscillatoria lauterbornii* *Phormidium autumnale*

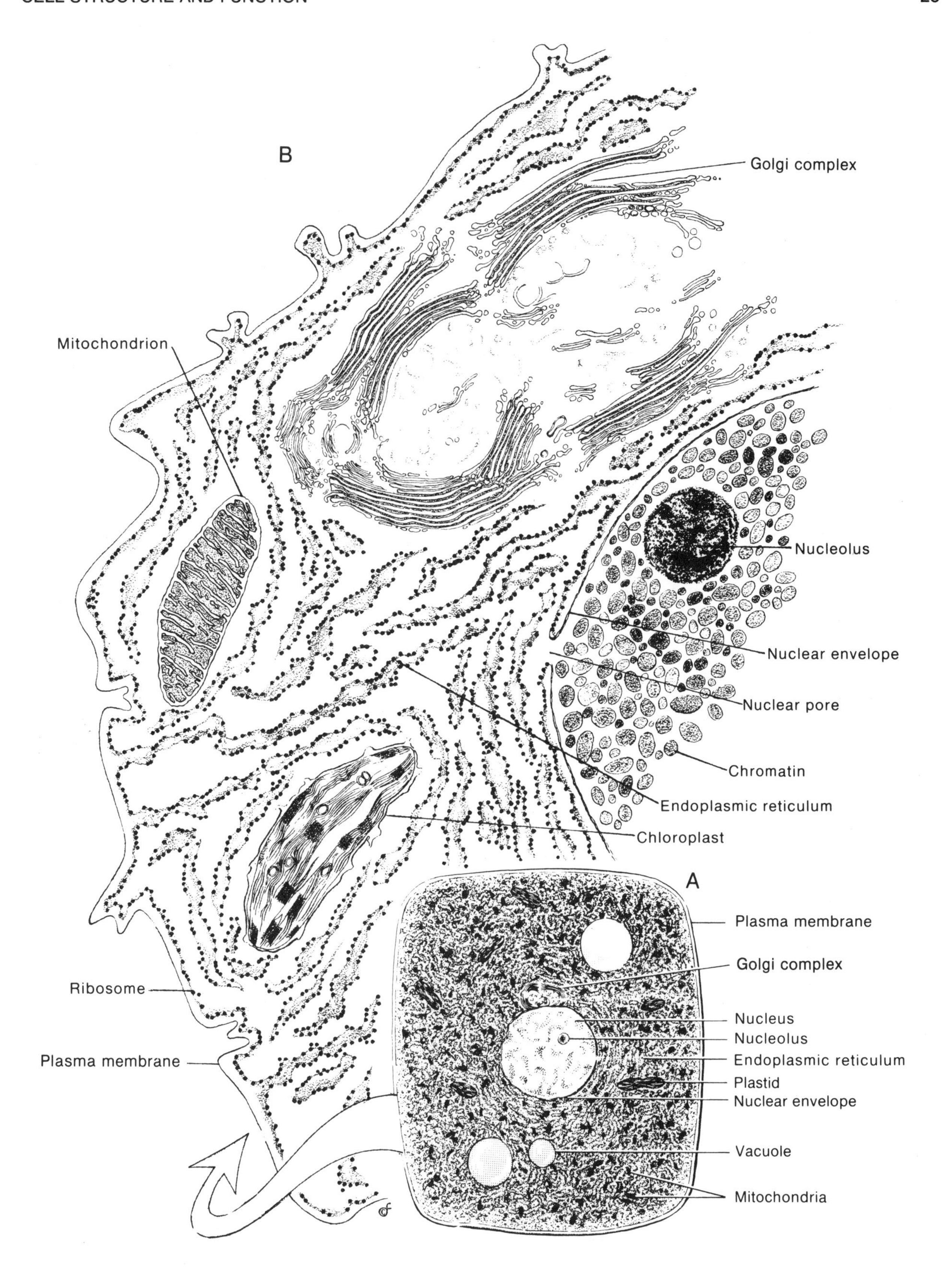

FIG. 2-4
Some of the intracellular components found in cells at the (A) light and (B) electron microscopic levels.

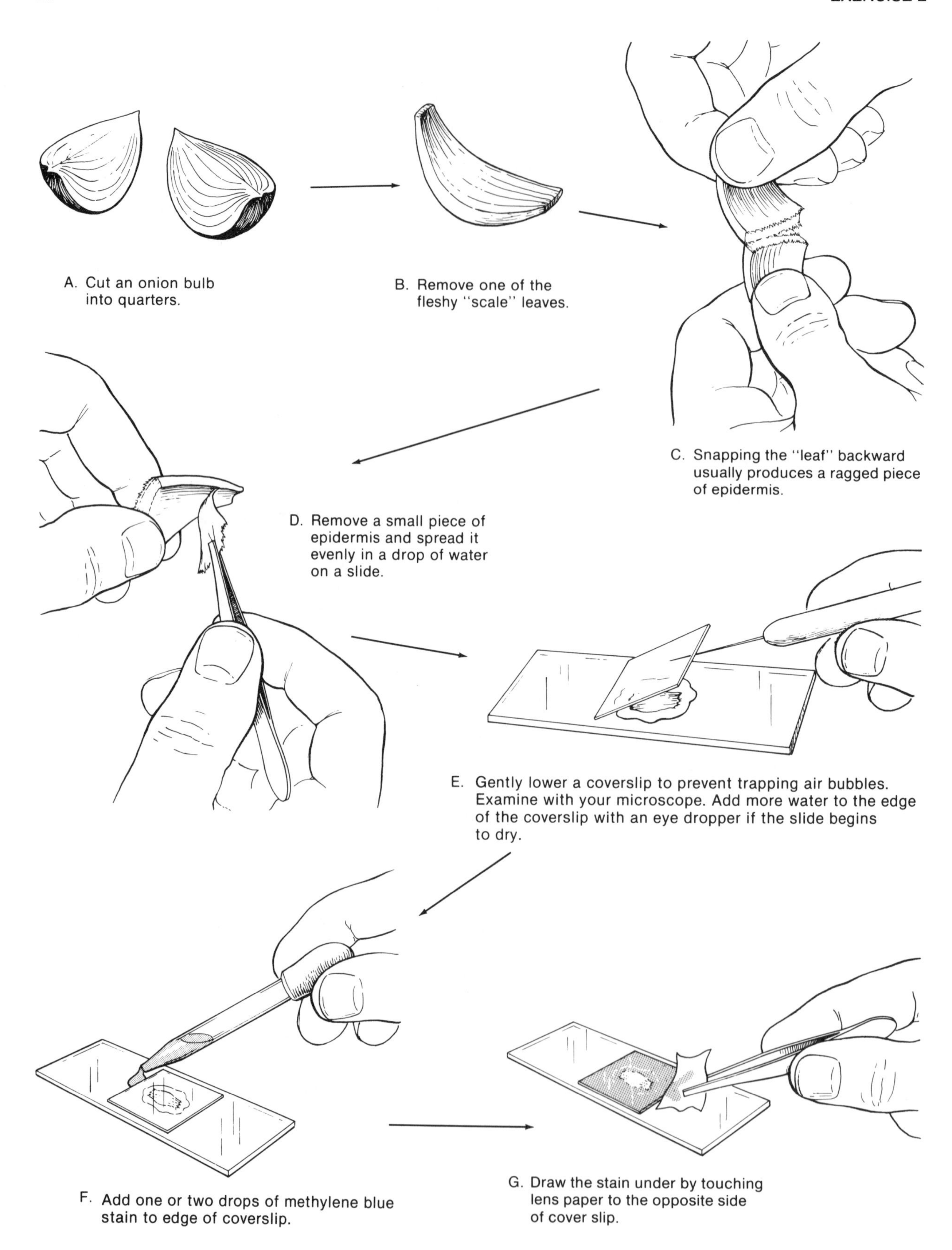

FIG. 2-5
Procedure for studying living onion cells.

spheres at the periphery of the cell. They should be blue in color when you first view your preparation. If they are not stained, place a strip of filter paper or other absorbent paper at the edge of the coverslip. Add a few drops of Janus Green stain at the opposite edge of the coverslip, and draw it through your preparation by absorbent paper, as shown in Fig. 2-5F, G. Stained mitochondria lose their color in about 5 minutes as a result of the action of a dehydrogenase enzyme located on their surface. How long did it require for the mitochondria in your preparation to lose their color?

____

2. Elodea Cells

In this exercise, you will examine cells from the leaf of an aquatic plant commonly called Elodea (Fig. 2-6). These cells are green because they contain a pigment called **chlorophyll.** In **photosynthesis,** this pigment absorbs light energy and converts it into chemical energy.

Remove a young leaf from the tip of the plant. Place the leaf in a drop of water on a slide, and add a coverslip. Examine the preparation with the low-power objective in position. Locate the nucleus, cytoplasm, and cell wall.

Examine a group of cells near the center of the leaf. Carefully switch to high power. Bring the cells into focus by using the fine adjustment, and try to determine the number of cell layers.

____

Note that the green pigment is located in small structures in the cytoplasm. These structures are called **chloroplasts.** As you examine the chloroplasts, you should see them moving in the cell, a phenomenon involving **cyclosis** (see Part B of this exercise).

The plant cell is enclosed by a nonliving cell wall and a cell membrane that is difficult to observe because it is pushed tightly against the inside of the cell wall. You can make this membrane easier to see, however, by placing the cell in a hypertonic saline solution (i.e., a solution that is more concentrated than the cytoplasm). Being hypertonic to the cytoplasm, the saline causes water to move out of the cell and the cell protoplast to shrink away from the cell wall. Under these conditions, you can readily see the plasma membrane, which covers the surface of the protoplast.

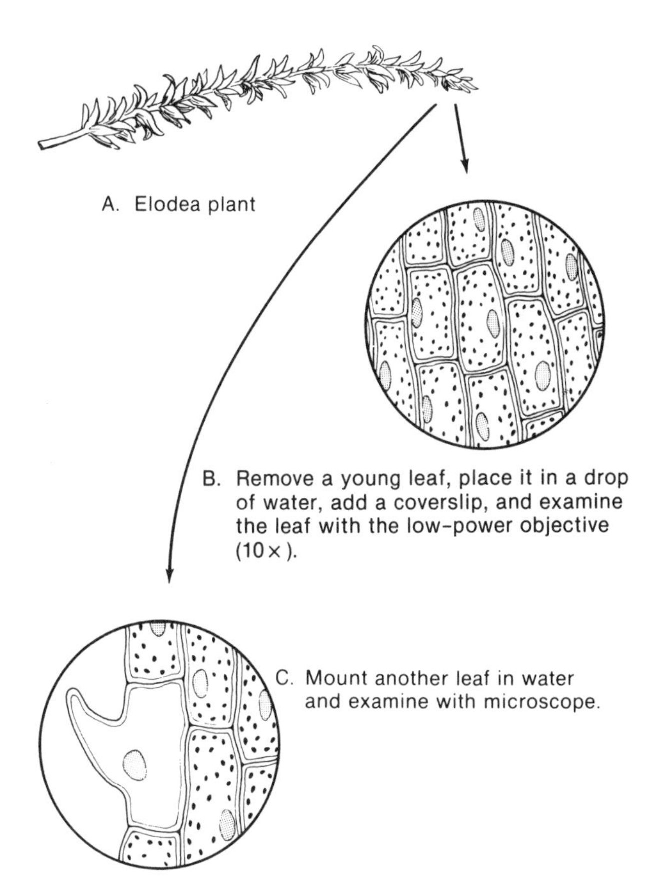

FIG. 2-6
Preparation of Elodea cells for microscopic examination.

Select another young Elodea leaf, mount it in a drop of water, and add a coverslip. Examine the preparation with the low-power objective. Along the edges of the leaf, locate "spine" cells (Fig. 2-6D). Switch to high power and study the cell.

Add one or two drops of a concentrated saline solution to one edge of the coverslip. Then touch the liquid on the opposite side of the coverslip with a piece of lens paper (or paper toweling) so that the paper draws up the liquid (Fig. 2-5F, G). Repeat this step two more times to be sure that the original water has been replaced by the saline. Examine the spine cell closely. Describe your observations, and account for the results you observed.

____

____

____

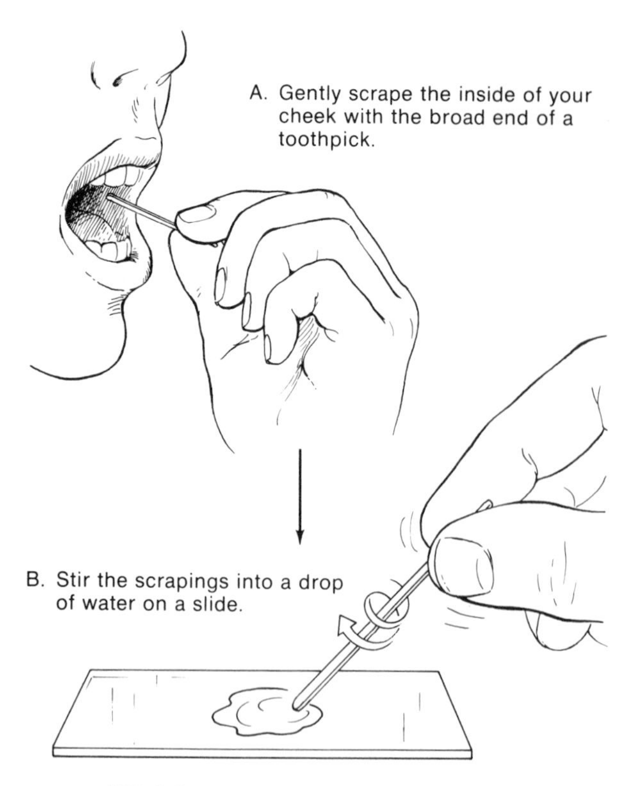

FIG. 2-7
Procedure for studying human epidermal cells.

3. Frog Epidermal Cells

On a clean glass slide, place a drop of water from a container in which frogs have been kept. The water will usually contain cells from the frog's skin, which is normally sloughed off.

Add a small drop of methylene blue stain and a coverslip, then examine microscopically. From what you see, in what way(s) are animal cells different from plant cells?

___

___

___

4. Human Epidermal Cells

Following the procedures outlined in Fig. 2-7, examine cells obtained from the epidermal lining of your inner cheek. Locate the cells under high power, and examine them carefully. What do these epidermal cells have in common with plant cells?

___

___

How are they different?

___

___

Some of the epidermal cells may have had their edges folded over. What does this indicate about the thickness of these cells?

___

___

Add a drop of methylene blue to the edge of the coverslip, and draw it under as shown in Fig. 2-5F, G. What structures in the cell have been stained by this dye?

___

5. Sperm Cells

On a clean glass slide, place a small drop of sperm suspension that has been prepared by removing the testes from a male frog and macerating (cutting) them into small pieces in water.

Add a coverslip, and examine microscopically. You may have to close the diaphragm to obtain more contrast. From what you see, in what way are these cells more specialized than the cheek cells?

___

___

___

C. CYTOPLASMIC STREAMING

In many cells, the cytoplasm can be seen flowing throughout the cell. This phenomenon is called **cytoplasmic streaming,** or **cyclosis.**

The phenomenon of cyclosis is perhaps best observed in the plasmodium of the slime mold *Physarum polycephalum.* The plasmodium, or plant body, is a multinucleate mass of protoplasm that lacks a cell wall. This organism is easily propagated during its dormant stage when it becomes a hard, crusty structure, called a **sclerotium.** To do this, pieces of sclerotia are placed on a moist substrate (e.g., agar) containing nutrients. In a short time, the organism begins to grow out over the surface. Channels of streaming cytoplasm become visible, usually after 72 hours of growth.

Examine a plasmodium of *Physarum* growing in a petri dish. Leaving the cover on, examine the organism carefully using a dissecting microscope. What seems to be unusual about cytoplasmic streaming in *Physarum?*

____

____

____

____

____

____

REFERENCES

Alberts, B., D. Bray, J. Lewis, M. Raff, K. Roberts, and J. D. Watson. 1983. *Molecular Biology of the Cell.* Garland.

Cairns, J. 1966. The Bacterial Chromosome. *Scientific American* 214(1):36–44 (Offprint 1030). *Scientific American* Offprints are available from W. H. Freeman and Company, 41 Madison Avenue, New York, 10010 and 20 Beaumont Street, Oxford OX1 2NQ, England. Please order by number.

DeRobertis, E. D. P., and E. M. F. DeRobertis, Jr. 1980. *Cell and Molecular Biology.* 7th ed. Saunders.

DeWitt, W. 1977. *Biology of the Cell: An Evolutionary Approach.* Saunders.

Green, D. E., and H. Baum. 1970. *Energy and the Mitochondrion.* Academic Press.

Holtzman, E., and A. B. Novikoff. 1984. *Cells and Organelles.* 3d ed. Saunders.

Karp, G. 1984. *Cell Biology.* 2d ed. McGraw-Hill.

Kennedy, D., ed. 1965. *The Living Cell: Readings from Scientific American.* W. H. Freeman and Company.

Staehelin, L. A., and B. E. Hull. 1978. Junctions Between Living Cells. *Scientific American* 238(5):140–152 (Offprint 1388).

Swanson, C. P., and P. L. Webster. 1985. *The Cell.* 5th ed. Prentice-Hall.

EXERCISE 3

Subcellular Structure and Function

A basic problem in microscopy is the limited resolution of light microscopes. For example, no subcellular structures (mitochondria, Golgi complex, or other organelles) that are closer together than 0.1 μm can be clearly seen through a light microscope. Cell membranes and other subcellular organelles are indeed smaller or closer together than this.

Some 40 years ago, it became apparent that microscopes could be built that would use electrons rather than light as a source of illumination. Theoretically the resolution of such an instrument could be at least 100,000 times greater than that of a light microscope since, under high voltage conditions, electrons have a wavelength of 0.005 nm as compared to the 500 nm wavelength of visible light. Thus, the development of the electron microscope has significantly enhanced our understanding of the subcellular structure of cells.

In Exercise 2, you were introduced to the differences between plant and animal cells as viewed with the light microscope, and you examined a number of different cells to become acquainted with a diversity of cell types. In this exercise, you will become familiar with transmission and scanning electron microscopes, the structure and function of some subcellular components, and how to determine the dimensions of objects seen with the electron microscope.

A. SUBCELLULAR ORGANIZATION

Modern microscopic procedures have shown that eukaryotic cells contain numerous highly specialized structures that carry out a variety of activities. These activities include acquiring and assimilating nutrients, eliminating wastes, synthesizing new cellular materials, moving and reproducing. All cells have an internal structure that includes organelles specialized to carry out these various functions. It is important to note that the cell is not a random assortment of parts but a highly structured and integrated entity (Fig. 3-1).

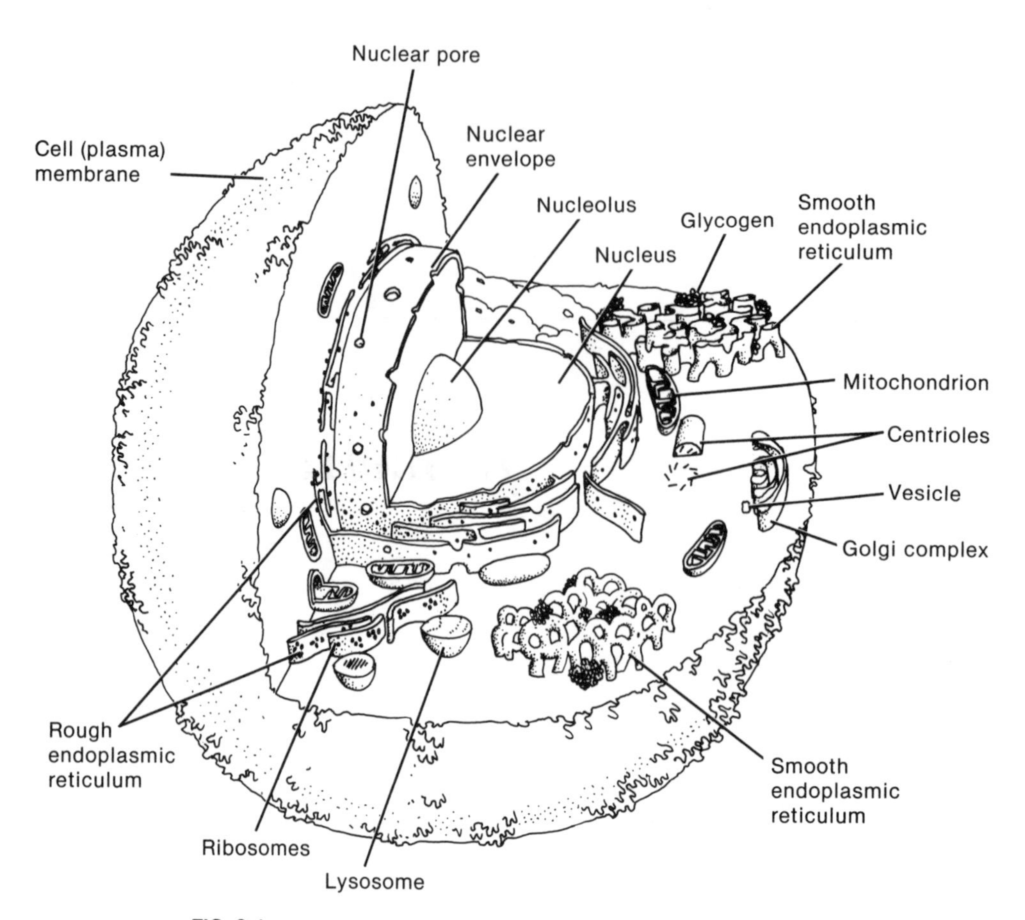

FIG. 3-1
Composite cell as interpreted from electron microscopic observations.

1. The Cell Membrane

Cells exist as separate and distinct entities because they are surrounded by a **cell membrane** that regulates the movement of materials into and out of the cell. The cell (or plasma) membrane, the structure of which cannot be seen through the light microscope, is approximately 6–9 nm thick and appears as a thin double line through the transmission electron microscope (Fig. 3-2A). In the **fluid mosaic model,** the primary framework of all membranes is provided by a lipid bilayer of cholesterol and phosphoglyceride molecules positioned so that their hydrophobic ends are facing inward as diagrammed in Fig. 3-2B. Embedded or attached to this lipid bilayer are a variety of proteins, which can move within the layer because the lipid layer is in a fluid state. In addition, short carbohydrate chains are attached to some of the protein and phosphoglyceride molecules that face the external environment; other proteins are attached to the side of the membrane facing the cytoplasm. All membranes found in the cell have the same basic structure, although there may be regional differences in the types of lipids, proteins, and carbohydrates they contain. These differences are important because they confer unique properties on different cells and organelles within the cell, which are correlated with differences in membrane function.

Cell membranes of prokaryotic cells have essentially the same structure as eukaryotic cells, except that some do not have cholesterol in the lipid bilayer.

2. Nucleus

The **nucleus,** in eukaryotic cells, is a large, usually spherical structure. It is bounded by two lipoprotein membranes that together constitute the **nuclear envelope** (Fig. 3-3). At frequent intervals the membranes are fused together to create pores through which materials pass between the nucleus and the cytoplasm.

The nucleus performs two critical functions. It carries the genetic information that determines the

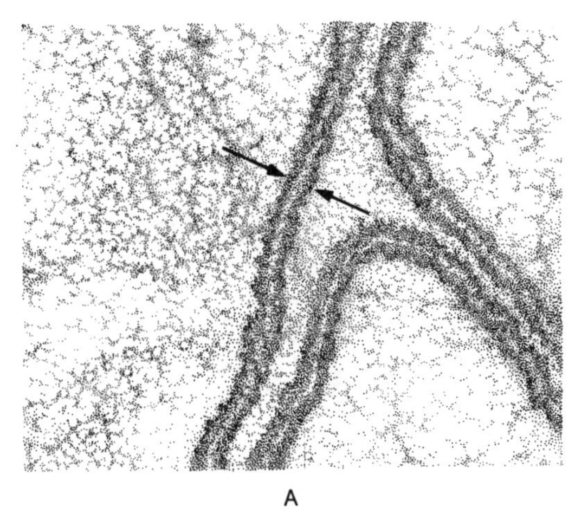

A

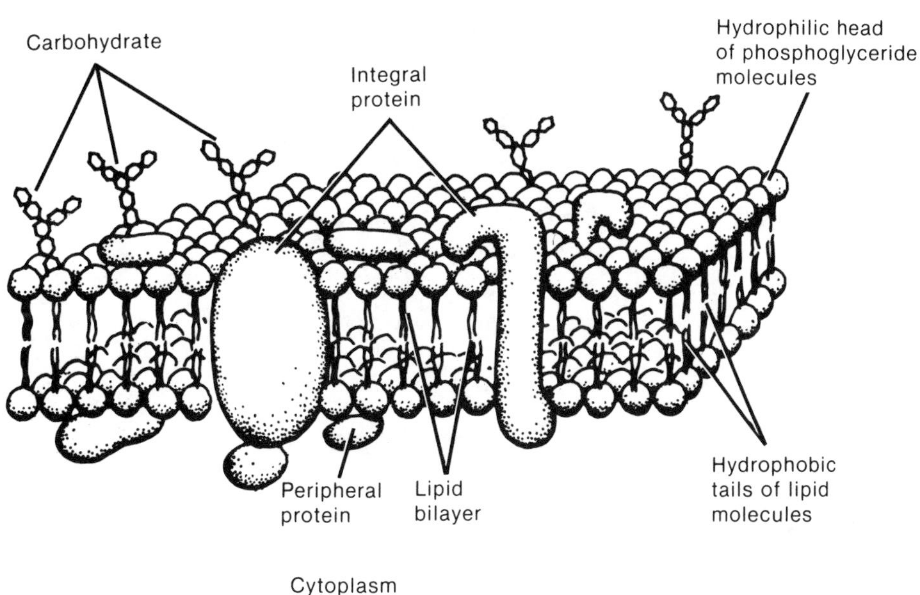

B

FIG. 3-2
(A) Transmission electron micrograph of plasma membrane consisting of two dark layers representing the hydrophilic phosphoglyceride heads of lipid molecules and a light layer representing the hydrophobic tails of lipid molecules. (B) Fluid mosiac model of the plasma membrane showing orientations of the hydrophilic and hydrophobic ends of phosphoglyceride molecules, integral proteins (those that penetrate the bilipid layer), peripheral proteins (those attached to surface of the lipid layer), and carbohydrate chains (that serve as receptor sites for hormones and antibodies).

direction a particular cell develops, and it continually influences the activities of the cell to ensure that the various complex molecules required by the cell are synthesized in the amount and kind needed.

3. Endoplasmic Reticulum and Ribosomes

Internal to the cell membrane, the cell contains a complex, three-dimensional canalicular system that extends throughout most of the cytoplasm. This is the **endoplasmic reticulum,** or the ER, as it is frequently called (Fig. 3-4).

The ER has the same membrane structure as the cell membrane and frequently is continuous with the cell and nuclear membranes.

The ER is made up of two types of membranes: rough and smooth. The rough ER is located in cells active in protein synthesis and involves structures called **ribosomes,** some of which are associated with the outer surfaces of the ER (Fig. 3-4). Large numbers of ribosomes, which are composed of pro-

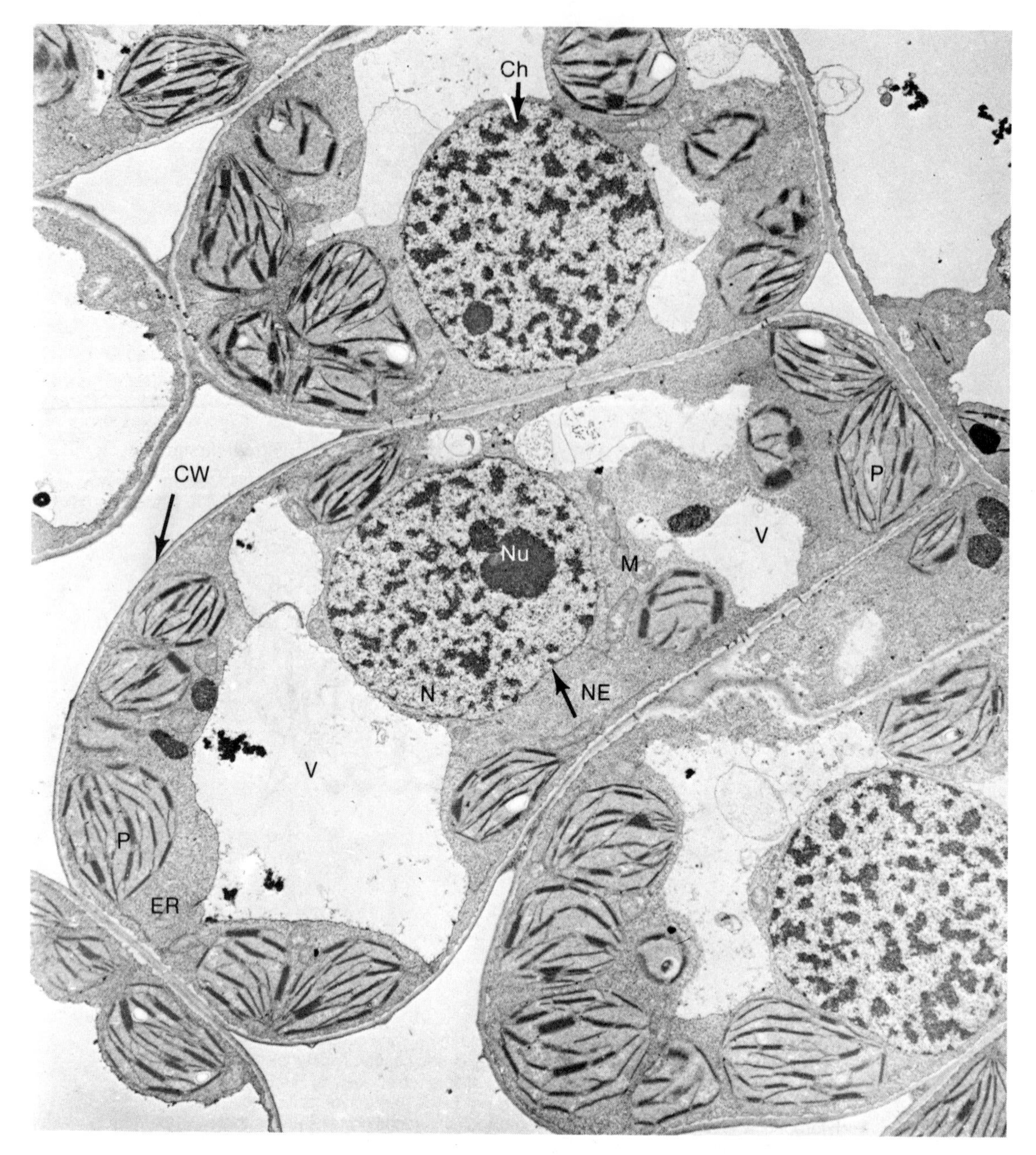

FIG. 3-3
Structure of a higher plant cell: N, nucleus; Nu, nucleolus; NE, nuclear envelope; Ch, chromatin; CW, cell wall; V, vacuole; P, plastids; M, mitochondrion; ER, endoplasmic reticulum. (Transmission electron micrograph courtesy of Dr. Eugene L. Vigil, University of Maryland.)

tein and ribonucleic acid (RNA), give the rough appearance to the ER. Smooth ER lacks the extensive accumulations of ribosomes.

It has been suggested that proteins, synthesized by the ribosomes of the rough ER, are released into the channels of the endoplasmic reticulum, stored, then transported to other parts of the cell or to the outside.

4. Mitochondria

Mitochondria are subcellular structures that participate in cell respiration and contain enzymes involved in the Krebs cycle. Fine structure (i.e., electron microscopic) analysis of these organelles shows them to consist of two membranes: an outer membrane and an inner membrane that invaginates

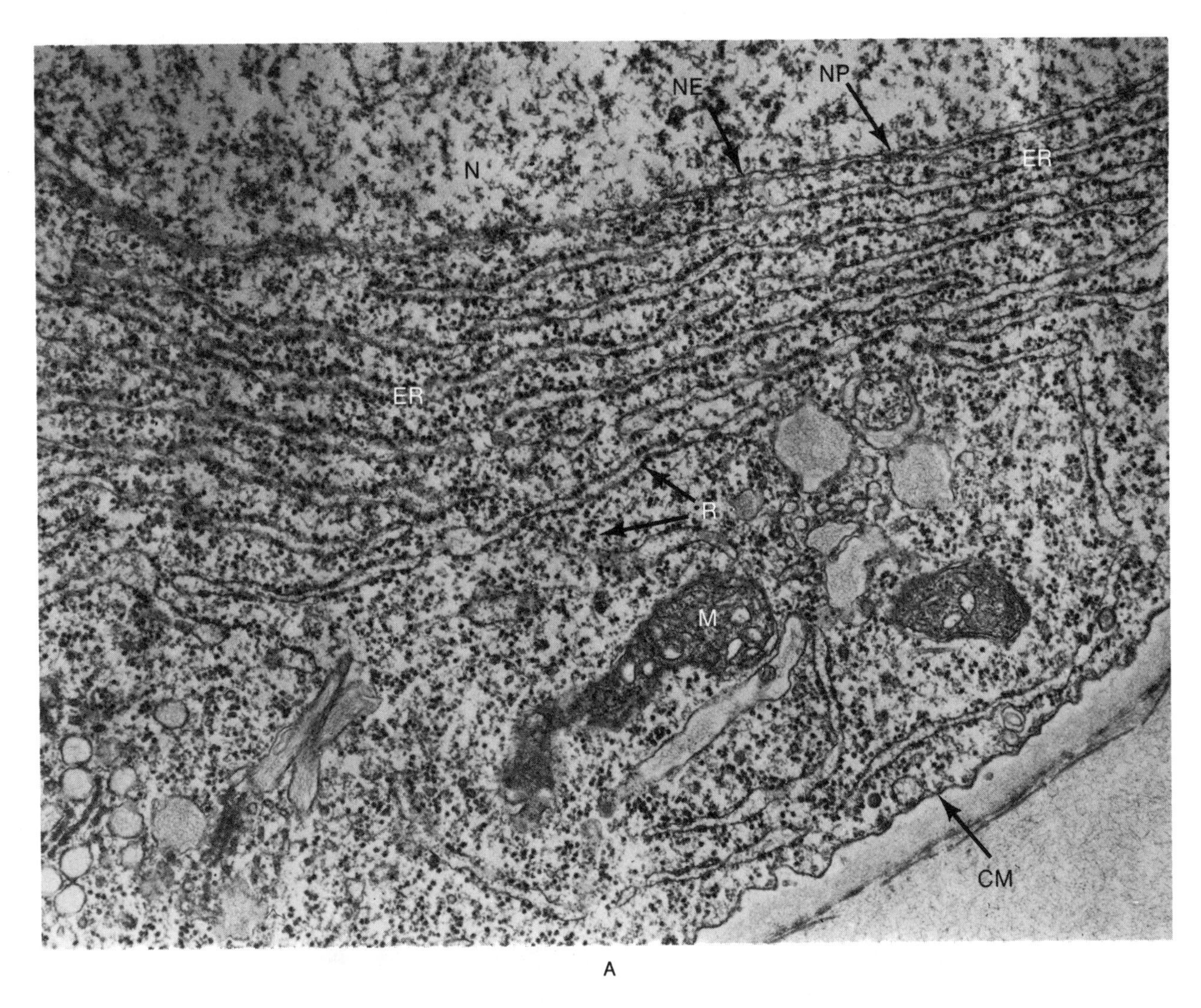

A

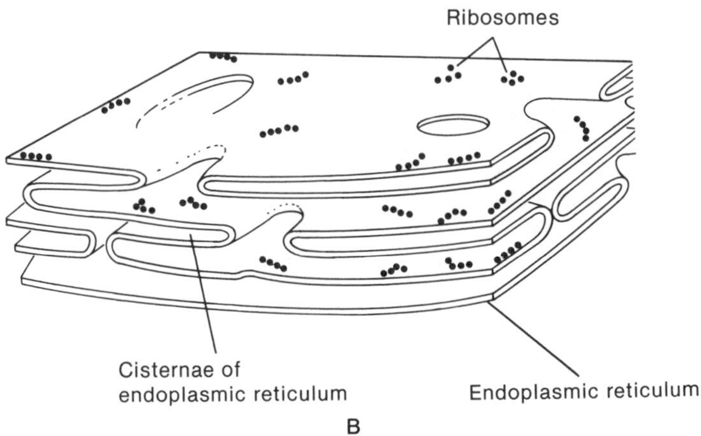

B

FIG. 3-4
(A) Transmission electron micrograph showing cytoplasmic organization of cells: N, nucleus; NP, nuclear pore; NE, nuclear envelope; ER, endoplasmic reticulum; M, mitochondrion; R, ribosome; CM, cell membrane. (Electron micrograph courtesy of Dr. Eugene L. Vigil, University of Maryland.) (B) The structural organization of the endoplasmic reticulum.

(folds in upon itself) to the interior to form **cristae** (Fig. 3-5). The space between the membranes is filled with a watery fluid. The inner and outer membranes have thousands of small particles attached to their surfaces. Those on the inner membrane are attached by small stalks. These particles are the elementary units that carry out the various chemical activities peculiar to the mitochondrion.

The stalkless particles on the outer surface of the outer membrane carry out the various oxidative reactions that provide electrons to the interior of the mitochondrion. The stalked particles on the surface

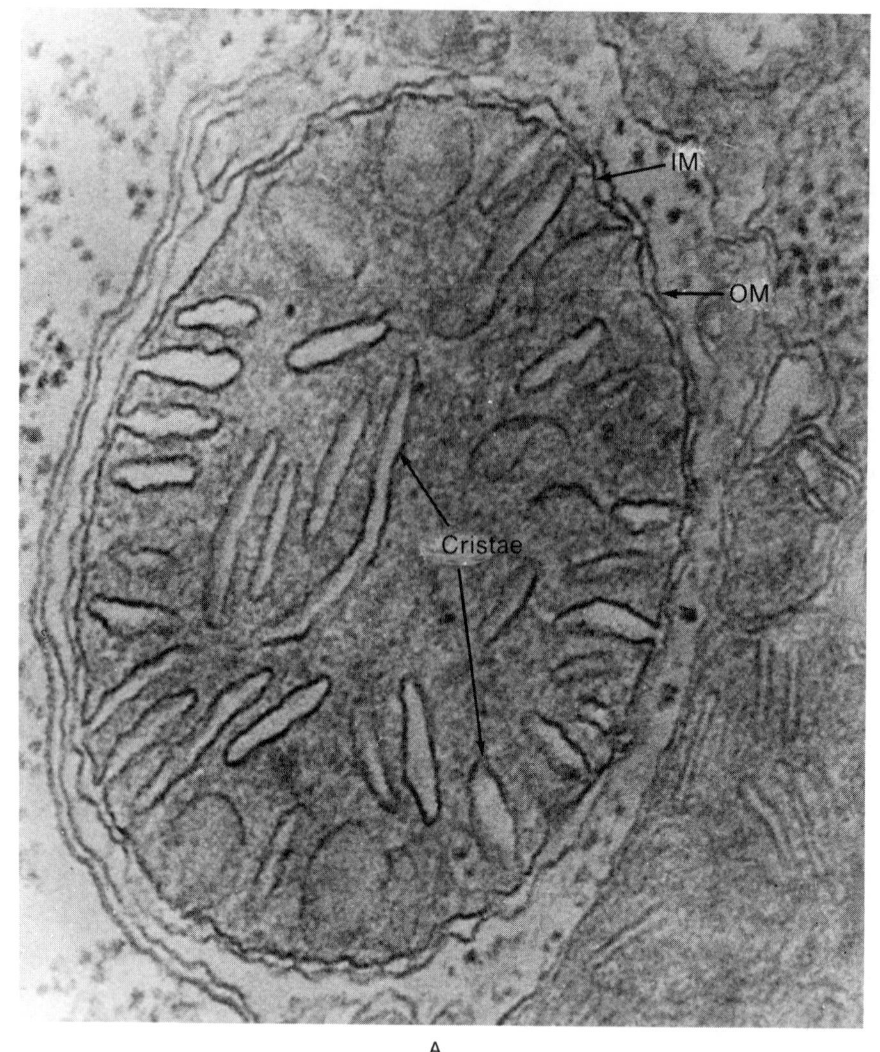

A

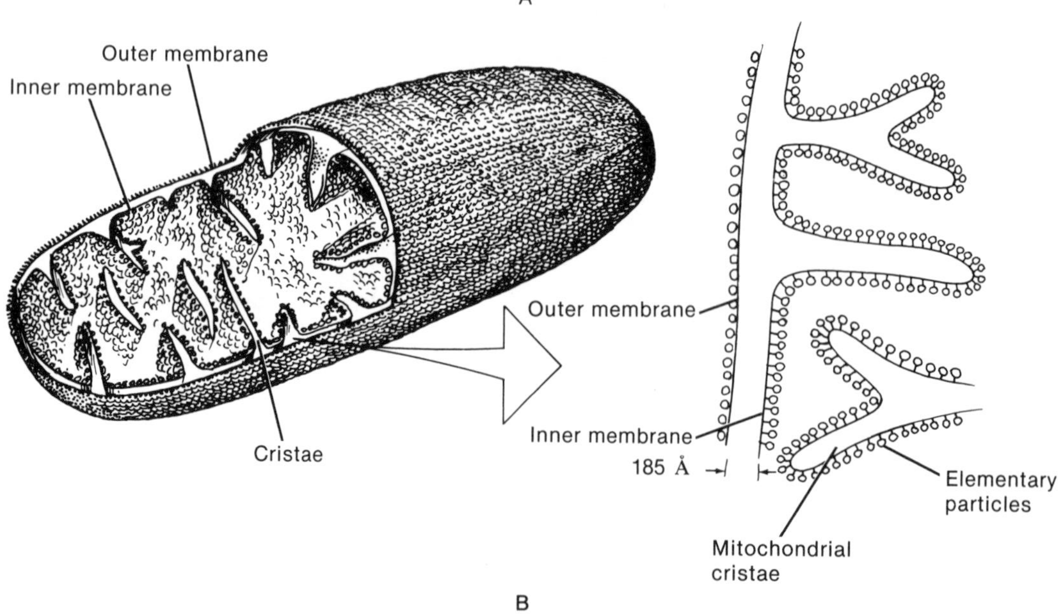

B

FIG. 3-5
(A) Fine structure of a mitochondrion: IM, inner membrane; OM, outer membrane; cristae. (B) The structural organization of mitochondria showing arrangement of elementary particles, stalked on inner membrane surface and stalkless on outer membrane surface.

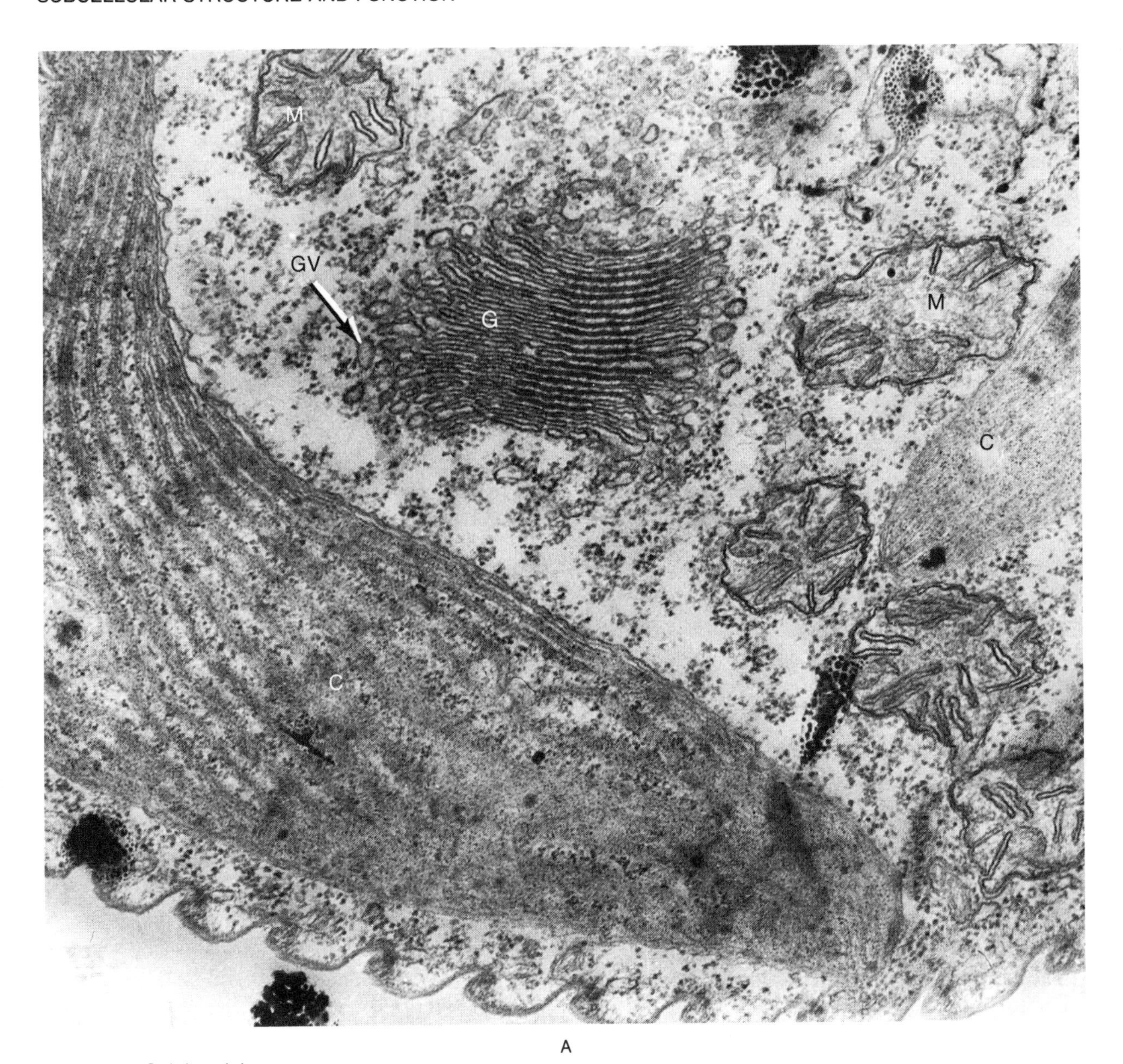

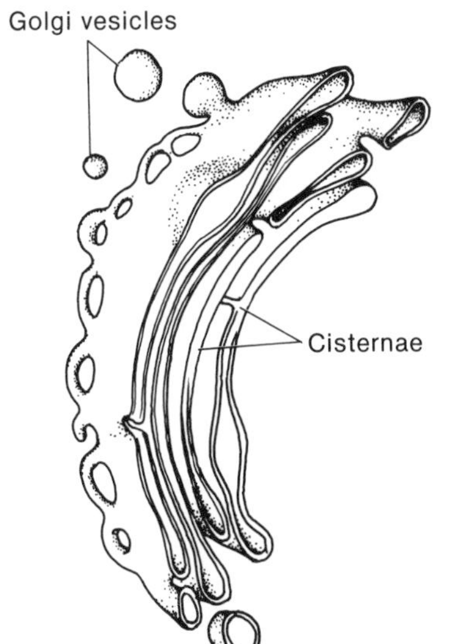

FIG. 3-6
(A) Transmission electron micrograph of the alga *Euglena* showing G, Golgi complex; GV, Golgi vesicle; M, mitochondrion; and C, chloroplast. (B) Diagram of the structural organization of the Golgi complex.

of the inner membrane are involved in the sequential transfer in electrons along a chain of molecules and ultimately synthesize adenosine triphosphate (ATP) through a process called **oxidative phosphorylation.**

5. Golgi Complex

The Golgi complex is another system of membranes found in both plant (called dictyosomes) and animal cells. As shown in Fig. 3-6, it consists of parallel flattened sacs, the **cisternae.** At the margins, the cisternae form spherical vesicles by a "pinching off" process. The Golgi complex receives vesicles formed from the endoplasmic reticulum, modifies the membranes of these vesicles, and further processes and distributes their contents to other parts of the cell, particularly the cell surface. The Golgi complex is the packaging and distribution center of the cell. Indeed, the final assembly of the various proteins and carbohydrates associated with the surface of cell membranes takes place in the Golgi complex.

B. TRANSMISSION ELECTRON MICROSCOPE

The transmission electron microscope (TEM) is shown in Fig. 3-7. The microscope proper consists of a tall, central column. The remainder consists of various electronic equipment including a beam detector and cathode ray viewing screen, vacuum pumps, and electrical and plumbing components.

The TEM is essentially a vertically positioned television tube; the electron gun (or source) is at the top and the viewing screen is at the bottom. Basically, the electrons emerge from the hot filament of an electron gun and are accelerated by high voltages down the tube to the viewing screen. The column must be evacuated of air so that the electrons pass without interference. Various magnets on the column act like the adjustment knobs on the light microscope to focus the streaming electrons onto the specimen being examined.

When the electrons strike the screen, it glows. If a specimen is properly prepared and placed in a holder in the column, an image of this specimen will appear on the screen as dark and light areas. The dark areas correspond to electron opaque regions of the specimen, since electrons are deflected and do not pass through the specimen. The light areas represent the electron transparent areas of the specimen. Normally, images that are viewed on the screen are observed by photographing them. This is done by moving the screen out of the way and allowing the electrons to strike a photographic film. This technique produces an image like that shown in Fig. 3-8 or 3-9. You can see the increase in resolution afforded by such an instrument. You must keep in mind, however, that TEM photographs are only taken of very thin sections of material. A structure that does not appear in the photograph may still be present in the cell; when the particular section was cut, the structure may just not have been in that section. This is somewhat similar to slicing hard salami, where not all slices will contain a peppercorn. This concept must be remembered when one is analyzing **electron micrographs** for various organelles associated with cells.

C. SCANNING ELECTRON MICROSCOPE

If we consider the transmission electron microscope as being analagous to the compound light microscope, then we can consider the scanning electron microscope (SEM) as the counterpart of the dissecting microscope. In preparing materials to view in the SEM, whole, not sectioned, materials are used because the image is formed by reflected rather than transmitted electrons. As shown in Fig. 3-10 and 3-11, the SEM images are three-dimensional. What advantages are provided by the SEM over the TEM?

____

____

____

D. INTERPRETATION AND MEASUREMENT IN ELECTRON MICROGRAPHS

1. Using Magnification of Micrograph

1. Examine the micrographs in Figs. 3-8 and 3-9. How many and what types of cells are visible?

____

____

____

Look for different kinds of structures in the cells. Find nuclei, mitochondria, endoplasmic reticulum, Golgi complex, and cell membranes. If you cannot

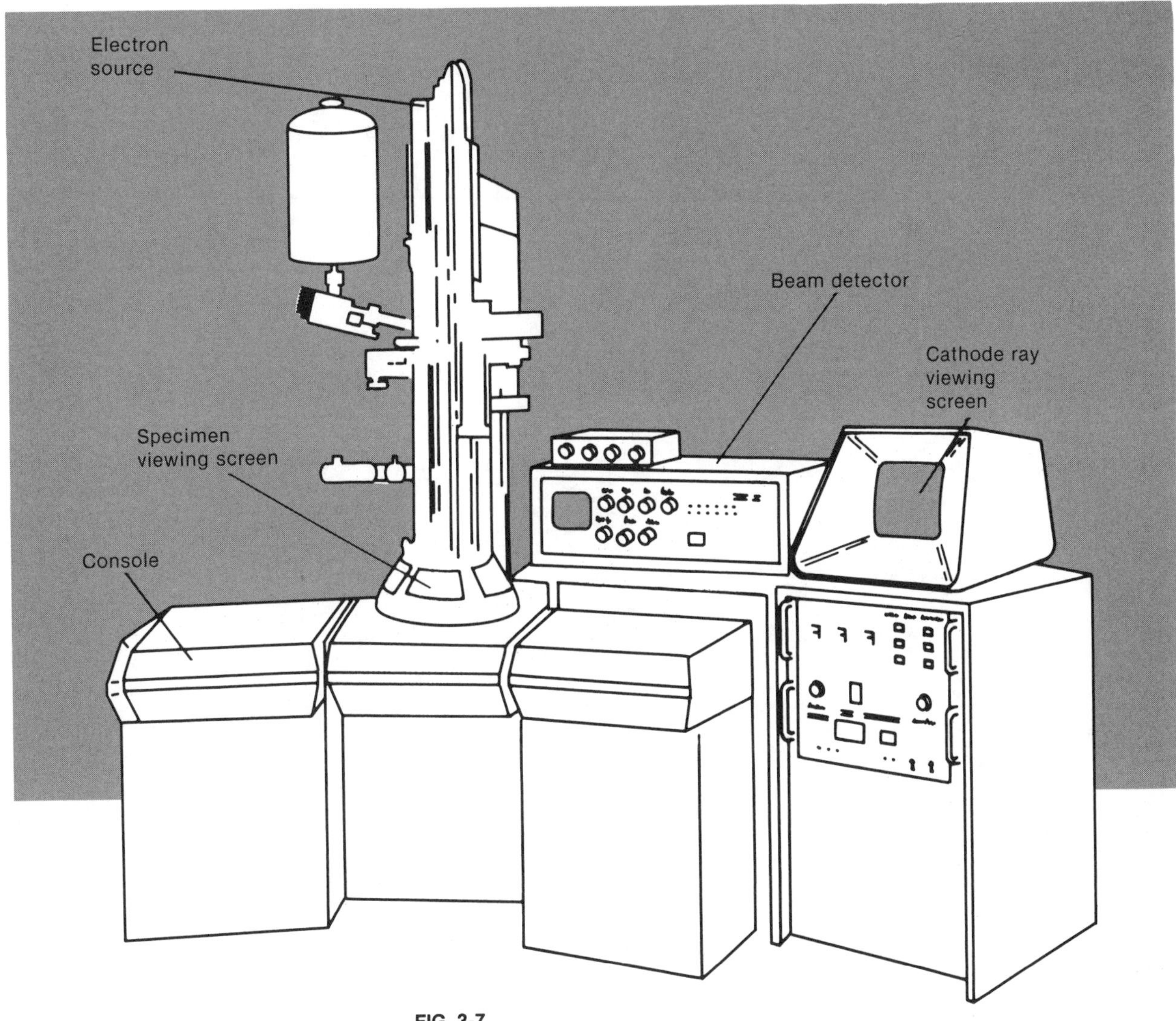

FIG. 3-7
Transmission electron microscope (TEM).

find a particular structure, can you assume that the cell doesn't contain it? Explain.

____

____

____

2. Note that Fig. 3-8 has a magnification of 11,390 × and Fig. 3-9, 8545 ×. In Fig. 3-8, measure the longest dimension of the macrophage in millimeters (mm) __________ and then convert this figure to micrometers (μm) __________. To determine the actual size of this cell use the following formula:

$$\text{Actual size} = \frac{\text{measured size of structure } (\mu\text{m})}{\text{magnification of micrograph}}$$

What is the longest dimension of the macrophage in micrometers?

____

If the macrophage in this micrograph was cut into sections 40 nm thick, how many sections would you get if you cut through the entire cell?

____

What advantages are there to examining more than one section of a cell or tissue?

____

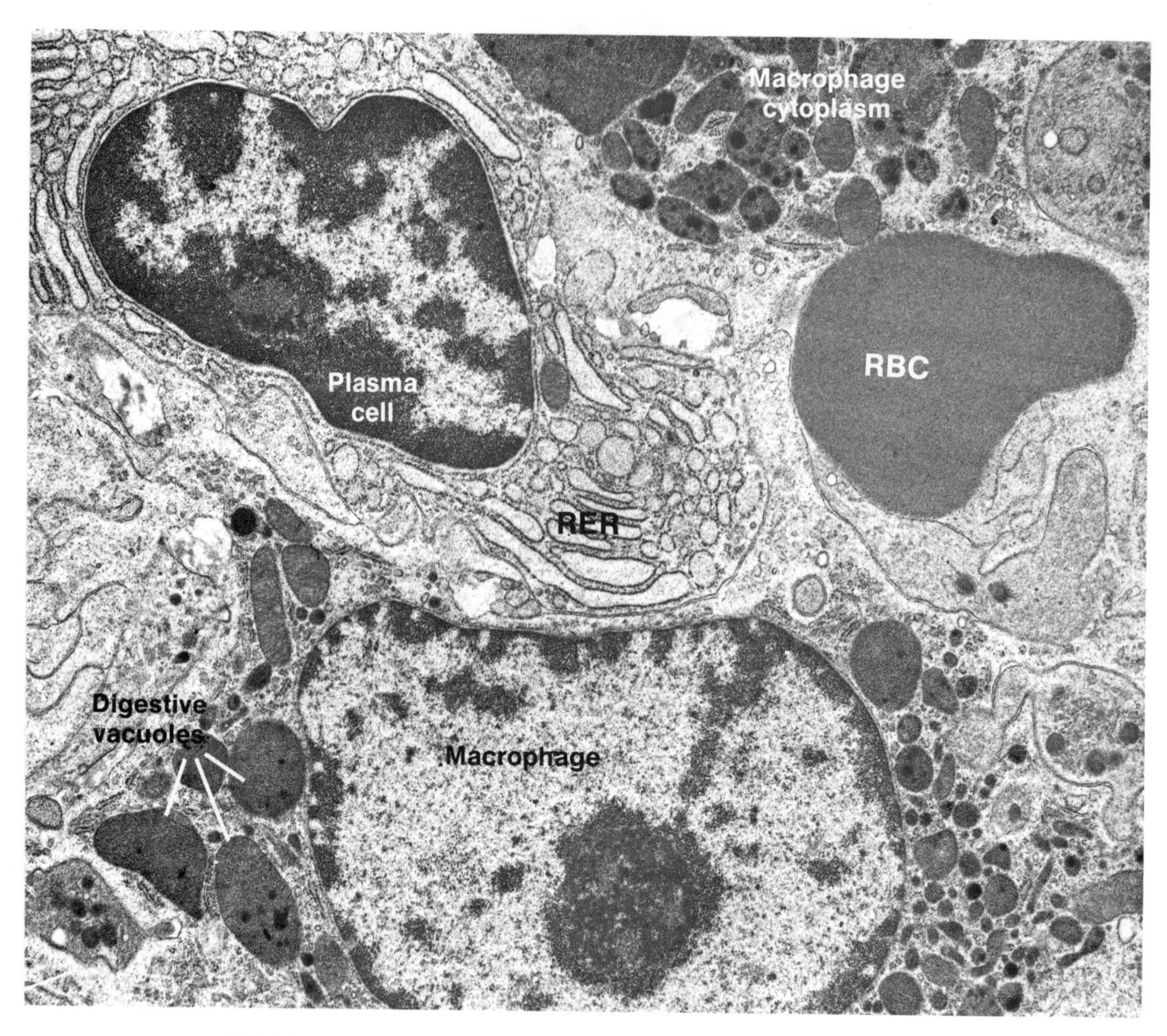

FIG. 3-8
Transmission electron micrograph showing a macrophage, plasma cell with rough endoplasmic reticulum (RER), and red blood cell (RBC). Digestive vacuoles in the macrophage function in hydrolysis of material taken into cell by phagocytosis (× 11,390). (From *Tissues and Organs: A Text-Atlas of Scanning Electron Microscopy* by R. G. Kessel and R. H. Kardon. W. H. Freeman and Company. Copyright © 1979.)

3. Measure the following cellular structures and determine their size relationships:

a. From Fig. 3-8
size of a plasma cell (longest dimension)

__________ μm

size of red blood cell (longest dimension)

__________ μm

thickness of a single endoplasmic reticulum

__________ μm

b. From Fig. 3-9
length of a mitochondrion[1]

__________ μm

[1] Why do you think some mitochondria in these pictures look round while others are elongated?

____

____

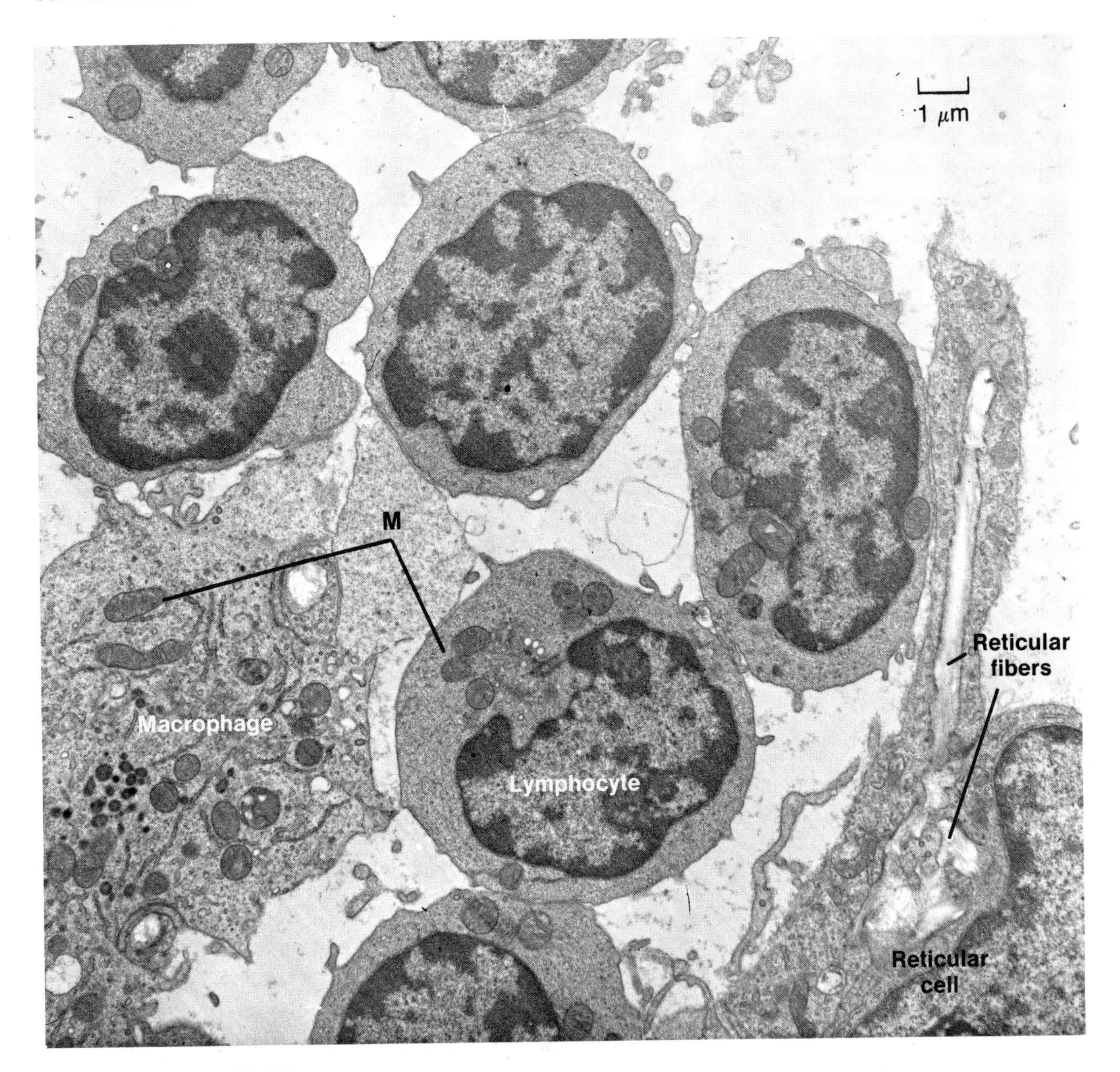

FIG. 3-9
Transmission electron micrograph through a lymph node showing several lymphocytes, macrophages with mitochondria (M), and a reticular cell with reticular fibers. (× 8545) (From *Tissues and Organs: A Text-Atlas of Scanning Electron Microscopy* by R. G. Kessel and R. H. Kardon. W. H. Freeman and Company. Copyright © 1979.)

size of lymphocyte nucleus

__________ μm

size of the plasma membrane

__________ μm

c. From Fig. 3-10
size of a bone lacuna (longest dimension)

__________ μm

size of Haversian canal __________ μm

size of canaliculus (length)

__________ μm

d. From Fig. 3-11
sizes of fungiform papillae

__________ μm; __________ μm

size of circumvallate papillae

__________ μm

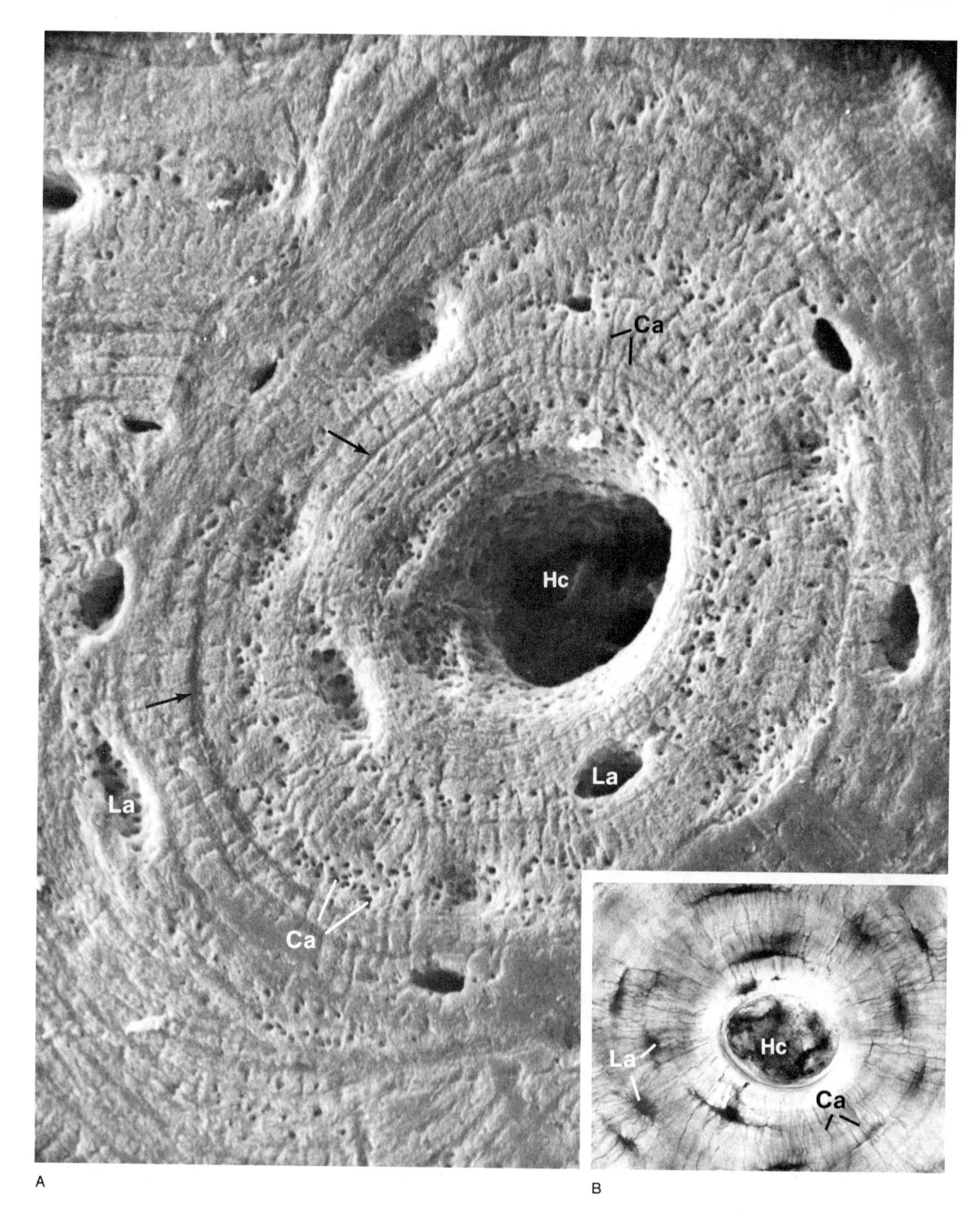

FIG. 3-10
(A) Scanning electron micrograph of compact bone. (× 1345). (B) Light microscope photograph of a similar specimen. La, lacunae or spaces in which bone cells, called osteocytes, are found; Ca, canaliculi (small channels between lacunae); Hc, Haversian canal containing blood vessels. (From *Tissues and Organs: A Text-Atlas of Scanning Electron Microscopy* by R. G. Kessel and R. H. Kardon. W. H. Freeman and Company. Copyright © 1979.)

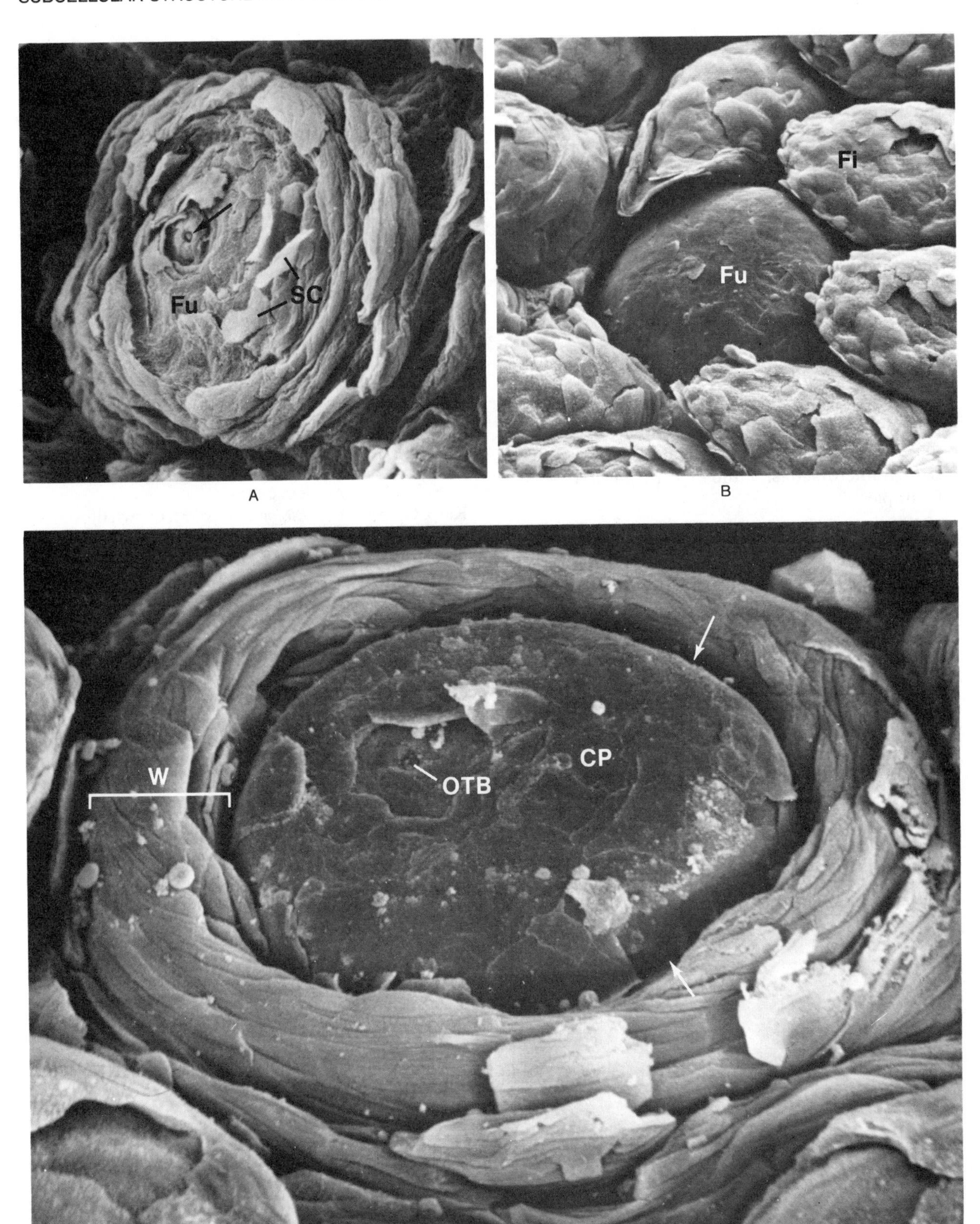

FIG. 3-11
Scanning electron micrographs of papillae, found on the surface of the tongue, that have taste buds associated with them. (A) and (B) are fungiform (Fu) papillae; (C) is a circumvallate (CP) papilla: OTB, opening of pore to taste bud; SC, squamous cells; Fi, filiform papillae; scale bar (W) = 1 μm. (Magnifications: (A) × 615; (B) × 245; (C) × 910.) (From *Tissues and Organs: A Text-Atlas of Scanning Electron Microscopy* by R. G. Kessel and R. H. Kardon. W. H. Freeman and Company. Copyright © 1979.)

2. Using Indicator Scale on Micrograph

In some electron micrographs, a scale indicating the length of 1 μm is often added to the photograph. You can use this indicator scale directly to measure sizes of subcellular structures or, alternatively, you can

1. Measure the size of this scale in millimeters and determine the decimal factor obtained from calculating the number of micrometers equal to 1 millimeter. Thus,

$$\text{Decimal factor} = \frac{1\ \mu\text{m}}{\text{length of indicator scale in mm}}$$

2. Then measure any object in the micrograph in millimeters, and multiply this number by the decimal factor obtained in (1). This will give you the size, in micrometers, of the object measured. Example:

a. length of an indicator scale = 10mm

$$\text{Decimal factor} = \frac{1\ \mu\text{m}}{10} = 0.10$$

b. length of mitochondrion (in millimeters) as measured in micrograph = 14 mm

$$14\ \text{mm} \times 0.10 = 1.4\ \mu\text{m (length of mitochondrion)}$$

Using the indicator scale in Fig. 3-9, measure the size of some of the cells, nuclei, and mitochondria.

REFERENCES

Alberts, B., D. Bray, J. Lewis, M. Raff, K. Roberts, and J. D. Watson. 1983. *Molecular Biology of the Cell.* Garland.

Capaldi, R. A. 1974. A Dynamic Model of Cell Membranes. *Scientific American* 230(3):26–33. (Offprint 1292). *Scientific American* Offprints are available from W. H. Freeman and Company, 41 Madison Avenue, New York 10010, and 20 Beaumont Street, Oxford OX1 2NQ, England. Please order by number.

DeRobertis, E. D. P., and E. M. F. DeRobertis, Jr. 1980. *Cell and Molecular Biology.* 7th ed. Saunders.

Everhart, T. E., and T. L. Hayes. 1972. The Scanning Electron Microscope. *Scientific American* 226:54–67.

Fawcett, D. W. 1981. *An Atlas of Fine Structure: The Cell.* 2d ed. Saunders.

Glauert, A. M. 1972. *Practical Methods in Electron Microscopy.* Vol. 1. Elsevier North-Holland.

Glauert, A. M. 1974. The High-Voltage Electron Microscope in Biology. *J. Cell Biol.* 63:717–748.

Goodenough, U. W., and R. P. Levine. 1970. The Genetic Activity of Mitochondria and Chloroplasts. *Scientific American* 223(5):22–29 (Offprint 1203).

Green, D. E., and H. Baum. 1970. *Energy and the Mitochondrion.* Academic Press.

Karp, G. 1984. *Cell Biology.* 2d ed. McGraw-Hill.

Kessel, R. G., and R. H. Kardon. 1979. *Tissues and Organs: A Text-Atlas of Scanning Electron Microscopy.* Freeman.

Lodish, H. F., and J. E. Rothman. 1979. The Assembly of Cell Membranes. *Scientific American* 249(1):48–63 (Offprint 1415).

Neutra, M., and C. P. Leblond. 1969. The Golgi Apparatus. *Scientific American* 220(2):100–107 (Offprint 1134).

Nomura, M. 1969. Ribosomes. *Scientific American* 221(4):28–35 (Offprint 1157).

Racker, E. 1968. The Membrane of the Mitochondrion. *Scientific American* 218(2):32–39 (Offprint 1101).

Staehelin, L. A., and B. E. Hull. 1978. Junctions Between Living Cells. *Scientific American* 238(5):140–152 (Offprint 1388).

Swanson, C. P., and P. L. Webster. 1985. *The Cell.* 5th ed. Prentice-Hall.

Wolfe, S. L. 1981. *Biology of the Cell.* 2d ed. Wadsworth.

EXERCISE 4

Cellular Reproduction

A multicellular organism typically begins life as a **zygote,** which is formed from the union of a sperm and an egg (the **gametes**). The egg and sperm, although usually unequal in size, give an equal number of chromosomes to the newly developing organism: each contributes the **haploid** number (half complement) of chromosomes. The zygote, therefore, contains a **diploid** number of chromosomes—a haploid set from each parent. For example, the diploid complement of the human zygote (and subsequently all other body cells) is 46 chromosomes. The haploid complement for the egg and sperm is 23 chromosomes. The reduction from the diploid condition to the haploid condition is brought about by a type of cell division called **meiosis,** which occurs during the formation of the gametes. After fertilization, the zygote, through repeated cell division, called **mitosis,** gives rise to all the cells that make up the organism.

In this exercise, you will compare and contrast mitosis and meiosis.

A. MITOSIS

The tremendous diversity of form and function that cells assume is even more remarkable when you consider that multicellular organisms begin life as a single fertilized cell, the **zygote.** This cell, through repeated divisions, gives rise to all the cells that make up the organism. The complex series of events that encompasses the life span of an actively dividing cell is termed **mitosis,** or the **mitotic cell cycle** (Fig. 4-1).

The process of mitosis is conveniently divided into four reasonably distinct stages: **prophase, metaphase, anaphase,** and **telophase.** The period between successive mitoses is called **interphase.** However, mitosis does not usually occur in well-defined stages but is a continuous process with each phase blending into the next. The actual replication (doubling) of chromosomes, including the synthesis of deoxyribonucleic acid (DNA), the main structural component of chromosomes, as well as the RNA and protein synthesis essential for mitosis, occurs during

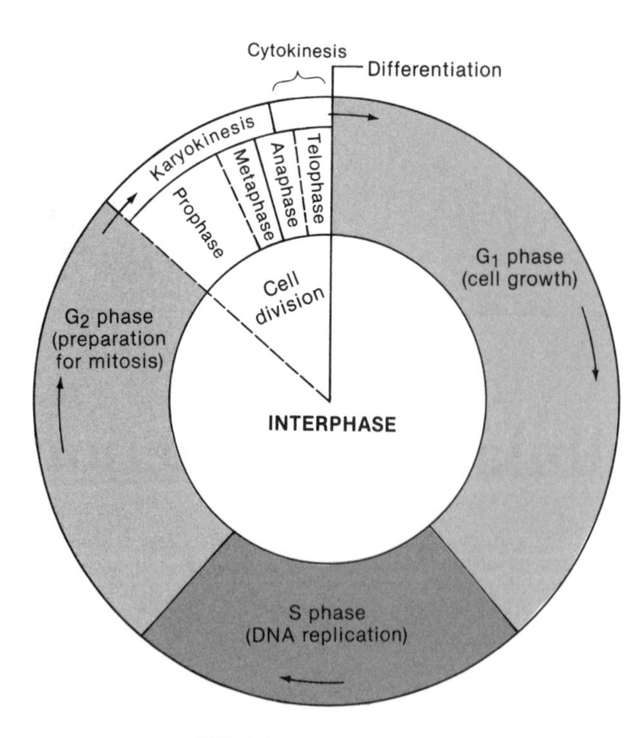

FIG. 4-1
Stages of the mitotic cycle.

interphase. However, the replication of DNA occurs only during a period of interphase called the **S** (for **synthesis**) **phase.** The doubling of DNA during the S phase provides a full complement of DNA for the daughter cells that will result from the next mitotic division. During interphase, there are also two non-mitotic and non-DNA-synthetic phases, called **G** (for **gap**) **phases.** The G phase preceding DNA is called G_1 and represents the period between the end of the telophase of one mitotic division and the beginning of the S phase of the next division. During G_1, a cell may follow a pathway leading to differentiation, rather than continue in active division. This possibility is indicated in Fig. 4-1 by the arrow. The period from the end of the S phase to the beginning of the prophase of the next mitosis is the G_2 phase. During this phase, the structures directly involved with mitosis, such as the spindle fibers, are assembled. The combination of G_1, S, G_2, and M (for mitosis) phases comprise the **cell,** or **mitotic, cycle.**

The time required for one mitotic cycle varies from tissue to tissue and from organism to organism. The actual mitotic process usually occupies only 10% of the total time of the cell cycle. The rest of the time is spent in interphase. It is important to distinguish between several parts of mitosis. Thus, chromosomal duplication and separation (i.e., nuclear division) is termed **karyokinesis.** Division of the cell body (i.e., cytoplasm and its organelles) is termed **cytokinesis.** Cytokinesis and karyokinesis need not occur together.

There are certain structural differences in plant and animal cells; the two vary to some extent also in the process of mitosis. However, in all organisms this process is essentially the same. It is the objective of this exercise to examine the essential steps in mitosis and to characterize the similarities and differences in this process between plant and animal cells.

1. Mitosis in Plant Cells

The onion root tip is one of the most widely used materials for the study of mitosis because it is available in quality, the preparations of the dividing cells are easily made, and the chromosomes are relatively large and few in number, hence easier to study than the cells of many other organisms. There are regions of active cell division in root tips; therefore, the chances are good that within such tissues one can identify every stage in mitosis.

Obtain a slide of onion root tips, and note a series of dark streaks on the slide. Each streak is a very thin longitudinal section through an onion root tip.

Place the slide on the stage of your microscope, and locate one of the sections under low power. It is often possible, under low power, to determine whether a given section shows good mitotic stages. Because each section is very thin, not all will be equally good for study. After your preliminary examination under low power, change to high power, being very careful not to break the slide. Keep in mind the sequence in which the different stages occur, but do not try to find them in sequence. Thus, if you happen to find an anaphase first, study it before proceeding to another stage. Chances are that most of the cells will be in interphase. The next largest number will be in prophase, and only a few will be seen in metaphase, anaphase, and telophase. This is because the cells remain in interphase and prophase longer than in the other stages.

a. Interphase

Erroneously said to be "resting," the interphase cell is actively respiring, undergoing protein and ribonucleic acid (RNA) synthesis, and replicating chromosomes (which includes the synthesis of deoxyribonucleic acid, DNA) in preparation for mitosis (Fig. 4-2A).

b. Prophase

During prophase, the chromosomes become distinguishable in the nucleus. The nuclear membrane breaks down, and the chromosomes become distributed throughout the cytoplasm (Fig. 4-2B, C). At

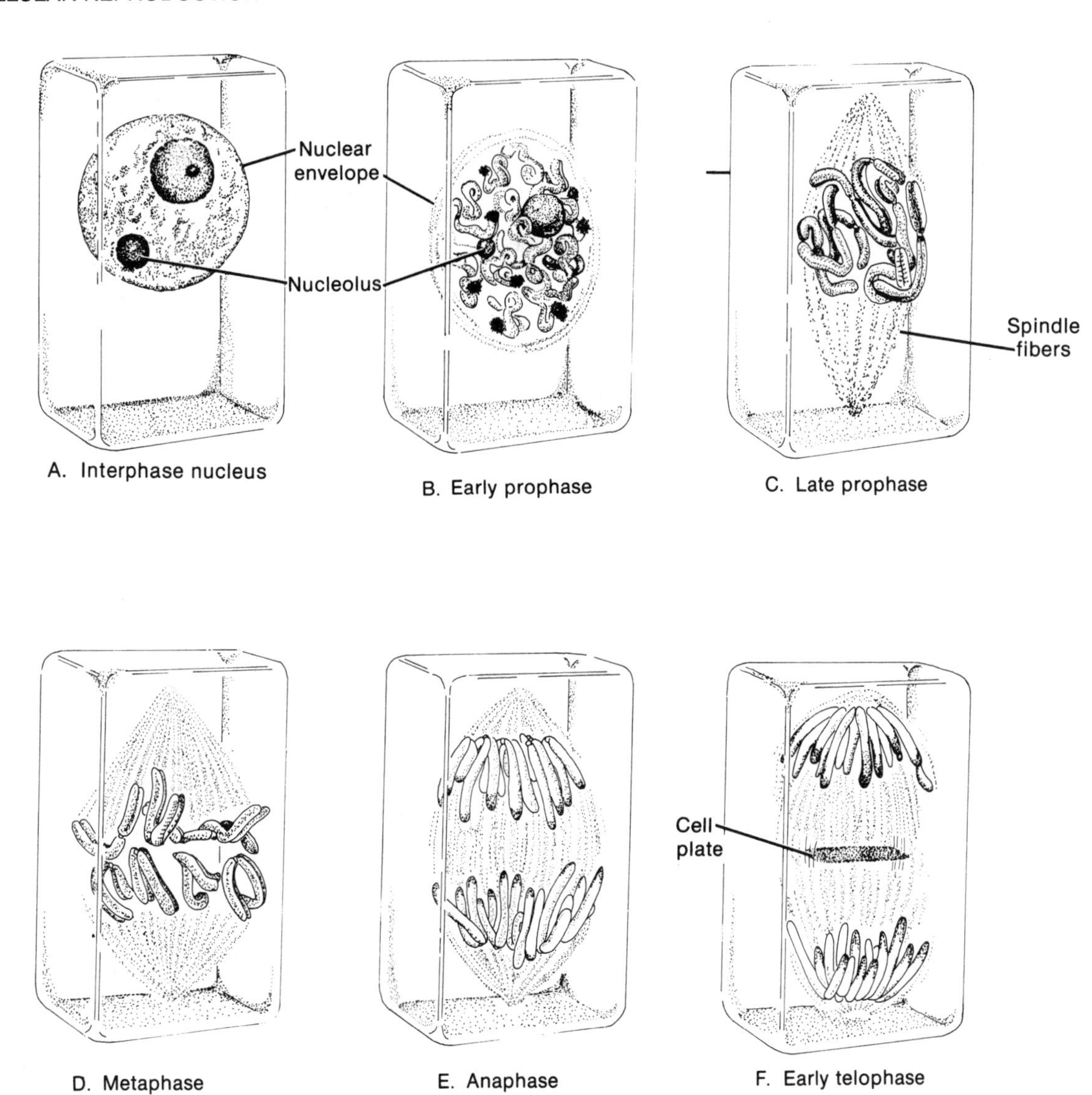

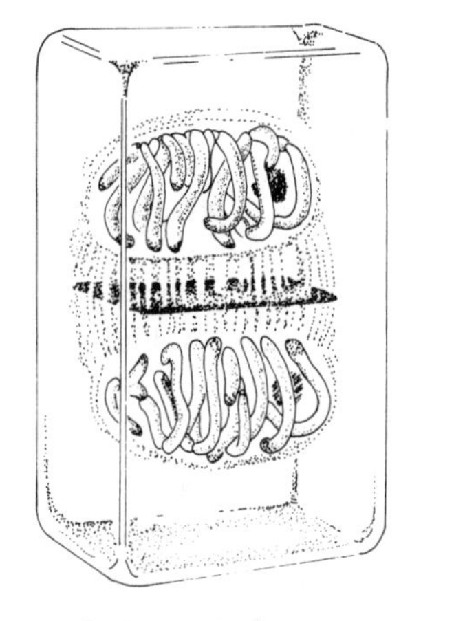

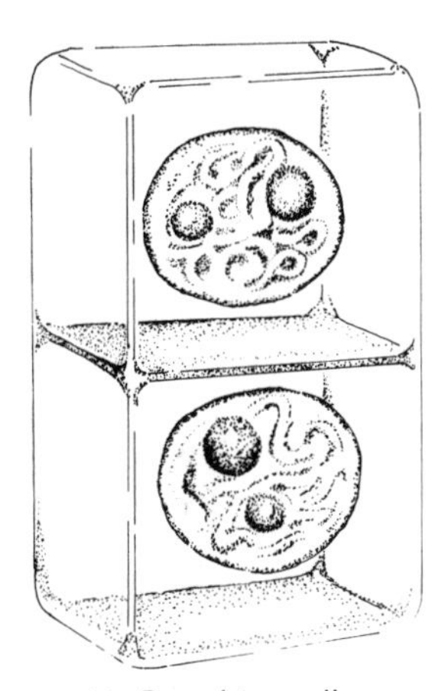

FIG. 4-2
Diagrammatic representation of the mitotic stages in the onion root tip.

this stage in the onion root tip, the chromosomes often appear as a coiled mass. These elongated chromosomes later become condensed into shorter chromosomes and the nuclear membrane disappears. Even at this early stage, each chromosome has doubled, although this will be difficult to see on the slides. Under very high magnifications, it is possible to see that each chromosome is composed of two separate strands, the sister **chromatids.** The chromatids, which are attached to each other in a region called the **centromere,** are identical copies made during DNA replication in the S phase of the previous interphase.

While the chromosomes are undergoing condensation, the apparatus that will sort them out and move them apart is being organized. This is the **spindle,** a system of microtubules that form between the poles of the nucleus. It has been suggested that they form by the aggregation of protein molecules. Under the electron microscope, these fibers appear as fine, straight, hollow microtubules (Fig. 4-3). Although they lengthen and then shorten during mitosis, they do not appear to get thicker or thinner. This suggests that they do not stretch or contract but that new material is added to the fiber or removed from it as the spindle changes shape. In animal cells, the poles of the spindle are attached to organelles called **centrioles.** No centrioles are found in onion cells or those of higher plants. In most cells, as the centrioles migrate to the poles, the nuclear envelope begins to disintegrate.

c. Metaphase

At metaphase, the short, thick, double-stranded chromosomes, each attached by its centromere to a spindle fiber, become arranged near the center of the cell in a plane at right angles to the long axis of the spindle fibers (Fig. 4-2D). The spindle microtubule attachment occurs so that one chromatid of each pair is connected to one pole of the spindle and its sister chromatid to the other pole. By the time the chromosomes are organized along the metaphase plate, the nuclear membrane has completely disintegrated.

d. Anaphase

At the beginning of anaphase, the two members of each of the previously doubled chromosomes separate and move to opposite ends of the cell (Fig. 4-2E). This stage can be recognized in the onion by the two groups of V-shaped chromosomes on opposite sides of the cell. The sharp end of the V is pointed toward the cell wall. The onion has 16 chromosomes; hence it is seldom possible to see all of them at one time.

Reduce the light by adjusting the diaphragm of your microscope, and see if you can find any spindle fibers near the center of the cell. They will appear as very fine lines between the two groups of chromosomes, but they are often not visible in a study of this kind. Anaphase ends when the two sets of single-stranded chromosomes have arrived at opposite ends of the cell.

e. Telophase

Karyokinesis is completed during telophase, and reorganization of the contents of the two daughter cells (cytokinesis) begins. It is often difficult to distinguish between late anaphase and early telophase in the cells of the onion root tip. During telophase, however, a **cell plate,** the first indication that cytokinesis is beginning, starts to form as a fine line across the center of the cell (Fig. 4-2F). When complete, the cell plate will divide the original cell into two daughter cells. In some cells, the plate will be indistinct. As telophase progresses, the nuclei begin to reorganize and the chromosomes uncoil, becoming longer and thinner, the nuclear membrane reforms, and the nucleoli reappear (Fig. 4-2G). Mitosis ends with the reassembly of two interphase nuclei, each with one complete set of single-stranded chromosomes (Fig. 4-2H).

The daughter cells resulting from mitotic division have the same number and kinds of chromosomes as the original cell from which they came. Thus, in the onion each daughter cell has 16 chromosomes, just as the original cell had. This is not as evident in the slides you are studying, because specially prepared slides and considerable experience are necessary before this many chromosomes can be accurately counted.

One difference in mitosis between most plant and animal cells should be pointed out at this time. When plant cells divide, a cell plate forms, as we saw in the onion root tip. Animal cells do not have a cell wall but are surrounded only by the plasma membrane. When most animal cells divide, a constriction is formed, by a process called **furrowing,** near the center of the original cell. This constriction, which goes completely around the cell, deepens until the cell cleaves into two daughter cells.

2. Mitosis in Animal Cells

Mitosis is readily observed in animal cells by examining a prepared slide of a whitefish blastula—the early developmental stage formed by successive mitotic divisions after the egg has been fertilized by the sperm.

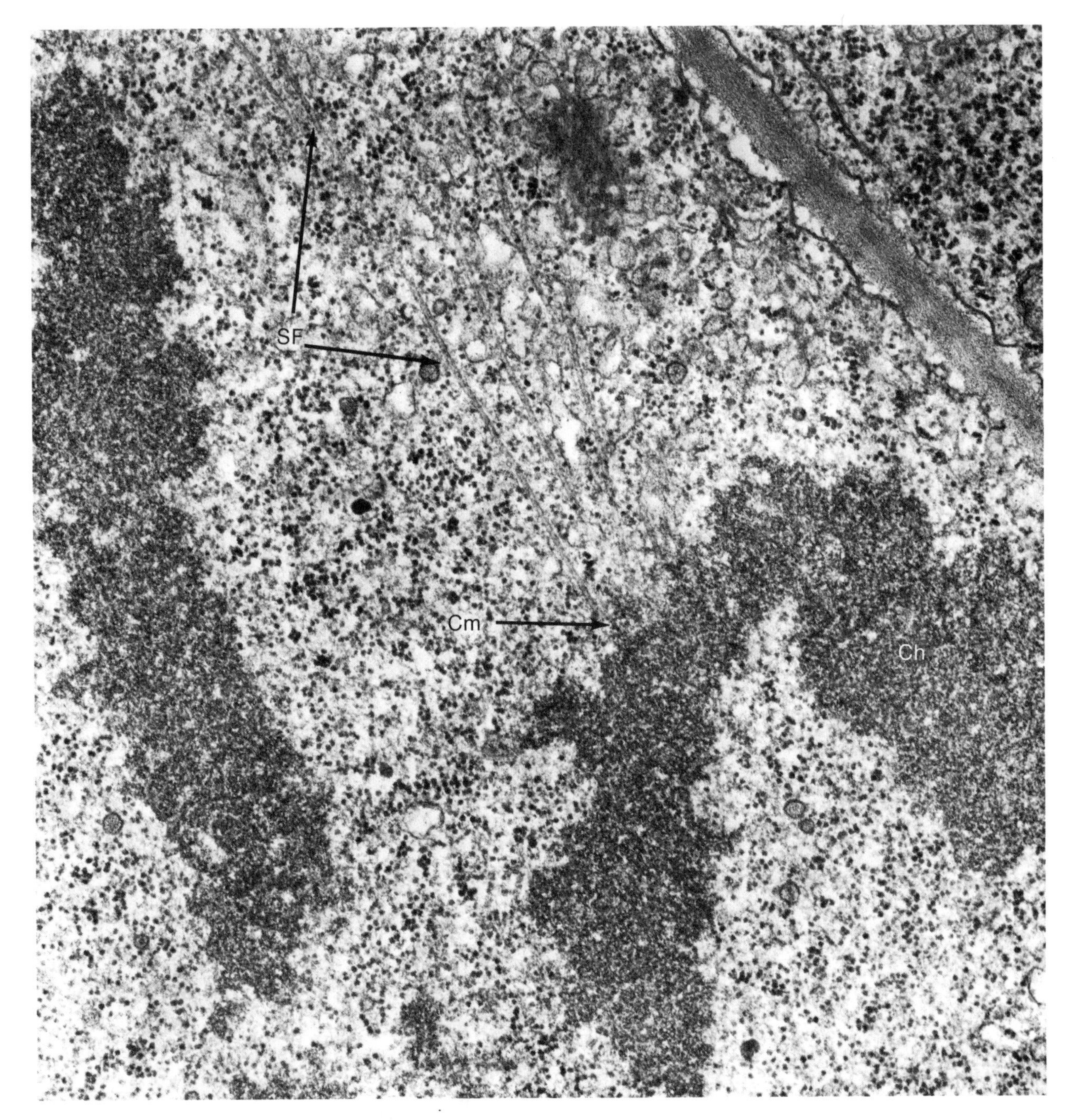

FIG. 4-3
Spindle fibers (SF) attached to a chromosome (Ch) at the centromere (Cm). Note the tubular structure of the spindle fibers. (Electron micrograph courtesy of Dr. Eugene L. Vigil, University of Maryland.)

a. Interphase

Locate interphase cells, which are characterized by having a distinct nucleus bounded by a nuclear envelope (Fig. 4-4A). The nucleolus should be readily identifiable. Immediately adjacent to the nuclear envelope is a cytoplasmic organelle sometimes referred to as the **cell center,** or **centrosome,** which contains the centrioles.

b. Prophase

Karyokinesis begins with prophase. The chromosomes of early prophase, which have already doubled longitudinally during the S phase of the previous interphase, appear as thin, coiled filaments (Fig. 4-4B). As prophase progresses, the chromosomes shorten and thicken (Fig. 4-C, D). Each doubled chromosome consists of two sister **chromatids,**

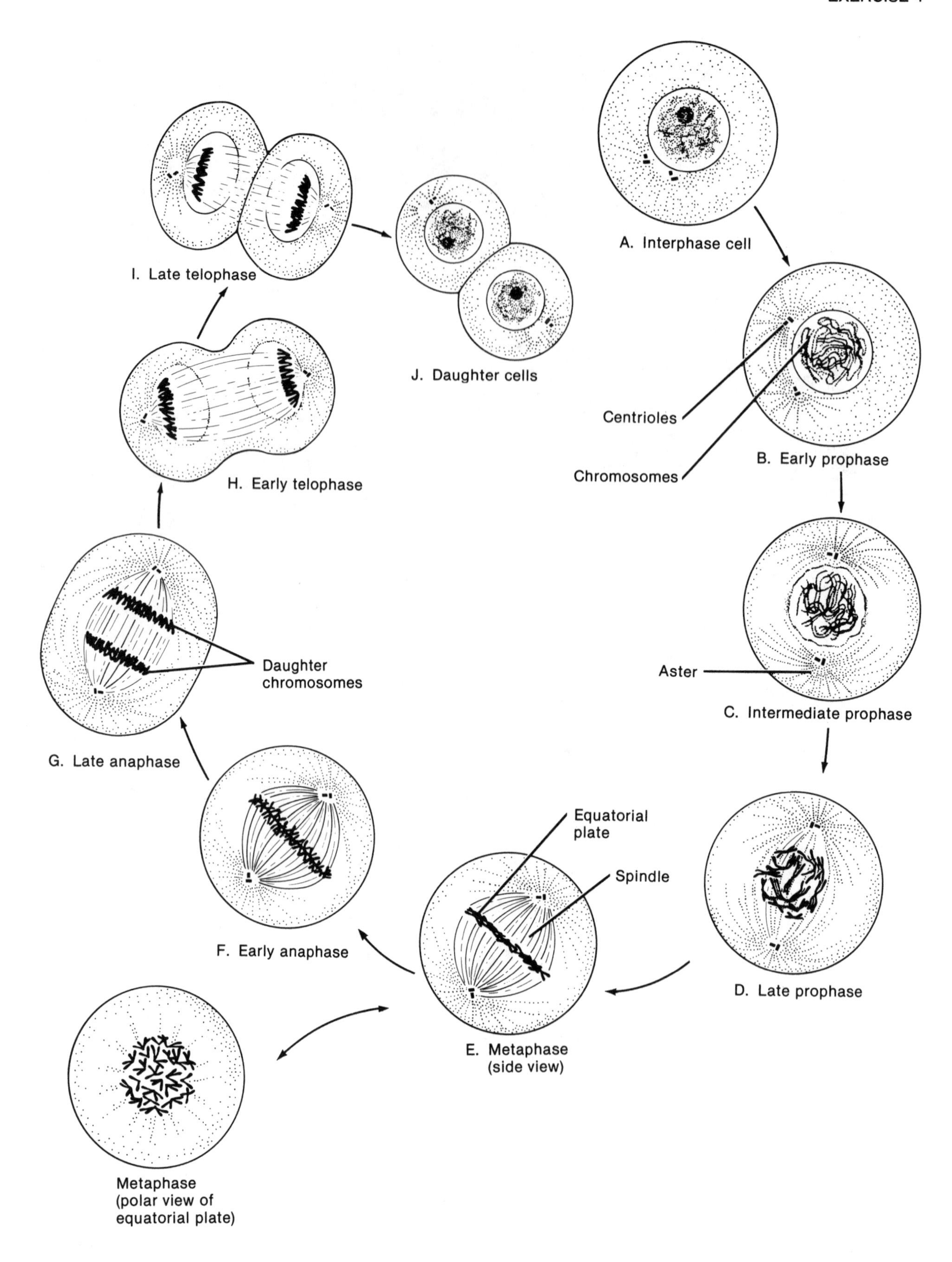

FIG. 4-4
Mitotic stages in the whitefish blastula.

which are joined together at a point called the **kinetochore** (an attachment point of the centromeres of each daughter chromatid). During prophase, the nuclear envelope becomes indistinct and the nucleolus disappears. Electron microscopic studies have shown that the nuclear envelope becomes part of the endoplasmic reticulum.

During prophase, two pairs of centrioles are organized within the centrosome. They begin to move apart, as if they repelled each other, migrating toward opposite poles of the cell (Fig. 4-4B, C). Fibers that radiate from each pair of centrioles like spokes on a wheel, then appear, forming a configuration known as an **aster.** The radiating fibers are called **astral rays.** The centrioles continue to migrate until they lie at opposite sides of the cell. Their positions mark the **poles** toward which the chromosomes will move. When the nuclear envelope breaks down, the region between the centrioles becomes clearly visible as a relatively transparent region called the **spindle.** The spindle consists of microtubules about 150 Å in diameter, which are so arranged as to form the spindle fibers. With the help of Fig. 4-4B–D, locate various stages of prophase on the slide.

c. Metaphase

During metaphase, the chromosomes move toward the central region of the spindle to form the metaphase, or **equatorial, plate** (Fig. 4-4E). The chromosomes are maneuvered into position by the spindle fibers (produced during prophase) that are attached to the centromere of each chromosome. Locate this phase of mitosis on your slide. Look for two views of the chromosomes: a side view and a polar view (Fig. 4-4E).

d. Anaphase

Anaphase begins when the centromeres of the doubled chromosomes divide and the chromatids separate and become **daughter chromosomes.** Following this stage, the daughter chromosomes begin moving to the polar regions of the cell (Fig. 4-4F, G). The completion of the movement to the poles by the chromosomes marks the beginning of telophase.

e. Telophase

During telophase, the spindle disappears, two daughter nuclei are organized, the nucleoli reappear, and the nuclear envelopes are reformed by the fusion of parts of the endoplasmic reticulum (Fig. 4-4H, I).

In late telophase, the cytoplasm has become deeply furrowed, or pinched in, and cytokinesis takes place. This results in two daughter cells having equivalent nuclear contents and equal amounts of cytoplasm (Fig. 4-4J).

B. MEIOSIS

Meiosis is an important biological event that not only maintains the chromosome number constant for most species of animals but also provides a means of genetic variability because of crossing over and the subsequent exchange of genetic material. In meiosis, immature or primordial germ cells undergo a "reduction" in the diploid number of chromosomes characteristic for the species and become mature gametes.

1. Meiosis in the Lily

Meiosis will be studied as it occurs in the development of mature pollen grains of the flowering plants. These pollen grains give rise to male gametes, which fuse with an egg to produce a zygote.

As you examine the series of slides in the meiotic sequence, refer to Fig. 4-5 to help locate the stages indicated in the following description.

a. Meiosis I

Examine a lily flower, and locate the **anthers,** or pollen sacs, which contain numerous pollen mother cells (Fig. 4-5A, B). Each of these cells may undergo meiosis and produce mature pollen grains. Next examine slides of a cross section through a young lily anther, and locate the pollen mother cells (Fig. 4-5C, D). The nuclei of these cells contain the **diploid number** of chromosomes. Many of the pollen mother cells are in the prophase of the first meiotic division. During this phase, **homologous** chromosomes, each composed of two **chromatids,** lie adjacent to one another **(synapse),** form **tetrads,** and exchange genetic components by a process called **crossing over.** What is the significance of this process?

____

____

This phenomenon is diagrammatically shown in Fig. 4-6.

Subsequent to synapsis the homologous chromosomes separate during anaphase, with one member moving to each pole of the division spindle. Because this is a separation of entire chromosomes, and not chromatids, the chromosome content of the cells at

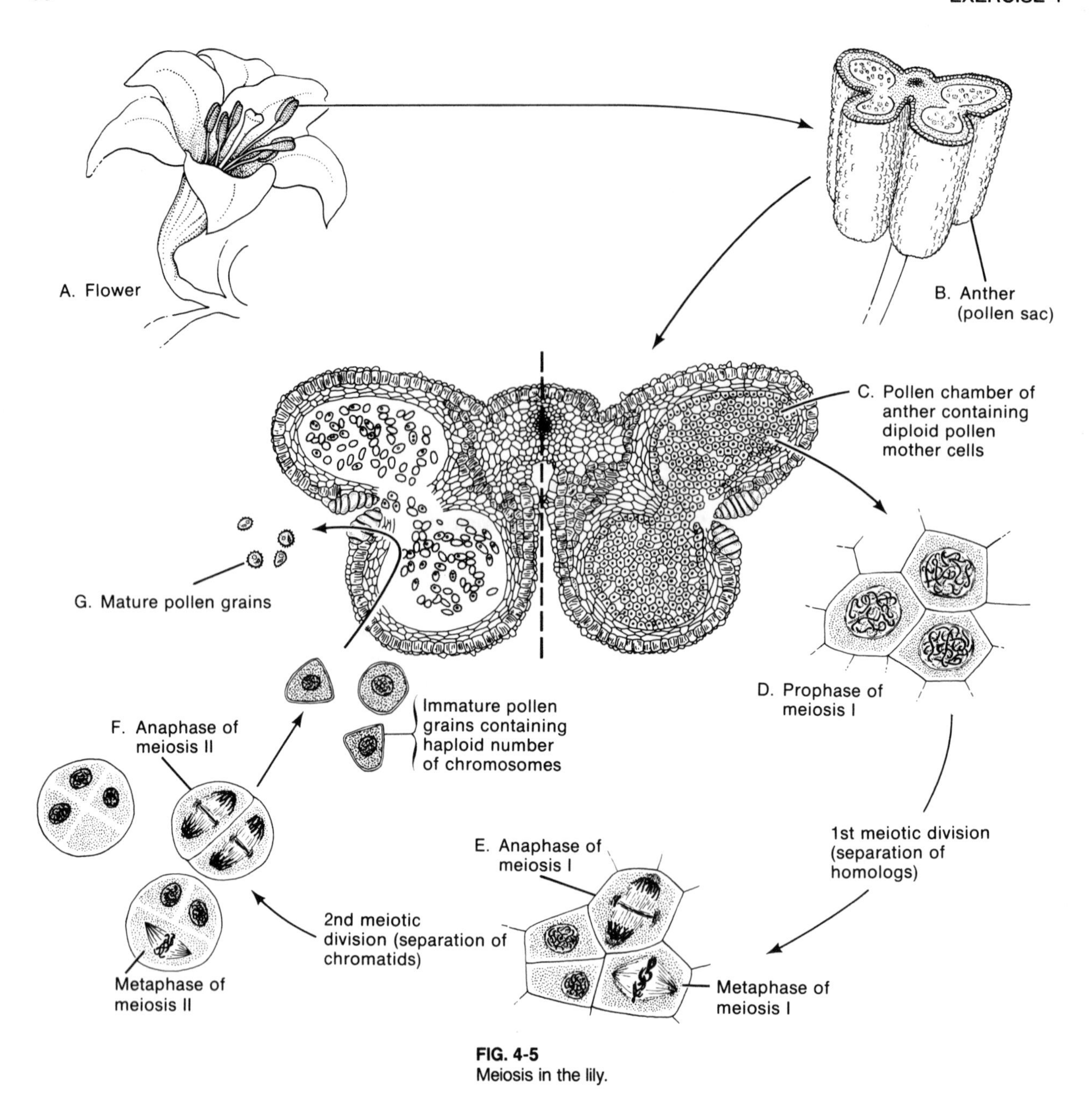

FIG. 4-5
Meiosis in the lily.

the end of meiosis I has been "reduced" from the diploid to the **haploid** condition. Examine slides of a lily anther showing separation of homologous chromosomes (Fig. 4-5D, E). The diploid number in lily is 24. What is the chromosome number following the first meiotic division?

____

b. Meiosis II

Examine slides of anthers in which the cells resulting from the first meiotic division are in anaphase of meiosis II (Fig. 4-5F). In these cells, the chromatids that make up each chromosome now separate and migrate to the spindle poles (Fig. 4-7). As in mitosis, the chromatids when separated from each other are called **daughter chromosomes.** At the spindle poles, each group of daughter chromosomes becomes enclosed in a nuclear membrane.

Cytokinesis follows the division of the nucleus. How many pollen grains are formed as a result of the two meiotic divisions of the pollen mother cells?

____

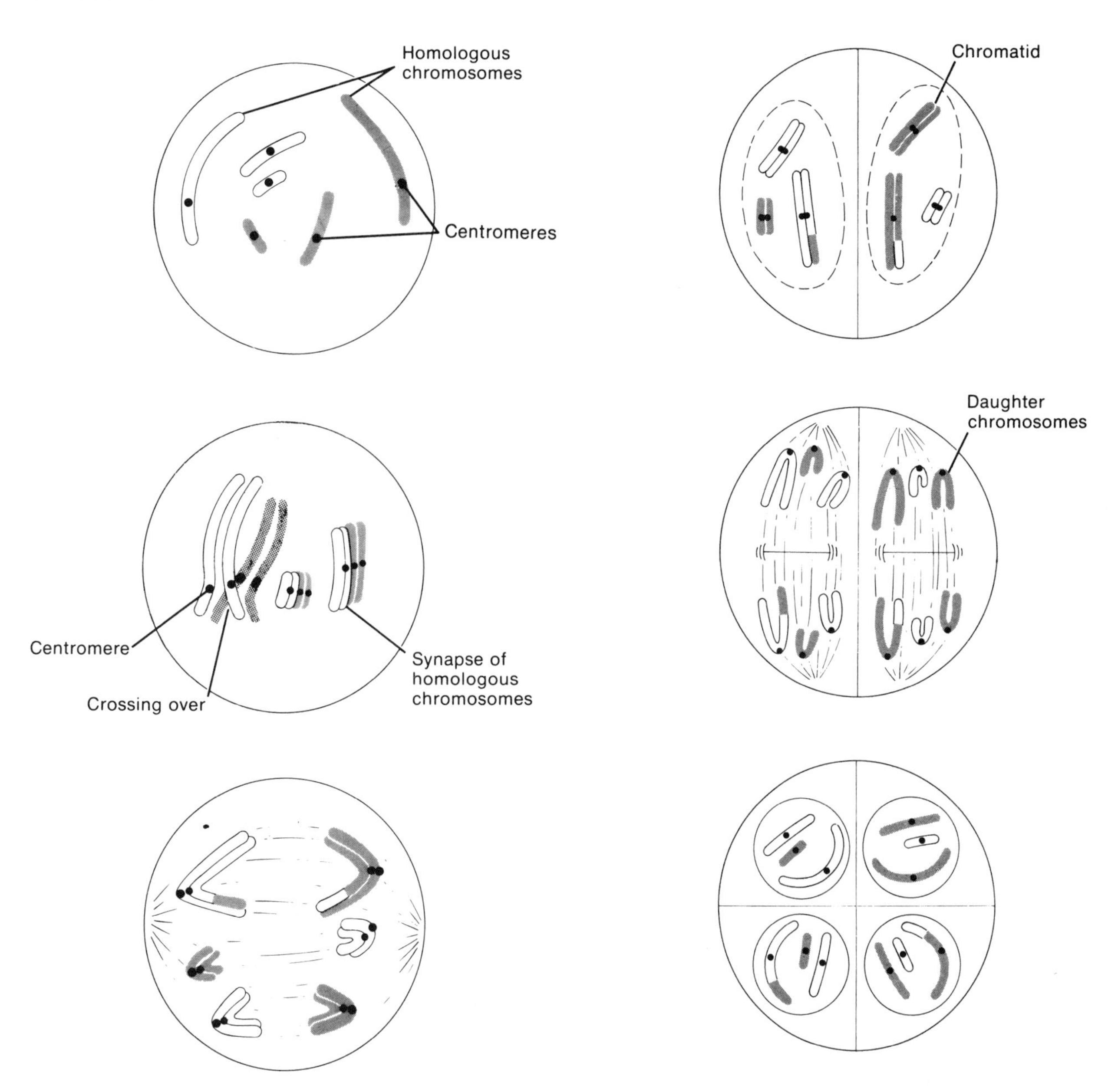

FIG. 4-6
First meiotic division, showing crossing over.

FIG. 4-7
Second meiotic division, showing the separation of the daughter chromosomes.

What is the chromosome complement of each pollen grain?

Examine lily anthers showing mature pollen grains (Fig. 4-5G).

2. Meiosis in *Ascaris*

The development of the gametes is called **gametogenesis.** During meiosis I, the diploid number of chromosomes is reduced to the haploid number. In males, gametogenesis occurs in the testes; in females, in the ovaries or oviducts. In this part of the exercise, the maturation of the egg **(oogenesis)** will be studied as it occurs in the parasitic roundworm *Ascaris.* Because the diploid number of chromosomes in *Ascaris* is only four, it is ideal for the study of this process.

The reproductive organ in *Ascaris* consists of a pair of long, highly coiled tubes that are regionally divided into the ovary, oviduct, and uterus (Fig. 4-8A). The "eggs," which are produced in the ovaries, pass into the oviducts, where they are fertilized by sperm.

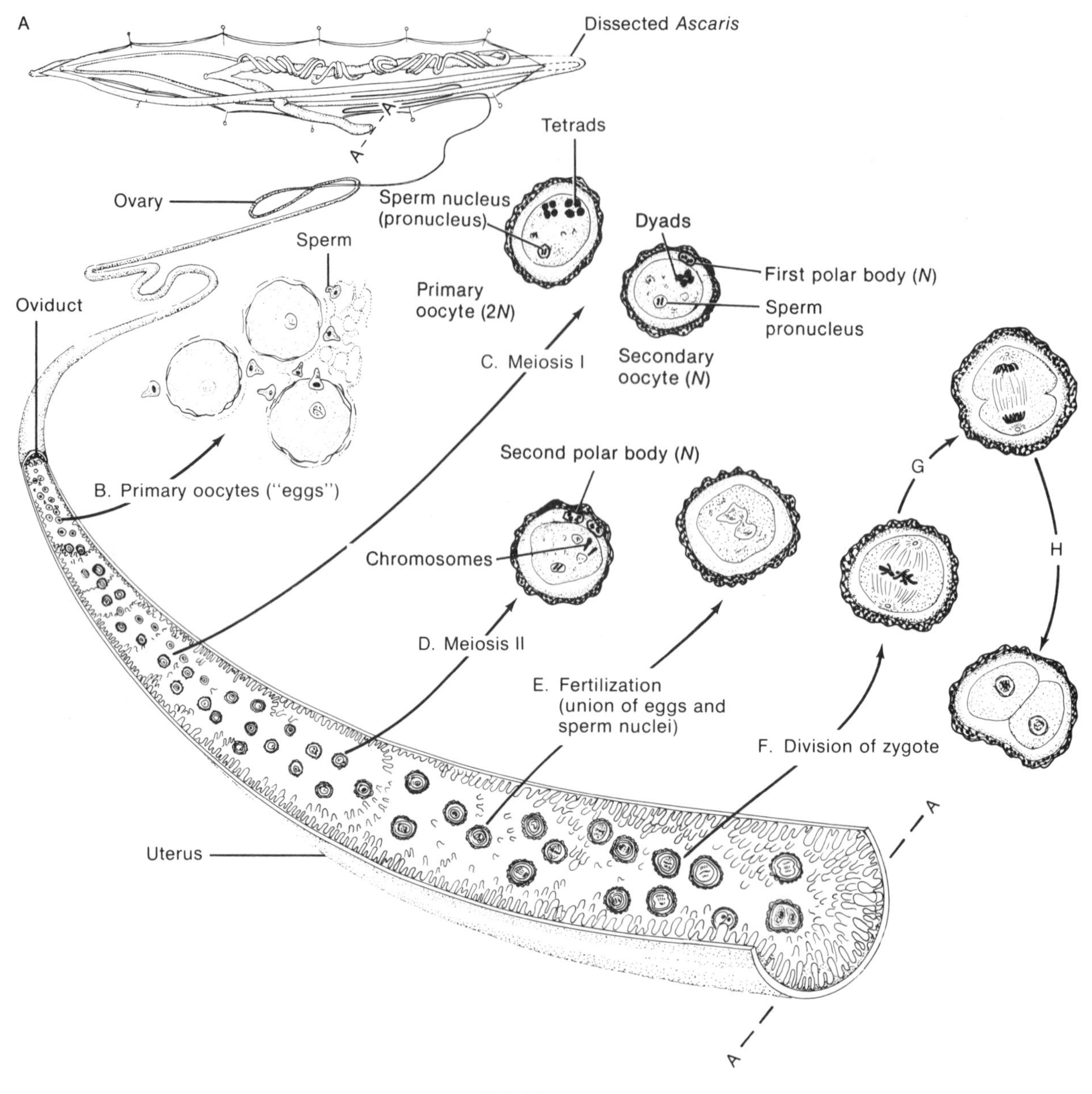

FIG. 4-8
Meiosis in *Ascaris.*

Examine slides containing the oviduct and uterus of *Ascaris.* Locate the oviduct, which characteristically has a large number of triangular sperm interspersed among numerous "eggs" (Fig. 4-8B). The eggs at this stage are still diploid, because in *Ascaris* the maturation process, or oogenesis, does not begin until after the egg has been penetrated by a sperm. Therefore, the term *egg* at this stage of development is not quite accurate. A more correct term would be **primary oocyte,** a cell that will undergo meiosis and produce the mature egg. On the slide, locate primary oocytes; some of them may have been penetrated by sperm. How many chromosomes are in the primary oocyte of *Ascaris?*

____

Shortly after fertilization, the homologous chromosomes of the primary oocyte duplicate to form chromatids and then, during prophase of meiosis I, come to lie adjacent to each other (synapse), forming **tetrads** (Fig. 4-8C). At this time, crossing over may occur, which results in an exchange of genetic components of homologous chromosomes (Fig. 4-6).

Locate primary oocytes in which synapsis has taken place. At anaphase of meiosis I, the members of homologous pairs of chromosomes, each **homolog** still consisting of a pair of chromatids, move to opposite poles (Fig. 4-8C). Since the centromeres do not separate, the sister chromatids remain together and are called **dyads.** At telophase of meiosis I, an unequal division of the cytoplasm occurs (cytokinesis) so that the first meiotic division produces one relatively large cell called a **secondary oocyte** and a small one called the **first polar body** (Fig. 4-8C). How many chromosomes are found in the first polar body of *Ascaris?*

How many are found in the secondary oocyte?

Locate the first polar body on the slide.

After an interphase, the second division begins when the homologous chromosomes (homologs), which separated during meiosis I, line up on the equator of a division spindle in metaphase of meiosis II (Fig. 4-8D). Each homolog is composed of two chromatids, which, at anaphase of meiosis II, separate and migrate to the poles. The second meiotic division of the secondary oocyte produces a second polar body and a large cell, which differentiates into the egg cell, or ovum. The first polar body produced in meiosis I may or may not go through a second meiotic division. Thus, when a diploid cell in the *Ascaris* ovary undergoes complete meiosis, only one mature ovum is produced; the polar bodies are essentially nonfunctional.

The unequal cytokinesis of oogenesis ensures that an unusually large supply of cytoplasm and stored food is allotted to the nonmotile ovum for use by the embryo that will develop from it. In fact, the ovum provides almost all the cytoplasm and initial food supply for the embryo. The tiny, highly motile sperm cell contributes essentially only its genetic material.

During the maturation of the egg, the sperm nucleus has been lying inactive in the cytoplasm. Following the second meiotic division, the egg and sperm nuclei unite **(fertilization)** and form a single cell called, a **zygote.** Locate eggs that show separate sperm and egg nuclei and eggs in which these two nuclei have fused (Fig. 4-8E). How many chromosomes does the zygote contain?

If you did not know the actual number of chromosomes in the zygote, how would you describe the chromosome content of this cell?

In the uterus, the zygote nucleus and cell soon divide and form two cells, each of which divides again until a multicellular embryo is formed (Fig. 4-8F–H). What type of division is this?

REFERENCES

DeRobertis, E. D. P., and E. M. F. DeRobertis, Jr. 1980. *Cell and Molecular Biology.* 7th ed. Holt/Saunders.

DuPraw, E. J. 1970. *DNA and Chromosomes.* Holt, Rinehart and Winston.

Karp, G. 1984. *Cell Biology.* 2d ed. McGraw-Hill.

Mazia, D. 1961. How Cells Divide. *Scientific American* 205(3):100–120 (Offprint 93). *Scientific American* Offprints are available from W. H. Freeman and Company, 41 Madison Avenue, New York 10010 and 20 Beaumont Street, Oxford OX1 2NG, England. Please order by number.

Mazia, D. 1974. The Cell Cycle. *Scientific American* 230(1):54–64 (Offprint 1288).

McFall, F. D., and D. G. Keith, 1969. Animal Mitosis. *Carolina Tips* 32:1–3.

Moens, P. B. 1973. Mechanisms of Chromosomes Synapsis at Meiotic Prophase. *International Review of Cytology* 35:117–134.

Rappaport, R., 1971. Cytokinesis in Animal Cells. *International Review of Cytology* 31:169–213.

Taylor, J. H. 1958 The Duplication of Chromosomes. *Scientific American* 198(6):36–42 (Offprint 60).

Wilson, G. B. 1966. *Cell Division and the Mitotic Cycle.* Reinhold.

Yeoman, M. M., ed. 1976. *Cell Division in Higher Plants.* Academic Press.

EXERCISE 5

Plant Anatomy: Roots and the Shoot

Growth and development are phenomena common to both plants and animals. As you would expect, plants and animals differ significantly in the changes that occur between the time of fertilization and the achievement of the characteristic form for any given plant or animal.

Development in plants is **indeterminate,** while development in animals is **determinate.** That is, in plants development continues throughout the life of the plant, with new organs, tissues, and cells being formed perpetually. For example, in a great redwood tree growing in California, cells and tissues near the base of the tree may be more than 2000 years old, while in the tips of its branches new cells, tissues, and organs continue to form and differentiate into their final structures. Theoretically, there is no limit to the growth of this tree. Conversely, the animal becomes complete (i.e., determined) early in its life. In a chicken, for example, differentiation of cells and tissues occurs early in embryonic development. After the animal matures, no further growth and development occurs. An exception is the development of gametes and blood cells, which are formed throughout the life of the organism.

Another major distinction between plants and animals is that cell division in plants occurs in localized regions called meristems, whereas in animals it is general throughout the immature organism. The three principal growth regions of postembryonic plants are (1) the **apical meristem,** which determines growth in the length of the stem and root and is instrumental in the production of leaves, flowers, and branches; (2) the **lateral meristem** consisting of the **vascular cambium,** which determines the growth in diameter of the stem and root; and (3) the **cork cambium,** which determines the production of cork—the protective outer covering of the stem and root.

In order to understand the growth and development of plants and to comprehend their complex structure and the interrelationships between structure and function, it is necessary to study their anatomy. In this exercise you will become familiar with the anatomy of the roots and stems of flowering plants.

A. ROOTS

Characteristically, the root is described as that part of the plant growing beneath the surface of the soil. The principal function of roots is to absorb water and soluble minerals and to transport these substances to the above-ground parts—the stem, leaves, flowers, and fruits. In addition, roots anchor the plant in the soil and store reserves of food for the plant, as in sweet potatoes, carrots, and turnips. Roots may also manufacture food. For example, aerial roots of orchids contain chlorophyll, thereby supplementing the leaves in the photosynthetic process.

Roots may be broadly classified by form into two groups: **tap roots** and **fibrous roots.** In tap roots, the main root becomes many times larger than the branch roots (those that arise from the main root) and penetrates some distance into the soil. In some cases, the tap root may be greater in diameter than the stem. Examine the tap root of a carrot plant for the presence of branch roots. In fibrous root systems, the primary root and the branch roots are approximately the same length and diameter. Examine the fibrous root system of a bean plant.

1. Root Apex

Obtain a germinating radish or grass seed. Mount it in a drop of water on a slide and examine the young root with a dissecting microscope. Locate a cone-shaped mass of cells covering the root tip. This **root cap** covers the apical meristem and protects it from damage as the root passes through the soil. By adjusting the light you may be able to observe the root apical meristem, which appears as a relatively dense, opaque region at the tip of the root. Somewhat behind the root tip, note the presence of **root hairs**—special absorbing cells on the root. These generally persist for a short time and then die. New root-hair cells continue to be formed near the root tip to replace those that are lost. In what way do root hairs increase the efficiency of absorption of water and minerals by the root?

____

____

____

Examine a prepared slide of a longitudinal section through an onion root tip and locate the structures just described (Fig. 5-1A, B).

2. Primary Tissues of the Root

Using a compound microscope, examine a prepared slide of a cross section of the buttercup *(Ranunculus)* root, cut through a region in which the cells have become differentiated (Fig. 5-1C, D). Three general regions are readily seen: the **epidermis,** the **cortex,** and the **vascular cylinder.** Starting from the outside, the **primary tissues** (those tissues having their origin from cells produced in the apical meristem) consist of an outer epidermis, which is a single layer of cells covering the outside of the root. In some instances, the epidermis dies and the outer cells of the cortex function as a special epidermal covering called the **hypodermis.** The cortex in most roots consists largely of thin-walled parenchymatous cells. Note the presence of starch grains in many of the cortical cells, and the large intercellular spaces formed where the cells abut one another.

Lining the cortex on the inside, and considered to be part of it, is the **endodermis.** It is represented by a single layer of cells separating the parenchyma of the cortex from the vascular cylinder. The endodermis, present in all roots, is believed to function in directing the flow of water from the cortex into the xylem of the vascular cylinder.

The vascular cylinder consists of several tissues lying internal to the cortex. The **pericycle,** a unicellular layer of cells adjacent to the endodermis, has the ability to become meristematic and to initiate the growth of lateral roots. The **xylem,** the primary water conducting tissue, is represented by three, four, or five radiating arms or plates. Alternating with the xylem arms are groups of **phloem** cells. The phloem conducts various organic molecules including products of photosynthesis and various hormones. This alternate arrangement is unique to roots and anatomically distinguishes roots from stems. Locate the region between the phloem and the xylem where the cambium will develop. The cambium is a lateral meristem responsible for growth in the diameter of a root. When the cambium divides, it gives rise to cells that will differentiate as xylem and phloem, which are then called **secondary** xylem and phloem, because they originated in meristems other than an apical meristem.

Obtain a carrot root. Using a sharp razor, cut a thin cross section from the root and place it on a slide. Add one or two drops of iodine and a coverslip. Examine microscopically. Iodine reacts with starch to form a deep blue-black color. Where is the starch located in the root?

____

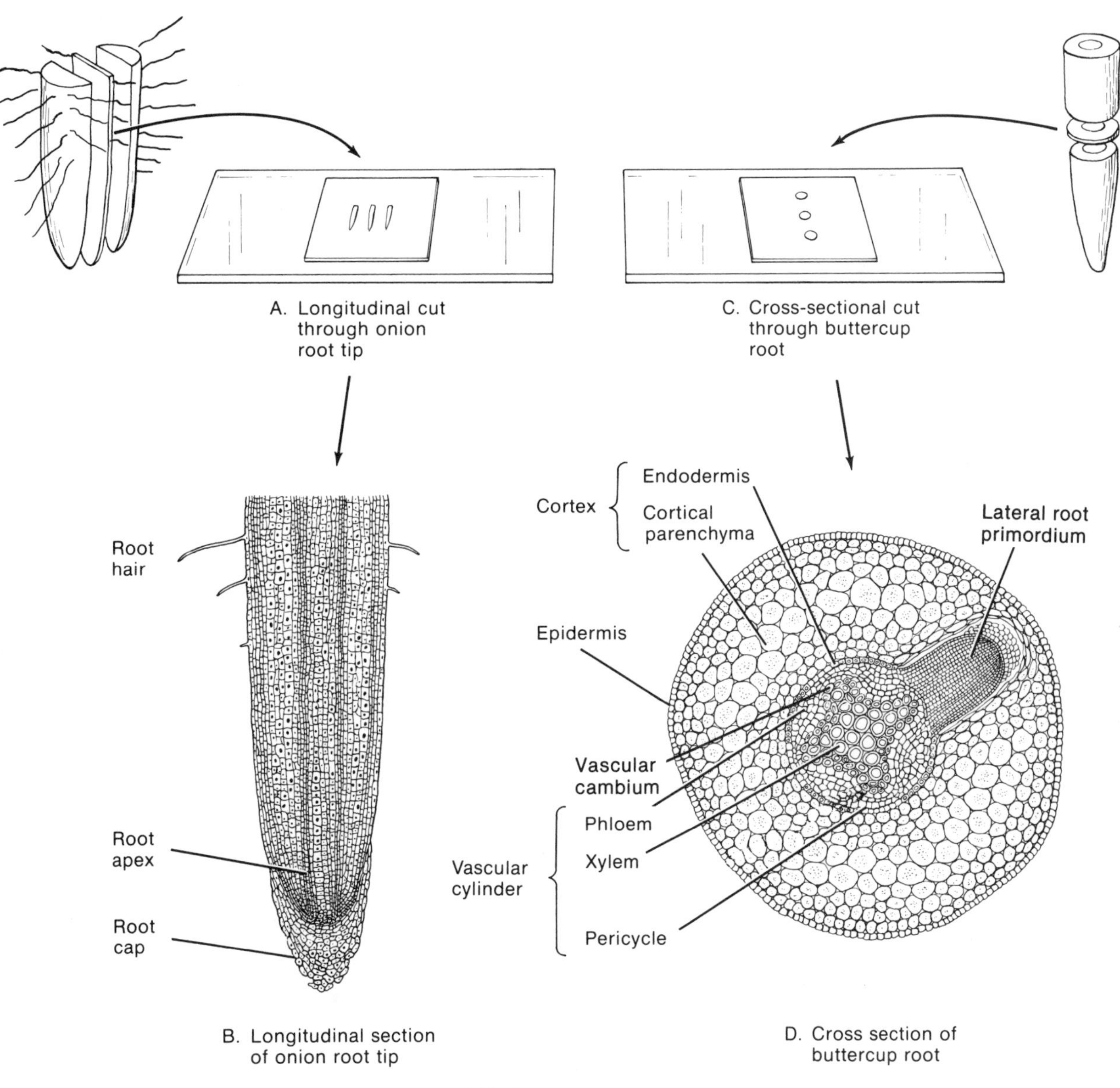

FIG. 5-1
Study of root anatomy.

Prepare a second cross section of carrot and stain with phloroglucinol-HC1. (CAUTION: *This solution contains concentrated hydrochloric acid.)* This chemical reacts with a substance called **lignin** in the xylem cells that make up the vascular tissue. How does the organization of vascular tissue of the carrot root differ from that of the *Ranunculus* root?

____

____

3. Lateral Roots

As noted, the pericycle may give rise to lateral (branch) roots. In lateral root formation, a new meristem is formed, complete with a root cap. As the young lateral root grows, it penetrates the cortex and epidermis to the outside, and continues growing. Examine demonstration slides of developing branch roots (Fig. 5-1D). Remove young roots from water lettuce plants *(Pistia),* if available, and examine developing branch roots.

4. Adventitious Roots

Examine stem cuttings taken from various plants (i.e., *Coleus,* willow, geranium) that have been rooted in sand or vermiculite (Fig. 5-2A). The roots that arise from such cut stems are called **adventitious** roots and generally result from regenerative processes in stem cuttings, or in some cases, from the "dedifferentiation" of cells in detached leaves of such plants. What does the term *adventitious* mean?

Do you agree with the statement that adventitious roots are not part of the normal developmental morphology of the plant as a whole? Explain.

Examine the adventitious **"prop"** roots of a corn plant. (Fig. 5-2B). Do these roots originate in the stem or root?

How would you determine this?

What is (are) the function(s) of prop roots?

If available, examine a microscopic section of a prop root. In what way(s) is it similar to the *Ranunculus* root previously examined?

In what way(s) is it different?

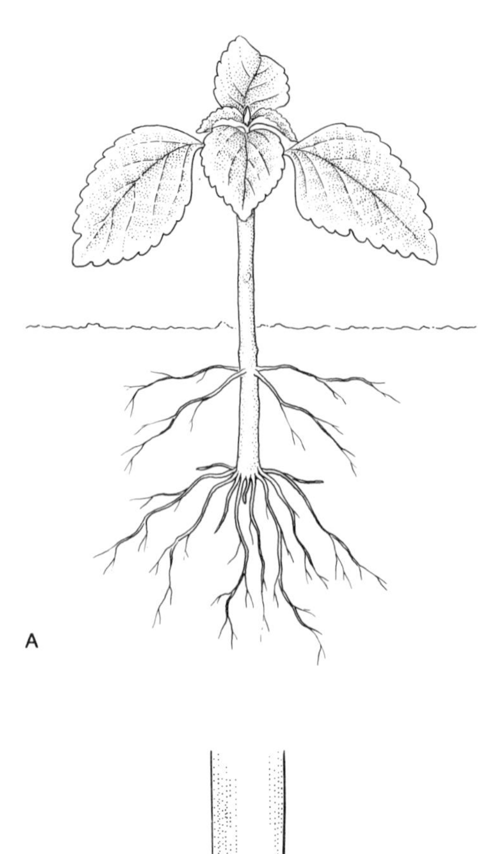

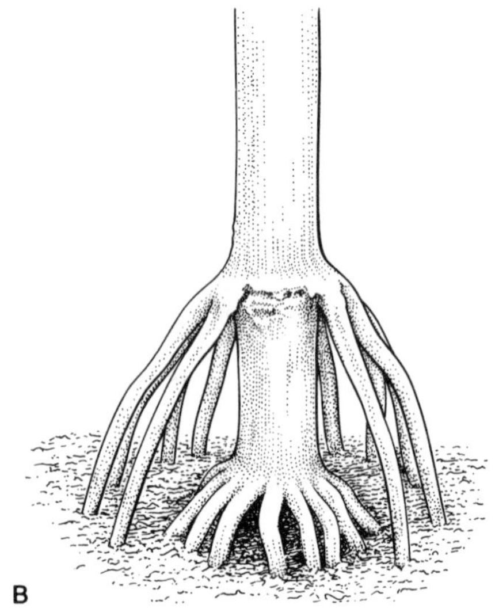

FIG. 5-2
Types of roots: (A) adventitious roots on stem cutting and (B) prop roots of corn.

B. THE SHOOT

The shoot consists of the stem and its leaves and is commonly aerial and upright. Stems are like roots in general structure: they have an epidermis, cortex, and vascular cylinder, although there are variations within this overall structure. Stems differ from roots in the fundamental vascular structure and in having lateral appendages called *leaves*.

Several basic functions are fulfilled by the stem. It is the framework supporting the leaves, flowers, and fruits. It provides for the transport of water and solutes absorbed by the roots and serves to transport and distribute the sugar manufactured by the leaves to the places where it is used or stored. Some stems are green and carry on a limited amount of photosynthesis. Many stems function as food storage areas —sometimes to the extent of being economically important (e.g., Idaho or Wisconsin potatoes and sugar cane).

Growth and development in the stem begins in the bud, a terminal or axillary structure consisting of a small mass of meristematic tissue, the **shoot apex,** or **apical meristem.**

1. The Bud and Other Structures Present on the Shoot

a. Classification of Buds

Buds may be classified in a variety of ways.

By content
- foliage buds
- floral buds
- mixed buds (i.e., flower and foliage primordia)

By position
- terminal buds
- axillary buds
- adventitious buds

By time of development
- active buds
- resting buds
- latent buds

By protective covering
- protected buds
- naked buds

b. Examination of the Shoot of a Woody Perennial

Examine branches from the hickory tree. Locate and label the following structures on Fig. 5–3.

Terminal bud: a bud located at the tip of the branch and involving the entire stem tip.

Axillary bud: a bud developed from a branch primordium and located in the axil of a leaf (above the leaf scar if the leaf has fallen from the twig).

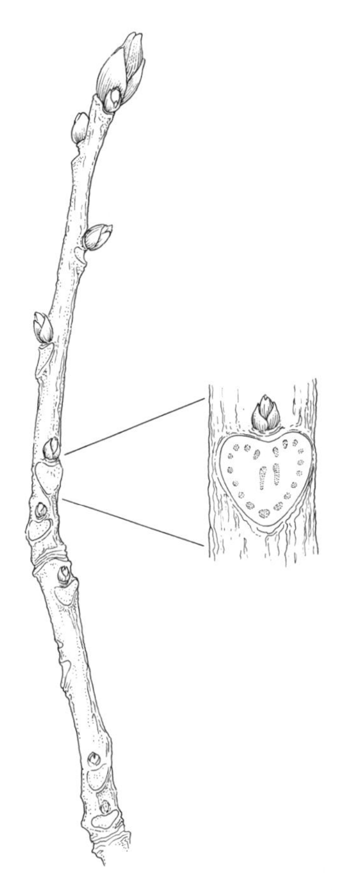

FIG. 5-3
External features of hickory branch.

Latent buds: small axillary buds that normally do not develop into branches the spring following their formation. They may develop into branches if an accident destroys the terminal or other axillary buds.

Leaf scar: a scar on the stem formed when the petiole separated from the stem at the time of leaf fall.

Vascular bundle scars: small spots located in the area of a leaf scar and indicating the number of vascular bundles that extended from the stem into the leaf. These vascular bundles were severed when the leaf fell.

Bud scales: protective scalelike leaves covering all the buds.

Clusters of bud scale scars: scars left at a point where there was formerly a bud. When a bud opens, the bud scales are shed, and each one leaves a scar. Because the bud scales are very close together, these scars will be found clustered together. The age of a twig can be determined by counting the sections of stem between successive clusters of bud scale scars.

For example, from the tip back to the first cluster of bud scale scars is the growth of the past season. From the first cluster back to the second cluster will be the growth of the season before that, and so on.

Growth produced during the past growing season.

Growth produced during the growing season preceding the past growing season.

Lenticels: small openings in the bark of the stem allowing for exchange of gases between the outside and inside of the stem. Loosely packed cells may protrude from these openings.

Floral branch scars: small, round scars directly above some of the leaf scars. These scars were produced when the floral branches broke from the stem and fell off.

Node: place where leaf is attached to stem.

Internode: region between nodes.

c. Examination of the Shoot Apex

A brief study of the shoot apex shows how new stem tissue and the leaves it supports are formed and become differentiated.

1. Dissect one of the larger buds on the twigs that have been given to you. Note the *bud scales* that enclose the bud. The bud scale represents a modification of what plant structures?

What makes up the bulk of the bud enclosed by the bud scales?

If possible, locate the extremely minute tip of the shoot within the bud. What is the shape of the actual stem shoot tip?

---

2. Examine a prepared slide of the *Coleus* shoot tip showing a longitudinal section through the middle of a bud (Fig. 5-4). Locate the shoot apical meristem, which is at the tip of the shoot and enclosed by the embryonic leaf primordia. Note how the leaves increase in size progressively as you move away from the shoot tip. Can you find evidence of cell division in the apical meristem? If not, can you suggest a reason for absence of divisions?

Are the cells in the leaf primordia all alike? If not, what evidence of differentiation do you find?

What do you find in the angle that each leaf makes with the stem?

---

2. Primary Tissues of a Stem

Flowering plants are divided into two major groups called the **dicotyledons** (dicots) and **monocotyledons** (monocots). The differences in the number of cotyledons (leaf-like food storage organs) in the seed—one in the monocots and two in the dicots—provide the most familiar distinction between these groups. Other differences are based on (1) the leaf venation pattern (parallel in monocots; reticulate in dicots); (2) The vascular bundle arrangement in the stem (peripheral and arranged in a cylinder in dicots; scattered in monocots); (3) the presence (in dicots) or the absence (in monocots) of a vascular cambium; and (4) the numerical arrangement of floral parts (three in monocots; four or five, or multiples thereof, in dicots). Although you will examine plants that exhibit these classical distinctions in the studies that follow, be aware that there are always exceptions to the rule.

a. Primary Tissues of a Dicot Stem

Examine various representatives of dicots such as *Coleus,* geranium, bean plants *(Phaseolus),* tomato plants *(Lycopersicon),* and any other provided by your instructor. Identify the following:

Terminal or apical bud: a bud located at the tip of the main stem or at the end of a branch. To what does this structure give rise?

Node: that place on the stem where leaves arise.

Internode: the region between two nodes. Stem elongation is primarily the result of elongation of the internodes.

Leaf: composed of two parts called the **blade** (the expanded flat portion) and the **petiole** (the stalk that attaches the blade to the stem). In some cases, the blade may be subdivided into numerous smaller segments, called **leaflets.**

Stipules: leaflike structures found near the base of the petiole. These are not always present.

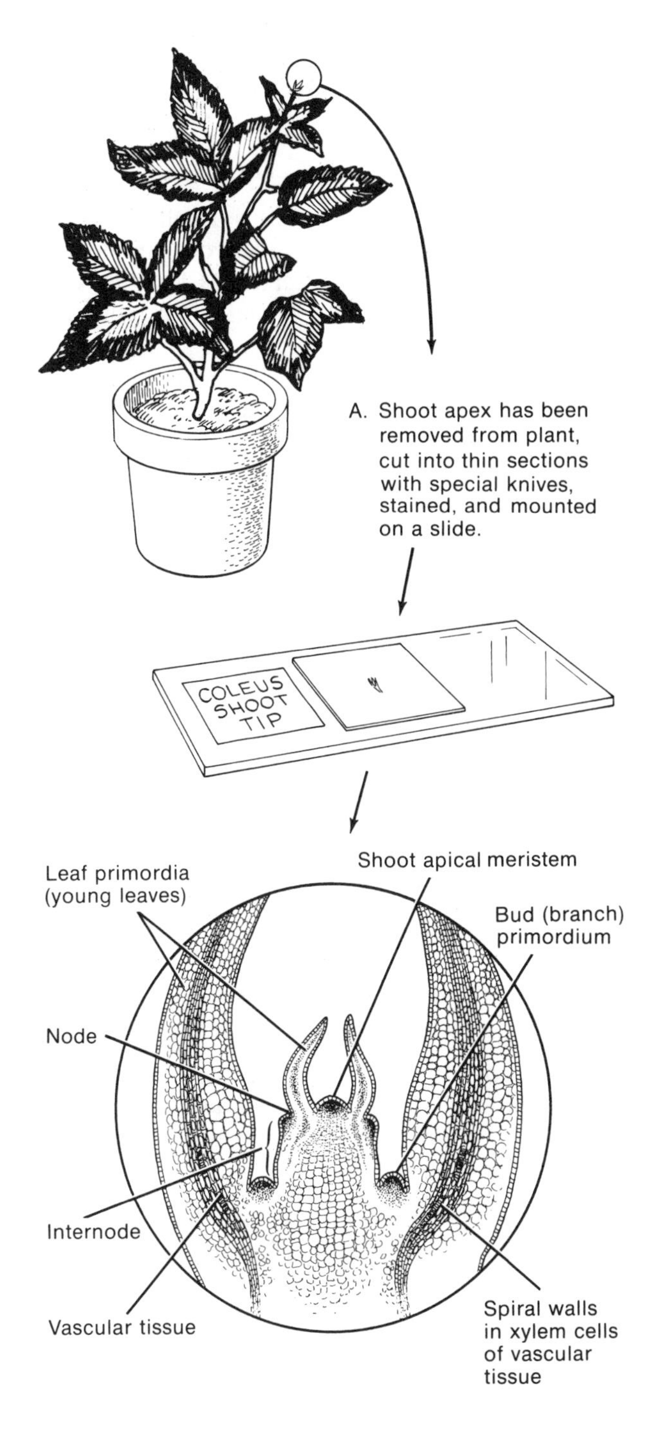

FIG. 5-4
Study of the *Coleus* shoot tip.

Veins: the vascular network located in the leaf blade. In dicots, this pattern is described as **net** or **reticulate venation.**

Axil: the angle between the upper side of the petiole and the stem.

Lateral or axillary bud: a bud formed in the axil of the leaf. Develops into a leafy branch or flower, or sometimes both. You examined these in Part B-1 of this exercise.

Obtain a portion of a geranium stem from the instructor. With a razor blade, make a transverse cut to obtain the thinnest section possible. Prepare a wet mount and examine it with the low power of the microscope. Locate the outer epidermal layer and the numerous hairs associated with it. Beneath the epidermis is a layer of tissue composed of parenchymatous cells. This is the cortex. Note the presence of numerous chloroplasts in the cells. Suggest a function for this tissue in the living plant.

___

___

The cortex is separated from the centrally located pith by a band of vascular tissue. Note that the vascular cylinder is not of uniform thickness but is composed of more or less separate vascular bundles. The vascular cylinder may be more readily observed if the tissue is stained with phloroglucinol-HC1. Mount the section in phloroglucinol solution. Wait about 1 minute and reexamine the section. How does the arrangement of the vascular tissue differ from that of the root.

___

For a detailed examination of the primary tissues, obtain a prepared slide of a cross section of the alfalfa *(Medicago)* stem (Fig. 5-5). The cells in the central part of the stem differentiate into pith cells. Encircling the pith are vascular strands, which appear in cross section as separate vascular bundles. A mature vascular bundle consists of three tissues: xylem, phloem, and cambium. The xylem is located on the inner side toward the pith. Adjacent to the xylem, appearing as small rectangular cells having the long axis at right angles to the radius of the stem, is the cambium. Division of cambial cells results in the formation of secondary tissues, which increase the diameter of the stem. How does the vascular arrangement in the stem contrast with that of the root?

___

The phloem lies immediately external to the cambium.

Capping each vascular bundle is a group of thick-walled fibers. Morphologically, these fibers are part

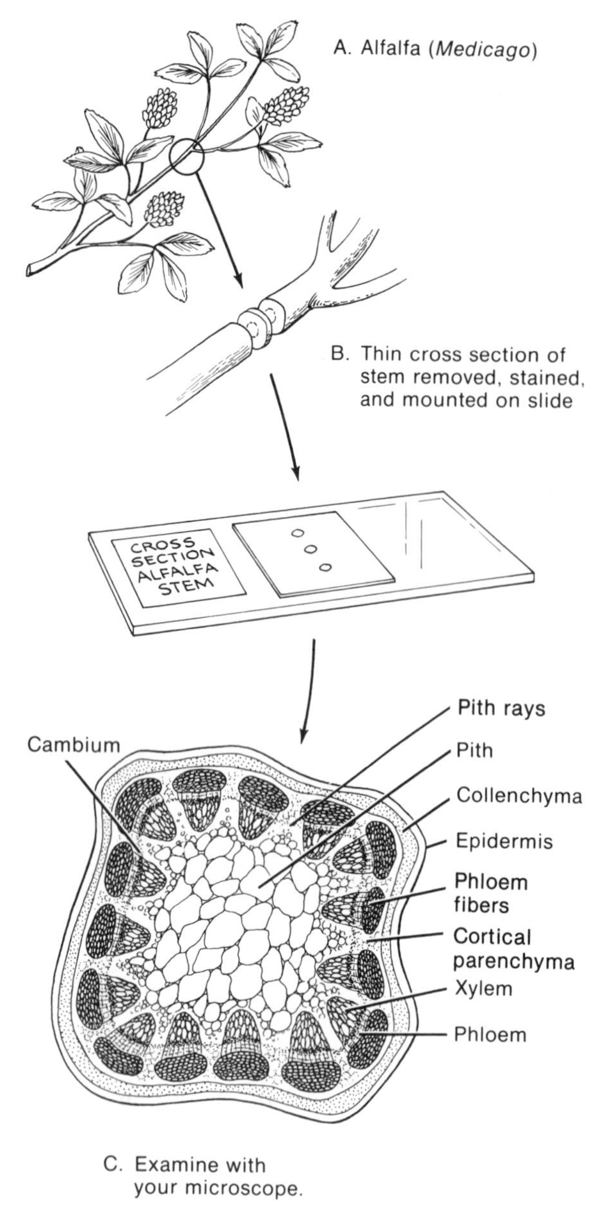

FIG. 5-5
Study of a dicot (alfalfa) stem anatomy.

of the primary phloem tissue and are properly called **phloem fibers.** The parenchymatous cells lying between the vascular bundles are **pith rays.**

The cortex lies immediately outside the phloem fibers and consists of two distinguishable tissues: **cortical parenchyma** and **collenchyma** (mechanical tissue). The cortical parenchyma consists of loosely packed parenchymatous cells, many of which contain chloroplasts. In many stems, the collenchyma may be found exterior to the cortical parenchyma and serves a supportive (strengthening) function. In alfalfa, there is considerable development of the collenchyma at the corners.

The outermost layer of cells constitutes the epidermis. How many cell layers thick is the epidermis?

____

The outer walls of epidermal cells usually have become thickened, and during their development a waxy substance **(cutin)** is secreted. Close the diaphragm on your microscope so that you can observe this layer of cutin. It will appear as a faint, pink, noncellular layer covering the epidermis. What effect does this waxy layer have on the loss of water from the stem?

____

____

____

What characteristics of dicots are exhibited by geranium and alfalfa?

____

____

____

b. Primary Tissues of a Monocot Stem

Examine various representatives of monocots such as corn *(Zea mays),* wandering Jew *(Zebrina pendula),* dumbcane *(Dieffenbachia),* lawn grass, oats, and any others provided by your instructor. Identify the following:

Leaf blade: generally narrow, parallel sided, and flat. Which of the demonstration plants fit this description?

____

Which do not?

____

Sheath: the base of the leaf blade, which may partially or completely enclose the stem.

Node: As in dicots, the region of the stem at which the leaf arises. In monocots, the nodes frequently are swollen and look jointed.

Veins: The vascular tissue in monocots generally runs parallel to the leaf margins. Thus the network is usually called **parallel venation.**

FIG. 5-6
Study of a monocot (corn) stem anatomy.

Examine a slide of a cross section of a young stem of corn *(Zea mays)* (Fig. 5-6). Answer the following questions:

	Yes	*No*
Are the vascular bundles located peripherally and in a ringlike pattern as they are in dicots?	___	___
Is there any kind of pattern evident with respect to the arrangement of vascular bundles?	___	___
If yes, describe the pattern. ________ ____________________		
Are the vascular bundles all of the same size?	___	___
If not, are they larger in the center or at the periphery? ____________		
Is there any pith?	___	___
Is there any cortical tissue?	___	___

Using the high-power objective lens, study the details of an individual vascular bundle. Locate the following, and label them in Fig. 5-6.

1. *Xylem:* The xylem tissue consists of xylem **vessels** and xylem **parenchyma.** Two or three of the vessels frequently are arranged to give the appearance of the eyes and nose of a face. Indeed, the vascular bundle sometimes looks like a skull. Located between the larger vessels are smaller vessels and thinner-walled xylem prenchyma. A large hole (which looks like a "mouth" in the skull) is created by cells that have been stretched during elongation of the stem and have collapsed.

2. *Phloem:* The phloem tissue, consisting of larger **sieve tubes** and small **companion cells,** is located in what could be described as the "forehead." In some sections, you may be able to see the perforated **sieve plate** of the sieve tube cell. What is the function of the sieve tubes and the companion cells?

____

3. *Fibrous sheath:* Thick-walled fibers **(sclerenchyma cells)** surround each vascular bundle. Suggest a function for this tissue.

____

3. Secondary Growth

In addition to the great diversity in structure among primary tissue systems in angiosperms, all dicots develop secondary tissues from a cambium and usually form an external tissue called **cork.** The extensive proliferation of secondary tissues results in the crushing and elimination of epidermal tissues, cortex, and primary xylem and phloem of the stem.

Because monocots typically lack cambial activity and the resulting secondary growth, this exercise will examine the secondary growth of a dicotyledenous plant.

Plants grow in diameter by the division and differentiation of cells produced in the lateral meristem or cambium. Tissues formed from cambial activity are designated as **secondary tissues.** Carefully examine a slide that contains a cross section of one-, two-, and three-year-old basswood stems. Label Fig. 5-7 as you locate each of the following:

a. Epidermis

The outermost portion of the cross section. There is a waxy layer, the **cuticle,** covering this outer layer of cells. In older stem sections, this cell layer may be disrupted and appear to be flaking away from the surface.

b. Periderm

This region is composed of

1. *Cork:* Cork cells are rectangular and are produced by the cork cambium. Walls of cork cells are impregnated with **suberin,** which makes the cells waterproof. Cork cells die soon after they are mature.

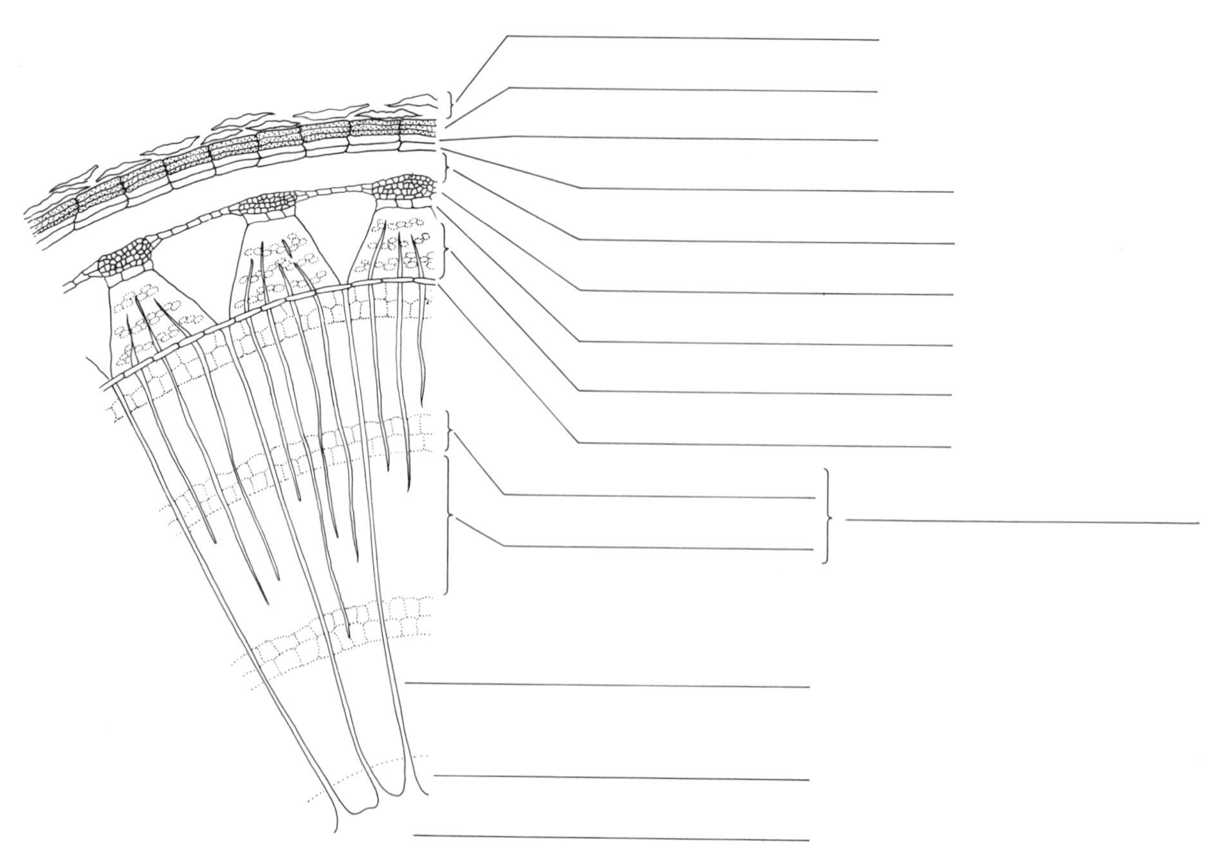

FIG. 5-7
Cross section of a 3-year-old basswood stem.

How do you account for their early death?

____

____

These cells are filled with **tannin** and are darkly stained.

2. *Cork cambium:* The cells of the cork cambium are shaped like the cork cells, but in the cork cambium cell nuclei are usually plainly visible.

3. *Cork parenchyma:* In addition to producing cork, the cork cambium produces a small layer of parenchyma cells located just inside the cork cambium layer . The walls of these cells are flattened on the radial and outer tangential sides, with the inner side being rounded like the adjacent mechanical tissue cells.

c. Cortex

This zone is composed of several layers of thick-walled cells—**mechanical tissue**—and, just within the mechanical tissue, an area of thin-walled parenchyma cells. Look for unique crystals in some of these cells.

d. Vascular Cylinder

1. *Phloem:* Recall from lectures and previous laboratory work the physiological function of the phloem.

   a. *Primary phloem:* The thick-walled phloem fibers are located as a cap over the remainder of the primary phloem. Examination of the thick-walled phloem with the high-power objective of your microscope will show the wall structure and reveal the very small "lumen" or cavity of the cell. Parenchyma cells are located at the outer end of the rays.

   b. *Secondary phloem:* The secondary phloem in the basswood stem is conspicuously banded with alternating layers of thick-walled cells **(phloem fibers)** and layers of thin-walled cells (the other types of phloem cells such as **sieve tubes, companion cells,** and **phloem parenchyma).** The thick-walled phloem fibers here are very similar in appearance to the thick-walled cells of the primary phloem.

2. *Vascular cambium:* The cambium consists of a layer of thin-walled cells between the xylem and phloem. Cells of the cambium are rather uniform in size and appear rectangular in cross section. The cells of the cambium differ from those of other tissues in that they continue to divide. The planes of division are largely tangential. The cambium layer does not increase in thickness, because the daughter cells produced mature into xylem and phloem cells. In the vascular bundle, therefore, the cells that flank the cambium have been produced by it. Although primary and secondary phloem differ from each other in origin, they are similar in structure and function and, in the cross section of the stem, cannot be distinguished from each other except by position. The same is true of primary and secondary xylem.

3. *Xylem*

a. *Secondary xylem:* The secondary xylem of woody plants normally shows a differentiation between the xylem cells formed early in the growing season **(spring wood)** and those xylem elements formed later in the growth period **(summer wood).** Those formed during the spring are larger and the amount of wall material is small in proportion to the total cell area. Those produced during the later part of the growing season are smaller and the walls are relatively thick. The zone of spring wood and summer wood produced during one growing season constitutes an **annual ring.** The annual ring shows up conspicuously because of the size difference between the small cells produced at the end of one growing season and the large ones formed when growth is resumed in the next growing season.

b. *Primary xylem:* The primary xylem is located in contact with the pith. Usually primary xylem cells are somewhat more irregularly arranged and are smaller in diameter than the secondary xylem cells. The primary xylem makes up only a small portion of the inner part of the first annual ring.

e. Pith

The pith occupies the central area of the stem. In the pith of the basswood stem, there are numerous cells containing tannin and mucilage material, which strongly absorbs the stains used in preparation of the slides. Near the outer edge of the pith, there is a ring of such darkly stained cells.

f. Rays

Some of the rays found in the stem extend from the pith outward to the cortex. As the stem grows in diameter, the cambium adds new parenchyma cells to these rays and also initiates new rays. These later-formed rays do not extend to the pith or outwardly to the cortex: they extend from the point in the xylem at which they originated outwardly through the secondary xylem and secondary phloem produced since the time of the origin. Sometimes the portions of rays found in the xylem, or wood, are called **"wood rays."**

REFERENCES

Cutler, D. 1978. *Applied Plant Anatomy.* Longman.

Cutter, E. 1978. *Plant Anatomy: Cells and Tissues.* Part 1. 2d ed. Arnold.

Cutter, E. 1982. *Plant Anatomy: Organs.* Part 2. 2d ed. Addison-Wesley.

Epstein, E. 1973. Roots. *Scientific American* 228:48–58 (Offprint 1271). *Scientific American* Offprints are available from W. H. Freeman and Company, 41 Madison Avenue, New York, NY 10010 and 20 Beaumont Street, Oxford OX1 2NQ, England. Please order by number.

Esau, D. 1977. *Anatomy of Seed Plants.* 2d ed. Wiley.

Fahn, S. 1982. *Plant Anatomy.* 3d ed. Pergamon Press.

Foster, A. S., and E. M. Gifford, Jr. 1974. *Comparative Morphology of Vascular Plants.* 2d ed. W. H. Freeman and Company.

Jensen, W. A., and F. B. Salisbury. 1984. *Botany.* 2d ed. Wadsworth.

Raven, P. H., R. F. Evert, and H. Curtis. 1981. *Biology of Plants.* 3d ed. Worth.

Saigo, R. H., and B. W. Saigo. 1983. *Botany: Principles and Applications.* Prentice-Hall.

Torrey, J. G., and D. T. Clarkson, eds. 1975. *The Development and Function of Roots.* Academic Press.

Weier, T. E., C. R. Stocking, and M. G. Barbour. 1982. *Botany: An Introduction to Plant Biology.* 6th ed. Wiley.

EXERCISE 6

Plant Anatomy: Leaves, Flowers, and Fruits

We have already discussed two plant organs in Exercise 5: the root, which functions chiefly as an absorbing organ, and the shoot, which serves to conduct water and soluble materials. A third organ, the leaf, is specialized to carry on photosynthesis; a physicochemical process that converts light energy to chemical energy. All forms of life, with the exception of some types of bacteria, depend on the process of photosynthesis at some point in their food chain.

Flowers are reproductive branches consisting of the stem axis and lateral appendages called **sepals, petals, stamens,** and **carpels.** The significance of the flower lies in its role in sexual reproduction and its contribution to the formation of the fruit.

A. LEAVES

Leaves consist basically of three parts: the **blade,** an expanded or flattened portion; the **petiole,** a thin, stemlike portion; and the **stipules,** small, paired, lobelike structures at the base of the petiole. The petiole and stipules may not be present in some plants.

Leaves may be classified according to the type of venation or the arrangement of the leaves on the stem. Thus, leaves are parallel-veined when the larger vascular strands traverse the leaf without apparent branching, and net-veined or reticulate when the main branches of the vascular system form a network. According to the position on the stem, leaves may be **alternate** or **opposite** (borne singly or in pairs at each node). Examine the plants on demonstration, and become familiar with the parts of the leaf and the simple classification just given.

1. Epidermis

Obtain a leaf from a bean or geranium plant. Remove a small piece of the lower epidermis by ripping the leaf as shown in Fig. 6-1A. Mount the thin, transparent piece of tissue in a drop of water on a slide, and then add a coverslip (Fig. 6-1B). Observe under low and high power. Note that many of the

A. Remove a leaf from a plant. Make a ragged tear in the leaf to obtain a piece of the epidermis.

B. Place epidermis in drop of water on slide and place a coverslip over it.

C

Stoma

Guard cell

Other epidermal cells

D

E

CROSS SECTION OF LILAC LEAF

F

Upper epidermis

Palisade tissue

Stoma

Guard cell

Bundle sheath

Intercellular space

Leaf mesophyll

Phloem

Xylem

Vascular tissue (vein)

FIG. 6-1
Study of leaf anatomy.

epidermal cells have an irregular shape. Scattered throughout the epidermis are openings called **stomata** (singular, **stoma**) (Fig. 6-1C). Each stoma is surrounded by two bean-shaped guard cells. Suggest a function for the stomata.

Which of the epidermal cells contain chloroplasts?

What is the function of the chlorophyll in these cells?

Where it is difficult to remove a piece of the epidermis you can use the following procedure to make a surface replica to show stomatal distribution or other surface features.

1. Place two leaves of the same type, one facing up and one facing down, on masking tape to hold them in position.
2. "Paint" a thin layer of Elmer's glue over the exposed surface.
3. Allow the glue to nearly dry. When it is peelable, remove the strip carefully from the surface of the leaf.
4. Mount the *replica strip* on a slide (without water or a coverslip) and examine through the microscope. What differences are observed between the living epidermal peel and the replica strip?

2. Internal Anatomy

Obtain a prepared slide of a cross section of a lilac *(Syringa)* leaf for examination of the internal tissues (Fig. 6-1F). Below the upper epidermis is a region of **palisade tissue,** consisting of one or more layers of elongated cells with the long axis of the cell perpendicular to the surface of the leaf. Palisade cells contain numerous chloroplasts. What is the function of this tissue?

Beneath the palisade tissue, locate a region of rounded, cells, which constitute the **spongy** tissue. The palisade and spongy tissue are collectively called **leaf mesophyll.** Note the presence of numerous intercellular spaces in the mesophyll. Locate the stomata. What is the relationship between the stomata and the intercellular spaces of the mesophyll tissue?

The veins of a leaf are vascular bundles (strands) that are continuous with the vascular tissues of the petiole and stem. A vein contains xylem and phloem and has the same cellular elements as the stem. What is the function of the **bundle sheath** surrounding the vascular strand?

Examine demonstration slides of a leaf from corn, a monocot. Describe any differences in the arrangement of tissues between this leaf and the net-veined lilac leaf.

Examine demonstration slides of cross sections of a pine needle leaf (Fig. 6-2). Note the heavily cutinized epidermal cells. Locate a layer of sclerenchyma just beneath the epidermis. Find depressions, or pits, in the epidermis. Where are the stomata located in the pine leaf?

How does the tree benefit by having the stomata located as they are?

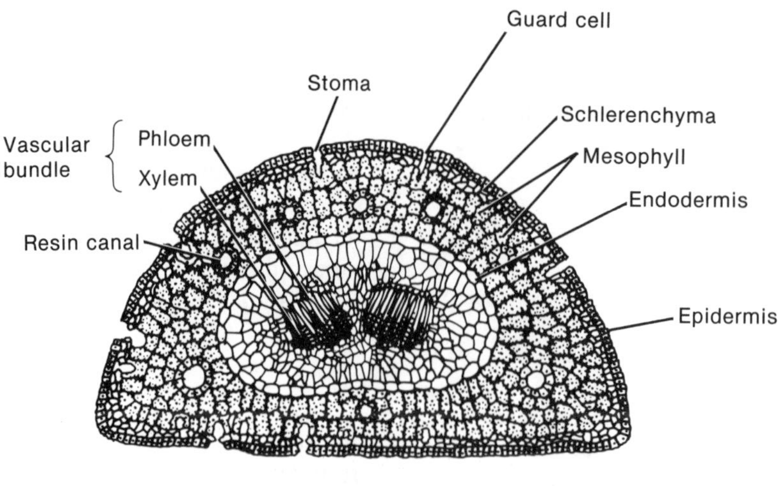

FIG. 6-2
Pine needle leaf.

Locate the central vascular cylinder surrounded by an endodermis. How many vascular bundles are present?

____

Examine cross sections of leaves of other species of pines. Describe any variation in structure among the various needles.

____

____

Examine the slides of the various leaves with a dissecting microscope. What correlation exists between the number of leaves in a cluster and the shape of an individual needle leaf?

____

____

____

____

3. Vascular System of the Leaf

The vascular system of the leaf consists of a series of veins that distribute water, minerals, and various plant growth regulators into the leaf through the xylem. It also transports photosynthetic products out of the leaf through the phloem. In order to appreciate the extensive nature of the leaf's vascular network, you can examine a leaf "cleared" of its pigmentation using hot lactic acid.

Obtain a monocot leaf and a dicot leaf cleared by immersion in hot lactic acid for several hours. Rinse the acid from the leaves with tap water. Examine the dicot leaf and note the variety of branching among the veins beginning with the prominent **midrib (primary vein)** and continuing down through the **secondary** and **tertiary veins** until the vein endings are located. Note that the finer veins divide the leaf mesophyll into small areas, called **areoles,** where the veins terminate. As a result, no mesophyll cell is very far from a vein.

Recall that in your observations of a cross section through a dicot leaf you observed that the larger veins were surrounded by a bundle sheath. Locate these bundle sheaths in the cleared leaf. Describe their distribution (i.e., are they found throughout the entire vascular network or associated with only certain of the veins?).

____

____

____

Compare the monocot leaf with the dicot leaf with respect to the following characteristics:

1. Presence of secondary and tertiary veins. _______________

____

____

2. Extent of bundle sheaths. ______________

____

____

____

4. Morphological Adaptations of Leaves to the Environment

Water is one of the basic raw materials of photosynthesis. It is the major component of plant tissues, making up 90% of the plant body. Water is the substance in which most materials enter and leave the cells of plants, and it is the solvent for the various biochemical reactions that occur in living cells.

The amount of water used by plants is far greater than that used by animalss of comparable weight. The reason for this is that a large amount of the water used by animals is recirculated in the form of blood plasma or tissue fluid. In plants, over 90% of the water taken in by the root system is evaporated into the air as water vapor. This process, which largely occurs through the leaves, is called **transpiration.** Consequently, plants not only have developed extensive and efficient transport systems but also have evolved numerous morphological adaptations, many of which involve the leaves, to conserve water. In this exercise, you will examine some of the modifications of plants and plant parts to obtain and retain water.

Many types of habitats can be found in nature with respect to water supply. These can be conveniently divided into **xeric, mesic,** and **hydric** habitats. The plants that are adapted for living in these habitats are called *xerophytes, mesophytes,* and *hydrophytes,* respectively.

Xerophytes include a number of species that live in habitats where the supply of water is physically or physiologically deficient (xeric habitat). **Mesophytes** inhabit regions of average or optimum water conditions (mesic habitat) and include the majority of wild and cultivated plants of the temperate regions. **Hydrophytes** form an extensive flora living on the surface of water or submerged at various depths (hydric habitat).

In this exercise, representative examples of hydrophytes and xerophytes will be studied, because the structural adaptations are more obvious in these groups. Following your examination of specimens characteristic of these groups, you will be asked to classify unknown specimens.

a. Hydrophytic Adaptations

The chief structural modifications exhibited by hydrophytes are an increase in leaf surface, the presence of air chambers, and a reduction in protective, supportive, and conductive tissues.

1. *Dissected leaves:* Examine a hydric habitat. Note that many of the plants have finely dissected leaves. How may this condition be advantageous to the plant?

____

____

2. *Air chambers:* The leaves and stems of many plants that are submerged in water have chambers filled with air. Examine prepared slides of cross sections of the leaf of *Potamogeton.* Locate these large air spaces, which are separated from each other by partitions of photosynthetic tissue. List several ways in which these air spaces may be of benefit in a hydric habitat.

____

____

3. *Supporting tissues:* Remove one of the hydrophytic plants from the water, and note how flaccid it becomes. This condition is due to the marked reduction of thick-walled supporting tissues or cells. Confirm this by examining prepared slides of *Potamogeton* stems and leaves. Why do submerged hydrophytes not need large amounts of supporting tissues?

____

4. *Vascular tissues:* Because aquatic plants are submerged in, or floating on, a nutrient solution, the structures necessary for absorption and transport of mineral nutrients and water are greatly reduced and in some cases absent. The greatest reduction occurs in the xylem. The phloem, although reduced in amount, is fairly well developed. Examine prepared slides of *Potamogeton* stems and leaves. Locate the vascular tissue, and note the absence of xylem. By what process does water enter aquatic plants?

____

____

5. *Protective tissues:* The epidermis of aerial plants has become modified to prevent or reduce desiccation. Under normal conditions, aquatic plants do not lose water through the epidermis, so the epidermis in hydrophytes is not a protective tissue. In these plants, nutrients and gases may be absorbed directly from the water. The cuticle overlying the epidermis is extremely thin and may be lacking. The epidermal cells usually contain chloroplasts and may form a considerable part of the photosynthetic tissue. Would you expect to find guard cells in submerged hydrophytes? If not, why not?

___

Where would they be located in floating hydrophytes?

___

Examine slides of *Potamogeton* and locate the modifications just indicated.

b. Xerophytic Adaptations

The lack of water that characterizes xeric habitats may be the result of various environmental conditions—for example, intense light, heat, and high-velocity winds. Xerophytes have evolved many adaptations to prevent desiccation of the plant when exposed to one, or any combination of these factors.

1. *Stomata:* The stomates function in the exchange of carbon dioxide and oxygen between the plant and the environment. When the stomates are open, however, water may also leave the plant, which harms the plant as a whole. Consequently, it is important to xerophytic plants to reduce the rate of transpiration. One way this is accomplished is by having the stomates situated below the level of the surrounding epidermal cells. Examine prepared slides of the cross section of a pine leaf and locate sunken stomates. How does this position of the stomates reduce the rate of transpiration?

___

2. *Protective tissues:* In contrast to the hydrophytes, the epidermis of xerophytes commonly has a thick layer of a waxy material called **cutin.** In addition, the walls of the epidermal cells may be highly lignified. How does this cut down on the amount of water lost from the plant?

___

___

___

Examine slides of pine leaf and locate the modifications mentioned here.

3. *Supporting tissues:* Xerophytes generally have a large proportion of supportive tissues that not only prevent water loss but also help support the stem or leaf. Examine a slide of pine leaf, and locate a layer of thick-walled tissue just below the epidermis. Why is it important that stems and leaves of aerial plants be supported in some manner?

___

4. *Leaf rolling:* During drying conditions, the leaves of many xerophytes—notably the xerophytic grasses—roll up tightly. Because the stomates in these plants are more numerous on the upper surface, what effect does this have on the rate of transpiration? Explain.

___

___

___

___

___

Examine a prepared slide of a corn leaf. In the upper epidermis, locate large, bulbous cells. These are **bulliform cells** (motor cells), which function in the rolling of the leaves in dry weather.

5. *Water storage:* Some xerophytes possess large amounts of water-storage tissue. These plants are called *fleshy xerophytes.* Examine specimens of this group. In some plants, the leaves may be fleshy **(leaf succulents).** In others, the stem is fleshy, and therefore the plants are called **stem succulents.** Note the absence, or greatly reduced number, of leaves on the stem succulents. What is the primary photosynthetic organ in these plants?

___

Cut a thin cross section of a leaf succulent (for example, *Aloe*), and mount it in a drop of water on a slide. Examine microscopically. Note the large amount of water-storage tissue that makes up the bulk of the leaf. What other xerophytic characteristics are present that you are able to observe microscopically?

____

____

____

____

____

c. Unknown Specimens

Examine slides of the species of plants listed in Table 6-1. Enter the information asked for in the table, and then decide in what habitat (xeric or hydric) each plant would be found. Your decision should be based on the information obtained from the examination of the various slides and characteristics cited in this study.

TABLE 6–1
Adaptations of vascular plants to the environment.

Species	Characteristics						Habitat and reason for choice (xeric, hydric)
	Air chambers	Supporting tissues	Protective tissues	Leaf modifications (stomata, motor cells, other)	Vascular tissue	Water storage tissue	
Potamogeton							
Yucca							
Myriophyllum							
Ammophila							
Acorus							
Typha							
Pinus							

B. FLOWERS

The roots, stems, and leaves develop and function so as to ensure the successful existence of the plant. Ultimately, however, the plant dies. It is through a constant succession of new individuals that the species survives. This is accomplished by a process called *reproduction.*

There are two main types of reproduction: **asexual** (or vegetative) and **sexual.** Flowering plants can reproduce asexually by numerous means. For example, in strawberry plants the tip of a branch may arch down, touch the soil, and develop leafy branches that root and form new plants. The most common method of reproduction, however, is by sexual means. To understand sexual reproduction in plants, it is necessary to study the anatomy of the flower.

Your instructor will provide you with two or three different flowers that you may use for this part of the exercise.

With the help of Fig. 6-3, locate the following parts of these flowers:

Sepals (which are modified leaves) are the outermost structures of the flower. They are typically green, although they may be other colors. In some cases, they may be absent.

Petals lie to the inside of the sepals and are often brightly colored. Both the sepals and petals are attached to the enlarged end of the branch—the **receptable.**

Carefully remove the sepals and petals. In the center of the flower, locate a stalklike structure (Fig. 6-3A, B). This is the female part of the flower: the **pistil.** Pistils are made up of one or more **carpels,** leaflike structures bearing seedlike structures called **ovules.** It is thought that, during evolution, the carpels rolled inward, enclosing the ovules. The pistil is composed of a swollen base, the **ovary,** and an elongated **style** that terminates in a **stigma.** A flower may contain more than one pistil.

The ovary contains one or more ovules. You can see these if you cut the ovary lengthwise and examine it with a dissecting microscope. One of the cells in the young, developing ovule undergoes cell division through meiosis, where the daughter cells formed have one-half (haploid) the number of chromosomes possessed by the parent cell. One of these haploid daughter cells develops into a microscopic haploid plant that will produce a female gamete, the egg cell (Fig. 6-3C, D, E). Why is it necessary that meiosis occur during the formation of the gametes?

____

____

____

Locate the stamens that surround the pistil (Fig. 6-3F). These are the male parts of the flower and consist of a terminal capsule—the **anther**—attached to a slender **filament.**

The anthers also contain cells that undergo meiosis to produce cells that eventually develop into microscopic gamete-producing plants, called **pollen grains** (Fig. 6-3G, H). Crush a small piece of an anther in a drop of water on a slide and add a coverslip. Examine it with your microscope and locate the pollen grains. If available, examine pollen from different plants. Note the diversity in size and in surface markings. During **pollination,** the pollen grains are transferred to the stigma—the sticky surface of the pistil. The pollen grain germinates, and the pollen tube grows through the style to the ovary and enters the ovule (Fig. 6-3I). It is known that many organisms, or parts of organisms, grow toward or away from various stimuli such as light, chemicals, and gravity. Suggest a mechanism that could direct the growth of the pollen tube toward the ovules.

____

During the growth of the pollen tube, two male gametes (sperm) are produced. These are released when the tube enters the ovule. One sperm fertilizes the egg cell and the other unites with the nucleus of another cell in the ovule. This second fertilization results in the formation of a special tissue—the **endosperm**—which functions as a nutrient tissue to the developing embryo.

C. FRUITS

Fertilization of the egg initiates extensive changes in the carpel. The ovary enlarges and develops into the fruit; in some cases, other floral parts also become part of the fruit. The ovules develop into seeds, which now contain the embryos of new plants (Fig. 6-3J, K).

The fruit protects and disperses the seed, which contains the next generation of the plant. Indeed, numerous and varied adaptations have evolved for obtaining maximal dispersion of the species. For example, some fruits have wings or similar structures that use the wind to disperse them. Fleshy fruits having hard, inedible, indigestable seeds may be eaten by animals. The seeds are later dispersed in the feces during the animals wanderings, which not only disperses the seed but also supplies the growing plant

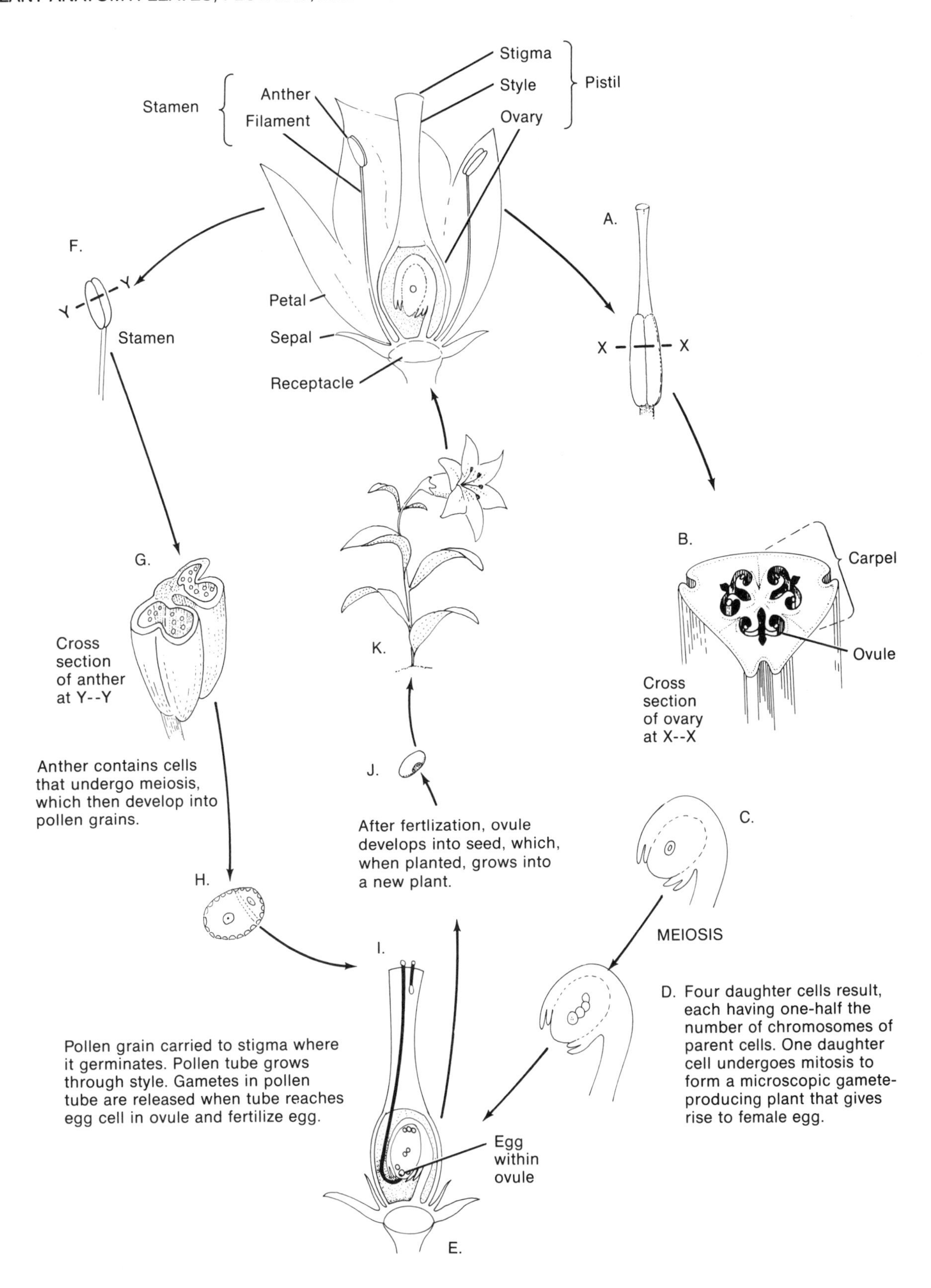

FIG. 6-3
Study of flower anatomy.

with nutrients. Spiny fruits may be inadvertently picked up by animals and carried along.

Structurally, the fruit consists of a mature ovary or cluster of ovaries. In some cases, other parts of the flower are modified and incorporated into the ovary and form part or all of the fruit. The seeds, usually located inside the fruit, develop from the ovules.

During the development of the fruit, the wall of the ovary, called the **pericarp,** usually thickens and becomes differentiated into three layers which may or may not be easy to distinguish visually depending upon the species. These three layers are called the **exocarp** (outer epidermal layer), the **mesocarp** (middle layer), and the **endocarp** (inner layer). As an example, in the peach the exocarp is the skin, the mesocarp consists of the fleshy part of the peach, and the stony pit is the endocarp. The seed, containing the embryo, is inside the pit.

1. Classification of Fruits

A simple classification of fruits follows.

1. **Simple fruits** (derived from a single ovary).
   a. Pericarp fleshy. Examples: drupe, pome, berry, pepo.
   b. Pericarp.
      (1) Indehiscent (does not split open when ripe). Examples: akene, nut, caryopsis.
      (2) Dehiscent (splits open when ripe). Examples: legume, silique, capsule, follicle.

2. **Aggregate fruits** (derived from numerous ovaries of a single flower that are scattered over a single receptacle and later unite to form a single fruit). Examples: strawberry, blackberry, raspberry.

3. **Multiple fruits** (derived from the ovaries of several flowers united into a single mass). Example: pineapple.

2. Some Common Types of Fruit

In this exercise, you will become familiar with several types of fruit. After examining them, fill in Table 6-2 using the information given in the classifi-

TABLE 6-2
Characteristics of some common fruits.

Name of plant	Name of fruit	Dry or fleshy	Dehiscent or indehiscent	Structure(s) other than ovary involved in fruit formation	Fruit type

cation scheme from Part 1 and the characteristics from your observations of each fruit type.

a. Legume (Fig. 6-4A)

Examine a bean or pea pod. Along how many sides does this fruit split open?

____

Remove a bean (or pea) seed and locate the

Hilum: a scar on the seed representing the point of attachment of the seed to the wall of the fruit.

Micropyle: a small opening adjacent to the hilum through which the pollen tube entered the ovule. It now serves to admit water to the seed as a preliminary step in the germination process.

Carefully split open the seed. The two fleshy halves, called **cotyledons,** are attached to the embryo, which consists of an embryonic root (**radicle**) and bud (**plumule**).

b. Follicle (Fig. 6-4B)

Examine a milkweed pod. Open it and remove some seeds. Suggest a function for the large number of "hairs" associated with each seed.

____

____

How does the pod of the follicle differ from that of the legume?

____

____

c. Akene (Fig. 6-4C)

Examine the fruit of the sunflower. Crack open the hard fruit coat and examine the seed. Remove the seed coats to observe the embryo. In other plants, such as dandelions and lettuce, the **akenes** are winged (Fig. 6-4D). Of what advantage is this to the species?

____

____

____

d. Samara (Fig. 6-4E)

Examine the winged fruit of the ash or maple tree. Let the fruit drop from a height of several feet. Explain how the response you observe is of value to the plant?

____

____

____

e. Nuts (Fig. 6-4F)

True nuts are represented by acorn and filberts as opposed to what you might call "commercial" nuts such as almonds or walnuts (which are the stones of drupes) or Brazil nuts (which are hard-walled seeds). In true nuts the pericarp is stony.

Obtain and examine an acorn. At its base locate a cuplike structure called an **involucre.** Remove the seed from the shell. Is this seed from a monocot or a dicot plant?

____

____

f. Berry (Fig. 6-4G)

Examine the fruit of a grape or tomato. Cut them in half and determine the arrangement of the seeds and the number of carpels that are present.

A modified berry, called a **hesperidum** (Fig. 6-4H) has a leathery rind (oranges and lemons). From what part of the flower is the pulp in this fruit derived?

____

g. Drupe (Fig. 6-4I)

Cherries, peaches, plums, and olives are examples of **drupes.** Remove the fleshy mesocarp from only one side of the fruit exposing the stone. Then remove the stone and crack it open to expose the seed. Usually only an abortive ovule may be present; occasionally two seeds may be found.

h. Aggregate Fruit (Fig. 6-4J)

Examine a strawberry in which the fleshy part of the fruit consists of the enlarged receptacle. The seedlike structures embedded in the "flesh" are small akenes.

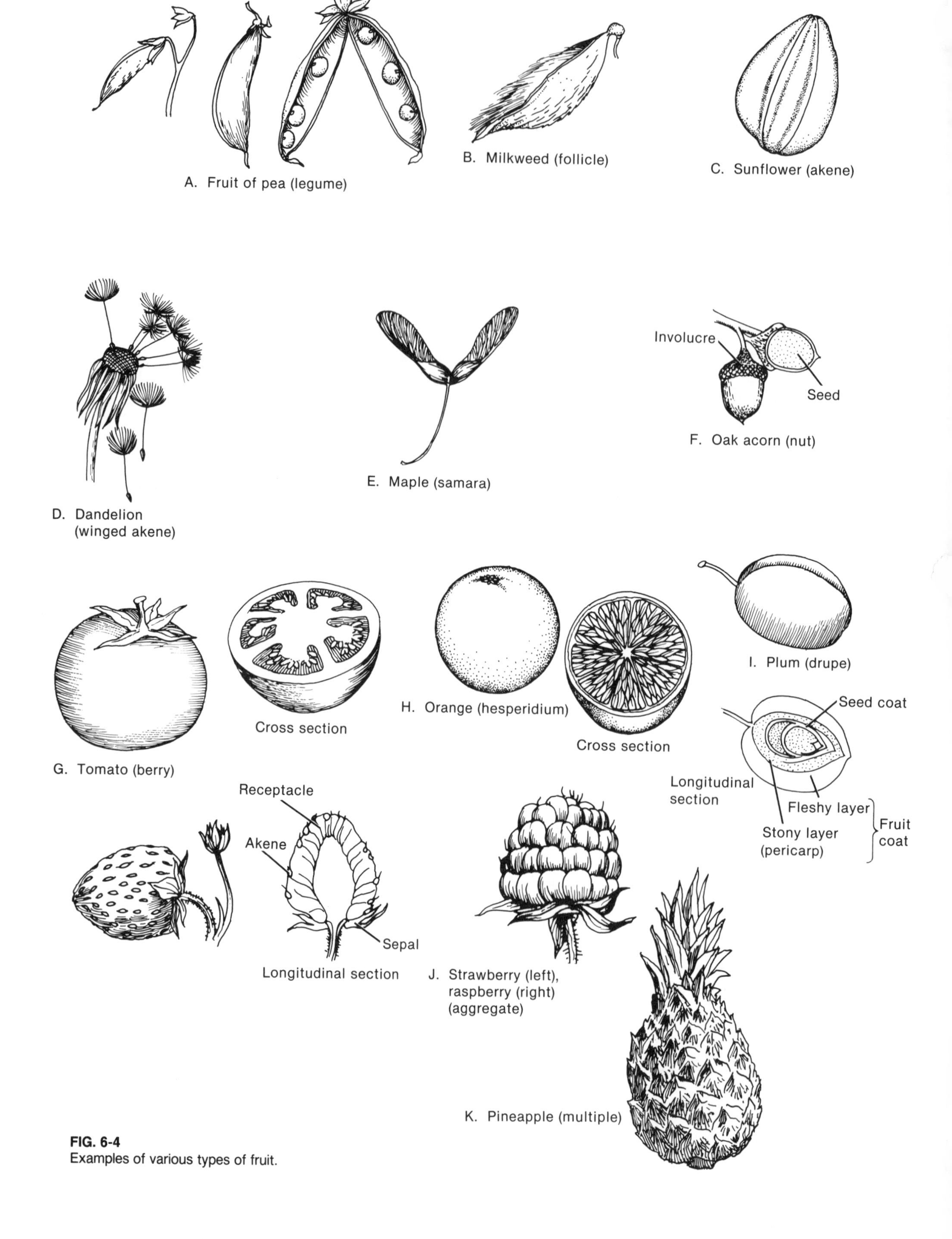

FIG. 6-4
Examples of various types of fruit.

i. Multiple Fruit (Fig. 6-4K)

Examine the fruit of a pineapple, which consists of an aggregation of individual fruits (each derived from a single flower) spirally arranged around a fleshy axis.

REFERENCES

Bold, H. C. 1980. *Morphology of Plants and Fungi.* 4th ed. Harper and Row.

Cutter, E. 1978. *Plant Anatomy: Cells and Tissues.* Part 1. 2d ed. Addison-Wesley.

Echlin, P. 1968. Pollen. *Scientific American* (Offprint 1105). *Scientific American* Offprints are available from W. H. Freeman and Company, 41 Madison Avenue, New York, 10010, and 20 Beaumont Street, Oxford OX1 2NG, England. Please order by number.

Esau, K. 1977. *Anatomy of Seed Plants.* 2d ed. Wiley.

Foster, A. S., and E. Gifford. 1974. *Comparative Morphology of Vascular Plants.* 2d ed. W. H. Freeman and Company.

Jensen, W. A., and F. B. Salisbury. 1984. *Botany.* 2d ed. Wadsworth.

Pijl, S. Van Der. 1982. *Principles of Dispersal in Higher Plants.* Springer-Verlag.

Raven, P. H., R. F. Evert, and H. Curtis. 1981. *Biology of Plants.* 3d ed. Worth.

Saigo, R. H., and B. W. Saigo. 1983. *Botany: Principles and Applications.* Prentice-Hall.

Stebbins, H. S. 1974. *Flowering Plants. Evolution Above the Species Level.* Belknap Press of Harvard University Press.

EXERCISE 7

Vertebrate Anatomy: External Anatomy, Skeleton, and Muscles

Frogs are probably the most commonly studied animals in general biology and zoology courses because they are easy to obtain and dissect and their structure clearly shows the major features of the organ systems characteristic of backboned animals, or vertebrates. Furthermore, the frog is one of the most widely used animals in vertebrate physiology and other experimental work. Vertebrates are of particular interest to us because we, too, are members of the phylum Vertebrata. It is therefore an animal with which you should be familiar.

Contemporary amphibians are grouped into three orders: frogs and toads, salamanders, and legless, wormshaped caecilians of the tropics. All reproduce in the water or in very moist places on land. Most go through an aquatic larval stage, commonly called the **tadpole.** This period of development is followed by a metamorphosis to a terrestrial adult. Adult amphibians have only a rudimentary ability to conserve body water, so they must live in damp habitats on land and must often return to water. As a result of their phylogenetic position between fishes and reptiles and their double mode of life in water and on land, amphibians have a mixture of aquatic and terrestrial attributes.

The most widespread of our North American frogs is the leopard frog, also known as the grass or meadow frog. Its scientific name is *Rana pipiens.* Another frog, which is often studied because of its large size, is the bullfrog, *Rana catesbeiana.*

(*Note:* In this exercise, which deals with vertebrate anatomy, refer to Appendix A for a discussion and glossary of the vocabulary used in the study of anatomy.)

A. EXTERNAL ANATOMY OF THE FROG

The body of a frog is divided into a **head**, which extends posteriorly to the shoulder region, and a **trunk** (Fig. 7-1). Fishes, which were ancestral to amphibians, have a powerful tail used in swimming. Aquatic tadpoles retain such a tail, but it is lost in

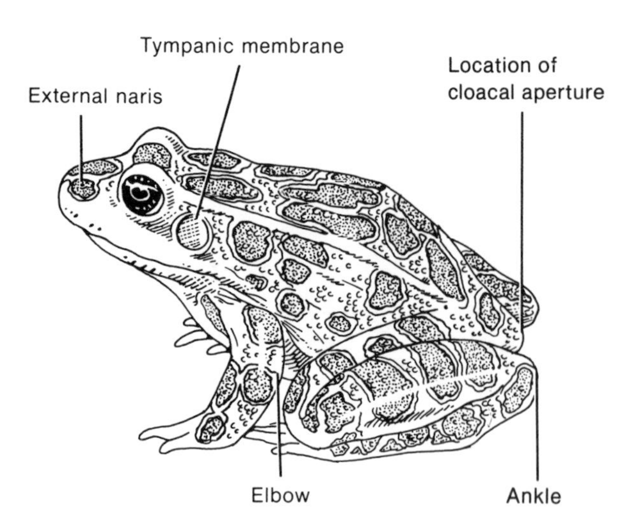

FIG. 7-1
Lateral view of a leopard frog, *Rana pipiens*. (From *Dissection of the Frog*, 2d ed., by Warren Walker, Jr. W. H. Freeman and Company. Copyright © 1981.)

the adult frog, which moves instead by means of its powerful hindlegs. The much smaller front legs are used primarily to keep the front of the body raised from the ground and in the male to clasp the female during mating in the spring. Notice that a distinct neck is absent. This is due to the retention of a characteristic of fishes, for which an independent motion of the head and trunk would be disadvantageous during swimming.

The common opening of the digestive and urogenital tracts, the **cloacal aperture**, is at the posterior end of the trunk just above the attachment of the hindlegs to the trunk. Through this opening are discharged undigested food wastes (feces), the liquid excretory waste (urine) from the kidneys, and the gametes (sperm or eggs) from the reproductive organs. The anus of human beings and other mammals, because it is the opening of just the digestive tract, is only partly comparable to the cloacal aperture.

A large **mouth**, a pair of **nostrils**, or **external nares** (singular **naris**), and the **eyes** will be recognized on the head. The upper **eyelid** is a simple fold of skin. The lower lid is a transparent **nictitating membrane** that can be drawn across the surface of the eyeball. The flat disc-shaped area posterior to each eye is the eardrum, or **tympanic membrane**. Although it is the same size in male and female leopard frogs, it is considerably larger in male than in female bullfrogs. This is one example of **sexual dimorphism** (structural differences between the sexes). Look carefully at the top of the head between the eyes, and you may see a small, light **brow spot**, which is about the diameter of a pin. This is a remnant of a light-sensitive eye that characterized primitive groups of fishes and amphibians.

The appendages of a frog have the same parts as your own arms and legs. In the front leg, or **pectoral appendage**, locate the upper arm, the elbow, the forearm, the wrist joint, and the hand. There are only four fingers. Although the finger closest to the body is comparable to our second digit, it is often called the *thumb* because it is stouter than the others. During the breeding season, the thumb in the male is particularly stout and darkly pigmented—another example of sexual dimorphism. In the hind limb, or **pelvic appendage**, locate the thigh, the knee, the shank, the ankle joint, and the elongated foot. Two elongated ankle bones lie within the proximal part of the foot: the distal part bears five toes with a conspicuous web between them. The first toe, the smallest and closest to the body, is comparable to our great toe. Technically it is called the **hallux**. The small spur at its base is the **prehallux**. The prehallux is much larger in toads, which use their hind feet in burrowing.

B. VERTEBRATE SKIN

All organisms have some sort of external covering. In one-celled organisms, this may be only a thin cell membrane; in higher organisms, the coverings are complex structures consisting of several layers of cells. The skin of vertebrates is a highly functional organ that provides physical protection, excludes disease organisms, acts as a means of water absorption, and in some animals serves an essential role in respiration.

Shortly after hatching, the tadpole develops gills, which also function in the absorption of oxygen. This **branchial respiration** lasts until the tadpole begins to change into a frog. In both the adult and tadpole stages oxygen is absorbed and carbon dioxide is eliminated through the skin. This **cutaneous respiration** occurs both in water and in air. During the winter, when the frog lies buried in the mud, the skin becomes the only respiratory organ. Adult frogs also absorb oxygen and give off carbon dioxide through the lining of the mouth (**buccal respiration**) and through the lungs (**pulmonary respiration**). Even in full-grown frogs, however, three or four times as much oxygen can be absorbed through the skin as through the lungs. In warm-blooded mammals, the skin helps to regulate the temperature of the body. The skin of some vertebrates has been modified into hair (mammals), feathers (birds), scales (fish), and teeth (sharks).

1. Amphibian Skin

Microscopically examine a stained slide of a cross section of frog skin (Fig. 7-2). The stains used are hematoxylin, which stains nuclei blue or purple, and eosin, which stains most cytoplasmic structures pink or shades of red.

The skin consists of two distinct regions. The outer **epidermis** is an epithelial layer, predominantly stained purple because it contains large numbers of nuclei. The **dermis** is predominantly pink and contains only a few nuclei. Rounded structures in the dermis are sections through glands. The dark layer between the epidermis and the dermis contains **chromatophores,** cells that contain pigment granules. Under the influence of light and hormones, these cells lighten or darken and, thus, regulate the protective coloration of the animal.

Note that the cells of the outermost layer of the epidermis are thin (squamous) and lie parallel to the surface; the deeper epithelial cells are more cuboidal. The cells in the basal (bottom) layer are usually columnar in shape—that is, they are longer in the direction perpendicular to the skin surface. Often, the cell membranes are not clearly visible, but the shapes of the nuclei give an idea of the shapes of the cells; if the nucleus is long in one direction, the cell is ordinarily long in the same direction. The epidermis, with its gradation from squamous to columnar cells in successive layers, is an example of **stratified squamous epithelium.** The basal, **germinative layer** continually produces new cells that are pushed out toward the surface, becoming steadily flatter and harder (more **cornified**) as new cells are formed in the germinative layer. Periodically, during the summer, a new layer forms beneath the old one and the old skin covering is molted, or sloughed off.

In the dermis, locate two types of glands: **mucous (slime) glands** and **poison glands.** These glands produce fluids that are secreted onto the surface of the epidermis through ducts. Both types of glands are lined by a layer of cuboidal secretory cells. The poison gland can be identified by its larger size and the granular material within the lumen of the gland. If a frog is roughly handled, the poison glands discharge a thick, whitish secretion that produces a burning sensation, protecting the animal, to a degree, from enemies. The smaller, mucous glands secrete a colorless, watery fluid that keeps the skin moist, glistening, and sticky. The surface openings of these glands can be widened or narrowed by the contraction of a **stoma** cell, which regulates the amount of mucus discharged. Examine a slide of frog skin. Find a stoma cell and note its star-shaped opening.

Just beneath the epidermis locate a prominent layer of pigment cells, or **chromatophores.** Grass frogs, usually protectively colored to resemble their surroundings, are green on the dorsal (back) and lateral surfaces and pale green to whitish on the ventral (belly) surface. The patterns of most amphibians are stable, but the colors of some undergo marked changes. The skin darkens when the pigment granules in the chromatophores spread out and cover other elements in the skin, and it lightens when they are concentrated. Changes in color result from both external conditions and internal states; a low temperature produces darkening, whereas a high temperature, drying of the skin, or increased light intensity causes the color to lighten.

The dermis contains a large number of small blood vessels that transport food and carbon dioxide to the skin and take back oxygen absorbed through surface layers from the air. Nerve and muscle fibers

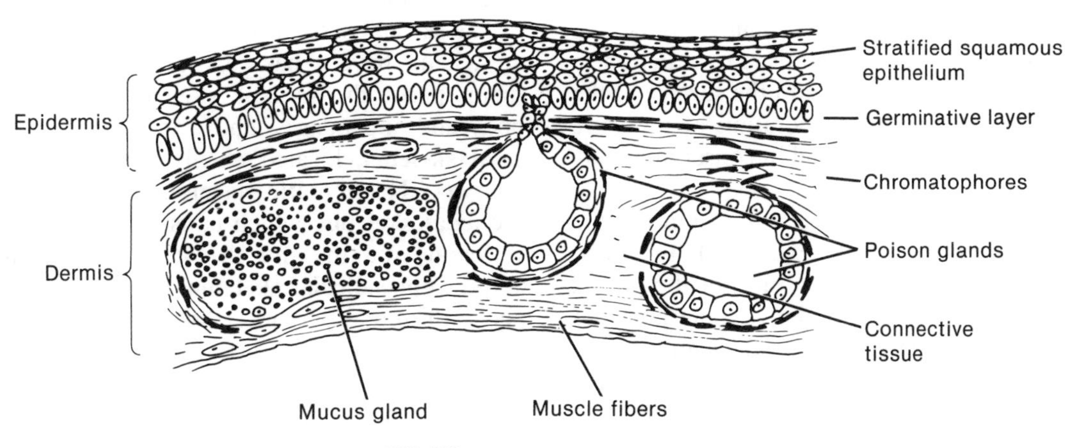

FIG. 7-2
Vertical section through frog's skin.

are also present in the dermis but will not be readily seen in your slides.

2. Mammalian Skin

Examine a slide of mammalian skin taken from the scalp. Compare and contrast it with the skin of the frog. The **epidermis** consists of stratified squamous epithelium, usually composed of several more layers of cells than that of the frog and having more flattened keratinized layers at the surface (Fig. 7-3A). Keratin is a tough proteinaceous material that serves several functions: (1) It is a relatively waterproof substance and therefore prevents water from entering the body through the skin, but, more importantly, it prevents water loss. (2) Because it is a tough material, it protects the underlying epithelial layers from damage through ordinary wear. It is particularly protective on the palms of the hands and soles of the feet.

The **dermis** is composed mainly of densely interwoven connective tissue. Note the abundance of small blood vessels. Suggest a mechanism by which the blood vessels regulate body temperature.

Chromatophores, when present, are usually confined to the basal layer of the epithelium. **Hair follicles** and **sweat** and **sebaceous (oil) glands** are located at various levels in the dermis. Locate a hair follicle, and notice that each hair arises in a tubular invagination of the epithelium. The root of the hair develops into the hair shaft, the free end of which protrudes beyond the surface of the skin. One or more sebaceous glands (Fig. 7-3B) are located in the dermis and open into the hair follicle. These glands secrete an oily substance, called **sebum,** onto the hair or on the surface of the epidermis. The function of sebum is to lubricate the hair and the surface of the skin. It may also prevent evaporation of moisture during cold weather and thus aid in conserving body heat.

Distributed along the surface of the skin are the openings of the sweat glands (Fig. 7-3C). The long, tubular, and highly convoluted ducts of the glands penetrate deeply into the underlying dermis and connect with the secretory parts of the gland.

Because of the highly coiled structure of a sweat gland, it is difficult to see an entire duct or gland in one section. Rather, you will see cross or oblique sections of the various parts (Fig. 7-3C).

The secretions produced by sweat glands flow onto the surface of the skin, where the surface is cooled by the process of evaporation. The activity of these glands is another body temperature regulating mechanism that is under the control of the nervous system.

The **subcutaneous** layer beneath the dermis is composed of fat cells (**adipose tissue**); the number of such cells present depend on the part of the body and the nutrition of the organism. The fat cells in your slide will look empty because the method of preparing the slide dissolved the fat droplet that occupied the greater part of each cell, leaving only a thin film of cytoplasm with its compressed nucleus. The fat cells are held together by fine, fibrous connective tissue. The subcutaneous layer is especially important in heat conservation by acting as an insulating layer.

C. VERTEBRATE SKELETON

One of the more significant evolutionary advances of the vertebrates over the invertebrates is the possession of an internal or **endoskeleton.** The external or **exoskeleton** of the invertebrates limits the ultimate size of an organism and in several cases is so heavy as to restrict the movement of some animals. The endoskeleton of vertebrates, on the other hand, permits relatively unrestricted development in size, as evidenced by such vertebrates as the whale and elephant.

The vertebrate skeleton, in addition to protecting and supporting the internal organs of the body, is a highly efficient supporting structure for the attachment of muscles. Furthermore, the flexibility due to the large number of separate bones, coupled with the strength and relative lightness of vertebrate skeletons, enables even the largest animals to have a great deal of mobility.

The skeletal systems of all animals having an endoskeleton have a basic pattern consisting of two major parts: the **somatic skeleton,** located in the body and appendages, and a less conspicuous **visceral skeleton,** located chiefly in the wall of the pharynx. The somatic skeleton is subdivided into the **axial skeleton,** which consists of the skull, the vertebral column, and the sternum (breastbone), and the **appendicular skeleton,** which consists of the pelvic (hip) and pectoral (shoulder) girdles and the limbs.

As you study the parts of the vertebrate skeleton, examine a mounted specimen of a frog skeleton and, if possible, a human skeleton to see both the similarities and differences between them. (Fig. 7-4).

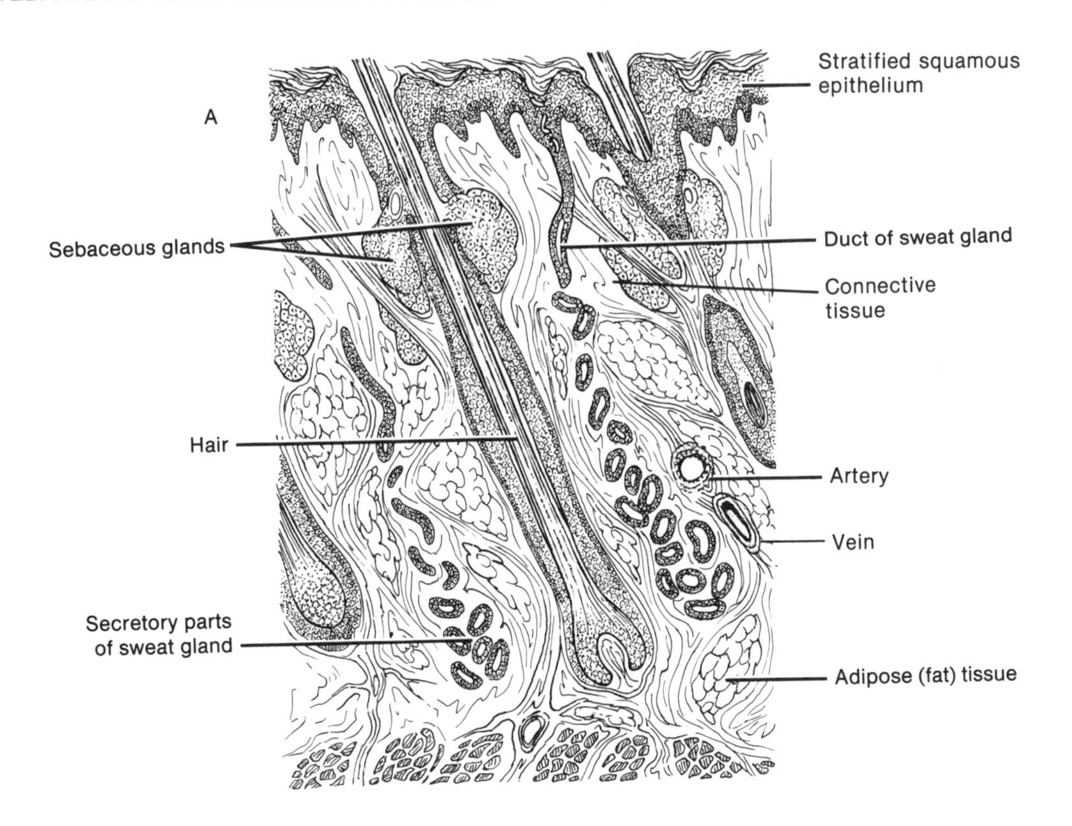

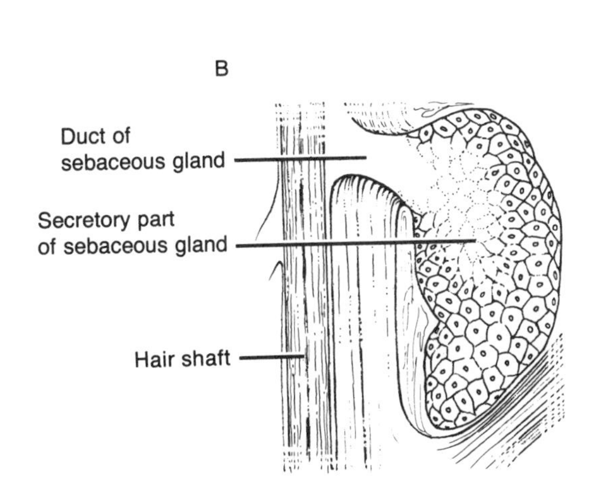

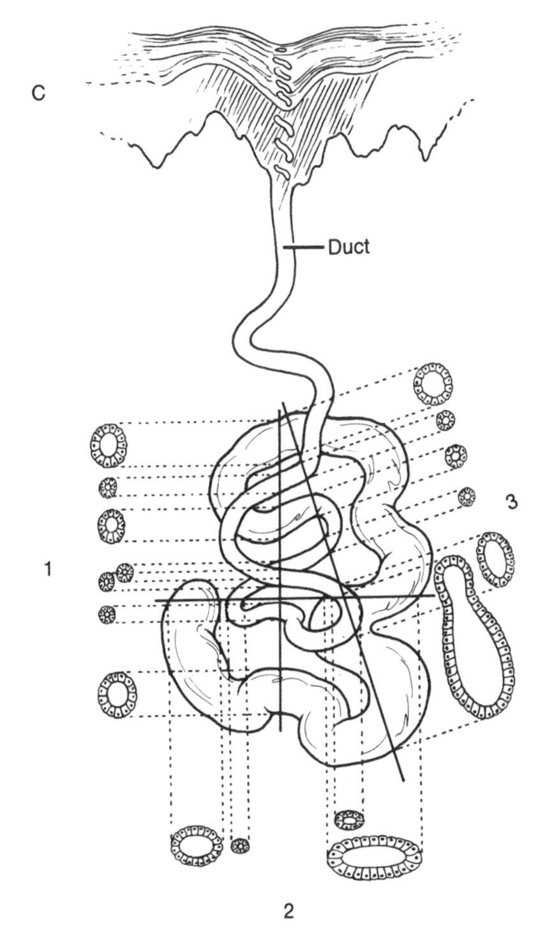

FIG. 7-3
(A) Vertical section of human scalp. (B) Sebaceous gland and adjacent hair follicle. (C) Sweat gland. Sections 1, 2, and 3 indicate the appearance, in cross and oblique section, of various parts of the gland as it would be seen in a tissue section. (After *Atlas of Human Histology,* 3d ed., by M. DiFiore. El Ateneo. Copyright © 1967.)

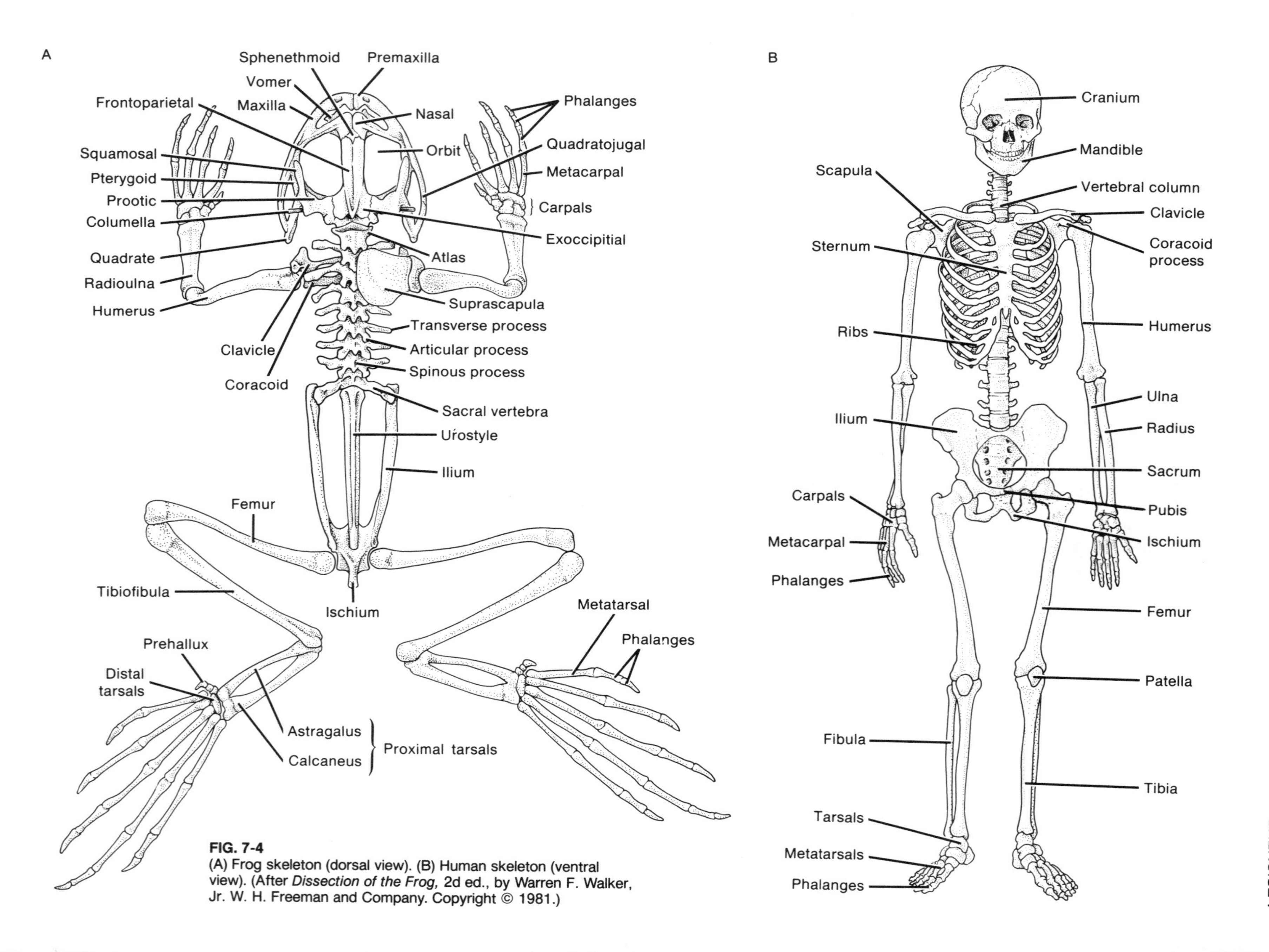

FIG. 7-4
(A) Frog skeleton (dorsal view). (B) Human skeleton (ventral view). (After *Dissection of the Frog,* 2d ed., by Warren F. Walker, Jr. W. H. Freeman and Company. Copyright © 1981.)

1. Axial Skeleton

a. Skull

The **cranium** is the part of the **skull** that contains the brain and the olfactory, optic, and auditory capsules for the organs of smell, sight, and hearing. In sharks and rays, the skull is cartilaginous. In higher vertebrates most of the cartilage has been replaced by bone. Primitive vertebrates tend to have a larger number of skull bones than do higher forms. Some fish have 180 skull bones; amphibians and reptiles, from 50 to 95; and mammals, 35 or less. Human beings have 29.

Examine the dorsal surface of a frog skull (Fig. 7-5). The skull is attached to the anterior end of the vertebral column. Locate the large jaws, the small brain case, or **cranium,** which is roofed mainly by the **frontoparietal** bones, the **nasal** bones (which cover the nasal capsules), the **prootics** (which house the inner ears), and the **exoccipital** bones (each of which has rounded **occipital condyles**). The two condyles fit depressions in the first vertebra, permitting slight movements of the head on the spinal column. Between the condyles is a large opening, the **foramen magnum,** through which the spinal cord passes. Between the prootics and the **maxilla** are the **squamosal** and **pterygoid** bones. These form the lateral borders of the cranium. The upper jaw consists of the small **premaxillary** bones in front and the larger **maxillary** bones that extend posteriorly to join with the pterygoid.

Now examine the ventral surface of the skull. Locate the **maxillary teeth** on the margin of the upper jaw and the **vomerine teeth** on the roof of the mouth (Fig. 7-6A). The **sphenethmoid** bone, which connects with the frontoparietal bone dorsally and the **parasphenoid** bone ventrally, make up the anterior part of the cranium.

The lower jaw (mandible) consists of a rodlike **Meckel's cartilage** encased by the **dentary** and **angulare** bones, which are united with the **quadratojugal bone** by means of the **quadrate cartilage** (Fig. 7-6B). Notice that the bones of the lower jaw do not have teeth.

b. Vertebral Column

The vertebrae vary greatly among different animals and among different regions of the vertebral column in the same animal. The vertebrae in fish are differentiated into trunk vertebrae and caudal vertebrae. In many of the other vertebrates, they are differentiated into neck (**cervical**), chest (**thoracic**), back (**lumbar**), pelvic (**sacral**), and tail (**caudal**) vertebrae. In birds, as well as in human beings, the caudal ver-

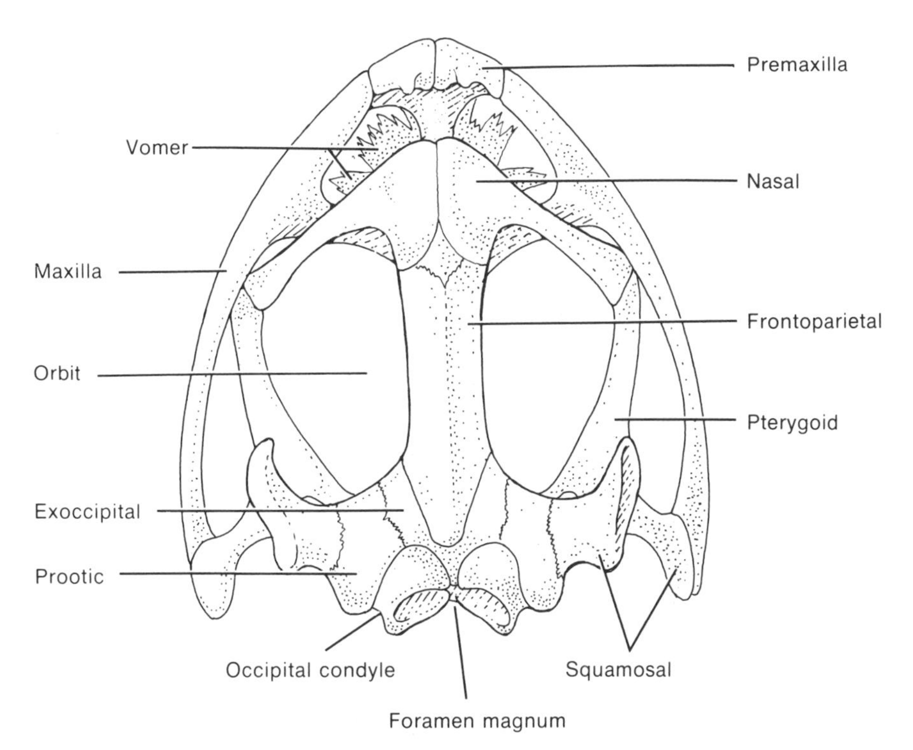

FIG. 7-5
Frog skull (dorsal view).

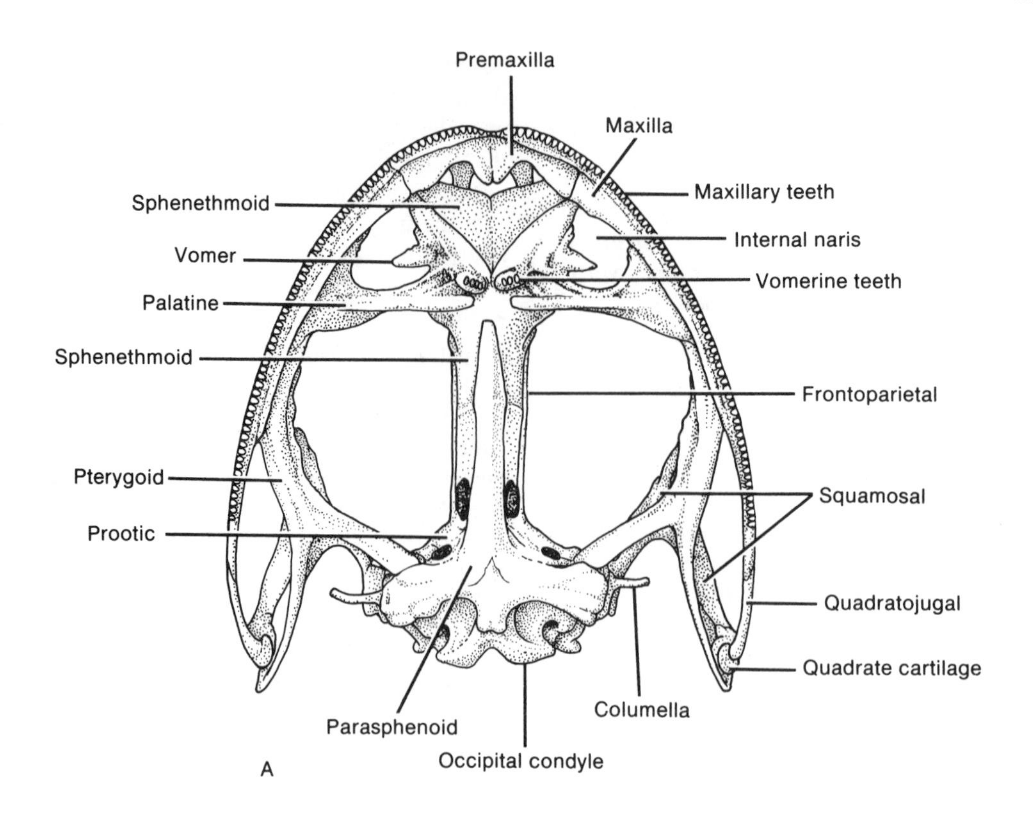

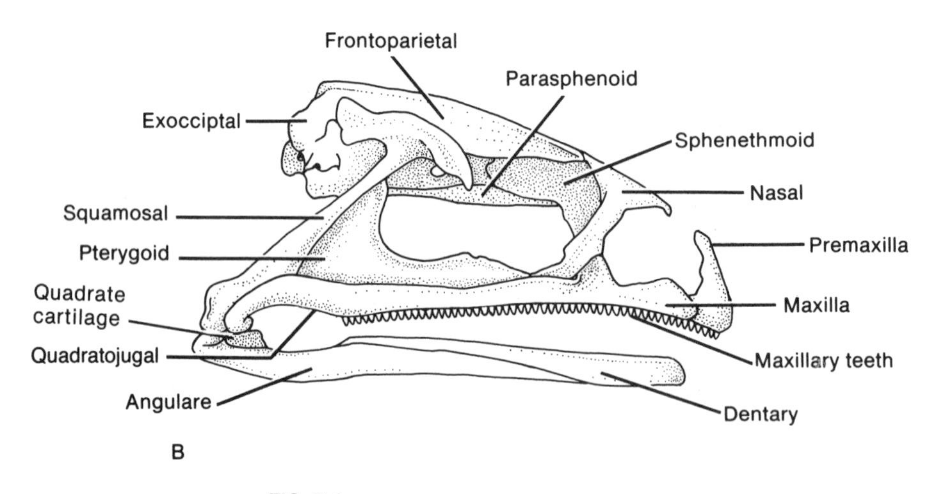

FIG. 7-6
Frog skull: (A) ventral view and (B) lateral view.

tebrae are reduced in number and size; and the sacral vertebrae are fused. The number of vertebrae varies for different animals. The python has the largest number—435. Human beings have 7 cervical, 12 thoracic, and 5 lumbar vertebrae, plus 5 that have fused to form the **sacrum** and **coccyx.** How many vertebrae does the frog have? ______________ The first cervical vertebra in the frog, the **atlas,** is modified for articulation with the skull. The vertebral column terminates in a long bone, the **urostyle.**

Note that the vertebrae are much alike. The lateral projections are called **transverse processes.** The ribs of higher vertebrates are attached to these processes. There are many variations in the ribs of the vertebrates. The basic plan seems to have been a pair of ribs for each vertebra from head to tail, but the tendency has been to reduce the number from the lower to the higher forms. Ribs, however, are not universal among vertebrates; many, including the frog, do not have them at all. Human beings have 12 pairs of ribs.

2. Visceral Skeleton

The visceral skeleton of very primitive fishes consists of a series of arches, of either cartilage or bone, embedded in the lateral wall and the floor of the pharynx. One is located at the base of each gill. Because terrestrial vertebrates do not have gills, their visceral skeleton is greatly reduced. Of what remains in terrestrial vertebrates, some components have become a part of the skull. The quadrate cartilage, Meckel's cartilage, and, in some vertebrates, the columella are of visceral origin, but most of these form the **hyoid apparatus,** a plate of cartilage on the floor of the pharynx at the base of the tongue forming a sling that supports the tongue. Arches of cartilage and bone around the pharynx extend from the hyoid apparatus toward the skull. Some of the visceral skeleton is found in the walls of the glottis, the entrance to the larynx, but the amount is very small and difficult to observe.

3. Appendicular Skeleton

Most vertebrate animals have some form of paired appendages, which are supported by **pectoral** (shoulder) and **pelvic** (hip) **girdles.** Among vertebrates, there are many modifications in the girdles, limbs, and digits, which enable the animals to meet the requirements of their special modes of life. Whatever the modification, in forms higher than the fish, the girdles and appendages have the same basic structure. Both the cartilaginous and bony fishes have pectoral and pelvic fins, which are supported by the pectoral and pelvic girdles, respectively. The single strong bones closest to the body in the legs of a frog are the **humerus** in the anterior limb (Fig. 7-7) and the **femur** in the posterior limb (Fig. 7-8). Distal (away from the body) to these are the **radioulna** (a fusion of the **radius** and **ulna**) in the forelimb and the **tibiofibula** (a fusion of the **tibia** and **fibula**) in the hindlimb. The humerus is attached to the pectoral girdle at the **glenoid fossa** by means of ligaments. The bone passing dorsally, the **scapula,** has a broad extension called the **suprascapula.** Ventrally, and posterior to the **clavicle** is the **coracoid.** At the point of junction of the two clavicles is the sternum, which continues anteriorly as the **omosternum** and **episternum** and posteriorly as the **mesosternum** and **xiphisternum.**

The femur attaches to the pelvic girdle (Fig. 7-8) in a socket, the **acetabulum,** which is formed by the fusion of three bones of the pelvic girdle, the **ilium, ischium,** and **pubis.** The feet and hands are built according to a common pattern, with a number of **carpal** (wrist) or **tarsal** (ankle) bones followed by a group of elongated hand (**metacarpal**) and foot (**metatarsal**) bones, and then the bones of the fingers or **phalanges** (toes). The **astragalus** (inside the foot) joins the **tibiofibula** to the **tarsal** bone.

D. VERTEBRATE MUSCULATURE

1. Joints

Before studying the muscles of the frog, it is advisable to learn something about the joints between bones, because these joints allow the muscles to bend, twist, and turn the various parts of the body. Become familiar with the following classification of joints before proceeding with the muscle dissection.

Ball-and-socket joints allow movement in any direction, including rotation, as in the shoulder and hip.

Condyloid joints allow movement in any direction except rotation, as between metacarpals and phalanges.

Hinge joints allow bones to bend in only one direction, as in the knee, and between flat surfaces of bones, as between the vertebrae.

Plane joints allow sliding movements between flat surfaces of bones, as between the vertebrae.

Radial joints permit rotation of one bone on another, as between the proximal ends (ends nearest the humerus) of the radius and ulna.

2. Muscle Terminology

Muscles function in several ways. They support the body by holding the bones in proper relationship to each other. They are responsible for the movement of the body as a whole, the passage of food down the digestive tract, the ventilation of the lungs, the circulation of the blood, and the movements of most of the materials in the body.

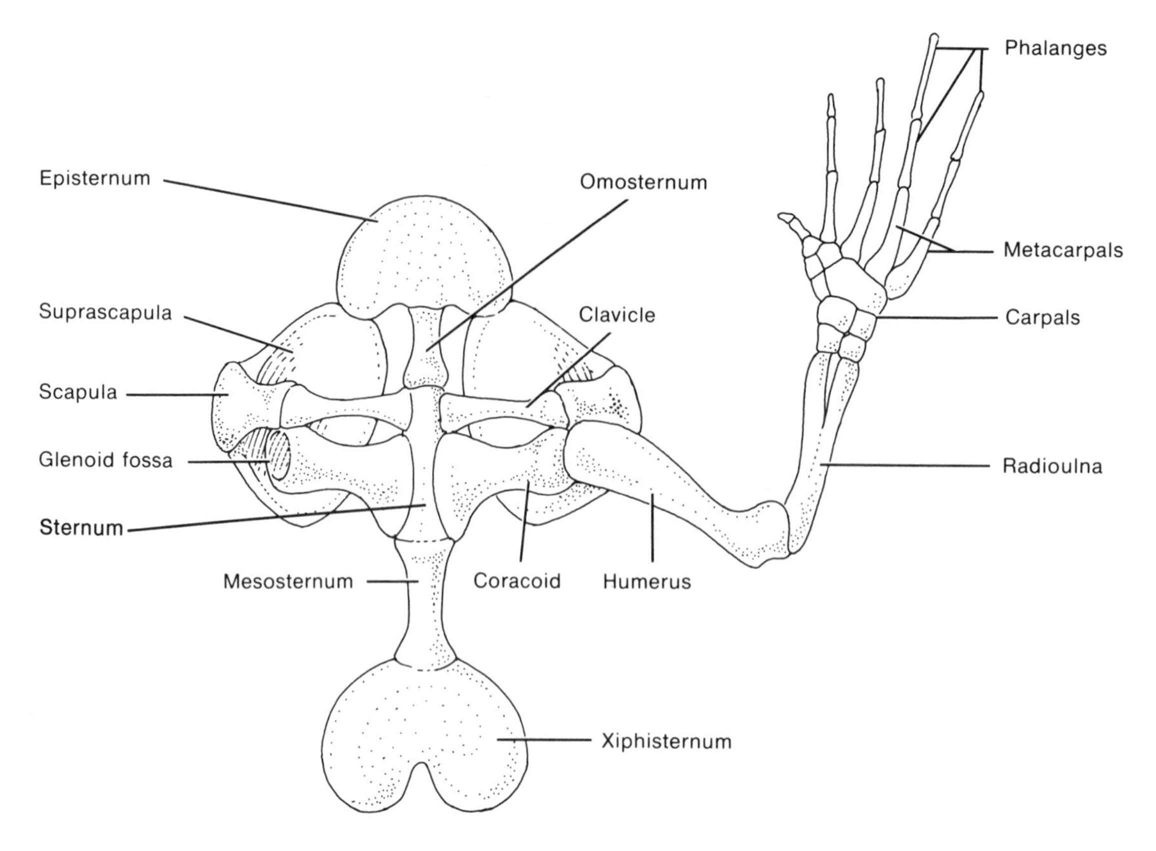

FIG. 7-7
Frog pectoral girdle (ventral view).

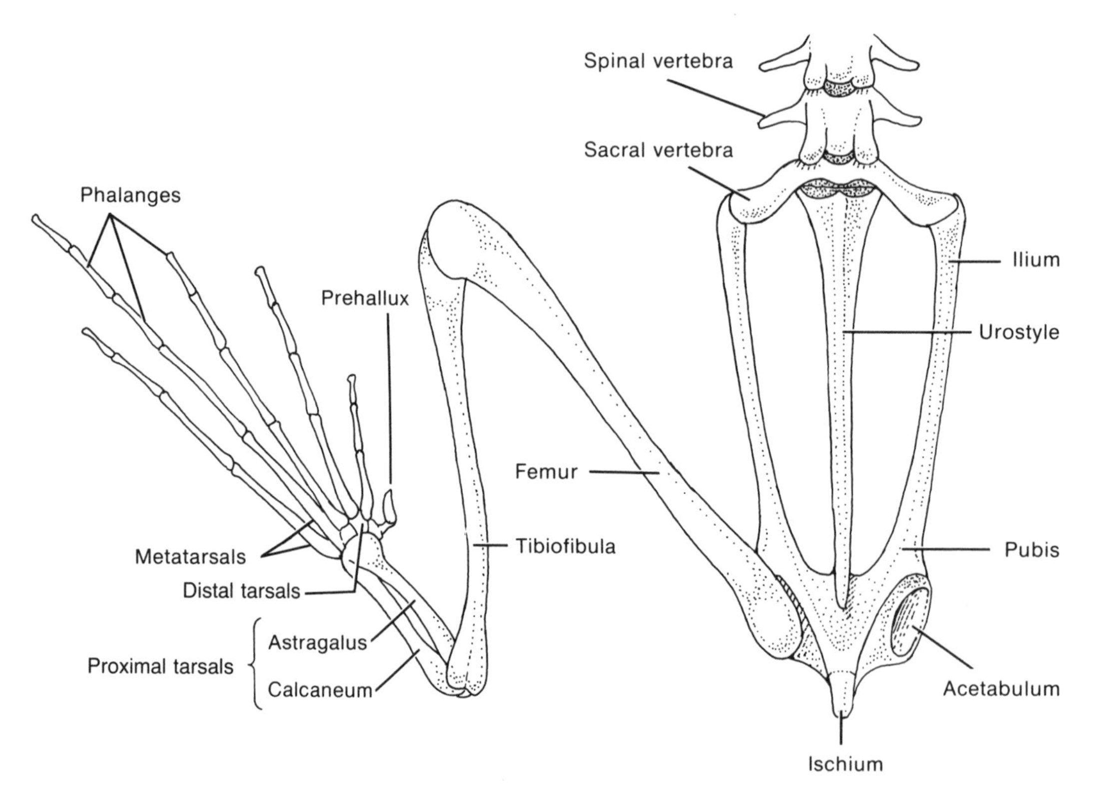

FIG. 7-8
Frog pelvic girdle (dorsal view).

Muscles can be attached to bone or other muscles by a cordlike, connective tissue **tendon** or by a broad, connective tissue sheet called an **aponeurosis.** One end of a muscle is usually fixed in position; this end is called the *origin.* Contraction of the muscle causes the other end, the *insertion,* to move, pulling a bone or other structure toward the fixed end. The part that lies between the two ends is the muscle **belly.**

The action of a muscle is to cause a contraction, or shortening in length, so that the structures to which it is attached are brought together. Muscles are commonly arranged in opposing groups, with one muscle moving in one direction and an opposing muscle moving in the opposite direction. The following list includes some of the common types of opposing muscles:

Flexors cause bending, by decreasing the angle between two structures (e.g., the biceps muscle bends the forearm toward the upper arm).

Extensors straighten or extend a part of the body (e.g., the triceps muscle extends the forearm away from the upper arm).

Adductors move a part towards the axis or midline of the body (e.g., the latissimus dorsi muscle draws the arm back against the body).

Abductors move a part away from the axis or midline of the body (e.g., the deltoid muscle draws the arm away from the body).

Depressors lower a part (e.g., the depressor mandibulae muscle moves the jaw down to open the mouth).

Levators raise or elevate a part (e.g., the masseter muscle raises the jaw to close the mouth).

Begin your dissection by cutting through the skin around the abdomen of the frog. Grasp the skin firmly, and pull it posteriorly over the legs, turning it inside out, until the abdomen and legs are completely free of skin. It will be necessary to remove sheets of tough connective tissue, called **fascia,** from some parts of the body. This tissue binds certain groups of muscles together. Separate the muscles by carefully dissecting away connective tissue from between them. **Do not try to cut muscles apart.** Observe the direction of the groups of muscle fibers, and then separate the muscle from the adjacent muscles by using a blunt probe to separate the fascia holding the muscles together. You can generally tell one muscle from another since the fibers of a given muscle, or major part of it, run in the same direction; those of an adjacent muscle run in a different direction.

3. Muscles of the Shoulder

The major muscles of the shoulder and forelimb can be identified by referring to Figs. 7-9 and 7-10. The **cucullaris,** the **dorsalis scapulae,** and the **latissimus dorsi** extend from the scapula, across the shoulder to the proximal end of the humerus (Fig. 7-9). All these muscles move the forelimb toward the body. On the ventral surface of the body, identify the **deltoid,** the **pectoralis,** and the **coracobrachialis** muscles, which extend from the pectoral girdle to the humerus (Fig. 7-10). These three muscles move the forelimb away from the body. The **coracoradialis,** another ventral shoulder muscle, extends from the girdle to the radioulna and is the major flexor of the forearm. The **anconeus,** situated on the dorsal surface of the humerus, is the major extensor of the forearm.

4. Muscles of the Forearm

Using Fig. 7-9, locate the major group of muscles that extends the hand and digits. The muscles in this group are the **extensor carpi radialis,** the **abductor indicis longus,** the **extensor digitorum communis longus,** and the **extensor carpi ulnaris** (Fig. 7-9). The flexor muscles that bend the forearm toward the body are the **flexor carpi radialis,** the **flexor carpi ulnaris,** and the **palmaris longus** (Fig. 7-10).

5. Muscles of the Trunk

The **cutaneous abdominis** on the dorsal surface of the trunk and the **cutaneous pectoris** on the ventral surface extend from the abdominal wall to the skin. It is believed that the contraction of these two muscles compresses certain lymph sacs and thus helps to circulate lymph fluid.

The **longissimus dorsi** and **iliolumbaris,** located on the dorsal side of the trunk (Fig. 7-9) serve to flex and extend the vertebral column. Located behind these two muscles is the **coccygeoiliacus** muscle, which is attached at one end to the urostyle and at the other to the ilium bones. These three muscles are thought to extend the urostyle and pelvic girdle when the frog is making strong leaps.

The ventral body wall of the frog is largely supported by the **rectus abdominis** (Fig. 7-10). The rest of the abdominal wall is covered on the outside by an **external oblique muscle** and an inner **transverse muscle,** whose fibers run nearly perpendicular to each other. These muscles support and tense the abdomen.

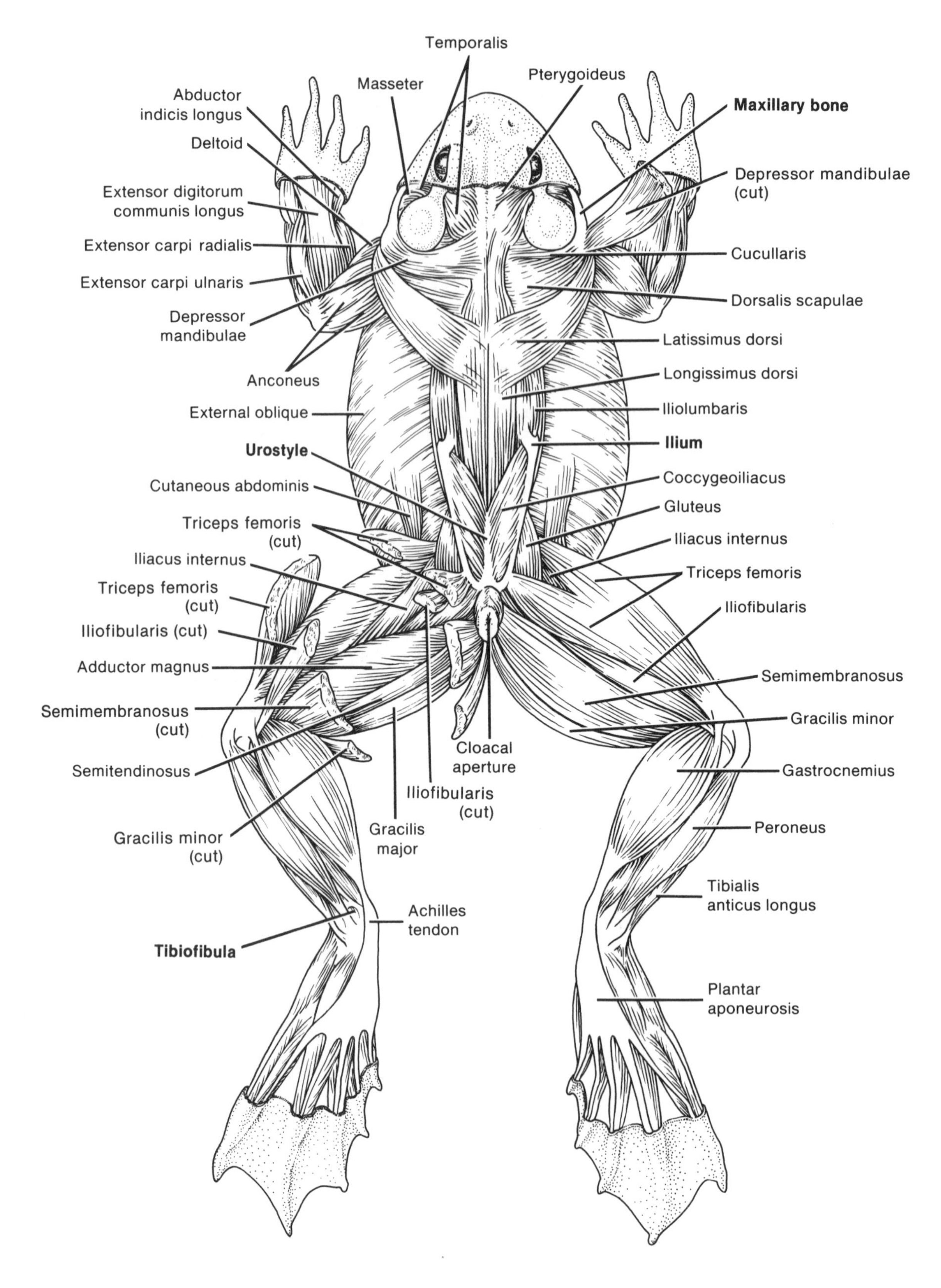

FIG. 7-9
Muscles of the frog as seen in a dorsal view: superficial muscles (right side of drawing), deeper muscles (left side). Skeletal parts are in boldface. (After *Dissection of the Frog,* 2d ed., by Warren F. Walker, Jr. W. H. Freeman and Company. Copyright © 1981.)

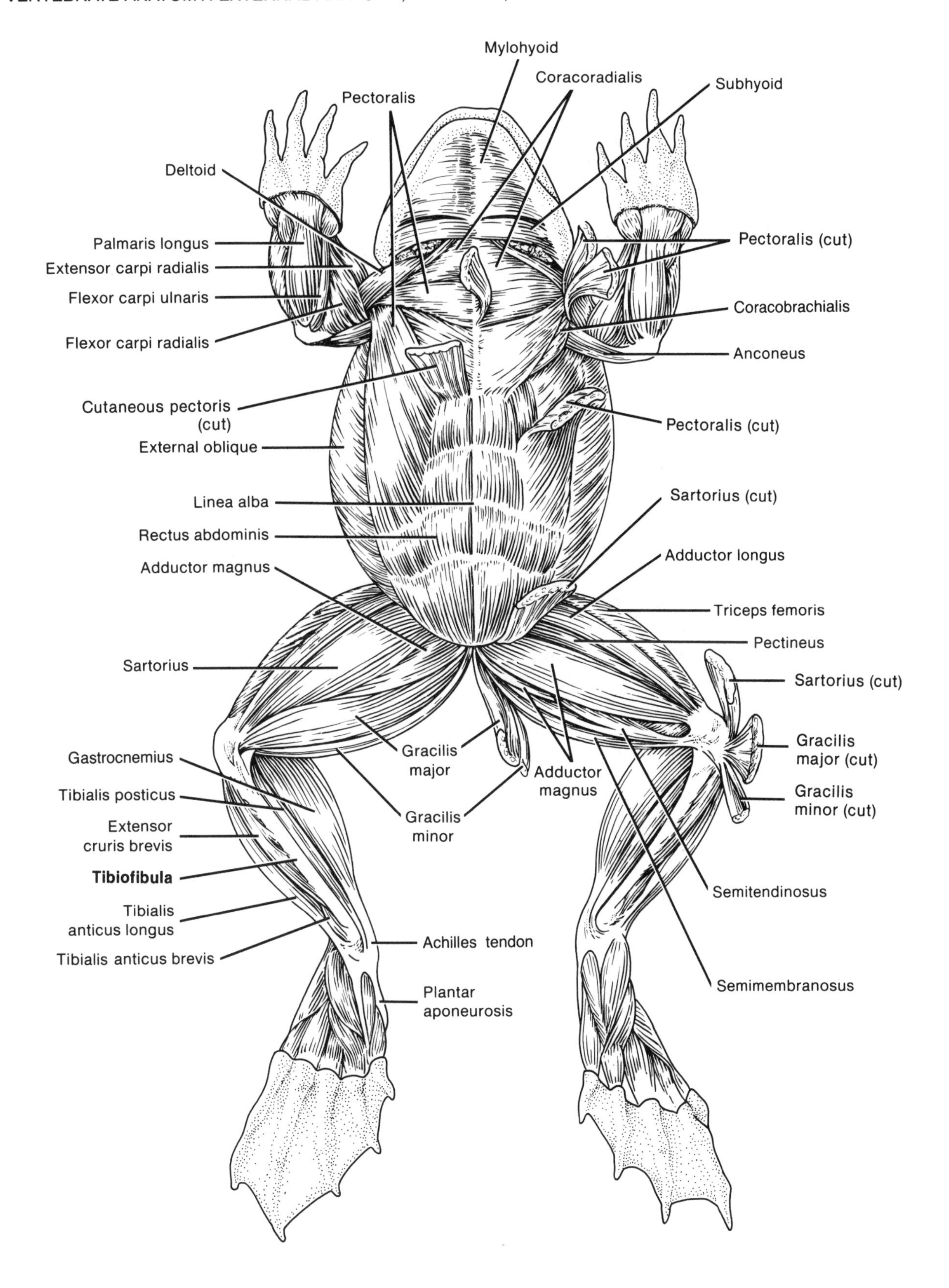

FIG. 7-10
Ventral view of the muscles: superficial muscles (left side of drawing); deeper muscles (right side). (After *Dissection of the Frog,* 2d ed., by Warren F. Walker, Jr. W. H. Freeman and Company. Copyright © 1981.)

6. Muscles of the Head

The **temporalis, pterygoideus,** and **masseter** muscles, located on the dorsal surface of the head (Fig. 7-9) elevate the mandible. Extending transversely on the ventral surface of the head are the **mylohyoid** and **subhyoid** muscles, forming the ventral surface of the lower jaw. These muscles raise the floor of the mouth and are therefore important in the swallowing and breathing movements of the mouth.

7. Muscles of the Thigh

Located on the anterior part of the thigh is the **triceps femoris** (Fig. 7-9), which extends the shank during jumping and swimming. The rest of the dorsal surface of the thigh is composed of the large **semimembranosus,** the **gracilis minor,** and a slender **iliofibularis** muscle (Fig. 7-9). The origins of all three of these muscles are on the bones of the pelvic girdle; their insertions are on the tibiofibula.

On the ventral surface of the thigh is the **sartorius** (Fig. 7-10); its origin is on the pubis and its insertion on the tibiofibula. This muscle flexes the thigh at the hip. The **adductor longus** and **adductor magnus** muscles lie on either side and partly beneath the sartorius. Their main function is to move the thigh toward the body. Lying posterior to the adductor magnus is the **gracilis major,** which flexes the knee and extends the thigh.

In order to observe the short muscles that extend from the pelvic girdle to the femur, it is first necessary to cut through some of the superficial thigh muscles as shown in Fig. 7-9. When this is done, you can identify the **gluteus** and **iliacus internus** (Fig. 7-9), which extend from the ilium bone of the pelvic girdle to the proximal end of the femur. These two muscles flex and rotate the thigh.

In order to observe the deep muscles on the ventral side of the thigh, cut the gracilis major, gracilis minor, and sartorius muscles as shown in Fig. 7-10. Locate the **pectineus** muscle between the adductor longus and adductor magnus. This muscle helps to move the thigh toward the body. Posterior to the adductor magnus is the **semitendinosus,** which acts both to reflect the shank and to extend the thigh.

8. Muscles of the Shank of the Hindlimb

The largest muscle of the shank is the **gastrocnemius.** This muscle arises by means of two large tendons from the distal end of the femur and ends in the Achilles tendon, which passes around the ankle and spreads into a sheet of connective tissue called the **plantar aponeurosis.** Its function is to flex the knee and toes. Anterior to the gastrocnemius, as viewed from the dorsal surface, are the **peroneus** and **tibialis anticus longus** muscles, both of which assist in foot extension.

From the ventral view of the shank, identify the **extensor cruris brevis** (Fig. 7-10), which inserts on the tibiofibula and is an extensor of the flank. The **tibialis posticus,** which arises from the shaft of the tibia and inserts on the astragalus bone of the foot, assists in foot extension. The **tibialis anticus brevis,** which arises from the distal part of the tibiofibula and inserts on the tarsal bones, is a flexor of the foot.

REFERENCES

Gilbert, S. G. 1965. *Pictorial Anatomy of the Frog.* University of Washington Press.

Romer, A. S., and T. S. Parsons. 1977. *The Vertebrate Body.* 5th ed. Holt/Saunders.

Underhill, R. A. 1980. *Laboratory Anatomy of the Frog.* 4th ed. Brown.

Walker, W. F. 1981. *Dissection of the Frog.* 2d ed. W. H. Freeman and Company.

Young, J. A. 1962. *The Life of Vertebrates.* 2d ed. Oxford University Press.

EXERCISE 8

Vertebrate Anatomy: Digestive, Respiratory, Circulatory, and Urogenital Systems

(*Note:* During this exercise, which deals with the digestive, respiratory, circulatory, and urogenital systems of the frog, refer to Appendix A for a discussion and glossary of the vocabulary used in the study of anatomy.)

A. DIGESTIVE SYSTEM

1. Oral (Buccal) Cavity

Place a frog on its back, and with scissors make a small cut at the angle of the jaw on each side so that the jaws can be opened widely (Fig. 8-1). The **oral (buccal) cavity** narrows into the **pharynx,** which connects to the **esophagus** (or gullet). Just anterior to the opening to the esophagus is the **glottis,** a short longitudinal slit in the floor of the pharynx (Fig. 8-2). The glottis opens for breathing but closes when food is being swallowed. Behind each eyeball and near the corners of the mouth are small openings to the **eustachian tubes,** each of which connects to the chamber of the middle ear beneath the tympanic membrane. In many species of frog, the males have openings into two **vocal sacs** located at the posterior corners of the oral cavity. The vocal sacs are inflated to amplify the croaking of the male frog, especially during the breeding season.

Locate the **maxillary teeth** on the margin of the upper jaw and the **vomerine teeth** on the roof of the mouth. Because the frog swallows its food whole, these sets of teeth are not used for chewing but primarily for holding the food.

Lying on the floor of the oral cavity and attached to the mouth anteriorly is a large tongue. The forked end of the tongue, covered with a sticky substance secreted by glands in the roof of the mouth, can be quickly thrust out of the mouth to capture small insects or other live prey. This substance makes the prey adhere to the tongue as it is retracted into the mouth.

Near the vomerine teeth are two small openings, the **internal nares** that lead to the **external nares** (nostrils) through which air passes to and from the oral cavity during breathing.

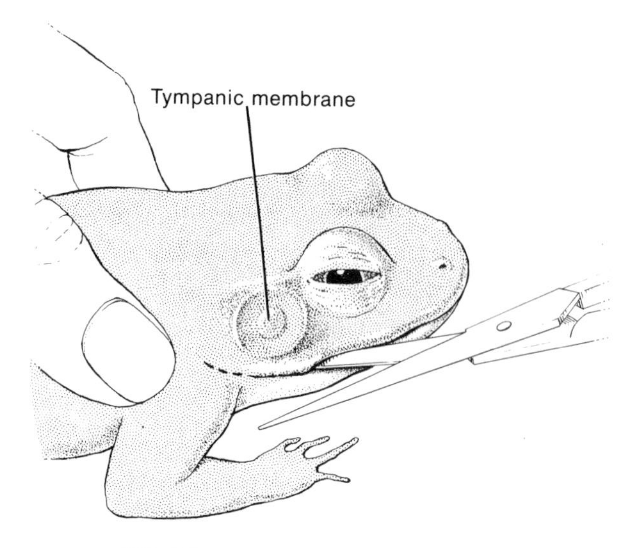

FIG. 8-1
Procedure for dissection of the frog mouth.

2. Body Cavity

Make a longitudinal incision through the abdominal wall (X–Y), just left of the midventral line, and extend it from the pectoral to the pelvic girdle (Fig. 8-3). Cut through the layers of muscle and lift the wall up. Now make transverse incisions (A–B and A′–B′) through the skin and muscle just anterior to each hindleg and posterior to each foreleg. Continue the cuts to the back, turning the flaps of the body wall back and pinning them to a dissecting pan. Lift the pectoral girdle and cut through the bone to expose the cavity that contains the heart.

The body cavity, or **coelom,** in which the internal organs **(viscera)** are located, is completely lined by a shiny layer of epithelium, called **parietal peritoneum.** The epithelium covering the viscera is called

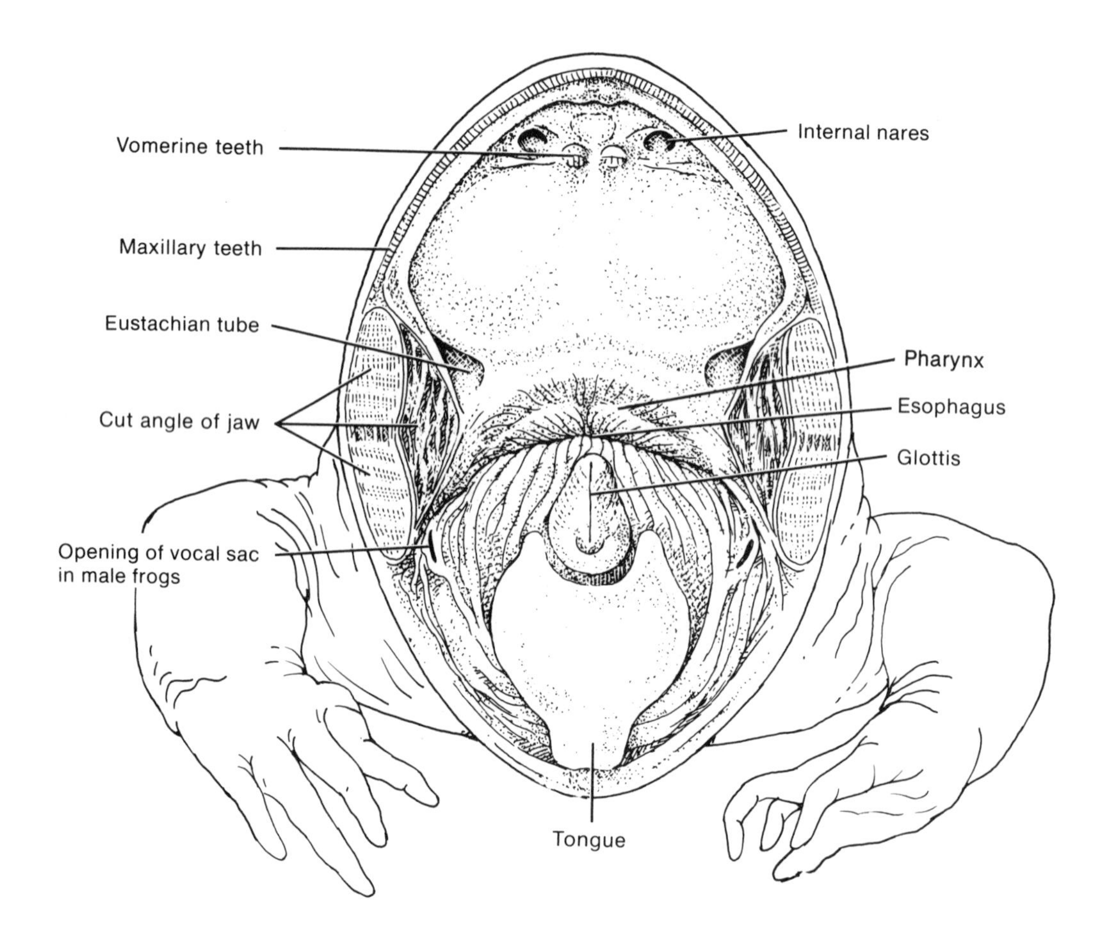

FIG. 8-2
Interior view of the oral (buccal) cavity of a male frog. The angles of the jaw have been cut so that the mouth can be opened widely. (After *Dissection of the Frog,* 2d ed., by Warren F. Walker, Jr. W. H. Freeman and Company. Copyright © 1981.)

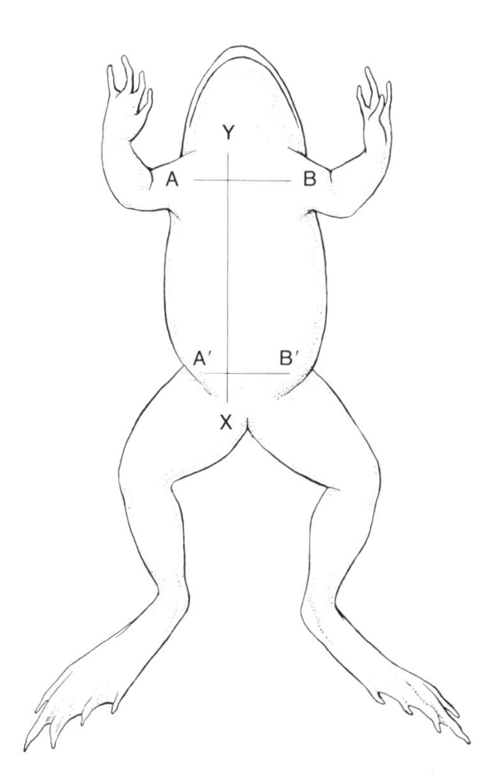

FIG. 8-3
Procedure for dissection of the abdominal body wall.

visceral peritoneum. Membranous sheets of this epithelium, called the **mesenteries,** extend from the dorsal body wall to the organs as well as between many of the organs. The mesenteries also serve as pathways for blood vessels and nerves as well as to limit the movement of various internal organs.

The coelom of the frog consists of two cavities: the **pleuroperitoneal cavity,** containing the lungs and most of the other viscera, and the **pericardial cavity,** containing the heart. The epithelial lining of the pericardial cavity is called the **pericardium.**

3. Visceral Organs

In beginning your dissection, first locate the **liver,** a large, trilobed organ occupying much of the anterior part of the coelom (Fig. 8-4). Locate the **stomach,** which lies dorsal to the liver and to your right as you view the body cavity. The stomach is connected to the pharynx by the short **esophagus.** It is a large, saclike organ in which food is stored and digestion begins. Use a scalpel to open the stomach lengthwise. Observe the longitudinal folds on the inner walls that help to grind the food. At the posterior end of the stomach is the **pyloric sphincter,** a band of muscle tissue that regulates the passage of partially digested food from the stomach into the small intestine.

The stomach connects to the slender, coiled small intestine, where digested food is absorbed. Note that the intestine is held in place by a dorsal fold of the peritoneum. The intestine is divided into a short **duodenum,** curving upward to the liver, and a much-coiled **jejunoileum,** leading into the **colon,** where water and ions are absorbed and undigested waste products (feces) stored before elimination through the **cloacal opening.**

Digestion is aided by secretions from the two largest digestive glands: the **liver** and **pancreas.** One function of the liver is to produce bile, an alkaline secretion that aids in the digestion of fatty materials. Bile is stored in the gallbladder until food enters the intestine, at which time the bile is released through a **common bile duct** into the duodenum. Locate the gallbladder. It is a small, dark sac attached to the dorsal surface of the liver.

Lying in the mesentery between the duodenum and the stomach is the other major digestive gland —the slender, irregular, and whitish **pancreas.** This gland secretes an alkaline digestive fluid that empties into the common bile duct. The pancreas has microscopic clumps of endocrine tissue, the **islets of Langerhans,** that produce the hormones insulin and glucagon, which regulate the levels of glucose in the blood.

Lift up the stomach and locate a small, dense, round structure—the **spleen**—on the left side of the mesentery supporting the intestine. The spleen is intimately related to the circulatory system, because in adult frogs it is an important site for the production of red and white blood cells and the storage and destruction of senescent (aging) red blood cells. It is also a major site for the production of antibodies (protein molecules produced, as a defense mechanism, in response to foreign organisms).

In the male frog, locate small, oval bodies, one on each side of the mesentery supporting the intestine; these are the **testes.** The fingerlike lobes attached to the testes are **fat bodies.** The **ovaries,** which are located in the same position in the female as the testes are in the male, vary greatly in size, depending on the female's reproductive state. Just before ovulation, they are filled with ripe eggs, taking up all the available space in the pleuroperitoneal cavity. Dorsal to each **gonad** (testis or ovary) is an elongated **kidney.** A long, coiled white tube—the **oviduct**—is lateral to each kidney in the female. Males may have a small vestigial oviduct. The bilobed sac lying ventral to the large intestine is the urinary **bladder.**

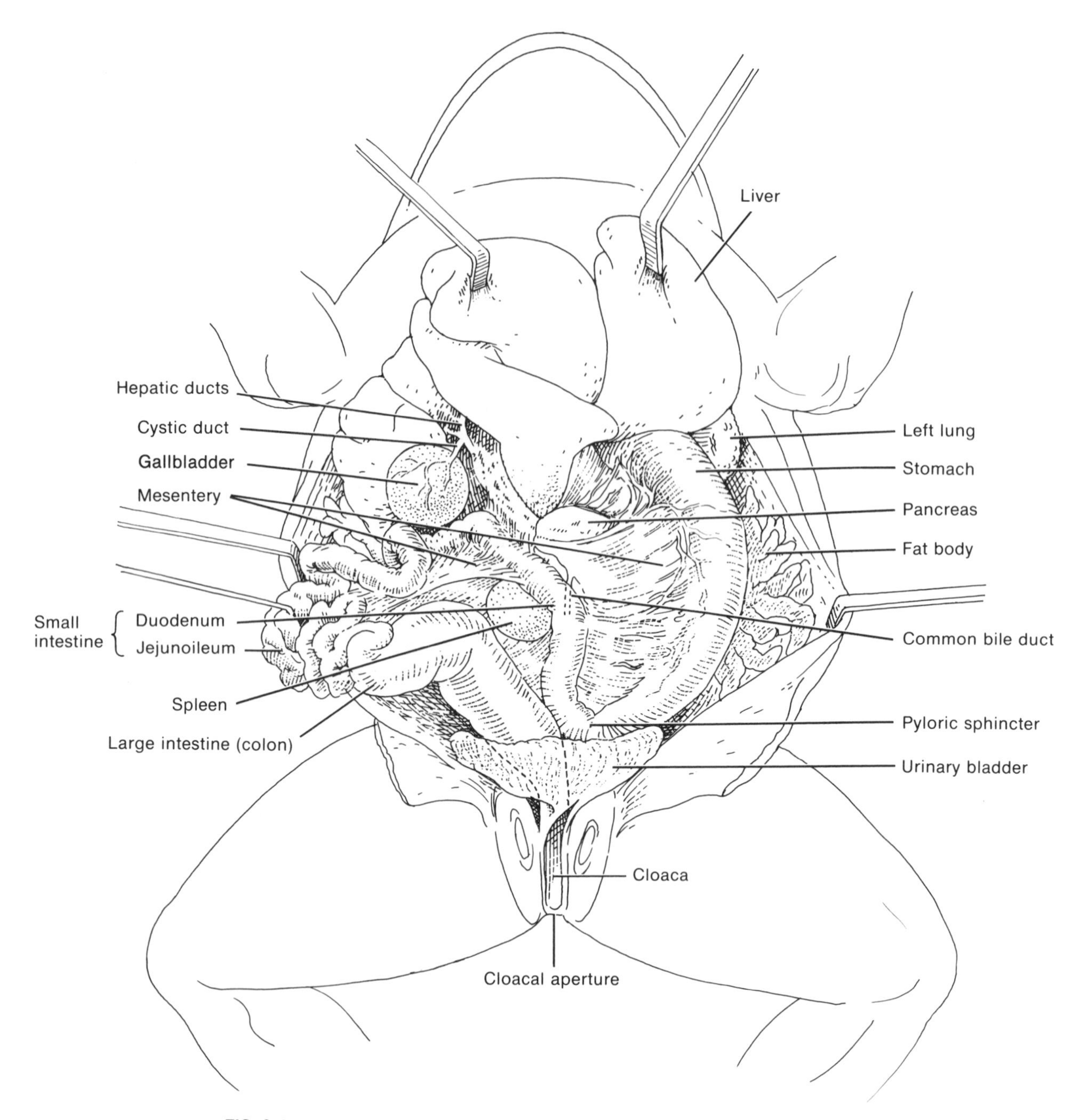

FIG. 8-4
Ventral view of the digestive tract and associated organs. The liver has been pulled forward and the pelvic region cut open. (After *Dissection of the Frog,* 2d ed., by Warren F. Walker, Jr. W. H. Freeman and Company. Copyright © 1981.)

B. RESPIRATORY SYSTEM

The primary functions of the respiratory system are to supply the tissues with oxygen and to eliminate excess carbon dioxide. In the frog, such gas exchanges take place not only in the lungs **(pulmonary respiration),** but also through the skin **(cutaneous respiration)** and the epithelial lining of the mouth and pharynx **(buccopharyngeal respiration).**

In pulmonary respiration, air passes through the nostrils (external nares) into the nasal cavity and then through the internal nares into the oral cavity. The external nares are then closed, the floor of the mouth is raised, and air is forced through the glottis into a short tube, the **larynx** (Fig. 8-5). The larynx is reinforced by cartilage and contains two elastic bands, the vocal cords, that vibrate and produce croaking sounds when air is forced vigorously from the lungs. The larynx is connected to each lung by a very short tube, the **bronchus.** In most preserved specimens, the lungs tend to be contracted by the preservative. An idea of the true size of the lungs can be gained by inserting a small glass tube into the glottis of a recently killed frog and then blowing air into them. Remove part of the lung, and slit it open to observe the network of partitions that divides the lung into many minute chambers called **alveoli.** These alveoli, with their blood capillaries, supply the extensive surface area necessary for gas exchange.

C. CIRCULATORY SYSTEM

The chief functions of the circulatory system are to transport (1) body fluids and nutrients to all of the cells and tissues; (2) oxygen and carbon dioxide between the respiratory organs and body tissues; (3) stored materials from place to place as needed; (4) organic wastes, water, and excess minerals to the excretory organs; (5) hormones from endocrine glands to the target organs; and (6) molecules (antibodies) and cells of defense (leukocytes) to areas in which they are needed.

The circulatory system in vertebrates includes a main pumping organ (the heart) and a system of vessels (arteries, veins, and capillaries) to conduct blood and lymph throughout the body.

1. Frog Heart

In order to study the structure of the frog heart, you will first have to cut away the pericardial membrane that encloses the heart. Using Fig. 8-6, identify the thin-walled left and right **auricles** and the thick-walled **ventricle.**

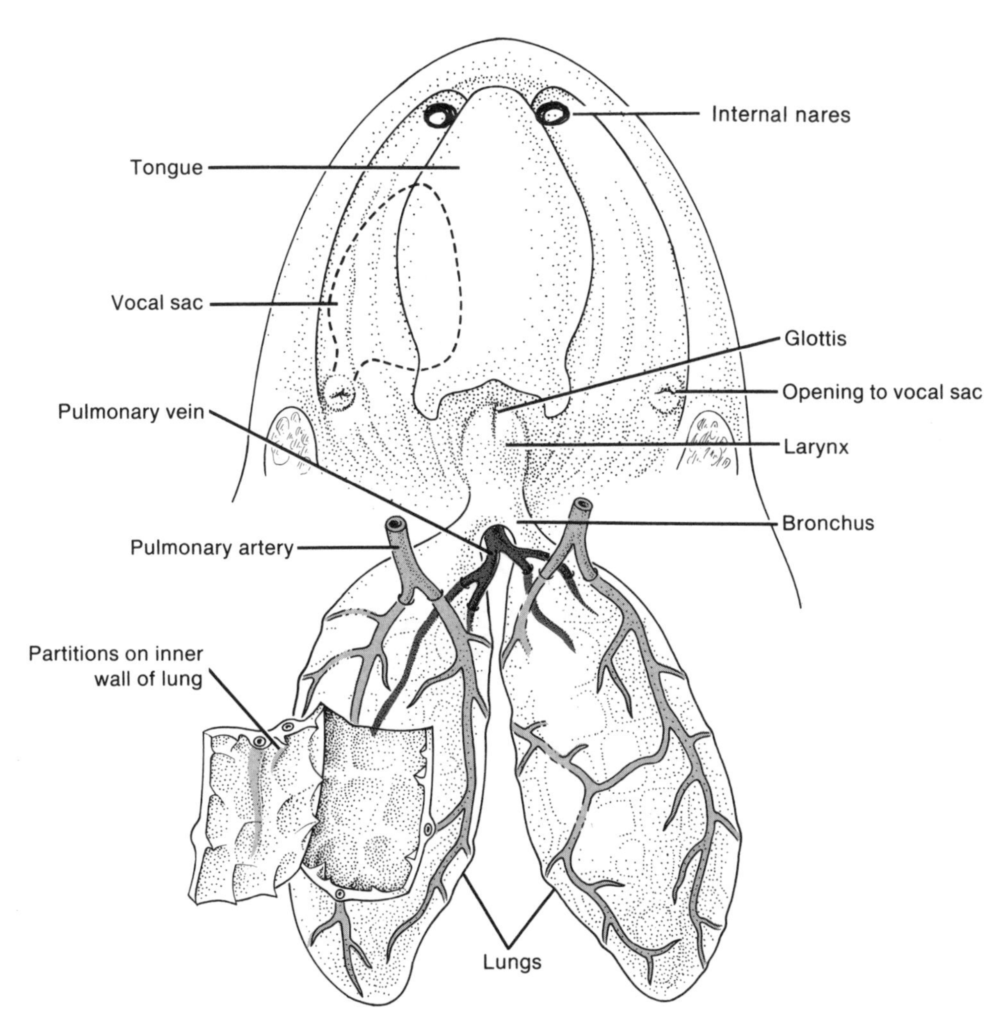

FIG. 8-5
Dorsal view of floor of mouth and respiratory system.

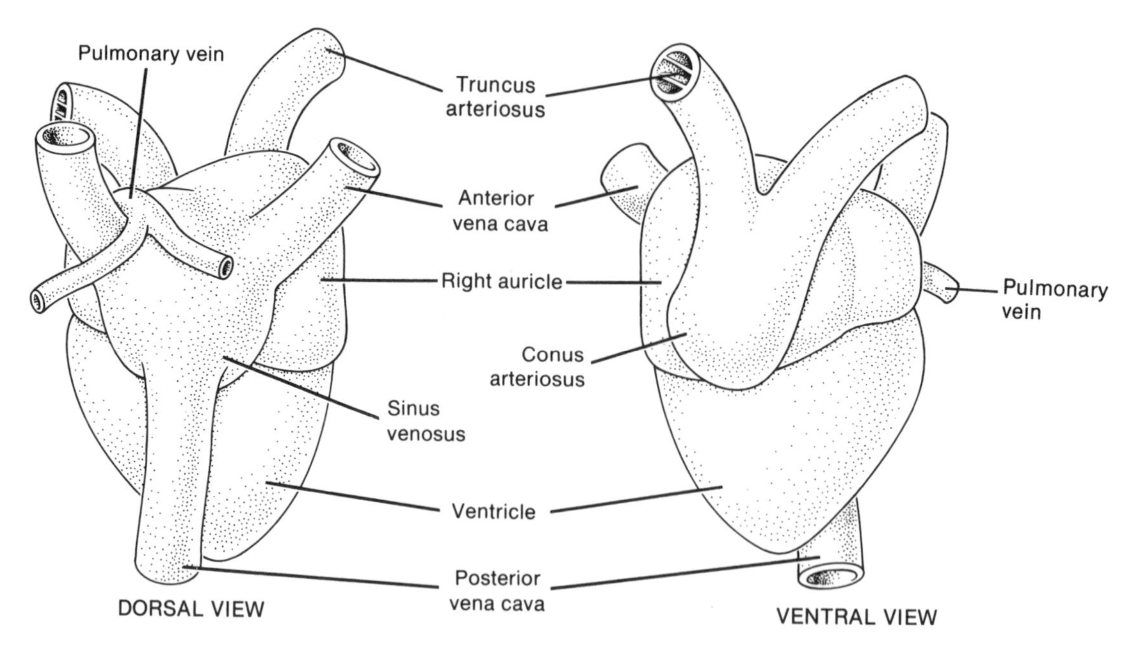

FIG. 8-6
Dorsal and ventral views of frog heart.

The frog heart is different from that of most other vertebrates because it has only one ventricle instead of two. Because the walls of the auricles are thinner than those of the ventricle, the blood within them makes them look darker. Arising from the base of the ventricle on its ventral side is a stout, cylindrical vessel, the **conus arteriosus.** This vessel receives the **posterior vena cava** and, at the two anterior apices, the left and right **anterior vena cavae.** Locate the **pulmonary veins,** which drain the lungs and enter the left atrium.

The action of the heart can be studied by examining a pithed frog that has been dissected to expose the beating heart. (Refer to Appendix E for the procedure of pithing a live frog.) As you observe the heart (and using Fig. 8-7), note that the sinus venosus contracts first and sends the nonoxygenated venous blood into the right auricle. Oxygenated blood from the lungs passes into the left auricle via the pulmonary veins. Then both auricles contract and pump their contents into the ventricle. Contraction of the ventricle and conus arteriosus next drives the blood to the lungs and the rest of the body. The blood is prevented from flowing back into the heart by means of the valves shown in Fig. 8-8. Because these valves are difficult to find in a dissection of the frog heart, you will next study the sheep heart, which is approximately the size of a human heart.

2. Sheep Heart

The sheep heart is enclosed in a tough, double-walled membranous sac, the **pericardium.** The inner surface of the pericardium is a membrane that faces the outer surface, or **epicardium,** of the heart. The **pericardial space** between these facing membranes contains a serous (watery) pericardial fluid. In removing the pericardium, observe that it is attached only around the base of the heart, where the large vessels emerge. The cavities of the heart are lined with **endocardium,** which is a type of epithelium (endothelium) as that lining all of the blood vessels. Between the epicardium and endocardium is the bulk of the heart musculature, the **myocardium.**

The right and left sides of the heart can be identified in two ways: (1) the pointed end, or **apex,** is entirely a part of the left ventricle; the division between the right and left ventricles is indicated superficially by a diagonal furrow containing the coronary blood vessels. (2) The left ventricle is firm and muscular to the touch, whereas the right ventricle feels soft and flabby. The auricles look like ears projecting from the base of the heart. The cavity within each auricle is referred to as an **atrium,** a receiving chamber for incoming blood returning from the body (right auricle) or lungs (left auricle).

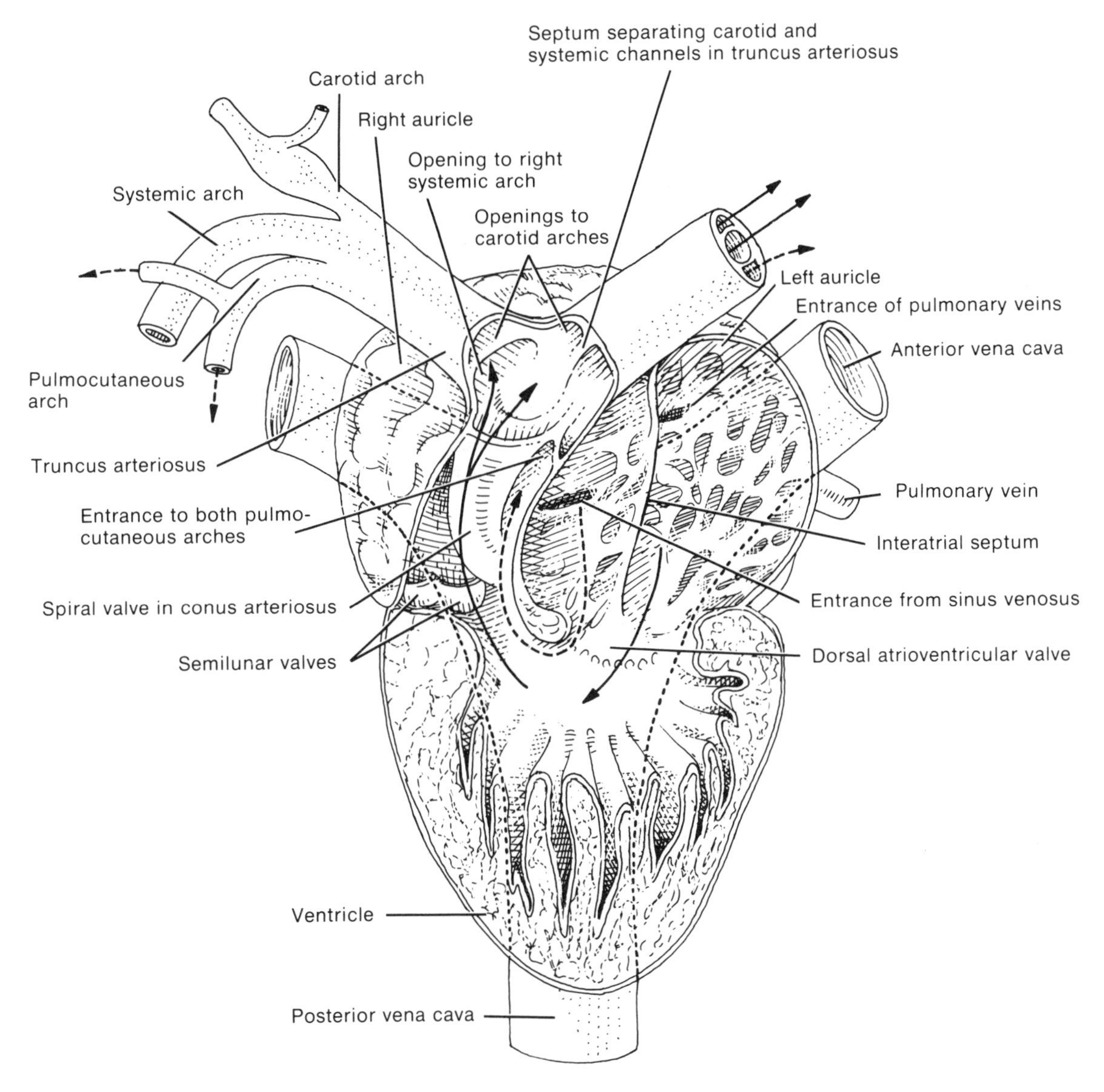

FIG. 8-7
Longitudinal section of a frog heart. The position of the sinus venosus on the dorsal surface of the heart is shown by broken lines. (After *Dissection of the Frog,* 2d ed., by Warren F. Walker, Jr. W. H. Freeman and Company. Copyright © 1981.)

1. Make a long incision through the **right auricle** in line with the superior vena cava. Lift the cut edges and observe the wide openings of the **superior** and **inferior venae cavae** as they enter the atrium (Fig. 8-8). Note the irregular bands of muscle lining the interior of the auricular wall, the **pectinate muscles.** Find the **coronary sinus,** which receives venous blood directly from the heart musculature and enters the auricle. Note that the partition separating the auricles is membranous rather than muscular. Locate the thinnest part of the wall between the two auricles. This area is the **fossa ovalis,** an oval depression marking the remants of an opening in the fetal heart through which blood was carried directly to the left side of the heart, thus bypassing the pulmonary circuit.

2. Carry the incision from the right auricle in a straight line through the lateral wall of the **right ventricle.** Note the three rounded flaps of membranous tissue suspended into the ventricle and held in place by tendinous cords. These flaps constitute the **tricuspid valve.** Study in detail its position, structural character, and attachments. If necessary, wash out both cavities. Note that pointed columns of ventricular muscle **(papillary muscles)** are continuous with the wall of the ventricle and with the strong fibrous cords **(chordae tendineae)** that extend to the edges of

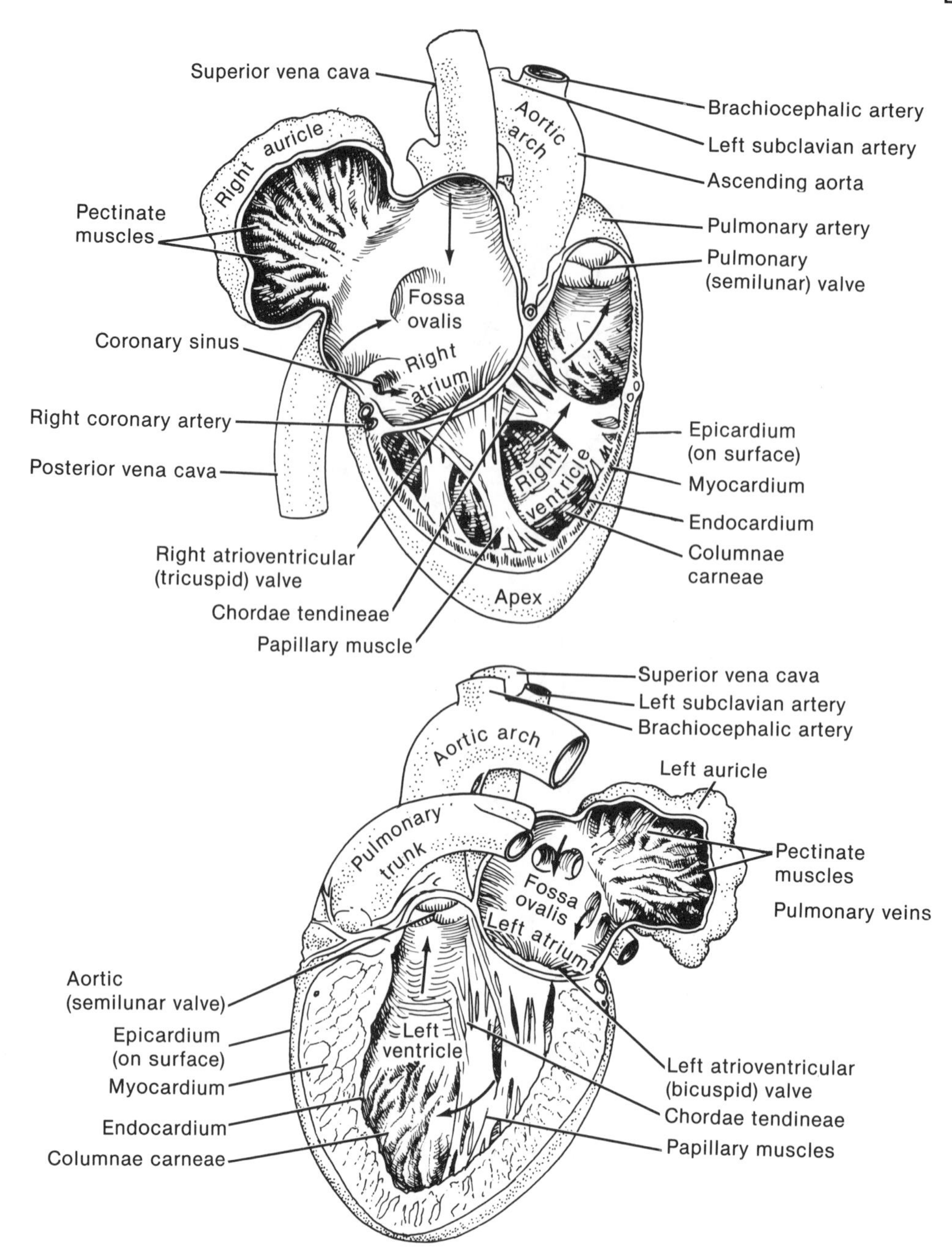

FIG. 8-8
Schematic longitudinal section of the mammalian heart.

the cusps, or flaps, of the valve. Observe the heavy muscular ridges within the ventricle, the **columnae carneae.** Find the exit for blood from the right ventricle, considering the tricuspid valve to be closed between auricle and ventricle. Carry an incision upward through the wall of this exit **(pulmonary artery),** and note that the mouth of the artery is surrounded by three membranous pockets (the pulmonary **semilunar valves**).

3. Open the **left auricle** in the same manner as the right one. Before cutting down through the ventricle, push your finger from the auricle into the ventricles and distend the auriculoventricular opening. Determine the number of openings draining into the auricle from the lungs. These are the openings of the **pulmonary veins.**

4. On the **left ventricle,** make an incision to expose the cavity. Study the details of the **mitral,** or **bicuspid** (two-part), **valve** between this auricle and ventricle. Explore the thickness of the septum between the two ventricles. Probe with your finger to find the outlet of the ventricle into the aorta. Open

the aorta to expose its semilunar valve. Find the openings into the two coronary arteries just above the valve—that is, within the pockets of the valve. Trace these to the walls of the heart. Observe the tough ligamentous connection, which is usually covered with a conspicuous pad of fat, between the pulmonary artery and aorta. This is a remnant of the vessel that connected the pulmonary artery and aorta in the embryo. If this vessel remains open in human beings after birth, oxygenated and unoxygenated blood mix, leading to a condition commonly referred to as "blue baby."

In studying the dissected heart, emphasis should be placed on the structural arrangement of the parts and their functional continuity. The mammalian heart is essentially a double pump, equipped with specially designed valves. It is the propelling force for two circuits of blood that do not mix in the heart. The circuit from the heart to the lungs and back is the **pulmonary circuit.** The circuit from the heart to all the body tissues and back is the **systemic circuit.** These circuits act concurrently and interdependently in that no more blood can be sent through the pulmonary circuit than is delivered to it by the systemic circuit.

3. Frog Arterial System

The arterial system consists of vessels that carry blood away from the heart to the capillaries and tissues. These vessels have thick, muscular walls that do not collapse when empty.

The conus arteriosus of the heart divides into two branches. Each branch, a **truncus arteriosus,** gives rise to three arteries (Fig. 8-9). The first of these is the **pulmocutaneous arch,** which divides into the **pulmonary artery** going to the lung and the **cutaneous artery** to the skin. This artery and its branches carry nonoxygenated blood.

The next artery to come off each truncus arteriosus is the **carotid arch,** which divides into the **external carotid,** leading to the ventral part of the head and the tongue, and the **internal carotid,** leading to the dorsal part of the head. The most highly oxygenated blood coming from the heart goes to the carotid arches. Note the small oval swellings, called the **carotid bodies,** at the junctions of the two carotids. The carotid bodies are **chemoreceptors** that are thought to be sensitive to changing levels of oxygen in the blood and which are involved in regulating blood pressure.

The final blood vessel to come off each truncus arteriosus is the **systemic (aortic) arch.** The two systemic arches bend dorsally around the esophagus and unite to form the large **dorsal aorta.** Before this union, several blood vessels are given off leading to the skull and vertebral column. The next blood vessel coming off the systemic arch is the **subclavian artery,** which gives off branches to the shoulder region and the forelimbs.

The first vessel arising from the aorta, near the junction of the systemic arches, is the **coeliaco-mesenteric artery.** This vessel divides into two branches, the **coeliac** and **anterior mesenteric arteries.** The coeliac further divides into the **hepatic artery,** going to the liver, and the **gastric arteries,** supplying the stomach. The mesenteric supplies branches to the spleen (**splenic**) and the large and small intestines. The dorsal aorta next gives off from four to six **urogenital** arteries to the kidneys, gonads, and fat bodies. The **posterior mesenteric artery** arises from the dorsal aorta to supply the colon. Next, the aorta divides into two **common iliac arteries,** which supply the hindlimbs. As you trace these vessels down the legs, you can observe the **epigastric arteries,** which extend to the urinary bladder, large intestine, and body wall; the **femorals,** which supply the hip and outer part of the thigh; and the **sciatics,** which serve the thigh, shank, and foot.

4. Frog Venous System

The vertebrate venous system, which consists of the vessels that carry blood back toward the heart, is somewhat more complex than the arterial system. Anteriorly, the head, forelimbs, and skin are drained by a pair of **anterior venae cavae,** which enter the heart by way of the sinus venosus (Fig. 8-10). Each anterior vena cava receives blood from three veins. The first of these is the **external jugular,** which returns blood from the tongue and floor of the mouth (**lingual**) and from the lower jaw (**mandibular**) vein. The **innominate** veins collect blood from the head by means of the **internal jugulars** and from the shoulder and back of the forelimb by means of the **subscapular veins.** The **subclavian** veins collect blood from the forelimbs by means of the **brachial veins** and from the muscles and skin on the lateral and dorsal part of the head and trunk by means of the **musculo-cutaneous veins.**

The single **posterior vena cava** enters the sinus venosus posteriorly and collects blood from the testes or ovaries through the **genital vein,** from the kidneys through from four to six pairs of **renal veins,** and from the liver by means of a pair of **hepatic veins.**

Veins that do not pass from organs or tissues directly to the heart but instead enter another organ

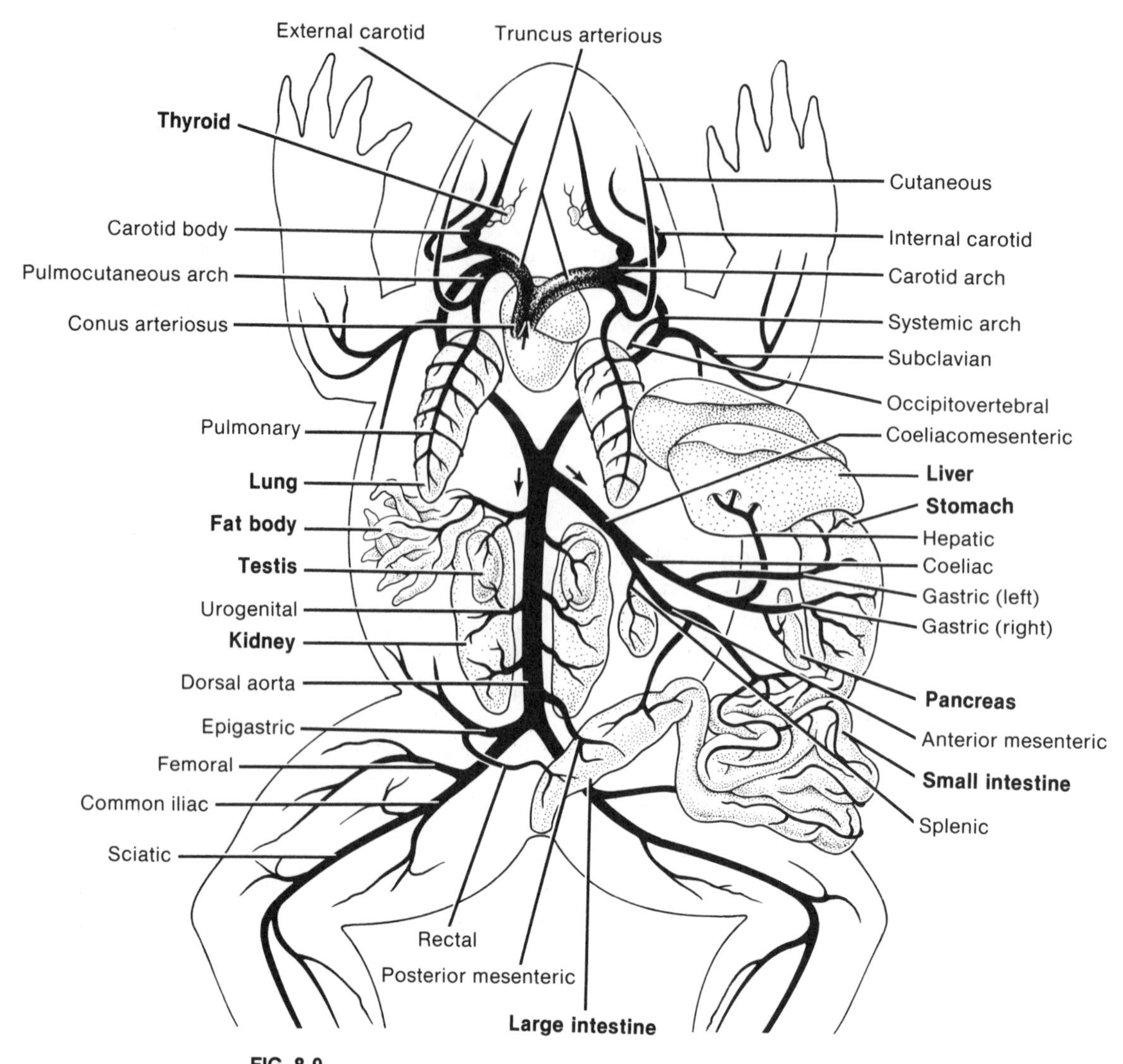

FIG. 8-9
Arterial system of the frog. Names of various organs are in boldface.

and subdivide within it to join its capillary system, are termed **portal veins** and constitute a **portal system.** In the frog, there are two such portal systems, the **renal** and the **hepatic portal systems.** The renal portal system is found only among lower vertebrates.

The renal portal system consists of a **renal portal vein,** which receives blood from the hindlimbs by means of the **sciatic, external iliac,** and **femoral veins** and from the body wall by means of the **dorsolumbar** vein. The renal portal vein carries blood to the dorsal border of the kidney.

The hepatic portal system consists of a large number of veins that carry blood into the liver from the stomach, intestine, spleen, and pancreas. Such blood is heavily laden with recently digested food products absorbed by these organs. As this blood passes through the liver before entering the main circulation, food products are removed and stored, and other substances are added to the blood. In addition, bacteria that may have entered the blood through the digestive tract are phagocytized by Kupffer cells found in the liver. The main vein draining into the liver is the **hepatic portal vein,** which collects blood from the spleen via the **splenic vein,** from the stomach via the **gastric vein,** from the small intestine via the **intestinal vein,** and from the large intestine via the **mesenteric vein.**

D. UROGENITAL SYSTEM

Because the urinary and reproductive system of the frog are closely connected, they are termed the **urogenital system.**

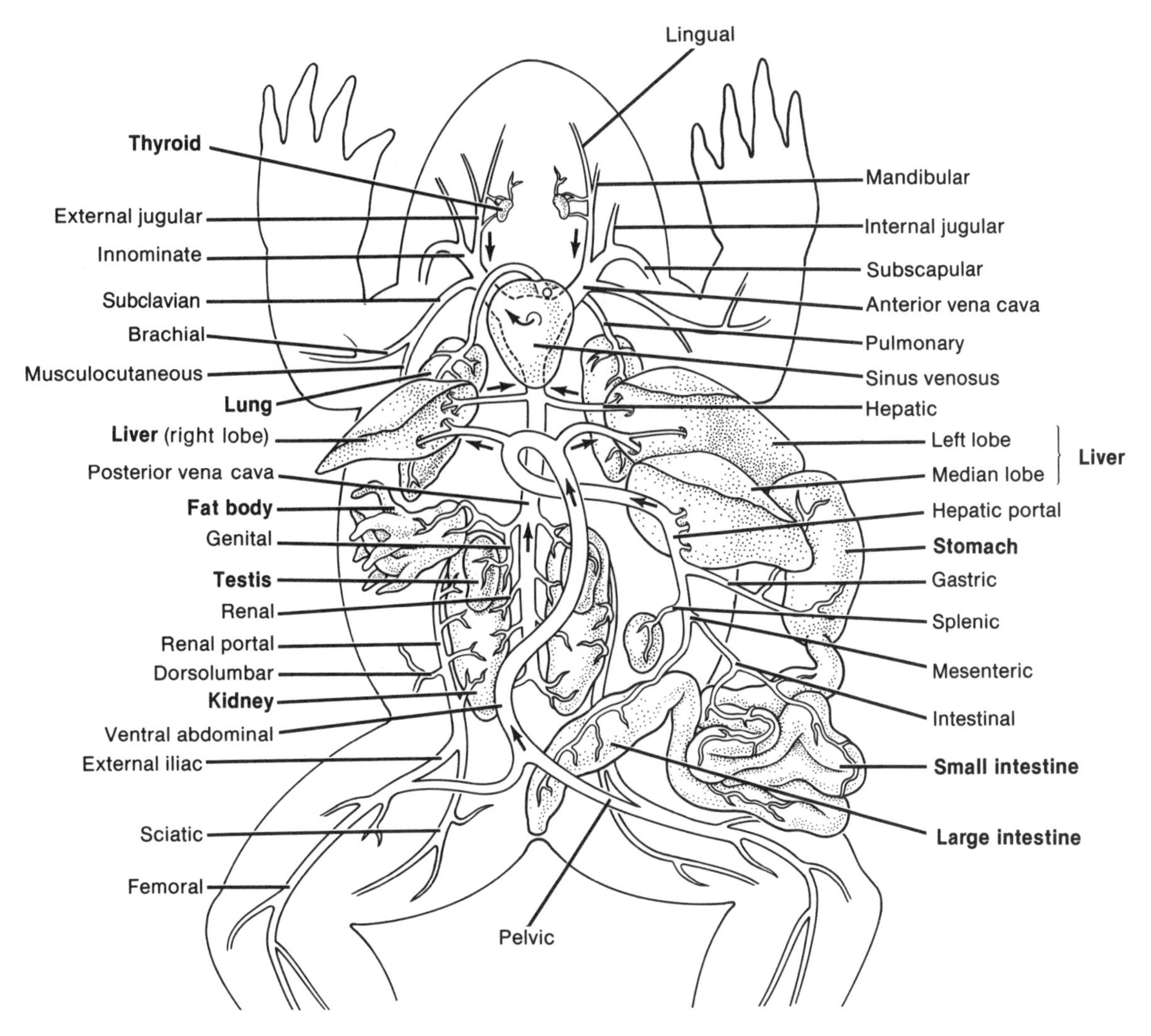

FIG. 8-10
Venous system of the frog. Names of various organs are in boldface.

1. Urinary System

The main organs of the urinary system are two elongate, reddish brown **kidneys** that lie close together on the dorsal wall at the posterior end of the body cavity. The urine from each kidney is drained by a **ureter,** which can be seen along the kidney's lateral edge. The ureters empty into the dorsal surface of the cloaca, and the urine must flow across the cloaca to enter the bilobed **urinary bladder,** which is attached to its ventral surface. The yellow or orange stripe running along the ventral side of each kidney is an adrenal gland, which secretes adrenaline into the blood. This hormone increases the heart rate, raises blood sugar levels, and helps the body react to stress conditions.

2. Female Reproductive System

The **ovaries** in most frogs are very conspicuous, occupying much of the body cavity. However, they vary greatly in size according to the season of the year. They are largest during the fall and winter, when they are filled with thousands of ripe eggs, and smallest after ovulation in the spring.

In order to identify the rest of the female reproductive system, dissect out one of the ovaries. Using Fig. 8-11, locate the **oviduct,** a highly looped duct extending forward to the anterior end of the body cavity. At the end of the oviduct is a funnel-shaped **ostium.** In the spring, eggs are released from the ovary into the body cavity and are carried to the ostium by currents created by movement of the cilia of

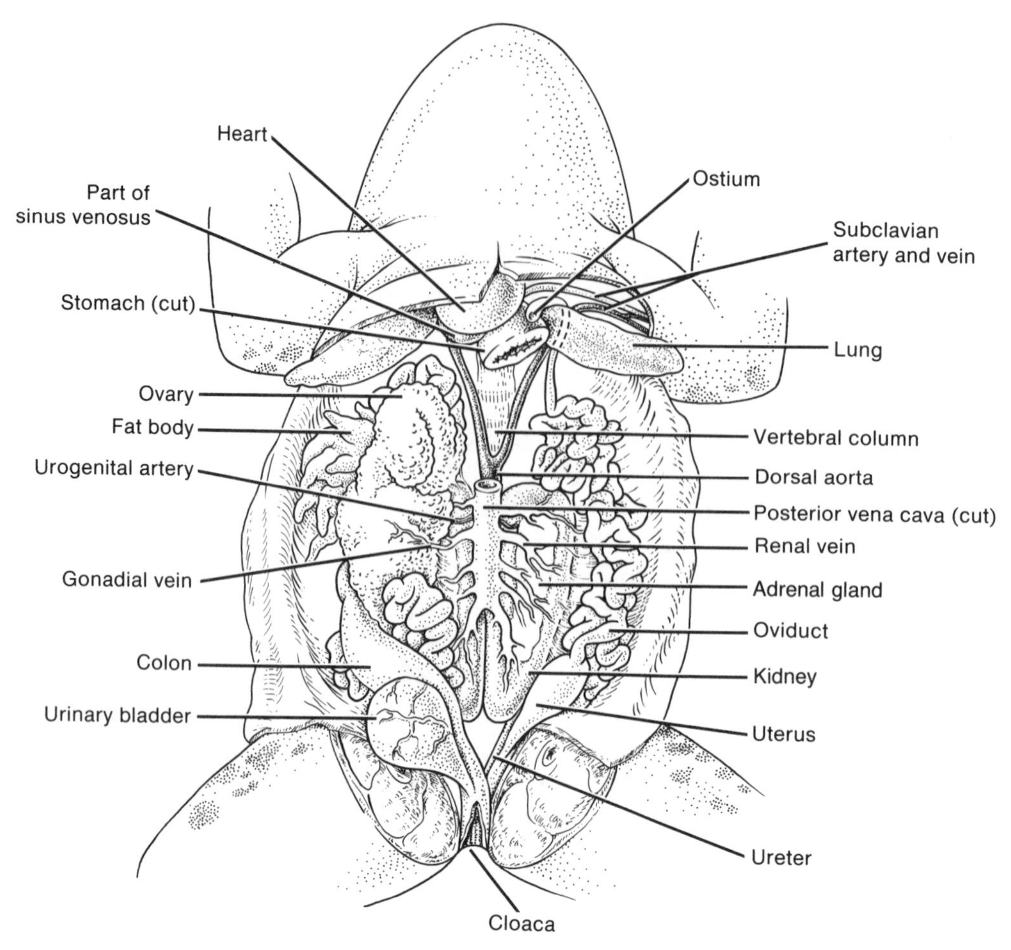

FIG. 8-11
Ventral view of the female urogenital system. One ovary and one fat body have been removed. Because this drawing is based on a specimen taken in midsummer, the ovaries and fat bodies have not reached their full size. (After *Dissection of the Frog,* 2d ed., by Warren F. Walker, Jr. W. H. Freeman and Company. Copyright © 1981.)

the coelomic epithelium. The posterior ends of the oviducts are enlarged to form the **uteri,** whose entrance to the cloaca is on its dorsal surface next to that of the ureters.

3. Male Reproductive System

Using a male specimen, locate the two oval, yellowish **testes,** each suspended from the dorsal abdominal wall by a mesentery (Fig. 8-12). Nearby are the bright yellow, finger-shaped masses of the **fat bodies.** These fat bodies are largest during the fall of the year. Some of the food stored in them is used during hibernation, and the rest is absorbed during the highly active spring breeding season.

Lift up one of the testes, and locate the **vasa efferentia,** the fine, threadlike tubes that connect the testes to the kidneys. Sperm produced by the testes are carried to the kidneys via the vasa efferentia and from the kidneys to the cloaca via the ureters, which also serve as the urinary ducts. It is not uncommon to find vestigial oviducts (**mesonephric ducts**) alongside the kidneys in male leopard frogs; these ducts should not be mistaken for the ureters.

REFERENCES

Gilbert, S. G. 1965. *Pictorial Anatomy of the Frog.* University of Washington Press.

Griffin, D. R. 1970. *Animal Structure and Function.* 2d ed. Holt, Rinehart and Winston.

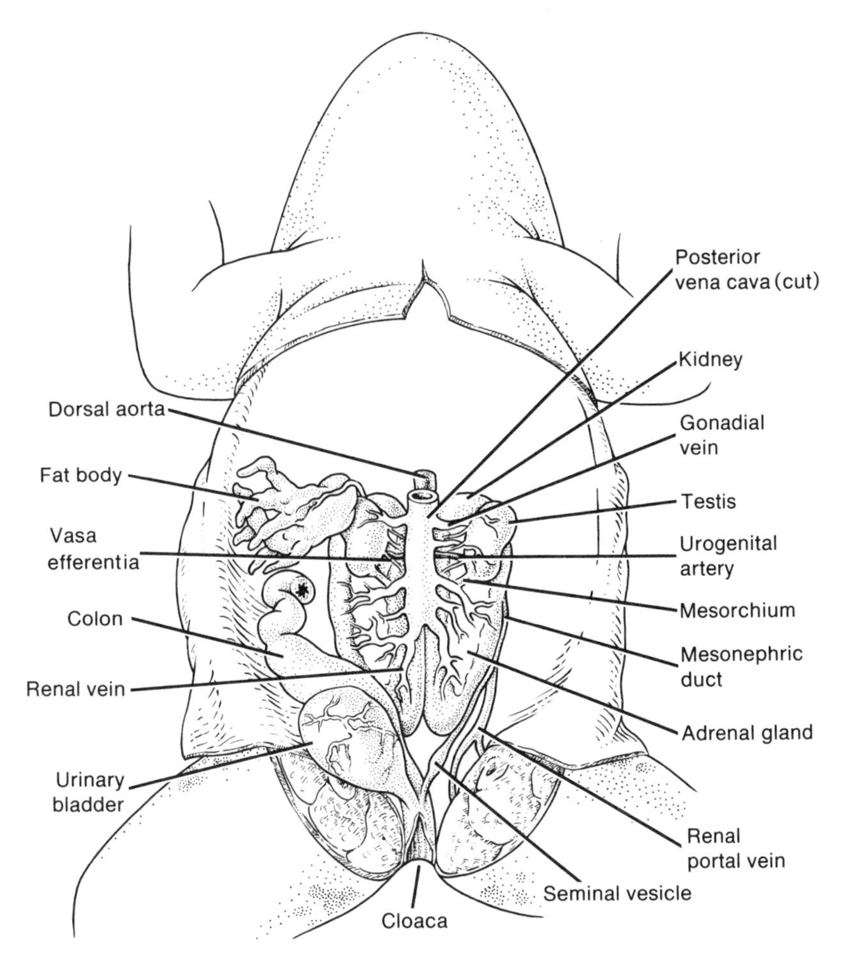

FIG. 8-12
Ventral view of the urogenital system of a male frog. One fat body has been removed. (After *Dissection of the Frog,* 2d ed., by Warren F. Walker, Jr. W. H. Freeman and Company. Copyright © 1981.)

Romer, A. S. and T. S. Parsons. 1977. *The Vertebrate Body.* 5th ed. Holt/Saunders.

Underhill, R. A. 1980. *Laboratory Anatomy of the Frog.* 4th ed. Brown.

Walker, W. F., Jr. 1981. *Dissection of the Frog.* 2d ed. W. H. Freeman and Company.

Wiggers, C. J. 1957. The Heart. *Scientific American* 196(5):74–87 (Offprint 62). *Scientific American* Offprints are available from W. H. Freeman and Company, 41 Madison Avenue, New York 10010, and 20 Beaumont Street, Oxford OX1 2NQ, England. Please order by number.

Wood, J. E. 1968. The Venous System. *Scientific American* 218(1):86–99.

Zweifach, B. W. 1959. The Microcirculation of Blood. *Scientific American* 200(1):54–74.

EXERCISE 9

Movement of Materials Through Cell Membranes

To perform their functions, cells must maintain a steady state (**homeostasis**) in the midst of an ever-changing environment. This constancy is maintained by regulating the movement of materials into and out of the cell. To achieve this control cells are enclosed by a complex membrane that can differentiate between different substances; slowing down the movement of some while allowing others to pass through. Since not all substances penetrate the membrane equally well, the membrane is said to be **differentially permeable.**

The external and internal environment of cells is an aqueous solution of dissolved inorganic and organic molecules. Movement of these molecules, both in the solution and through the cell membrane, is by a physical process called **diffusion**—a spontaneous process by which molecules move from a region in which they are highly concentrated to a region in which their concentration is lower.

As an example of diffusion, consider a colored dye dissolved in water. If a drop of the dye solution is placed in the water, it will soon become dispersed throughout the water as a result of the random movements and kinetic energy of the solute and solvent molecules. In this type of diffusion, called **passive diffusion,** movement is caused by the kinetic energy of the diffusing molecules. No outside energy is required for passive diffusion to occur.

In contrast, **active transport** is a type of diffusion in which solute molecules may move *against a concentration gradient.* For example, it has been shown that human erythrocytes have almost 30 times more potassium than the surrounding blood plasma. These cells accumulate potassium by an active process that requires energy.

Another kind of diffusion that occurs in biological systems is **osmosis.** Simply defined, osmosis is the diffusion of water through a differentially permeable membrane from a region of higher concentration to a region in which its concentration is lower. This phenomenon is diagrammatically shown in Fig. 9-1.

Diffusion and osmosis, although basically phenomena resulting from the kinetic activity of molecules, are affected by a number of other factors such

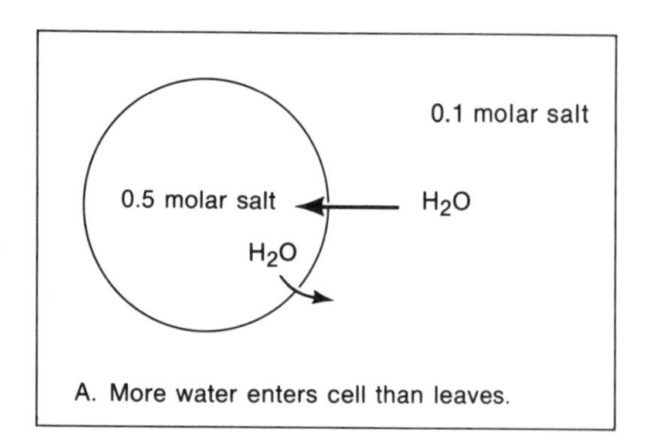

A. More water enters cell than leaves.

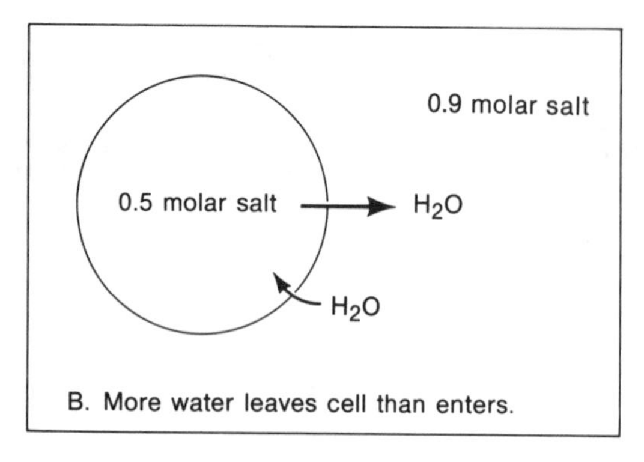

B. More water leaves cell than enters.

FIG. 9-1
Osmosis.

as temperature, the physical nature of the medium in which diffusion is occurring, the molecular weight of the diffusing substance, and the lipoid solubility of the solute. In this study, you will become familiar with diffusion and osmosis and examine some of the factors regulating these processes.

A. DIFFUSION

The kinetic energy of molecules causes them to move from regions in which their concentration is high to regions in which it is lower until they become uniformly distributed throughout the space available to them. Thus, a gas set free in a room is in time equally distributed throughout the room. A crystal of salt put into a glass of water dissolves, and the molecules of which it is composed become uniformly distributed throughout the water. This movement of gases and dissolved substances from a region of high concentration to one of low concentration is termed *diffusion.* Diffusion can thus be defined as *the net movement of the molecules of a substance, under their own kinetic energy, from a region of greater to one of lesser molecular activity of that particular substance.* The rate at which particles will diffuse depends on several factors, such as their size, their concentration, and the permeability of the dispersion medium.

1. Diffusion of a Gas in a Gas

A striking demonstration of diffusion is the movement of gases through air. An apparatus to demonstrate such diffusion is shown in Fig. 9-2.

Saturate a piece of absorbent cotton on one side with ammonium hydroxide (NH_4OH). Saturate another piece of cotton with hydrochloric acid (HCl). *(CAUTION: Handle chemicals carefully. Open only one reagent bottle at a time.)* Place the pieces of cotton *simultaneously* in opposite ends of the glass tubing as shown in Fig. 9-2. Why simultaneously?

Ammonium hydroxide and hydrochloric acid react to form ammonium chloride (NH_4Cl, a cloudy, white precipitate) and water. The equation for this reaction is as follows:

$$NH_4OH + HCl \rightarrow NH_4Cl + H_2O$$

Describe what happens when these two gases meet in the tube?

The molecular weights of the ammonium (NH_4^+) and chlorine (Cl^-) ions are 18 and 35.5, respectively. At which end of the tube does the reaction occur?

What, if any, relationship is there between the molecular weights of these gases and their rates of diffusion?

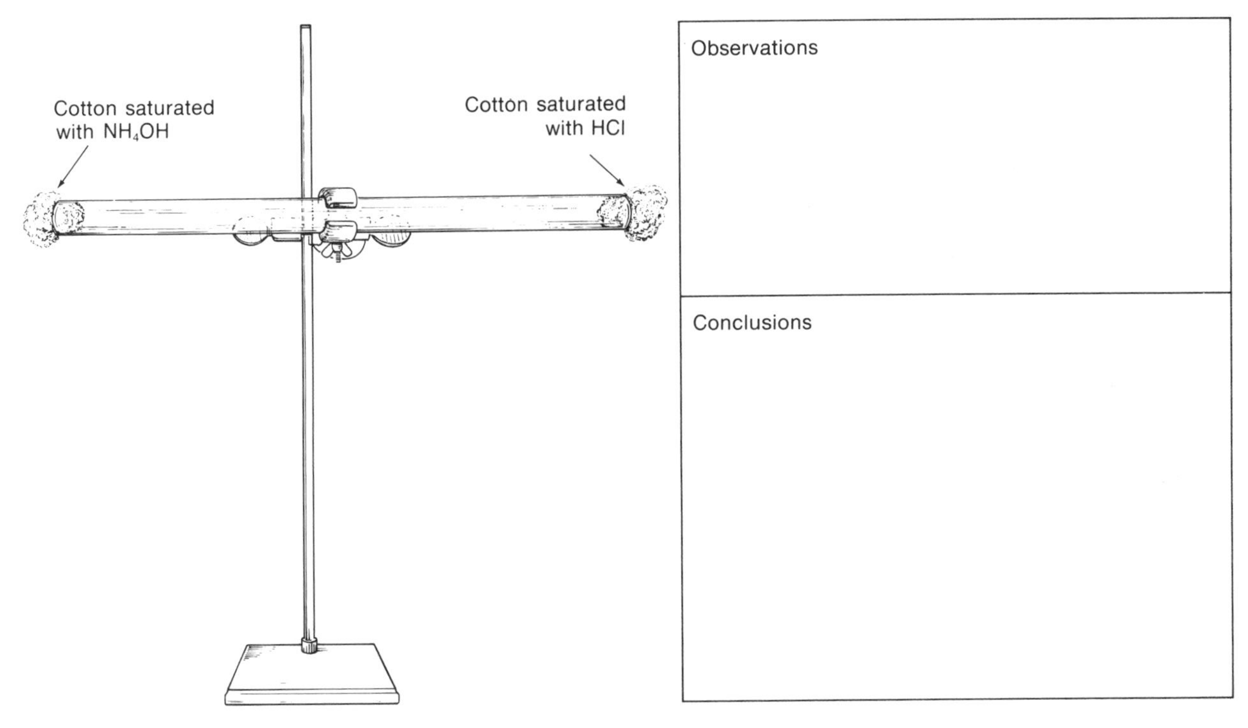

FIG. 9-2
Apparatus for studying gas diffusion.

2. Diffusion of a Liquid in a Solid

The procedure used in this exercise is a modification of a rather simple but useful and informative technique used to characterize the complex relationships between antigens (those substances that induce an immune response) and antibodies (those specific protein molecules that are produced by the body in response to exposure to antigens). This technique has widespread clinical application, although it is being replaced by more sensitive procedures.

Basically, the procedure consists of pouring agar, a gellike material obtained from certain seaweeds, into a petri dish and then punching circular wells close to one another in the gel. The liquid substances that are to be characterized are poured into the wells and allowed to diffuse radially until they meet and react to form a line of precipitate. In this exercise, you will use a procedure, called double immunodiffusion or the **Ouchterlony technique,** to study the effect of molecular weight on diffusion.

Your instructor will give you a disposable petri dish containing agar (Fig. 9-3A). Using a No. 5 cork borer, punch four holes in the agar as shown in Fig. 9-3B and C. Fill each of the holes uniformly with a small amount of 1 *N* solutions of sodium chloride (NaCl), potassium bromide (KBr), potassium ferricyanide ($K_3Fe[Cn]_6$), or silver nitrate ($AgNo_3$). The approximate molecular weights of each of the migrating groups formed when these substances are placed in solution are chloride anion (Cl), 35; bromide anion (Br), 80; ferricyanide anion ($Fe(CN)_6$), 212; nitrate anion (NO_3), 62. Periodically examine the petri dishes and record your observations in Fig. 9-4.

From this study, what can you conclude about the relationship between the rate of diffusion of a molecule and its molecular weight?

____

____

____

B. DIALYSIS

Essentially, **dialysis** is diffusion through a differentially permeable membrane that separates small molecules from large molecules. The principle of

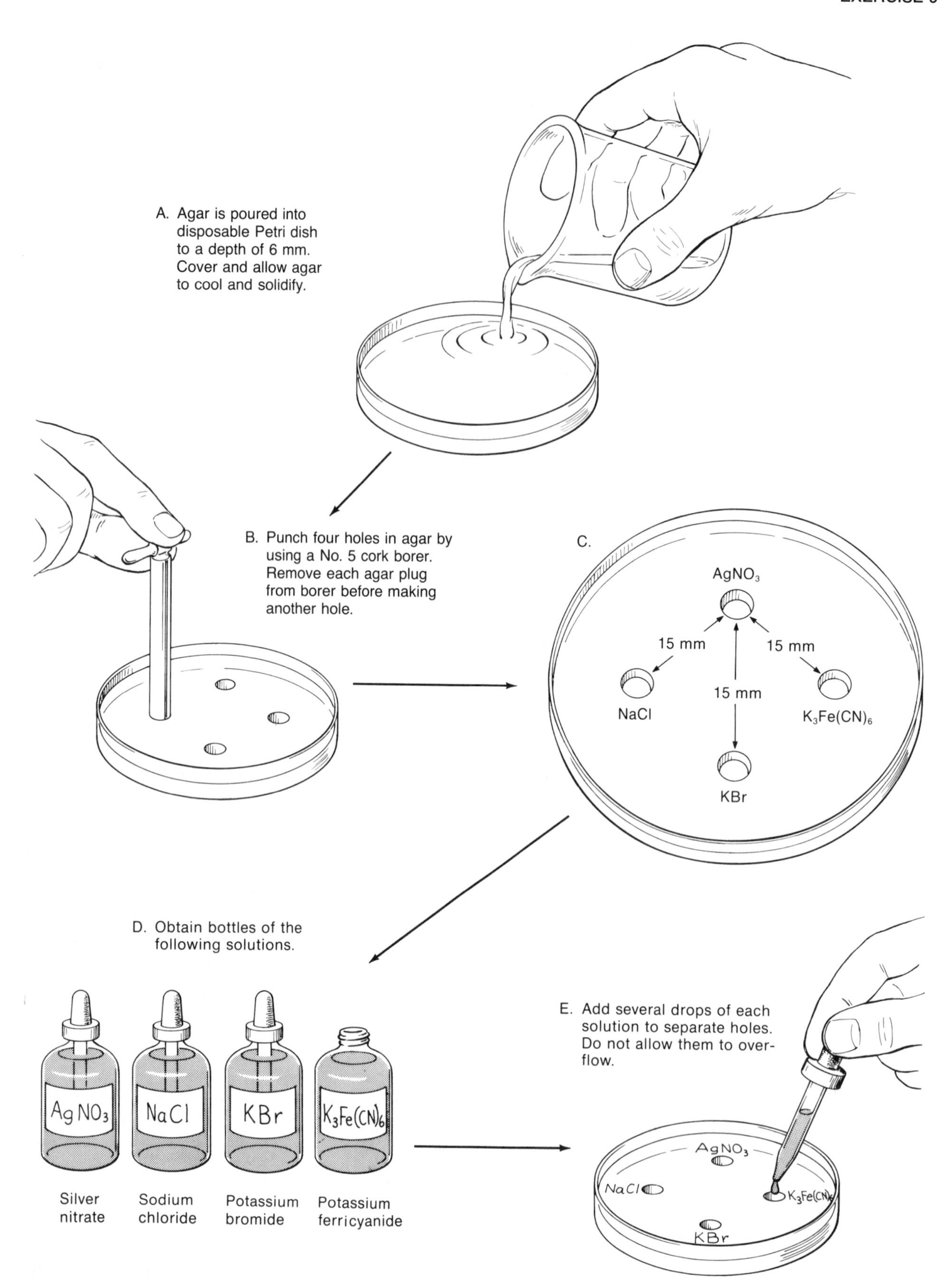

FIG. 9-3
Procedure for determining the effect of molecular weight on diffusion.

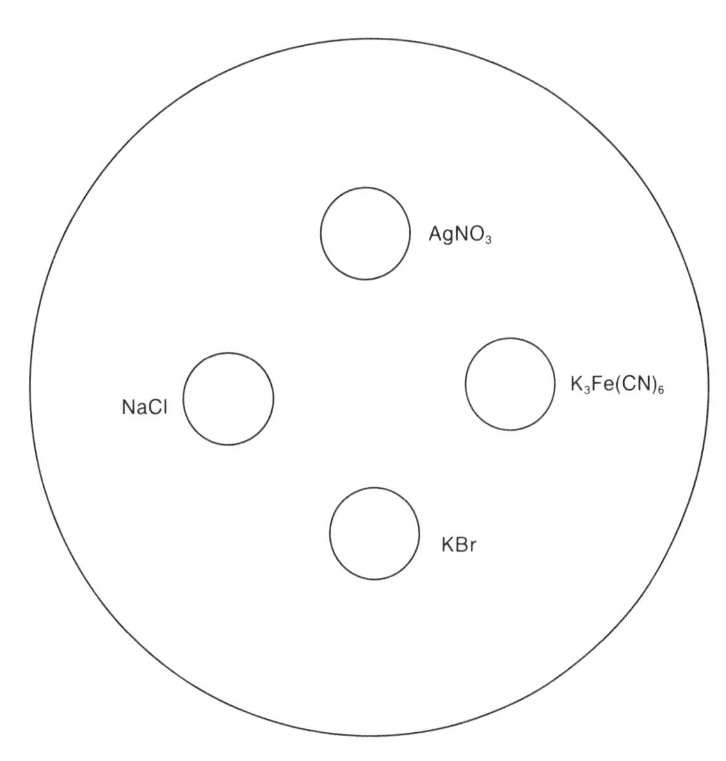

FIG. 9-4
Effects of molecular weight on diffusion.

dialysis is used in artificial kidney machines, where the patient's blood is passed through a tube of **dialyzing membrane** outside of the body; this artificial membrane takes the place of the damaged or defective kidneys (or the lack of kidneys). As the blood moves through the membranous tube, small particle waste products move, by diffusion, from the blood into a solution surrounding the membrane. The purified blood is then returned to the body.

The concept of dialysis will be illustrated by removing chloride ions from a solution of starch and sodium chloride. In order to detect the presence of these compounds in solution, you must be familiar with simple tests used to identify sodium chloride and starch.

1. Tests for Chloride Ions and Starch

1. Fill six numbered test tubes with 5 ml of each of the solutions shown in Fig. 9-5.

2. To tube numbers 1, 3, and 5, add two to three drops of silver nitrate ($AgNo_3$).

3. To tube numbers 2, 4 and 6, add two to three drops of iodine solution.

4. Mix the contents of each tube by swirling and record your observations in Table 9-1.

2. Dialysis of a Starch/ Sodium Chloride Mixture

Following the procedure shown in Fig. 9-6, fill a dialyzing membrane with about 15 ml of a solution containing starch and sodium chloride. Seal the dialysis bag as shown and place in a 250 ml beaker filled with distilled water. Why would you *not* use tap water?

____

____

____

____

How would you determine if NaCl *or* starch had diffused through the membrane into the distilled water?

____

____

____

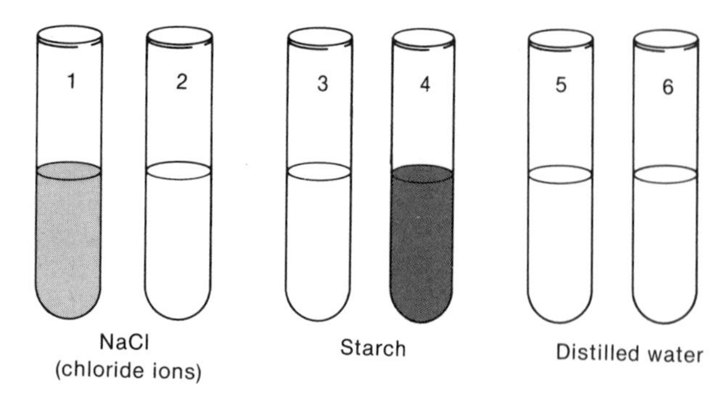

FIG. 9-5
Tests for chloride ions and starch molecules.

Carry out the tests you have described and explain the results you obtained.

____

____

C. OSMOSIS

1. Hemolysis

The cell membrane of **erythrocytes** (red blood cells) is freely permeable to water but relatively impermeable to salts. Thus, if red blood cells are placed in an **isotonic** saline solution (i.e., one that has the same salt concentration as found in plasma and cytoplasm—0.85% NaCl), the cell will retain its normal shape and size. Why?

____

____

____

If red blood cells are placed in a **hypotonic** saline solution (i.e., one that has a lower salt concentration

TABLE 9–1
Test for chloride ions and starch.

Test solution	Reagent	
	Silver nitrate	Iodine
	Observations	
Sodium chloride (chloride ions)		
Starch		
Distilled water		
Conclusions:		

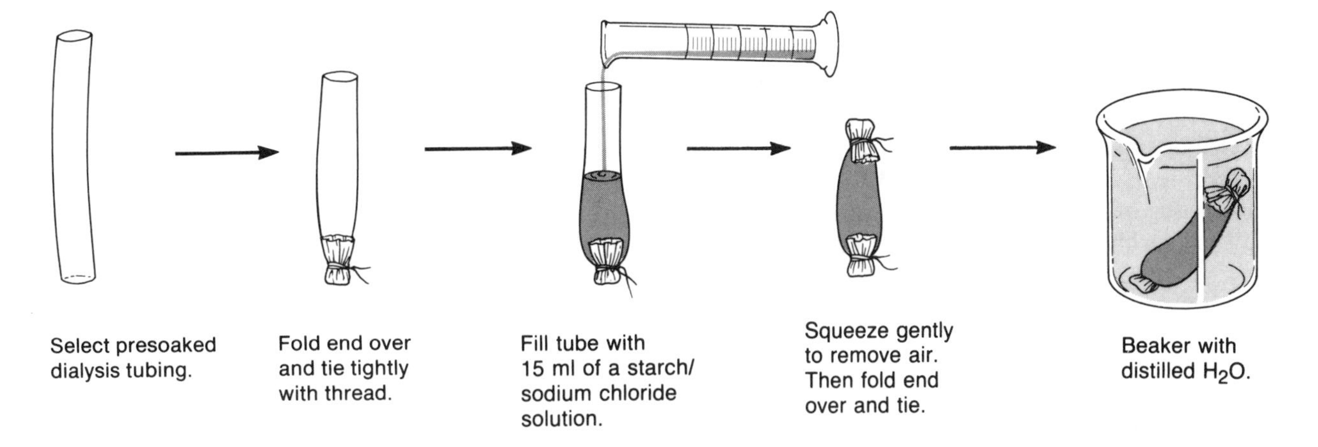

FIG. 9-6
Procedure for dialysis of starch/NaCl mixture.

than found in plasma or cytoplasm), water will enter the cells more rapidly than it leaves and, as a consequence, the red blood cells will swell and ultimately burst, releasing the hemoglobin. This phenomenon is called **hemolysis.** Red blood cells placed in a **hypertonic** saline solution (i.e., one that has a higher salt concentration than found in plasma or cytoplasm) will shrink and appear to have a bumpy, irregular outline. These cells are said to be **crenated.**

To demonstrate the changes to red blood cells under the conditions just described carry out the following procedure:

1. Add a small drop of 0.85% NaCl to a clean glass slide.

2. If whole blood is available in the lab, use it and proceed to step 4. Otherwise, sterilize the tip of your index finger with 70% alcohol or an alcohol pad. Allow the alcohol to evaporate.

3. Prick your finger with a sterile disposable hemolet and gently squeeze a small drop of blood into the saline on the slide. Add a coverslip.

4. Examine the blood using the high-power (43 ×) lens of your microscope. Observe a region where the red blood cells are not too dense. Note the size and shape of these normal cells. Draw a few of these cells in Fig. 9-7.

5. Next, add two to three drops of 5% NaCl (hypertonic solution) to one edge of the coverslip. Continue to observe the blood cells and observe the changes that occur as the more concentrated saline solution reaches them. Record your observations in Fig. 9-7.

6. To a second slide, add a drop of distilled water and a drop of blood. Add a coverslip and observe the cells in this hypotonic solution for several minutes. Record any changes that occur in Fig. 9-7.

Isotonic	Hypotonic	Hypertonic

FIG. 9-7
Appearance of red blood cells in isotonic, hypotonic, and hypertonic saline solutions.

The following is a practical application of the changes that may occur in the tonicity of plasma and/or tissue fluid: one of several symptoms of diabetes mellitus (sugar diabetes) is extreme thirst caused by the decreased production of insulin by the endocrine tissue of the pancreas. What brings about this feeling of thirst? (Refer to any medical physiology text for the answer to this question.)

____

____

____

____

2. Effect of Solute Concentration on the Rate of Osmosis

The rate at which osmosis occurs (i.e., the rate at which water moves into or out of cells) is a function of the tonicity of the cytoplasm of the cell or of the extracellular fluid. In this exercise, you will use an artificial membrane to measure the effect, on osmosis, of varying the tonicity of the fluid within this differentially permeable membrane.

Obtain five dialysis bags, which will function as artificial, differentially permeable membranes. Each bag should be tied at one end by folding the end over and tying with a thread (Fig. 9-8). Fill these bags as follows:

Bag 1. 15 ml of tap water

Bag 2. 15 ml of 20% sucrose solution

Bag 3. 15 ml of 40% sucrose solution

Bag 4. 15 ml of 60% sucrose solution

Bag 5. 15 ml of tap water

As each bag is filled, remove the air by gently squeezing the bottom end of the bag to bring the liquid to the top of it. Press the sides of the bag together so that air does not reenter. Fold the end of the bag over about 5 cm (2 inches) and tie securely with a thread. Wipe each bag dry and weigh each separately to the nearest 0.5 g. Record the weights in Table 9-2 at zero time.

Place bags 1, 2, 3, and 4 in separate beakers of water, and place bag 5 in a beaker of 60% sucrose solution.

At 15-minute intervals (i.e., after 15, 30, 45, 60, and 75 minutes) remove the bags from the beakers, carefully wipe off all excess water, and again weigh each bag separately. Record the data in Table 9-2. Plot the changes in weight (i.e., Δ wt) of each bag against time in Fig. 9-9. What relationship (if any) is there between the concentration of sucrose and the rate of osmosis? How do you account for the differences observed?

____

____

____

____

____

____

TABLE 9–2
Osmosis data.

Time (minutes)	Tubing number, weight, and change in weight (Δ wt)*									
	1		2		3		4		5	
0	weight	Δ wt	weight	Δ wt	weight	Δ wt	weight	Δ wt	weight	Δ wt
15										
30										
45										
60										
75										

* Change in weight (Δ wt) may be recorded as differences (+) or (–) between each reading, or as differences for each reading from 0 time.

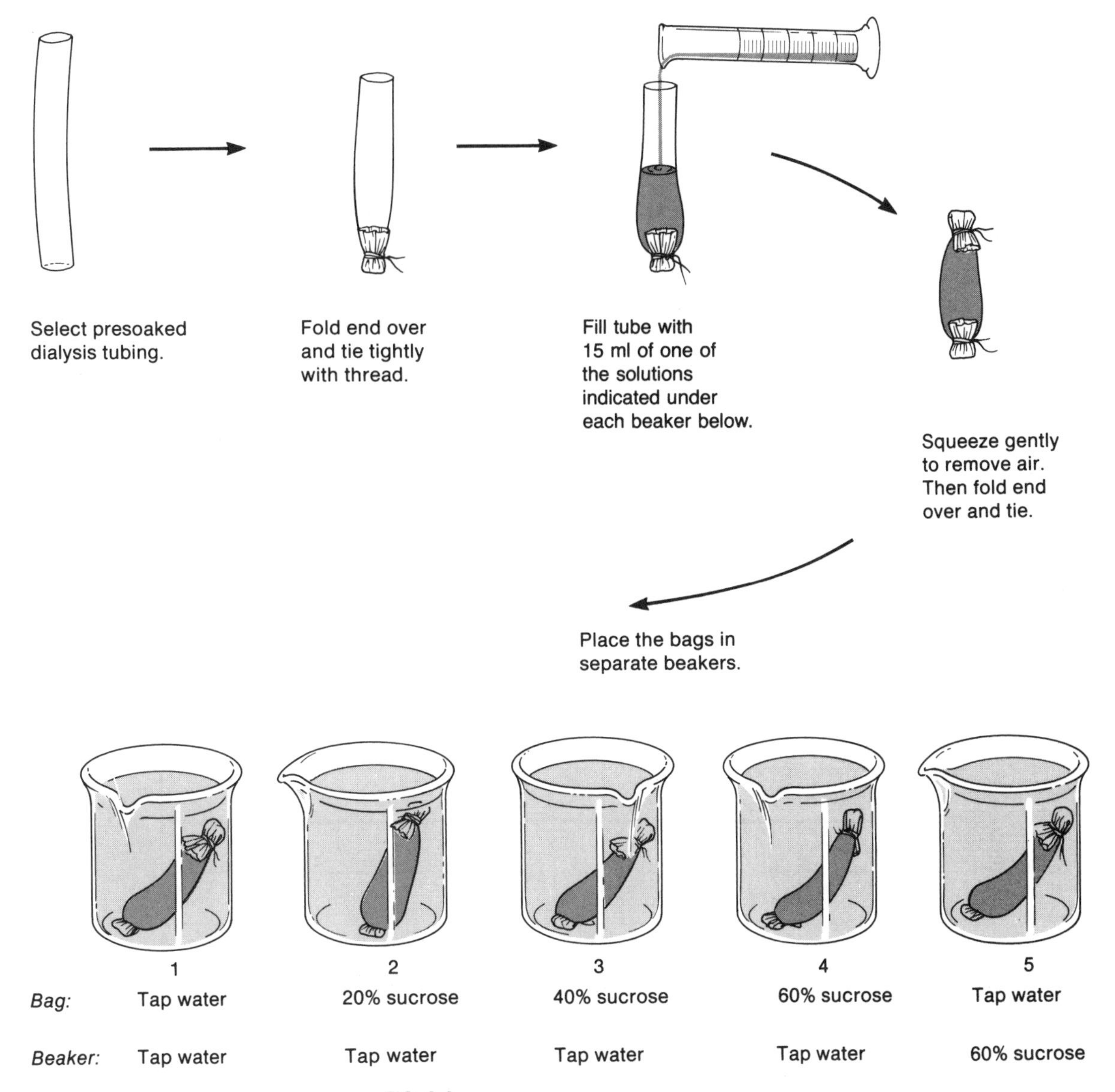

FIG. 9-8
Procedure for measuring rate of osmosis.

3. Effect of Ionization of Molecules on Osmosis

In this exercise, we want to examine electrolytes (molecules that ionize or dissociate) and nonelectrolytes (those that do not) in terms of their osmotic effects on red blood cells. Recall from Part C-1 of this exercise that red blood cells **hemolyze** when exposed to a solution that is hypotonic to the cells. Further, a dilute solution of red blood cells transmits very little light and appears turbid when examined with the eye. On the other hand, if red blood cells are hemolyzed, the solution in the tube becomes so transparent that a printed page placed behind the tube can be easily read. For this reason, you can use the rate at which a suspension of red blood cells becomes clear due to hemolysis to examine the effects of various factors on the rate of diffusion and osmosis.

Since the osmotic effect exerted by a solute is proportional to the concentration of the solute (in terms of the numbers of ions or molecules in solution), it is meaningless to express solute concentration merely in terms of mass (e.g., 25 g/liter). Rather, you express concentration in terms of the number of particles (ions or molecules) in solution. The term normally used for this is **osmole.**

One osmole represents the number of particles in 1 gram of undissociated solute (i.e., a nonelectrolyte). For example, 180 g of glucose is equal to 1 osmole of glucose because glucose does not dissociate

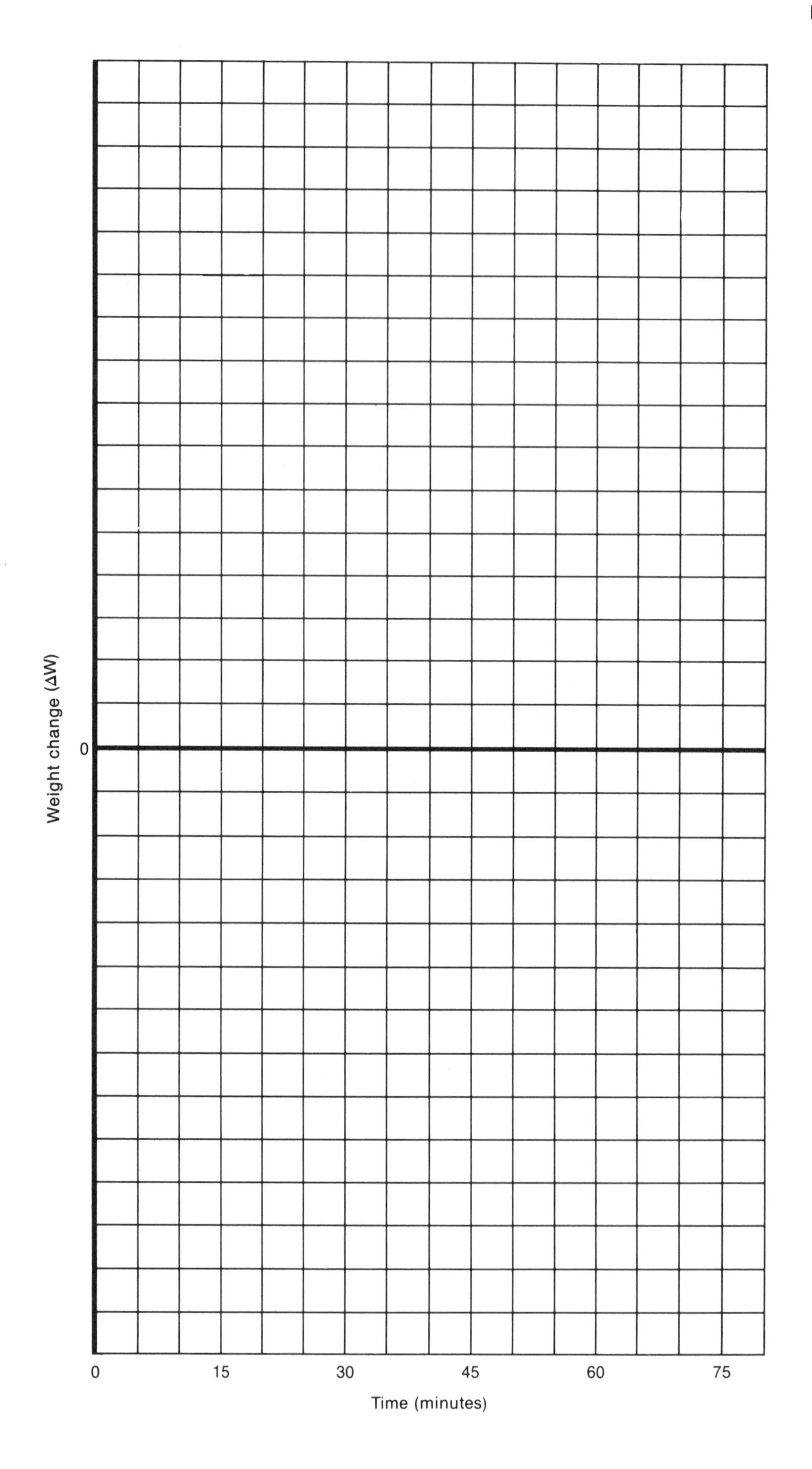

FIG. 9-9
Osmosis data.

in solution. On the other hand, 58.5 g of sodium chloride (NaCl) is equivalent to 2 osmoles, since NaCl dissociates into sodium and chloride ions. In the case of NaCl, the number of osmotically active particles is twice as great as that of the undissociated glucose molecule.

In defining osmotically active solutions, the term **osmolarity** is used. A 1 **osmolar** solution has 1 osmole of solute dissolved in 1 liter of water. For example, a 1 osmolar solution of glucose has 180 g of glucose dissolved in 1000 ml of water and a 0.1 osmolar solution has 18 g in 1000 ml of water.

Procedure

1. Set up two rows of test tubes corresponding to the various osmolar solutions of glucose and NaCl shown in Table 9-3.

2. To each tube add and immediately mix two drops of a red blood cell suspension (Fig. 9-10).

3. Set the tubes aside for 30 minutes and then examine each tube for hemolysis by holding the tube flat against a printed sheet of paper. Hemolysis is complete when the print is plainly readable through the tube.

4. In Table 9-3, indicate by an "X" in which tube(s) hemolysis has occurred.

5. Determine which tube contains the osmolar concentration that is **isotonic** to the red blood cells. In doing this, assume that the tube immediately preceding the one(s) in which hemolysis had occurred is isotonic. Is this a valid assumption? Explain.

____

____

____

Is the tube identified exactly isotonic in terms of its concentration? Explain.

____

____

____

____

Are the "isotonic" osmolar concentrations of glucose and NaCl the same?

____

If not, what concentrations of glucose and NaCl did you determine to be "isotonic"?

Glucose: ____________ NaC1: ____________

Explain any difference in isotonic osmolar concentration between glucose and NaCl.

____

____

____

____

____

____

____

TABLE 9-3
Determination of isotonic molar concentrations of glucose and sodium chloride.

Hemolysis solutions	Tube number and osmolarity						
	1 $\frac{1}{6}$ osmole	2 $\frac{1}{8}$ osmole	3 $\frac{1}{10}$ osmole	4 $\frac{1}{12}$ osmole	5 $\frac{1}{14}$ osmole	6 $\frac{1}{16}$ osmole	7 $\frac{1}{18}$ osmole
	(Indicate the occurrence of hemolysis by X)						
Glucose							
Sodium chloride							

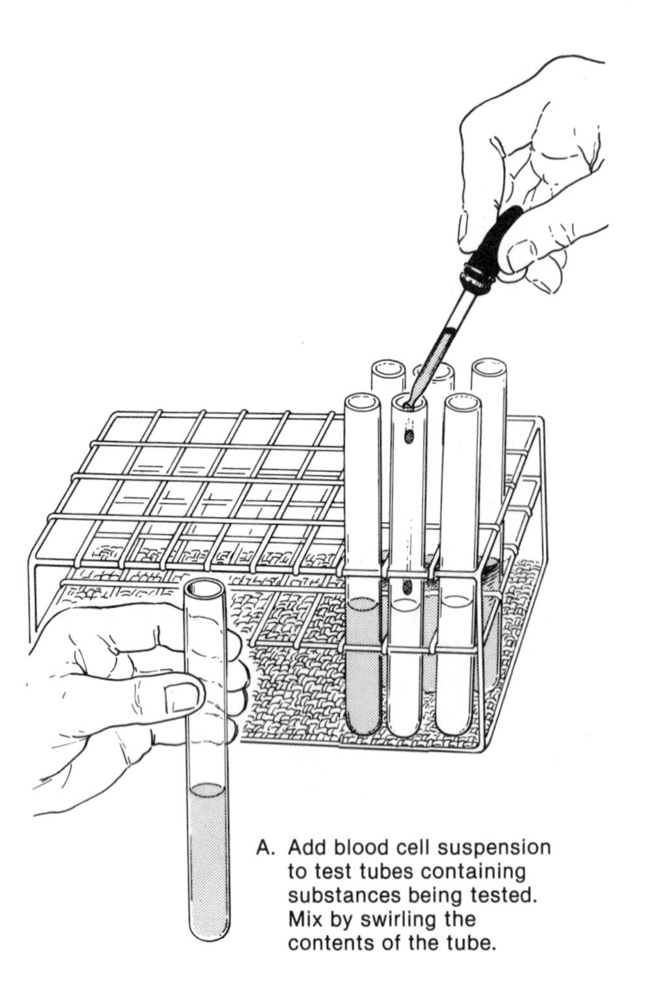

A. Add blood cell suspension to test tubes containing substances being tested. Mix by swirling the contents of the tube.

B. Hold each tube flat against a sheet of printed paper. Hemolysis is complete when the print is plainly readable through the tube.

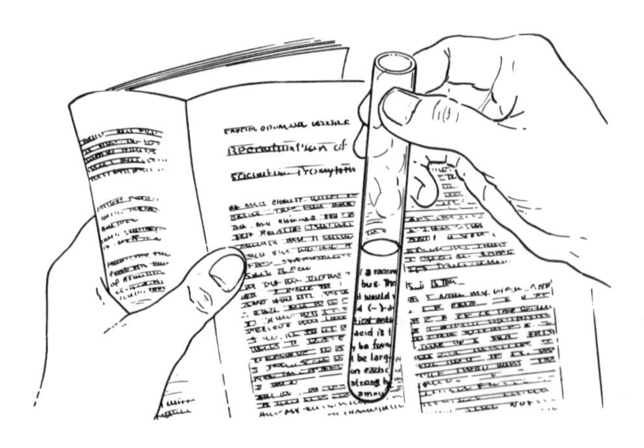

FIG. 9-10
Procedure for determining hemolysis time.

With the data you have obtained it is possible to calculate the degree of dissociation of NaCl using the formula

$$i = 1 + (k - 1)a$$

where

$$i = \frac{\text{isotonic osmolar concentration of glucose}}{\text{isotonic osmolar concentration of NaCl}}$$

k = number of ions from each molecule of NaCl

a = degree of dissociation (to obtain percent dissociation multiply by 100)

What is the percent dissociation of NaCl in your study?

____

What is the percent dissociation of NaCl obtained by several other students?

____

Explain any differences in the amount of dissociation of NaCl that was obtained by the different students.

____

____

____

____

____

____

REFERENCES

Capaldi, R. A. 1974. A Dynamic Model of Cell Membranes. *Scientific American* 230(3):26–33 (Offprint 1292). *Scientific American* Offprints are available from W. H. Freeman and Company, 41 Madison Avenue, New York 10010, and 20 Beaumont Street, Oxford OX1 2NQ, England. Please order by number.

DeRobertis, E. D. P., F. A. Saez, and E. M. F. DeRobertis. 1980. *Cell Biology.* 7th ed. Holt.

Finian, J. B., R. Coleman, and R. H. Mitchell. 1978. *Membranes and Their Cellular Functions.* 2d ed. Wiley.

Giese, A. C. 1979. *Cell Physiology.* 5th ed. Holt/Saunders.

Holter, H. 1961. How Things Get into Cells. *Scientific American* 205(3):167–180 (Offprint 96).

Karp, G. 1984. *Cell Biology.* 2d ed. McGraw-Hill.

Holtzman, E., and A. B. Novikoff. 1984. *Cells and Organelles.* 3d ed. Holt, Rinehart and Winston.

Robertson, J. D. 1962. The Membrane of the Living Cell. *Scientific American* 206(4):64–72 (Offprint 151).

Rustad, R. C. 1961. Pinocytosis. *Scientific American* 204(4):120–130.

Solomon, A. K. 1960. Pores in the Cell Membrane. *Scientific American* 203(6):146–156 (Offprint 76).

Solomon, A. K. 1962. Pumps in the Living Cell. *Scientific American* 207(2):100–108.

Swanson, C. P., and P. L. Webster. 1985. *The Cell.* 5th ed. Prentice-Hall.

Young, J. H., and L. Brubaker. 1963. A Technique for Demonstrating Diffusion in a Gel. *Turtox News* 41:274–276.

EXERCISE 10

Biologically Important Molecules: Proteins, Carbohydrates, Lipids, and Nucleic Acids

The bulk of the "dry matter" of cells consists of carbon, oxygen, nitrogen, and hydrogen organized into large molecules of four main types: proteins, carbohydrates, lipids (fats), and nucleic acids. These molecules often combine with each other to make even larger molecules of importance to the cell.

A. PROTEINS

Proteins are large molecules that range in molecular weight from about 5000 (insulin) to 40 million (tobacco mosaic virus protein). They are constructed of long chains of nitrogen-containing molecules known as **amino acids,** which are linked by **peptide bonds.** In protein molecules, the successive amino acid molecules are bonded together between the carboxyl and amino groups of adjacent amino acids to form long, unbranched molecules. Such bonds are commonly called **peptide bonds.** The structures resulting from the formation of peptide bonds are called dipeptides, tripeptides, or polypeptides, depending on the number of amino acids involved. The individual amino acids are called **residues.** Protein molecules exhibit an unlimited variety of sizes, configurations, and kinds of physical properties because of

- the large number of amino acids that enter into making a single protein molecule (thousands in many cases)
- the almost infinite number of combinations of different amino acids that can be formed
- the reactivity of side groups of the individual amino acids

For example, a human being is composed of thousands, possibly hundreds of thousands, of different proteins. Each protein has a special function, and by its unique chemical nature each is specifically suited to that function.

Proteins have no competitors in the diversity of roles they play in biological systems. They make up a significant portion of the structure of cells and form a large part of most cellular membranes and organelles (for example, mitochondria, ribosomes, spindle fibers, and chromosomes). In addition to their structural role, proteins act as biological catalysts (enzymes), regulate cellular and tissue functions (hormones), and fight disease (antibodies).

Because of their biological significance, it is important to know something about the chemistry of proteins. In the first part of this exercise, you will become familiar with several tests used to identify the presence of protein molecules and to determine the amino acid composition of casein, a milk protein.

1. Qualitative Tests to Detect Proteins

The free terminal amino ($-NH_2$) and carboxyl ($-COOH$) groups of proteins undergo the same kinds of chemical reactions as those of the amino and carboxyl groups of individual amino acids. This reaction is widely used for the detection and quantitative estimation of proteins in procedures used to separate complex mixtures of molecules.

a. Ninhydrin Reaction

The reaction of ninhydrin with amino acids is of particular importance for the detection and qualitative estimation of amino acids. Ninhydrin is a powerful oxidizing agent that removes the amino groups of amino acids. In the process, the reaction liberates ammonia, carbon dioxide, the corresponding aldehyde, and a reduced form of ninhydrin. The ammonia then reacts with an additional mole of ninhydrin and the reduced ninhydrin to form a purple substance. The appearance of the purple color is a positive test for protein.

Add 3.0 ml of distilled water to one test tube and 3.0 ml of 0.1% egg albumin solution to a second tube. Add solid sodium acetate to each test tube (scoop-type spatula loaded to a depth of 1 inch). Add eight drops of ninhydrin to each tube. Heat for 3 minutes in a boiling water bath, and cool. Record your observations in Table 10-1.

b. Sakaguchi Test

Alkaline solutions of proteins that contain the amino acid arginine react with α-naphthol and sodium hypobromite to produce an intense red color. This color disappears rapidly unless stabilized by the addition of urea. Thus, the Sakaguchi test is a useful tool for the detection of proteins containing arginine.

Pipet 3.0 ml of distilled water into Test Tube 1; 3.0 ml of a 1% albumin solution into Test Tube 2; and 3.0 ml of a 0.1% arginine solution into Test Tube 3. Add 1.0 ml of 10 *N* sodium hydroxide to each tube, followed by 1.0 ml of a 0.02% α-naphthol solution. Add two drops of sodium hypobromite to Tube 1, followed immediately (within 10 seconds) by 1.0 ml of a 40% urea solution. Repeat this procedure on Tubes 2 and 3. Record your observations in Table 10-1.

c. Pauly Test

When the amino acids tyrosine and/or histidine are present in a protein hydrolysate (product of enzymatic hydrolysis of protein), they react in alkaline solution with sulfanilic acid to give an intense red color. No other amino acids react. Thus, this test is useful as a confirmation of the presence of histidine and/or tyrosine.

TABLE 10-1
Qualitative chemical reactions of amino acids and proteins.

Reagents tested	Test	Observations
Distilled H_2O	Ninhydrin	
0.1% egg albumin		
Distilled H_2O	Sakaguchi	
1.0% egg albumin		
0.1% arginine		
Distilled H_2O	Pauly	
10 mg/ml tyrosine		
10 mg/ml glycine		
10 mg/ml histidine		

Pipet 2.0 ml of histidine (10 mg/ml) into Tube 1; 2.0 ml of tyrosine (10 mg/ml) into Tube 2; 2.0 ml of glycine (10 mg/ml) into Tube 3; and 2.0 ml of distilled water into Tube 4. Add 1.0 ml of sulfanilic acid reagent and 1 ml of 5% sodium nitrite to each tube. Mix and let stand for 30 minutes. Add 3.0 ml of 20% sodium carbonate to each tube, and mix. Record your observations in Table 10-1.

2. Quantitative Chemical Determination of Protein

Proteins may be assayed quantitatively using the **biuret reaction,** which is based upon a chemical interaction between biuret and copper sulfate in alkaline solution.

Biuret, a simple molecule prepared from urea, contains what may be regarded as two peptide bonds and, thus, is structurally similar to simple peptides. When treated with copper sulfate, this molecule turns an intense purple as a result of the reaction between the copper ions and the peptide bonds. Proteins give a particularly strong biuret reaction because they contain a large number of peptide bonds. You may use the biuret reaction to determine the concentration of proteins, because peptide bonds occur with approximately the same frequency per gram of material for most proteins.

In this experiment, you will determine the "unknown" concentration of two protein solutions by colorimetrically measuring the intensity of the color produced in the biuret reaction as compared with the color produced by a known concentration of the protein bovine serum albumin (BSA). Refer to Appendices B and C for instructions on the use of a colorimeter and a discussion of spectrophotometry.

1. Prepare a set of five test tubes each containing 5 ml of increasing standard solutions of bovine serum albumin, as shown in Table 10-2. Also prepare two test tubes containing 5 ml each of two unknown concentrations of protein solutions.

2. Add 2.5 ml of biuret reagent to each of the tubes, and mix thoroughly by rotating the tubes between the palms of your hands. The color will be fully developed in 30 minutes and will be stable for at least an hour. While waiting for the color to develop, standardize the colorimeter using Tube 1, which is the "blank" containing 5 ml of distilled water and 2.5 ml of biuret reagent. Set the instrument at a wavelength of 540 nm. After "blanking" the instrument, determine the percent transmittance (%*T*) for Tubes 2–5. Record your readings in Table 10-2. Convert the %*T* into absorbance *(A)* for Tubes 1–6 using Table 10-3, and then plot your data in Fig. 10-1 to obtain a **standard curve** (concentration) for BSA. Using this curve, determine the concentrations of your unknown protein solutions.

3. Chromatographic Separation of Amino Acids

A variety of physical procedures are available for the isolation and purification of proteins and amino acids. One of the simplest and most elegant techniques for the separation of the components of a mixture of molecules is **chromatography.** This technique permits the separation of a mixture of substances by separating the components between two

TABLE 10-2
Protocol for quantitative determination of protein.

Tube number	Protein		Biuret reagent (ml)	%*T*	A_{540} nm
	Volume (ml)	Concentration (μg/ml)			
1 (Control)	5	0 (H_2O)	2.5	100	0.0
2	5	250	2.5		
3	5	500	2.5		
4	5	1000	2.5		
5	5	2000	2.5		
No. 1 unknown	5		2.5		
No. 2 unknown	5		2.5		

TABLE 10–3
Conversion of percent transmittance (%*T*) into absorbance (*A*).

%*T*	Absorbance (*A*)				%*T*	Absorbance (*A*)			
		1 (.25)	2 (.50)	3 (.75)			1 (.25)	2 (.50)	3 (.75)
1	2.000	1.903	1.824	1.757	51	.2924	.2903	.2882	.2861
2	1.699	1.648	1.602	1.561	52	.2840	.2819	.2798	.2777
3	1.523	1.488	1.456	1.426	53	.2756	.2736	.2716	.2696
4	1.398	1.372	1.347	1.323	54	.2676	.2656	.2636	.2616
5	1.301	1.280	1.260	1.240	55	.2596	.2577	.2557	.2537
6	1.222	1.204	1.187	1.171	56	.2518	.2499	.2480	.2460
7	1.155	1.140	1.126	1.112	57	.2441	.2422	.2403	.2384
8	1.097	1.083	1.071	1.059	58	.2366	.2347	.2328	.2310
9	1.046	1.034	1.022	1.011	59	.2291	.2273	.2255	.2236
10	1.000	.989	.979	.969	60	.2218	.2200	.2182	.2164
11	.959	.949	.939	.930	61	.2147	.2129	.2111	.2093
12	.921	.912	.903	.894	62	.2076	.2059	.2041	.2024
13	.886	.878	.870	.862	63	.2007	.1990	.1973	.1956
14	.854	.846	.838	.831	64	.1939	.1922	.1905	.1888
15	.824	.817	.810	.803	65	.1871	.1855	.1838	.1821
16	.796	.789	.782	.776	66	.1805	.1788	.1772	.1756
17	.770	.763	.757	.751	67	.1739	.1723	.1707	.1691
18	.745	.739	.733	.727	68	.1675	.1659	.1643	.1627
19	.721	.716	.710	.704	69	.1612	.1596	.1580	.1565
20	.699	.694	.688	.683	70	.1549	.1534	.1518	.1503
21	.678	.673	.668	.663	71	.1487	.1472	.1457	.1442
22	.658	.653	.648	.643	72	.1427	.1412	.1397	.1382
23	.638	.634	.629	.624	73	.1367	.1352	.1337	.1322
24	.620	.615	.611	.606	74	.1308	.1293	.1278	.1264
25	.602	.598	.594	.589	75	.1249	.1235	.1221	.1206
26	.585	.581	.577	.573	76	.1192	.1177	.1163	.1149
27	.569	.565	.561	.557	77	.1135	.1121	.1107	.1083
28	.553	.549	.545	.542	78	.1079	.1065	.1051	.1037
29	.538	.534	.530	.527	79	.1024	.1010	.0996	.0982
30	.532	.520	.516	.512	80	.0969	.0955	.0942	.0928
31	.509	.505	.502	.498	81	.0915	.0901	.0888	.0875
32	.495	.491	.488	.485	82	.0862	.0848	.0835	.0822
33	.482	.478	.475	.472	83	.0809	.0796	.0783	.0770
34	.469	.465	.462	.459	84	.0757	.0744	.0731	.0719
35	.456	.453	.450	.447	85	.0706	.0693	.0680	.0667
36	.444	.441	.438	.435	86	.0655	.0642	.0630	.0617
37	.432	.429	.426	.423	87	.0605	.0593	.0580	.0568
38	.420	.417	.414	.412	88	.0555	.0543	.0531	.0518
39	.409	.406	.403	.401	89	.0505	.0494	.0482	.0470
40	.398	.395	.392	.390	90	.0458	.0446	.0434	.0422
41	.387	.385	.382	.380	91	.0410	.0398	.0386	.0374
42	.377	.374	.372	.369	92	.0362	.0351	.0339	.0327
43	.367	.364	.362	.359	93	.0315	.0304	.0292	.0281
44	.357	.354	.352	.349	94	.0269	.0257	.0246	.0235
45	.347	.344	.342	.340	95	.0223	.0212	.0200	.0188
46	.337	.335	.332	.330	96	.0177	.0166	.0155	.0144
47	.328	.325	.323	.321	97	.0132	.0121	.0110	.0099
48	.319	.317	.314	.312	98	.0088	.0077	.0066	.0055
49	.310	.308	.305	.303	99	.0044	.0033	.0022	.0011
50	.301	.299	.297	.295	100	.0000	.0000	.0000	.0000

Note: Intermediate values may be arrived at by using the .25, .50, and .75 columns. For example, if %*T* equals 85, the absorbance equals .0706; if %*T* equals 85.75, the absorbance equals .0667.

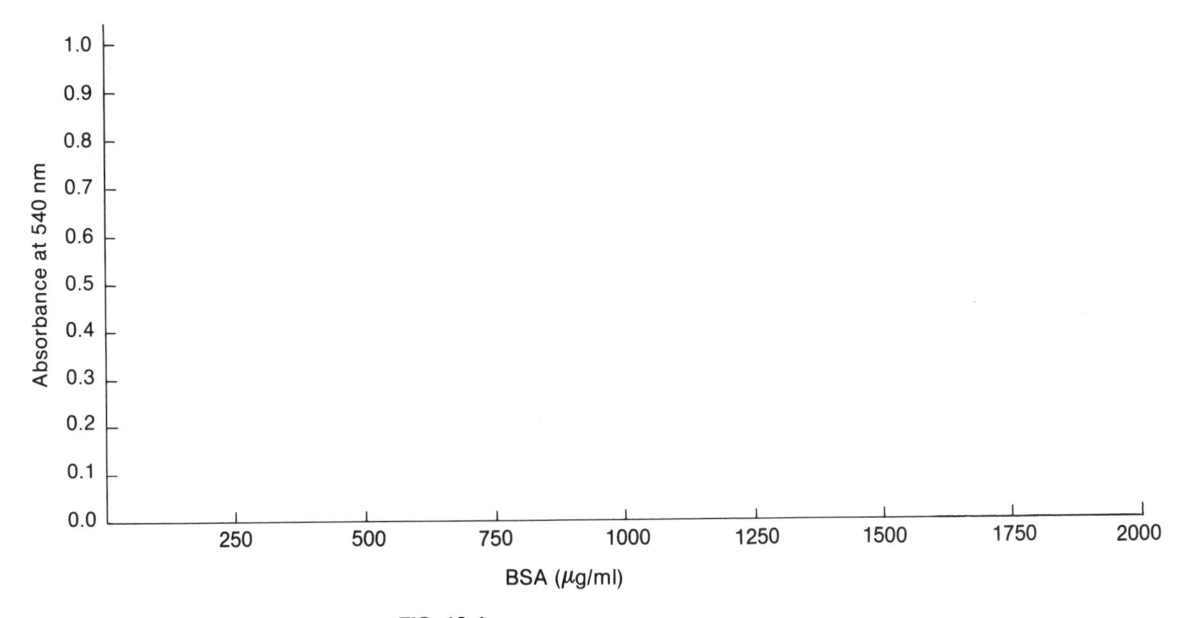

FIG. 10-1
Bovine serum albumin (BSA) standard curve.

phases. In **column chromatography,** these phases are a solid or **stationary phase,** such as calcium carbonate or sephadex, and a **mobile phase,** a liquid solvent. In **paper chromatography,** the paper acts as the stationary phase and the solvent as the mobile phase. **Thin-layer chromatography** (TLC) is a recent adaptation of chromatographic methods, designed to increase the resolution of component separations. The name of the process is derived from the thin, uniform layer of an **adsorbent,** such as silica gel or alumina, spread on a flat glass plate and treated in a manner similar to a paper strip. The thinness of the adsorbent layer requires less material and results in a very rapid migration of the components in a mixture. (See Appendix D for a more detailed discussion of chromatograhy.)

In this study, you will use thin-layer chromatography to separate and identify the amino acid residues of a hydrolysate of casein, a milk protein.

If possible, precoated silica gel chromatographic plates should be purchased for this exercise. Otherwise, the following procedure can be used to prepare your own slides coated with silica gel.

1. Holding a clean glass slide by the sides, immerse it as far as possible into a jar containing silica gel. Move the slide back and forth several times. Stop, then carefully lift the slide *straight up* (Fig. 10-2).

2. Allow the slide to air-dry for several minutes. (*Note:* The white, silica gel coat is very fragile. Do not damage the surface.)

3. Select the side of the slide having the smoothest surface. Then, remove the silica gel from the other side by wiping with a paper towel. (*Note:* Avoid excessive handling of the slide, because your hands may contaminate it with amino acids. Touch it only at the edges.)

4. Lay the slide down with the coated side up. "Spot" the coated surface at two points approximately 12 mm apart and 6 mm from the bottom. To Spot 1, add a small drop of casein hydrolysate. To Spot 2, add a known amino acid supplied by your instructor. Other students in the class will be given other known amino acids.

5. Allow the spots to dry. Then carefully place the slide in a chromatographic developing jar and cover. When the **solvent front** (i.e., the leading edge of the solvent) has moved to about 6 or 12 mm from the top of the slide (30–45 minutes), remove the slide from the jar. Allow the slide to dry for 4–5 minutes.

6. In a hood or other well-ventilated area, cautiously spray the surface of the slide with ninhydrin. *Do not inhale the fumes or get any spray in your eyes.* Allow the slide to dry for 2–3 minutes, then heat the slide as directed by your instructor for 2–3 minutes.

The amino acids in the hydrolysate will react with the ninhydrin and will appear as colored spots on the slide. The ninhydrin test yields purple colors with most amino acids and a yellow spot with the

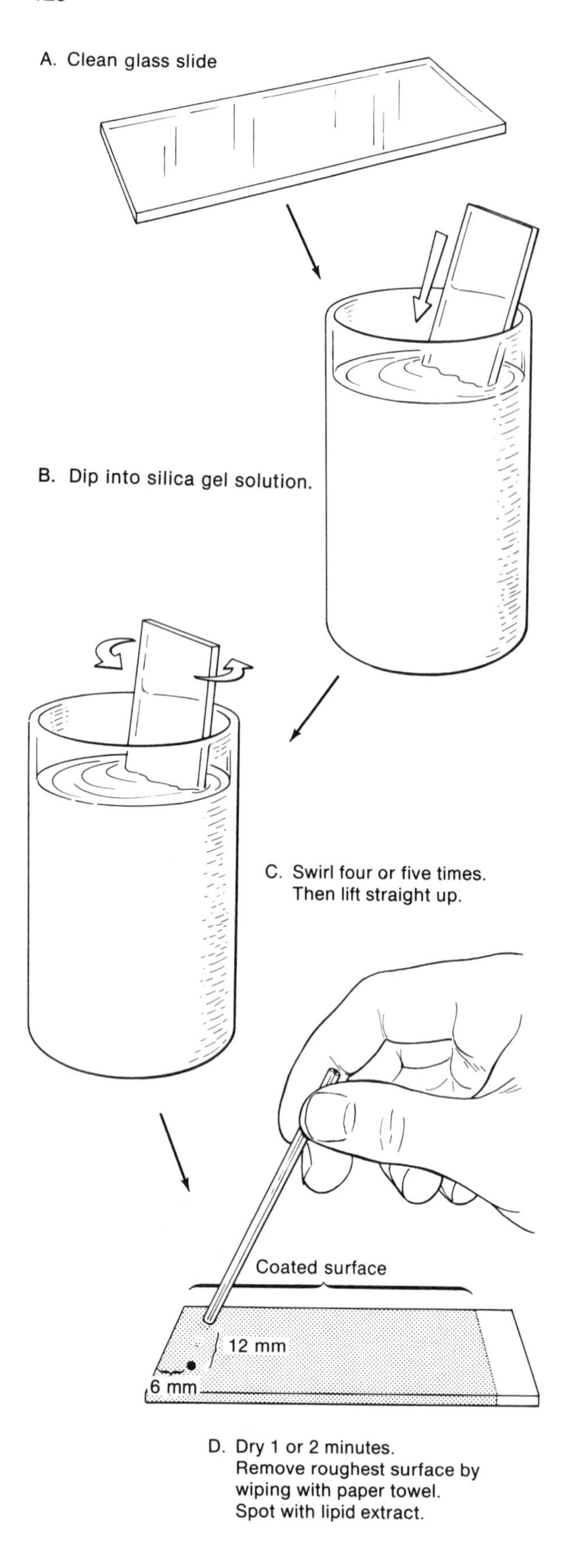

FIG. 10-2
Preparation of plate for thin-layer chromatography.

amino acid proline. Record the color of the spots on your chromatogram in Table 10-4.

Roughly estimate the center of each amino acid spot and measure the distance it has traveled up the slide from its point of application. This distance, divided by the total distance traveled by the solvent from the origin line, is known as the R_f **value.** Two substances having the same R_f are probably identical and thus this value can be used to identify specific amino acids separated from a mixture.

Record your observations, as well as those of other students who were given known amino acids other than your own, in Table 10-4.

B. CARBOHYDRATES

Carbohydrates, formed in the process of photosynthesis, play a key role in making captured energy available to the cell. They also supply important carbon "skeletons" (carbon atoms linked together) for the synthesis of other biologically important molecules.

The term *carbohydrate* means "hydrate of carbon." This name is used because it included many compounds that contain atoms of hydrogen and oxygen in the same proportion as in water—two hydrogens to one oxygen. Thus, a carbohydrate can be described by the general formula $C(H_2O)n$; n represents the number of $C(H_2O)$ units that make up carbohydrates, ranging from relatively simple molecules, called *sugars,* to the complex molecules of starch and cellulose.

Most carbohydrates have a basic unit of 6 carbon atoms (sugars), which are linked together in various ways. On the basis of the number of these 6-carbon compounds that are linked together to make longer units, the carbohydrates are divided into three classes: **monosaccharides,** which consist of a single 6-carbon molecule (e.g., glucose); **disaccharides,** which consist of two single sugar molecules linked together (sucrose is made up of glucose linked to fructose); and **polysaccharides,** which consist of three or more sugar molecules linked together (starch and glycogen are long chains of glucose molecules). Starch and glycogen are storage forms of carbohydrates; starch is usually produced in plants, and glycogen in animals.

Carbohydrates may be identified by color reactions with specific reagents. These tests can be used to determine the amount as well as the kind of carbohydrate by measuring the variations in color obtained with different concentrations of reagent. Most of the procedures require heating the carbohydrate and reagent together in a hot water bath. It is usually best to carry out a procedure on all the carbohydrates in the same water bath at the same time so that a direct comparison can be made of the relative behavior of all the carbohydrates in a given test.

In this exercise, you will become familiar with some of the more common tests for detecting the presence of specific types of carbohydrates. You will be asked to identify the carbohydrates composition of an "unknown" solution containing one or more carbohydrates. Only the carbohydrates you test will be found in the unknown. Each carbohydrate in the solution will be at a concentration of 6%. About one-half of each unknown solution should be diluted to give a 1% solution, which will be used in most of the tests. *The unknown should always be run simultaneously with the known carbohydrates.*

1. Tests for Reducing Sugars

A carbohydrate with a free or a potentially free aldehyde ($-\overset{O}{C}\diagdown_{H}$) or ketone ($>C=O$) group is a reducing sugar and, in a solution of sufficiently high pH, can reduce weak oxidizing agents such as cupric, silver, or ferricyanide ions. For example, Cu^{2+} ions react with glucose to form a colored precipitate of cuprous oxide. The color of the precipitate will range from green to reddish brown, depending on the quantity of the reducing sugar present.

$$\text{Glucose} + Cu(OH)_2 \xrightarrow{\text{heat}} \underset{\text{Colored}}{Cu_2O} + H_2O + \text{oxidized glucose}$$

a. Benedict's Test

Benedict's solution contains sodium bicarbonate, sodium citrate, and copper sulfate. If combined with a reducing sugar (such as glucose or fructose) and heated, the divalent copper ion (Cu^{2+}) of copper sul-

TABLE 10–4
Chromatography of amino acids.

Amino acid	Distance solvent moved	Distance spot moved	Color of spot	R_f value
Aspartic acid				
Glutamic acid				
Methionine				
Proline				
Tyrosine				
Histidine				
Alanine				
Lysine				
Slide No.				
Slide No.				
Slide No.				
Slide No.				
Slide No.				
Slide No.				
Slide No.				
Slide No.				

fate ($CuSO_4$) is reduced to the monovalent copper ion (Cu^+) of cuprous oxide (Cu_2O), which forms a precipitate.

Place 5.0 ml of Benedict's reagent in a test tube, and add eight drops of a 1% solution of the carbohydrate to be tested. Heat the contents for 2 minutes over a Bunsen burner (do not boil) and allow to cool to room temperature. Or place the test tube in a boiling water bath for 3 minutes, and allow to cool to room temperature. When several sugars are tested at one time, the latter method is preferred.

Record the color and amount of precipitate formed for each of the carbohydrates tested in Table 10-5. Use a plus sign (+) to indicate a small amount of precipitate, a double plus (++) for a moderate amount, and a triple plus (+++) for a large amount.

b. Barfoed's Test

Barfoed's test is used to distinguish between monosaccharides and disaccharides. The reagent is similar to Benedict's reagent except that it is slightly acidic, having a pH of about 4.5. At this pH, disaccharides, when heated for 2 minutes, *will not* reduce Cu^{2+} to Cu_2O, whereas monosaccharides *will* reduce the Cu^{2+}. Longer heating of disaccharides may lead to some reduction because of the formation of monosaccharides by hydrolysis. Therefore, it is essential that all the sugars be treated in exactly the same way and that the time of appearance of a precipitate be noted.

For each carbohydrate to be tested, put 5 ml of Barfoed's reagent in a test tube and add 0.5 ml of a 1% solution of the carbohydrate. Mix the solutions well. Place the test tubes in a boiling water bath, and heat for 2 minutes. A positive test for monosaccharides is the appearance of a red precipitate of cuprous oxide within 1 or 2 minutes. Record the results in Table 10-5, noting the time of appearance of the precipitate.

TABLE 10-5
Qualitative carbohydrate tests.

Carbohydrates	Tests		
	Benedict's	Barfoed's	Phenylhydrazine
Glucose			
Fructose			
Galactose			
Mannose			
Xylose			
Lactose			
Maltose			
Sucrose			
Starch			
Unknown			

CHO–CHOH–CHOH–CHOH–CHOH–CH_2OH (Mannose) + C_6H_5–NH–NH_2 (Phenylhydrazine) ⟶ HC=N–NH–C_6H_5–CHOH–CHOH–CHOH–CHOH–CH_2OH (Phenylhydrazone) + H_2O

FIG. 10-3
Reaction of phenylhydrazine with mannose.

c. Reactions with Phenylhydrazine

Reducing sugars react with phenylhydrazine to form phenylhydrazones. This reaction takes place at room temperature. All of the reducing sugars, except mannose, form phenylhydrazones that are soluble in H_2O. Mannose forms an insoluble phenylhydrazone (Fig. 10-3). Thus, this test is specific for the detection of mannose.

For each carbohydrate to be tested, put about 6 mm of solid phenylhydrazine reagent into a test tube. (CAUTION: *Phenylhydrazine is poisonous. Do not spill or get on skin.*) To each test tube, add 2 ml of a 1% solution of one of the carbohydrates to be tested. Mix until the solid has dissolved, and allow to stand at room temperature 10–15 minutes. The formation of a white, crystalline precipitate indicates mannose. Record your results in Table 10-5.

d. Unknown Carbohydrates

Using the tests in Part B, determine the carbohydrate content of the "unknown" that you have been given to analyze. Confirm your results with your instructor.

C. LIPIDS

Lipids are a group of fatty or oily substances, classified together because they are insoluble in water and soluble in the so-called fat solvents (e.g., ether, ace-

$$\underset{\text{Lipid}}{\begin{matrix} H_2C{-}O{-}\overset{O}{\overset{\|}{C}}{-}R^1 \\ | \\ H{-}C{-}O{-}\overset{O}{\overset{\|}{C}}{-}R^2 \\ | \\ H_2C{-}O{-}\overset{O}{\overset{\|}{C}}{-}R^3 \end{matrix}} \xrightarrow[\text{Hydrolysis}]{3H_2O} \underset{\text{Glycerol}}{\begin{matrix} CH_2OH \\ | \\ CHOH \\ | \\ CH_2OH \end{matrix}} + \underset{\text{Three fatty acids}}{\left\{\begin{matrix} R^1{-}\overset{O}{\overset{\|}{C}}{-}OH \\ R^2{-}\overset{O}{\overset{\|}{C}}{-}OH \\ R^3{-}\overset{O}{\overset{\|}{C}}{-}OH \end{matrix}\right.}$$

FIG. 10-4
Hydrolysis of a lipid.

tone, carbon tetrachloride). The simplest lipids (butter, coconut oil, and animal and plant fats) are composed of carbon, hydrogen, and oxygen and, on hydrolysis, yield glycerol and fatty acids (Fig. 10-4). They have a higher proportion of carbon-hydrogen bonds than carbohydrates and consequently release a larger amount of energy on oxidation than do other organic substances. Fats, for example, release about twice the calories released by an equal amount of carbohydrates. These molecules also form a large part of most cellular membranes, and control the movements of other lipids or lipid-soluble materials in and out of cells.

In this exercise, you will separate a lipid extract into its component fatty acids using thin-layer chromatography. If possible, use precoated silica gel chromatographic plates for this exercise. Otherwise, use the following procedure to prepare your own slides coated with silica gel.

1. Holding a clean slide by its sides, immerse it as far as possible into a jar containing silica gel. Move the slide back and forth several times and then carefully lift the slide *straight up* (Fig. 10-2).

2. Allow the slide to air-dry for several minutes. (*Note:* The white, silica gel coat is very fragile. Do not damage the surface.)

3. Select the side of the slide with the smoothest surface. Then, by wiping with a paper towel, remove the silica gel from the other side. (*Note:* Avoid excessive handling of the slide; your hands may contaminate it with lipids. Touch it only at the edges.)

4. Lay the slide down with the coated side up. "Spot" the hydrolyzed lipid extract on the coated surface about 6 mm from the bottom. Allow the spot to dry. Then carefully place the slide in a chromatographic developing jar and cover. When the solvent front (i.e., the leading edge of the solvent) has moved to about 6 or 12 mm from the top of the slide (30–45 minutes), remove the slide from the jar. Allow the slide to dry from 4–5 minutes.

5. Place the slide in a jar containing iodine crystals, cover, and leave until brownish spots appear on the surface of the silica gel.

Only fatty acids migrate in the solvent used. Other components present in the extract either do not absorb onto the silica gel or do not migrate and thus remain at the origin. For example, triacylglycerols (also called triglycerides) will be found at the edge of the solvent front while phosphoglycerides remain at the origin. Locate these components on your slide.

D. NUCLEIC ACIDS

Nucleic acids are so named because they were found in the nuclei of fish sperm by Miescher about 60 years ago. It has now been shown that there are two types of nucleic acids in all cells: **deoxyribonucleic acid** (DNA) and **ribonucleic acid** (RNA). DNA is composed of (1) the purine nitrogen bases adenine and guanine, (2) the pyrimidine nitrogen bases cytosine and thymine, (3) the pentose sugar deoxyribose, and (4) phosphoric acid. RNA is composed of essentially the same structural units except that the pyrimidine uracil is present instead of thymine and ribose is present instead of deoxyribose. DNA is found predominantly in the nucleus, whereas RNA is most abundant in the cytoplasm but is also found in the nucleoli of the nucleus.

The presence of the sugar ribose in RNA and deoxyribose in DNA can be used to identify and differentiate these nucleic acids. The colorimetric measurement of the green color that is obtained when ribose reacts with Bial's orcinol reagent can be used as a quantitative assay of ribose and thereby of the RNA from which it has been hydrolyzed. Deoxyribose can be quantitatively determined by measuring the blue color that is formed when it reacts with Dische diphenylamine reagent.

The orcinol and diphenylamine reagents may be applied directly to RNA and DNA solutions because the strong acids in these reagents hydrolyze the purine nucleotides to sugars, bases, and phosphoric acid. The sugars react with the proper reagent to yield color. The sugars attached to pyrimidine bases (in pyrimidine nucleotides) do not react under these conditions because the bond linking the sugar to the pyrimidine base is resistant to hydrolysis under the conditions used. Since purine and pyrimidine nucleotides are present in a ratio of approximately 1:1 in nucleic acids, about half of the total sugar present in a sample is measured under these conditions. The amount of sugar in a sample of purified nucleic acid is such that the molar ratio, sugar : phosphorus or sugar : base, is nearly 1.

In this exercise, you will prepare extracts of DNA and RNA from bovine spleen tissue. The Bial's and Dische reactions will be used to measure the amount of RNA and DNA in your extracts by colorimetric comparison with known amounts of these nucleic acids.

1. Extraction of DNA

Because the amount of DNA in most cells is rather small, it is important to select a tissue or organ that contains cells with a high nucleus-to-cytoplasm ratio (i.e., large nuclei surrounded by a relatively small amount of cytoplasm). Human lymphocytes are ideal sources for DNA extraction. Therefore, lymphoid tissues, such as spleen and thymus, are routinely used, because they contain large numbers of lymphoid cells.

In this experiment you will use frozen bovine (cow) spleen or thymus that was obtained fresh from a slaughter house. (*Note:* In doing this extraction every effort should be made to complete the entire extraction procedure in one laboratory period. If the extraction must be discontinued every effort should be made to complete the first six steps. At this point, the extract may be put into a flask, labeled with your name, and frozen until you can complete steps 7–10.)

1. Your instructor will give you frozen cubes of bovine spleen or thymus tissue. Carefully weigh out 15 g of the tissues. Return the excess to your instructor.

2. Pour 150 ml of a cold (4°C) citrate buffer solution (pH 7.2–7.4) into a chilled Waring or similar type blender. Start the blender and add the frozen cubes of tissue one at a time, blending until each tissue cube has been thoroughly homogenized. This procedure is an excellent method for breaking the cell and nuclear membranes to release their cytoplasmic and nuclear contents. Continue homogenizing for 30–60 seconds after the last cube has been added.

Citrate is added to the buffer to inhibit the activity of intracellular DNA-hydrolyzing enzymes (DNAases), which are released from disrupted lysosomes during the homogenization process. These enzymes require magnesium ions (Mg^{2+}) for their activity. Because citrate has a strong affinity for Mg^{2+}, it binds these ions and prevents the DNAases from inactivating the DNA in the course of extraction. Carrying out the extraction procedure at 0–4°C also retards the activity of these enzymes.

3. Pour the homogenate into centrifuge tubes and centrifuge for 15 minutes at $4000 \times G$ in a refrigerated centrifuge. The homogenate is centrifuged to separate cells that have not been broken, debris, and deoxyribonucleoprotein (DNP), which is insoluble in the citrate buffer used in the extraction procedure. RNA, on the other hand, is soluble under these conditions and will be found in the supernatant fluid.

4. Carefully decant the supernatant, measure its volume, and freeze for later use in the extraction of RNA (Part 3). The sediment or pellet at the bottom of your centrifuge tube contains the DNP from the bovine spleen.

5. Pour enough cold citrate buffer into your centrifuge tubes to fill them about half full. Stopper the tubes and shake well for several minutes to break up the pellet. When the pellet is thoroughly dispersed recentrifuge for 15 minutes at $4000 \times G$.

6. Pour off the supernatant and discard. Pour cold 2.6 *M* sodium chloride (NaCl) (about 15% concentration) into the centrifuge tubes until they are about half full. Break up the pellets with a glass rod, stopper the tubes, and shake them vigorously for several minutes to get the DNP into solution, since it is soluble in 2.6 *M* NaCl. The NaCl also dissociates the proteins (protamines and histones) from the DNP to yield free DNA. The proteins form a fine precipitate that can be removed at high speeds, leaving the DNA dissolved in the supernatant.

In order to dissolve all of the DNA present, pour the suspension into the blender and homogenize for 1 minute. If the suspension is quite thick, add a little more of the 2.6 *M* NaCl. If any lumps remain, homogenize for another 30 seconds. If necessary, the contents of your centrifuge tubes can be poured into a 250-ml beaker and the suspension mixed on a magnetic stirrer for 10 minutes.

(*Note:* At this point, the extraction procedure may be stopped and the extract frozen until the next lab. If you have sufficient time left (about 1 hour) continue with step 7.)

7. Transfer your preparation back into centrifuge tubes and centrifuge at $20{,}000 \times G$ for 20 minutes to sediment the protein (thaw first if previously frozen).

8. The supernatant, containing the dissolved DNA, should then be poured into a 1000-ml beaker. The DNA can now be precipitated by *slowly* adding approximately two volumes of 95% ethyl alcohol

(ethanol) so that it forms a layer over the supernatant fluid. A mass of white fibrous material will form at the interface of the DNA solution and the ethanol. Using a glass rod, stir the precipitate so as to force the alcohol into the supernatant containing the DNA. As you do this the precipitate will become wound around the glass rod. Continue this procedure until you can no longer see precipitate being added to the rod. This precipitate is the alcohol insoluble form of DNA.

9. Transfer the precipitated DNA to a 250-ml flask. Add 200 ml of distilled water, stopper the flask, and shake this DNA mixture vigorously. In a few minutes, the DNA should dissolve to form a viscous and colorless solution. If necessary, complete dissolving of DNA by use of a magnetic stirrer.

10. Transfer 5 ml of this DNA solution to a test tube and give the rest to your instructor. Using the 5-ml sample, you can measure the concentration of DNA in your preparation by means of the Dische Diphenylanime Reaction.

2. DNA Detection by the Dische Diphenylamine Reaction

The presence of deoxyribose in DNA can be used to characterize and differentiate this nucleic acid from RNA, which contains the sugar ribose. Furthermore, the concentration of DNA can be quantitatively determined by measuring the blue color that is formed when deoxyribose reacts with Dische diphenylamine reagent. In the presence of acid, deoxyribose is converted into a molecule which then binds to diphenylamine to form a complex that turns a blue color. The intensity of the color is proportional to the concentration of dissolved nucleic acid.

1. Prepare a test tube rack containing six test tubes numbered 1–6. Tubes 1–4 will be used to prepare a standard DNA concentration curve, Tube 5 will serve as a "blank" control, and Tube 6 will contain the extracted DNA solution of unknown concentration.

2. Dissolve exactly 5 mg of commercially prepared DNA in 5 ml of distilled water. This will be the stock DNA solution containing 1 mg of DNA per milliliter.

3. Pipet 2 ml of the stock DNA solution into Tubes 1 and 2. Pipet 2 ml of distilled water into Tubes 2, 3, 4, and 5. Mix the contents of Tube 2 thoroughly and then transfer 2 ml to Tube 3. Thoroughly mix the contents of Tube 3 and transfer 2 ml to Tube 4. Mix well and then discard 2 ml from Tube 4. Pipet 2 ml of your unknown DNA solution in Tube 6. All six tubes should now contain 2 ml of the solutions described in Table 10-6.

4. Pipet 4 ml of the Dische diphenylamine reagent into each of the six test tubes and mix thoroughly. Place all tubes in a boiling water bath for 10 minutes. While these tubes are being heated, prepare an ice bath by placing crushed ice in a 500-ml beaker and adding water until the beaker is about two-thirds full. After heating for 10 minutes, transfer the six tubes to the ice bath and agitate them gently for 5 minutes to cool the contents rapidly.

5. Turn on the colorimeter and allow 5 minutes for the instrument to warm up. (See Appendix B for instructions on the use of the Bausch & Lomb Spectronic 20 colorimeter and Appendix C for a discussion of spectrophotometry.) Check to make sure the sample holder is empty. Adjust the dial so that it reads 0% transmittance (%*T*) at a wavelength of 500 nm. Place the "blank" tube (Tube 5) in the tube holder, close the cover, and adjust the light control until the dial reads 100% transmittance. Remove the tube, then determine the percent transmittance of Tubes 1, 2, 3, and 4. Convert these readings into absorbance *(A)*, using Table 10-3. Record these data in Table 10-6.

6. Prepare a standard DNA curve (Fig. 10-5) by plotting the absorbance of Tubes 1, 2, 3, and 4 against the known concentrations of DNA in each tube. Using this standard curve, determine the concentration of DNA in your unknown preparation.

---

3. Extraction of RNA from Bovine Spleen

There are several methods for extracting RNA from tissues. The technique used in this experiment is not as sophisticated as some of the procedures but is ad-

TABLE 10–6
Protocol for quantitative determination of DNA.

Tube no.	Contents	%*T*	A_{500} nm
1	DNA (1 mg/ml)		
2	DNA (0.5 mg/ml)		
3	DNA (0.25 mg/ml)		
4	DNA (0.125 mg/ml)		
5	Water		
6	Unknown DNA preparation		

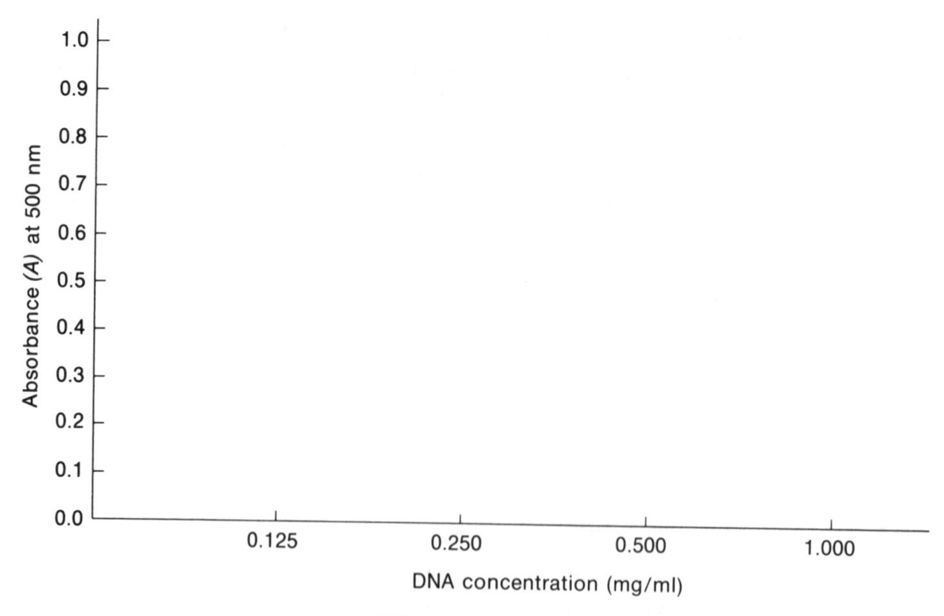

FIG. 10-5
DNA standard curve.

equate to prepare RNA for our purpose. The RNA preparation is somewhat impure and the yield is lower than that obtained by other methods.

1. Thaw out the frozen supernatant obtained in step 4 of the DNA extraction procedure. Add an equal volume of cold (4°C) 30% trichloroacetic acid (TCA), stir gently, and allow to stand for 5 minutes. CAUTION: *TCA is a strong acid. Use extreme care in handling.*

2. Centrifuge at 2000 × *G* for 5 minutes to collect the precipitate that forms. Pour off and discard the supernatant. Add enough cold (4°C) acetone to half fill the tube and stir to resuspend the pellet. Recentrifuge at 2000 × *G* for 5 minutes, discard the supernatant, and again add cold acetone. Recentrifuge, discard the supernatant, and this time add room temperature acetone. Recentrifuge, then discard the supernatant and retain the precipitate.

3. Dry the precipitate in air or under a hood until a fine powder appears. This powder contains a mixture of both RNA and proteins. Suspend this powder in approximately 10 ml of 10% sodium chloride in a test tube.

4. Cover the test tube with a loose cap and place it in a boiling water bath for 40 minutes. If the volume of liquid in the test tube decreases due to evaporation, add distilled water to restore the original volume.

5. Cool the contents of the test tube to room temperature and then centrifuge the suspension at 2000 × *G* for 10 minutes. Collect the supernatant, which contains dissolved RNA, and discard the precipitate, which contains the proteins.

6. Add 2 volumes of absolute ethyl alcohol to the supernatant and place the tube in an ice bath for 5 minutes. Collect the precipitate containing the RNA by centrifuging at 3000 × *G* for 10 minutes. Discard the supernatant. Wash the precipitate by adding acetone and stirring for several minutes. Centrifuge at 2000 × *G* for 10 minutes and discard the supernatant.

7. Place the precipitate in a beaker and allow to air-dry to obtain your final RNA preparation. This material may be frozen for use at a later time.

TABLE 10-7
Protocol for quantitative determination of RNA.

Tube no.	Contents	%*T*	A_{660} nm
1	RNA (0.166 mg/ml)		
2	RNA (0.083 mg/ml)		
3	RNA (0.042 mg/ml)		
4	RNA (0.021 mg/ml)		
5	Water		
6	Unknown RNA preparation		

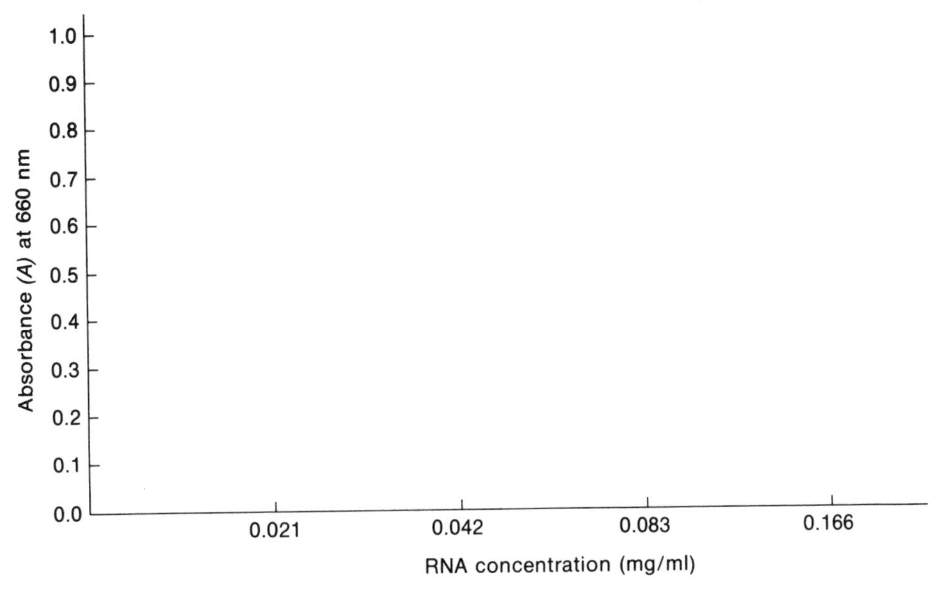

FIG. 10-6
RNA standard curve.

4. RNA Detection by the Orcinol Reaction

1. Prepare a test tube rack containing six test tubes numbered 1–6. Tubes 1–4 will be used to prepare a standard RNA concentration curve, Tube 5 will serve as a "blank", and Tube 6 will contain the extracted RNA solution of unknown concentration.

2. Dissolve 1 mg of commercially prepared yeast RNA in 6 ml of distilled water. The dissolution of RNA can be aided by adding several drops of 0.1 *N* HCl. This will be your stock RNA solution containing 0.166 mg of RNA per milliliter.

3. Pipet 3 ml of distilled water into Tubes 2, 3, 4, and 5. Pipet 3 ml of the stock RNA solution into Tubes 1 and 2. Mix the contents of Tube 2 thoroughly and then pipet 3 ml of this solution into Tube 3. Mix the contents thoroughly and transfer 3 ml to Tube 4. Mix the contents and discard 3 ml of solution from Tube 4. Into Tube 6, pipet 3 ml of your extracted RNA preparation.

4. Pipet 6 ml of the acid-orcinol reagent and 0.4 ml of the alcohol-orcinol reagent into each of the six test tubes. Place all tubes in a boiling water bath for 20 minutes. Then cool all of the tubes by immersing them in an ice bath.

5. Determine the absorbance (*A*) for each of the tubes at 660 nm following the procedure used for the DNA measurements. Record these data in Table 10-7. Prepare a standard RNA curve (Fig. 10-6) and from this determine the concentration of RNA in the spleen extraction.

____

REFERENCES

Alberts, B., D. Bray, J. Lewis, M. Raff, K. Roberts, and J. D. Watson. 1983. *Molecular Biology of the Cell.* Garland.

Bailey, J. S. 1967. *Techniques of Protein Chemistry.* 2d ed. Elsevier.

Dickerson, R. E., and I. Geis. 1985. *The Structure and Action of Proteins.* 2d ed. Harper & Row.

Friefelder, D. 1983. *Molecular Biology.* Van Nostrand Reinhold.

Gurr, A. I., and A. T. James. 1976. *Lipid Biochemistry: An Introduction.* 2d ed. Halstead Press.

Koshland, D. E. 1973. Protein Shape and Biological Control. *Scientific American* 229(4):52–64 (Offprint 1280). *Scientific American* Offprints are available from W. H. Freeman and Company, 41 Madison Avenue, New York 10010, and 20 Beaumont Street, Oxford OX1 2NQ, England. Please order by number.

Lehninger, A. L. 1982. *Principles of Biochemistry.* Worth.

Stryer, L. 1981. *Biochemistry.* 2d ed. W. H. Freeman and Company.

Swanson, C. P., and P. L. Webster. 1985. *The Cell.* 5th ed. Prentice-Hall.

EXERCISE 11

Photosynthesis

The living world, with few exceptions, operates at the expense of the energy captured by the photosynthetic machinery of green plants. From the products of photosynthesis and from a small number of inorganic compounds available in the environment, living organisms are able to build up the numerous complex molecules that contribute to their cellular structure or that in other ways are essential to their existence. Furthermore, the ultimate source of the energy expended by living organisms is the converted energy of sunlight that is trapped within the newly synthesized organic molecules during photosynthesis.

Classically, the reaction taking place in photosynthesis is given as follows:

$$\underset{\text{Carbon dioxide}}{6CO_2} + \underset{\text{Water}}{12H_2O} \xrightarrow[\text{chlorophyll}]{\text{light}} \underset{\text{Sugar}}{C_6H_{12}O_6} + \underset{\text{Oxygen}}{6O_2} + \underset{\text{Water}}{6H_2O}$$

This equation suggests that carbohydrate synthesis is the central feature in this process. Photosynthesis, however, is not a single-step reaction, as might be indicated by this equation. It is a complex process involving the interaction of many compounds. The large number of individual reactions can be divided into two groups: (1) the light, or photochemical, reactions in which light is required; and (2) the so-called dark, or biosynthetic, reactions, so named because they do not require light for the reactions to proceed.

In the **photochemical (light) reactions,** radiant energy is used for two purposes. First, light is used to split water molecules into oxygen and hydrogen. The hydrogen is then transferred to $NADP^+$ (nicotinamide adenine dinucleotide phosphate) to form NADPH, which in turn transfers hydrogen to other molecules.

Second, light energy absorbed by chlorophyll is converted into chemical energy, which is stored in the molecule ATP (adenosine triphosphate). This conversion occurs in the chloroplast and consists of the transport of electrons from "excited" chlorophyll through a series of acceptor molecules (including the cytochromes) that constitute an electron-transport system. Thus, mitochondria are not the only cytoplasmic structures capable of generating ATP. This phenomenon of light-dependent generation of ATP has been called **photophosphorylation** to differentiate it from **oxidative phosphorylation,** which occurs in mitochondria.

Thus, the light reactions result in the formation of NADPH, ATP, and the release of oxygen. In the "dark" reactions, NADPH and ATP are then used to reduce CO_2 to carbohydrate.

In this exercise, you will examine the role of light, carbon dioxide, and chloroplast pigments in the photosynthetic process.

A. ROLE OF LIGHT

1. Necessity of Light for Photosynthesis

There is no practical method in introductory courses to determine precisely the amount of sugar or oxygen produced during photosynthesis. However, it has been shown experimentally that during photosynthesis much of the sugar produced in the leaves is rapidly condensed into starch. Although starch is not a direct product of photosynthesis, because it is a condensation product of the glucose produced during photosynthesis we can use its synthesis as indirect evidence of photosynthetic activity.

Your instructor will supply geranium plants that have been kept in a dark place for 48 hours. Test a leaf for the presence of starch by the following procedure (Fig. 11-1A–E).

1. Hydrolyze the cell walls by boiling in a water bath for several minutes, then remove the pigment by putting the leaf in hot alcohol.

2. Transfer the leaf to a petri dish containing iodine. If starch is present, the leaf will turn a deep bluish black.

Similarly, test for the presence of starch in plants that were continuously exposed to light for 48 hours.

From your observations, what conclusions can be made about the necessity of light for photosynthesis?

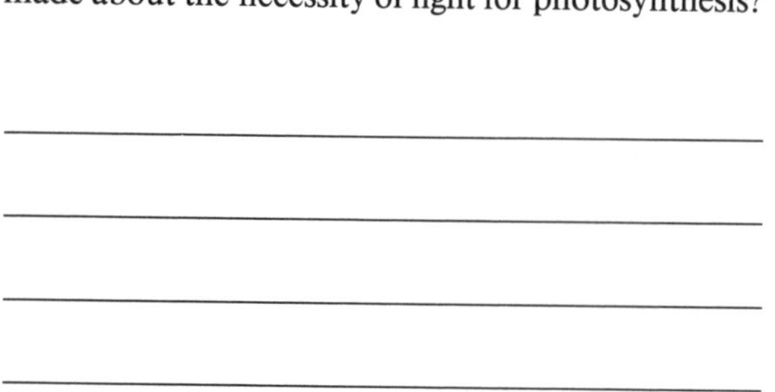

A. Remove a leaf from a plant that has been kept in the dark for 48 hours and one from a plant that has been exposed to sunlight for 48 hours.

B. Place leaves in boiling water bath for several minutes.

C. Place leaves in beaker of hot alcohol and heat until pigment is removed.

D. Place leaves in dish containing iodine for several minutes.

E. Remove leaves. If starch is present, leaves will become deep bluish black.

FIG. 11-1
Procedure for determining the necessity of light for photosynthesis.

2. Effect of Light Intensity on the Rate of Photosynthesis

The intensity of sunlight striking the earth's surface varies from hour to hour, from day to day, and from season to season. Because oxygen is a byproduct of photosynthesis, oxygen liberation may be used in designing an experiment to measure the effect of variations in light intensity on photosynthesis. In this study, light intensity is varied by placing an Elodea sprig at varying distances from a constant light source.

You can use either of the following methods. Method 1 is a semiquantitative procedure in which the amount of oxygen produced is equated to the amount of water displaced in a pipet attached to a photosynthesizing sprig of Elodea. In Method 2, changes in photosynthetic rate are measured as changes in the amount of oxygen produced as bubbles, a large number of bubbles indicates greater photosynthetic activity than a small number of bubbles.

a. Method 1

1. Select a 1-ml pipet graduated in hundredths. Turn it upside down, and place a short piece of rubber tubing over the delivery end (Fig. 11-2).

2. Select an undamaged sprig of Elodea about 15 cm in length. Insert it upside down into a large test tube filled with a 0.25% solution of sodium bicarbonate ($NaHCO_3$). This solution is a source of CO_2 for photosynthesis. Before completely submerging the plant, cut 2–3 mm from the end of the stem opposite the growing point with a sharp razor blade, being careful not to crush the stem. If there are any leaves within a few millimeters of the cut end, remove them with forceps.

3. Swab the rubber tubing on the pipet with cotton moistened with 70% ethyl alcohol. Allow to dry. Aspirate the sodium bicarbonate solution into the pipet until it is full. Hold your finger over the end of the rubber tubing to prevent the water column from dropping and attach a clamp over the tubing as shown in Fig. 11-2B and C.

4. Next, position the pipet gently over the cut end of the Elodea, and clamp in place on a ring stand, as shown in Fig. 11-2C and D. Keep the pipet and Elodea below the level of the water.

Obtain a reflector containing a 200-watt bulb and a container of cool water, and set them in the position shown in Fig. 11-2D. Why is the water container used in this system?

____

____

With the Elodea plant at a distance of 50 cm, turn on the lamp and allow the system to equilibrate for 7–10 minutes. Why?

____

____

Determine the total amount of oxygen given off by the plant during a 10-minute period, *by determining the amount of water in the pipet that is displaced during this time.* Fig. 11-3 illustrates how this is done.

Determine the amount of oxygen produced at distances of 50, 30, and 10 cm from the light source. Enter your results in Table 11-1, and plot your data in Fig. 11-14.

b. Method 2

Arrange materials as in Method 1. It is not necessary here, however, to use a graduated pipet. Any glass tubing that will fit closely over the cut end of the Elodea will do. Set the tube at the 50-cm distance, and allow the system to equilibrate for 7–10 minutes. Then determine the rate of photosynthesis by

TABLE 11-1
Effects of light intensity on photosynthetic activity.

Distance from light (cm)	Volume of oxygen (ml)	Average bubble count
10		
30		
50		

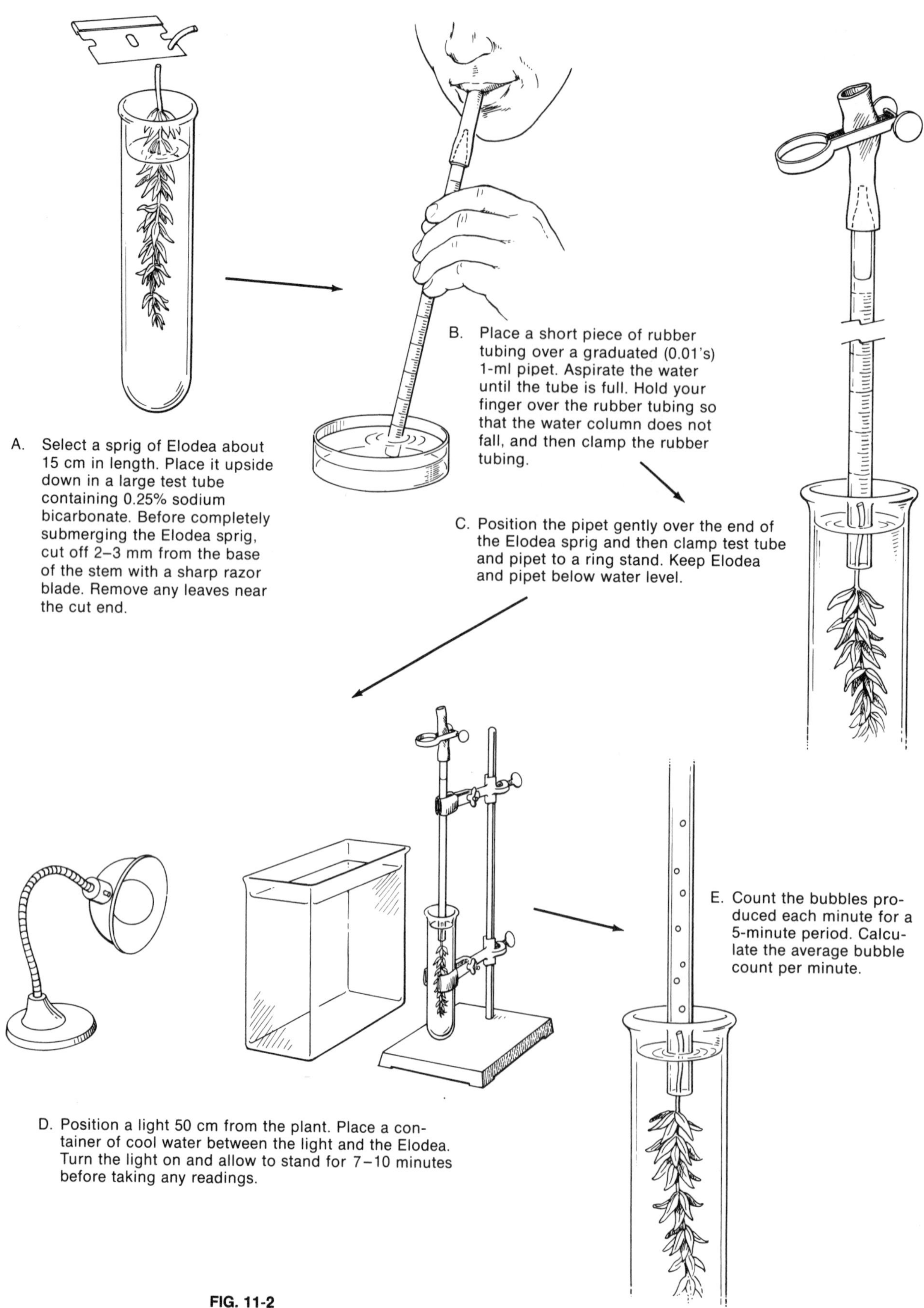

FIG. 11-2
Procedure for determining the effect of light intensity on photosynthesis.

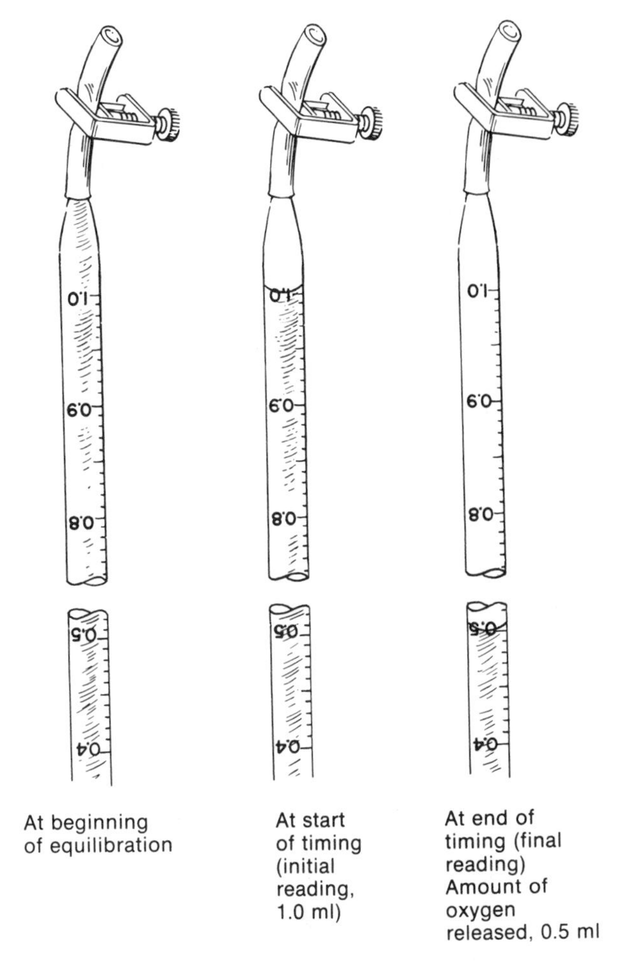

FIG. 11-3
Method of measuring oxygen evolved during photosynthesis.

counting the bubbles produced each minute for a 5-minute period. Calculate the *average* number of bubbles per minute, and record your data in Table 11-1. Move the tube to the 30-cm distance, allow the system to equilibrate, and calculate the average number of bubbles. Repeat this procedure for the 10-cm distance. Plot your data in Fig. 11-4.

As light intensity increases, does the rate of photosynthesis (as measured in oxygen production) increase along with it? If not, what does this suggest?

____

____

____

____

B. ROLE OF CARBON DIOXIDE

1. Necessity of CO_2 for Photosynthesis

To determine the necessity of CO_2 in photosynthesis, the apparatus in Fig. 11-5D will be used.

Select a leaf from a geranium plant that has been in the dark for 24 hours. Test the leaf for the presence of starch by immersing it in hot alcohol until it loses its green color, then place it in a petri dish containing iodine (Fig. 11-5A, B). Return the plant to the dark while performing the starch test. If the test results in a strong positive starch reaction, select another plant and test the leaves until a negative or very weak starch reaction is obtained. Why is this step necessary?

____

____

____

Select another leaf from the plant giving the negative starch reaction, and place the leaf in the jar as shown in Fig. 11-5D.

Place a 200-watt shielded lamp near the setup, but not close enough to heat the jar. (Better results may be obtained if this setup can be placed in direct sunlight.)

What is the "control" for this experiment?

____

____

Run the control simultaneously with the "experimental" setup. This experiment should be allowed to run for about 24 hours, after which time you should remove the leaf and test it for photosynthetic activity in terms of starch production.

After setting up your experiment, examine the demonstration arranged by your instructor (Fig. 11-6).

Why is there a beaker of $Ba(OH)_2$ in the bell jar?

____

____

What "control" would be needed for this experiment?

____

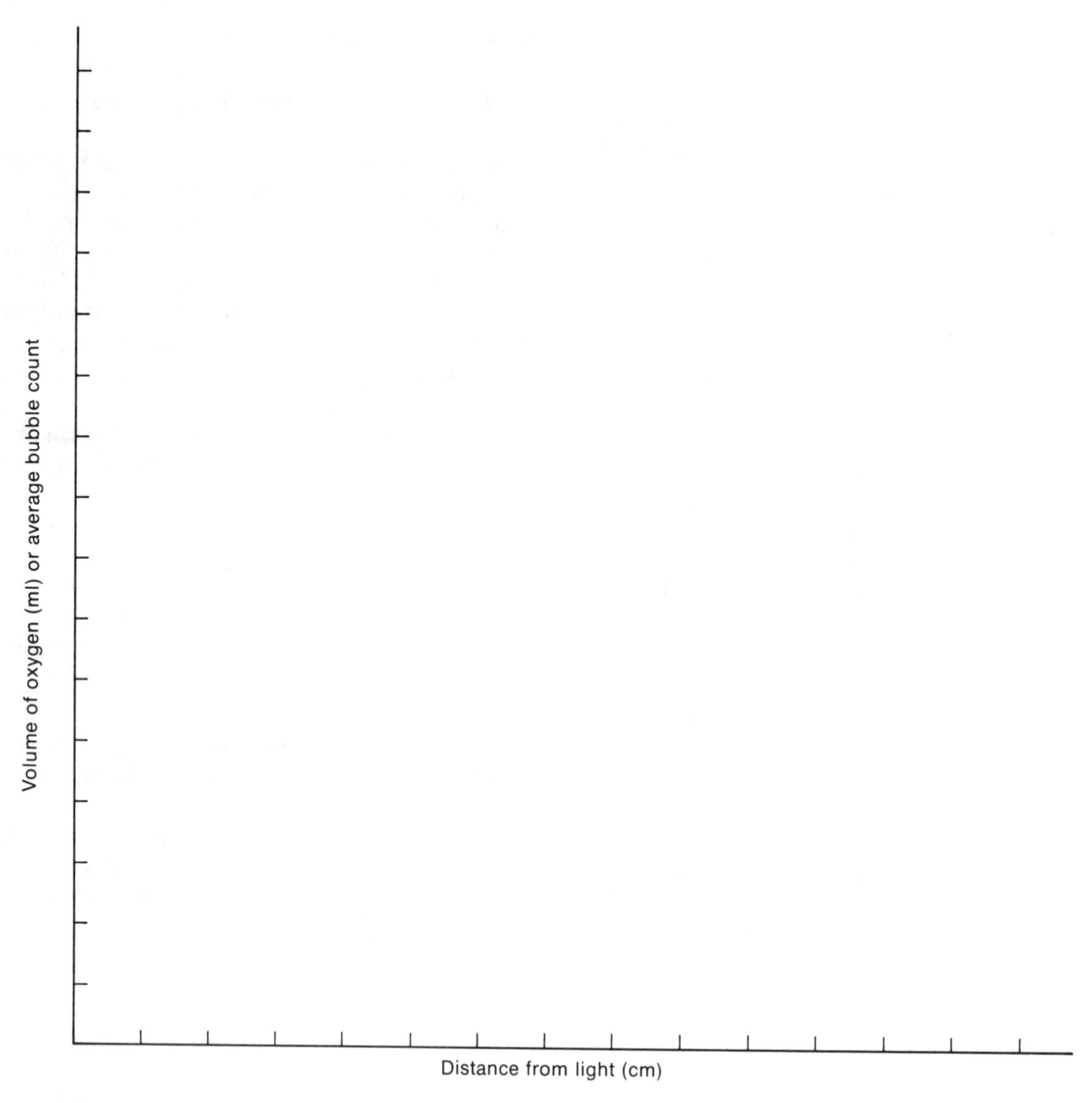

FIG. 11-4
Effect of light intensity on photosynthetic rate.

To save time, your instructor has tested the leaves of the "experimental" and "control" plants for photosynthetic activity. Under which condition is the starch test negative?

____

____

If the results of your experiment do not agree with those of your instructor's, suggest reasons for this difference.

____

____

____

2. Uptake of CO_2 by Aquatic Plants

That CO_2 is used during photosynthesis can be demonstrated by placing an Elodea plant in a test tube containing a chemical indicator that will change color in the presence or absence of CO_2. Phenol red is a chemical indicator that is red in an alkaline solution and yellow in an acid solution. Using this information, devise and run an adequately controlled

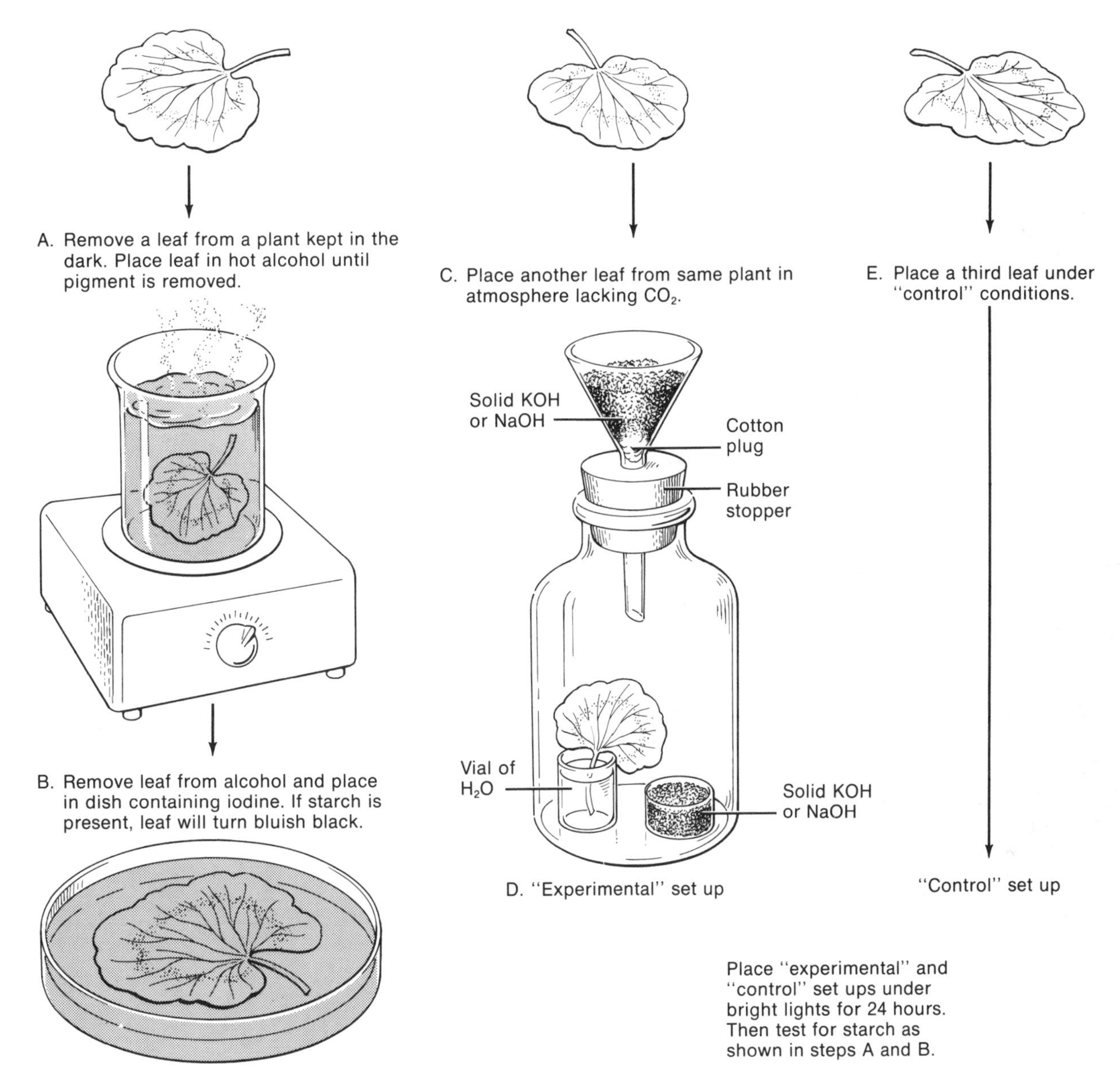

FIG. 11-5
Procedure for determining the necessity of CO_2 for photosynthesis.

experiment showing (1) that CO_2 is taken up by Elodea and (2) that light affects the plant's ability to take up CO_2.

C. ROLE OF CHLOROPLAST PIGMENTS

The second law of thermodynamics states that living organisms and cells create and maintain their steady-state condition at the expense of their surrounding environment, which tends to become disordered and random **(entropy).** To maintain this steady state, a constant source of energy must be available. Energy available to cells comes in two forms: radiant (light) energy from the sun and potential energy stored in chemical bonds.

Light energy, which is a type of electromagnetic radiation (Fig. 11-7), must first be transformed into chemical (bond) energy before it can be used by the living cell. This transformation takes place in green plant cells. Because only absorbed light can transfer its energy, it seems apparent that the colored components of plant cells act to absorb visible light. Those substances that have the ability to absorb light selectively are called *pigments.* In this exercise, you will determine experimentally the necessity for chloro-

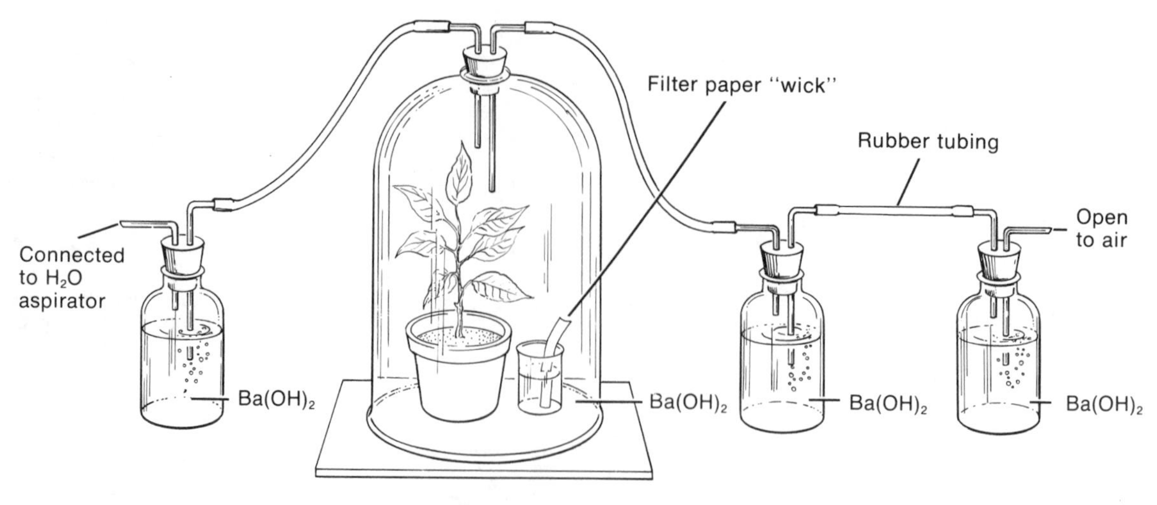

FIG. 11-6
Necessity of CO_2 for photosynthesis.

phyll in photosynthesis, the nature of the green color of plants, and the absorption spectrum of a chloroplast pigment solution.

1. Necessity of Chlorophyll for Photosynthesis

Select a leaf from a variegated *Coleus* and a silver-leafed geranium plant. In Row 1 of Table 11-2, draw an outline of each leaf, showing the distribution of the different pigments. The obvious pigments will be the "green" chlorophyll and the "red" anthocyanin in the *Coleus* leaf.

Place the leaves in a beaker of cold water for several minutes. Remove the leaves, and record your observations (by drawings or written comments) in Row 2 of Table 11-2.

Transfer the leaves to a beaker of boiling water for several minutes. Remove, and record any changes observed in Table 11-2. Account for the differences observed between Rows 2 and 3 in Table 11-2.

____

____

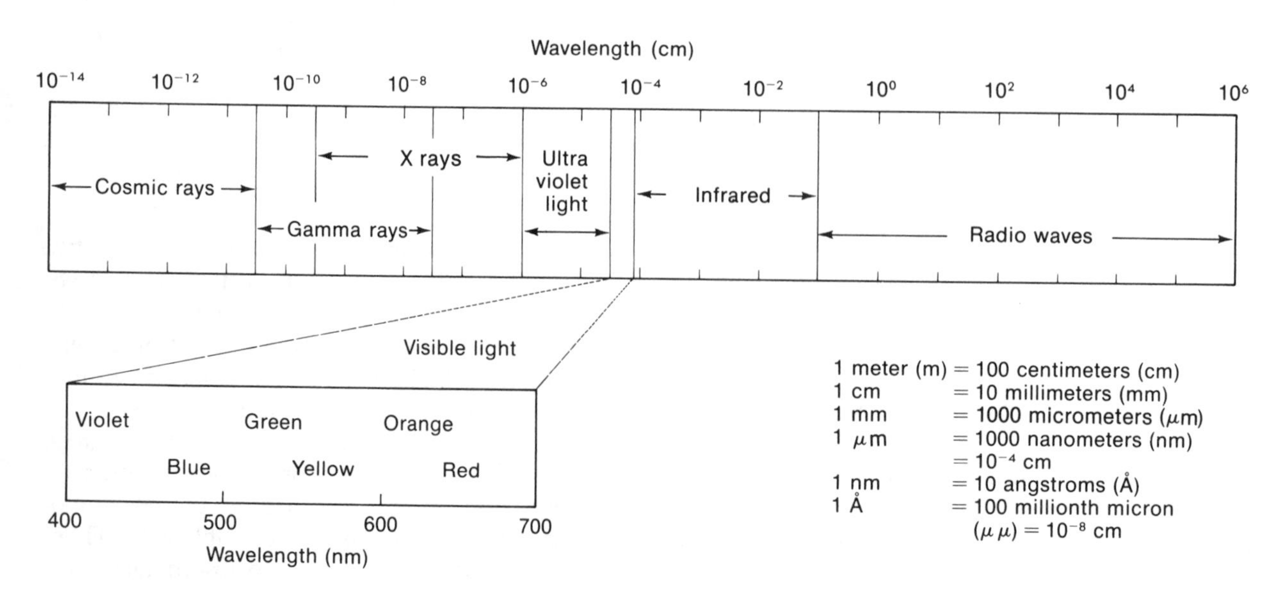

FIG. 11-7
The electromagnetic spectrum.

Next, place the leaves in hot alcohol. (CAUTION: *Use a hot water bath to heat the alcohol.*) After several minutes, the leaves will become whitish in color. At this point, transfer them to a petri dish containing iodine. Swirl the dish gently. Outline the distribution of starch in each leaf in Row 4 of Table 11-2. How do these experiments demonstrate the necessity for chlorophyll in photosynthesis?

2. Isolation and Characterization of Chloroplast Pigments

Complex mixtures of chemical substances may be separated by chromatography. The separation of the mixture is based on variations in solubility among the constituents of the mixture in different solvents. (See Appendix D for a discussion of the principles of

TABLE 11-2
The role of chlorophyll in photosynthesis.

Row	Treatment	*Coleus*	Silver-leafed geranium
1	None		
2	Cold H_2O for several minutes		
3	Boiling H_2O for several minutes		
4	Hot alcohol for several minutes. Place in dish with iodine.		

chromatography.) In this exercise, you will use the techniques of paper and thin-layer chromatography to analyze the pigment composition of chlorophyll.

a. Paper Chromatography

In paper chromatography, filter paper is usually used to separate different kinds of mixtures. The substance to be chromatographed is placed at one end of the paper. This end is then immersed in a solvent that separates the components of the mixture as it passes upward through the spot to the top of the paper. After drying, you can directly observe those materials that are separated if they are colored, or you can make them visible using various spray reagents.

1. Prepare a "chlorophyll" extract by grinding two or three spinach leaves in 5 ml of acetone (Fig. 11-8A). Adding a small quantity of quartz sand will make the grinding easier.

2. Using a small paint brush, apply a narrow strip of chlorophyll extract to the filter paper (Fig. 11-8B). Dry thoroughly by blowing on the paper or waving it in the air. Apply the extract five or six more times. Let it dry thoroughly after each application.

3. Place the strip in a test tube containing benzene-petroleum ether as shown in Fig. 11-8. Examine the chromatogram for the next several minutes. How long does it take for the solvent to reach the top of the paper?

____

Describe any separation that occurs.

____

____

____

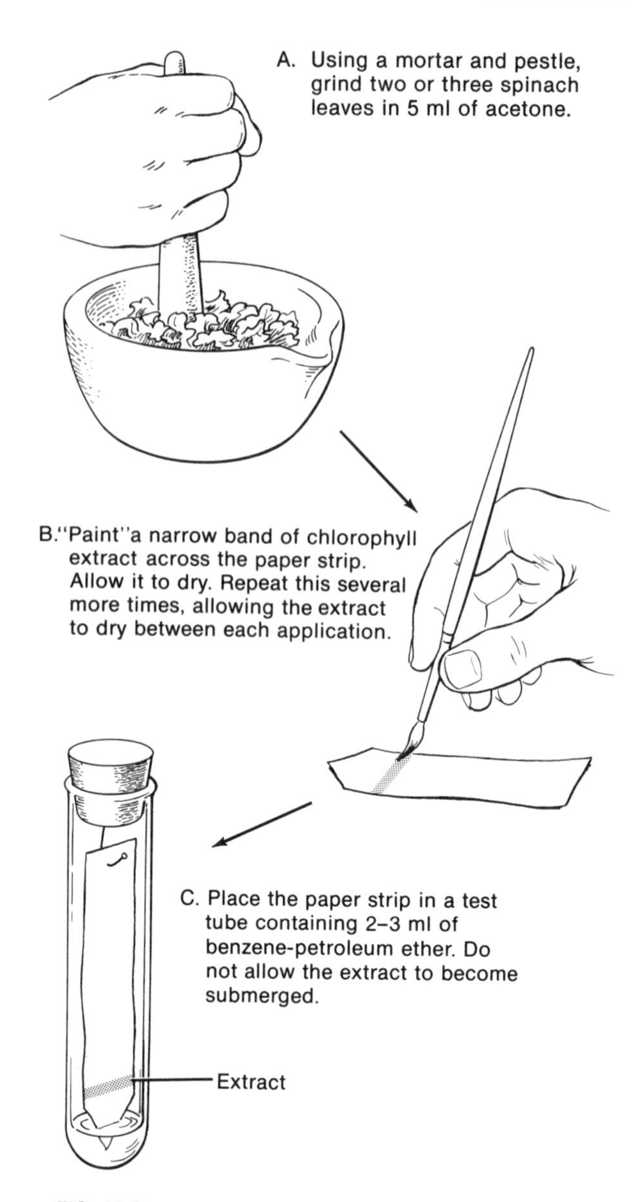

FIG. 11-8
Procedure for separating chlorophyll pigments using paper chromatography.

b. Thin-Layer Chromatography

In thin-layer chromatography, a gellike material is thinly applied to a support made of glass, aluminum, plastic, or cardboard. The gel can be prepared in several ways. In this exercise, you can use commercially prepared sheets, consisting of silica gel on acetate sheet backing, to assure uniformity of adsorbent thickness. They have the added advantage of being able to be cut to desired shape and size.

A unique advantage of using thin-layer chromatography is the speed at which separation occurs. Whereas paper chromatography can require up to 24 hours to separate a complex chemical mixture, the same separation using thin-layer methods can be done in an hour.

1. Place about 0.5 cm of solvent (isooctane-acetone-diethyl ether, 2:1:1) into a chromatographic jar. Cover the jar to allow the interior to become saturated with the fumes of the solvent (Fig. 11-9).

2. Using a capillary hematocrit tube, apply several drops of the previously prepared chloroplast extract about 2 cm from the bottom of a strip of

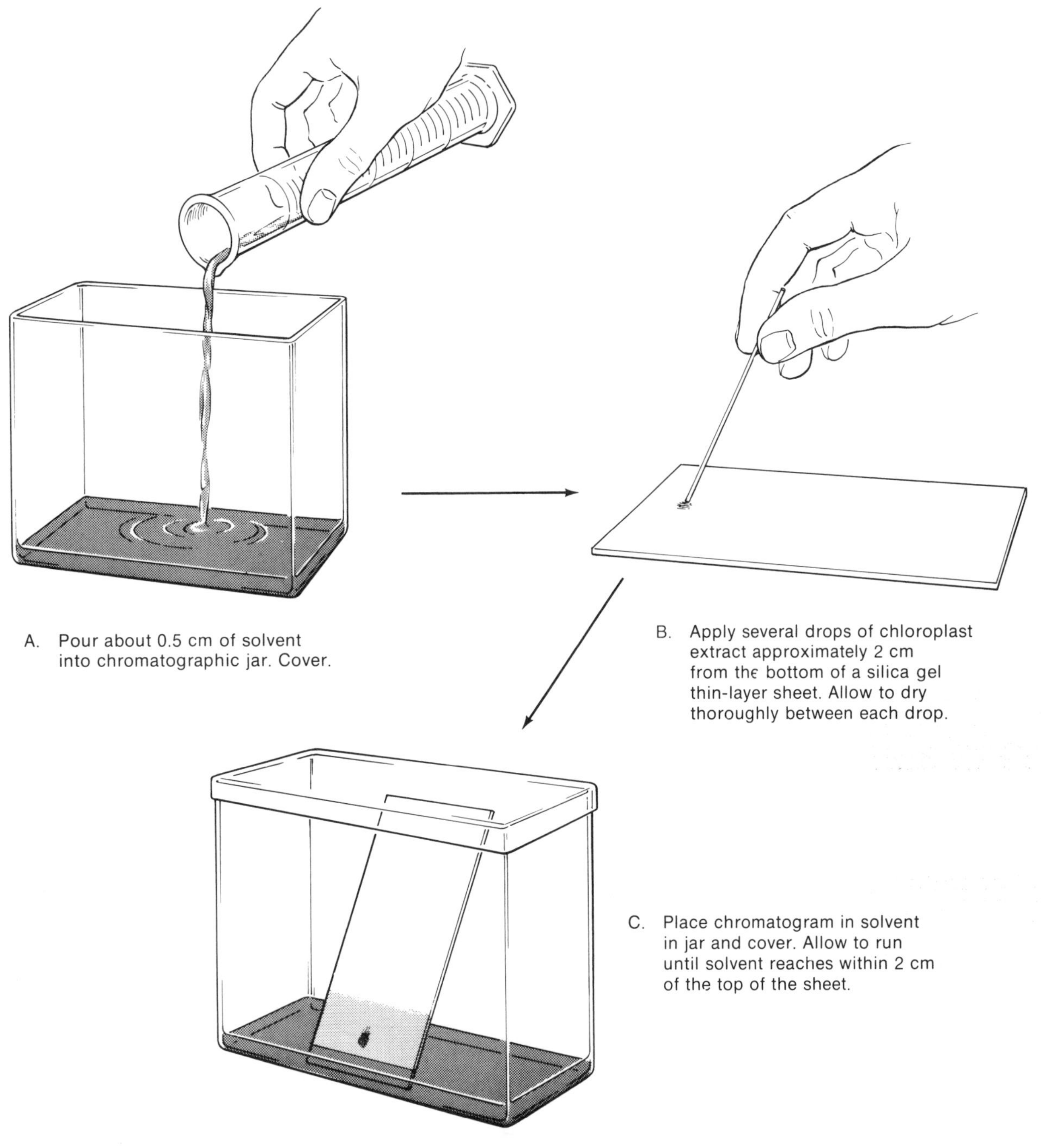

FIG. 11-9
Procedure for thin-layer chromatography of chloroplast pigments.

thin-layer silica gel. Try to keep the spot about 3–4 mm in diameter. Dry thoroughly between applications.

3. Insert the spotted sheet into the jar and cover. Allow the test to run until solvent reaches within 2 cm of the top of the sheet.

Approximately how long does it take to separate the pigments using the thin-layer as compared with the paper chromatographic method?

____

3. Absorption Spectra of Chloroplast Pigments

The part of the visible spectrum that is absorbed by the chloroplast pigments can be determined by using an instrument that disperses visible light into its component colors or wavelengths (Fig. 11-10). The disappearance from the spectrum of various colors, or wavelengths, as a result of passing light through a pigment solution, indicates that those wavelengths were absorbed by the pigments. A graph of the ability of a substance to absorb light against the wavelength of light is called an **absorption spectrum.** The absorption spectrum for a hypothetical substance is shown in Fig. 11-11.

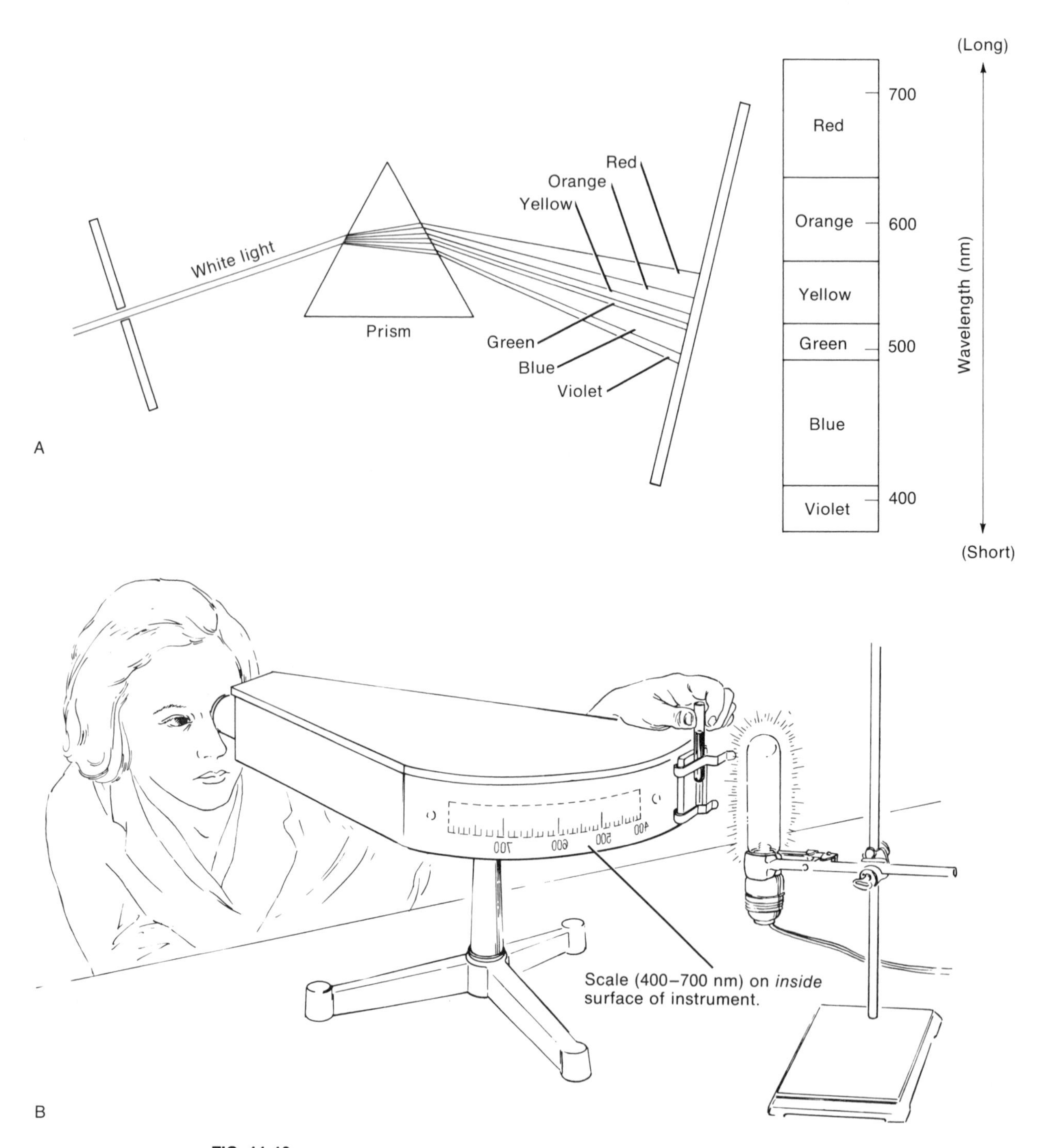

FIG. 11-10
(A) Dispersion of white light by a prism. (B) Determination of the absorption spectrum of chloroplast pigments by a spectroscope.

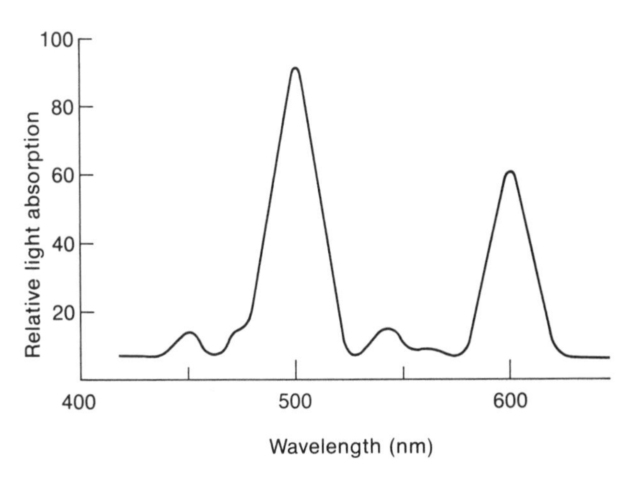

FIG. 11-11
A sample absorption spectrum.

What wavelengths are strongly absorbed by this hypothetical substance?

What wavelengths are weakly absorbed?

In this exercise, you will attempt to determine the absorption spectrum of a chloroplast extract using two different methods.

a. Method 1: Spectroscopic Determination

Pipet a sample of the extract available in the laboratory into a small test tube. Your instructor will help you use the instrument for determining the absorption spectrum for your extract. In the particular instrument used here, two spectra will be projected onto a scale (400–700 nm) on the back, inside surface of the instrument. The upper reference spectrum shows the various colors (wavelengths) of light. The lower sample spectrum results from passage of the light through the sample. In the chart below indicate the wavelengths of light that are absorbed by the chlorophyll extract.

b. Method 2: Spectrophotometric Determination

In this method, a Bausch & Lomb Spectronic 20 spectrophotometer will be used to determine the absorption spectrum of the chloroplast pigments more accurately. (Refer to Appendixes B and C for a description of the theory and mechanics of using this instrument.)

1. Beginning at a wavelength of 400 nm, standardize the instrument with the acetone-ethanol solvent used to extract the chloroplast pigments. Why is this solvent used to standardize the spectrophotometer?

---

2. Place the tube containing the extract into the sample holder, and determine the percent transmittance (%*T*).

3. Remove the sample, and reset the wavelength control to 425 nm. Restandardize the instrument to 0% and 100% transmittance, and then determine the percent transmittance of the sample.

4. Repeat the above procedure at 25-nm intervals. It will be necessary, however, to insert an accessory red filter and red-sensitive phototube for determinations above 625 nm.

5. Convert the percent transmittance (%*T*) into absorbance *(A)* using Table 11-3, and then plot your data in Fig. 11-12. At what wavelength(s) does the chlorophyll extract absorb maximally?

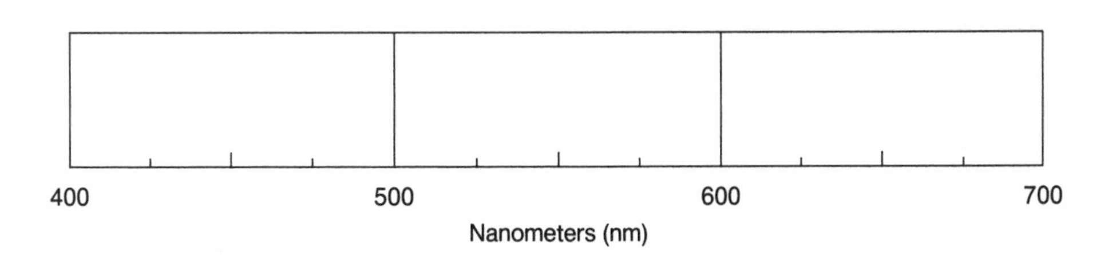

TABLE 11-3
Conversion of percent transmittance (%*T*) to absorbance (*A*).

%*T*	Absorbance (*A*)				%*T*	Absorbance (*A*)			
		1 (.25)	2 (.50)	3 (.75)			1 (.25)	2 (.50)	3 (.75)
1	2.000	1.903	1.824	1.757	51	.2924	.2903	.2882	.2861
2	1.699	1.648	1.602	1.561	52	.2840	.2819	.2798	.2777
3	1.523	1.488	1.456	1.426	53	.2756	.2736	.2716	.2696
4	1.398	1.372	1.347	1.323	54	.2676	.2656	.2636	.2616
5	1.301	1.280	1.260	1.240	55	.2596	.2577	.2557	.2537
6	1.222	1.204	1.187	1.171	56	.2518	.2499	.2480	.2460
7	1.155	1.140	1.126	1.112	57	.2441	.2422	.2403	.2384
8	1.097	1.083	1.071	1.059	58	.2366	.2347	.2328	.2310
9	1.046	1.034	1.022	1.011	59	.2291	.2273	.2255	.2236
10	1.000	.989	.979	.969	60	.2218	.2200	.2182	.2164
11	.959	.949	.939	.930	61	.2147	.2129	.2111	.2093
12	.921	.912	.903	.894	62	.2076	.2059	.2041	.2024
13	.886	.878	.870	.862	63	.2007	.1990	.1973	.1956
14	.854	.846	.838	.831	64	.1939	.1922	.1905	.1888
15	.824	.817	.810	.803	65	.1871	.1855	.1838	.1821
16	.796	.789	.782	.776	66	.1805	.1788	.1772	.1756
17	.770	.763	.757	.751	67	.1739	.1723	.1707	.1691
18	.745	.739	.733	.727	68	.1675	.1659	.1643	.1627
19	.721	.716	.710	.704	69	.1612	.1596	.1580	.1565
20	.699	.694	.688	.683	70	.1549	.1534	.1518	.1503
21	.678	.673	.668	.663	71	.1487	.1472	.1457	.1442
22	.658	.653	.648	.643	72	.1427	.1412	.1397	.1382
23	.638	.634	.629	.624	73	.1367	.1352	.1337	.1322
24	.620	.615	.611	.606	74	.1308	.1293	.1278	.1264
25	.602	.598	.594	.589	75	.1249	.1235	.1221	.1206
26	.585	.581	.577	.573	76	.1192	.1177	.1163	.1149
27	.569	.565	.561	.557	77	.1135	.1121	.1107	.1093
28	.553	.549	.545	.542	78	.1079	.1065	.1051	.1037
29	.538	.534	.530	.527	79	.1024	.1010	.0996	.0982
30	.532	.520	.516	.512	80	.0969	.0955	.0942	.0928
31	.509	.505	.502	.498	81	.0915	.0901	.0888	.0875
32	.495	.491	.488	.485	82	.0862	.0848	.0835	.0822
33	.482	.478	.475	.472	83	.0809	.0796	.0783	.0770
34	.469	.465	.462	.459	84	.0757	.0744	.0731	.0718
35	.456	.453	.450	.447	85	.0706	.0693	.0680	.0667
36	.444	.441	.438	.435	86	.0655	.0642	.0630	.0617
37	.432	.429	.426	.423	87	.0605	.0593	.0580	.0568
38	.420	.417	.414	.412	88	.0555	.0543	.0531	.0518
39	.409	.406	.403	.401	89	.0505	.0494	.0482	.0470
40	.398	.395	.392	.390	90	.0458	.0446	.0434	.0422
41	.387	.385	.382	.380	91	.0410	.0398	.0386	.0374
42	.377	.374	.372	.369	92	.0362	.0351	.0339	.0327
43	.367	.364	.362	.359	93	.0315	.0304	.0292	.0281
44	.357	.354	.352	.349	94	.0269	.0257	.0246	.0235
45	.347	.344	.342	.340	95	.0223	.0212	.0200	.0188
46	.337	.335	.332	.330	96	.0177	.0166	.0155	.0144
47	.328	.325	.323	.321	97	.0132	.0121	.0110	.0099
48	.319	.317	.314	.312	98	.0088	.0077	.0066	.0055
49	.310	.308	.305	.303	99	.0044	.0033	.0022	.0011
50	.301	.299	.297	.295	100	.0000	.0000	.0000	.0000

Note: Intermediate values may be arrived at by using the .25, .50, and .75 columns. For example, if %*T* equals 85, the absorbance equals .0706; if %*T* equals 85.75, the absorbance equals .0667.

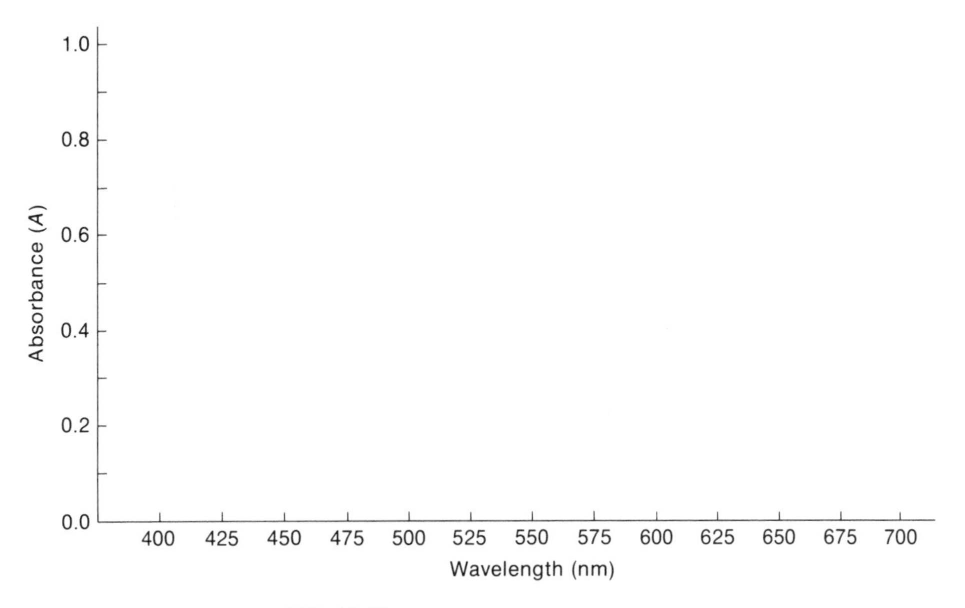

FIG. 11-12
Absorption spectrum of chloroplast extract.

In those plants having chlorophyll as the predominant pigment, why do the leaves appear green?

____

____

____

Because the chloroplast pigments consist of both chlorophylls and carotenoids, you cannot tell from the absorption spectrum which pigments are absorbing which wavelengths. How could you determine this?

____

____

____

REFERENCES

Alberts, B., D. Bray, J. Lewis, M. Raff, K. Roberts, and J. D. Watson. 1983. *Molecular Biology of the Cell.* Garland.

Allamong, B. D., and T. R. Mertens. 1976. *Energy for Life: Photosynthesis and Respiration* (Self-Teaching Text). Wiley.

Björkman, O., and J. Berry, 1973. High-Efficiency Photosynthesis. *Scientific American* 229:80–93 (Offprint 1281). *Scientific American* Offprints are available from W. H. Freeman and Company, 41 Madison Avenue, New York 10010, and 20 Beaumont Street, Oxford OX1 2NQ, England. Please order by number.

Govindjee and R. Govindjee. 1974. The Primary Events of Photosynthesis. *Scientific American* 231:68–82 (Offprint 1310).

Jensen, W. A., and F. B. Salisbury. 1984. *Botany.* 2d ed. Wadsworth.

Lehninger, A. 1961. How Cells Transform Energy. *Scientific American* 205:62–73 (Offprint 91).

Lehninger, A. L. 1971. *Bioenergetics: The Molecular Basis of Biological Energy Transformation.* 2d ed. Benjamin/Cummings.

Levine, R. P. 1969. The Mechanism of Photosynthesis. *Scientific American* 221(6):58–70 (Offprint 1163).

Levitt, J. 1974. *Introduction to Plant Physiology.* 2d ed. Mosby.

Rabinowitch, E. I., and Govindjee. 1965. The Role of Chlorophyll in Photosynthesis. *Scientific American* 213(1):74–83 (Offprint 1016).

Rabinowitch, E., and Govindjee. 1969. *Photosynthesis.* Wiley.

Stryer, L. 1981. *Biochemistry.* 2d ed. W. H. Freeman and Company.

EXERCISE 12

Carbohydrate Metabolism

In the broadest sense, metabolism includes all those events that occur in living cells. One of the more important results of these events is the formation of new cells. Concomitant with new cell formation is the active synthesis and use of carbohydrates, proteins, lipids, and other complex organic molecules. In this exercise, we will confine our studies to carbohydrate metabolism and, specifically, to the hydrolysis and synthesis of starch.

Starch is a polysaccharide composed of glucose molecules linked together as shown in Fig. 12-1. Starches are frequently stored by plants and animals (as glycogen) for use as a potentially available energy source. The breakdown of starch into its component "sugar units" is accomplished by hydrolysis. The important sugar unit we are concerned with is glucose because (1) it is a key compound in cellular metabolism (recall that is is synthesized from CO_2 and H_2O by photosynthesizing plants), and (2) it contains energy (found in the chemical bonds holding the molecule together) that can be released to do the work of the cell.

A. STARCH HYDROLYSIS

Starch hydrolysis is accomplished by amylases, enzymes that cleave the starch molecule into smaller and smaller subunits until maltose, a reducing sugar, is obtained. This sequence is shown in Fig. 12-2.

Maltose is enzymatically converted into glucose, which may then be shunted into the glycolytic and Krebs cycles, where it is further broken down into carbon dioxide, water, and energy. This process, called *cellular respiration,* will be considered in Exercise 13.

In the experiments that follow, the rate of hydrolysis of starch by salivary amylase will be measured colorimetrically under various conditions of temperature, pH, and enzyme and substrate concentrations. In these experiments, you will test for the presence of starch by adding several drops of iodine to the sample. If starch is present, the solution will turn a deep blue-black.

As hydrolysis proceeds, the amount of starch in the sample will gradually be reduced. This will be

FIG. 12-1
Structure of a starch molecule.

reflected in the color of the sample when iodine is added, so that if you test the sample at various intervals after adding the enzyme, you will observe a graded series of colors from deep blue (starch present) to red (partial hydrolysis) to the color of iodine. Thus, you may *qualitatively* measure starch hydrolysis by measuring the color changes. In this exercise, you will *quantitatively* measure the rate of hydrolysis by measuring the amount of light that the sample absorbs when placed in a colorimeter (refer to Appendix B for use of the Bausch and Lomb Spectronic 20 colorimeter and Appendix C for the principles of spectrophotometry).

Before beginning the experiment, carry out the following preliminary procedures to obtain the enzyme salivary amylase and standardize the colorimeter.

1. Collect 10 ml of saliva in a clean test tube. (You can stimulate the flow of saliva by chewing a small piece of paraffin.) Filter the saliva through a double layer of cheesecloth into a small beaker. This is the *stock enzyme solution.*

2. Turn on the colorimeter, and allow the instrument to warm up for 5 minutes. Check the meter needle to make sure it records 0% transmittance at a

FIG. 12-2
Action of amylase on starch.

wavelength of 560 nanometers (nm) with no test tube in the holder.

3. Prepare an iodine control tube by adding three drops of iodine to 3 ml of water in a colorimeter tube. Place the iodine control in the colorimeter, close the cover, and adjust the light control until the needle records 100% transmittance. Why is this step required?

1. Effect of Substrate Concentration on Activity of Salivary Amylase

In this exercise, you will determine the effect of altering the amount of substrate (starch) available to the enzyme.

1. Prepare a test-tube rack containing four rows of 10 colorimeter tubes each. Label the first tube in each row 1:2, 1:4, 1:8, and 1:16, corresponding to their respective dilutions of starch.

2. Label four Erlenmeyer flasks (125 ml) to correspond to the dilutions indicated, and add 50 ml of distilled water to each flask. To the first flask, add 50 ml of starch solution. Mix thoroughly, and transfer 50 ml of this mixture to the second flask. Mix thoroughly, and transfer 50 ml of the mixture to the third flask. Mix thoroughly and transfer 50 ml of the mixture to the last flask. Mix thoroughly. Remove and discard 50 ml of the last mixture.

3. Using a 1-ml pipet, add 0.2 ml of the stock enzyme solution to each flask, note the time, and mix thoroughly by swirling each flask. Two minutes after adding the enzyme, transfer 3 ml of the contents of each flask to the corresponding colorimeter tube. Add three drops of iodine to the tubes to determine the presence of starch. Mix thoroughly and then place each tube in the colorimeter. Determine the percent transmittance, and record this value in Table 12-1. Repeat this procedure at 2-minute intervals for each flask in the series. (*Note:* For these experiments, a reading of 90% *T* or higher will indicate that hydrolysis is complete.) Plot your data in Fig. 12-3. Interpret your results in terms of the amount of substrate available to a constant amount of enzyme.

____

____

____

____

2. Effect of Enzyme Concentration on Activity of Salivary Amylase

1. Prepare a test-tube rack containing four rows of 10 colorimeter tubes each. Label the first tube in each row 1:5, 1:25, 1:125, and 1:625 to correspond to the enzyme dilutions that will be used in this experiment.

2. Next, set up four Erlenmeyer flasks (125 ml), each containing 20 ml of distilled water. Label each flask to correspond to the dilution series. Add 5 ml of the enzyme solution to the first flask. Mix thoroughly, and transfer 5 ml to the second flask. Repeat this procedure for the third and fourth flasks, thoroughly mixing each flask before each transfer.

3. Next add 25 ml of starch solution to each flask, starting with the first flask. Note the time.

TABLE 12–1
Effect of substrate concentration on the activity of salivary amylase.

Time	Percent Transmittance			
	Substrate dilution			
	1:2	1:4	1:8	1:16

Conclusion:

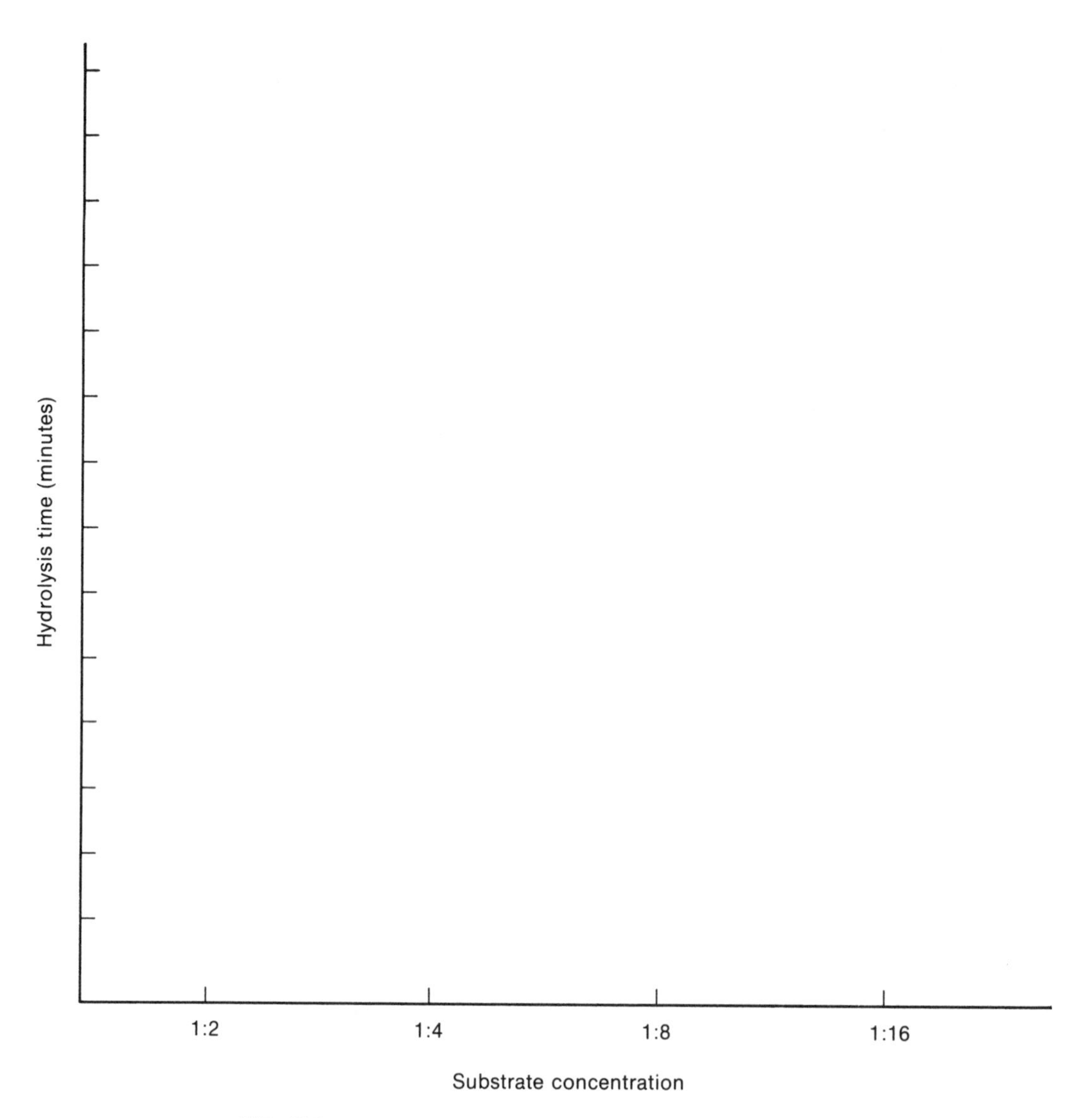

FIG. 12-3
Effect of substrate concentration on the activity of salivary amylase.

4. Two minutes after adding the starch, remove 3 ml of solution from each flask, transfer it to the first colorimeter tube in each series, and immediately test for starch by adding three drops of iodine to each tube. Determine the percent transmittance for each tube, and record your data in Table 12-2.

5. Repeat this procedure at 2-minute intervals for the remaining tubes in each series.

6. Plot your data in Fig. 12-4. Interpret your results.

3. Effect of pH on Activity of Salivary Amylase

Note: For this study, use the optimum enzyme and substrate concentrations from Parts 1 and 2.

1. Prepare a test-tube rack containing four rows of 10 colorimeter tubes each. Label the first tube in each row to correspond to pH values of 5, 6, 7, and 8.5, respectively. To each of four Erlenmeyer flasks (125 ml), add 25 ml of buffer solutions at a pH of 5.0, 6.0, 7.0, and 8.5, respectively. Into each flask, pipet 0.5 ml of enzyme solution. Mix thoroughly by swirling each flask. Beginning with the tube of lowest pH, add 25 ml of starch solution to each flask. Note the time. Thoroughly mix the contents immediately after adding the starch.

2. Two minutes after adding the starch to the first flask, quickly transfer 3 ml from each flask to the first tube in each corresponding series. Test for the presence of starch, using three drops of iodine. Record colorimeter readings as quickly as possible after the addition of iodine.

3. Repeat this procedure at 2-minute intervals for the remaining tubes of each series.

4. Record these times in Table 12-3, and plot your data in Fig. 12-5.

What is the optimum pH for the activity of salivary amylase?

____

Are your results consistent with your knowledge of where this enzyme functions in the body? Explain.

____

____

____

TABLE 12-2
Effect of enzyme concentration on the activity of salivary amylase.

Time	Percent Transmittance			
	Enzyme dilution			
	1:5	1:25	1:125	1:625
Conclusion:				

4. Effect of Temperature on Activity of Salivary Amylase

Note: For this study, use the enzyme and substrate concentrations (from Parts 1 and 2) that gave the optimum results.

1. Into each of five Erlenmeyer flasks (125 ml), pour 50 ml of the starch solution. Immerse these containers in large beakers of water or water baths adjusted to temperatures of 5°, 15°, 30°, 45°, and 70°C. This may be accomplished by adding ice water or hot water to the beakers. The temperatures in the beakers should be maintained throughout the experiment and should not vary more than ±3°C.

2. While waiting for the temperature to equilibrate in the beakers, prepare a test-tube rack containing five rows of 10 or 12 colorimeter tubes each.

TABLE 12-3
Effect of pH on the activity of salivary amylase.

Time	Percent Transmittance			
	pH			
	5	6	7	8.5
Conclusion:				

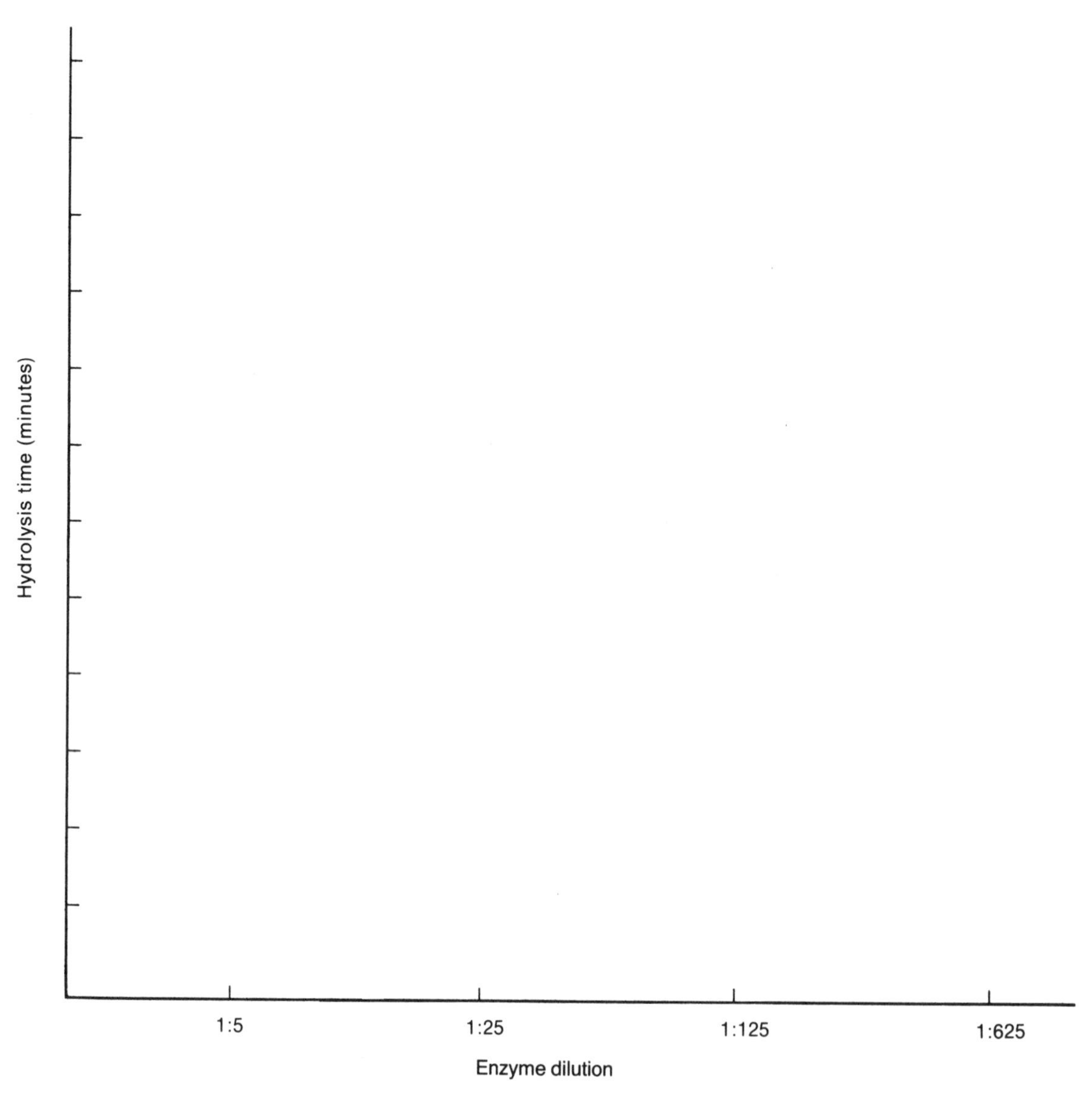

FIG. 12-4
Effect of enzyme concentration on the activity of salivary amylase.

Label the first tube in each row to correspond to the five temperature values used.

3. When the starch solution in the flasks has reached the temperature of the water bath (verify this by placing a thermometer in the starch solution), add 1 ml of enzyme solution to each flask, starting with that at 5°C. Note the time you add the enzyme to each flask. Mix each flask thoroughly as soon as the enzyme has been added, and put it back in the water bath.

4. Two minutes after the addition of the enzyme, quickly transfer 3 ml of the contents of each flask to the corresponding first colorimeter tube in each series. Immediately test for the presence of starch by adding three drops of iodine to each colorimeter tube. Mix thoroughly. (At the 70°C temperature, cool the tube with tap water before adding iodine.) Repeat this procedure at 2-minute intervals for the remaining tubes of each series. Be prepared to test for starch at longer, or shorter, intervals if necessary.

5. Record this time in the appropriate space in Table 12-4. Plot your data in Fig. 12-6. What is the effect of temperature on the activity of salivary amylase?

____

____

____

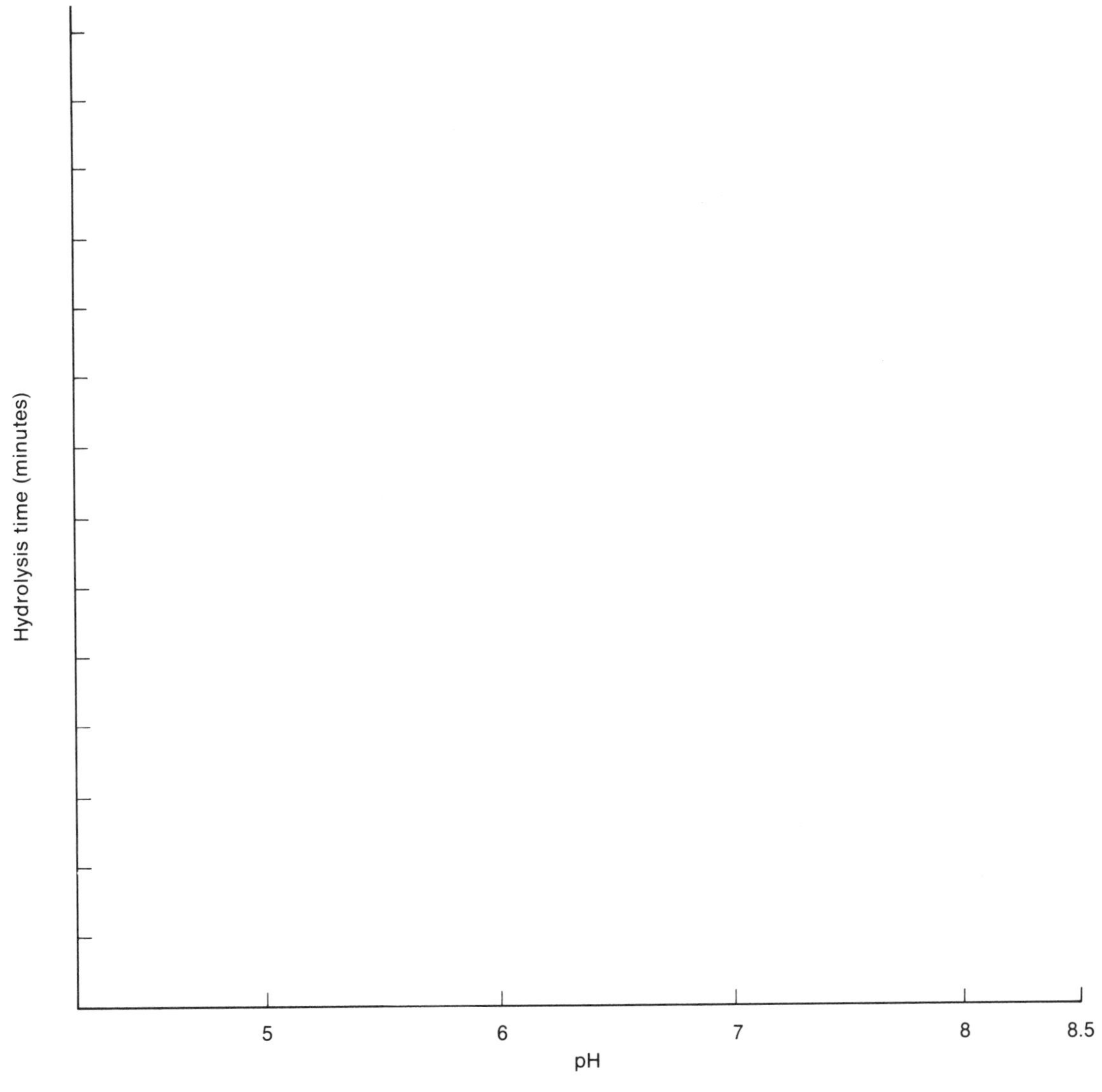

FIG. 12-5
Effect of pH on the activity of salivary amylase.

B. ENZYMATIC SYNTHESIS OF STARCH

In the first part of this exercise, you examined the hydrolysis of starch by a class of enzymes called **amylases.** However, other enzymes can also hydrolyze starch. These enzymes, called **phosphorylases,** degrade glycogen or starch more rapidly than any known amylase.

The activity of the two classes of enzymes differs in a number of ways. The differences in their rates have already been mentioned. A second difference is that amylases cleave this polysaccharide into maltose units and require a second enzyme, called **maltase,** to complete the conversion to glucose. Phosphorylases cleave polysaccharides into glucose-phosphate units, which are further hydrolyzed to glucose and phosphoric acid through the activity of the enzyme **phosphatase.**

The most important difference in these enzymes, however, lies in the reversibility of the phosphorylase reaction, which goes readily in either direction; the amylase reaction is almost irreversible (Fig. 12-7). Explain the difference in the reversibility of these two reactions.

____

____

____

____

TABLE 12–4
Effect of temperature on the activity of salivary amylase.

Time	Percent transmittance				
	Temperature (°C)				
	5	15	30	45	70
Conclusion:					

Use the following procedure to isolate the phosphorylase enzyme from fresh potatoes and then synthesize starch using this enzyme. In this experiment, enzyme activity will cause an increasing amount of starch to form from the reaction of glucose-1-phosphate and phosphorylase. You can monitor the starch formed by periodically adding samples of the reaction mixture to iodine. As starch is formed, the color, upon addition of iodine, will progress from that of the iodine solution (no starch present) to blue-black (indicating the presence of starch).

1. Cut a peeled potato into quarters. Place the pieces in a blender, add 65 ml of water, and homogenize until a slurry is formed. Pour the slurry through four layers of cheesecloth into a 125-ml beaker to remove any large pieces of potato that remain.

2. Place a piece of Whatman #1 filter paper on a Buchner funnel and wet the paper. Attach rubber tubing from the flask to the water aspirator and turn on the water faucet (Fig. 12-8).

3. Pour the potato filtrate into the funnel. After filtration is complete, remove the tubing from the aspirator and turn the faucet off.

4. Pour the filtrate into a beaker, and place the beaker in a 50°C water bath. Amylases are denatured at 50°C. Why is it necessary to destroy these enzymes in your preparation?

____

____

5. After 5 minutes, pour 25 ml of the heated filtrate into a beaker, and slowly add 5 g of ammonium sulfate crystals. Stir until the crystals are dissolved and a brown precipitate of amylase forms.

TABLE 12–5
Enzymatic synthesis of starch by phosphorylase.

Time (minutes)	Color intensity
0	
1	
2	
3	
4	
5	
6	

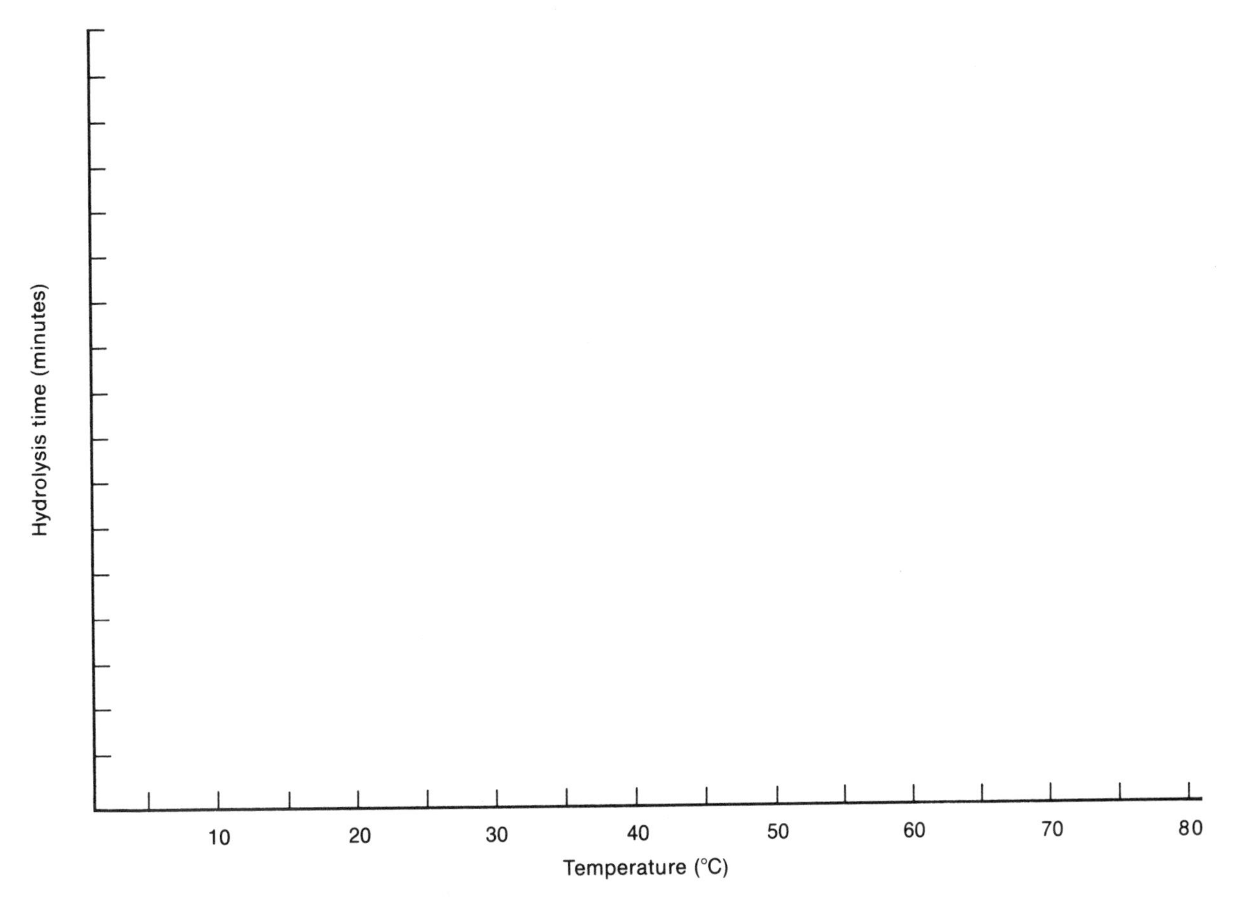

FIG. 12-6
Effect of temperature on the activity of salivary amylase.

FIG. 12-7
Action of phosphorylase. (From *Experimental Biochemistry*, 2d ed., by John M. Clark, Jr., and Robert L. Switzer. W. H. Freeman and Company. Copyright © 1977.)

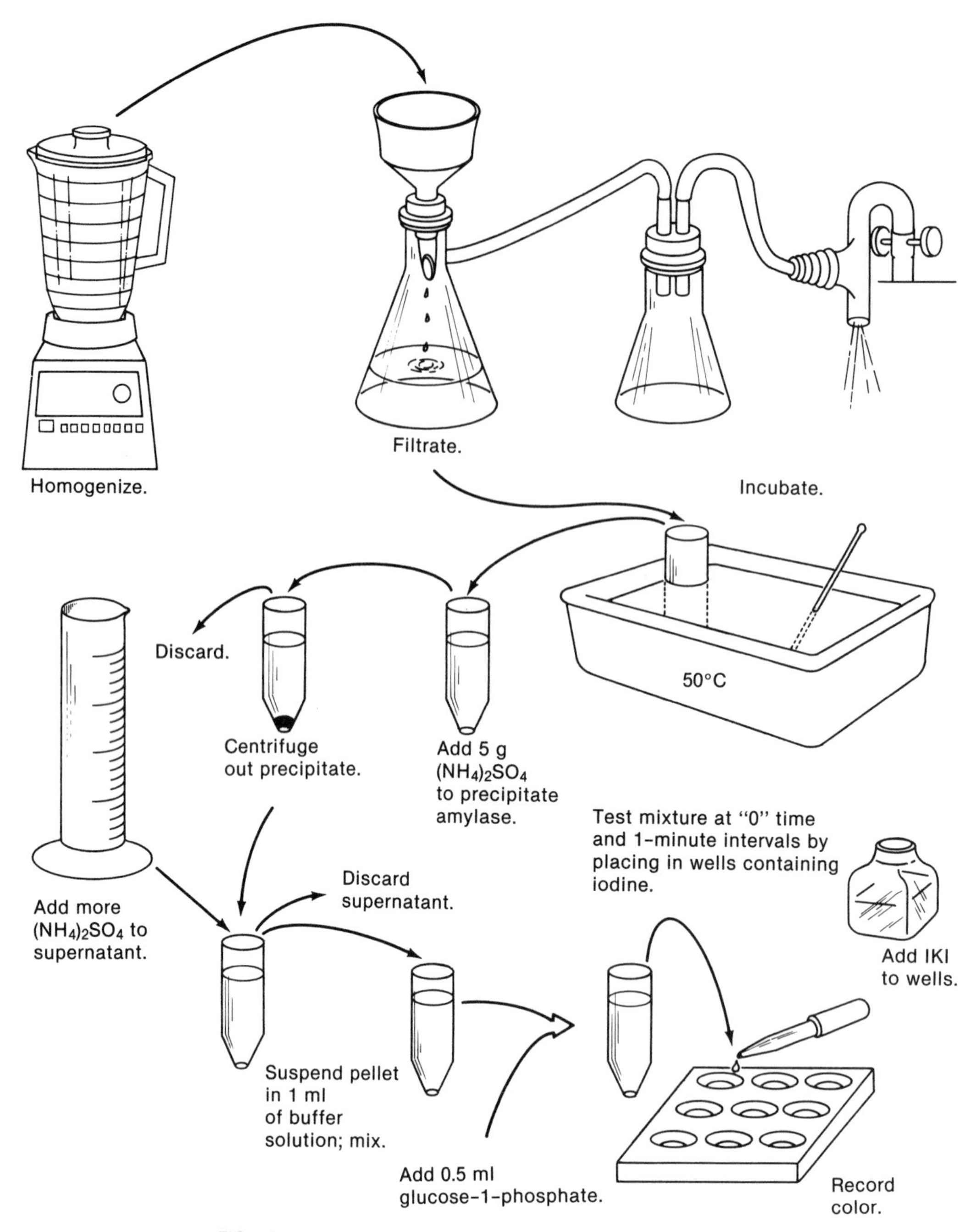

FIG. 12-8
Enzymatic synthesis of starch using the enzyme phosphorylase.

6. With a wax pencil mark a conical centrifuge tube 10 cm from the bottom and pour the amylase filtrate into the tube up to this mark.

7. Place two large beakers on the plates of a beam balance, and add water to the lightest of the two beakers until both beakers are balanced. Place your filled conical centrifuge tube in one of the beakers; have another student place his in the other beaker. Add water to the centrifuge tubes as necessary to balance them. Place the balanced centrifuge tubes directly opposite one another in the head of a clinical centrifuge. Centrifuge these preparations for 5 minutes at full speed.

8. Remove the centrifuge tubes and pour the supernatant liquid into a graduated cylinder. The pellet may not be packed too tightly so it is best to pour

the supernatant with a quick sudden motion. Discard the pellet by washing it out of the centrifuge tube.

9. Add water to the supernatant to bring the volume up to 25 ml. Add 4 g of ammonium sulfate crystals to the supernatant, stirring until all crystals dissolve and another brown precipitate forms. *This is the precipitate you want to keep because it is mostly the enzyme starch phosphorylase.*

10. Pour the mixture into a conical centrifuge tube and repeat the balancing procedures outlined in step 7. Centrifuge at full speed for 5 minutes. Remove the centrifuge tube after the centrifuge has stopped, and discard the supernatant.

11. Add 1 ml of the buffer solution to the pellet and stir.

12. Obtain a spot plate and place one drop of IKI solution in one of the wells. Add 0.5 ml of glucose-1-phosphate solution to the resuspended pellet in the tube and swirl to mix. Note the time ____________.

13. Immediately after mixing, remove some of the mixture with a Pasteur pipet and add three drops to the drop of IKI in the well of the spot plate. Record the color changes in Table 12-5. Place another drop of IKI in the next well on the spot plate.

14. One minute after the addition of the glucose-1-phosphate, remove some of the mixture with a pipet and add three drops to the drop of IKI in the well of the spot plate. Record the color changes in Table 12-5.

15. Continue to make 1-minute readings for a total of 6 minutes. How long did it take for starch to begin to be synthesized in measureable quantities?

____

____

____

Indicate three ways to speed up this reaction.

____

____

____

____

REFERENCES

Alberts, B., D. Bray, J. Lewis, M. Raff, K. Roberts, and J. D. Watson. 1983. *Molecular Biology of the Cell.* Garland.

Guthrie, R. D., and J. Honeyman. 1968. *An Introduction to the Chemistry of Carbohydrates.* Clarendon Press.

Lehninger, A. L. 1971. *Bioenergetics.* 2d ed. Worth.

Lehninger, A. L. 1960. Energy Transformation in the Cell. *Scientific American* 202(5):102–114 (Offprint 69). *Scientific American* Offprints are available from W. H. Freeman and Company, 41 Madison Avenue, New York 10010, and 20 Beaumont Street, Oxford OX1 2NQ, England. Please order by number.

McElroy, W. D. 1971. *Cellular Physiology and Biochemistry.* 3d ed. Prentice-Hall.

Pigman, W., ed. 1972. *The Carbohydrates.* Vol. 1a. Academic Press.

Stryer, L. 1981. *Biochemistry.* 2d ed. W. H. Freeman and Company.

EXERCISE 13

Cellular Respiration

All plants and animals require a continuous supply of energy for the performance of vital activities. This is obtained in the form of the chemical bond energy contained in the nutrient molecules. The source of the chemical bond energy is the solar energy used in the photosynthetic activities of green plants. The conversion of the chemical bond energy into a readily usable form—such as the high-energy phosphate bond of adenosine triphosphate (ATP)—and its expenditure are by means of processes collectively referred to as **respiration.** Broadly speaking, respiratory processes can be classified as either **aerobic** or **anaerobic.** Anaerobic respiration undoubtedly evolved first, and still plays an important role in many of the metabolic activities of plants and animals. Muscular work, for example, depends in large measure on anaerobic processes. Aerobic respiration, which requires oxygen, is not only a more recent biochemical innovation (in the evolutionary sense) but also a more efficient one in terms of energy recovery from nutrient molecules. The following diagram summarizes the many steps in the respiration of the simple 6-carbon sugar, glucose, under anaerobic and aerobic conditions:

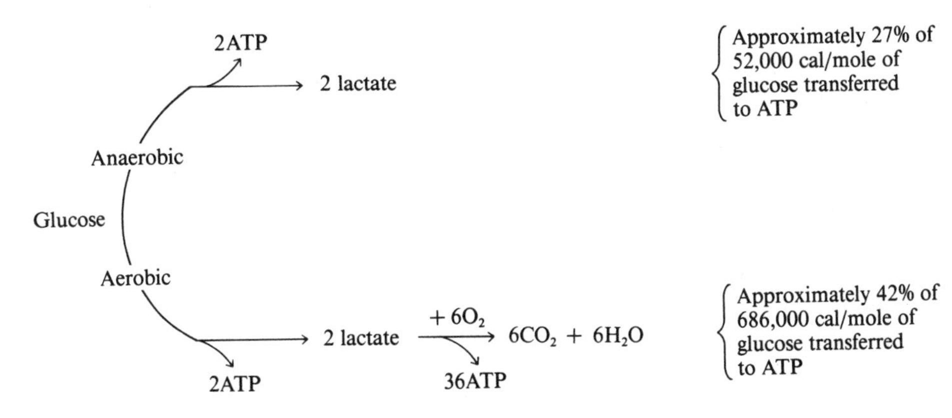

Aerobic respiration yields much more energy than does anaerobic respiration and results in the complete breakdown of glucose to carbon dioxide and water, whereas anaerobic respiration generally leads to endproducts, such as organic acids and alcohols, that may be toxic.

The glucose molecule is broken down and its energy released in three stages involving **glycolysis,** the **Krebs cycle,** and the **electron-transport chain** (Fig. 13-1).

During glycolysis, the 6-carbon sugar molecule is converted, through a complex series of enzyme-catalyzed reactions, into two 3-carbon molecules of pyruvic acid. In the presence of oxygen (aerobic respiration), pyruvic acid enters another series of enzyme-catalyzed reactions called the Krebs cycle. All of these reactions take place in the mitochondria. In the Krebs cycle, energy in the form of ATP is generated. This is accomplished by a series of oxidation-reduction reactions that take place in the electron-transport chain. In this process, hydrogen atoms that have been removed from Krebs cycle compounds are split into positively charged protons and negatively charged, *high-energy* electrons. The electrons are transported through a series of molecules that are alternately oxidized (lose electrons) and reduced (gain electrons). During these oxidation-reduction reactions, some of the energy of the electrons is incorporated into ATP. Finally, at the end of the chain the free protons combine with the now *low-energy* electrons and oxygen to form water.

The respiration taking place in an organism can be demonstrated in several ways: by measuring the energy given off in the form of heat, the amount of glucose used, the amount of oxygen consumed, and the amount of carbon dioxide released. In this exercise, you will study the external signs of respiration by indirectly determining the amount of oxygen consumed.

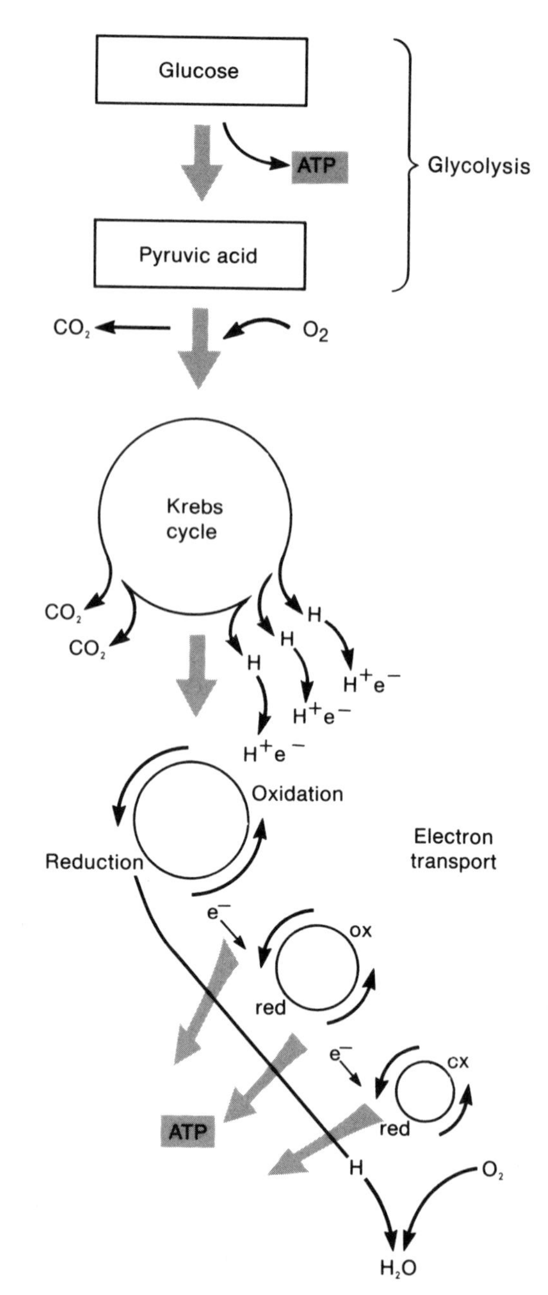

FIG. 13-1
Energy, in the form of ATP, produced during glycolysis and respiration.

A. AEROBIC RESPIRATION

Many methods used in the study of respiration depend on measuring changes in the volume or pressure of CO_2 or O_2. Any variation in the volume or pressure within a closed system in which an organism is respiring represents the net difference between oxygen consumption (which would decrease pressure and volume in the closed container) and carbon dioxide production (which would increase pressure and volume). However, if the carbon dioxide produced is absorbed in some way, any changes could be attributed to oxygen consumption.

A simple respirometer that can be used to detect changes in gas pressure and volume is shown in Fig. 13-2. This equipment consists of two vessels that can be closed to the outside. Respiring material is placed in one of the containers, along with potassium hydroxide (KOH), an agent used to absorb carbon dioxide. Because gas volume is influenced by such physical factors as atmospheric pressure and temperature, the second container (identical except for the living material) is employed as a compensation

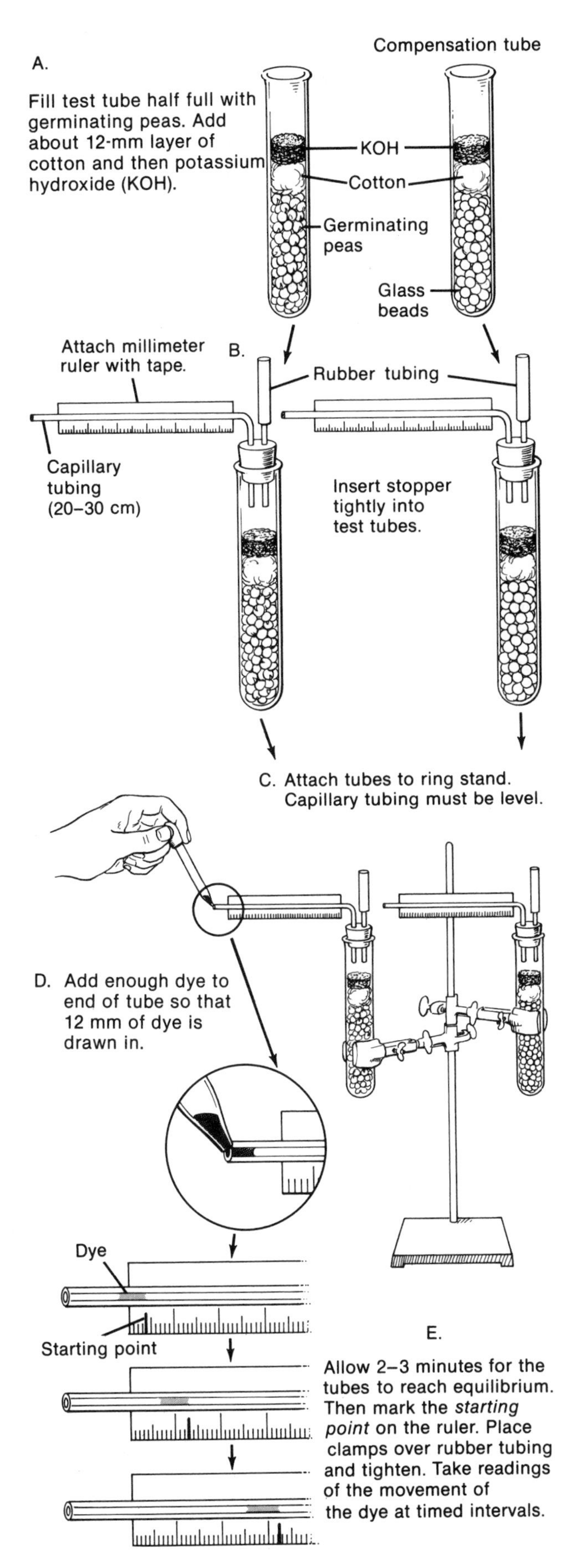

chamber. Be sure to take into consideration any changes in the volume of gas in the compensation chamber when evaluating changes occurring in the respiration chamber.

In this exercise, you will determine the respiratory rate of germinating peas in terms of oxygen intake.

1. Following the diagram in Fig. 13-2A, fill the respiration test tube half full with germinating peas.

2. Place a loose wad of cotton over the peas. Place about 12 mm of potassium hydroxide (KOH) pellets over the cotton. The cotton keeps the KOH from contacting the living seeds; it should not be packed tightly. KOH is a substance that will remove the CO_2 from the atmosphere of the tube as fast as it is given off by respiring peas. Why is it necessary to remove the CO_2 from the tube?

____

____

____

3. Prepare the compensation tube in the same way, except use glass beads in place of the peas. Why is it necessary to place the inert glass beads into the compensation tube?

____

____

4. Insert a rubber stopper, with attached capillary tubing, firmly into each tube (Fig. 13-2B).

5. Place the tubes in a vertical position by clamping to a ring stand (Fig. 13-2C). Using an eyedropper, add enough dye to the end of each capillary tube so that about 12 mm of the dye will be drawn into the tube (Fig. 13-2D).

6. After allowing 2–3 minutes for the gas pressures to reach equilibrium, note the position of the inner end of the dye column on the millimeter scale (Fig. 13-2E). Record this initial reading in Table 13-1. Then attach pinch clamps to the rubber tubing on each test tube. (*Note:* Because the respirometer is very sensitive to volume changes due to heat, keep it away from any heat sources such as lamps and hot plates.)

FIG. 13-2
Procedure for measuring oxygen consumption in germinating peas.

7. Take readings of the location of the column at 1-minute intervals for the next 5 minutes. Record your data in Table 13-1. (*Note:* If the movement of the dye is fairly rapid, you must be prepared to take your readings at shorter intervals—20 or 30 seconds—or the column may reach the bent portion of the tube before you have enough readings to complete your data.) The dye column can be returned to the outer end of the tube by opening the pinch clamp and tilting the capillary tube.

Why should the dye move toward the respiration chamber and not away?

___

___

Under what circumstances might the dye move away from the chamber?

___

___

___

8. Repeat the procedure to determine the effects of temperature on respiration and record your data in Table 13-1. Students should select different temperatures in order to reflect a variety of temperature effects.

9. Plot the data tabulated in Table 13-1 in Figs. 13-3 and 13-4. Identify each line on the graph with the appropriate label.

1. Respiration in Animals

Many factors affect respiration. Among them are the general state of health and the degree of physical activity of the organism. The activity of certain hormones also markedly affects respiration. As part of this study, you will attempt to determine what effect temperature has on the respiration of **poikilothermic** (cold-blooded) and **homeothermic** (warm-blooded) animals.

Poikilothermic animals lack the internal mechanisms necessary to regulate body temperature. As a consequence, their respiratory rates tend to be *directly* related to the temperature of their environments. In contrast, homeothermic animals have the ability to maintain their internal temperatures and, therefore, are *less affected* by temperature changes in their environments.

In this exercise, you will measure the volume of oxygen consumed during respiration by using a respirometer that consists in part of a **manometer.** In its simplest form, a manometer is a U-tube that is partly filled with fluid. The manometer measures the difference in the pressures on the two sides of the U-tube. For example, when the pressures on both sides of the tube are equal, the fluid levels in both sides are equal (Fig. 13-5A). However, when the pressure on one side is greater than that on the other side, the fluid level is lower in the side having the greater pressure (Figs. 13-5B, C).

In the respirometer used in this exercise, the manometer is connected to two jars. These jars can be closed off from the pressure of the outside atmosphere with pinch clamps attached to the vent tubes.

TABLE 13-1
Respiration data.

Time	Respirometer readings (in mm)					
	Germinating peas (Room temp: ______ °C)			Germinating peas (______ °C)		
	Respiration tube (1)	Compensation tube (2)	Corrected data (1 minus 2)	Respiration tube (1)	Compensation tube (2)	Corrected data (1 minus 2)

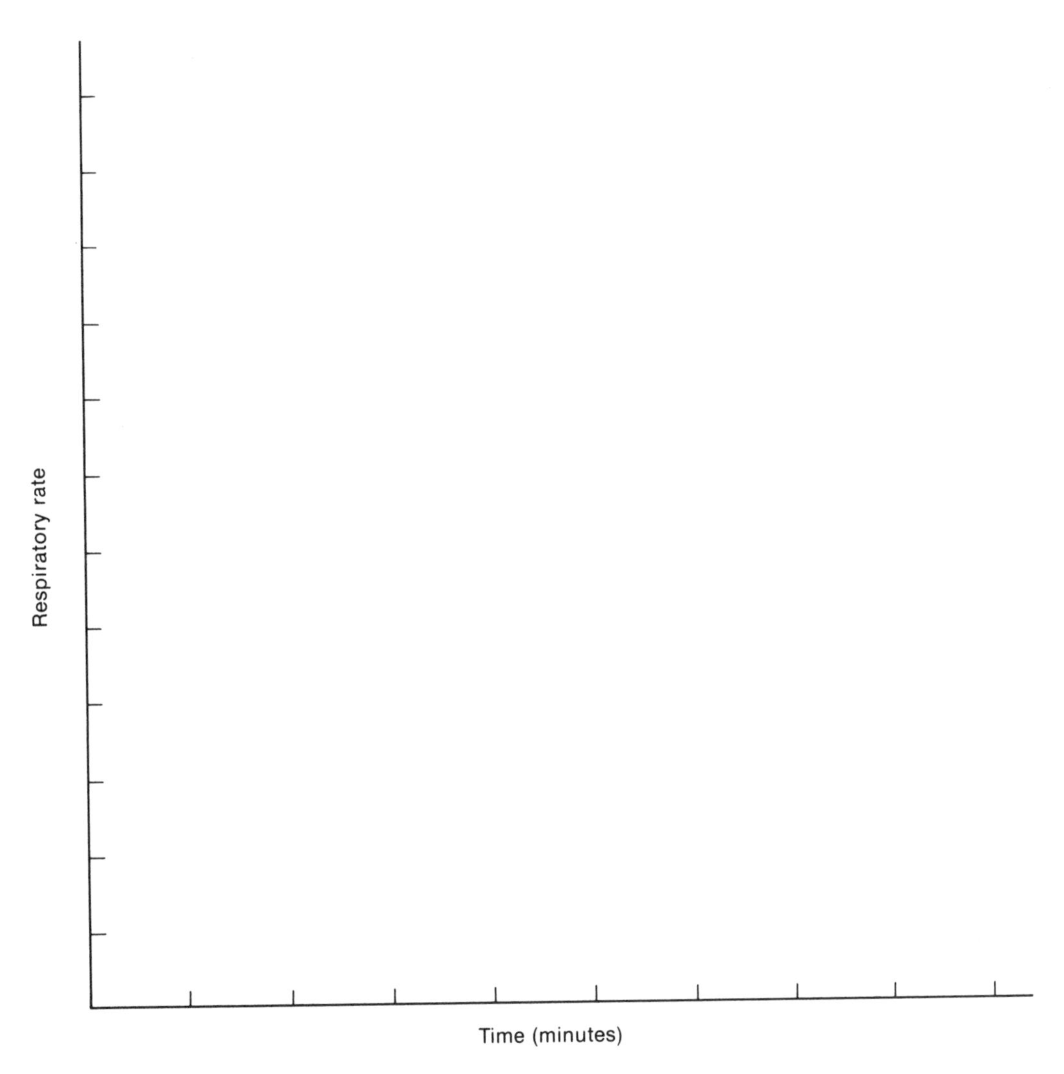

FIG. 13-3
Respiratory rate of germinating peas at room temperature.

Any pressure changes that develop in these jars will be revealed by the vertical movement of the fluid in the manometer tube. These changes can be measured if one arm of the manometer consists of a measuring device—in this case, a pipet.

If a live animal is placed in one of the jars and the vent tubes are sealed, then the exchange of respiratory gases (O_2 and CO_2) will be restricted to the inside of the jar. The oxygen in the jar containing the animal will be used up and will be replaced by an equal amount of carbon dioxide. Under these conditions, there will be little or no pressure change in the jar. If KOH, a compound that absorbs carbon dioxide as it is exhaled, is placed in the jar, then the carbon dioxide will be removed from the atmosphere in the jar. Consequently, the total amount of gas in the jar containing the animal will be lowered. This results in a decrease in the pressure in the jar containing the animal. Why?

____

____

As a result, the fluid in the manometer tube moves. In which direction?

____

The empty jar attached to the other side of the manometer tube serves as a temperature–volume control. Initially, its atmosphere is identical with

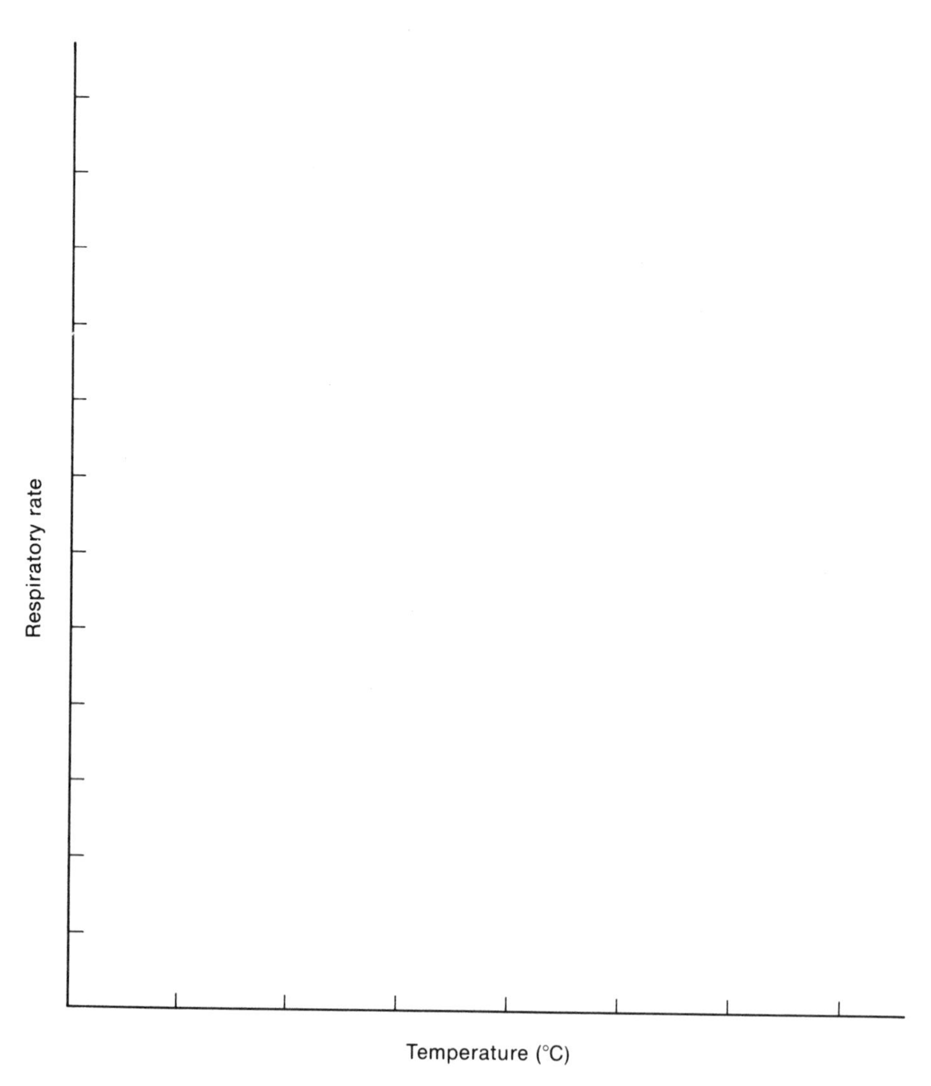

FIG. 13-4
Effect of temperature on respiratory rate.

that of the jar containing the animal. Because no respiration is occurring, there will be no change in the gas pressure in the empty jar. Keeping both jars at the same temperature, ensures that any changes in the height of the manometer fluid are due solely to pressure changes brought about by the respiratory activities of the animal in the respiration chambers.

A variety of animals are available in the laboratory. You are to measure their respiratory rates and, using the data obtained, determine whether the animal you have tested is poikilothermic or homeothermic.

Your instructor will determine the temperature each team of two students is to use in these experiments. Refer to Fig. 13-6A, B, C for the procedure to use to obtain that temperature. Before beginning this experiment, weigh the animal to the nearest gram. Record the animal's weight in Table 13-2. Because different animals and different temperatures are being used by different teams, the data collected should be exchanged with other teams and entered in Table 13-3.

After the animal and the water bath are at the desired temperature, allow the setup to stand for 5–10 minutes with the vent tube open. Then attach a pinch clamp to each vent tube. Record in Table 13-2 the initial position of the fluid (in milliliters) in the pipet arm of the manometer (Fig. 13-6D).

After 5 minutes, or sooner if the column of fluid moves rapidly, record the new position of the manometer fluid in Table 13-2. (*Note:* The time interval used will depend on the respiratory rate of the animal under study. Make it long enough so that the water column rises approximately one-third the height of the pipet. It should be in whole minutes to make succeeding calculations easier.) After complet-

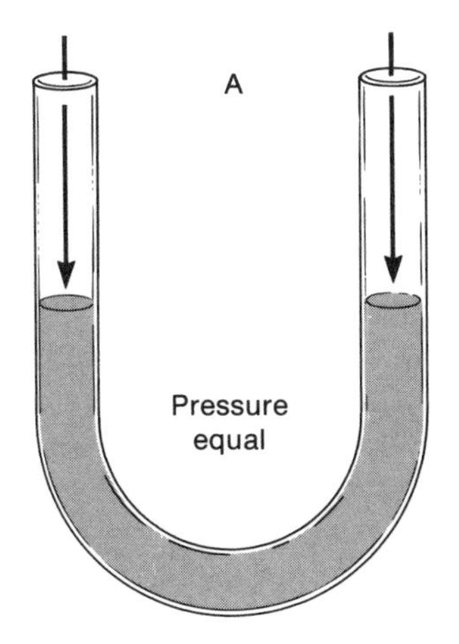

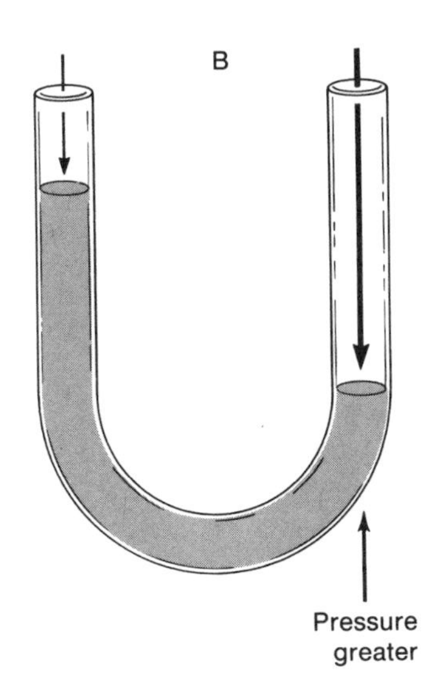

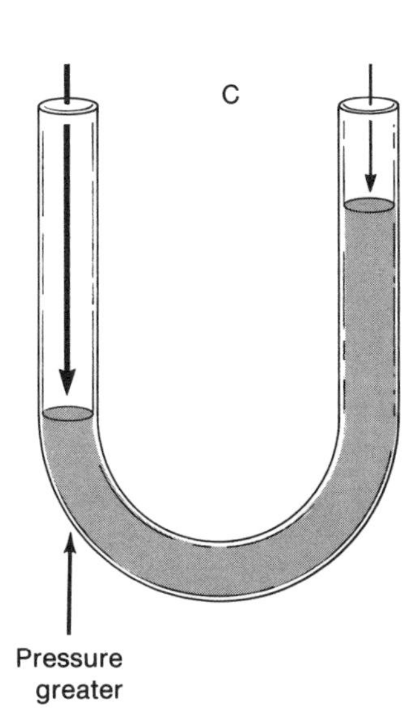

FIG. 13-5
Operation of a simple manometer. (After *Investigations of Cells and Organisms*, by P. Abramoff and R. G. Thomson, Prentice-Hall, 1968.)

ing one such reading, remove the pinch clamps from the vent tubes. This will allow the fluid in the manometer to equalize. Replace the pinch clamp, and repeat the procedure twice to obtain two additional manometer readings. Calculate the average value for the three readings, then determine how much oxygen was used per hour. For example,

$$\frac{12 \text{ ml } O_2}{\overset{}{\cancel{6}}\ \text{m}\cancel{\text{in}}_{1}} \times \frac{\overset{10}{\cancel{60}}\ \text{m}\cancel{\text{in}}}{1 \text{ hr}} = \frac{120 \text{ ml } O_2}{1 \text{ hr}}$$

$$= 120 \text{ ml } O_2/\text{hr}$$

Determine the respiratory rate of each animal in terms of the milliliters of O_2 used per kilogram per hour as follows:

$$\frac{\text{ml } O_2/\text{hr}}{\text{wt in g}/1000} = \frac{120 \text{ ml } O_2/\text{hr}}{120 \text{ g}/1000}$$

$$= \frac{120 \text{ ml } O_2/\text{hr}}{0.12 \text{ kg}}$$

$$= 1000 \text{ ml } O_2/\text{kg/hr}$$

Record your data in Tables 13-2 and 13-3.

From your data, and from the data of other students using the same type of animal, determine whether the animal you used was homeothermic or poikilothermic. Explain.

____

____

____

____

There are several different gases in air. How could you prove that the gas being used in the tube was actually oxygen?

____

____

____

B. BIOLOGICAL OXIDATION

As described in the first part of this exercise, oxidation-reduction reactions are important in releasing the energy in glucose and incorporating this energy into ATP molecules. Iron-containing compounds

TABLE 13-2
Calculation of respiratory activity.

Treatment	Trial	Initial manometer reading (ml)	Final manometer reading (ml)	Volume of O_2 used (ml)	Time interval (min)	Calculated average ml of O_2/hr	Calculated average respiratory rate (ml of O_2/kg/hr)
Room temperature ______°C	1						
	2						
	3						
High temperature ______°C	1						
	2						
	3						
Low temperature ______°C	1						
	2						
	3						

Type of animal ______________________ Weight (kg) ______________

Poikilothermic ______________ Homeothermic ______________

called **cytochromes** take part in these reactions, which are carried out in the mitochondria. In these reactions, oxidation takes place in the presence of oxygen. An example of an oxidation-reduction reaction is shown in Reaction 1, in which $A \cdot H_2$ is the hydrogen donor and B is called the hydrogen acceptor. Thus every oxidation must be accompanied by a simultaneous reduction. The energy required for the removal of hydrogens in oxidation reactions is supplied by the accompanying reduction. In order for this type of reaction to proceed, enzymes called **dehydrogenases** are required.

$$A \cdot \boxed{H_2} + B \xrightarrow{\text{Dehydrogenases}} A + B \cdot \boxed{H_2} \qquad (1)$$

Specific dehydrogenases called **oxidases** use oxygen as a hydrogen acceptor. Most oxidases contain a metal, such as copper, iron, or zinc, and a riboflavin-containing complex. The transfer of hydrogen from the substrate to oxygen by the oxidase usually results in the formation of hydrogen peroxide (H_2O_2)

TABLE 13-3
Respiratory data on other animals.

Type of animal	Temperature treatment	Respiratory rate (ml of O_2/kg/hr)	Respiratory rate (ml of O_2/kg/hr)	Temperature treatment	Type of animal

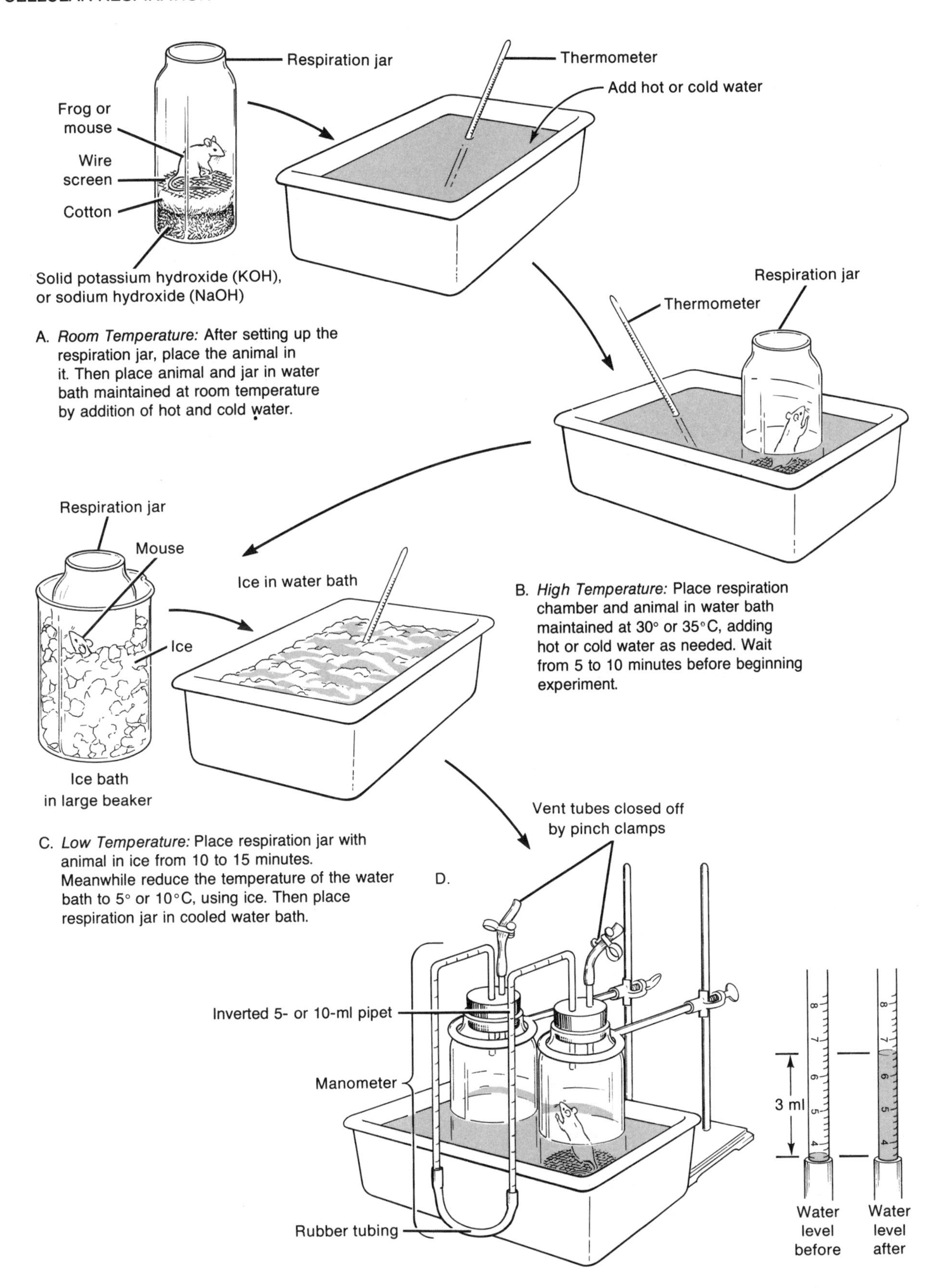

FIG. 13-6
Procedure for determining respiration rate of an animal. (After *Investigations of Cells and Organisms,* by P. Abramoff and R. G. Thomson. Prentice-Hall, 1968.)

(see Reaction 2). Although H_2O_2 is toxic to the tissues, its toxic effect is prevented by two important enzymes: **peroxidase** and **catalase.**

Substrate

$$A{\cdot}\boxed{H_2} \xrightarrow{\text{Oxidase}} A \quad O_2 \rightarrow \boxed{H_2}O_2 \xrightarrow{\text{Peroxidase}} O_2 \quad B \rightarrow B{\cdot}\boxed{H_2} \qquad (2)$$

Hydrogen peroxide

Peroxidase, as shown in Reaction 2, splits off the oxygen, which then acts to accept hydrogen from another substrate molecule. During the same reaction, another molecule (B in Reaction 2) picks up the H_2 released from H_2O_2 and is reduced. Thus, the toxic potential of hydrogen peroxide has been removed. Since oxidation-reduction reactions are important at the level of cellular respiration, it is important that you become familiar with these types of reactions. In this exercise, you will examine an oxidation-reduction reaction in which peroxidase is involved.

Guaiacol is oxidized to a colored product by H_2O_2 in the presence of peroxidase (Reaction 3).

The reaction can be followed colorimetrically by measuring the amount of light absorbed by oxidized guaiacol. (Refer to Appendix B for the use of the Bausch & Lomb Spectronic 20 colorimeter and Appendix C for the principles of spectrophotometry.)

$$\text{Guaiacol (reduced)} + H_2O_2 \xrightarrow{\text{peroxidase}} \text{Guaiacol (oxidized colored complex)} + 2H_2O \qquad (3)$$

Guaiacol (reduced) Guaiacol (oxidized colored complex)

1. Peel a turnip and cut it into small cubes. Place the cubes in a blender and grind for 30 seconds. Or, place the cubes in a mortar containing silica sand, and grind to pulp with a pestle (Fig. 13-7).

2. Filter the homogenate through a double layer of cheesecloth into a beaker. Squeeze out as much of the liquid as you can. Dilute 1 ml of the juice to 200 ml with distilled water.

3. Using a wavelength of 500 nm, blank the colorimeter with a tube containing 0.01 ml of guaiacol, 0.2 ml of 0.9% H_2O_2, and 9.8 ml water.

4. Add 0.01 ml of guaiacol, together with 0.2 ml of 0.9% H_2O_2 and 4.7 ml of distilled water, to a test tube. Put 1.0 ml of the diluted turnip extract and 4 ml of water into a separate colorimeter tube. Pour the guaiacol–H_2O_2 solution into the colorimeter tube. To mix, quickly pour the mixture back and forth between the empty test tube and the colorimeter tube.

5. Wipe the colorimeter tube clean and immediately place it in the Spectronic 20 colorimeter. Start the stopwatch, and read the percent transmittance. Take readings every 20 seconds, and record the data in Table 13-4. Plot these readings against time in Fig. 13-8. Repeat the determination, using one-half and then twice as much extract. Also measure the effect of extract that has been placed in a bath of boiling water for several minutes and then cooled. Then test the effect of adding 0.5 ml of 0.01 *M* sodium fluoride (CAUTION: *Poison!*) to the reaction mixture.

From these data, what can you conclude about peroxidase activity in turnip tissue?

____

____

TABLE 13-4
Peroxidase activity in turnip tissue.

Time (seconds)	Percent transmittance				
	1 ml of turnip extract	0.5 ml of turnip extract	2.0 ml of turnip extract	1 ml of boiled turnip extract	1 ml of extract + 0.5 ml NaF
20					
40					
60					
80					
100					
120					

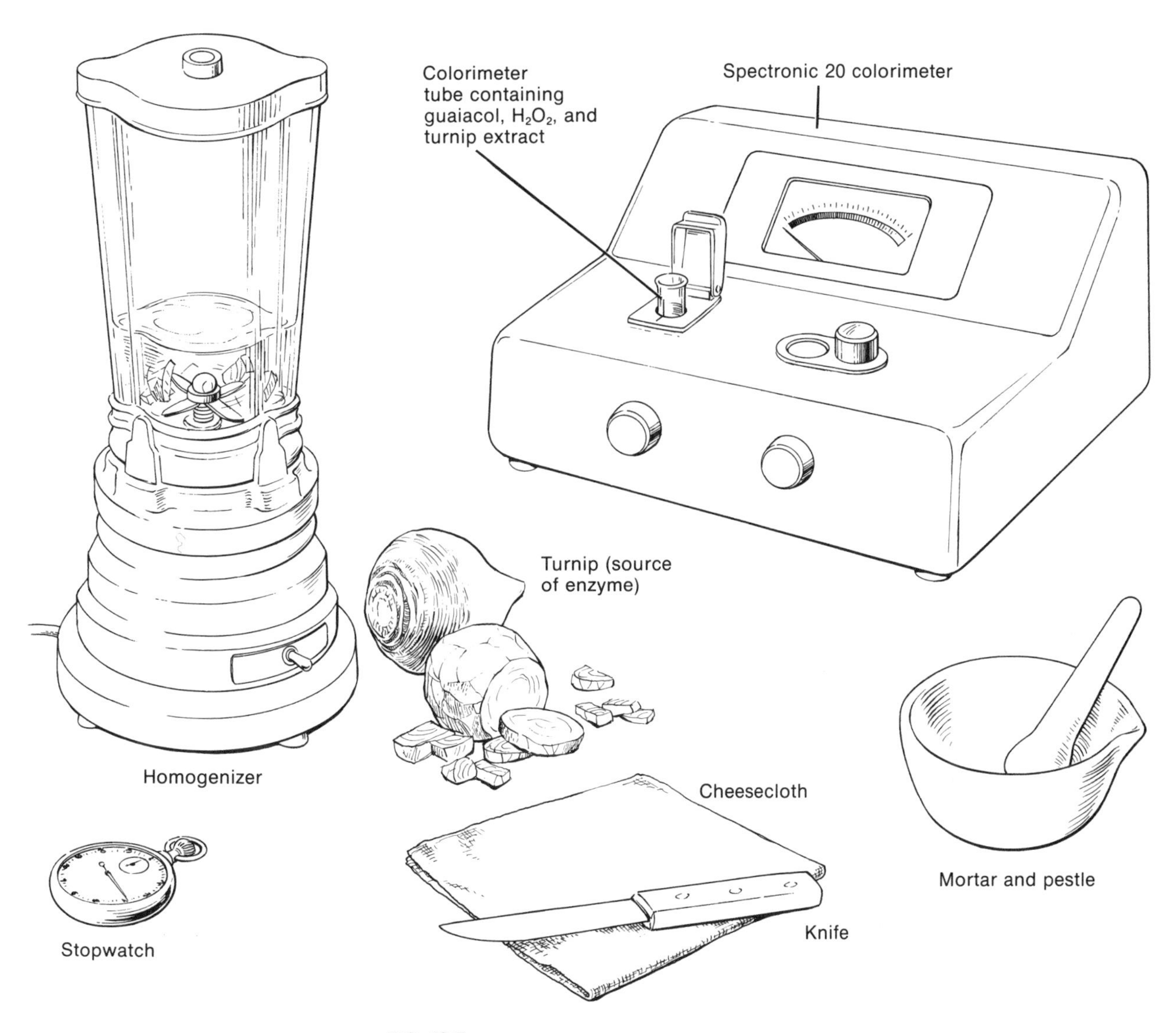

FIG. 13-7
Apparatus for extracting the enzyme peroxidase.

____

Fill in the blanks in Reaction 4. Indicate hydrogen donor, hydrogen receptor, the reaction catalyzed by peroxidase, and the reactions catalyzed by dehydrogenases.

In oxidation-reduction reactions, what component of the hydrogen atom is actually being shunted through the various carrier molecules?

____

$$\begin{array}{l} ___ \leftarrow \quad \rightarrow \text{Guaiacol}\cdot\boxed{H_2} \quad\quad O_2 \leftarrow \quad \rightarrow ___ \\ \text{A}\cdot\boxed{H_2} \quad\quad ______ \leftarrow \quad \rightarrow ___ \quad\quad \text{B} \leftarrow \end{array} \qquad (4)$$

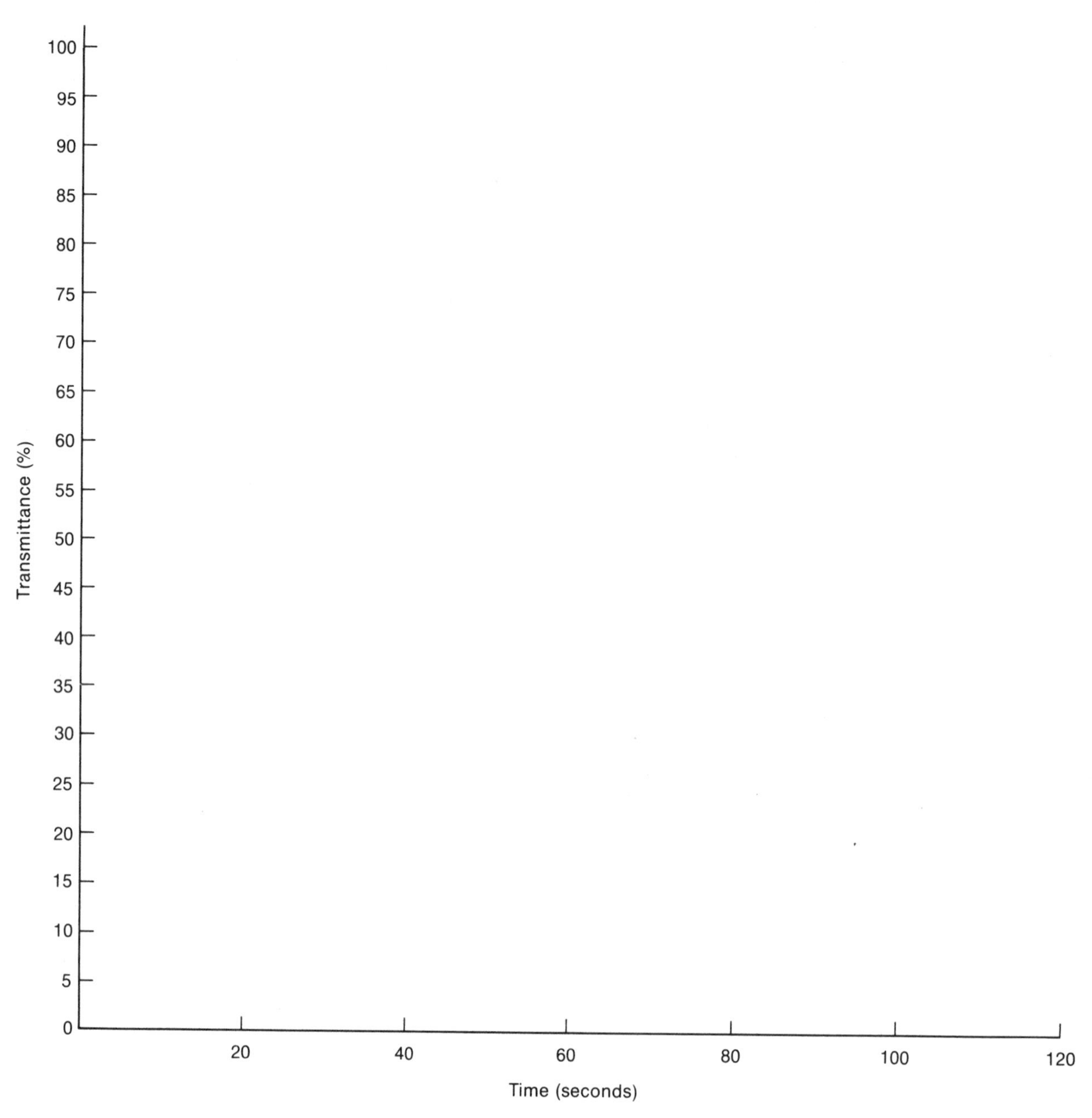

FIG. 13-8
Peroxidase activity in turnip tissue.

REFERENCES

Alberts, B., D. Bray, J. Lewis, M. Raff, K. Roberts, and J. D. Watson. 1983. *Molecular Biology of the Cell.* Garland.

Allamong, B. D., and T. R. Mertens. 1976. *Energy for Life: Photosynthesis and Respiration* (Self-Teaching Text). Wiley.

Green, D. E. 1964. The Mitochondrion. *Scientific American* 210(1):63–74.

Hinkle, P. C., and R. E. McCarty. 1978. How Cells Make ATP. *Scientific American* 238:104–123 (Offprint 1383). *Scientific American* Offprints are available from W. H. Freeman and Company, 41 Madison Avenue, New York 10010, and 20 Beaumont Street, Oxford OX1 2NQ, England. Please order by number.

Karp, G. 1984. *Cell Biology.* 2d ed. McGraw-Hill.

Lehninger, A. L. 1960. Energy Transformation in the Cell. *Scientific American* 202(5):102–114 (Offprint 69).

Lehninger, A. L. 1961. How Cells Transform Energy. *Scientific American* 205(3):62–73 (Offprint 91).

Racker, E. 1968. The Membrane of the Mitochondrion. *Scientific American* 218(2):32–39 (Offprint 1101).

Stephenson, W. K. 1978. *Concepts in Cell Biology: A Self-Instruction Text.* Wiley.

Stryer, L. 1981. *Biochemistry.* 2d ed. W. H. Freeman and Company.

EXERCISE 14

Transport in Biological Systems

The movement of water, metabolic wastes, food, hormones, minerals, gases, and many other materials within the organism is extremely important to higher plants and animals. Almost every metabolic activity of the organism depends ultimately on the exchange of materials between cells and their environment and between the organism and its environment.

Nutrient molecules must be absorbed, body wastes removed, and gases exchanged to support cellular respiration. A variety of substances, including such diversified materials as chemical regulators (hormones) and defensive materials (antibodies), must be exchanged if the organism is to function coherently. Basically, exchange depends on physical processes such as diffusion, osmosis, and pinocytosis and on biochemical processes such as active transport. Because these phenomena can affect rapid movement over short distances, very small organisms (i.e., protozoa) or organisms with most of their cells in contact with the environment (i.e., sponges) rely almost entirely on them to achieve exchange between their cells and the surrounding environment (Fig. 14-1A).

Larger and more complex organisms require supplementary systems. With only a small fraction of their cellular mass in direct contact with the environment, exchanges based solely on diffusion would not occur rapidly enough to meet their needs. In larger organisms, cellular exchanges take place in an internal environment (the extracellular fluid system), and a way must be developed to transport these fluids throughout the organism. Exchange between the organism and the environment than takes place through the body wall or some specialized part of it. These mechanisms constitute the transport system of plants and the circulatory systems of animals.

A. TRANSPORT IN ANIMALS

Circulatory systems in animals vary enormously in complexity. In the simplest cases, movement of extracellular fluids is achieved by the compressive

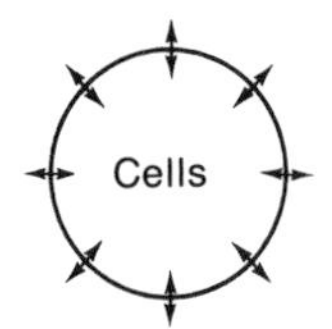

A. Direct exchange between cells and environment

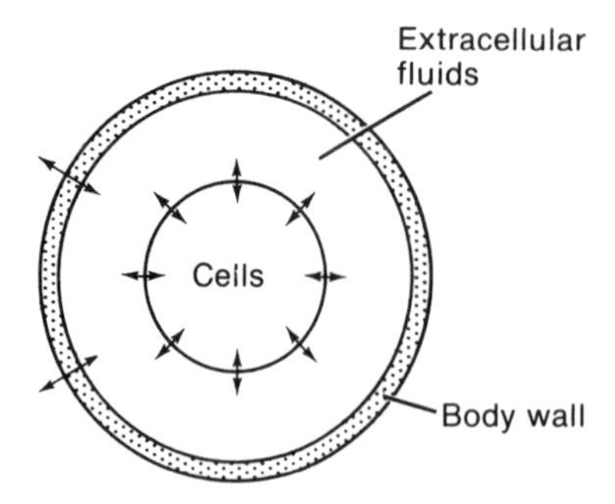

B. Circulation of extracellular fluids in the absence of a defined circulatory system

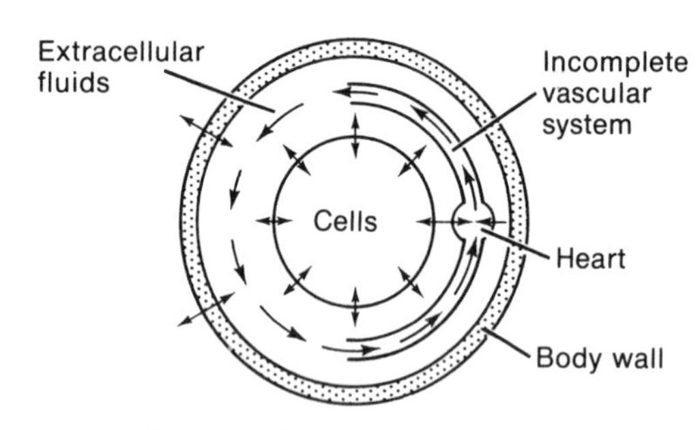

C. Open circulatory system

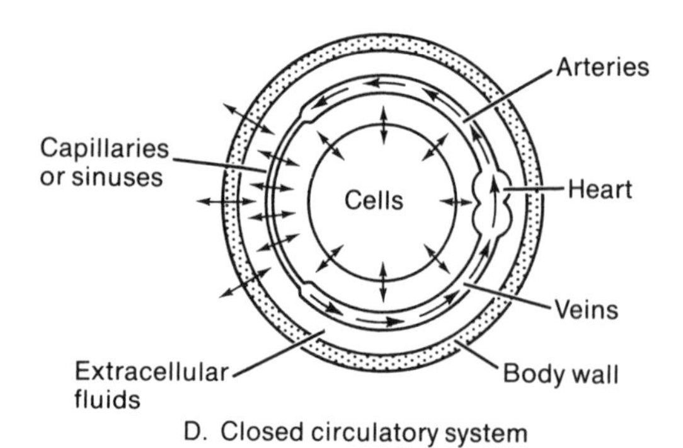

D. Closed circulatory system

FIG. 14-1
Types of exchange and circulatory systems.

action of body wall muscles or by ciliary activity (Fig. 14-1B).

Those only slightly more complex, however, have some type of pumplike heart. The pump may consist simply of a group of pulsatile vessels, as in the earthworm, or it may take the form of a simple membranous chamber, as in insects. In many molluscs and vertebrates, the heart is a complex, chambered structure. The evolution of a heart is normally associated with the development of vessels that direct the flow of extracellular fluids to specific areas of the body. In insects and several other invertebrates, the system of such vessels is incomplete in that the fluids carried by the vessels empty into irregular cavities and return to the heart through the body spaces (Fig. 14-1C). Such circulatory systems are said to be "open," and there is no distinction between the fluids within the heart and vessels and those outside.

In vertebrates, the circulatory system is completely closed and consists of a pump and a continuous series of vessels. The blood carried in these vessels is different from the extracellular fluid in many respects (Fig. 14-1D). Blood flow takes place within this tubular system, and exchange between the blood and other extracellular fluids takes place through the smallest vessels (capillaries) and irregular cavities (sinusoids).

In the early 1600s, William Harvey, an English physician, discovered the pattern of blood circulation. He subsequently concluded that the blood carries nutrients, oxygen, and various other materials to the body tissues, and carbon dioxide and other waste products away from the tissues. Harvey, however, never saw one of the more important parts of the circulatory system—the capillaries. These blood vessels are important because it is in the capillaries that materials are exchanged between the circulatory system and the cells. In this exercise, you will examine the circulation of blood through the capillaries in a fish tail *or* the web of a frog's foot.

1. Capillary Circulation

a. In a Fish Tail

Obtain a small (5–10 cm) goldfish from the aquarium in your laboratory (Fig. 14-2). Wrap the fish (except for the mouth and tail) in dripping wet cotton. *(Note: Do not allow the cotton around the fish to dry out.)* Place it in the bottom half of a petri dish (Fig. 14-2B, C). If the fish moves around too much, place it in a container of chloretone solution to anesthetize it. After a minute or two, the fish will roll

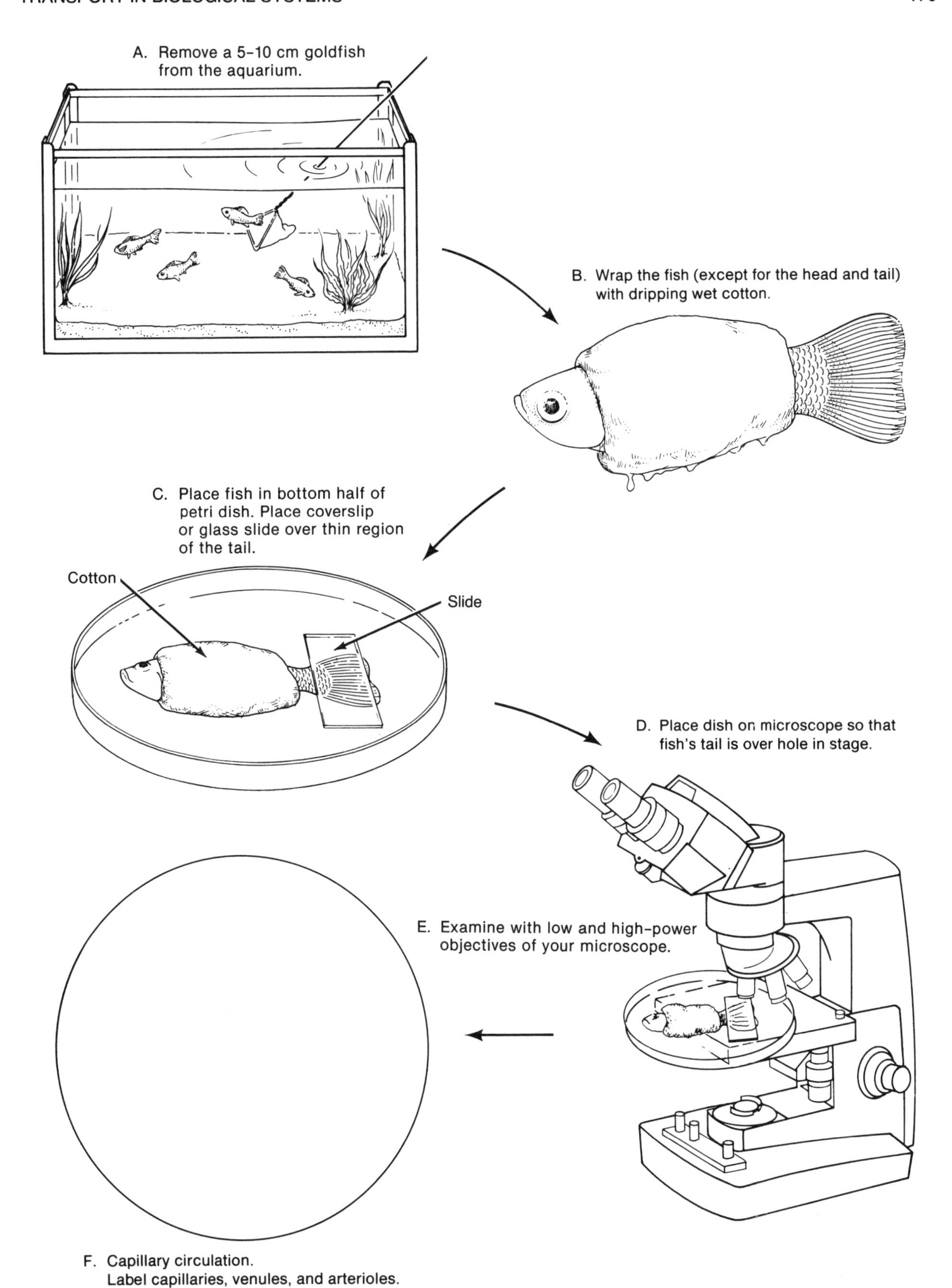

FIG. 14-2
Examination of capillary circulation in the goldfish tail. (After *Investigations of Cells and Organisms,* by P. Abramoff and R. G. Thomson. Prentice-Hall, 1968.)

over on its side and will remain anesthetized for about 1 hour.

Place a coverslip (or glass slide) over a thin region of the tail. Then position the petri dish on your microscope so that the fish's tail is over the hole in the stage (Fig. 14-2D). Focus on the tail with the low-power objective of your microscope. Note the movement of blood through the blood vessels. The smallest vessels you see are the capillaries. They are just wide enough to permit the passage of a single file of blood cells. What is the advantage of having red blood cells pass through the capillaries in single file?

Change to high power to examine the capillaries and blood cells more closely.

Blood enters the capillaries from small arteries called **arterioles.** Trace a capillary back to an arteriole. Does the blood flow more rapidly in the arterioles or in the capillaries?

Follow a capillary in the direction in which the blood is flowing. It will join a slightly larger blood vessel called a **venule.** Is the rate at which the blood flows greater in the venules or in the capillaries?

What is the relation between the diameters of the red blood cells and the capillaries?

What significance might this relation have with regard to exchange of gases and other products?

Does the red cell seem to be a rigid or a flexible structure?

What are the advantages and disadvantages of this?

Why are exchanges between the surrounding tissues and the blood more likely to occur in the capillaries than in the arteries or veins?

Record your observations of capillary circulation with a drawing in Fig. 14-2F.

b. In the Web of a Frog's Foot

Carefully wrap a live frog in a wet towel tightly enough to keep it from moving very much. Leave one foot exposed. Now spread and pin the web of the foot over a hole in a balsa board. Keep the skin of the frog moist with tap water. By supporting the whole preparation on the stage of your microscope, the actual circulation of cells in the blood vessels can be seen. Select a very small field and observe carefully the size of the vessels, the thickness of their walls, and velocity of blood flow. As you examine the preparation, keep in mind the three comparisons that you wish to make between arterioles, capillaries, and venules: (1) relative size, (2) comparative speed of flow of blood, (3) amount of pulsation. Stroke the web lightly in the field of view. Record your observations.

Blood enters the capillaries from small arteries called *arterioles.* Trace a capillary back to an arteriole. Does the blood flow more rapidly in the arterioles or in the capillaries?

Follow a capillary in the direction in which the blood is flowing. Is the flow of blood greater in the venules or in the capillaries?

What is the relation between the diameters of the red blood cells and the capillaries?

____

____

What significance might this relationship have with regard to exchange processes?

____

____

____

Does the red blood cell appear to be a rigid or a flexible structure?

____

What are the advantages and disadvantages of this?

____

____

Why are exchanges between the surrounding tissues and the blood more likely to occur in the capillaries than in the arteries or veins?

____

____

____

____

____

2. Factors Affecting Capillary Circulation

The flow of blood is affected by a variety of factors, most of which operate to ensure that blood is delivered at adequate rates to various sections of the body. The effects on capillary circulation of ethyl alcohol, nicotine, and lactic acid are given in Table 14-1. Determine the effects of temperature and adrenaline on the capillary circulation of the frog mesentery or fish tail, following the directions given below.

a. Temperature

Carefully examine the capillary network at room temperature to determine the normal blood flow. Then, using an eye dropper, apply several drops of chilled frog or fish Ringer's solution (isotonic salt solution) to the surface of the capillary network. Does the blood flow rate increase or decrease?

____

Record your results in Table 14-1. Rinse off the tissue using warm (room temperature) Ringer's solution, then apply Ringer's solution heated to 40°C. Record your results in Table 14-1.

b. Adrenaline

Wash the tissue with warm Ringer's solution. Then place one drop of 1:50,000 adrenaline solution on the capillary bed. Record your observations in Table 14-1. Again rinse the tissue with warm Ringer's solution.

From the data recorded in Table 14-1, answer the

TABLE 14-1
Factors that affect blood flow in capillaries.

Regulator	Diameter decreases (constricts) Blood flow rate decreases	Diameter increases (enlarges) Blood flow rate increases
Ethyl alcohol	+	(−)
Nicotine	(−)	+
Lactic acid	(−)	+
Adrenaline		
Temperature	Cold Ringer's:	Warm Ringer's:
Animal used:		

following questions regarding the effect of these factors on peripheral circulation.

Why is the temperature of the fingers of a nonsmoker higher than that of those who smoke?

It is known that the amount of lactic acid in muscle tissue increases during exercise. Why, then, would you expect more oxygen to be delivered to the muscle tissue of a person who is exercising than to that of a person who is resting?

Suggest a reason why an intoxicated person might more easily die from exposure to cold than a sober person.

Adrenaline is released into the bloodstream at times of stress. Why might a person who is frightened become pale rather than flushed and red?

3. Factors Affecting Heartbeat

Anesthetize the frog that was used in Part A1b, and cut from A to B as shown in Fig. 14-3A. (*Note:* If

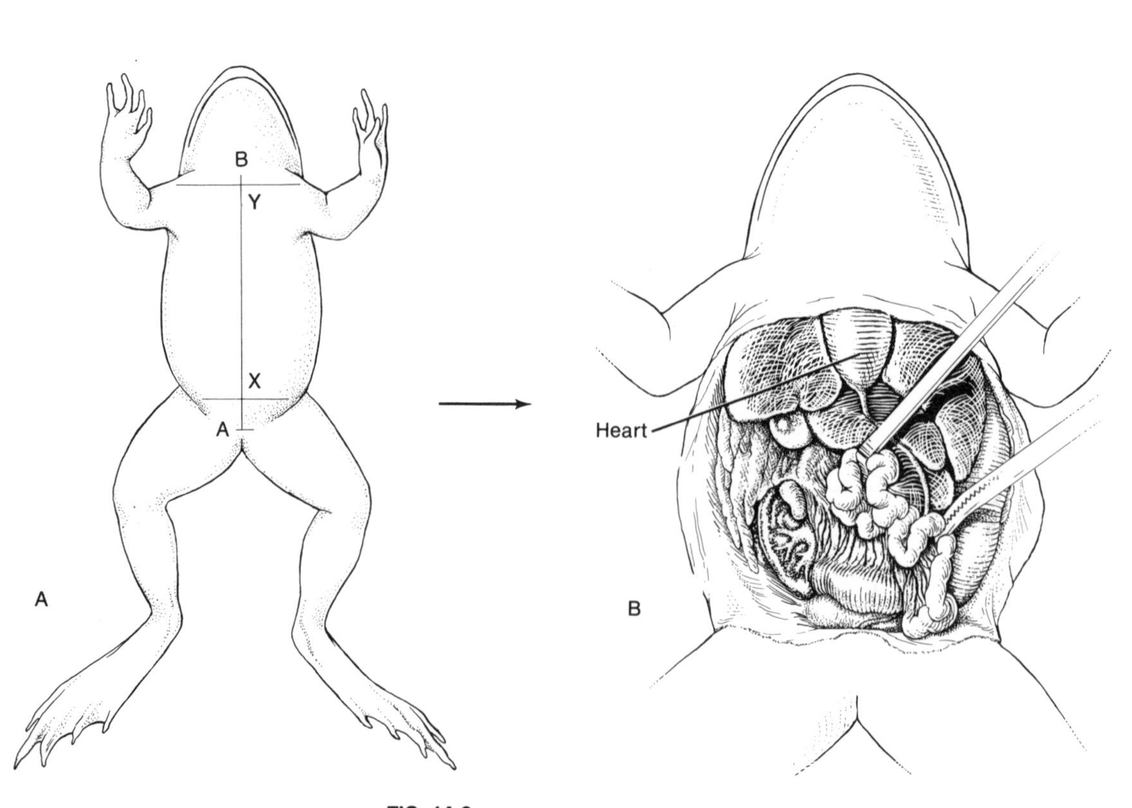

FIG. 14-3
Dissection of frog to expose the heart.

you used a fish in Part 1, your instructor will give you an anesthetized frog.) Extend the incision forward to the shoulder girdle. Lift upward and forward with each cut to avoid damaging the underlying organs. Cut through the bones of the shoulder girdle. Then, cut laterally at points X and Y and pull back the skin to expose the internal organs (Fig. 14-3B). Use clean filter paper to sponge up any blood in the area.

Note that the heart is enclosed in a membrane, the **pericardium.** Observe the volume of the heart during its diastolic (expansion) and systolic (contraction) phases. Carefully slit the pericardium, and note any changes in volume during **diastole** and **systole.** There are two anterior chambers and a posterior third chamber; the anterior chambers are the **auricles.** The single posterior chamber is the **ventricle;** note any differences in the texture of its wall relative to those of the atria. Observe the groove marking the boundary between the auricles and ventricle. This is the **atrioventricular groove.** Observe the **truncus arteriosus,** a thick-walled tube that arises dorsally from the ventricle and divides into two **aortic arches.** With fine-nosed forceps, lift the ventricle by its apical tip and reflect it forward. Note the narrow **V**-shaped **sinus venosus,** which opens into the right auricle.

Finally, observe the main veins, which return blood to the heart. The right and left **anterior venae cavae,** or caval veins, enter the sinus venosus from the anterior part of the body. A single **posterior vena cava,** or caval vein, empties into the posterior part of the sinus venosus.

Carefully observe the sequence of contraction of all parts of the heart and the veins and truncus arteriosus. List the order in which these parts contract.

____

____

Using the second hand of a watch, or using a stopwatch, determine the number of contractions per minute made by the auricles, ventricle, truncus arteriosus, and sinus venosus.

Left auricle	________/minute
Right auricle	________/minute
Ventricle	________/minute
Truncus arteriosus	________/minute
Sinus venosus	________/minute

Are the rates equal or different? What can you conclude from this?

____

____

____

____

____

Observe and record the rate of contraction of the heart: ________/minute. Decerebrate the frog as outlined in Appendix E, Part C. After 5 minutes, observe and record the rate of contraction again: ________/minute. Destroy the midbrain and hindbrain areas by pithing the animal as described in Appendix E, Part B. After 5 minutes, observe and record the rate of contraction: ________/minute.

Finally, insert a dissecting needle into the part of the vertebral column containing the spinal cord, and complete the destruction of the central nervous system by forcing the needle down as far as it will go. After a recovery period of 5 minutes, observe and record the rate of contraction of the heart: ________/minute.

Do your observations suggest that the central nervous system plays a role in governing heart action?

____

If so, which part of the central nervous system seems to be the most important?

____

Did the heart cease beating with complete destruction of the central nervous system?

____

What conclusions are suggested by this?

____

____

____

Chill several glass or metal rods in a beaker of ice water, and apply the rods one at a time to the ventricle. When it has cooled, record the contraction rates of the auricles, ventricle, and sinus venosus in Table 14-2. Restore the heart to room temperature with frog Ringer's solution. Repeat, cooling the atria and then determining the contraction rates as before. Again warm the heart. Finally, cool the sinus venosus, and determine the rate of contraction of each

TABLE 14-2
Effects of differential cooling on heart contraction.

Portion cooled	Rate (contractions per minute)		
	Ventricle	Auricles	Sinus venosus
Ventricle			
Auricles			
Sinus venosus			

chamber. (It will be necessary to tip up the ventricle to reach the sinus venosus.)

Do your observations suggest that one part of the heart has a greater effect on heart rate than the other parts?

____

How do you account for this?

____

____

____

____

To examine this aspect of heart control further, remove the heart from the body by lifting the ventricular apex and cutting through the anterior and posterior vena cavae and the right and left aortae. Be careful not to stretch the heart, and do not damage the chambers, particularly the sinus venosus. In addition, be careful not to puncture the gallbladder, a small, dark green sac lying close to the heart near the liver.

Place the heart in a watch glass containing Ringer's solution at room temperature. After 2–3 minutes, the heart should begin to contract again. Does this support your earlier conclusions? Explain.

____

____

____

____

Determine the rate of contraction of the whole isolated heart: ________/minute. Carefully separate the sinus venosus from the right auricle, without damaging either one. After 2–3 minutes, determine the contraction rate of the sinus venosus and of the atrioventricular part: sinus venosus: ________/minute; atrioventricular: ________/minute. As a last step, separate the auricles from the ventricle. It may be necessary to wait a few minutes for the ventricle to resume contractions. If it shows no sign of contracting after 5–10 minutes, stimulate it by touching it with a dissecting needle. After it resumes beating, determine the contraction rate of each segment: sinus venosus: ________/minute; auricles: ________/minute; ventricle: ________/minute.

Summarize your conclusions regarding the control of heart activity in the frog.

____

____

____

____

____

____

____

____

4. Function of Hemoglobin

Almost all of the oxygen carried in the blood is transported in combination with hemoglobin. In vertebrate animals, hemoglobin is contained within red blood cells. It consists of a protein called **globin** to which is attached a prosthetic group called **heme.** Heme is a complex consisting of protoporphyrin and ferrous iron (FE^{2+}).

To function, hemoglobin depends on its ability to reversibly combine with oxygen:

$$\underset{\text{Hemoglobin}}{Hb} + O_2 \rightleftharpoons \underset{\text{Oxyhemoglobin}}{HbO_2}$$

Hemoglobin combines with oxygen in the lungs, where the oxygen concentration is high. Oxygen is then released in the tissues, where the oxygen concentration is low. The iron of the heme group is the site of the reversible reactions involving oxygen. Part of the carbon dioxide released from the tissues by cellular respiration is then transported by the hemoglobin to the lungs, where carbon dioxide levels are low, and is exchanged for oxygen. Carbon dioxide is transported in combination with the free amino groups of the protein (globin) part of the hemoglobin molecule.

a. Absorption Spectrum of Hemoglobin

Hemoglobin possesses a distinct absorption spectrum, as does each of its derivatives (i.e., oxyhemoglobin, deoxyhemoglobin, carboxyhemoglobin). In this exercise, you will determine the spectrum of hemoglobin, as well as those of some of its alternate forms. The part of the visible spectrum that is absorbed by hemoglobin can be determined by using a **spectroscope,** which disperses visible light into its component colors, or wavelengths (Fig. 14-4). The disappearance from the spectrum of various colors, or wavelengths, as a result of passing light through a pigment solution, indicates that those wavelengths were absorbed by the pigments. The absorption spectra of hemoglobin and its derivatives will be visible as shadows in the yellow and green regions. These are **absorption bands.** The wavelength at which the shadow is darkest is the wavelength of maximum absorption (A_{max}).

1. *Oxyhemoglobin:* A major physiological function of hemoglobin resides in its ability to react reversibly with oxygen. Under optimal conditions 1 g of hemoglobin will combine with 1.36 g of oxygen. Oxyhemoglobin is quite stable as long as sufficient oxygen is present to prevent the reverse reaction from occurring.

To determine the absorption spectrum of oxyhemoglobin, pipet a sample of defibrinated blood (diluted 1:100) into a small test tube. Your instructor will demonstrate the use of the instrument for determining the absorption spectrum of your sample. In the particular instrument referred to here, two spectra are projected onto a scale (400–700 nm) on the back, inside surface of the instrument. The upper reference spectrum shows the colors (wavelengths) of light. The lower spectrum results from passage of the light through the sample. In the first chart below, indicate the wavelengths of light that are absorbed by oxyhemoglobin.

2. *Reduced hemoglobin:* To the test tube you have just examined, add a small amount of sodium dithionite (sodium hydrosulfite), a reducing agent. Reexamine the blood and determine its absorption spectrum; draw it in the chart in Fig. 14-5. You

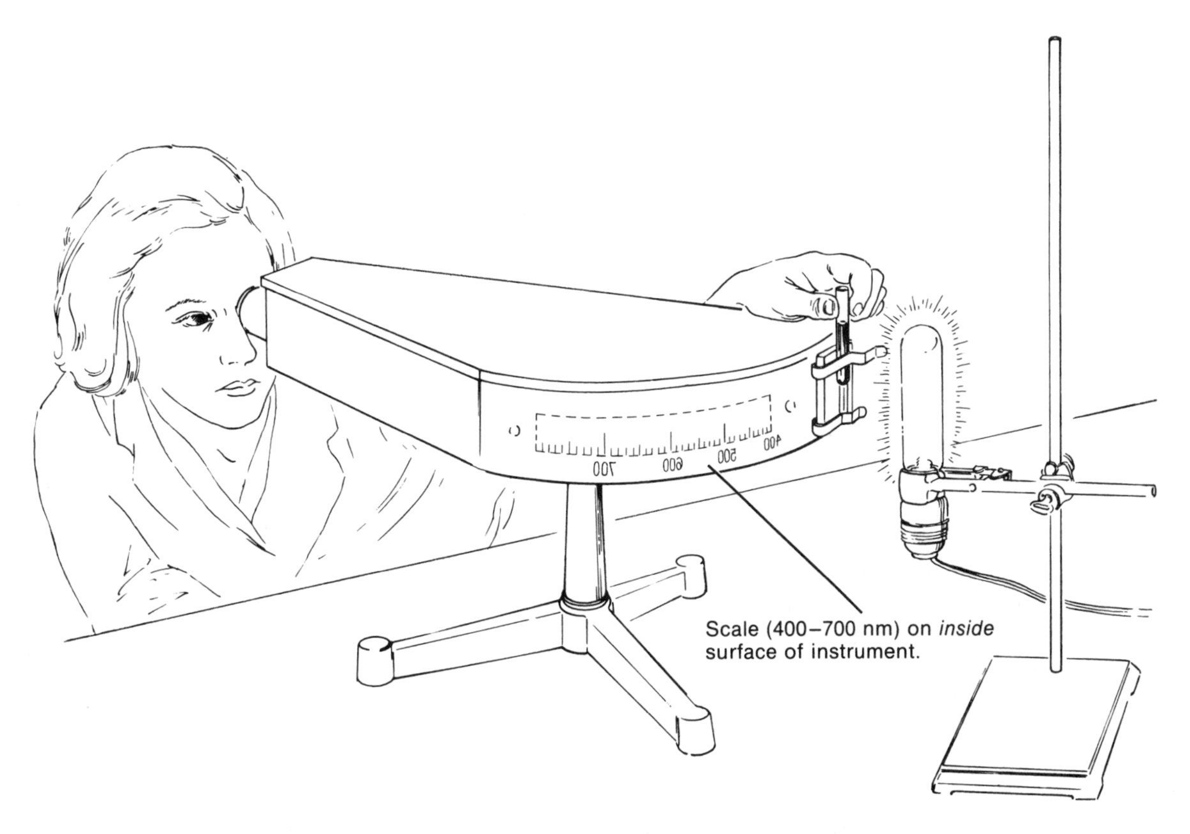

FIG. 14-4
Spectroscope for demonstrating the absorption spectrum of hemoglobin.

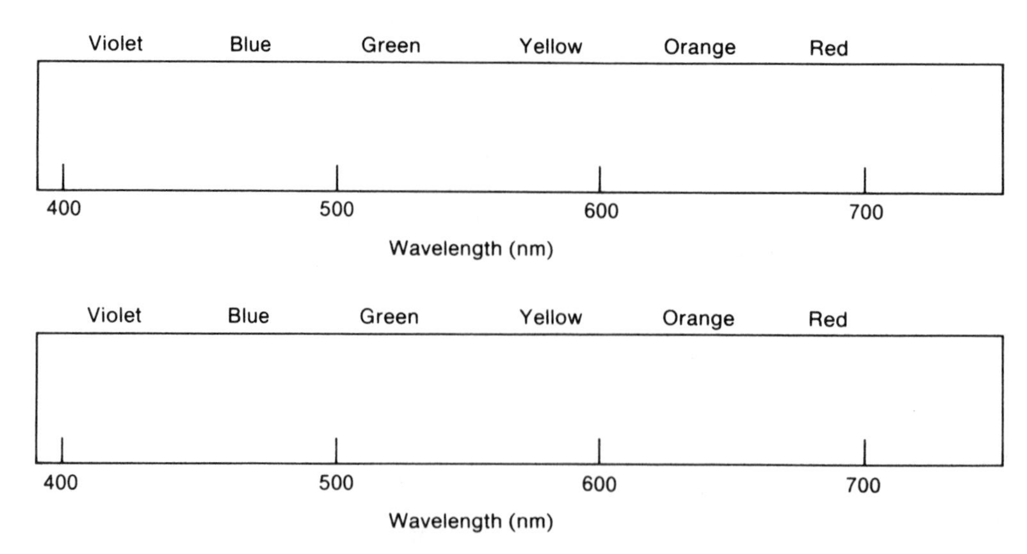

FIG. 14-5
The absorption spectrum of carboxyhemoglobin.

should see that the two bands of oxyhemoglobin have been replaced by a single broad band.

Shake the tube vigorously, and reexamine the absorption spectrum. How do you account for what you observed?

____

____

____

____

____

____

____

3. *Carboxyhemoglobin:* In addition to its ability to combine with oxygen, hemoglobin will also react with a variety of other substances such as hydrogen sulfide, ferricyanide, nitric oxide, and carbon monoxide (CO). The product in most cases is a colored compound that can also be characterized by its absorption spectrum. The combination of hemoglobin and carbon monoxide is of particular interest because carbon monoxide is found in city air, in the exhaust of motor vehicles, and in industrial gases and because it has about 200 times as much affinity for hemoglobin as does oxygen. In addition, the formation of carboxyhemoglobin (HbCO) prevents hemoglobin from combining with oxygen, probably because the two gases compete for the same spot in the hemoglobin molecule. The reaction between oxyhemoglobin and carbon monoxide is

$$\underset{\text{Oxy-hemoglobin}}{HbO_2} + \underset{\text{Carbon monoxide}}{CO} \rightleftharpoons \underset{\text{Carboxy-hemoglobin}}{HbCO} + \underset{\text{Oxygen}}{O_2}$$

Examine a sample of blood through which carbon monoxide has been bubbled. Determine its absorption spectrum and draw it on Fig. 14-5.

b. Hemoglobin Crystals

You can crystallize hemoglobin and its derivatives if you treat them with acid and gently heat them. Although the shapes of the crystals are very characteristic and can be used as a reliable and delicate test for the presence of blood, they cannot be readily distinguished from one animal species to another.

To observe the presence of hemoglobin crystals,

1. Place a drop of blood in the center of a clean glass slide, and allow it to air-dry.

2. Cover the dried blood with one or two drops of glacial acetic acid, and mix thoroughly with a toothpick.

3. Add a coverglass. (*Note:* The entire area under the coverglass should be filled with acid.)

4. Heat the slide *gently* over a low flame until bubbles are observed in the acid.

5. Allow the slide to cool, examine it microscopically, and locate the hemoglobin crystals, which are

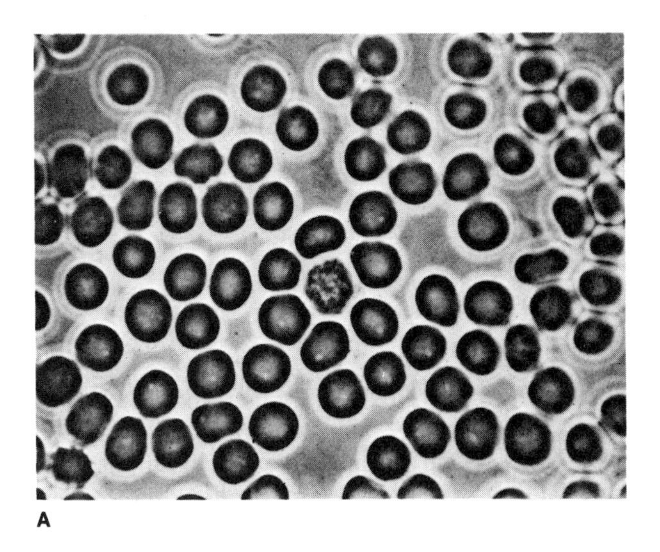

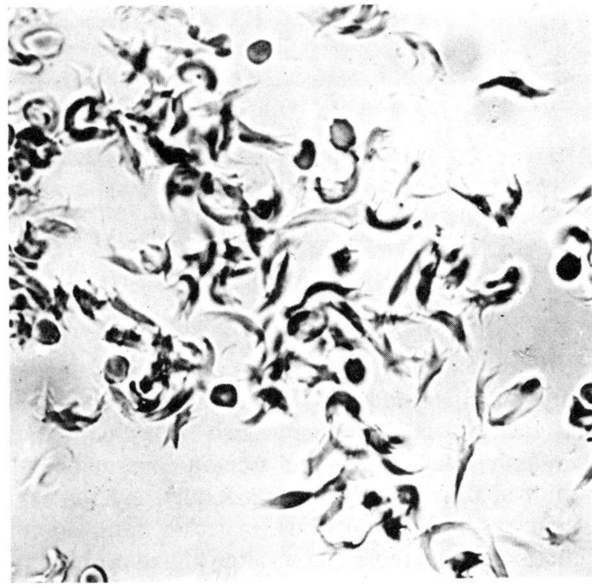

FIG. 14-6
(A) Normal red blood cells and (B) red blood cells of a person who has sickle-cell anemia. (Photomicrographs by Anthony C. Allison.)

generally rhomboid in shape, although they may have other forms.

c. Abnormal Hemoglobin (Sickle Cell)

Sickle-cell anemia was first discovered by J. B. Herrick in 1910. In this inherited disease (predominantly, but not exclusively, found in blacks and Latin Americans), a defective hemoglobin molecule (hemoglobin S) causes the red blood cells to assume an elongated, sickled shape. Not only do these cells have a low oxygen-carrying capacity, but their abnormal shapes also slow the blood flow and mass the cells in the capillaries, causing anoxia (lack of oxygen) in the tissues. In some cases, tissue may be destroyed. This disease is particularly harmful to normally sedentary people, who exercise vigorously only occasionally. Examine microscope slides of normal and sickled red blood cells. Compare them with the normal red blood cells and the sickled cells shown in Fig. 14-6.

In sickle-cell anemia, the protein (globin) part of the normal hemoglobin molecule (Hb A) has been slightly altered by a mutation of a gene. When the abnormal genes (Hb^S) are inherited from both parents, the resulting homozygous condition (Hb^S Hb^S) causes a severe disability known as **sickle-cell disease.** When only one abnormal gene is inherited, resulting in the heterozygous condition (Hb^A Hb^S), the offspring is said to possess **sickle-cell trait,** a condition considerably less severe. Sickle trait also produces a resistance to malaria.

The two types of hemoglobin, Hb A and Hb S, can be distinguished from each other by electrophoresis, a procedure in which proteins migrate along an electrical gradient. The difference in the charges of each hemoglobin causes them to separate from each other (Fig. 14-7). This procedure can be used to distinguish between people whose hemoglobin is normal, those having sickle-cell disease, and those having sickle-cell trait.

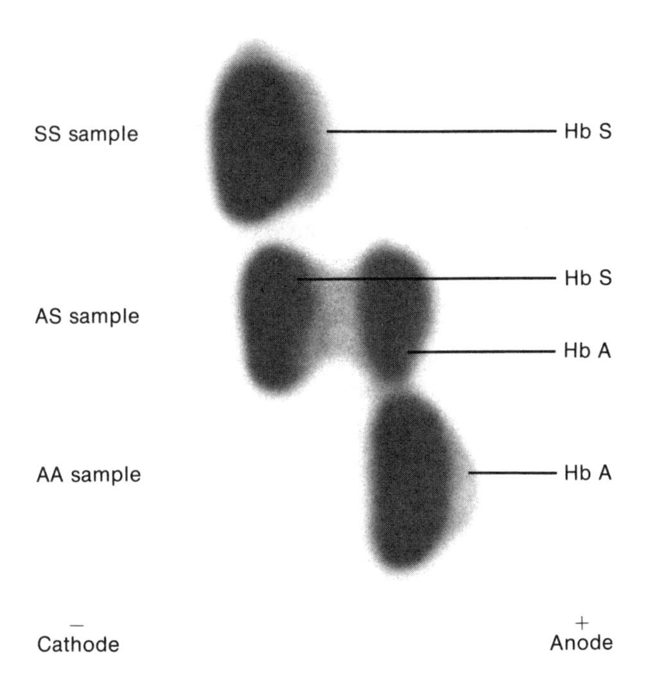

FIG. 14-7
An example of electrophoretic separation of hemoglobins: SS and AS contain abnormal hemoglobin S; AA is normal.

Normal ($Hb^A Hb^A$)	leucine—threonine—proline—**glutamic acid**—glutamic acid—lysine leucine—threonine—proline—**glutamic acid**—glutamic acid—lysine
Sickle disease ($Hb^S Hb^S$)	leucine—threonine—proline— **valine** —glutamic acid—lysine leucine—threonine—proline— **valine** —glutamic acid—lysine
Sickle trait ($Hb^A Hb^S$)	leucine—threonine—proline—**glutamic acid**—glutamic acid—lysine leucine—threonine—proline— **valine** —glutamic acid—lysine

FIG. 14-8
Sequences of amino acids in peptide of beta chains of hemoglobin A and hemoglobin S.

Recall that all proteins consist of chains of smaller units called amino acids. The hemoglobin molecule has four polypeptide chains: two beta chains and two alpha chains. Each alpha chain consists of 141 amino acids in a known specific sequence. Each beta chain has 146 amino acids. In sickle hemoglobin, glutamic acid is replaced by valine at one position in each chain (Fig. 14-8).

Until recently there was no effective treatment for this debilitating disease. However, intravenous treatment with glucose and urea, which blocks and reverses sickling, has shown some promise.

B. TRANSPORT IN PLANTS

Water is one of the basic raw materials of photosynthesis. It is the major component of plant tissues, making up 90% of the plant body. Water is the substance in which most materials enter and leave the cells of plants, and it is the solvent for the various biochemical reactions that occur in living cells.

The amount of water used by plants is far greater than that used by animals of comparable weight. The reason for this is that a large amount of the water used by an animal is recirculated in the form of blood plasma or tissue fluid. In plants, over 90% of the water taken in by the root system is evaporated into the air as water vapor. This process, which largely occurs through the leaves, is called **transpiration.** Consequently, plants not only have developed extensive and efficient transport systems but also have evolved numerous morphological adaptations to conserve water. In this exercise, you will become familiar with some of the factors that regulate transpiration.

1. Effect of Environmental Factors on Transpiration Rate

1. Cut a branch from a geranium or *Coleus* plant and insert it into a rubber stopper as shown in Fig. 14-9. Place the branch and stopper in a potometer flask that was previously filled with distilled water. Insert the stopper slowly to avoid creating bubbles. If this is done properly, water will be forced out of the end of the capillary tubing. However, when the pressure on the stopper is released, the fluid in the capillary tube will tend to recede. If this should occur, fill a 5-ml syringe with water and insert the needle into the rubber coupling between the flask and the capillary tubing. Slowly inject water until it comes out of the end of the tubing. If the apparatus has been properly set up, the water in the tube will begin to recede slowly. The rate at which the meniscus moves along the tubing is a measure of the rate of water uptake by the branch and may be used as a measure of the rate of transpiration. Determine the transpiration rate by recording the distance the meniscus moves each minute for a period of 10 minutes. If the meniscus goes beyond the graduated scale at the right, it may be returned to the zero mark by injecting water into the rubber coupling as previously described. Record your results in Table 14-3. Plot your data in Fig. 14-10.

2. Design and perform experiments to show the effects (if any) of light intensity, air movement, and humidity and the role of the leaves in the process of transpiration. Record the experimental conditions and your results in Table 14-3. Plot your data in Fig. 14-10. What would be the control for these various experiments?

What is the reason for covering the potometer flask with aluminum foil?

All aerial parts of plants may lose water by transpiration. In most herbaceous and woody plants, a

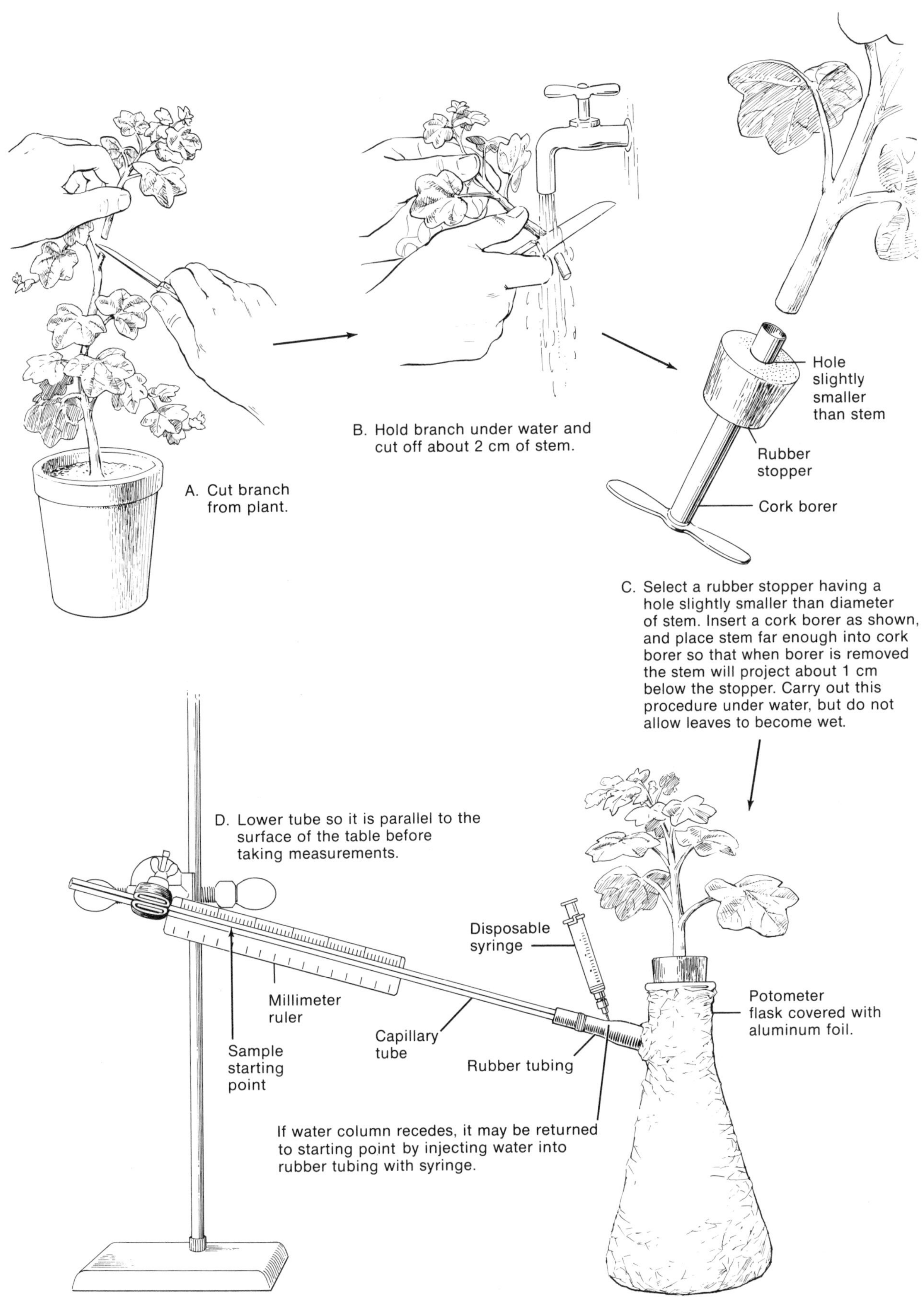

FIG. 14-9
Procedure for determining the rate of transpiration.

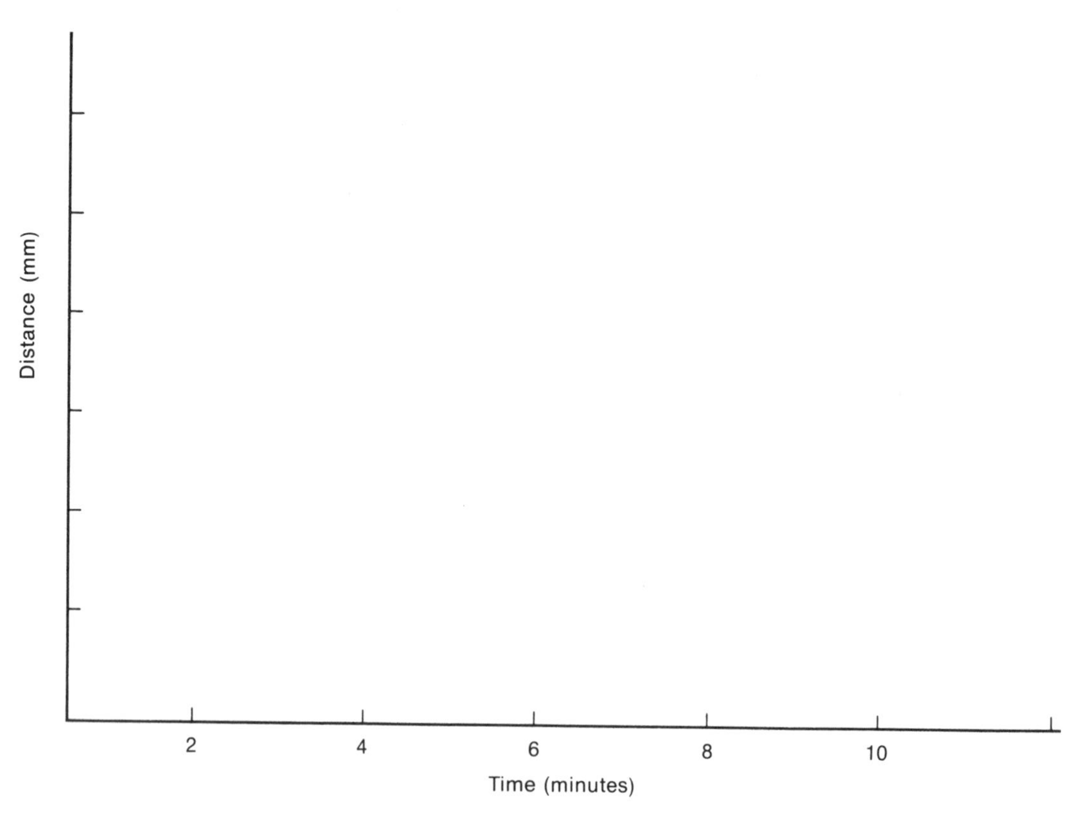

FIG. 14-10
Comparison of transpiration rates under various environmental conditions.

large proportion of the water is lost through openings called **stomates.** Discuss the effects of the various environmental factors as they affect transpiration in terms of their effects on the activity of the stomates.

Why is it important that ornamental evergreen plants be thoroughly watered before winter sets in?

TABLE 14-3
Potometer data.

Experimental conditions	Time (minutes)									
	1	2	3	4	5	6	7	8	9	10
	Distance (mm)									
Control										
Light intensity										
Air movement										
Humidity										
Effect of leaves										

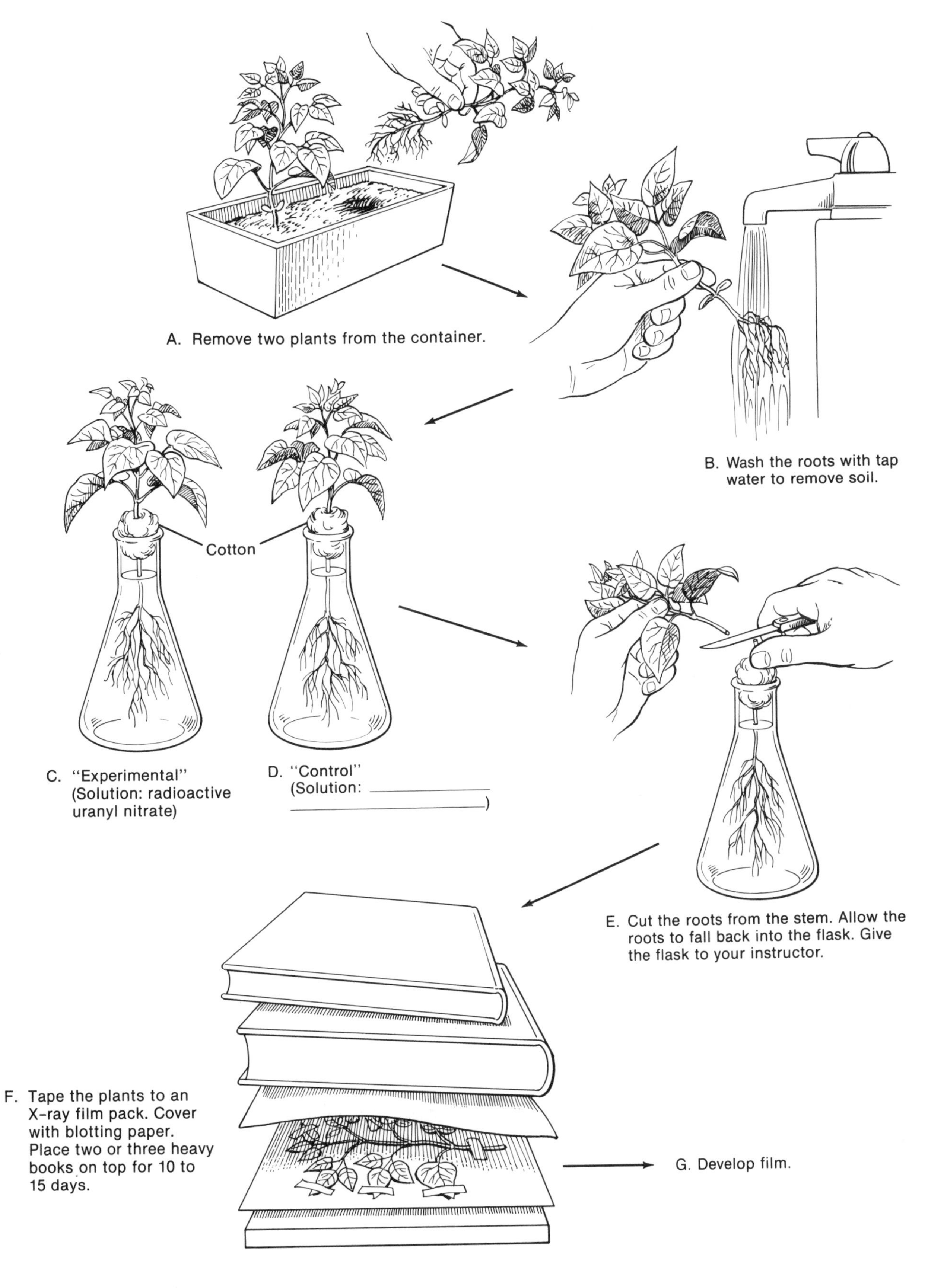

FIG. 14-11
Procedure for preparing an autoradiograph.

How does the process of transpiration benefit the plant?

____

____

____

____

2. Movement of Minerals

The study of relationships between structure and function in plants and animals has been aided by the use of radioactive substances called **radioisotopes** (see Appendix H). Radioisotopes can be used to trace the pathways of substances during their movement throughout an organism. When used in this way, they are called **tracers.** For example, biologists using radioactive carbon dioxide ($^{14}CO_2$) have been able to identify the various intermediate compounds with which the carbon becomes associated during the photosynthetic process.

In this part of the exercise, a radioactive compound, uranyl nitrate, will be used to determine if the minerals that are absorbed by roots are distributed throughout the plant.

Carefully remove two young bean seedlings from their container (Fig. 14-11A). Try not to damage the roots. Wash the roots with tap water (Fig. 14-11B). Place one of the plants in a flask containing a solution of uranyl nitrate so that the roots are completely submerged (Fig. 14-11C). Label this flask "experimental." The second plant should be set up as your control (Fig. 14-11D). What solution should be put in the control flask?

____

Cover both flasks with aluminum foil, and place them on a window ledge in the light for 1–2 hours. Next lift the plants out of the flasks just high enough so that the root can be cut off at the base of the stem. Allow the roots to fall back into the flasks (Fig. 14-11E). Return the containers to your instructor for disposal. Place each plant on a separate X-ray film pack. Spread the leaves so they do not overlap, and tape the plants to the film packs (Fig. 14-11F). Mark each film pack with your name and the date, and indicate whether it is the "experimental" or the "control" plant. Now place each film pack between two pieces of blotting paper (or paper towels). Add a heavy weight (two or three heavy books), and allow to remain this way for 10–15 days. At the end of this time (they may be left for a longer period), develop each film according to the directions given by your instructor. Discard the plants and blotting paper in the container provided for this purpose.

The radioactive uranyl nitrate is emitting high-energy particles. When these strike the emulsion of the film, latent images are formed, which, when treated with photographic developing solution, are reduced to metallic silver. This produces a black spot on the film. In this way you will be able to show the distribution of the radioactive uranyl nitrate in the plant. This procedure is called **autoradiography.** The picture you obtain is an **autoradiograph.**

REFERENCES

Allison, A. C. 1956. Sickle Cells and Evolution. *Scientific American* 195:87–94 (Offprint 1065). *Scientific American* Offprints are available from W. H. Freeman and Company, 41 Madison Avenue, New York 10010, and 20 Beaumont Street, Oxford OX1 2NQ, England. Please order by number.

Baker, D. A., 1978. *Transport Phenomena in Plants.* Wiley.

Biddulph, S., and O. Biddulph. 1959. The Circulatory System of Plants. *Scientific American* 200:44–49 (Offprint 53).

Guyton, A.C. 1981. *Testbook of Medical Physiology.* 6th ed. Saunders.

Hoar, W. S. 1983. *General and Comparative Physiology.* 3d ed. Prentice-Hall.

Jensen, W. A., and F. B. Salisbury. 1984. *Botany.* 2d ed. Wadsworth.

Lassen, N. A., D. H. Ingvar, and E. Skinhøj. 1978. Brain Function and Blood Flow. *Scientific American* 239:62–71 (Offprint 1410).

Lerner, I. M., and W. J. Libby. 1976. *Heredity, Evolution, and Society.* 2d ed. W. H. Freeman and Company.

Macey, R. I. 1975. *Human Physiology.* 2d ed. Prentice-Hall.

Perutz, M. F. 1978. Hemoglobin Structure and Respiratory Transport. *Scientific American* 239:92–125 (Offprint 1413).

Raven, P. H., R. F. Evert, and H. Curtis. 1981. *Biology of Plants.* 3d ed. Worth.

Zimmerman, M. H. 1963. How Sap Moves in Trees. *Scientific American* 208:132–142 (Offprint 154).

EXERCISE 15

Biological Coordination in Plants

As organisms become more complex and individual cells become more specialized, the various activities within the organism must be coordinated. Furthermore, organisms need effective ways to respond to physical and chemical changes in their environment. Thus, all organisms must become adapted to receive various internal and external stimuli and to respond to these stimuli in a coordinated manner.

Although higher plants do not have nervous systems, the coordination of many activities is necessary if the plant is to grow and develop. Two types of mechanisms bring about coordination in plants. One, consisting of a system of chemical messengers, directs the cells of the plant to carry out different functions. This system includes the various plant hormones and other growth regulators. A second coordinating system consists of various physical forces; for example, radiant energy, electrical gradients, gas exchange gradients, and metabolic or pressure differences throughout the plant. The endproduct of these controls is a highly complex plant body with localized areas of growth, special regions of photosynthetic activity, and the ability to translocate and accumulate the necessary raw materials for reproduction, growth, and differentiation to assure the continuation of the species.

A. PHOTOTAXIS

The ability to respond to a stimulus (a chemical or physical change in the environment) is a striking characteristic of living organisms. The response of the organism to any stimulus serves to coordinate the activities of the organism to maintain its equilibrium with the environment. In this exercise, you will examine the effects of various wavelengths of light in orientating movements of the green alga, *Euglena.*

Following the directions outlined in Fig. 15-1, expose a tube of *Euglena* to various wavelengths of light. The *Euglena* will tend to concentrate and adhere to the sides of the tube, in the area of the wavelength to which they are most sensitive. At the conclusion of the exercise, carefully decant off the

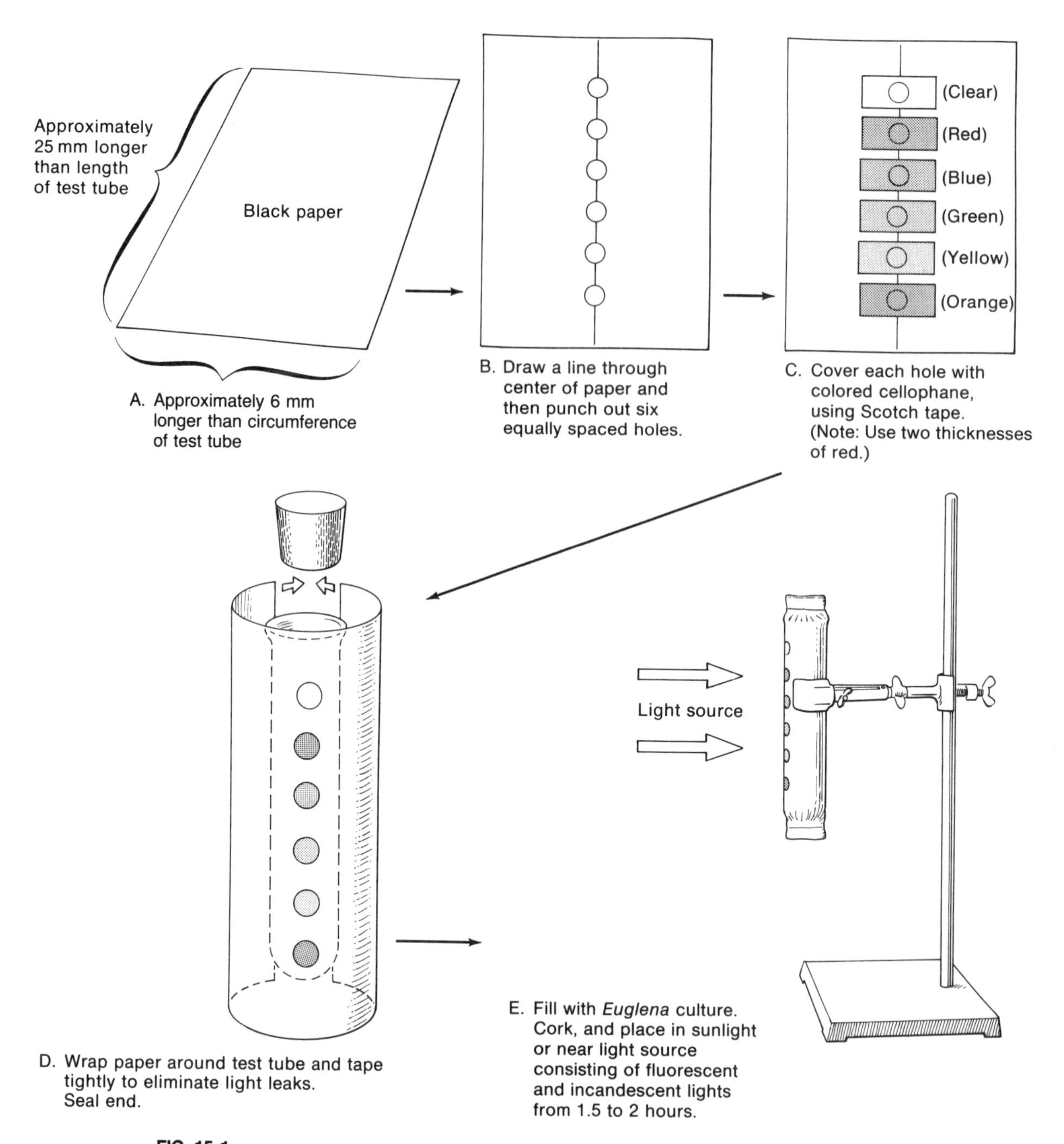

FIG. 15-1
Procedure for studying the effect of different wavelengths of light on the movements of *Euglena*.

liquid from the tube, and remove the black paper sleeve. Record your observations in Fig. 15-2 by shading the appropriate circles to represent the density of *Euglena* found associated with each colored filter. What wavelength of light attracted the heaviest concentration of *Euglena?*

____

Prepare a wet mount of *Euglena,* and locate a red-pigmented "eye-spot" near the flagella. Suggest a role for this structure in the response just studied.

____

____

How does the response observed help *Euglena* to survive?

B. GEOTROPISM

The effect of gravity on the direction of growth in various plant parts is called **geotropism.** Stems generally grow in a direction opposite the force of gravity. Is this negative or positive geotropism?

Characterize the response of roots in regard to their reaction to gravity.

In the following parts of this exercise, special attention will be given to the geotropic response of *Coleus* or *Iresine* shoots. For convenience, each team will be assigned one aspect of the problem. Because the response of the plant usually requires several hours, set the experiment aside for examination during the next laboratory period. In Table 15-1, record the results you expect and compare these with the actual results.

Filter

Clear

Red

Blue

Green

Yellow

Orange

Record density of *Euglena* at each wavelength by shading as indicated below:

None Moderate

Light Heavy

FIG. 15-2
Effect of wavelengths of light on the orienting movements of *Euglena.*

TABLE 15-1
Data on the effect of gravity on stem orientation.

Results	Stem position		
	Vertical	Horizontal	Inverted
Expected			
Observed			
Conclusions:			

1. What is the effect of gravity on stem orientation? Remove three *Coleus* or *Iresine* branches, keeping their surfaces moist during the cutting operation. Set them up as indicated in Fig. 15-3. What is the "control" in this experiment?

____

____

____

2. Does defoliation affect the geotropic response of *Coleus or Iresine* shoots? Design and perform an experiment to answer this question. Complete Table 15-2.

3. Does the shoot apex exert any effect on the geotropic response of *Coleus* or *Iresine?* Design and perform an experiment to answer this question. Complete Table 15-3.

4. Will replacing the shoot apex with auxin have any effect on the geotropic response of *Coleus* or *Iresine?* Design and perform an experiment to answer this question. Complete Table 15-4.

C. CHEMOTROPISM

The response of nonmotile cells to chemical stimulation, in which growth occurs toward or away from the chemical stimulus, is called **chemotropism.** In this part of the exercise, you will study the chemotropic response of lily pollen tubes to various materials (pistils, auxin, and gibberellic acid). Petri dishes containing a nutrient medium will be used in these experiments. Your instructor will give you the pollen

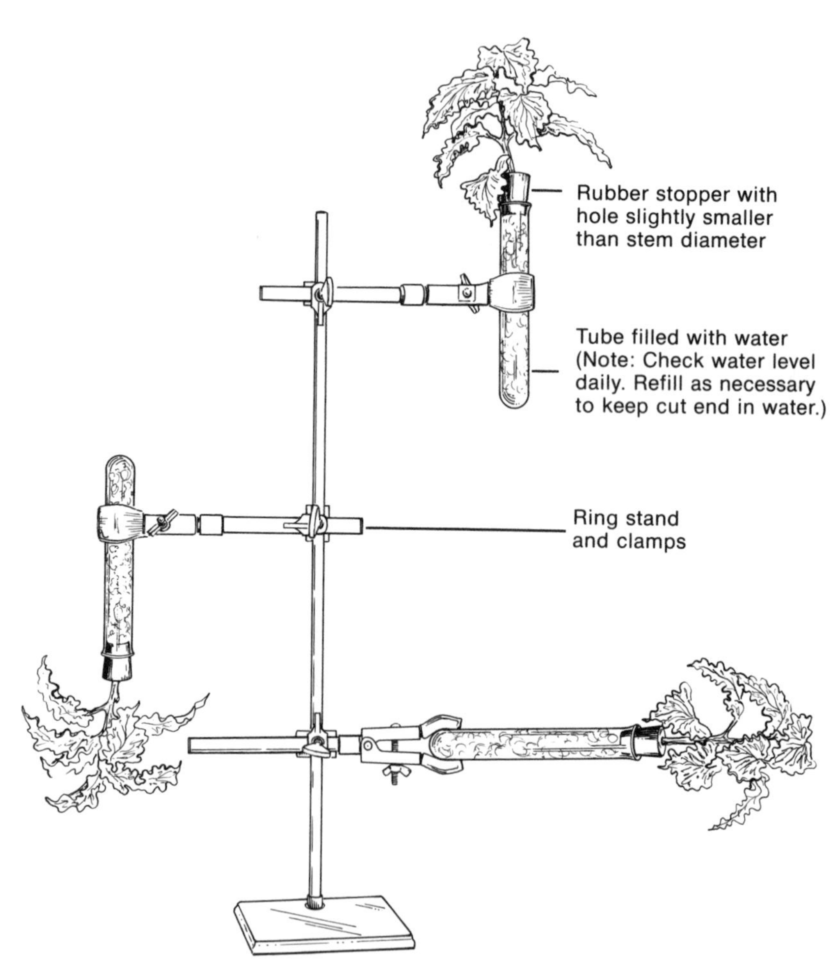

FIG. 15-3
Experimental setup to show effect of gravity on growth.

TABLE 15-2
Data on the effect of defoliation on the geotropic response.

Procedure: 1. Experimental treatment 2. Control treatment
Expected results:
Observed results:
Conclusions:

TABLE 15-3
Data on the effect of shoot apex on the geotropic response.

Procedure: 1. Experimental treatment   2. Control treatment
Expected results:
Observed results:
Conclusions:

TABLE 15-4
Data on the effect of replacing the shoot apex with auxin on the geotropic response.

Procedure: 1. Experimental treatment 2. Control treatment
Expected results:
Observed results:
Conclusions:

and the materials to be tested for chemotropic activity. Place a small piece of material to be tested on the surface of the nutrient medium and position the pollen around it, using a camel-hair brush and a binocular microscope. For best results, distribute the pollen in small clumps within 2 or 3 mm of the material to be tested. To test the chemotropic activity of liquids, cut a well in the medium with a cork borer, place the liquid in the well, and position the pollen around the rim.

1. Is chemotropic activity exhibited by the lily pistil? Obtain petri dishes, pollen, and lily pistils from your instructor. Determine if chemotropic activity is exhibited by all parts of the pistil; design and perform an experiment to answer this question. Test for chemotropic activity as described. What controls should be used?

What criteria would you use to determine if positive or negative chemotropism is indicated?

Complete Table 15-5.

2. Is the chemotropic activity of pistils altered by heat? Design and perform an experiment to answer this question. Complete Table 15-6.

3. Is chemotropic activity exhibited by auxin or gibberellic acid? Design and perform an experiment to test the chemotropic activity of each of these known plant growth substances. It will be necessary to know the solvents for each substance so that adequate controls can be established.

If one (or both) of these substances shows chemotropic activity, determine its optimum concentration by testing with a series of concentrations—for example, 1, 10, 50, and 100 parts per million (ppm). Complete Table 15-7.

D. REGULATION OF ENZYME ACTIVITY BY LIGHT

Numerous developmental phenomena in plants are regulated by light. Among these are the germination of seeds, the appearance of flowers, and the ripening of fruit. However, these light-controlled processes are separate from photosynthesis; instead of chlorophyll, a pigment called **phytochrome** perceives the light stimulus and initiates the developmental process. An interesting phytochrome-mediated event is the appearance of the enzyme inorganic pyrophosphatase in corn leaves when the plant's phytochrome system is activated by light.

This exercise uses two groups of 6-day-old corn seedlings. The "control" group has been kept in darkness since the seeds germinated. The "experimental" group has been kept in darkness for 4 days, exposed to incandescent light for 10 minutes, then returned to darkness. To assay whether or not pyrophosphatase is present, you follow the procedure given below, in which pyrophosphate is hydrolyzed to orthophosphate through the action of pyrophosphatase.

$$\underset{\text{Pyrophosphate}}{O{=}\overset{\overset{\displaystyle O^-}{|}}{\underset{\underset{\displaystyle O^-}{|}}{P}}{-}O{-}\overset{\overset{\displaystyle O^-}{|}}{\underset{\underset{\displaystyle O^-}{|}}{P}}{=}O} \xrightarrow[\text{pyrophosphatase}]{\text{water: } H \;\; OH} \underset{\text{Orthophosphate}}{O{=}\overset{\overset{\displaystyle O^-}{|}}{\underset{\underset{\displaystyle O^-}{|}}{P}}{-}O\,H + OH{-}\overset{\overset{\displaystyle O^-}{|}}{\underset{\underset{\displaystyle O^-}{|}}{P}}{=}O}$$

The appearance of orthaphosphate is then measured spectrophotometrically.

1. Remove the leaves from five seedlings in each of the "control" and "experimental" groups.

2. Grind each sample of leaves in a mortar and pestle using approximately 1 g silica sand and 10 ml 0.05 *M* Tris buffer at pH 7.5.

3. Using filter paper, filter each sample into separate test tubes, label the tubes "control" or "experimental" and place them into a beaker containing ice.

4. Number eight test tubes 1–8 and set them up as indicated in Table 15-8. Mix the contents of the

TABLE 15-5
Data on the chemotropic activity of lily pistils.

Procedure: 1. Experimental treatment 2. Control treatment
Results:
Conclusions:

TABLE 15-6
Data on the effect of heat on the chemotropic activity of lily pistils.

Procedure: 1. Experimental treatment 2. Control treatment
Results:
Conclusions:

TABLE 15-7
Data on the chemotropic activity of auxin and gibberellic acid.

Procedure: 1. Experimental treatment 2. Control treatment
Results:
Conclusions:

TABLE 15-8
Assay for pyrophosphatase.

Solution added	Tube number and volume of solution added (ml)							
	1	2	3	4	5	6	7	8
0.001 *M* pyrophosphate	0.6	0.6	0.6	—	0.6	0.6	0.6	—
0.5 *M* $MgC1_2$	0.2	0.2	0.2	0.2	0.2	0.2	0.2	0.2
Control sample:								
Undiluted	0.2	—	—	0.2	—	—	—	—
Diluted 1:10*	—	0.2	—	—	—	—	—	—
Diluted 1:50	—	—	0.2	—	—	—	—	—
Experimental sample:								
Undiluted	—	—	—	—	0.2	—	—	0.2
Diluted 1:10	—	—	—	—	—	0.2	—	—
Diluted 1:50	—	—	—	—	—	—	0.2	—

* Dilute with Tris buffer solution.

tubes thoroughly and allow each tube to incubate for 2 minutes.

5. To each tube add 2.5 ml of "Reagent S," which stops the chemical reaction in the tube. Allow the tubes to stand for 10 minutes.

6. Blank the spectrophotometer at 620 nm and then record the absorbance for each tube in Table 15-9. (*Note:* Refer to Appendix B for the use of the Bausch and Lomb Spectronic 20 colorimeter and Appendix C for the principles of spectrophotometry.")

Discuss the evidence for phytochrome-mediated synthesis of inorganic pyrophosphatase.

____

____

____

____

____

____

____

The phytochrome system is affected by red and far-red light. Briefly discuss the effect that each of these wavelengths has on phytochrome.

____

____

____

____

TABLE 15-9
Data for the pyrophosphatase assay.

Tube number and contents	Absorbance (620 nm)
1. 0.001 *M* pyrophosphate/0.5 *M* $MgC1_2$: undiluted control sample	______
2. 0.001 *M* pyrophosphate/0.5 *M* $MgC1_2$: 1:10 dilution control sample	______
3. 0.001 *M* pyrophosphate/1.5 *M* $MgC1_2$: 1:50 dilution control sample	______
4. 0.5 *M* $MgC1_2$: undiluted control sample	______
5. 0.001 *M* pyrophosphate/0.5 *M* $MgC1_2$: undiluted experimental sample	______
6. 0.001 *M* pyrophosphate/0.5 *M* $MgC1_2$: 1:10 dilution experimental sample	______
7. 0.001 *M* pyrophosphate/0.5 *M* $MgC1_2$: 1:50 dilution experimental sample	______
8. 0.5 *M* $MgC1_2$: undiluted experimental sample	______

If indeed red and far-red light are necessary to the phytochrome system, how can incandescent light be used in this exercise to trigger the production of pyrophosphatase by the activation of phytochrome? Could fluorescent light have been used?

What "control," other than leaving one group of plants in the dark, could you use to provide additional support for the hypothesis that the pigment phytochrome and not chlorophyll is involved in the synthesis and appearance of the enzyme pyrophosphatase in leaves exposed to light? Explain.

REFERENCES

Baron, W. M. M. 1979. *Organization in Plants.* 3d ed. Wiley.

Butler, L., and V. Bennett. 1969. Phytochrome Control of Maize Leaf Inorganic Pyrophosphatase. *Plant Physiology* 44:1285.

Galston, A. W. 1980. *The Life of the Green Plant.* 3d ed. Prentice-Hall.

Galston, A. W., and P. J. Davies. 1970. *Control Mechanisms in Plant Development.* Prentice-Hall.

Jensen, W. A., and F. B. Salisbury. 1984. *Botany.* 2d ed. Saunders.

Machlis, L., and J. G. Torrey. 1956. *Plants in Action: A Laboratory Manual of Plant Physiology.* W. H. Freeman and Company.

Salisbury, F. B. 1957. Plant Growth Substances. *Scientific American* 196:125–134 (Offprint 110). *Scientific American* Offprints are available from W. H. Freeman and Company, 41 Madison Avenue, New York 10010, and 20 Beaumont Street, Oxford OX1 2NQ, England. Please order by number.

Van Overbeek, J. 1968. The Control of Plant Growth. *Scientific American* 219:75–81 (Offprint 1111).

EXERCISE 16

Biological Coordination in Animals

An entire animal has functional capabilities that exceed those of its constituent cells, tissues, organs, and organ systems. This observation is the basis for the generalization that "the organism is more than a simple sum of its parts." Each subunit of the body makes a specific contribution, but the proper functioning of the whole organism depends on the integration of a wide spectrum of individual processes into a coherent unit. Such integration in turn depends on the exchange of information among parts of the body, and the nervous system and endocrine organs play a central role in this vital process. Although structurally quite different, they exert their effects in an essentially similar fashion—through the production and release of molecules that modify cellular activity. The effects produced by the nervous system are normally quite localized in that the materials released at nerve endings (such as acetylcholine or norepinephrine) act on specific targets: other nerve cells, muscle cells, and secretory cells. The effects of the endocrine glands, on the other hand, are more generalized in that their secretions (hormones) are distributed throughout the body by the circulatory system. Specific hormonal effects are achieved by the varying sensitivity of different types of cells to particular hormones. In a general sense, the nervous system contributes to organismic coordination by providing rapid and highly specific but short-term controls, whereas long-term modifications of physiological activity are normally the province of the endocrine system.

The nervous system is divided into the **central nervous system,** composed of the brain and spinal cord, the **peripheral nervous system,** consisting of the cranial and spinal nerves (those nerves originating in the brain and spinal cord, respectively), and the **autonomic nervous system,** which regulates those parts of the body under involuntary control (i.e., intestines and blood vessels). In this exercise, you will become familiar with the anatomy and physiology of the nervous system of the frog.

A. THE BRAIN

Carefully remove the skin from the top of the head of a frog. Then remove any muscles present on the posterior part of the skull. The brain occupies that part of the skull extending from between the eyes to the foramen magnum. Using a scalpel (or fine scissors) and forceps, carefully remove the thin bone on the roof and lateral margins of the skull and expose the brain and anterior part of the spinal cord. Take care in doing this, because brain tissue is fragile and easily damaged.

Note that the brain is covered by a layer of membranous tissue, called the **meninges.** The outer nonvascular meningeal membrane is called the **dura mater;** the inner layer, called the **pia mater,** is highly vascularized and is tightly bound to the surface of the brain. Between the two is the **arachnoid membrane.** In human beings, inflammation of the meninges is called **meningitis** and can affect the brain or the spinal cord, or both. It can be caused by bacteria, fungi, or viruses. One of the more common agents of this disease is the bacterium *Neisseria meningitidis,* frequently called *meningococcus.* Carefully remove these meninges to obtain an unobstructed view of the surface features of the brain.

1. Features of the Dorsal Surface

The brain is composed of five regions. In the frog, these regions are in a linear arrangement. The most anterior region, the **telencephalon,** is subdivided into the **olfactory bulbs** and the **cerebral hemispheres** (Fig. 16-1A). Locate the **olfactory nerves,** which lead from the external nares to the olfactory bulbs. These nerves carry odor stimuli to the olfactory bulbs and cerebral hemispheres. The cerebral hemispheres are predominantly olfactory in the frog, but in human beings they involve reasoning, memory, foresight, emotion, personality, intelligence, and speech mechanisms.

Immediately posterior to the cerebral hemispheres is the **diencephalon,** the second region in the linear arrangement of the brain. Locate the **pineal gland** protruding from the surface. The function of the pineal gland is unknown, although it does contain light-sensitive and secretory cells.

Just anterior to the pineal gland, the roof of the diencephalon is quite thin and vascularized. Careful removal of this membranous roof reveals a cavity in the brain, the **third ventricle.** Just lateral to the third ventricle, the walls of the diencephalon make up the **thalamus.**

The third region, the **mesencephalon** (or midbrain), is posterior to the diencephalon and is chiefly composed of the **optic lobes.** These structures receive sensory impulses from the eyes, as well as from many other sense organs. It is thought that the optic lobes carry out many of the functions associated with the cerebrum in higher vertebrates.

The **cerebellum,** located just behind the optic lobes, constitutes the major part of the fourth region of the brain, the **metencephalon.** In the frog, as in all vertebrates, this region functions in the coordination of such motor activities as walking, twisting, and kicking.

The fifth region, the **myelencephalon** consists of the **medulla oblongata,** which continues into the spinal cord. The medulla contains control centers regulating breathing, heart rate, visceral activity, and swallowing. The roof of the medulla covers the fourth ventricle of the brain. If you carefully remove the membranous roof, you should see the **choroid plexus,** a series of vascular folds of tissue, attached to its inner surface.

2. Features of the Ventral Surface

The brain must be removed from the cranium in order to view the various regions and attachments of nerves on the ventral and lateral surfaces. To do this, first place the frog in a tray of water (to minimize damage) and then sever the olfactory nerves and the medulla oblongata, leaving as much of these structures attached to the brain as possible. Then, as you carefully lift the brain, cut the various cranial nerves (see Fig. 16-1B) extending from the ventral surface.

As you lift out the brain, look for a small bulbous structure projecting from the ventral surface in the region of the optic lobes. This structure, called the **hypophysis** (pituitary gland), lies in a small depression in the bone on the floor of the skull. You may have to chip some bone away to remove this structure.

With the aid of Fig. 16-1, locate the optic nerves and the hypothalamus. The **optic nerves** cross as they reach the brain and form the **optic chiasma** from which they continue as optic tracts to the optic lobes. Thus the optic nerve from the right eye stimulates the sight centers of the left side of the brain and vice versa.

The **hypothalamus** makes up the floor of the diencephalon just posterior to the optic chiasma. Extending from the hypothalamus is the hypophysis (if you do not find it on the brain, it may have broken off and still be in the bone pocket). The hypothalamus in the frog and other vertebrates regulates the **autonomic nervous system,** which controls such activities as blood pressure, salt balance, water balance, peristaltic movements of the intestines, and secretions of the pituitary gland.

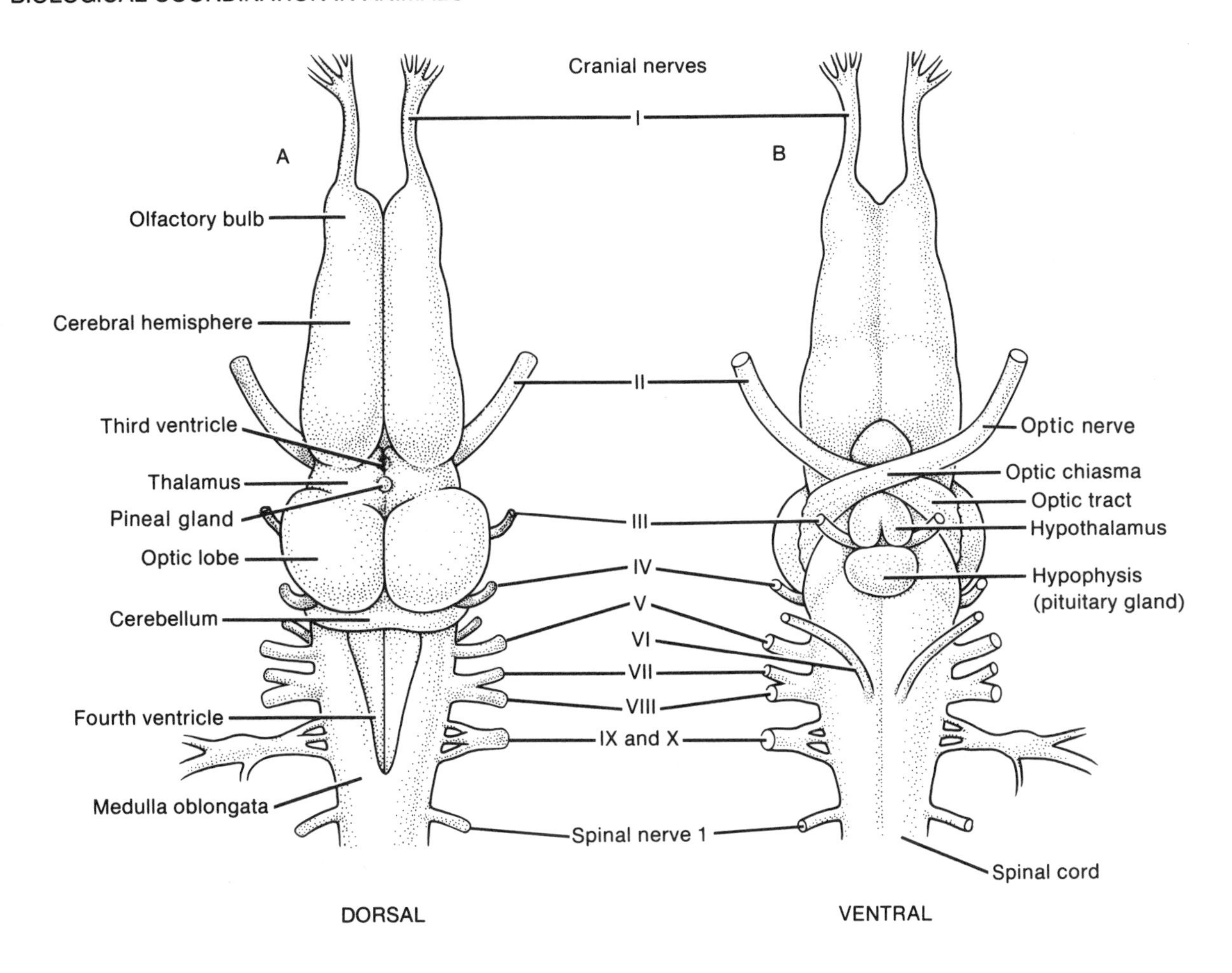

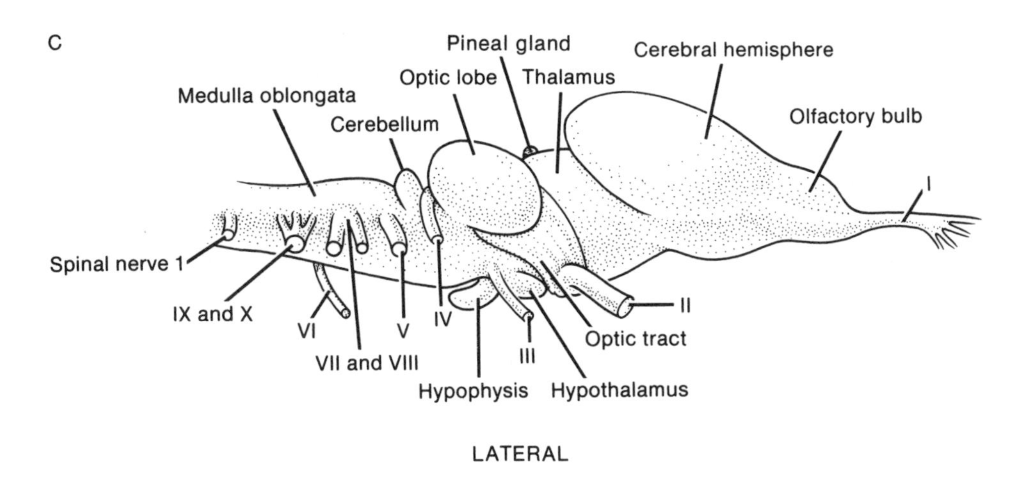

FIG. 16-1
The frog brain and the stumps of the cranial nerves: (A) dorsal, (B) ventral, and (C) lateral views.

The pituitary gland, often called the *master gland,* secretes hormones that affect growth, activities of the adrenal and thyroid glands, and the production of hormones in the testes and ovaries. In humans, the pituitary has posterior **(neurohypophysis)** and anterior **(adenohypophysis)** lobes. A description of the hormones secreted by these lobes and their functions in humans follows.

a. Hormones of the Anterior Lobe

In human beings, the adenohypophysis produces several hormones, called **tropic** hormones, because they act largely on other endocrine glands.

Somatotropin, or **growth hormone,** promotes growth (particularly of bones and muscles) by increasing protein synthesis until puberty. *Hypersecretion* of somatotropin (due to a pituitary tumor, for example) results in giantism in children and acromegaly in adults. *Hyposecretion* results in dwarfism. Secretion of growth hormone is regulated by a somatotropin-releasing factor (SRF) secreted by the hypothalamus.

Adrenocorticotropic hormone (ACTH) stimulates the outer cortical part of the adrenal gland to produce its hormones (cortisone, aldosterone, and others). Production of ACTH is curtailed when the levels of adrenal cortical hormones are adequate. Conversely, when these levels decrease, the pituitary is stimulated to produce ACTH.

Thyroid-stimulating hormone (TSH, or **thyrotropin)** stimulates the thyroid gland to produce thyroxine and triiodothyroxine, which regulate the rate of metabolism, the level of calcium in the blood, and the level of phosphate in bone.

Follicle-stimulating hormone (FSH) stimulates the growth of the ovarian follicles in women and the seminiferous tubules in the testes during spermatogenesis in men. Pituitary hormones that affect the gonads are called **gonadotropic hormones.**

After FSH has acted, **luteinizing hormone (LH)** stimulates the development of ovarian follicles and the corpus luteum, ovulation, and the production of estrogens (which are important in the development of secondary sex characteristics and regulation of the menstrual cycle). The hypothalamus secretes a luteinizing hormone-releasing factor (LRF), which is thought to regulate the production of LH.

The **interstitial cell-stimulating hormone (ICSH)** is produced only in men. This hormone promotes the secretion of **testosterone** by the interstitial cells of the testes. Testosterone affects secondary sex characteristics of men.

Lactogenic hormone (prolactin) promotes the production of milk by the mammary glands. There is evidence that the hypothalamus produces a prolactin-inhibiting factor (PIF) that suppresses prolactin production until it is needed.

Melanocyte-stimulating hormone (MSH) stimulates the deposit of the pigment melanin in the skin after exposure to sunlight. A condition similar to melanin pigmentation is produced if the adrenal cortex atrophies, as in Addison's disease.

b. Hormones of the Posterior Lobe

Oxytocin causes the smooth muscles of the uterus to contract during childbirth and stimulates smooth muscles of the mammary glands during nursing. This hormone is secreted by the hypothalamus and then moves into the pituitary before it is released.

The **antidiuretic hormone (ADH,** or **vasopressin)** promotes reabsorption of water by the kidney. Like oxytocin, ADH is secreted by the hypothalamus when osmoreceptors within that gland detect an increase in the osmotic pressure of the blood following a loss of water. ADH thus conserves water by decreasing urine output.

Interestingly, the vertebrate brain has also been shown to produce morphinelike substances called **endorphins.** These substances bind to specific receptors in the brain, as do other opiates. When this happens, nerve impulses decrease in the particular circuit mediated by the receptors. This is of some significance because these endorphins, and the receptor sites, are believed to act as natural **analgesics** (pain relievers).

B. THE CRANIAL NERVES

Cranial nerves are those nerves that are attached to the brain. Using Fig. 16-1, locate as many of their stumps as you can. Although there are 10 pairs of cranial nerves in the frog (identified by Roman numerals in Fig. 16-1), some are very small and may not be found. A description of each nerve follows:

I **Olfactory.** Sensory in function. Has nerve endings in the lining of the nasal cavity. Sensation is interpreted as odor in the cerebral hemispheres.

II **Optic.** Sensory in function. Has nerve endings in the retina of the eye. Stimulates vision centers in optic lobes.

III **Oculomotor.** Primarily a motor nerve that innervates the muscles that move the eyeballs. A branch of this nerve regulates the ciliary body of the eye, which in turn regulates the size of the pupil. There are also a few sensory fibers in the nerve that provide information on the intensity of muscle contractions during eye movement.

IV **Trochlear.** A motor nerve that sends impulses to the muscles of the eyes.

V **Trigeminal.** Motor and sensory in function. Receives sensory stimuli from the skin of the head and the lining of the mouth. Carries motor stimuli to the region of the jaw.

VI **Abducens.** Primarily a motor nerve that innervates the muscles of the eyes.

VII **Facial.** Motor and sensory in function. It carries sensory stimuli from the taste buds, lining of the mouth cavity, and the tongue. Motor impulses control the hyoid apparatus, a Y-shaped skeletal component in the region of the larynx and tongue. The hyoid apparatus anchors the tongue, provides for the attachment of some muscles of the larynx, and is the attachment for muscles used in swallowing. In human beings, this nerve is sometimes called the "nerve of facial expression" because it supplies motor fibers to all muscles that affect expression.

VIII **Acoustic.** Sensory in function. Sometimes called *auditory nerve,* it functions in equilibrium and hearing.

IX **Glossopharyngeal.** Sensory and motor in function. Sensory fibers return stimuli from the tongue, taste buds, and lining of the buccal (cheek) cavity. Motor fibers innervate some throat muscles.

X **Vagus.** Has both sensory and motor fibers affecting the heart, lungs, laryngotracheal chamber, bronchi, esophagus, stomach, intestine, and many of the abdominal organs.

Human beings have two pairs of nerves in addition to the aforementioned ones: spinal accessory nerves (XI) and hypoglossal cranial nerves (XII). One branch of the vagus nerve in the frog has the function of the human spinal accessory nerve; it supplies motor fibers to the shoulder muscles. Part of the olfactory nerve in the frog serves the same function as that of the human hypoglossal nerve; it supplies motor fibers to the tongue.

To help you remember the order in which the 10 cranial nerves in the frog, plus the 2 additional nerves in humans, are arranged, you might want to use the following rhyme, in which the first letter of each word corresponds to the first letter of the name of a nerve:

On Old Olympus' Towering Top,
A Finn And German Vended Some Hops.

C. THE SPINAL CORD AND SPINAL NERVES

The spinal cord continues posteriorly from the brain into the vertebral column (Fig. 16-2). To expose the spinal cord, remove the skin and muscle from the dorsal surface of the frog. Then insert fine-pointed scissors into the vertebral column, lateral to the midline, and carefully cut through the neural arch of the first vertebra. Repeat this on the other side, and then lift off the top of the vertebra. Repeat this procedure for all remaining vertebrae (Fig. 16-2). Locate the various spinal nerves along the length of the spinal cord. Note that they pass through an opening in the vertebral column, called the **intervertebral foramina.** Observe that the spinal cord is uniformly cylindrical except where the nerves to the fore- and hindlimbs emerge. There the spinal cord enlarges into the **brachial** (fore) and **lumbar** (hind) enlargements. These swellings are to accommodate the large numbers of neurons that send branches to the muscles of the limbs.

If you carefully cut away the bone and muscle surrounding one of the nerves, you will see that it, like the others, branches as it enters the spinal cord. The upper branch, usually swollen, is called the *dorsal root* and contains neurons of sensory nerves. Motor neurons leave the spinal cord through the ventral root. The frog has 10 pairs of spinal nerves. Using a dissecting microscope, find one of each pair. A description of each nerve follows:

Spinal nerve 1. Usually destroyed during dissection. Emerges between the first and second vertebra. Innervates the muscles in the buccopharyngeal cavity.

Spinal nerve 2. One of the largest nerves. With branches of the first and third spinal nerves, it forms the **brachial plexus.** Supplies muscles of the shoulder and arm.

Spinal nerves 3, 4, 5, and **6.** Innervate the muscles and skin of the abdomen.

Spinal nerves 7, 8, and **9.** Together, these nerves form the **sciatic plexus,** sending branches to the skin and muscles of the posterior part of the abdomen, the pelvic region, and the muscles of the hind leg. The largest nerve, the sciatic nerve, runs parallel to the sciatic artery and vein into the thigh.

Spinal nerve 10. A very small nerve that leaves the urostyle through a small opening. It supplies the urinary bladder and cloaca.

D. SPINAL CORD ANATOMY

Study a stained slide of a cross section of mammalian spinal cord passing through the dorsal root ganglia. You will notice that it is not a perfect circle but has notches in its circumference (Fig. 16-3). The deepest notch is the **ventral fissure.** Rotate your slide so that this fissure is toward you.

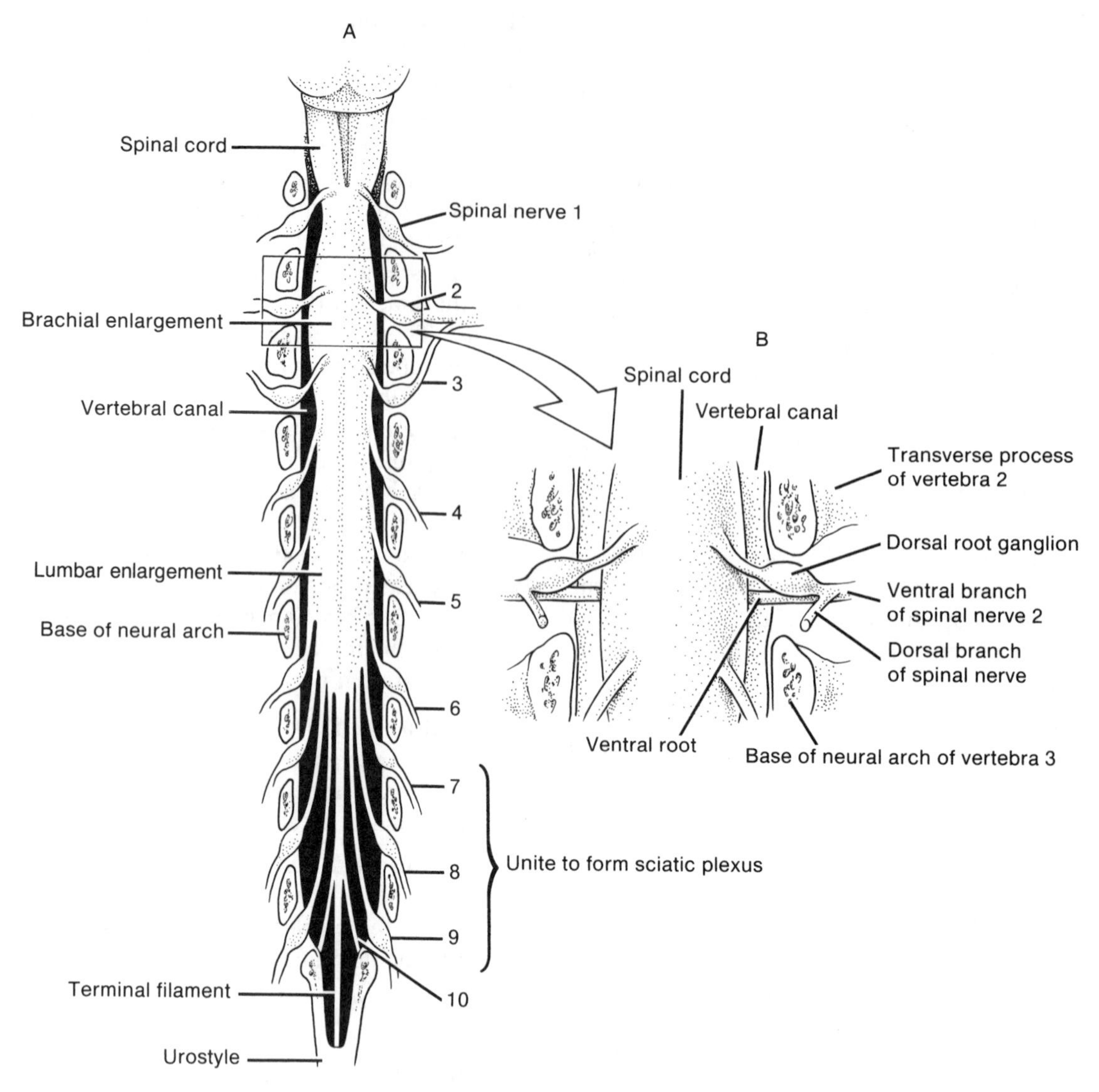

FIG. 16-2
(A) Dorsal view of the spinal cord and the origins of the 10 spinal nerves. (B) Enlargement of the origin of the second spinal nerve.

The brain and spinal cord are surrounded by membranous coverings called **meninges.** there are three such membranes in mammals: an outer dura mater next to the skull and vertebral bones, an inner pia mater fitting tightly to the brain and spinal cord, and a median arachnoid membrane.

In the center of the cord, locate a small, clear space, the **central canal.** This canal runs the length of the cord and, at its anterior end, is continuous with the fourth ventricle of the medulla oblongata. It is filled with **cerebrospinal fluid,** as are all the cavities within and around the central nervous system. Looking at the cross section of the cord, you will notice an H-shaped area of **gray matter** occupying much of the central region. The gray matter contains many cell bodies and looks gray in fresh tissue. In the outer area of the cord is the more uniformly colored **white matter,** containing fewer nuclei. The white matter is made up mostly of fibers that appear white in fresh tissue, owing to the glistening white sheaths around them. In your slide, the white matter has a granular appearance because almost all of its fibers run lengthwise and thus are cut transversely in the slide, each fiber end appearing as a dot.

The gray matter shows some differentiation. In the lower corners of the H-shaped area are extralarge cells—the cell bodies of **efferent (motor) neurons.** Because their processes (cytoplasmic extensions) are normally very irregular, they are seldom visible in their entirety in a thin section. The axons of these

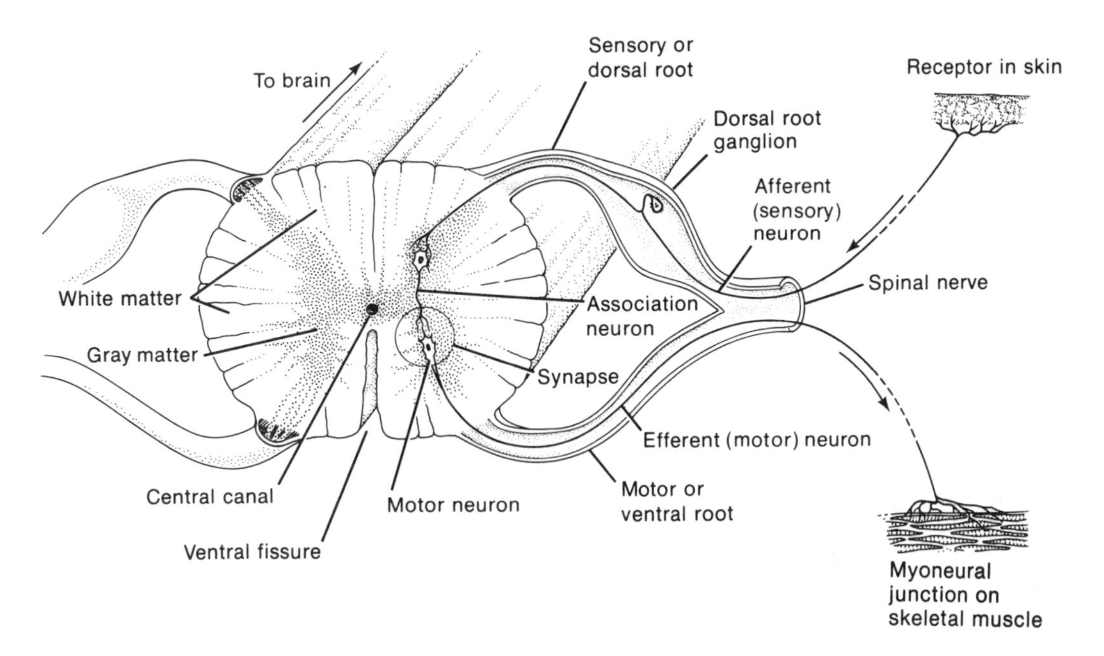

FIG. 16-3
Diagrammatic representation of the relationship of spinal cord, spinal nerves, and the simple reflex arc.

large neurons run out through the ventral root into the spinal nerve.

Besides the large motor neurons, numerous other cells will be found. The smallest belong to a special type of supporting cell characteristic of the central nervous system; these cells are not neurons. The intermediate-sized cells are called **association (adjustor) neurons** because they take impulses from the incoming sensory fibers and relay them to other adjustors, or up to the brain by long fibers in the cord, or directly to the motor neurons. Because the axons of adjustor neurons often branch, they can transmit an impulse to a number of other cells and these, in turn, to several more. Thus, in the end, hundreds of cells may be stimulated by one entering impulse.

E. REFLEX ARCS

Each spinal nerve is attached to the spinal cord by two connections, a **sensory (dorsal)** and a **motor (ventral) root** (Fig. 16-3). The dorsal root has an enlargement called the **dorsal root ganglion,** containing numerous cell bodies. The dorsal root contains only those fibers that carry impulses toward the spinal cord; the impulses that originate in sense organs from all parts (internal and external) of the body. The ventral root contains no cell bodies.

The process (projection) transmitting the nerve impulse from the sense organ to the nerve-cell body is the **dendrite.** Another process, the **axon,** transmits the impulse from the cell body and, in this case, to the dorsal region of the gray matter of the spinal cord. The neuron that carries impulses into the spinal cord from the sense organs—for example, skin and stomach smooth muscle—is called an **afferent (sensory) neuron.**

The axon of the afferent neuron usually synapses with an **association (adjustor) neuron** found within the gray matter of the spinal cord. The parts of the association neuron are similar to those of the afferent neuron, but it has several short dendrites and a short axon that ends close to the ventral region of the gray matter.

If the reaction produced is of a simple type, the axon of the association neuron synapses with the dendrites of an **efferent (motor) neuron.** The cell body, as well as the short dendrites of motor neurons, is found in the ventral part of the gray matter. A long axon proceeds from the cell body of the motor neuron and passes ventrally out of the spinal cord into the ventral root; it then enters the spinal nerve. It passes through the spinal nerve and terminates in an effector organ, which, in this case, is always a striated, skeletal muscle. The pathway that the impulse takes from the sense organ through the afferent, association, and efferent neurons to the effector is known as a **reflex arc.**

Reflex arcs vary considerably in their complexity. For example, among very primitive animals, **direct effector** systems are found in which a nerve cell responds directly to stimulation and transmits directly to an effector (Fig. 16-4A). However, in vertebrates, a specialized sensory organ detects stimuli and activates nerve cells. The nervous pathway from sense organ to effector unit may consist of a sensory-motor neuron (Fig. 16-4B), or separate sensory and motor nerve cells (Fig. 16-4C), or one or more association neurons may be interpolated to increase the number of circuits possible (Fig. 16-4D).

A nervous reflex can be defined as a coordinated response by effectors to a stimulus applied to a receptor, mediated through two or more neurons. Much animal behavior can be interpreted in terms of the activity of reflex circuits in the spinal cord, but this activity is modified by influences from higher centers, to the extent that the brain has achieved physiological dominance.

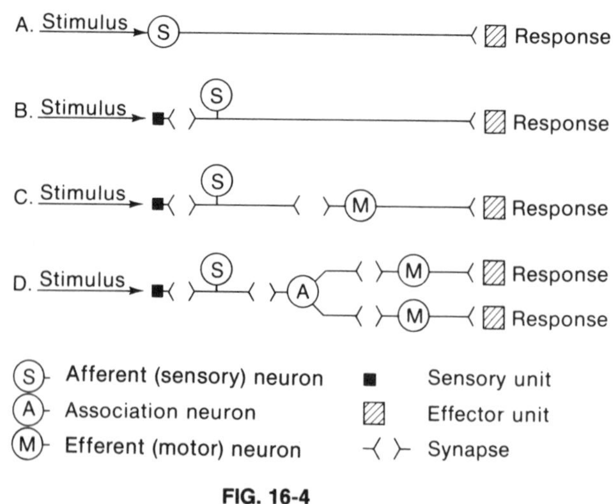

FIG. 16-4
Typical reflex arcs.

1. Reflexes in the Living Frog

a. Blinking

Hold a live frog securely in one hand by grasping the hind legs. Place a blunt probe or a glass rod close to one eye, and observe that the frog blinks. In fact, the response at times is so strong that the eye is pulled into the throat region. Note the "lids" covering the eyes.

b. "Scratch" Reflex

Grasp the frog's head and forelimbs securely in one hand. Wash off any mucus present on the back of the frog. With forceps, place a small (6-mm-square) piece of filter paper moistened with dilute acetic acid on the lower surface of the back. Notice how the hind leg attempts to brush off the irritant. This is the "scratch" reflex. A stronger acid would produce a more violent response, with the animal trying to use both legs to brush off the irritating substance. Wash off the acid with water.

c. Other Reflexes

Lay the frog on its back, and stroke the throat and belly regions. When first stroking a male (a male has a large, swollen "thumb"), watch the distinct clasping reflex in the forelimbs. Notice the quieting, almost hypnotic, effect on the frog as stroking is continued.

2. Spinal Reflexes in the Decerebrate Frog

The frog is an excellent animal for the study of **spinal reflexes** because of the brevity of the period of spinal shock following destruction of the brain, the small blood loss in decerebration, and the adequacy of the cutaneous respiration for supporting metabolism. (See Appendix E for frog pithing and decerebration techniques.)

Holding the frog firmly in your left hand, quickly sever the spinal cord at the neck (on a line just posterior to the eardrums) with a probe. Then, using scissors, cut off the head, but leave the lower jaw. Wipe off the blood, and suspend the frog from a ring stand with the hook through the lower jaw, with the frog's back toward you. Wash off the mucous covering, but remember to keep the skin moist. If spinal shock has occurred, the animal does not respond. Wait 5 or 10 minutes for recovery before proceeding. Such a preparation is known as a **decerebrate frog.** Because it lacks a brain, the frog is completely without consciousness. Does the scratching reflex remain in this "decerebrate frog"?

____

Pinch the toe of one foot with a forceps. What happens?

____

____

What does the other foot do?

____

Allow the frog to recover for a minute or so and then stimulate the toe of the other foot in a similar manner. Now pinch the toe of one foot, holding it tightly with forceps so that it cannot be withdrawn. Describe what happens.

How do you think the nerve impulse reached the unstimulated leg?

F. STIMULATION OF MUSCLE THROUGH NERVE

Cells are, in general, subject to stimulation from a variety of impulses. In studying the different agencies that can cause a muscle to react, remember that the muscle cell differs only in degree, not in kind, in the way it responds to a stimulus. Cells specialized to perform other functions may respond to the same forms of stimulation that muscle cells do, but in a different way. A nerve will transmit an impulse in response to an adequate stimulus. If the nerve is in physiological connection with an excitable muscle, the muscle will contract on receiving that impulse. Adequate stimulation of the nerve is attained when the muscle that it innervates contracts. In this part of the exercise, we will examine the effects of several factors on the nerve impulse, using an isolated gastrocnemius **muscle-nerve preparation.**

To obtain your muscle-nerve preparation, kill a frog by pithing both the brain and spinal cord (see Appendix E). Remove the skin from the legs. This can be easily done if the skin is cut completely around the "thigh" of the frog leg. The entire skin of the leg can then be pulled off in the same way that you would strip a glove from your hand (Fig. 16-5). Place the animal, dorsal side up, on moist filter paper. Keep the animal moist with physiological saline. Note the large "calf" muscle of the lower leg. This the gastrocnemious muscle, the one to be used. Run a blunt probe, or the end of a forceps, between this muscle and the long bone of the leg, noting how easily it is freed in the region between the two ends.

Notice the fairly well marked groove running along the upper leg. This groove marks the position of the sciatic nerve, which innervates the gastrocnemius muscle and which must be freed from the rest of the upper leg. With two pairs of forceps, pull the muscles apart on either side of the groove. If you do this carefully, you will expose the faintly yellowish sciatic nerve along with reddish blood vessels. Without touching or pulling the nerve, remove or pull aside the muscles so that the nerve is exposed from a point near the knee up to the hip. Then, with a forceps, grasp the hip end of the nerve and, cutting across the nerve on the hip side of the forceps, gradually free the nerve of connective tissue down to a point near the knee joint. This is the most delicate operation of the whole dissection. You should not stretch the nerve or touch it with metal *except where it is held at the hip end.* Lay the free nerve down on top of the lower leg. Using scissors, cut through the middle of the upper leg. Cut through everything, including bone and muscle, but be sure that the nerve is laid back so you do not cut through it. Sever the tendon at the lower end of the gastrocnemius muscle, where it is attached to the foot. Cut the lower leg bone near the knee. Your muscle-nerve preparation is now ready to use. Place it on filter paper moistened with physiological saline, or, if you are not ready to use it at once, place it in a beaker of the solution.

1. Effect of Mechanical Stimulation

Tap the nerve lightly near its distal (cut) end with a small glass rod. How does the muscle respond?

Repeat the procedure with a harder tap. Does the muscle respond more fully than the first time?

Pinch the nerve sharply with the forceps about 0.5 cm from its cut end. Is the response any greater?

Pinch the nerve approximately 1 cm from the cut end. What is the result?

Account for the results you observed.

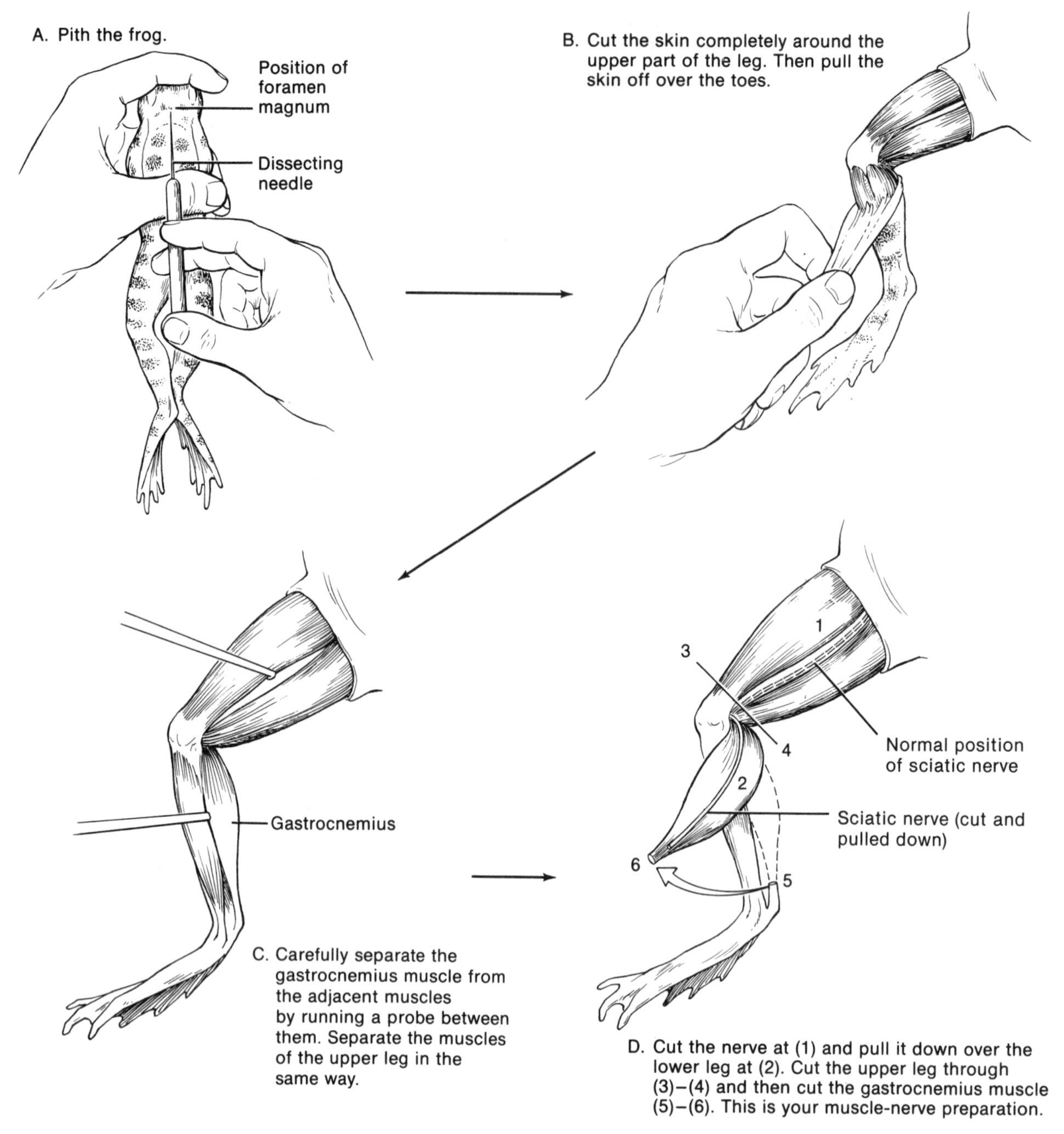

FIG. 16-5
Frog muscle-nerve preparation.

2. Effect of Thermal Stimulation

Touch the nerve with the pointed end of a small glass rod so gently that you evoke no response. Heat the end of the rod until it is rather warm to the touch, and again touch the nerve as gently as before. What is the result?

____

3. Effect of Chemical Stimulation

Cut off the part of the nerve that has already been used. Put a drop of 1 *N* HCl on the newly cut end of the nerve. Describe the response, if any.

____

Wash off the HCl from the nerve.

4. Effect of Electrical Stimulation

When two different metals are connected and their free ends are put in contact with a salt solution, a current of electricity will flow as it does in a battery. Scrape the ends of a piece of iron and copper wire (12–15 cm in length) with a knife or an emery cloth. Twist them together tightly at one end to make a V-shape. Making sure that the nerve-muscle preparation is moistened with physiological saline solution, touch the uninjured part of the nerve with the free ends of the wires. Repeat several times. Describe the response observed.

____

____

Place one end of the V-shape on a nerve and the other end on a muscle. What is the result?

____

____

____

Reverse the positions of the wires. What difference in the response, if any, is observed?

____

Repeat the foregoing procedures, stimulating the muscle directly. Describe the response.

____

____

____

Contrast the direct electrical stimulation of the muscle with the response obtained when the nerve is electrically stimulated.

____

____

Would you say that nerve tissue is a better conductor than muscle? Explain.

____

____

REFERENCES

Baker, P. F. 1966. The Nerve Axon. *Scientific American* 214:74–82 (Offprint 1038). *Scientific American* Offprints are available from W. H. Freeman and Company, 41 Madison Avenue, New York 10010, and 20 Beaumont Street, Oxford OX1 2NQ, England. Please order by number.

Guyton, A. C. 1981. *Textbook of Medical Physical Microbiology.* 6th ed. Saunders.

Katz, B. 1966. *Nerve, Muscle, and Synapse.* McGraw-Hill.

Levi-Montalcini, R., and P. Calissano. 1979. The Nerve-Growth Factor. *Scientific American* 240:68–77 (Offprint 1430).

Luria, A. R. 1970. The Functional Organization of the Brain. *Scientific American* 222:66–78 (Offprint 526).

Macey, R. I. 1975. *Human Physiology.* 2d ed. Prentice-Hall.

Marrazzi, A. S. 1957. Messengers of the Nervous System. *Scientific American* 196(2):86–94.

Merton, P. A. 1972. How We Control the Contraction of Our Muscles. *Scientific American* 226:30–37 (Offprint 1249).

Patterson, P. H., D. D. Potter, and E. J. Furshpan. 1978. The Chemical Differention of Nerve Cells. *Scientific American* 239:50–59 (Offprint 1393).

Schmidt, R. F. 1978. *Fundamentals of Neurophysiology.* 2d ed. Springer-Verlag.

Schmidt-Nielsen, K. 1970. *Animal Physiology.* 3d ed. Prentice-Hall.

Shepherd, G. M. 1983. *Neurophysiology.* Oxford University Press.

Stent, G. 1972. Cellular Communication. *Scientific American* 227:42–51 (Offprint 1257).

Storer, T. I., R. L. Usinger, R. C. Stebbins, and J. W. Nybakken. 1979. *General Zoology.* 6th ed. McGraw-Hill.

Vander, A. J., J. H. Sherman, and D. Luciano. 1980. *Human Physiology: The Mechanism of Body Function.* 3d ed. McGraw-Hill.

EXERCISE 17

Mendelian Genetics

Like many subjects, genetics can be studied through the use of lectures and books without laboratory work. However, the personal experience of performing experimental procedures, collecting and analyzing data, and drawing conclusions gives insights and appreciation of a discipline that cannot be achieved by simply reading about the subject.

Modern genetics, and indeed much of contemporary evolutionary theory, has as its basis the foundations of experimental evidence established by Gregor Mendel. His genius in studying the inheritance of single and multiple characteristics of the sweet pea led to the discovery of his principles (now laws) of **segregation** and **independent assortment**. From these simple beginnings evolved the studies leading to the characterization of DNA ("the stuff of life") and to the forefront of modern molecular biology and studies involving recombinant DNA.

In this exercise, you will become familiar with basic studies involving Mendelian genetics; that is, the laws of segregation and independent assortment as observed through studies involving monohybrid and dihybrid crosses. For demonstrating Mendel's laws, a variety of organisms may be used. We will study the fruit fly, *Drosophila,* and maize (corn), because these are the two eukaryotic organisms for which substantial genetic knowledge has accumulated. Before beginning these studies, read Appendix J for a discussion of Mendel's laws of inheritance.

A. GENETIC STUDIES USING *DROSOPHILA*

The most widely used organism for genetics studies is the common vinegar, or fruit fly, *Drosophila melanogaster.* This fly is easily cultured, and its generation time is only 9 or 10 days at normal room temperature (25°C). Because *Drosophila* is small, the cultures occupy little space; it is therefore a convenient and inexpensive organism with which to work.

Drosophila is well understood genetically and is therefore used more than any other organism for laboratory experiments concerning the basic principles of Mendelian genetics. Wild-type (normal) and mutant strains of *Drosophila* can be easily obtained for instructional purposes. An enormous number of spontaneous mutations have been found, and many others have been induced by radiation, thus making *Drosophila* highly suitable for the investigation of genetic crosses.

Before performing genetics experiments with *Drosophila,* refer to Appendix J to learn some basic facts about the biology and culture of this organism.

1. Examination of Wild-Type *Drosophila*

In this part of the exercise, you will become familiar with the characteristics of the wild type and some of the common mutants of *Drosophila.* Then you will be given several flies whose mutant traits you are to identify.

Obtain a vial of wild-type fruit flies from your instructor and etherize them according to the procedure outlined in Fig. 17-1. When all of the flies have stopped moving, turn them onto a white card and carefully examine them with a dissecting microscope or hand lens. Become familiar with male and female characteristics by recording your observations in Table 17-1.

Obtain from your instructor a numbered vial containing a mixture of the following mutant flies:

Vestigial: a wing mutant characterized by highly reduced, withered wings.

White: a mutant whose eyes look white.

Bar: an eye mutant in which the number of facets of the eye is reduced, resulting in the appearance of a "bar" down the middle of the eye.

Black: a mutant whose body is black.

Other mutants

Etherize the flies and determine the nature of the mutant flies in your vial. Record your observations in Table 17-2.

2. Experimental Genetic Crosses

In making these genetic crosses, you will use methods essentially similar to those used by scientists doing research on *Drosophila.* You will be supplied with pedigreed stocks, maintained in the laboratory, that carry mutant genes. Flies in these cultures tend to breed true as long as they mate among themselves. Very rarely, however, new hereditary variations (i.e., spontaneous new mutations) do occur, and you should be on the lookout for these.

For these studies, flies having readily recognizable differences (traits) will be used. Names and symbols used to identify the mutant genes carried by the flies are those derived by research workers. Breeding stocks will be conveniently designated by symbols indicating the particular mutant gene(s) carried. Flies that exhibit traits that may be considered standard or normal are designated as **wild:** a plus sign (+) indicates wild type with reference to any gene. A lowercase letter indicates that the mutant allele is recessive to the wild-type allele. The symbol *e,* for example, represents the recessive mutant allele for ebony body color and $e+$ the dominant allele for the wild-type gray body. Homozygous ebony flies are symbolized *ee* and homozygous wild-type, e^+e^+.

When making a cross between two varieties of fruit flies, only those characters in which the parent

TABLE 17-1
Comparison of male and female fruit flies.

Characteristic observed	Comparison	
	Male (♂)	Female (♀)
Which is relatively larger in overall size?		
What is the difference in banding on the abdomen?		
What is the shape of the tip of the abdomen?		
Are sex combs present or absent?		

TABLE 17-2
Unknown mutant files.

Mutation	Sex

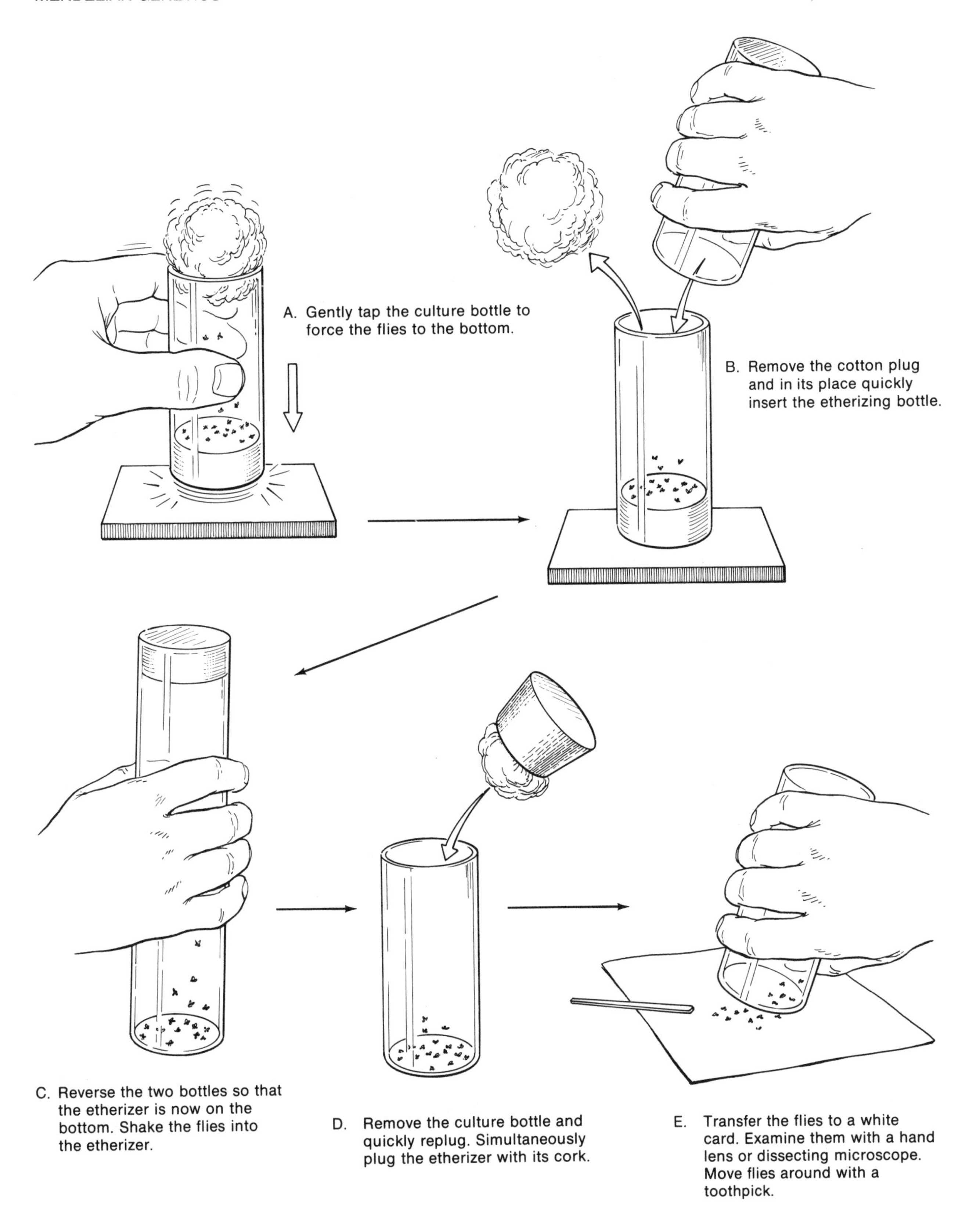

FIG. 17-1
Procedure for etherizing fruit flies.

flies differ should be considered. For example, in crossing flies having vestigial wings and those having sepia eyes, only wing shape and eye color should be observed carefully.

B. MENDEL'S LAW OF SEGREGATION

According to Mendel's first law, the law of segregation, genes do not blend, but behave as independent units. They pass intact from one generation to the next, in which they may or may not produce visible traits, depending on their dominant characteristics. Furthermore, genes segregate at random, thus producing predictable ratios of traits in the offspring.

1. In *Drosophila*

In this experiment, you will make several crosses to demonstrate Mendel's law of segregation—either crosses given in the following table or others suggested by your instructor. If time is limited, your instructor may have prepared the necessary matings for you.

Parent (P_1)		
Female (♀)	×	Male (♂)
1. Sepia eye color *(se)*	×	Wild-type (red eye color)
2. Dumpy wings *(dp)*	×	Wild-type (long wings)
3. Vestigial wings *(vg)*	×	Wild-type (long wings)
4. Ebony body color *(e)*	×	Wild-type (gray-brown body color)

However, if you are to make the necessary crosses, your instructor will supply you with appropriate fly cultures. As you make these crosses, complete the information indicated in Table 17-3. Isolate an etherized male of one variety and an etherized virgin female of the other, using the procedures described in Appendix J. While holding a culture bottle on its side, place these flies in the bottle. Be sure to add some dry yeast granules or yeast suspension to the medium *before* introducing the flies. Keep the bottle on its side until the flies have recovered from etherization, to prevent them from becoming stuck in the medium. After 7 or 8 days, remove the parent flies. Flies of the first filial generation (F_1) will begin to appear about 10 days after mating. After several F_1 flies have appeared, etherize them and examine under low-power magnification. Separate the flies by sex and record the phenotypes in Table 17-3.

To establish an F_2 generation, select three F_1 flies of each sex to be mated and place them in a fresh bottle of medium, using the mating procedure described for the first mating. It is not necessary that the F_1 female flies be virgins for this mating. Why?

____

____

Remove the F_1 flies in 7 or 8 days. About 14 days after mating, etherize the F_2 flies, separate them by sex, and record the phenotypes in Table 17-3. Record the results of the F_2 cross you made by entering the genotypes of the parent, F_1, and F_2 generations.

From this data, indicate the F_2 phenotypic ratio you obtained.

____

What F_2 genotypic ratio was observed?

____

Which trait is dominant in your crosses?

____

2. Chi-Square Analysis of *Drosophila* Data

In general, when you carry out crosses involving one or more pairs of alleles, the results are fairly predictable. For example, if you carry out the following crosses you expect the typical 3:1 monohybrid ratio in the F_2 generation:

P_1: $AA \times aa$
F_1: Aa
F_1: (self-fertilized P_1's for F_2's) $Aa \times Aa$
F_2: AA Aa aa

1 2 1 Genotypic ratio
3 1 Phenotypic ratio

The same expectation could be made for a dihybrid cross where the expected 9:3:3:1 phenotypic ratio would be expressed.

The ratios just given are the *expected* ratios. In natural populations, the *expected values* rarely are identical to the *experimental* or *observed* values you get when counting the numbers of individuals expressing the characteristic in which you are interested.

The purpose of the **chi-square test** is to determine whether the experimentally obtained data (i.e., the *observed* data) are a good fit for, or satisfactory ap-

TABLE 17-3
Monohybrid crosses in *Drosophila*.

Student name	P_1 female ×	P_1 male
Date P_1 mated	Date P_1 removed	
Phenotype of F_1 females	Phenotype of F_1 males	
F_1 female ×	F_1 male	Date F_1 mated
Date F_1 removed		Date F_2 examined

F_2 males		F_2 females	
F_2 phenotypes	Number	F_2 phenotypes	Number
Total =		Total =	

	Male	Female
Genotype of P_1		
Genotype of F_1		
Genotypes of F_2		

proximation of, the *expected* data for a given ratio. In other words, this test determines whether any deviations from the expected values are due to something other than chance.

Record the data obtained in Part B1 in Table 17-4. Carry out a chi-square analysis of the data (read Appendix F for a description of how to do a chi-square analysis).

Are the observed values you obtained small enough to be within the limits expected by chance alone?

If not, what factors, other than chance, could account for the larger deviations of the observed values from the values you expected to obtain?

3. In Maize (Corn)

You will be given F_1 corn seeds that, when planted, will grow into F_2 plants that will exhibit an obvious deficiency. Your instructor will give you planting and watering instructions. It takes about 7–10 days for the seedlings to emerge from the soil. When they are 50–75 mm tall, count the two different types of seedlings in your own tray and in those of the rest of the class. You may use whatever symbols you feel are appropriate to distinguish one allele from the other. Set up appropriate tables, properly labeled, to tabulate your data.

What is the deficiency expressed by one of the alleles?

What is the F_2 ratio obtained?

Is the ratio you obtained in your tray the same or different from that obtained when you count the plants of the entire class? Explain any differences.

What phenotypes are expressed in the F_2 generation?

What genotypes are expressed in the F_2 generation?

What would be the phenotype and genotype of the F_1 generation?

What would be the phenotypes and genotypes for the original parents (i.e., the P_1 generation)?

Would it be possible to have had sexually mature P_1's for each of the traits expressed in the F_2 generation? Explain.

TABLE 17-4
Data for chi-square analysis of monohybrid cross in *Drosophila*.

Phenotype	Genotype	Observed (O)	Expected (E)	(O − E)	$(O - E)^2$	$(O - E)^2/E$
Totals						
Conclusion:						

Analyze your results using the chi-square test (see Appendix F). Are your experimental values consistent with what you would expect to obtain? If not, suggest why the deviation from the expected value could have occurred.

C. MENDEL'S LAW OF INDEPENDENT ASSORTMENT

Mendel's second law was derived from his studies involving crosses dealing with two gene pairs. He demonstrated that, whether you are dealing with one, two, three, or more alleles, each acts independently of the others and is not changed in transmission from one generation to the next. However, we now know that this is valid only if the genes involved are located on different chromosomes, as was the case in all of Mendel's studies. Thus, Mendel always obtained the classical 9:3:3:1 dihybrid ratio. Subsequent studies have shown other ratios for dihybrid crosses in which the genes are linked, that is, located on the same chromosomes. In the space below, give examples of other dihybrid ratios (i.e., 9:7) and the circumstances under which they would be obtained (i.e., linkage).

1. In *Drosophila*

The gene for ebony body color (*e*) is located on the third chromosome and the gene for vestigial wings (*vg*) is located on the second chromosome. Reciprocal crosses should be made from these stocks as follows:

vestigial ♂ $(vgvg; e^+e^+)$ × ebony ♀ $(vg^+vg^+; ee)$

and

vestigial ♀ $(vgvg; e^+e^+)$ × ebony ♂ $(vg^+vg^+; ee)$

Note: Make certain only virgin females are used in these crosses. Why?

After about 8 days, remove the parental stock and discard them. When the F_1 flies emerge, count them and determine the characteristic(s) which they express. Record the data you obtain in Table 17-5 or 17-6. What do you predict the F_1 phenotype and genotype to be?

Mate the F_1 males and F_1 females in new bottles. Again, after 8 days, remove and discard all F_1 flies. When the F_2 generation flies have emerged, count them and determine the numbers and ratios for the characteristics being studied. Record this information in Table 17-5 or 17-6.

Also, fill in Table 17-7 and then carry out a chi-square analysis to determine if your experimental values are consistent with what you would theoretically expect to obtain. (See Appendix F for a discussion of chi-square analysis.)

2. In Maize (Corn)

Your instructor will provide you with F_2 maize seeds removed from an ear of corn similar to the one shown in Fig. 17-2.

Determine the four different phenotypes being expressed. Count the number of kernels represented by each phenotype and record them in the space below. Select the symbols you feel would represent the phenotypes being expressed.

Allele	Phenotype

TABLE 17-5
Dihybrid cross in *Drosophila*.

______ Student name	______ P_1 female ×	______ P_1 male
______ Date P_1 mated	______ Date P_1 removed	
______ Phenotype of F_1 females	______ Phenotype of F_1 males	
______ F_1 female ×	______ F_1 male	______ Date F_1 mated
______ Date F_1 removed		______ Date F_2 examined

F_2 males		F_2 females	
F_2 phenotypes	Number	F_2 phenotypes	Number
Total =		Total =	

	Male	Female
Genotype of P_1		
Genotype of F_1		
Genotypes of F_2		

Which characteristics are recessive?

______

Which characteristics are dominant?

______

Give the phenotype ______ and genotype ______ of the F_1 generation.

Give the phenotype and genotype of the original P_1, homozygous parents.

______

Complete Table 17-8 and carry out a chi-square analysis to determine if the experimenta values are

TABLE 17-6
Dihybrid cross in *Drosophila* (reciprocal cross).

Student name	P_1 female ×	P_1 male
Date P_1 mated	Date P_1 removed	
Phenotype of F_1 females	Phenotype of F_1 males	
F_1 female ×	F_1 male	Date F_1 mated
Date F_1 removed		Date F_2 examined

F_2 males		F_2 females	
F_2 phenotypes	Number	F_2 phenotypes	Number
Total =		Total =	

	Male	Female
Genotype of P_1		
Genotype of F_1		
Genotypes of F_2		

consistent with what you would expect to obtain in this dihybrid cross.

D. INHERITANCE OF HUMAN BLOOD GROUPS

The discovery of human blood groups was reported by Dr. Karl Landsteiner in 1900. This important discovery, for which Landsteiner received a Nobel Prize, led to the establishment of four blood groups: O, A, B, and AB. The basis for these groups in the presence of **antigens,** which are molecules that form part of the surface coating of red blood cells. These antigens stimulate lymphocytes to produce a class of globular protein molecules called **antibodies.** These antibodies combine with the red blood cell antigens to cause agglutination (clumping) and subsequent elimination from the circulation by macrophages.

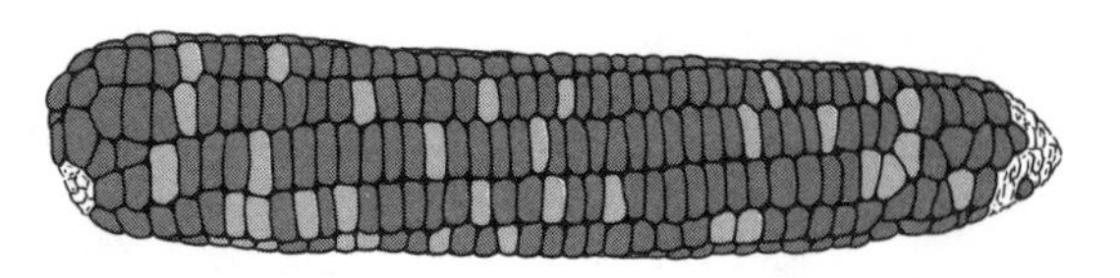

FIG. 17-2
Example of a dihybrid cross in maize. Typical 9:3:3:1 ratio expressed for genes located on different chromosomes.

The two antigens discovered by Landsteiner are glycoproteins and are designated A and B. If you have Type A blood, then the surface of your red blood cells has the specific Glycoprotein A; if you have Type B, then you have Glycoprotein B on your blood cells; if you have Type AB, your blood cells have both types of antigens; and if you have Type O you have neither A nor B antigens on your red blood cells. Furthermore, an organism's serum contains antibodies against whichever antigen is not present on its red cells. Thus human beings with Type A blood have antibodies against Type B red cells in their blood. Similarly, Type B people have antibodies against A, Type O have antibodies against both A and B, and Type AB have neither antibody in their blood.

If a patient having blood type A is given a transfusion of Type B or AB blood, the anti-B antibodies in his or her blood will agglutinate the Type B or AB donor red blood cells almost immediately, with severe—perhaps fatal—consequences. In practice, blood types are cross-matched before transfusion. Thus a person who has Type A can be safely transfused with Type O cells because those who have Type A carry no antibodies against O. Similarly, Type O can be transfused to a person who has Type B or AB. For this reason, those who have Type O blood are known as **universal donors.** In practice, it has been found best to use a donor of the same blood type as the recipient, and the universal donor is used only in an emergency.

TABLE 17-7
Data for chi-square analysis of dihybrid cross in *Drosophila.*

Phenotype	Genotype	Observed *(O)*	Expected *(E)*	$(O - E)$	$(O - E)^2$	$(O - E)^2/E$
Totals						
Conclusion:						

TABLE 17-8
Data for chi-square analysis of dihybrid cross in maize.

Phenotype	Genotype	Observed *(O)*	Expected *(E)*	$(O - E)$	$(O - E)^2$	$(O - E)^2/E$
Totals						
Conclusion:						

If you have Type AB red blood cells, you are a **universal recipient** because you have no naturally occurring antibodies to Type A or B red cells, and therefore neither A nor B antigens will be foreign to your blood.

The degree of agglutination depends largely on the concentration of antibodies in the serum. Thus, when blood is transfused, the antigen of the donor and the antibody of the recipient are all that matter because the antibodies in the blood of the donor are extensively diluted by the blood of the recipient and thus are essentially ineffective in agglutinating the recipient's blood.

The antigens that determine the four blood groups are the result of the expression of three alleles for phenotypes O, B, and A. Alleles for phenotypes A and B are apparently dominant over that for O. The distribution of the antigens and antibodies in the human blood groups is shown in Table 17-9.

Many blood groups have been identified in addition to the O, A, B, and AB series. These include the MNS, Pp, Lutheran, Kell, Lewis, Duffy, Kidd, and the Rh series—the last being very important in certain pregnancies. If a woman is Rh negative (i.e., she lacks the Rh antigens on her red blood cells) and her husband carries the Rh antigens on his cells, then some of the children born to these parents may be Rh positive. Should any of the blood of an Rh- positive fetus cross the placenta, antibodies against the Rh antigens on the fetal red blood cells will form in the woman. Some of these antibodies will then recross the placenta into the fetus, where they will react with the Rh-positive red cells, producing anemia and jaundice. The anemic condition causes the fetus to produce immature red blood cells (**erythroblasts**) at a high rate, leading to the condition known as **hemolytic disease of the newborn.**

To determine your own blood type, it is necessary to have only two reagents: one containing anti-A antibodies and the other containing anti-B antibodies. Each of the two antibodies should agglutinate the red blood cells containing the corresponding antigen. Table 17-10 shows the results that will be obtained if cells of the four different blood types are used. A plus sign (+) indicates agglutination of cells and a minus sign (−) indicates no agglutination.

Draw three circles with a wax pencil on a clean microscope slide. Label one circle "A" (for anti-A), one "B" (for anti-B), and the third "C" (for control). Add a drop of anti-A serum (colored blue for identification purposes) to circle A, and a drop of anti-B (yellow) to circle B.

Add a drop of 0.9% saline to circle C (Fig. 17-3). Why is this control needed?

____

____

Apply alcohol to the tip of the finger to be pricked. Prick that fingertip with a sterile, disposable lancet supplied by the instructor. Squeeze the finger and wipe off the first drop of blood. Continue to squeeze the finger, and apply a large drop of blood to

TABLE 17-9
Human blood groups.

Phenotype	Genotype	Antigen found on surface of red blood cells	Antibody found in serum
O	OO	none	anti-A and anti-B
A	AA or AO	A	anti-B
B	BB or BO	B	anti-A
AB	AB	A and B	none

TABLE 17-10
Determination of human blood groups.

Unknown red blood cells from individual	Serum used and reaction		Phenotype
	Anti-A	Anti-B	
1	+	−	A
2	+	+	AB
3	−	−	O
4	−	+	B

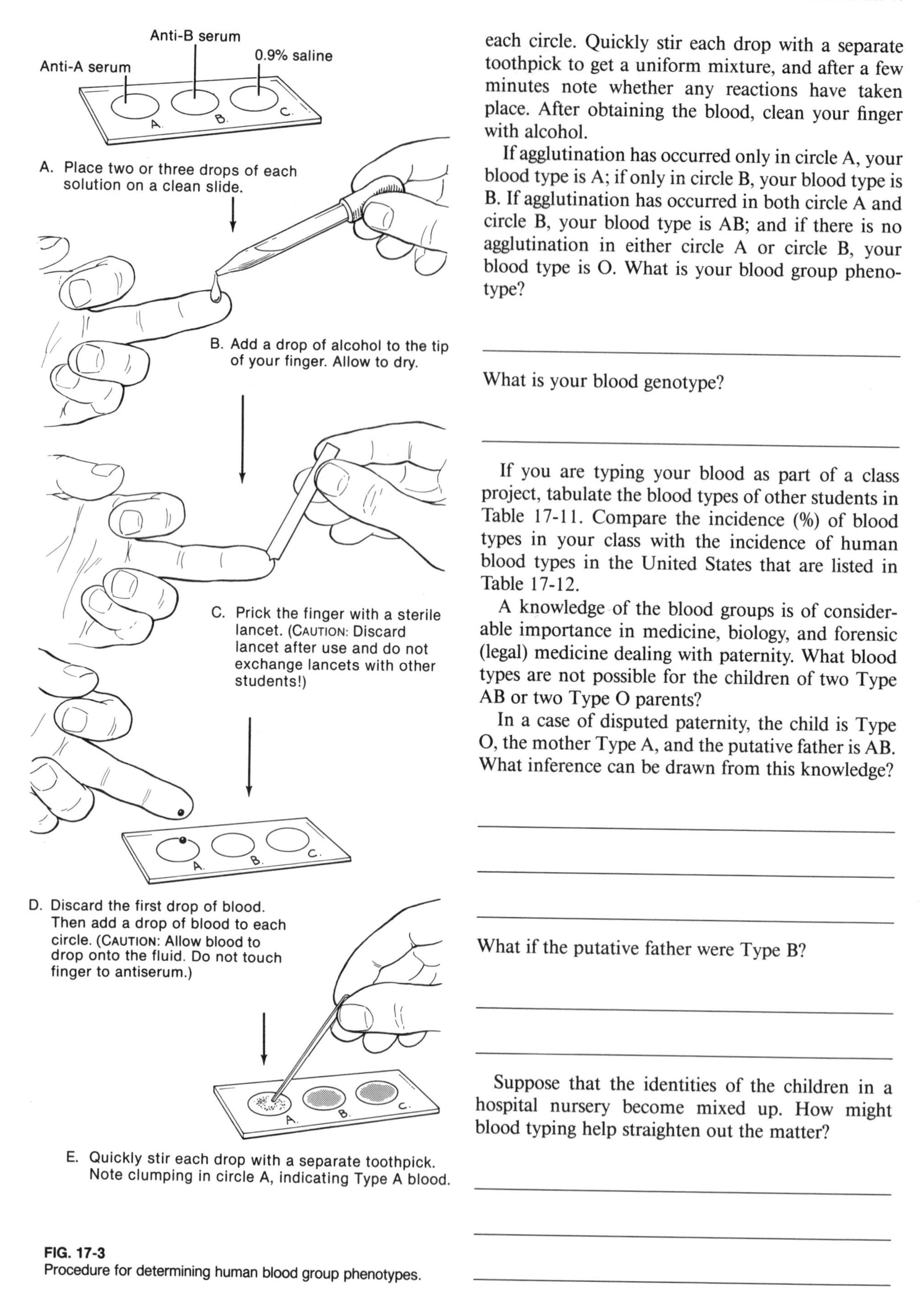

FIG. 17-3
Procedure for determining human blood group phenotypes.

each circle. Quickly stir each drop with a separate toothpick to get a uniform mixture, and after a few minutes note whether any reactions have taken place. After obtaining the blood, clean your finger with alcohol.

If agglutination has occurred only in circle A, your blood type is A; if only in circle B, your blood type is B. If agglutination has occurred in both circle A and circle B, your blood type is AB; and if there is no agglutination in either circle A or circle B, your blood type is O. What is your blood group phenotype?

____

What is your blood genotype?

____

If you are typing your blood as part of a class project, tabulate the blood types of other students in Table 17-11. Compare the incidence (%) of blood types in your class with the incidence of human blood types in the United States that are listed in Table 17-12.

A knowledge of the blood groups is of considerable importance in medicine, biology, and forensic (legal) medicine dealing with paternity. What blood types are not possible for the children of two Type AB or two Type O parents?

In a case of disputed paternity, the child is Type O, the mother Type A, and the putative father is AB. What inference can be drawn from this knowledge?

____

____

____

What if the putative father were Type B?

____

____

Suppose that the identities of the children in a hospital nursery become mixed up. How might blood typing help straighten out the matter?

____

____

____

TABLE 17-11
Summary of class blood types.

Blood Type	Number of students with this blood type	Incidence (%)
A		
B		
AB		
O		

TABLE 17-12
Incidence of blood types in several populations in the United States.

Type	Incidence (%) among			
	Whites	Blacks	Chinese	American Indians
O	45	48	36	23
A	41	27	28	76
B	10	21	23	1
AB	4	4	13	1

REFERENCES

Crow, J. F. 1979. Genes that Violate Mendel's Rules. *Scientific American* 240:134–146 (Offprint 1418). *Scientific American* Offprints are available from W. H. Freeman and Company, 41 Madison Avenue, New York 10010, and 20 Beaumont Street, Oxford OX1 2NQ, England. Please order by number.

Demerec, M., and B. P. Kaufman. 1969. *Drosophila Guide: Introduction to the Genetics and Cytology of Drosophila.* 8th ed. Carnegie Institution.

Flagg, R. O. 1971. *Drosophila Manual.* Carolina Biological Supply Co.

Goodenough, U. 1984. *Genetics.* 3d ed. Saunders.

Lerner, I. M., and W. J. Libby. 1976. *Heredity, Evolution, and Society.* 2d ed. W. H. Freeman and Company.

Singer, S. 1985. *Human Genetics.* 2d ed. W. H. Freeman and Company.

Srb, A. M., R. D. Owen, and R. S. Edgar, eds. 1970. *Facets of Genetics: Readings from Scientific American.* W. H. Freeman and Company.

Stern, C. 1973. *Principles of Human Genetics.* 3d ed. W. H. Freeman and Company.

Strickberger, M. W. 1962. *Experiments in Genetics with Drosophilia.* Wiley.

EXERCISE 18

Molecular Genetics

The science of genetics has passed through several phases during the past 85 years. Each period was initiated by some important event.

The first period began with the rediscovery of Mendel's papers in 1900. The results of this period demonstrated the universal application of the laws of heredity; that is, the transmission of heritable characteristics is brought about by the same mechanism that regulates Mendelian segregation.

The second period, beginning around 1910, saw the introduction of a new "tool" to be used in genetic research—the fruit fly, *Drosophila melanogaster*. From this period came the experimental evidence that genes are located in linear order on the chromosomes.

Further studies on the structure of chromosomes and the discovery that radiation (such as X rays) caused **mutations** (a potentially transmissible change in a gene or chromosome) marked the beginning of the third period of activity. This information was developed further during the succeeding years.

The fourth period introduced microorganisms in genetic studies and provided evidence that the genes regulated the activities of cells by controlling the production of enzymes.

The formulation of the Watson-Crick model of the molecular structure of deoxyribonucleic acid **(DNA)** opened the present era of molecular genetics. DNA is unique in three respects. First, it is a very large molecule, having a certain uniformity of size, rigidity, and shape. Despite this uniformity, however, it has almost infinite internal variety that gives it the complexity required for carrying information. The second characteristic of DNA is its capacity to make copies of itself with remarkable exactness; that is, such a molecule can **replicate** itself. The third characteristic is its ability to transmit information to other parts of the cell. The behavior of the cell reflects the information transmitted.

The past several years have shown that this genetic information (carried in the chemical structure of DNA) consists of a four-letter alphabet of which the "words" and "sentences" of the genetic material and the role it plays in determining the structure of proteins can be regarded as one of the greatest discoveries of all time.

A. CHROMATOGRAPHIC SEPARATION OF *DROSOPHILA* EYE PIGMENTS

In 1941, George W. Beadle and Edward L. Tatum presented experimental evidence supporting the concept that genes control the chemical activities of the cell by controlling the production of biochemical catalysts called *enzymes.* Enzymes, in turn, catalyze the numerous chemical reactions taking place in cells that are ultimately expressed morphologically, physiologically, biochemically, or in the behavior of an adult organism.

Eye pigmentation in the fruit fly *Drosophila melanogaster* is under genetic control. The "normal" eye color of this insect is correlated with the presence of a characteristic series of substances called **pteridines** (derived from the Greek word *pteron,* meaning wing). This term was chosen because the first substances in this class of compounds were extracted from butterfly wings. These compounds are readily separated by chromatography and, if viewed under ultraviolet light, produce seven distinctive fluorescent patterns in wild-type *Drosophila* (Fig. 18-1). Pteridines are present in many invertebrates, in certain pigment cells of amphibians and fishes, and in plants where they may participate in photosynthesis. The pathway of pteridine synthesis is not well understood; however, certain steps have been identified:

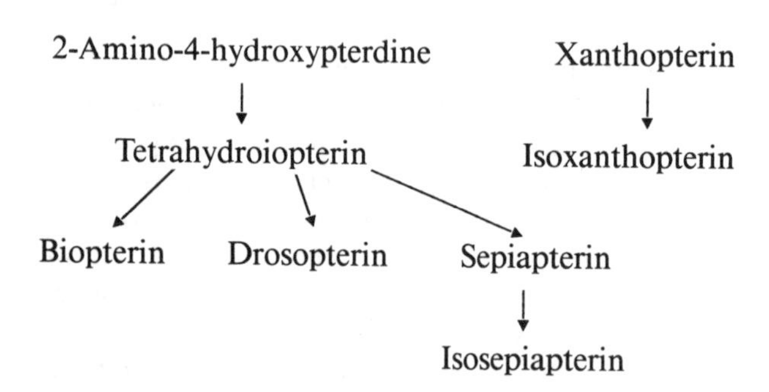

Flies with mutant eye colors have pteridine patterns that are distinctly different from that of the wild-type flies. Certain "normal" wild-type pteridines may be completely missing, whereas others may be present in abnormally large quantities.

Recessive mutations (e.g., X-linked *w,* white eyes) affect the pteridine profiles of wild-type heterozygotes. Although a wild-type heterozygote may appear to have normal eye color, it can be shown chromatographically to be not fully normal at all.

Two mutants that have been independently isolated have dull, reddish brown eyes in contrast with the bright red of normal flies. The two genes producing this trait have been found to reside in two locations in the *Drosophila* genome. One of them, termed *maroonlike (mal),* is located on the first (or sex) chromosome (Chromosome I), whereas the other, termed *rosy (ry),* is located on the third chromosome (Chromosome III). Both of these genetic loci must be wild-type in order for normal wild-type pigmentation to occur; thus flies that are $mal/mal{:}ry^{+}/ry$ have the same phenotype as flies that are genotypically $mal/mal{:}ry^{+}/ry^{-}$. Enzymatic analysis has revealed that both mutant types are deficient in xanthine dehydrogenase, the enzyme that converts 2-amino-4-hydroxypteridine into isoxanthopterin. As a result of the enzymatic deficiency, mutant flies accumulate 2-amino-4-hydroxypteridine, whereas the wild-type flies have little or no substrate but do contain considerable quantities of the product isoxanthopterin. This particular example supports the generalization that genes produce their effects through the action enzymes.

In this exercise, you will analyze chromatographically the pteridines in the normal (wild-type) fruit fly and compare the wild type with several eye-color mutants to establish that the eye-color mutations are accompanied by differences in pteridine patterns. (See Appendix D for a discussion of chromatography.)

1. On a 20-by-20 cm silica gel thin layer chromatography plate, draw a pencil line parallel to, and about 25 mm from, one edge (Fig. 18-2). Lightly pencil in several evenly spaced dots along the origin line. The number of dots will depend on the number of flies supplied by your instructor. Handle the silica gel plate as little as possible because fingerprints may interfere with separation.

2. Obtain three wild-type flies, three each of the eye-color mutants rosy *(ry)* and maroonlike *(mal),* and three each of one or more of the following mutants:

Sepia *(se)*	Cinnabar *(cn)*
Brown *(bw)*	Vermillion *(v)*
Plum *(Pm)*	Eosin (w^{e})
Scarlet *(st)*	Apricot (w^{a})
	White *(w)*

Because sex differences exist with respect to the pteridines, the following analyses should be made using flies of the same sex. Choose adult males or females.

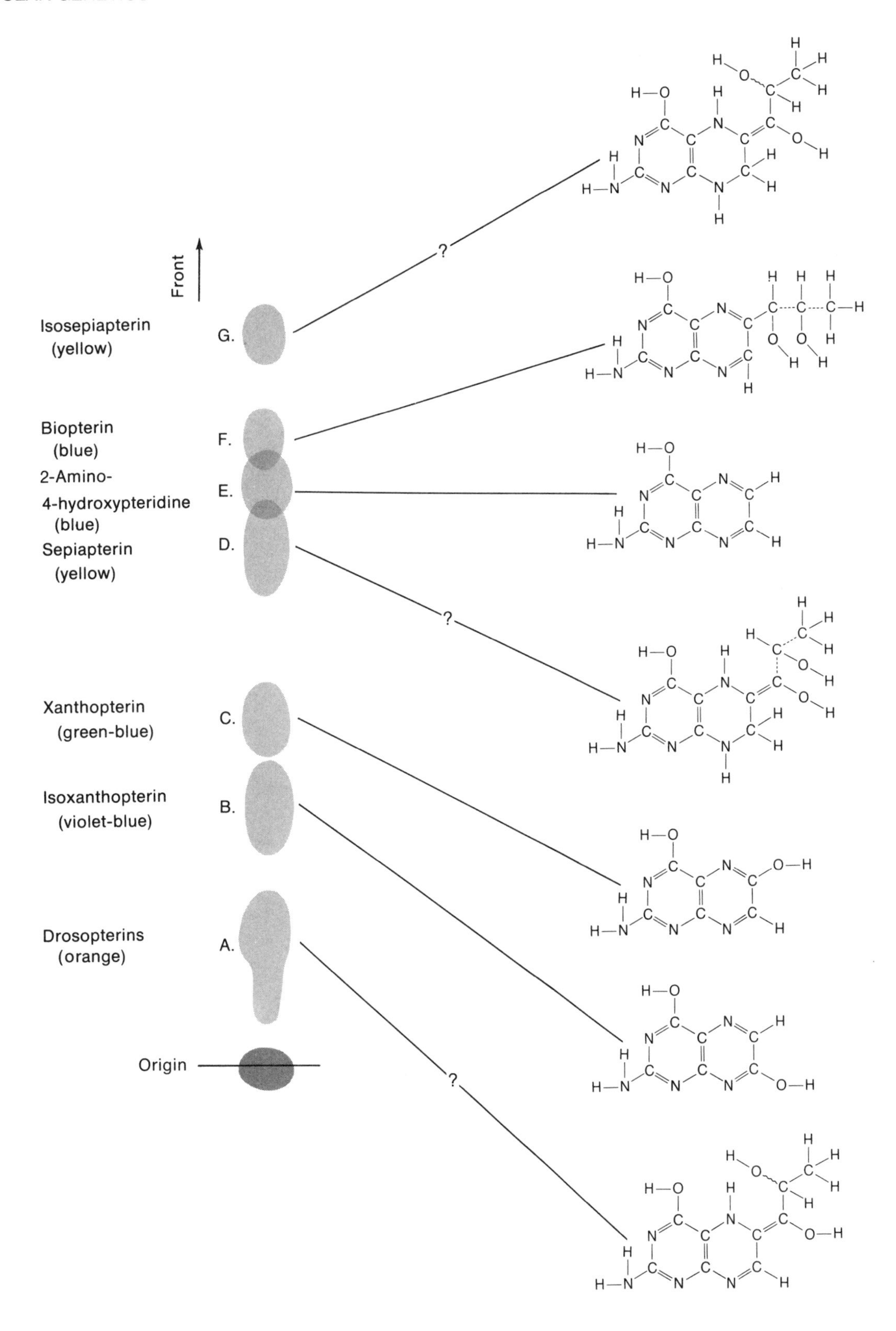

FIG. 18-1
Pteridines of the wild-type fruit fly. The chemical structures were worked out by Max Viscontini, Institute of Organic Chemistry, University of Zurich. (From *Fractionating the Fruit Fly,* by Ernst Hadorn. Copyright © 1962 by Scientific American, Inc. All rights reserved.)

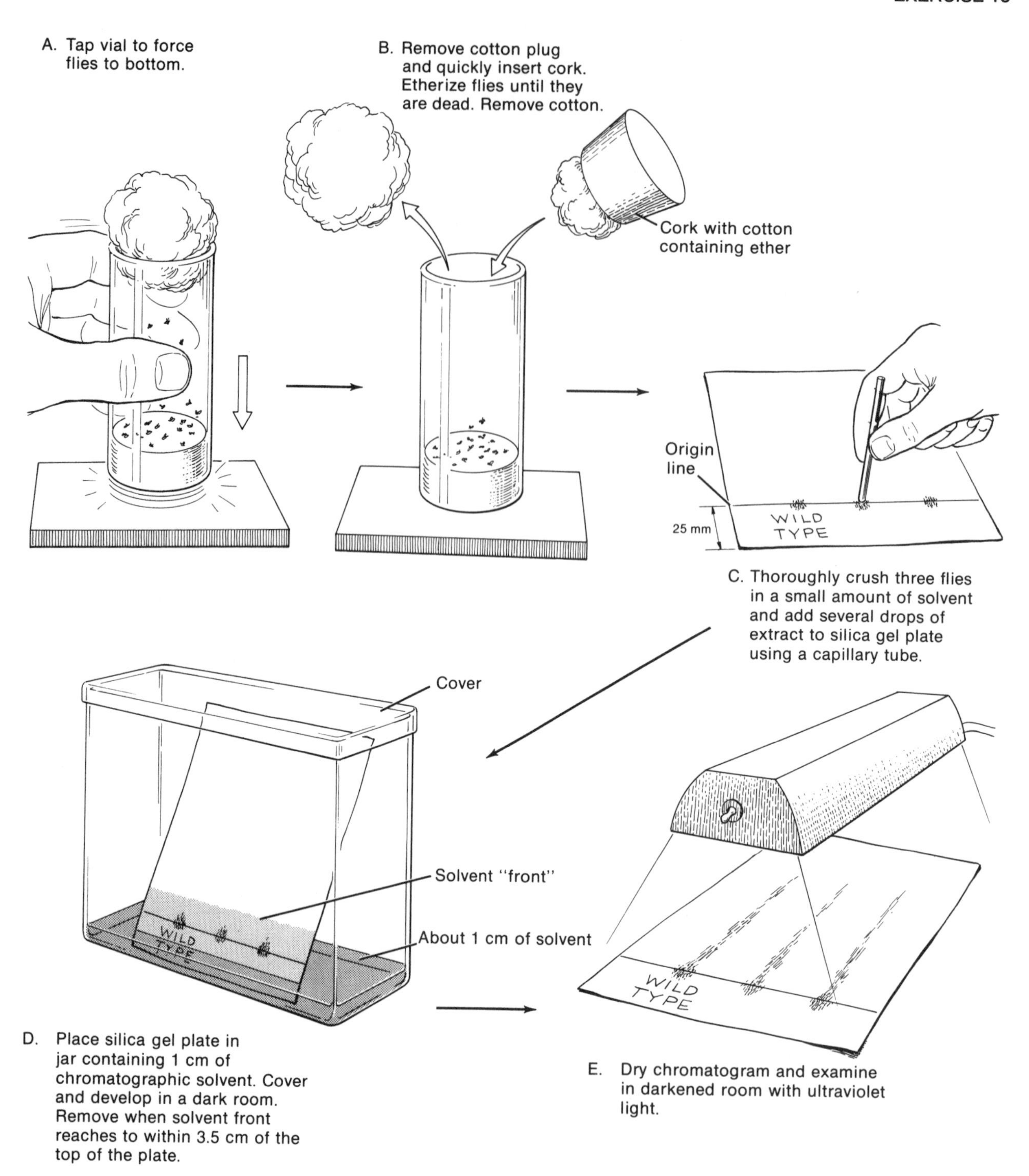

FIG. 18-2
Procedure for chromatographic separation of pteridines.

The wild-type fruit fly has dark red eyes, a tannish, bristle-covered body, and a long, straight pair of wings reaching just beyond the tip of the abdomen (Fig. 18-3). The adult males are differentiated from females by their somewhat smaller bodies; a rounded, heavily pigmented abdomen; and "sex comb" (a tuft of bristles) on the forelegs. The abdomen of the female is somewhat pointed, is traversed by several dark stripes, and may have a terminal tuft of short bristles.

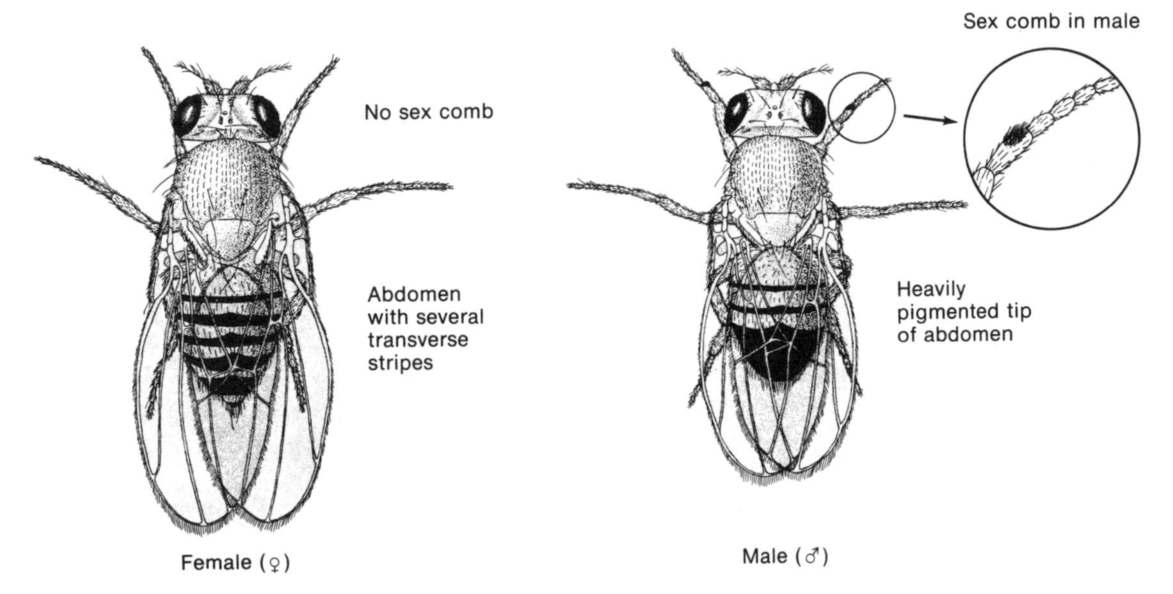

FIG. 18-3
The adult fruit fly, *Drosophila melanogaster.*

3. Etherize three wild-type flies and place them in a vial containing about 0.25 ml of the chromatographic solvent. Crush the flies with a glass rod to dissolve the pigments. Using a capillary tube, add several drops of the extract to the silica gel, as shown in Fig. 18-2. Allow the spot to dry between applications. Be careful not to disturb the silica gel coating any more than necessary. Label for identification. Repeat this procedure for each of the eye-color mutants. What is one control that should be used?

____

Add this control to the thin-layer sheet.

4. Allow the spots to dry for several minutes. Then place the silica gel sheet in the solvent in the chromatographic jar (Fig. 18-2). Because the pteridines are light sensitive, you should develop the chromatograms in a dark room (or cover the jar with foil).

5. Allow the chromatogram to develop until the solvent front reaches to within 3.5 cm of the top of the sheet.

6. Remove the plate from the tank, mark the solvent front with a pencil, and air-dry for several minutes. Examine the sheet, using ultraviolet (UV) light of "long" wavelengths (360 nm). (CAUTION: *Do not look directly into the UV lamp and wear goggles to protect your eyes from UV reflection from the laboratory table.*) Note the fluorescent colors of the various pteridines (Fig. 18-1). Outline each spot with a pencil. The pteridines are listed in Table 18-1 in the order in which they are separated on the chromatogram, with the pteridine listed last being found at the bottom of the chromatogram, nearest to the origin. Indicate which pteridines are present in the flies you examined by checking the appropriate boxes in Table 18-1.

Calculate the R_f values, and record them in Table 18-2 for each of the pteridines isolated on your chromatogram, using the procedure outlined in Appendix D.

Does your chromatogram of sepia-eyed flies show any pteridines present in excess of those found in wild-type? __________. If so, what pigments?

____

What pteridines did you or your classmates observe in male and female rosy and maroonlike eye-color mutant flies?

____

____

____

____

Explain any differences you may have observed.

____

____

____

____

Discuss the eye-color mutations in terms of the genetic control of enzyme synthesis and activity.

____

____

____

____

____

If time permits, you might consider using this technique to answer the following questions.

1. Do adult male and female flies have the same pteridine patterns?

2. Do the patterns of pteridine synthesis change during the development of the fly from egg to the adult?

3. Will there be pigment differences in body-color mutants similar to those seen in eye-color mutants?

B. INDUCTION OF MUTATION BY ULTRAVIOLET LIGHT

The capacity for mutation is inherent in the genetic material of all living organisms, as well as in viruses. Although mutations may occur spontaneously, the frequency with which mutations occur can be increased by a variety of agents called **mutagens.** One of the more commonly used mutagenic agents is ultraviolet light. Ultraviolet light produces its effects (at least in part) by causing breaks in chromosomes that result in the loss or rearrangement of parts.

In this experiment, ultraviolet radiation will be used to study the mutation rate of the antibiotic-producing mold *Penicillium.* This study will be limited to such obvious mutations as shape and pigmentation of the colony and the effects on growth rate.

1. Working in pairs, obtain two petri dishes containing nutrient medium. Label one "Control" and

TABLE 18-1
Distribution of pteridines in the wild-type *Drosophila* and in eye-color mutants.

Pteridine (color)	Fruit fly eye type					
		Mutants				
	Wild type Sex: ___	Rosy Sex: ___	Maroonlike Sex: ___	______ Sex: ___	______ Sex: ___	______ Sex: ___
Isosepiapterin (yellow)						
Biopterin (blue)						
2-Amino-4-hydroxypteridine (blue)						
Sepiapterin (yellow)						
Xanthopterin (green-blue)						
Isoxanthopterin (violet-blue)						
Drosopterins (orange)						

TABLE 18-2
Calculation of R_f values for pteridine pigments of wild-type *Drosophila melanogaster.*

Pigment	Distance from base line to center of spot	Distance from base line to solvent front	R_f values
Isosepiapterin			
Biopterin			
2-Amino-4-hydroxypteridine			
Sepiapterin			
Xanthopterin			
Isoxanthopterin			
Drosopterins			

the other "UV," and add you names, date, and laboratory section number.

2. Add 1 ml of a *Penicillium* spore suspension to each of the plates as follows (Fig. 18-4).

a. Gently shake the flask containing the spores to make the contents uniformly suspended.

b. Fill the pipet in the flask to the 1-ml mark. Add the spore suspension to the surface of the agar in the control dish. Then, holding the agar plate at eye level, tilt the plate, allowing the suspension to run to the edge. (CAUTION: *Do not tilt it so far that the spore suspension runs over the lip of the plate.*) Then hold the plate horizontally and gently jerk it toward you. Repeat this several times, rotating the plate part-way between "jerks." In this way, the spore suspension will cover the entire surface of the agar.

c. Repeat steps a and b for the UV plate.

3. Set the plate aside for $\frac{3}{4}$–1 hour, so that the spores settle onto the agar. Do not move the plates during this period.

4. Irradiate the UV plate as follows (Fig. 18-4).

a. Turn the UV lamp on several minutes before irradiation.

b. Place the plate, cover on, under the UV lamp.

c. Remove the cover and expose the plate to the UV light for 35–40 seconds. Replace the cover and remove the plate.

5. Incubate the plates for 1 week at 28°C. Then count the number of *Penicillium* colonies that have grown on the control and irradiated plates. (Each colony arises from one spore.) Record your data in Table 18-3.

6. Carefully study the appearance of the colonies on the control plate. These are the normal or wild-type colonies. They should be bluish-black in color, with a narrow white fringe. If available, examine color transparencies of wild-type colonies to aid you in your identification of the normal condition. If any of the colonies on the control plate deviate from the wild-type (for example, in colony size, margin width, color, or surface appearance), consider them to be mutants. Would you expect to find mutant colonies on the control plate? Explain.

____

____

____

TABLE 18-3
Effects of ultraviolet radiation on *Penicillium.*

Penicillium	Class data		Team data	
	Control	UV	Control	UV
Total number of colonies				
Number of mutant colonies				
Percentage of mutant colonies				
Percentage of spores surviving irradiation				

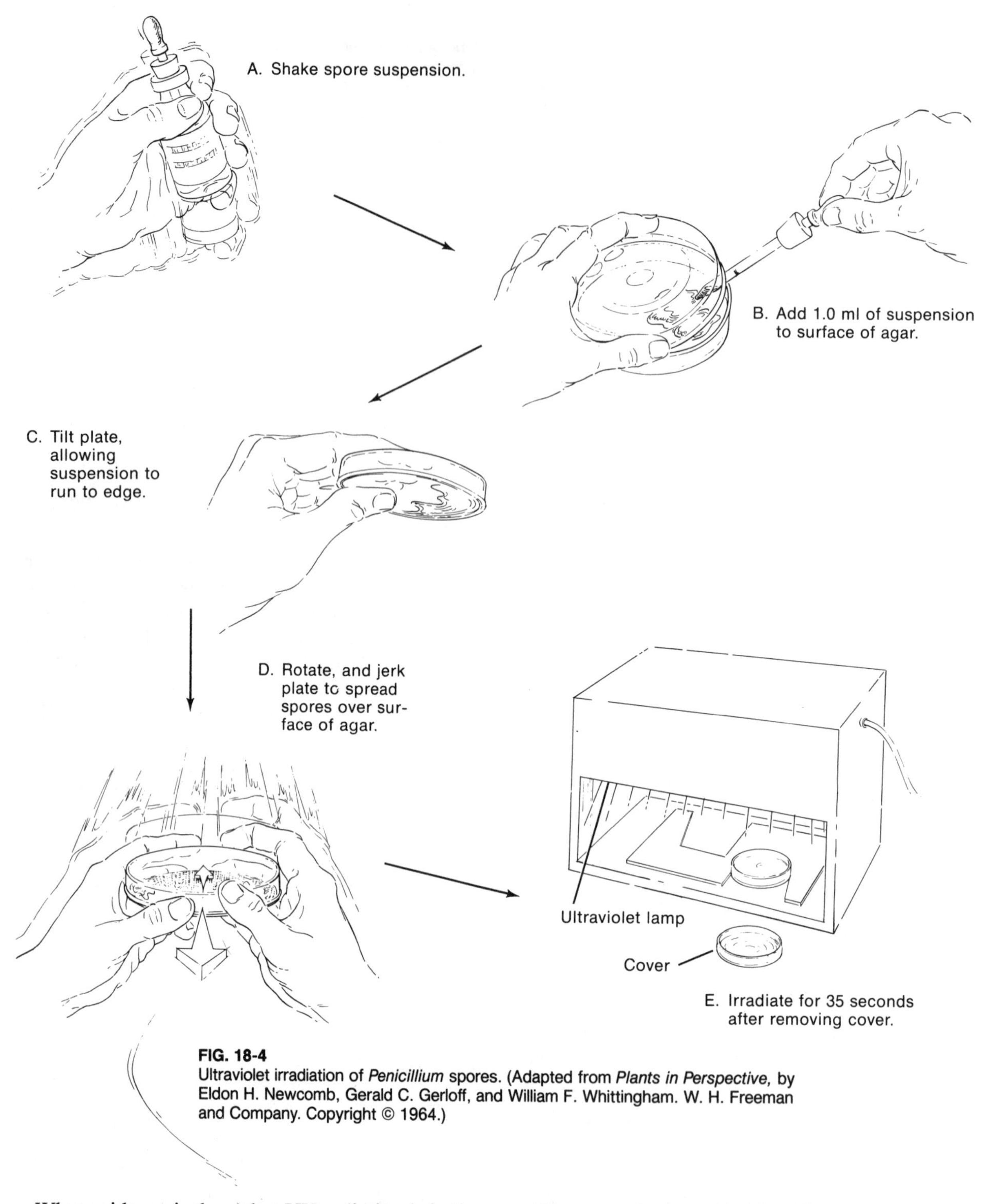

FIG. 18-4
Ultraviolet irradiation of *Penicillium* spores. (Adapted from *Plants in Perspective,* by Eldon H. Newcomb, Gerald C. Gerloff, and William F. Whittingham. W. H. Freeman and Company. Copyright © 1964.)

What evidence is there that UV radiation is indiscriminate in its ability to induce mutations?

____

____

____

The mutations studied in this exercise were expressed as gross visible changes. Is it possible that some of the wild-type colonies are carrying UV-induced mutations? Explain.

____

____

____

____

____

____

C. CHROMOSOME MORPHOLOGY

1. Salivary Gland Chromosomes in *Drosophila*

The giant salivary gland chromosomes of *Drosophila* and other flies afford an excellent opportunity to study the components of the nucleus in relation to chromosomal structure. Although it is true that these cells are highly specialized and that certain aspects may not readily apply to other tissues, they are like most other cells in terms of their major nucleoprotein components. They are unlike many other tissues, however, in that from the time the first gland primordium is formed in the embryo the cells do not divide. The larval salivary gland thus contains a constant number of cells from before the egg is hatched until the gland degenerates during formation of the puparium. The gland grows entirely through cell enlargement, and, as the cell enlarges, the nucleus and its chromosomes also enlarge.

Giant salivary gland chromosomes occur primarily in dipteran flies. The chromosomes attain their maximum size just before pupation. The salivary gland undergoes degeneration (histolysis) in *Drosophila* during the formation of the puparium, although in other species (e.g., *Sciara*) gland degeneration occurs about 30 hours after pupation. An important physiological function of these glands is the secretion of "silk" or a similar protein, which can be used to spin a web or a cocoon or to attach the puparium to a substratum. Salivary glands also secrete digestive enzymes during the active ingestion phase of larval growth and are well developed in some dipteran larvae that do not appear to secrete silk. Large, multi-stranded, or **polytene,** chromosomes are not confined to the dipteran salivary gland, but are also present in cells of the gut epithelium, in Malpighian tubules of the larva, and in the foot pads of adult flies. Chromosomes of these tissues are usually not as large as those of the salivary gland. The micromorphology of chromosome structure is ideally studied in giant chromosomes of the salivary gland.

In this exercise, you will study the morphology of salivary gland chromosomes from a slide that you will prepare from the fruit fly *Drosophila melanogaster.*

1. Examine a culture and locate wormlike larvae. Select one of the larger, slower-moving ones, preferably one that is crawling up the side of the culture container. Gently remove it with forceps and place it on a glass slide lightly moistened with saline solution. Examine it with your dissecting microscope. Note that it has a blunt rear end and a pointed head end that contains black mouth parts. To obtain the salivary glands, the head end must be dissected away from the rest of the body (Fig. 18-5).

2. With a finely pointed needle (which has been dipped into saline solution), pierce the head as close to the anterior end as possible. The larva wriggles, so you will probably have to make several attempts before you are successful.

3. After the head has been secured, grasp the rear end with a pair of finely pointed forceps and with one smooth, quick motion at the anterior end stretch the larva until the mouth parts are torn off and pulled onto the moist slide. All of this can be done while observing the operation with the dissecting microscope.

4. The salivary glands, when pulled out, will probably be accompanied by the digestive tract and fat bodies. Add a drop of saline to the slide. When you are sure that you have the salivary glands, remove the digestive tube and any other extraneous parts with a sharp razor blade or scalpel. Discard these parts, leaving only the salivary glands on the slide.

5. The salivary glands are now ready to be stained. Remove the excess saline solution from the slide by soaking it up with a small piece of paper toweling or filter paper. Try not to touch the glands with the paper because they might adhere to it. Cover the glands with a drop of aceto-orcein stain.

6. After staining for 5 minutes, carefully place a plastic cover slip over the glands. Then place a small piece of paper toweling over the coverslip and press down hard with your thumb. Examine the preparation with your compound microscope. If properly squashed, the cells of the gland will be separated from each other. You should be able to see a nucleus in most of the cells.

7. If your preparation has been adequately stained, you should be able to see banding along the length of the chromosomes. The bands may be better observed if the chromosomes are released from the nuclei. To do this, firmly tap the coverslip with a pencil eraser several times. Examination under the

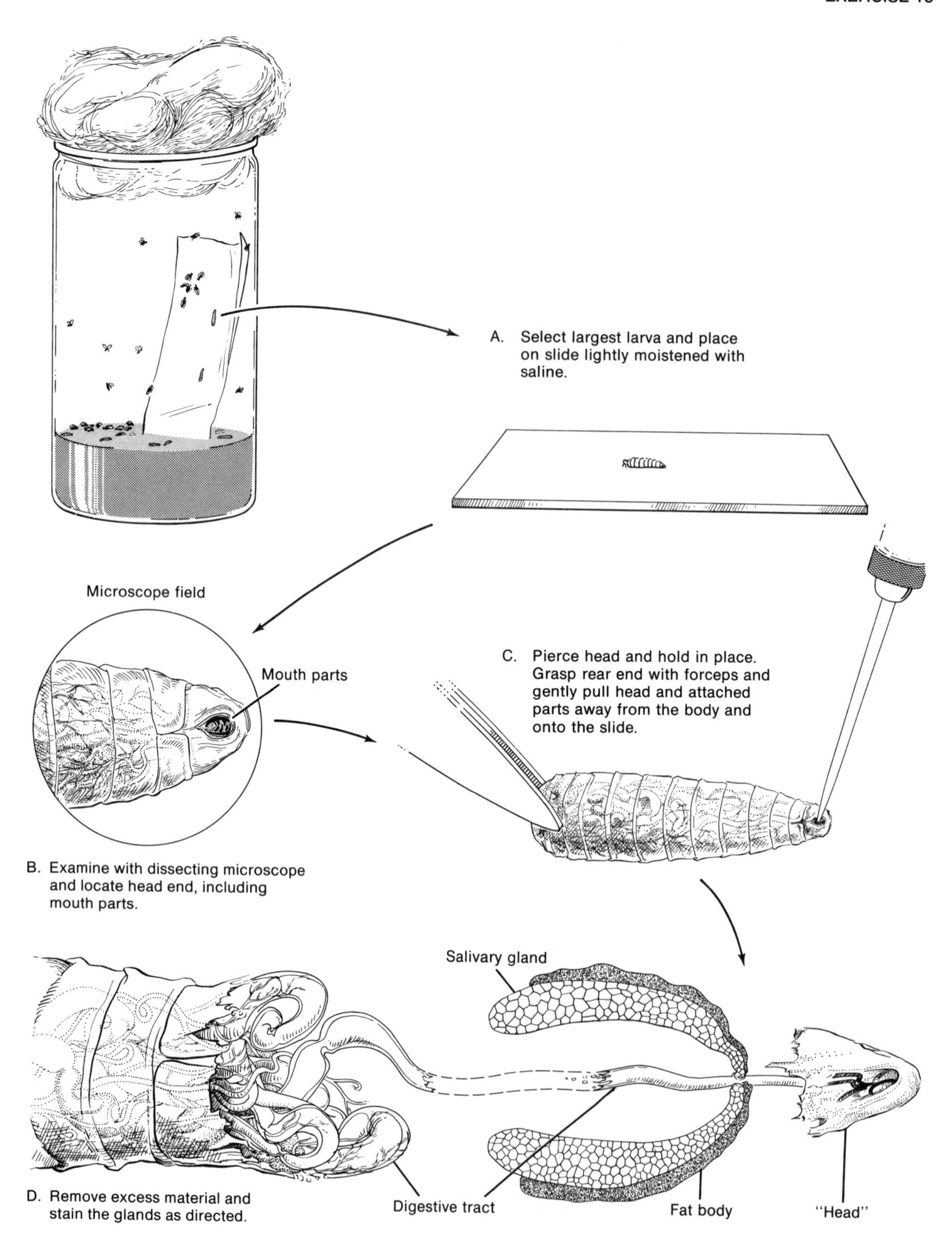

FIG. 18-5
Procedure for removing *Drosophila* salivary glands.

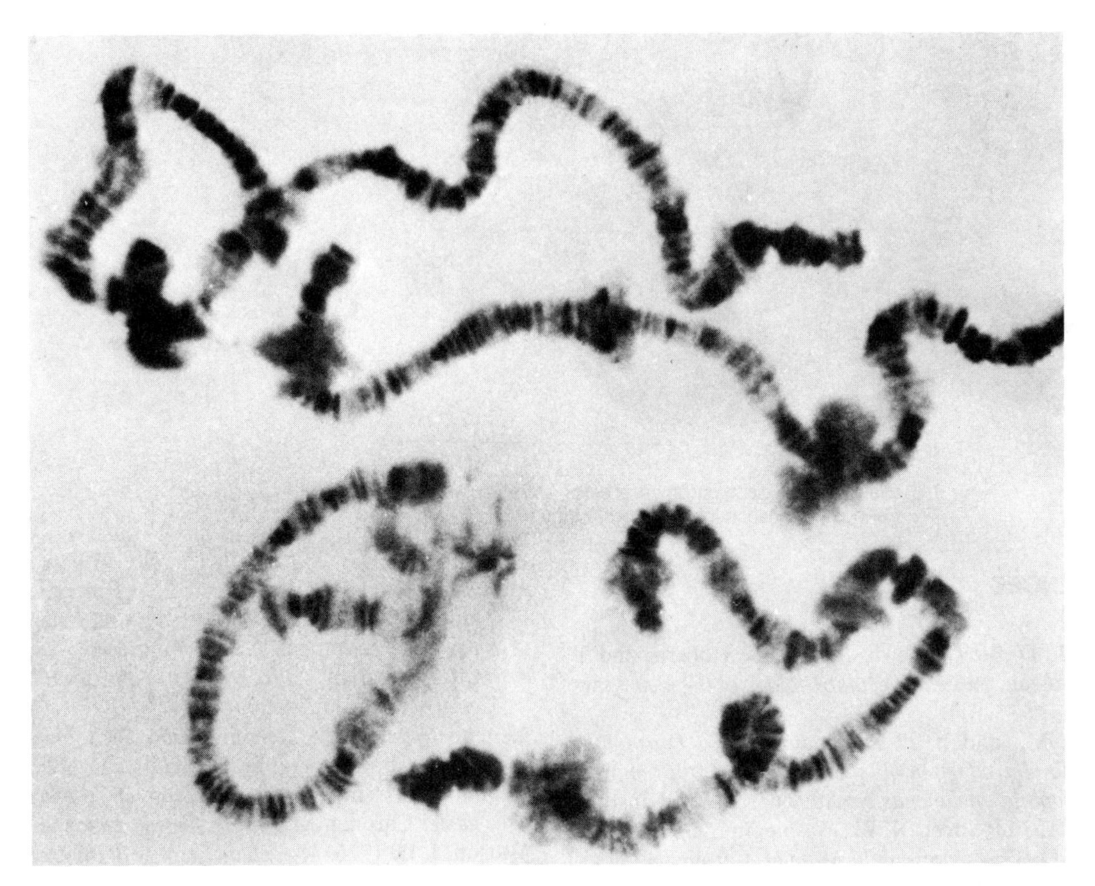

FIG. 18-6
Salivary gland chromosomes of *Sciara coprophila,* showing distinct longitudinal differentiation. (Photomicrograph courtesy of Dr. Ellen M. Rasch, East Tennessee State University.)

high power of the microscope should now show the chromosomes separated and the bands sharply stained. Compare your preparation with that of the *Sciara* chromosomes shown in Fig. 18-6.

2. Chromosome "Puffs"

In your studies of mitosis and meiosis, the chromosomes appeared as heavily stained bodies that apparently lacked distinct morphology. However, when you examined the giant chromosomes found in the salivary glands of *Drosophila* larvae (and of certain other flies), you became aware of a great deal of structural detail that is not apparent in "ordinary" chromosomes.

1. Examine Fig. 18-7. This is a photograph of the giant salivary gland chromosomes of the "midge" fly, *Chironomus.* (If they are available, supplement your studies with specially prepared slides of these chromosomes.) Note the darkly stained bands that are separated from each other by lighter areas.

These bands are areas of DNA and show up when these chromosomes are stained by Feulgen's reagent, a dye that stains DNA.

2. In the center of the photograph, note the appearance of a chromosome "puff." It has been shown that the pattern of "puffing" changes during the development of this organism. Recent experimental evidence suggests that these puff regions are sites of special gene activity, where a special "message-carrying" chemical is formed.

It has further been shown that the hormone involved in the insect molting process can induce puffing. This hormone, called **ecdysone,** appears to act on specific genes, which in turn generate a "message" that is transferred to the cell proper, where it initiates the synthesis of specific protein. It has also been demonstrated that one of the plant growth regulators—**gibberellin**—will accelerate molting and that ecdysone, when used in one of the standard gibberellin bioassays, brings about a growth response similar to that of gibberellin.

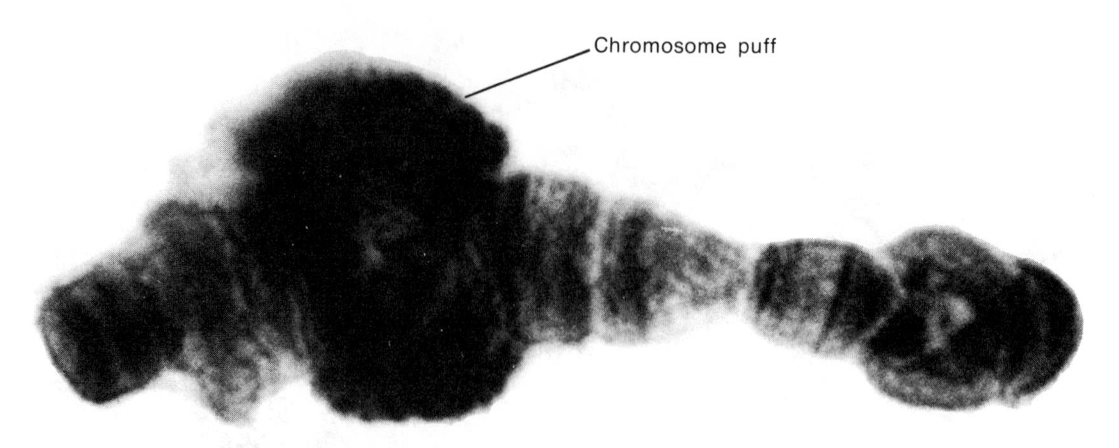

FIG. 18-7
Giant salivary gland chromosomes of *Chironomus*. (Photomicrograph by Claus Pelling, Max Planck Institute for Biology.)

REFERENCES

Alberts, B., D. Bray, J. Lewis, M. Raff, K. Roberts, and J. D. Watson. 1983. *Molecular Biology of the Cell*. Garland.

Demerec, M., and B. P. Kaufmann. 1969. *Drosophila Guide: Introduction to the Genetics & Cytology of Drosophila melanogaster*. 8th ed. Carnegie Institution, 1530 P Street, N. W., Washington, D. C.

Flagg, R. O. 1971. *Drosophila Manual*. Carolina Biological Supply Company.

Hadorn, E., and H. K. Mitchell. 1951. Properties of Mutants of *Drosophila melanogaster* and Changes During Development as Revealed by Paper Chromatography. *Proceedings of the National Academy of Sciences* 37:650–665.

Leitenberg, M., and E. L. Stokes. 1964. *Drosophila melanogaster* Chromatography, I. *Turtox News* 42:226–229.

Leitenberg, M., and E. L. Stokes. 1964. *Drosophila melanogaster* Chromatography, II. *Turtox News* 42:258–260.

Watson, J. D. 1976. *Molecular Biology of the Gene*. 3d ed. Benjamin.

EXERCISE 19

Fertilization and Early Development of the Sea Urchin

The development of a fertilized egg into the complex, interdependent systems of tissues and organs that make up an adult animal is one of the more fascinating studies in developmental biology. In this exercise and the two that follow, the patterns of early development in an echinoderm (the sea urchin), an amphibian (the frog), and a bird (the chick) will be studied.

An important factor that determines the type of development an embryo undergoes is the amount of yolk in the egg. Not only the amount, but the position and distribution of the yolk will markedly affect the patterns of cleavage and subsequent events in early development. In the more primitive eggs, the amount of yolk is small and is evenly distributed throughout the cytoplasm. This type of egg, called **isolecithal,** is found in the sea urchin. Another type of egg has a large amount of yolk that is displaced toward the vegetal pole, the cytoplasm being concentrated at the animal pole. This type of egg, called **telolecithal,** is found in the frog and chick.

In this exercise, you will observe fertilization and the early stages of embryonic development of a marine echinoderm, the sea urchin. This animal, related to the starfish, can be induced to shed its gametes either by injecting potassium chloride into the body cavity or by stimulating it with a weak electric current. Each female will lay approximately one billion eggs and each male will eject several billion sperm.

A. FERTILIZATION

Before the laboratory meeting, your instructor will have obtained living eggs and sperm from female and male sea urchins. The procedure used is shown in Fig. 19-1.

Obtain a depression slide and coverslip as shown in Fig. 19-1. With a clean pipet, transfer a drop of the egg suspension to the slide and examine microscopically.

Inject 2.0 ml of 0.5 *M* KCl into body cavity through soft membrane surrounding mouth.

♀

Mouth (oral surface)

♂

Place oral surface down on clean glass plate until animal begins to shed gametes from gonopores on aboral surface.

Aboral surface

Invert ♀ over beaker containing 25 ml of sea water and allow eggs to drain into water. When eggs have settled, pour off the water. Add fresh water and repeat the washing procedure three times.

Invert ♂ over dry Petri dish to collect sperm.

Cover and store sperm in a cool, dry place until used.

Prepare dilute egg suspension by adding five drops of concentrated egg solution to graduate cylinder and bring to 100-ml volume with sea water. Eggs will last 5 or 6 hours at 25°C.

To use: Dilute one or two drops of dry sperm with 10 ml of sea water. The sperm must then be used within 20 or 30 minutes.

Add one drop eggs and one drop sperm.

This

or this

Wax to hold coverslip away from slide

Depression slide

FIG. 19-1
Preparation of sea urchin egg and sperm suspensions for observing fertilization and early development.

Add one drop of dilute sperm suspension to the eggs on the slide, and examine immediately under high power of the compound microscope. Note the time so that you will know when to expect the cleavage stages (see Table 19-1). The rate of development varies with temperature; therefore, the eggs should be kept at temperatures not to exceed 25 °C. How do the sperm respond to the eggs?

What might be the cause of this response?

Soon after a sperm penetrates an egg, a **fertilization membrane** begins to lift from the egg's surface and completely surrounds the egg. The entire process—from the time the sperm touches the egg to the formation of the fertilization membrane—takes 2–5 minutes.

The first cleavage does not occur until 50–70 minutes after the formation of the fertilization membrane. The changes taking place within the egg during this time are difficult to observe in living material. After the sperm enters the cell, the male nucleus migrates toward the female nucleus and unites with it, reestablishing the diploid number of chromosomes. Subsequently, the chromatin is organized into chromosomes, and the first nuclear division, followed by cell division, occurs.

B. CLEAVAGE

The process by which the fertilized egg develops into a multicellular embryo is called **cleavage.** During this process, complex events occur within the nucleus and cytoplasm. You will only be concerned with the more readily observable external changes.

While waiting for the first division, examine prepared slides of sea urchin or starfish eggs. How can you tell if an egg on a slide is fertilized or unfertilized?

Next, prepare a slide of the living sperm. By reducing the light, you may be able to observe the flagellum by which the sperm moves. If, at this point, the fertilized eggs have not reached the first cleavage stage, continue with Part C.

From your study of meiosis, you will recall that, in the development of the egg, polar bodies are formed during the first and second meiotic divisions. In the sea urchin, the first cleavage begins in the region where the polar bodies are formed: the **animal pole.** If you carefully observe the egg, you should notice that it begins to elongate slightly before it divides. The second cleavage occurs from approximately 15–20 minutes after the first cleavage. Describe the plane of this division with respect to the first cleavage.

Describe the plane of the third cleavage with respect to the first two cleavages.

If your preparation does not attain this stage of development, examine fertilized eggs that were prepared earlier by your instructor or examine your slide later in the day.

TABLE 19-1
Approximate time sequence for the development of fertilized sea urchin eggs.

Stage	Time
Formation of fertilization membrane	2–5 minutes
First cleavage	50–70 minutes
Second cleavage	78–107 minutes
Third cleavage	103–145 minutes
Blastula	6 hours
Hatching of blastula	7–10 hours
Gastrula	12–20 hours
Pluteus larvae	24–48 hours

C. LATER STAGES OF DEVELOPMENT

Before the eight-cell stage is reached, the cleavages are uniform. Subsequent to the eight-cell stage, the four cells of the vegetal pole divide unequally, resulting in four large cells (**macromeres**) and four small cells (**micromeres**). Then the four cells of the animal pole divide into eight cells of equal size (**mesomeres**) (Fig. 19-2).

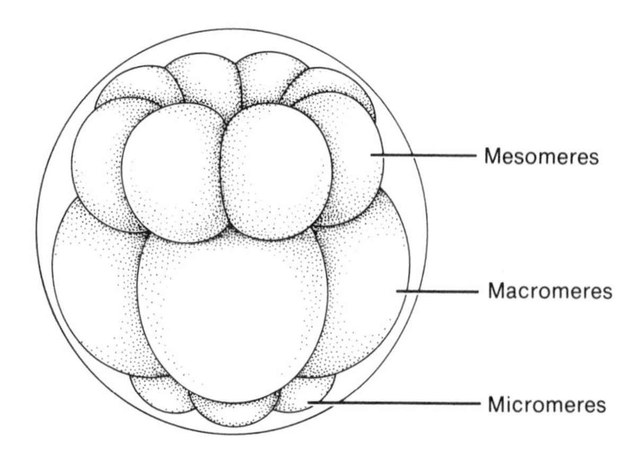

FIG. 19-2
Unequal cell division in the sea urchin (fourth division).

1. Blastula

As the cells of the embryo undergo additional cleavage, a spherical mass containing a fluid-filled central cavity, the **blastocoel,** is formed. With further cleavages, the cavity increases in size until the embryo, now called a **blastula,** consists of several hundred cells that become arranged in the form of a single-layered hollow ball. At this stage, the embryo begins to rotate within the fertilization membrane. Approximately 8 hours after fertilization, the bastula breaks through the membrane by secreting a "hatching enzyme," which apparently digests the fertilization membrane. If you are unable to observe a living blastula, (Fig. 19-3A), examine slides of this stage in the sea urchin or in related forms such as the starfish.

2. Gastrula

For several hours after "hatching," the blastula swims actively. About 15 hours after fertilization, the single-layered blastula changes into a double-layered **gastrula** (Fig. 19-3C). In the sea urchin, this process begins when cells at the lower, or vegetal, pole begin to pulsate intensely, particularly on the inner surfaces facing the blastocoel. This causes some of the cells to be forced into the blastocoel forming a slight indentation called the **blastopore.** These cells then send out long cytoplasmic extensions that become attached to the opposite wall near the animal pole. Contraction of these cytoplasmic strands results in a deepening of the blastopore; ultimately forming an invagination that will become the **archenteron,** or primitive gut, of the embryo. The blastocoel remains, but is soon invaded by other cells. The invagination continues forward to the animal pole, where it will eventually meet the opposite wall. At this junction the mouth will form. The opening of the blastopore will become the anus in the adult.

At the conclusion of gastrulation, the embryo consists of an outer presumptive ectoderm and an inner presumptive endoderm that lines the archenteron. A third layer, the mesoderm, will develop between the ectoderm and endoderm from the proliferation of cells arising in the endodermal layer (Fig. 19-3D). These embryonic germ layers develop into the various tissue and organ systems of the adult animal. The skin, sense organs, and nervous system arise from the ectoderm. The muscles, skeletal elements, and blood originate in the mesoderm. The endoderm gives rise to the alimentary canal and its various derivatives: the pharynx, esophagus, stomach, liver, intestine, pancreas, and endocrine glands.

Examine living gastrulae or slides of this stage of development.

3. Pluteus Larva

In the later stages in the development of the gastrula, the skeletal elements begin to form as pairs of rod-like structures on either side of what will become the anus. Twenty-four hours after fertilization, the larva is well formed. Examine slides or living specimens of the pluteus larva (Fig. 19-4A). When 2 or $2\frac{1}{2}$ months

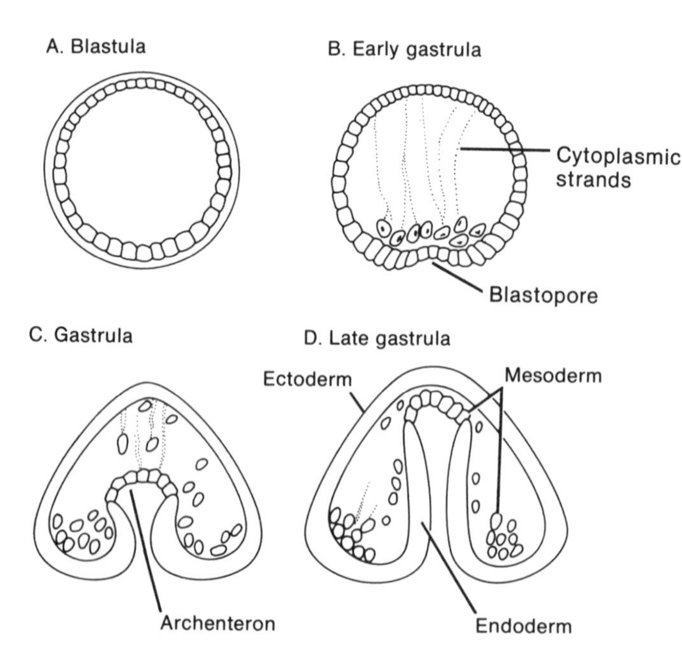

FIG. 19-3
Blastula and gastrula of the sea urchin.

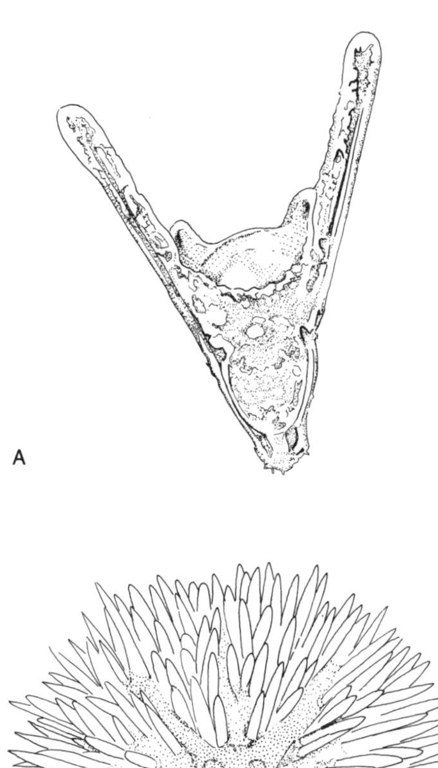

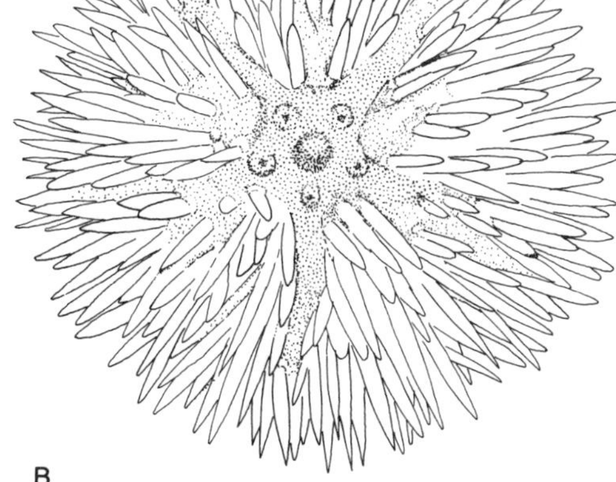

FIG. 19-4
Sea urchin: (A) pluteus larva and (B) adult.

old, the larva metamorphoses into the adult sea urchin (Fig. 19-4B). Examine specimens of the adult animal. Echinoderms are characterized by having numerous "tube feet." Where on the sea urchin are these feet located?

____

What seems to be their function?

____

____

____

____

D. ARTIFICIAL PARTHENOGENESIS

Eggs of sea urchins, and of numerous other organisms, have the ability to develop **parthenogenetically** (from the Greek words *parthenos,* for "virgin," and *genesis,* "birth"); that is, without being fertilized by sperm. Sea urchins, in particular, are very susceptible to the action of external substances. For example, various concentrations of electrolytes such as magnesium chloride ($MgCl_2$) and potassium chloride (KCl), as well as nonelectrolytes such as urea and sucrose, induce parthenogenetic cleavage in sea urchin eggs. In the following exercise, you will attempt to determine if hypertonic seawater will induce cleavage in nonfertilized eggs.

1. Prepare petri dishes containing the following concentrations of seawater:

   Normal concentration
   1.5 × normal
   2 × normal
   4 × normal

2. Transfer samples of washed, unfertilized eggs (as prepared in Part A) to each of the dishes.

3. After treatment for 5 minutes, wash the eggs in four changes of normal seawater and then leave them in the fourth change.

4. Examine the eggs for the presence of fertilization membranes, which appear more slowly in parthenogenetically induced eggs than in normal eggs. Check further development, noting particularly if the development of any of the eggs is arrested at a given cleavage stage.

E. DEVELOPMENT OF EMBRYOS FROM ISOLATED BLASTOMERES

In 1892, Hans Dreisch separated sea urchin blastomeres (cells) in early embryonic stages and followed the development of these isolated cells. He was able to do this because, immediately after fertilization, the fertilization membrane is quite fragile and can be removed by vigorously shaking the test tube containing the eggs. Without the membrane, the two cells can then be separated. You will attempt to duplicate this procedure in this study.

1. Place a sample of unfertilized eggs into a test tube half full of seawater. Add a drop of sperm suspension, and seal with a clean rubber stopper. Mix the contents by gently inverting the test tube once or twice.

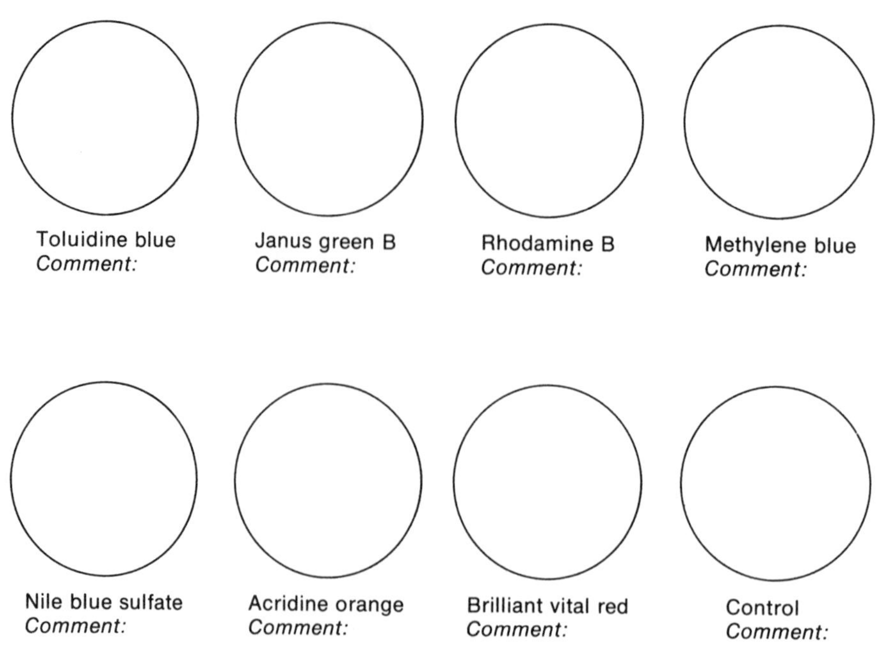

FIG. 19-5
Vital staining of sea urchin eggs.

2. After 1 minute, vigorously shake the test tube for 2 minutes. This should rupture a large number of fertilization membranes.

3. Transfer the contents of the test tube to a petri dish. After an hour, examine the eggs for the appearance of the first cleavage. If cleavage has begun, watch carefully until the first cleavage is just completed. Then, using a clean pipet, vigorously squirt seawater into the culture several times. Then allow the eggs to settle to the bottom and examine microscopically for isolated blastomere pairs.

4. Transfer the isolated blastomeres to finger bowls of seawater and observe development for the next several days.

Compare "normal" development with that of isolated "half embryos." Consider such things as rate of development and size of pluteus larvae.

____

____

____

____

____

____

F. VITAL STAINING OF SEA URCHIN EGGS

Many chemical and structural components of cells can be "visualized" through the use of vital stains, which do not kill the cells in the course of staining. Several of these stains are listed, together with the cellular constituents with which they react:

Toluidine blue stains mucopolysaccharides pink and nucleic acids (mostly RNA) blue.

Janus green B, Rhodamine B, and Methylene blue stain mitochondria.

Nile blue sulfate stains phospholipids.

Acridine orange stains DNA yellow and RNA deep orange (these are fluorescent colors under ultraviolet).

Brilliant vital red stains acid and neutral proteins.

Using these stains and the following procedure (you may have to modify the staining time), treat samples of unfertilized sea urchin eggs and locate the various constituents just listed.

1. Using a marking crayon, label the depressions in a spot plate to correspond to the various stains listed above. Use one depression as a control.

2. Add 1 ml of seawater and four or five unfertilized sea urchin eggs to each depression.

3. Add one drop of stain to each depression. Add nothing to the control.

4. Stain from 1–2 hours. Then carefully pipet off the staining solution, and add fresh seawater. Examine the eggs microscopically, and "describe" your results by drawing them in Fig. 19-5.

REFERENCES

Balinsky, B. I. 1981. *An Introduction to Embryology.* 5th ed. Saunders College/Holt, Rinehart and Winston.

Browder, L. W. 1984. *Developmental Biology.* 2d ed. Saunders College/Holt, Rinehart and Winston.

Costello, D. P., M. E. Davidson, A. Eggers, M. H. Fox, and C. Henley. 1957. *Methods for Obtaining and Handling Marine Eggs.* Marine Biological Laboratory, Woods Hole, Massachusetts.

Harvey, E. B. 1954. Electrical Method of Determining the Sex of Sea Urchins. *Nature* (London) 176:86.

Harvey, E. B. 1956. *The American Arbacia and Other Sea Urchins.* Princeton University Press.

Karp, G., and N. J. Berrill. 1981. *Development.* 2d ed. McGraw-Hill.

Slack, J. M. 1983. *From Egg to Embryo: Determinative Events in Early Development.* Cambridge University Press.

Trinkaus, J. P. 1984. *Cells into Organs.* 2d ed. Prentice-Hall.

EXERCISE 20

Fertilization and Early Development of the Frog

The eggs of most frogs form during the summer months when the frogs are feeding heavily. By fall, the eggs are fully developed, but they are usually retained in the ovaries until the following spring, at which time they are released en masse. The release of the eggs from the ovaries is called **ovulation.** The eggs subsequently pass into the oviducts and are then passed out of the body.

Ovulation normally results from stimulation of the ovaries by gonadotropic hormones produced in the pituitary gland. A mature female frog can be induced to ovulate earlier than usual by injecting whole pituitary glands into the body cavity. The number of glands needed varies with the time of year and with the sex of the frog supplying the glands. Because male pituitaries are about one-half as potent as female pituitaries, twice the number of glands must be used. Table 20-1 outlines the number of female pituitaries needed at various times of the year to induce ovulation.

In this exercise, you will induce ovulation in mature female frogs and prepare a frog sperm suspension. You will use the sperm suspension to fertilize the eggs so that you can then examine and become familiar with the patterns of cleavage and early development in the frog embryo.

Your instructor will demonstrate how to remove and inject the pituitary glands. To have frogs that are ovulating at the time of the laboratory meeting, the female frogs should be injected 48 hours ahead of time. (*Note:* A powdered extract is available that can be hydrated and injected as an alternate to pituitaries from living frogs.)

A. ARTIFICIAL INDUCTION OF OVULATION

1. Anesthetize the number of frogs necessary to obtain the quantity of pituitaries shown in Table 20-1 by placing them in a sealed container in which a wad of cotton saturated (but not dripping) with ether has been suspended. Use a fume hood or a well-ventilated room to perform this operation. Shake the container and, when the frogs fail to move, re-

move them and dissect out the glands as illustrated in Fig. 20-1. Place the glands in a dish containing Holtfreter's solution (an optimal chemical solution for the maintenance of amphibian eggs and embryos).

2. When the required number of pituitaries has been collected, remove the needle from a 2-ml syringe and draw the glands into the barrel. Bring the volume to 1 ml, and attach an 18-gauge needle. Hold the syringe with the needle down, so that the glands settle into the mouth of the needle. If any pituitaries stick to the sides of the barrel, tap the syringe until all of them collect on the bottom.

3. Select the female to be injected and, holding her as shown in Fig. 20-2, insert the needle into the body cavity. Keep the needle parallel to the body wall so as not to injure any internal organs. When the needle is deeply inserted (check to see if the pituitaries are at the needle end of the syringe), quickly inject the pituitary glands. Leave the needle in position for a minute and then slowly remove it, pinching the skin to prevent any loss of fluid or glands.

4. Draw more Holtfreter's solution into the syringe to determine if any pituitary material is lodged in the needle. If so, eject all but about 0.5 ml of solution and then inject the preparation containing the pituitary material.

5. Place the frog in a covered container with about 50 ml of water, and put in a cool room at about 68°F (20°C). Check for ovulation after 24 hours by gently squeezing the abdomen, as shown in Fig. 20-3.

If eggs come out of the cloaca, ovulation has taken place. The frog may then be placed in a container of fresh water and kept in a refrigerator until needed. Eggs will remain viable in the oviduct 3–4 days if the animal is kept at a temperature ranging from 10–15°C. If ovulation has not occurred, put the frog in fresh water and try again 24 hours later.

B. PREPARATION OF SPERM SUSPENSION

This should be done about 30 minutes before the laboratory meeting so that active sperm will be avail-

TABLE 20-1
Number of female pituitaries needed to induce ovulation in frogs.

September–December	January–February	March–April
5–6	3–5	1–3

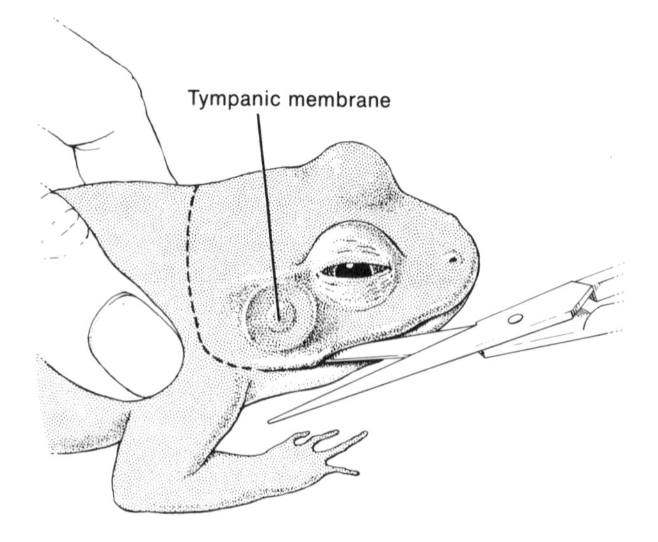

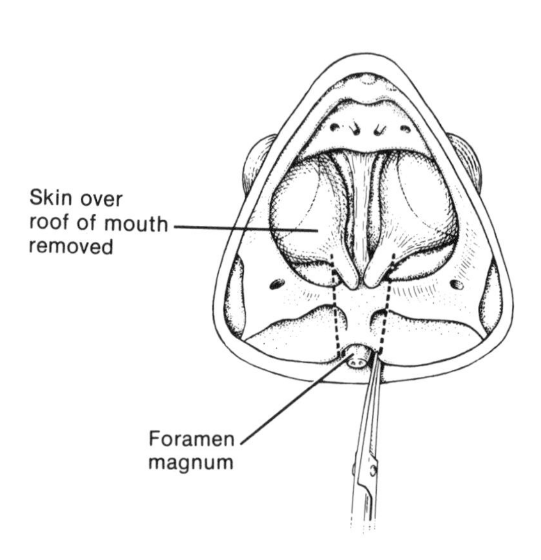

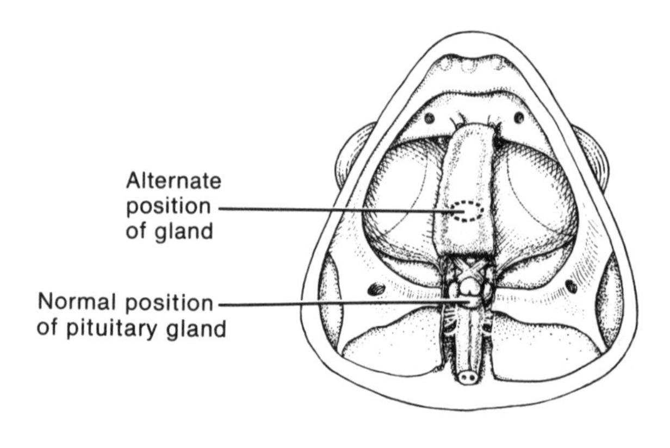

FIG. 20-1
Procedure for obtaining pituitaries.

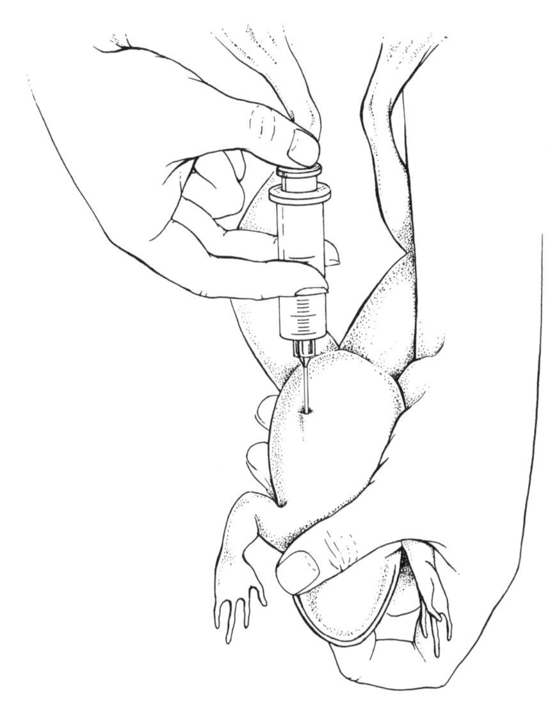

FIG. 20-2
Method of injecting pituitary preparation.

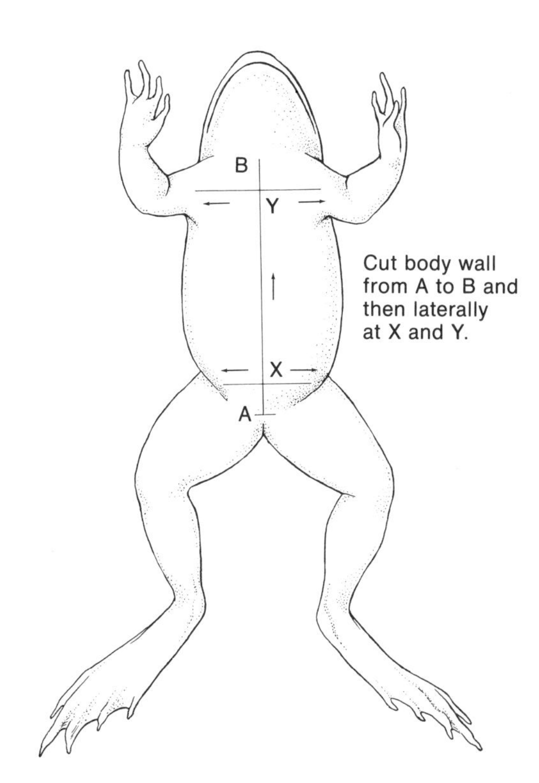

able at the time the eggs are stripped from the female.

1. Prepare a petri dish containing 20 ml of 10% Holtfreter's solution.

2. Pith a mature male frog (see Appendix E for this procedure) and remove the paired testes (Fig. 20-4). Clean away any adhering blood and tissue, and, using the blunt end of a clean probe, thoroughly macerate the testes in the Holtfreter's solution until a milky suspension is obtained.

3. Set the sperm suspension aside for 15–20 minutes to allow the sperm to become motile.

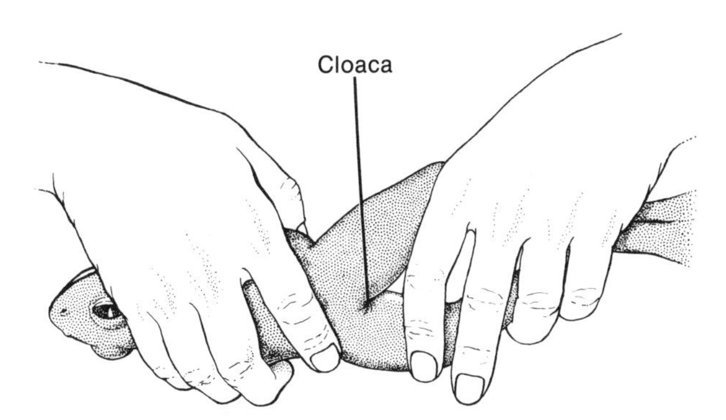

FIG. 20-3
Method of holding frog to test for ovulation.

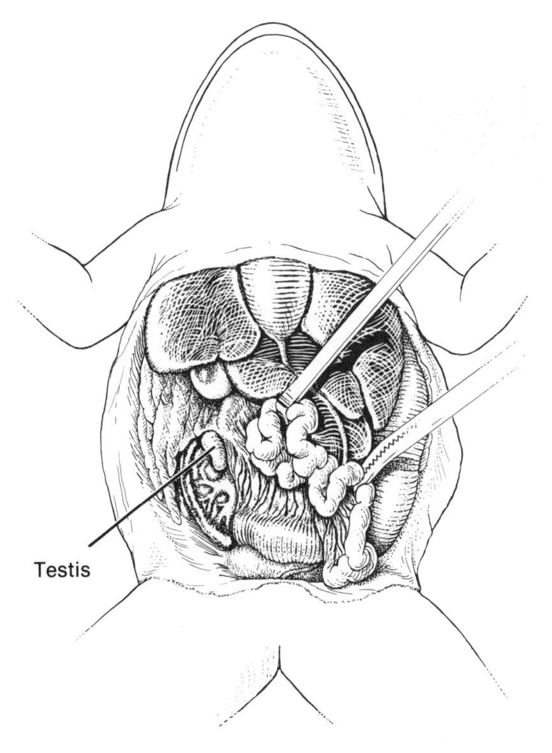

FIG. 20-4
Procedure for removing testes.

4. Pipet a drop of the sperm suspension onto a glass slide and examine the sperm under the high power of your microscope. Describe the shape of the sperm.

____

What is the physical basis for the motility of the sperm?

____

____

C. FERTILIZATION

Divide the motile sperm suspension prepared earlier among two or three *clean* petri dishes so that the bottom of each dish is just covered. Holding the female frog as shown in Fig. 20-3, strip the eggs directly into the sperm suspension. Line the egg masses in rows or in a spiral so that all the eggs are in contact with the sperm. *Do not place them in a heaping pile in the sperm.* Shake the dishes gently. Why?

____

Note the random orientation of the eggs. Some will have the pigmented animal pole uppermost; in others the lightcolored vegetal pole will face up.

When the eggs are shed, the first polar body has been formed. The formation of the second polar body can be observed shortly after fertilization. To see this process, remove two or three eggs after they have been in the sperm suspension for several minutes. Remove their jelly coats (Fig. 20-5), place them in a Syracuse dish, and cover completely with Holtfreter's solution. Examine the eggs with a dissecting microscope. Adjust the light so that it strikes the egg surface at an oblique angle. With careful focusing, you will see a lighter, circular area in the animal pole. By adjusting the light, you may observe a pit in this light area (Fig. 20-6). This is where the second polar body will be expelled. Carefully observe the formation of this polar body. It may sometimes be seen more readily if the egg is rotated so that the pit area is at right angles to your field of vision.

After the remaining eggs have been in the sperm suspension for 10 minutes, flood them with 10% Holtfreter's solution. Rinse and completely cover the eggs with fresh Holtfreter's solution. The first cleavage will occur in about 2 hours. During this interval, go on to Part D or E as directed by your instructor.

D. UNFERTILIZED FROG EGGS

From the injected female frog, gently squeeze about a dozen eggs into a petri dish containing Holtfreter's solution. The solution should completely cover the eggs. Note that the eggs are clustered together when they first leave the cloaca but tend to separate from each other in the water. The eggs separate because the jelly that surrounds each egg swells considerably as the protein in the jelly absorbs water. Why is the swelling important in the development of the eggs?

____

____

Place one or two eggs on a glass slide, cover with water, and examine them with the low power of your microscope. How many jelly layers are present?

____

Examine the remaining eggs, using a dissecting microscope. Adjust the spot lamp so the light is striking the eggs at about a 45° angle. How does the pigmentation of the frog egg differ from that of the sea urchin egg?

____

____

____

Remove as much jelly as possible from two or three eggs using the procedure shown in Fig. 20-5. Place them in a Syracuse dish animal pole up, completely cover with Holtfreter's solution, and examine with the highest power available on your dissecting microscope. Adjust the spot lamp for maximum light. The dark color of the egg is due to granules of melanin pigment. Although the egg may seem to be inactive, close observation of the cytoplasm will reveal a great deal of movement of the pigment granules. What might be causing this activity?

____

____

____

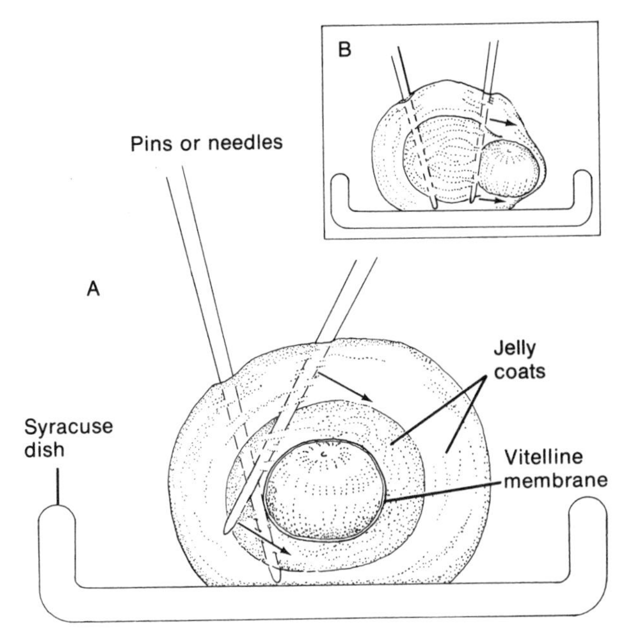

FIG. 20-5
Procedure for stripping jelly coats from frog eggs.

E. CLEAVAGE STAGES

Examine the fertilized eggs in your dish. When an egg is fertilized, it secretes a proteinaceous material. As this proteinaceous material begins to absorb water, it forces away the **vitelline membrane** that had been tightly bound to the egg surface. The yolk-laden vegetal pole, being heavier, shifts downward, so that the animal pole is oriented upward. Approximately what percentage of the eggs in your dish have been fertilized?

In nature, why is it advantageous to the development of the egg to have the pigmented area uppermost?

Examine the eggs with a dissecting microscope for the appearance of the first cleavage, resulting in two cells. The phenomenon of cleavage is the partitioning of the egg cell into a large number of smaller cells with no increase in the size of the mass. (Be careful not to confuse degenerating eggs with normally dividing eggs; a broken, mottled surface is a sign that the egg is dead.) If your eggs have not reached the first cleavage stage, obtain eggs that were fertilized before the class meeting. Continue to examine the eggs periodically. Where does the first cleavage furrow begin?

Note: Continued reference to Fig. 20-7 (on the next two pages) will help you in identifying stages of normal development. The stages were originally numbered by Waldo Shumway in 1940. If embryos are allowed to grow at either of the temperatures indicated, various developmental stages can be made available when needed.

F. LATER STAGES OF DEVELOPMENT

1. Blastula

The process of cleavage terminates with the formation of the blastula (Fig. 20-7), although the organ-

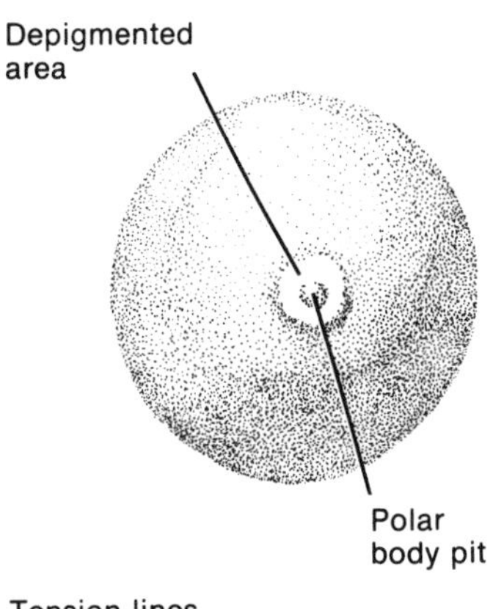

FIG. 20-6
Location of second polar body.

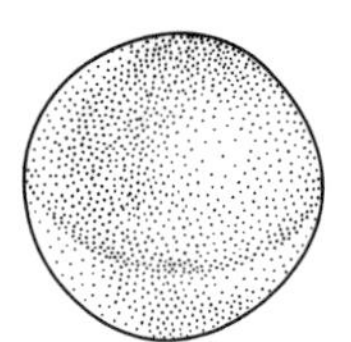
1. Unfertilized

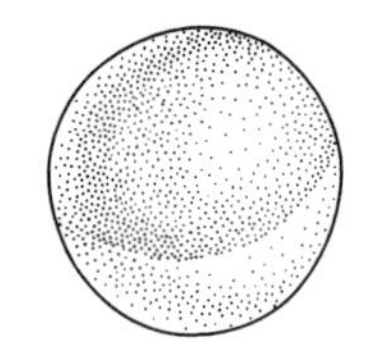
2. Gray crescent

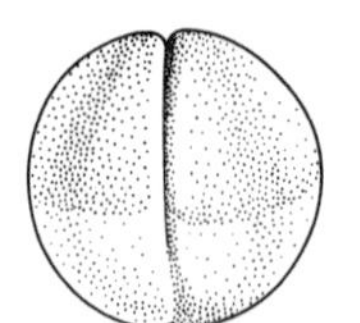
3. Two cells

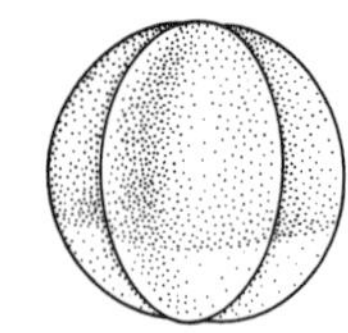
4. Four cells

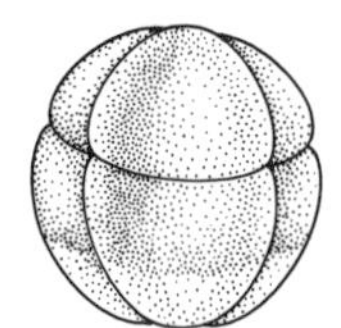
5. Eight cells

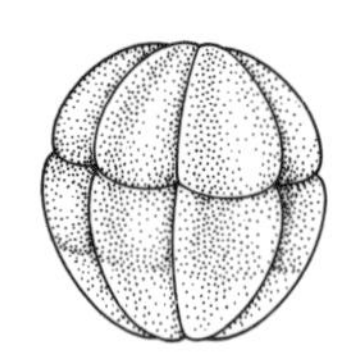
6. Sixteen cells

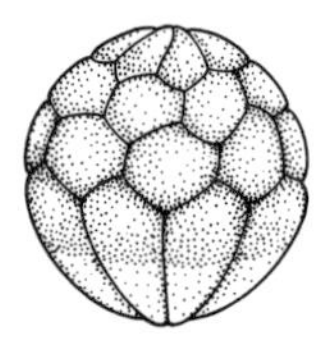
7. Thirty-two cells

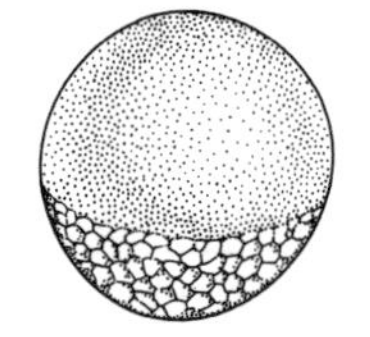
8. Early blastula

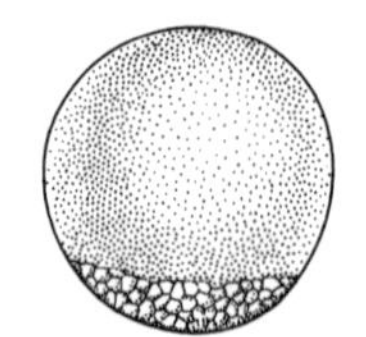
9. Late blastula

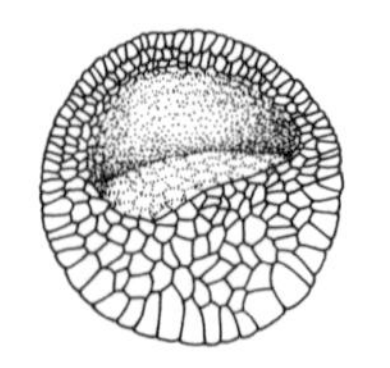
9A. Cross section of blastula

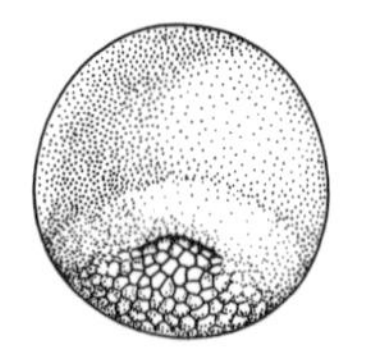
10. Dorsal lip

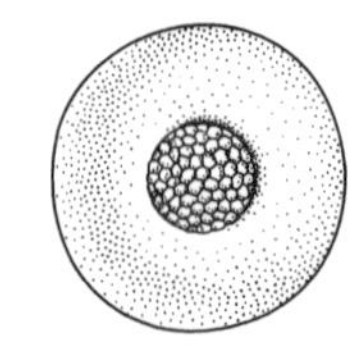
11. Mid-gastrula

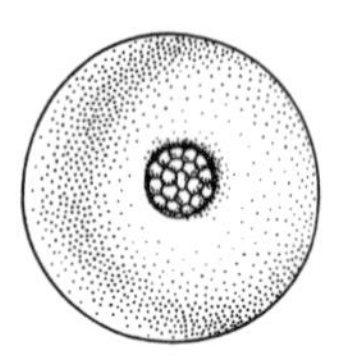
12. Late gastrula

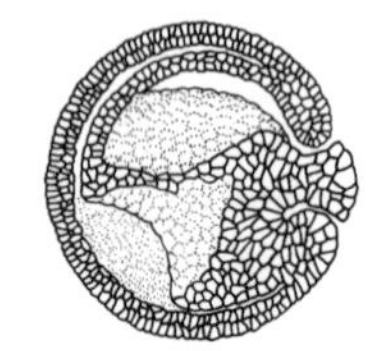
12A. Cross section of gastrula

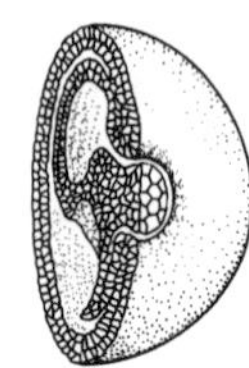
12B. Half gastrula

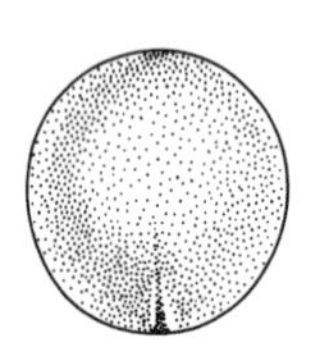
13. Neural plate

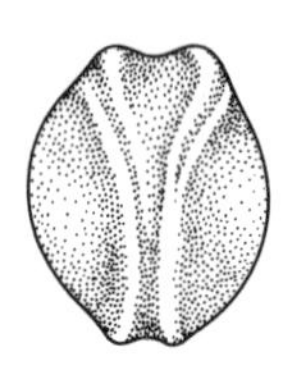
14. Neural folds

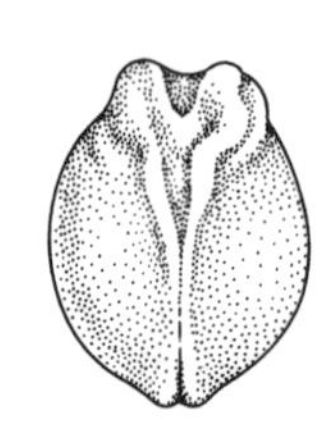
15. Rotation

Shumway stage	Description	Age in hours	
		18°C	25°C
1	Unfertilized	0.0	0.0
2	Gray crescent	1.0	0.5
3	Two cells	3.5	2.5
4	Four cells	4.5	3.5
5	Eight cells	5.5	4.5
6	Sixteen cells	6.5	5.5
7	Thirty-two cells	7.5	6.5
8	Early blastula	16.0	11.0
9	Late blastula	21.0	14.0
10	Dorsal lip	26.0	17.0
11	Mid-gastrula	34.0	20.0
12	Late gastrula	42.0	32.0
13	Neural plate	50	40
14	Neural folds	62	48
15	Rotation	67	52
16	Neural tube	72	56
17	Tail bud	84	66
18	Muscular response	96	76
19	Heart beat	118	96
20	Gill circulation	140	120
21	Mouth open	162	138
22	Tail fin circulation	192	156
23	Opercular fold	216	180
24	Operculum closed on right	240	210
25	Operculum complete	284	240

FIG. 20-7
Normal embryonic developmental stages of the frog from unfertilized egg through tadpole (after Shumway, 1940).

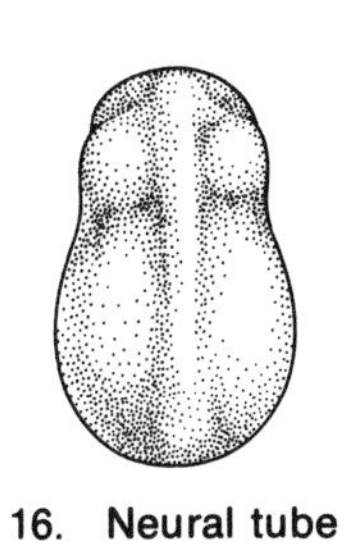

16. Neural tube

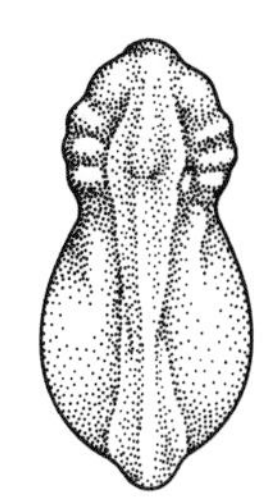

17. Tail bud

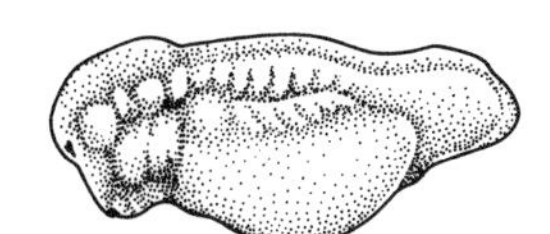

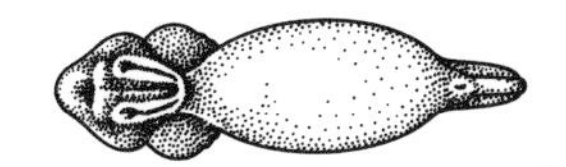

18. Muscular response

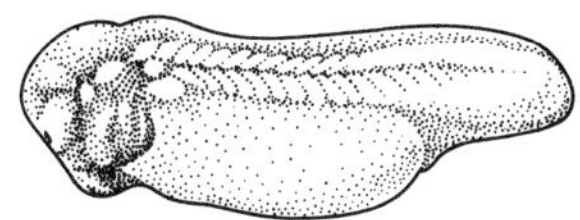

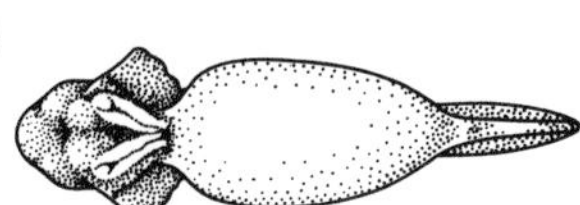

19. Heart beat

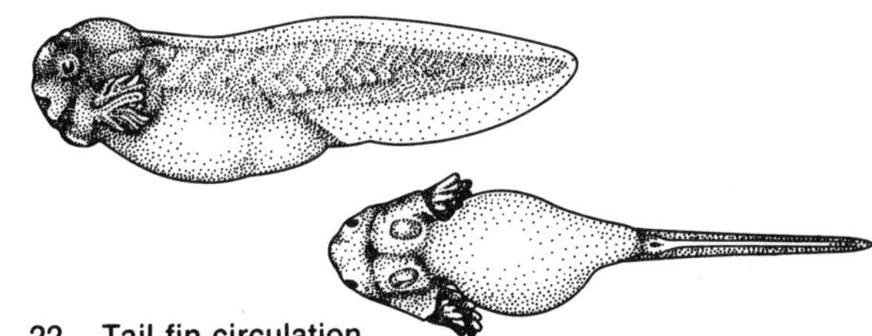

20. Gill circulation and hatching

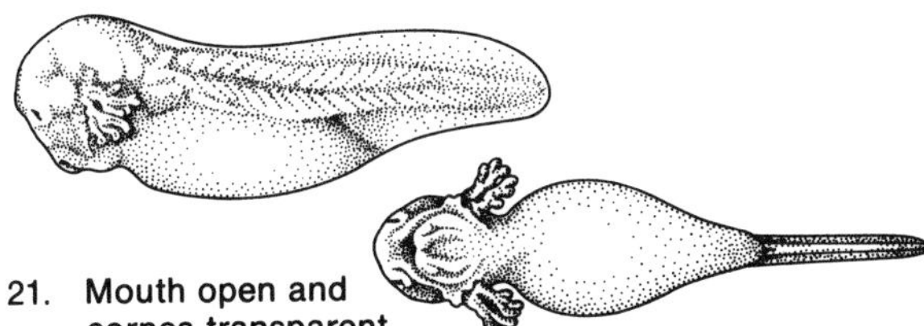

21. Mouth open and cornea transparent

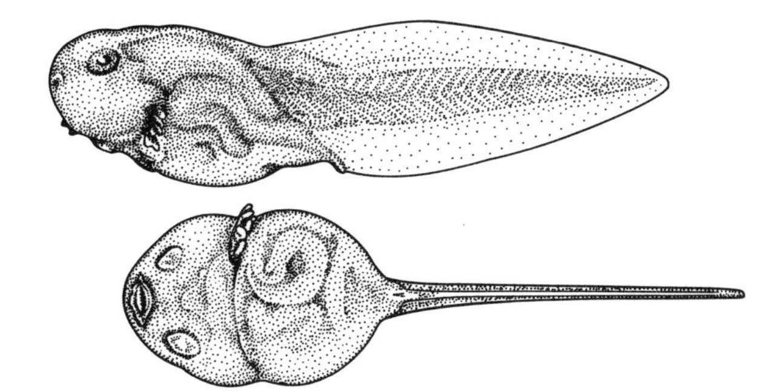

22. Tail fin circulation

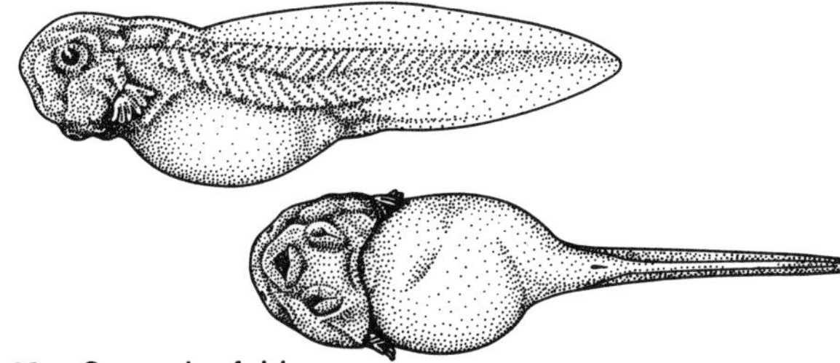

23. Opercular fold and teeth

24. Operculum closed on right

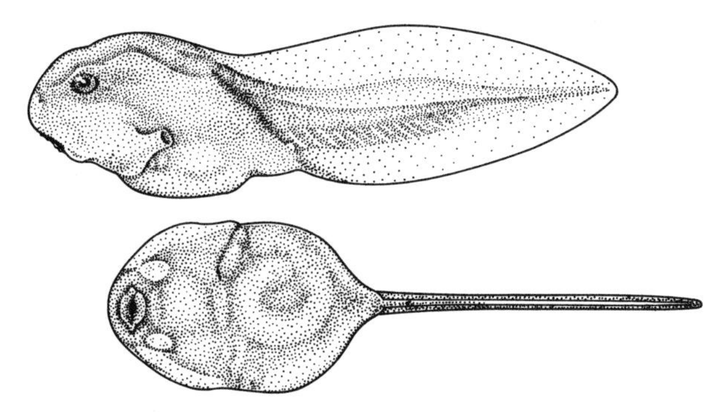

25. Operculum complete

ism is still undergoing cellular division. Different regions of the blastula have different prospective fates: that is, the cells of different areas will ultimately contribute to the skin, skeleton, nervous tissue, and so on (Fig. 20-8). How could you follow the movement of various cells as they proceed through subsequent developmental stages in the living embryo?

____

____

____

In commercially prepared models, these presumptive areas (called *ectoderm, mesoderm,* and *endoderm)* are arbitrarily colored blue, red, and yellow, respectively. A specialized region of mesoderm, called *chordamesoderm,* is colored green. From your readings and the instructor's discussion, give examples of tissues that have their origin in the following:

Presumptive mesoderm: ______________

Presumptive endoderm: ______________

Presumptive ectoderm: ______________

Chordamesoderm: ______________

At this stage of development, the presumptive areas are associated with the *outside* of the embryo. However, in the mature animal the muscles, alimentary tract, nervous tissue, and other tissues and organs that arise from the areas are situated *inside.* The process of getting these areas of cells to the interior is called *gastrulation.*

2. Gastrulation

Patterns of gastrulation vary. In the sea urchin, one side of the hollow blastula invaginates, forming a two-layer cup; later a third tissue layer, the mesoderm, arises between the inner and outer layers. In the frog, gastrulation is accomplished in a quite different manner. From the use of models, Fig. 20-9, and your instructor's discussion, answer the following questions.

What is epiboly?

____

What is the first visible indication that gastrulation has started?

____

____

____

What is the blastopore?

____

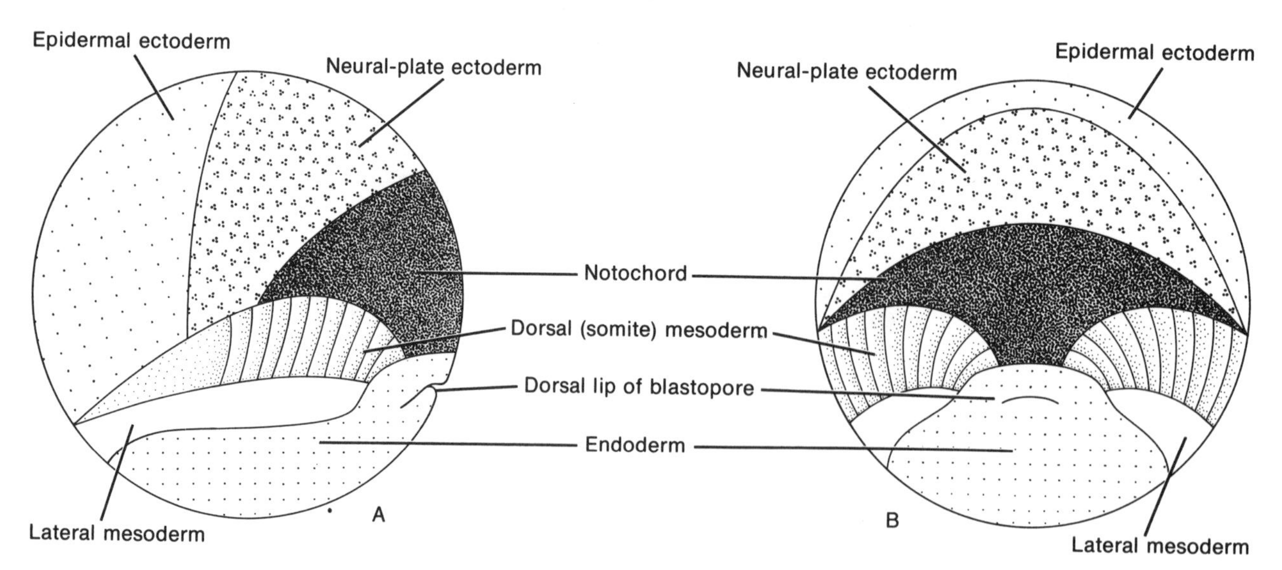

FIG. 20-8
Fate map of the frog embryo: (A) lateral view and (B) frontal view of the blastula.

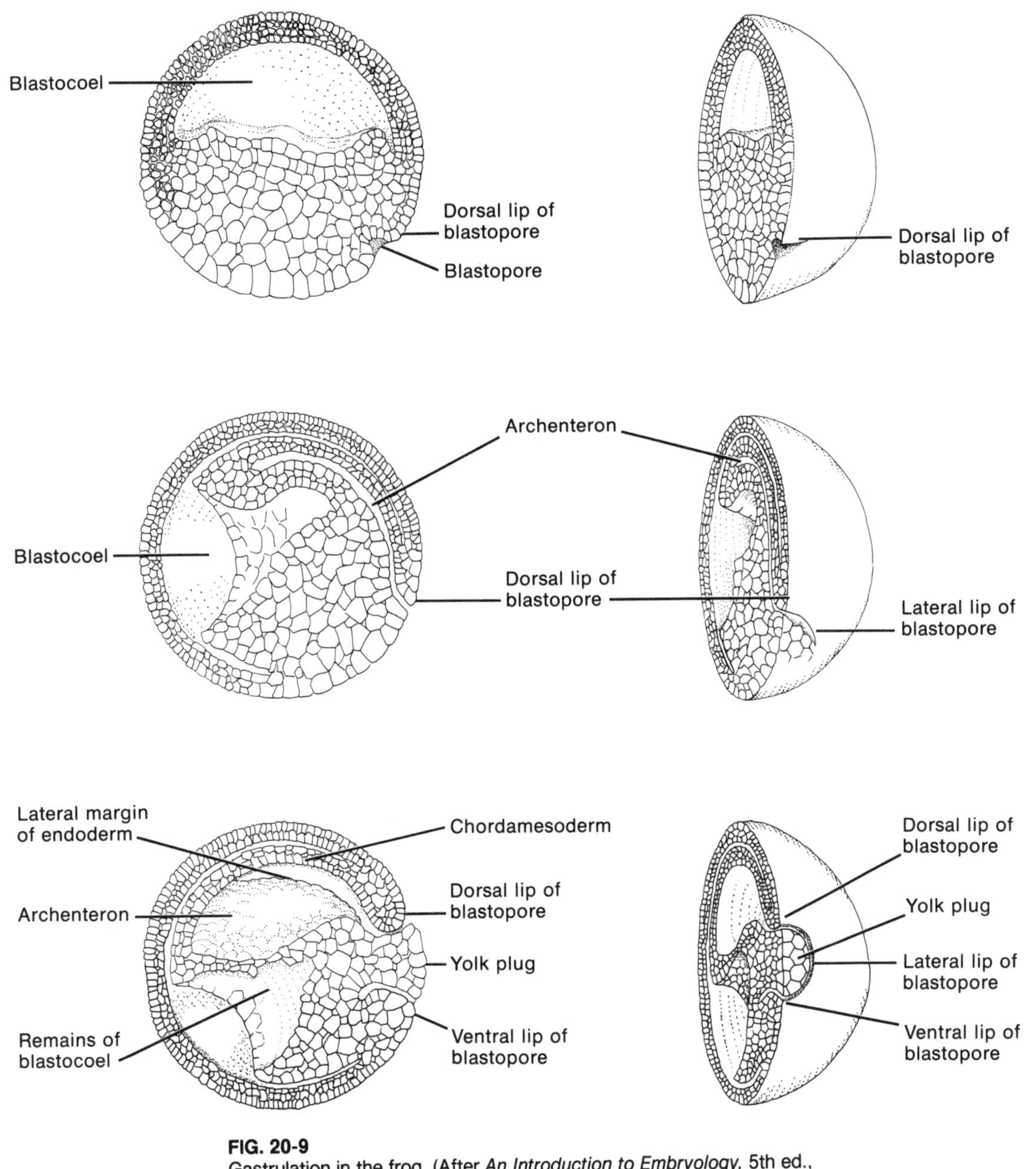

FIG. 20-9
Gastrulation in the frog. (After *An Introduction to Embryology,* 5th ed., by B. I. Balinsky. Holt, 1981.)

By what morphogenetic movement is the blastopore formed?

____

What role in development does the dorsal lip of the blastopore play?

____

____

____

What new cavity is formed during gastrulation?

____

What does this cavity ultimately become in the adult?

____

____

The chordamesoderm, which will form the notochord, plays what other important role in the development of the frog?

___

___

3. Formation of the Neural Tube

When gastrulation is completed, the embryo is completely covered, except for the yolk plug, by ectoderm. The endoderm and mesoderm have been moved inside. Soon, two ectodermal ridges will form on the dorsal surface of the gastrula. These grow upward and toward each other, eventually fusing in the midline. When the edges fuse, a tube of ectoderm will be formed, running from the anterior to the posterior end of the embryo. This tube, covered by an outer ectoderm layer, will develop into the brain and spinal cord. Follow these events, using Fig. 20-7 (Stages 13–16) and models. If available, examine living embryos at this stage of development.

4. Development of the Tadpole

The neural tube is formed as a result of the outfolding and fusion of the ectoderm. Most other tissue-forming processes occurring in later development consist of foldings, invaginations, evaginations, or cell movements of the various germ layers. For example, the digestive glands and lungs develop as evaginations of the endoderm at various stages. The eye develops, in part, from an evagination of the brain. Examine these morphogenetic movements on the models. List any others you find.

___

___

___

___

___

___

___

___

G. EXPERIMENTAL STUDIES ON AMPHIBIAN EMBRYOS

In the early stages of developmental biology, attention centered on describing the various orderly events occurring during the normal development of organisms. A wealth of detailed information was gained from these direct observations. However, studies of the developing organism raises a multiplicity of questions: How is growth initiated? Why does growth stop? Why do cells differentiate? What regulates form? In order to answer these questions embryologists moved from observation and description to experimental studies designed to shed light on these and other developmental phenomena.

In this exercise you will use the amphibian embryo to study two developmental phenomena: (1) initiation of development of the frog egg by artificial means and (2) the reaggregation of the dissociated cells of a blastula.

1. Parthenogenesis

In the normal sequence of development, a mature egg is penetrated by a sperm. As a result, the egg receives a set of the genetic characteristics from the male; it is also stimulated to develop. Because the sperm physically breaks through the surface of the egg, we might ask if artificially breaking the surface by using a glass needle would also initiate development. This process, whether it occurs artificially or naturally (as it does in the eggs of honeybees, with males developing from unfertilized eggs), is called **parthenogenesis** (from the Greek words *parthenos,* meaning "virgin," and *genesis,* meaning "birth"). To answer this question, the procedure given below should be followed.

1. To be sure that all instruments are clean and free of sperm, wash them in 70% alcohol, then rinse with distilled water and dry.

2. Strip the eggs from the ovulating frog in a single row along the length of a glass slide (Fig. 20-10A.). Prepare at least 10 slides in this way. Set each slide on a Syracuse dish containing water, and place them under a jar as shown in Fig. 20-10B.

3. Obtain a second female frog that has been segregated from male frogs for several days. Why is this necessary?

___

___

Pith the frog (see Appendix E for this procedure) and dissect it to expose the heart (Fig. 20-10C). Cut off the tip of the heart and let the blood flow into the body cavity so that it mixes with the coelomic fluid.

4. Dissect out a piece of muscle from the abdominal wall and dip it into the mixture of blood and coelomic fluid. Remove five slides from the moist chamber and streak the eggs with this fluid (Fig. 20-10D). Using a dissecting microscope, prick each egg, slightly off center in the animal pole, with a fine glass needle.

5. Place the slides of eggs in petri dishes containing 10% Holtfreter's solution so that the eggs are covered by the fluid. Cover the dishes, label them "Experimental Groups," and keep in a cool place at about 68°F (20°C).

6. Describe the control group that should be tested.

____

____

____

Use the remaining eggs for your controls.

7. Examine the eggs after 2 hours, and periodically thereafter. If cleavage occurs, is the pattern of cleavage similar to that found in normally fertilized eggs?

____

Describe the developmental timing and stages of parthenogenetically induced eggs in comparison with those of normal eggs as shown in Fig. 20-7.

____

____

If no cleavage at all occurs, does this prove that sperm are necessary for development? Explain.

____

____

____

____

2. Reaggregation of Embryonic Cells

Numerous studies have centered on the behavior of cells isolated from their normal relationships in a tissue, using various dissociation procedures. For example, in 1907, Wilson cut up sponges and squeezed the pieces through cheesecloth to separate individual cells. These isolated cells actively moved about for awhile and then aggregated into clumps that subsequently organized themselves into functional miniature sponges. Similar studies have now been done with amphibian, avian, and mammalian species. Moscona carried out an interesting one in which chick cartilage cells and chick liver cells were isolated and then mixed together. Upon reaggregation the cells sorted into discrete liver and cartilage aggregations without intermingling the cell types.

In this study you will attempt to dissociate the cells of a frog blastula and determine the pattern of reaggregation.

1. Obtain several frog embryos in a late blastula or early gastrula stage (Fig. 20-7, stages 9 or 10). Using a fine forceps, transfer one of the embryos to 70% ethyl alcohol *for not more than 5 seconds.* This will reduce the number of bacteria on the jelly layer without damaging the embryo. Rinse off the alcohol by immersing the embryo in sterile Steinberg's solution (a physiological solution for frog's eggs) and then place it in a separate container. (*Note:* Carry out these procedures on several eggs to obtain adequate results.)

2. Using a dissecting microscope and clean watchmaker's forceps, remove the jelly coats and vitelline membrane from the embryo. Then, using a special pipet supplied by your instructor, transfer the embryo to a small petri dish containing 2–3 ml of dissociating medium. This medium lacks magnesium and calcium ions (important in intercellular binding) and contains ethylene diamine tetraacetate (EDTA), a compound that binds and removes any calcium or magnesium from the intercellular "cement" holding the cells together. Thus, in a short time the cells of the embryo should begin to dissociate. You can speed up the process by gently aspirating the embryo in and out of the pipet, or by gentle manipulation with the forceps.

3. With a clean Pasteur pipet, transfer the cells to fresh Steinberg's solution to "wash" the cells free of any adhering dissociating solution. After a few minutes, transfer the washed cells to a flat-bottomed well-type slide containing "aggregating" solution (Steinberg's solution supplemented with serum or egg albumin). Transfer just enough cells to sparsely scatter over the floor of the culture vessel.

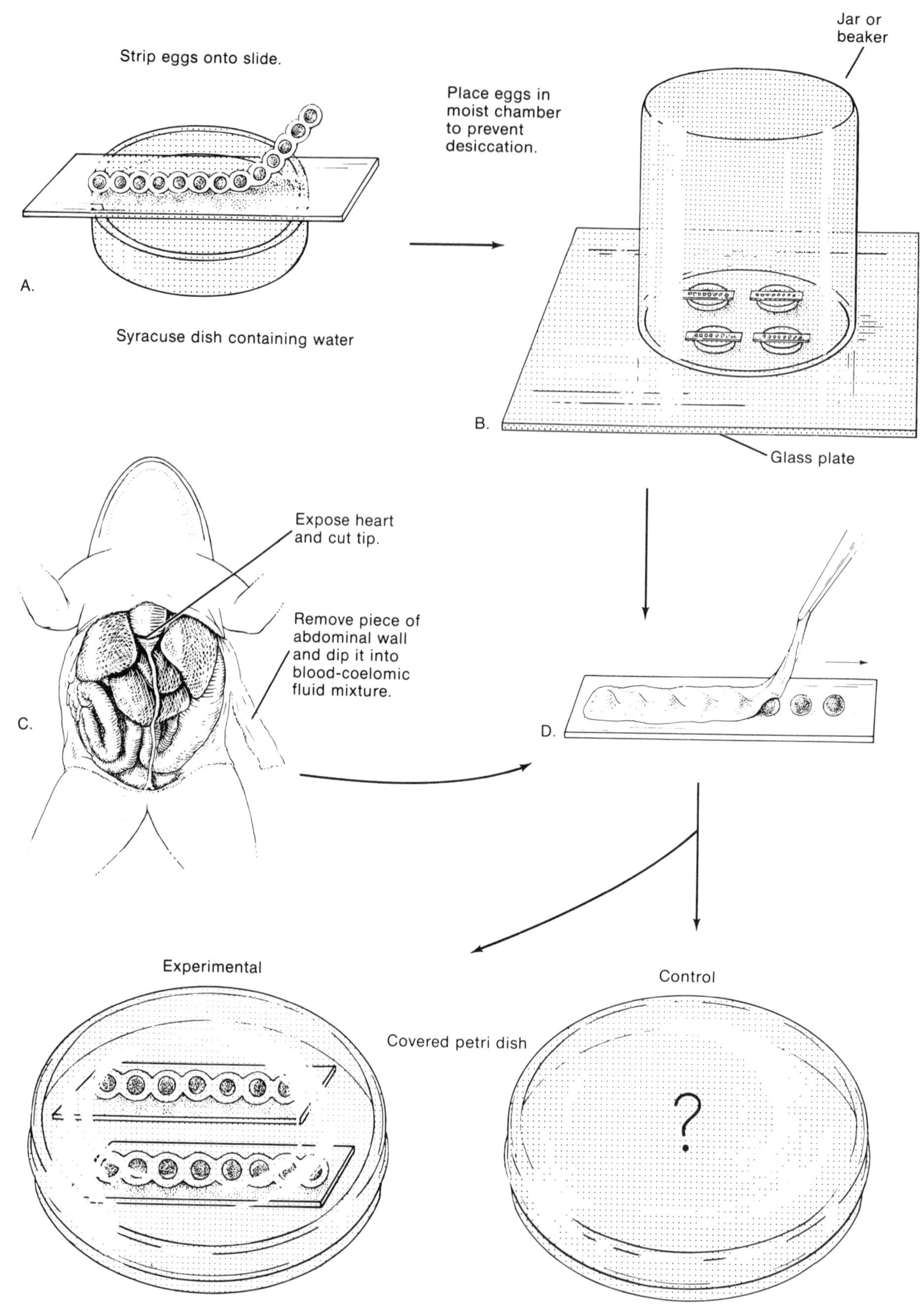

FIG. 20-10
Procedure for inducing parthenogenetic development of frog eggs.

4. Add a thin coat of vaseline to the rim of a coverslip and moisten the entire coverslip by dipping it into Steinberg's solution. Place the coverslip over the well of the slide so that the vaseline makes contact with the glass. This procedure will retard evaporation and reduce fogging of the coverslip during subsequent observations.

5. Using the scanning objective (4 ×–5 ×) of a compound microscope, focus on the cells. Disturb the cultures as little as possible during subsequent observations. (CAUTION: To prevent overheating the embryos, turn the microscope lamp off when you are not observing the cells.)

6. Observe the cells carefully throughout the next few hours and periodically for several days thereafter. The cultures should be kept at room temperature and the microscope covered with a plastic bag.

Record your observations. Look for such events as cyclosis (internal cytoplasmic streaming), protrusions from the cell surface, cellular movements, and so forth. You might consider recording your observations in the form of annotated drawings. For the purpose of this experiment, consider aggregation to be completed upon the formation of rather smooth, dark clusters of cells.

REFERENCES

Balinsky, B. I. 1981. *An Introduction to Embryology.* 5th ed. Saunders College/Holt, Rinehart and Winston.

Bradfield, B. 1967. The Use of Egg White as a Protein Additive in Medium for Embryonic Frog Cell Reaggregation. *Proceedings of the South Dakota Academy of Science* 46:259–260.

Browder, L. W. 1980. *Developmental Biology.* Saunders College/Holt, Rinehart and Winston.

Bryant, P. J., S. V. Bryant, and V. French. 1977. Biological Regeneration and Pattern Formation. *Scientific American* 237:66–81 (Offprint 1363). *Scientific American* Offprints are available from W. H. Freeman and Company, 41 Madison Avenue, New York 10010, and 20 Beaumont Street, Oxford OX1 2NQ, England. Please order by number.

Epel, D. 1977. The Program of Fertilization. *Scientific American* 237:128–138 (Offprint 1372).

Hamburger, V. 1960. *A Manual of Experimental Embryology.* rev. ed. University of Chicago Press.

Jones, K. W., and T. R. Elsdale, 1963. The Culture of Small Aggregates of Amphibian Embryonic Cells in Vitro. *Journal of Embryological and Experimental Morphology* 11:135–154.

Karp. G., and N. J. Berrill. 1981. *Development.* 2d ed. McGraw-Hill.

Moscona, A. 1961. How Cells Associate. *Scientific American* 205(3):142–162 (Offprint 95).

Rugh, R. 1951. *The Frog: Its Reproduction and Development.* McGraw-Hill.

Rugh, R. 1962. *Experimental Embryology: Techniques and Procedures.* 3d ed. Burgess.

Shumway, W. 1940. Stages in the Normal Development of *Rana pipiens.* I. External Form. *Anatomical Record* 78:139–147.

Slack, J. M. 1983. *From Egg to Embryo: Determinative Events in Early Development.* Cambridge University Press.

Trinkaus, J. P. 1984. *Cells into Organs.* 2d ed. Prentice-Hall.

Wolper, L. 1978. Pattern Formation in Biological Development. *Scientific American* 239:154–164 (Offprint 1409).

EXERCISE 21

Early Development of the Chick

The chick embryo has been used so extensively in developmental studies that it is considered to be one of the most intensely studied of all organisms. The interest in the chick embryo is due to its ready availability and its marked similarity in early embryonic development to the mammalian embryo.

In this exercise, you will become familiar with the early developmental features of the chick embryo. However, many of the very early events (i.e., cleavage, formation of the blastula, and gastrulation) are not observed as readily as similar stages in the frog embryo. One reason for this lies in the unique organization of the chick "egg." The egg (popularly called the *yolk*) is surrounded by various accessory coverings secreted by the female reproductive tract.

A. FERTILIZATION AND EGG LAYING

Direct observations of fertilization are very difficult in birds and mammals because the process takes place inside the body of the female. It is known that when the ovum is liberated from the ovary it begins to undergo certain changes that can be characterized as aging. These changes progress rapidly to a point at which the ovum, although technically still alive, can no longer be fertilized. If, however, the ovum is fertilized soon after liberation, these aging processes are checked. Fertilization in the chicken takes place just as the ovum is entering the oviduct (Fig. 21-1).

The accessory coverings are secreted around the ovum in the course of its subsequent passage toward the cloaca. In the part of the oviduct adjacent to the ovary, a mass of stringy **albumen** is produced. This adheres to the **vitelline membrane** and projects beyond it in two masses extending in both directions along the oviduct. Because of the spirally arranged folds in the oviduct wall, the ovum rotates as it moves toward the cloaca. This rotation twists the albumen to form spiral strands at either end of the yolk, known as the **chalazae.** Additional albumen is added in concentric rings as the ovum moves down the first half of the oviduct. These layers of albumen can readily be seen in boiled eggs, because the heat causes the albumen to coagulate.

The **shell membranes,** which consist of matted or-

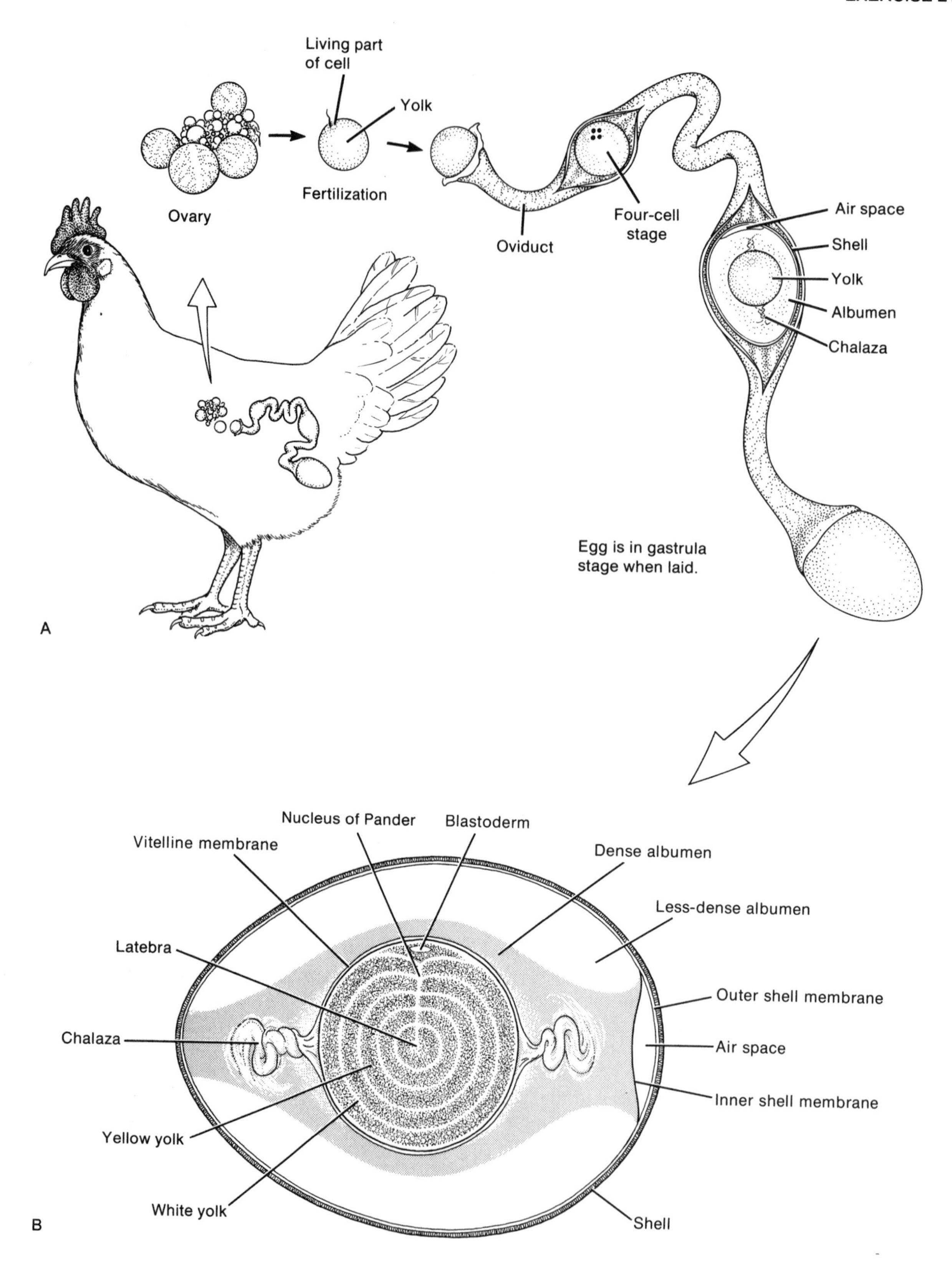

FIG. 21-1
(A) Stages of fertilization and early development in the hen oviduct. (B) Structure of the egg when it is laid.

ganic fibers, are added farther along in the oviduct. The **shell** is secreted as the egg passes through the shell gland (the uterus). The entire passage of the ovum—from its discharge from the ovary until it reaches the end of the oviduct and is ready for laying—takes about 22 hours.

To examine the internal organization of the chicken egg, remove the shell from a hard-boiled egg, and then carefully remove the coagulated egg white surrounding the yolk. On the surface of the yolk, locate a small, lighter circular area. With a sharp razor blade, cut the yolk into two equal parts through this area. Close examination of the yolk will reveal that it consists of **white yolk** and **yellow yolk,** which differ in the size and shape of their granules, thus giving this difference in color (Fig. 21-1). The granules and globules in white yolk are generally smaller and less uniform in appearance than those in the yellow yolk. These concentric layers of yolk are believed to indicate the daily accumulations of cytoplasm in the final stages of the formation of the ovum. The principal accumulation of white yolk lies in a central flask-shaped area, the **latebra,** which extends toward the blastoderm and flares out under it into a mass known as the **nucleus of Pander.**

The albumen, except for the chalazae, is nearly homogeneous in appearance. Two layers of shell membranes lie in contact with the albumen everywhere except at the large end of the egg, where the inner and outer membranes are separated to form an air space. This space is believed to form after the egg has been laid because evaporation of water causes the egg to contract as it becomes cooler than it was inside the hen. The size of the air space increases as eggs are kept over time owing to continued evaporation of water from the egg. This fact is used in the familiar method of testing the freshness of eggs by "floating" them.

The egg shell is composed largely of calcareous salts and is porous, which allows the exchange of gases between the embryo and the outside air.

On the surface of the yolk, locate a small, whitish spot—the **blastoderm.** This is the living part of the egg. The remainder is composed of inert food material and water. Normally the blastoderm faces up.

After the egg has been fertilized (as it passes through the upper region of the oviduct), cleavage of the blastoderm quickly occurs. The mitotic divisions of cleavage proceed as the egg passes through the oviduct, including the part that contains the shell gland. Cleavage in the chick is termed *partial,* or *discoidal,* because the highly irregular cleavage divisions are restricted to the blastoderm (Fig. 21-2), whereas the huge mass of yolk does not divide. As cleavage continues, the blastoderm becomes several

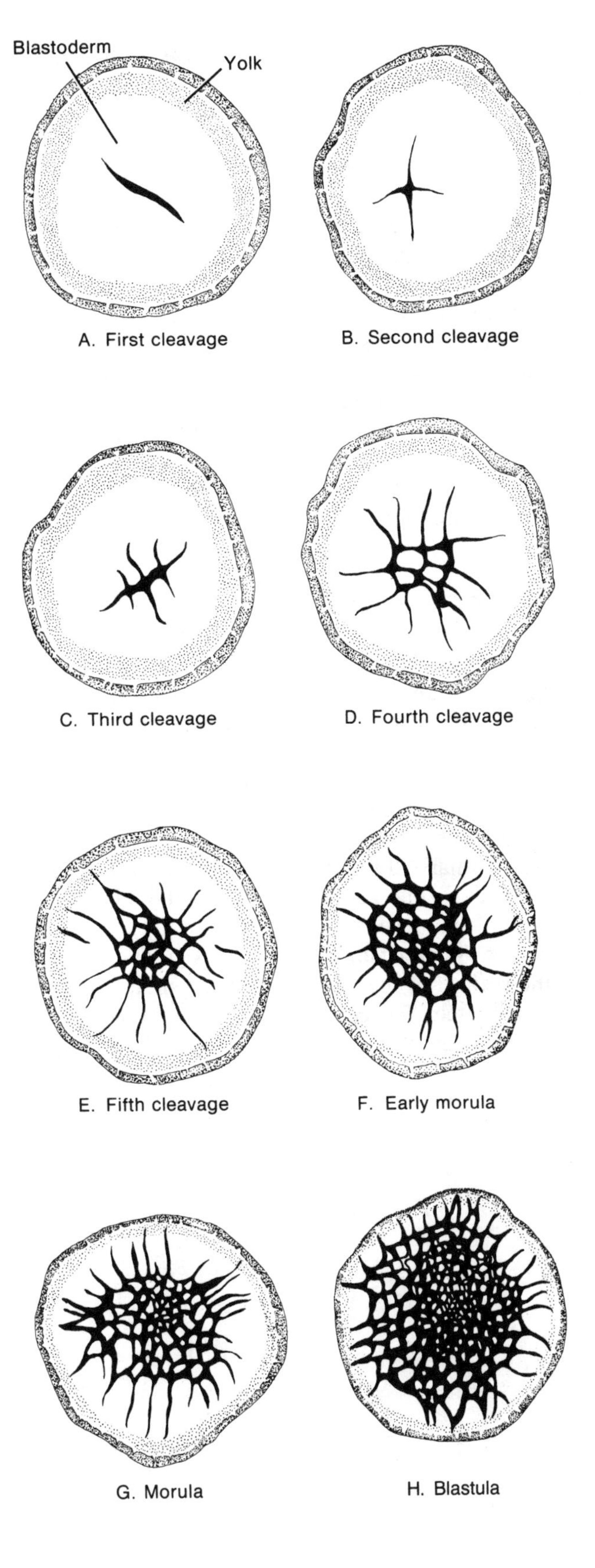

FIG. 21-2
Successive cleavages as observed looking down at the surface of the blastoderm.

cell layers thick and a fluid-filled **subgerminal space** develops beneath it (Fig. 21-3A).

B. BLASTULATION

The blastula begins to form when some of the cells of the blastoderm separate and move toward the subgerminal space (Fig. 21-3B). This movement leads to the formation of a two-layered structure, the two layers being separated by a space—the **blastocoel.** The newly formed and rather loosely organized lower layer is called the **hypoblast.** The upper **epiblast** contains cells that are arranged in an orderly epithelial (sheetlike) fashion. The developmental process of separation, by which the hypoblast and epiblast are formed is called **delamination.**

C. GASTRULATION

Following delamination (which is arrested when the egg is laid and only resumes when the egg is incubated), the cells of the epiblast reorganize themselves and move to new positions where they have different developmental fates; that is, the cells will ultimately contribute to the skin, skeleton, nervous system, and other tissues. The developmental roles of these cells are summarized in the **fate map** shown in Fig. 21-4.

At this time, a thickening of the blastoderm occurs that results in the formation of a narrow groove—the **primitive streak.** This groove traverses the surface of the pearshaped blastoderm from approximately the center to the tapered edge (Fig. 21-5). Cells of the epiblast then begin moving to-

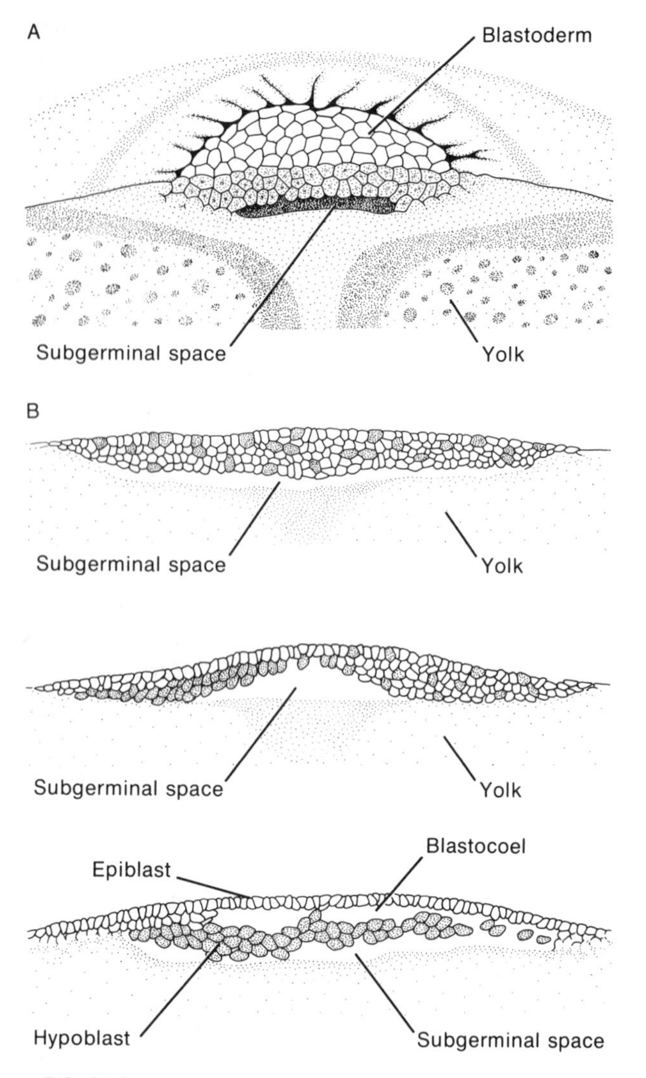

FIG. 21-3
(A) Cross section of the blastoderm. (B) Delamination of the blastoderm to form the epiblast and hypoblast.

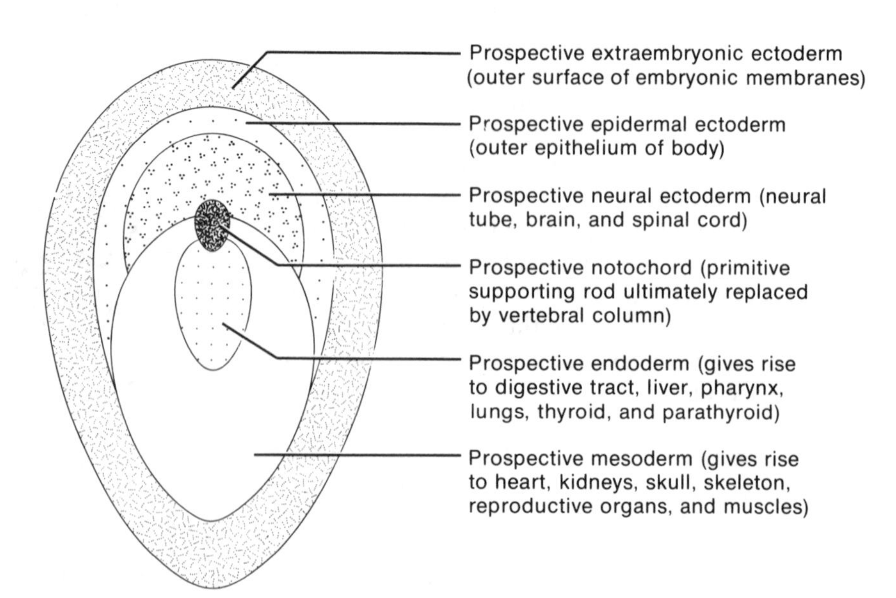

FIG. 21-4
Fate map of chick embryo epiblast.

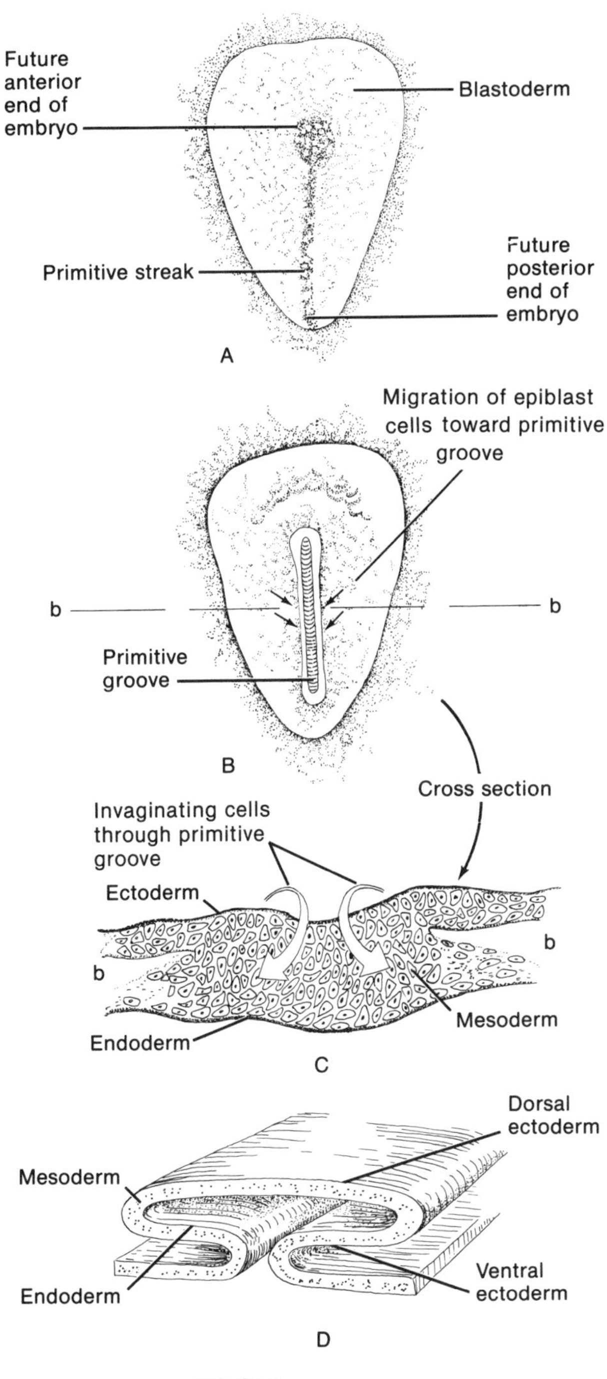

FIG. 21-5
Gastrulation in the chick.

ward, down, into, and then away from the primitive streak (Fig. 21-5B, C) to form a **mesodermal** layer and an embryonic **endodermal** layer. The remaining surface layer becomes the **ectoderm.**

When the mesoderm is fully established, gastrulation is complete. Eventually, the surface ectoderm will be folded under the embryo. When this is completed (Fig. 21-5D), the ectoderm will no longer be a single sheet of cells, but a complete outer covering.

D. LATER STAGES OF DEVELOPMENT

Like the frog embryo, the chick embryo passes through a number of recognizable morphological stages of development. These are usually identified by the number of hours of incubation (at 38°C) required to reach that stage of development. Stages after the appearance of blocklike cell masses of mesoderm, called **somites,** at about 21 hours are also often identified by the number of somites present.

In this exercise you will study two stages of chick development: the 33-hour stage (12–15 somites) and the 72-hour stage (36 somites). Your study will include examination both of living embryos at these two stages and of demonstration slides with cross sections cut through various regions of the 33-hour chick embryo.

1. The Living 33-Hour (12–15 Somites) Chick Embryo

Approximately 33 hours before the laboratory meeting, fertilized eggs were placed in an incubator by your instructor.

For observation of the living embryo, crack an egg, carefully remove the shell, and place the egg in a finger bowl containing warm chick Ringer's solution; the unbroken yolk with its developing embryo will float in the saline solution (Fig. 21-6A–C). (If only a white spot is found on the surface of the yolk, the egg has failed to develop. If this is the case, use another egg.) Remove the blastoderm from the yolk mass as shown in Fig. 21-6D–F. Examine the embryo, using a dissecting microscope. Supplement your observations of the living embryo with prepared slides and models available in the laboratory (Fig. 21-7).

At this stage of development, the blastoderm has become a thin disc about 20 mm across. Surrounding the embryo is a transparent region—the **area pellucida.** Outside of this is the **area opaca,** in which the blood vessels are beginning to develop. One of the most striking features of this stage is the presence of the "heart," which can be seen on the right side of the embryo. The heart originates as separate masses of cells on either side of the embryonic axis. These heart-forming regions move to the midline and eventually fuse (Fig. 21-8). If extra unincubated fertilized eggs are available, you might incubate them for 26 hours and then remove the embryo as shown

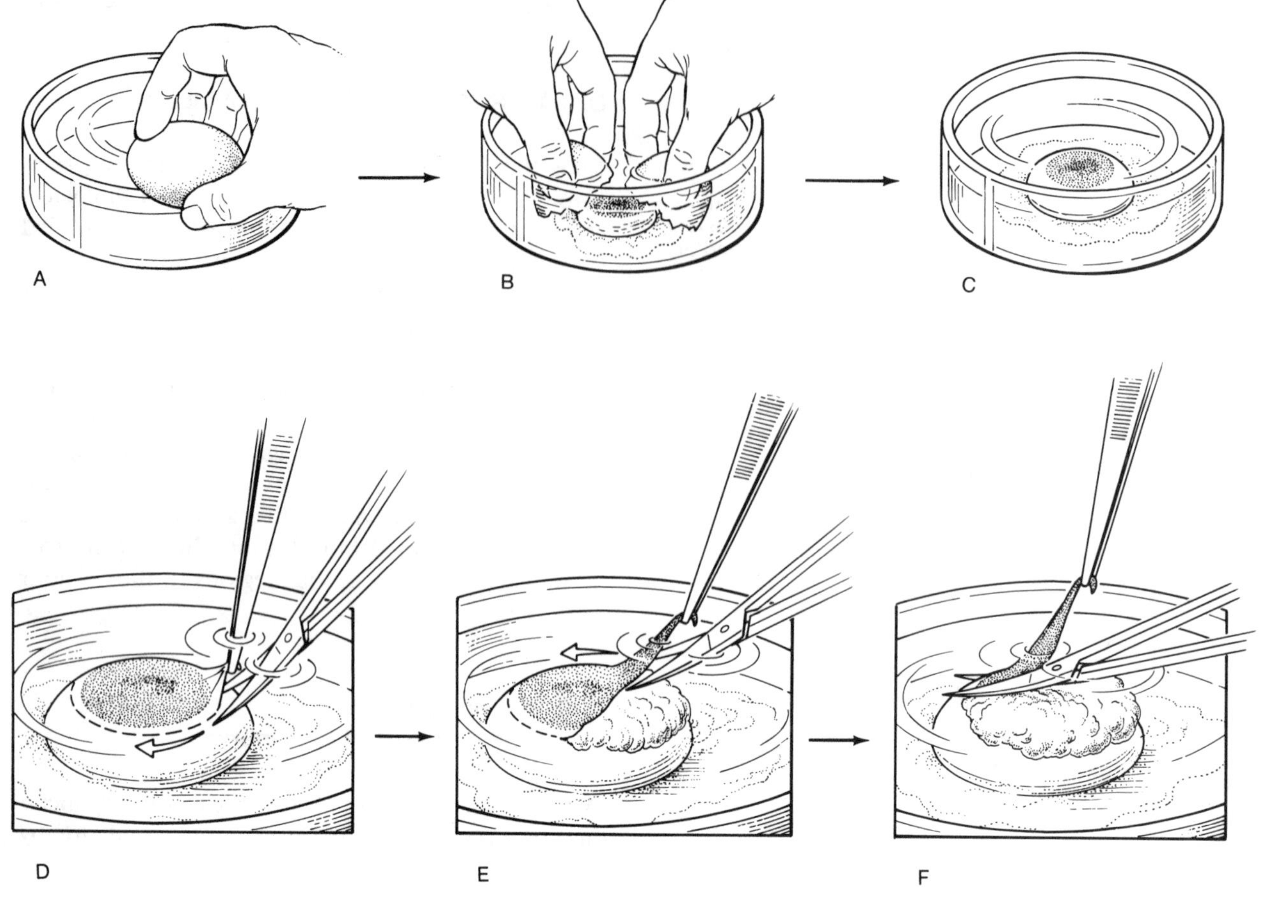

FIG. 21-6
Procedure for opening egg and cutting blastoderm from yolk.

in Fig. 21-6. After about 26 hours, the heart mass can be seen to beat, even though it has not yet formed a heart.

A pair of **vitelline veins** enters the heart at its posterior end. What is the function of these veins?

____

____

A single artery, the **ventral aorta,** leaves the heart anteriorly.

The brain region is divided into three parts: the anterior **prosencephalon (forebrain);** the **mesencephalon (midbrain);** and the **rhombencephalon (hindbrain),** which continues for the length of the embryo as the **neural tube.** The lateral evaginations of the forebrain are the **optic vesicles.**

The **notochord,** which is ventral to the neural tube, can be seen in the midline of the embryo, extending anteriorly to the forebrain as a solid rod. Lying on either side of the neural tube are 12–15 **somites.** These somites will form the vertebral column and the musculature on the dorsal side, and they may contribute to the ribs and musculature of the body wall.

2. Serial Sections of the 33-Hour Embryo

Study demonstration slides with cross sections cut through various regions of the 33-hour embryo and refer to Fig. 21-9. The approximate region from which each section shown in this figure was taken is indicated by the letters A through H in Fig. 21-7. The relationships of various body layers to one another and their differentiation into specific organs and tissues of the embryo can be reconstructed through a study of these sections.

Section A is through the prosencephalon and optic vesicles, which will produce major parts of the eyes. The head region is free and is separated from

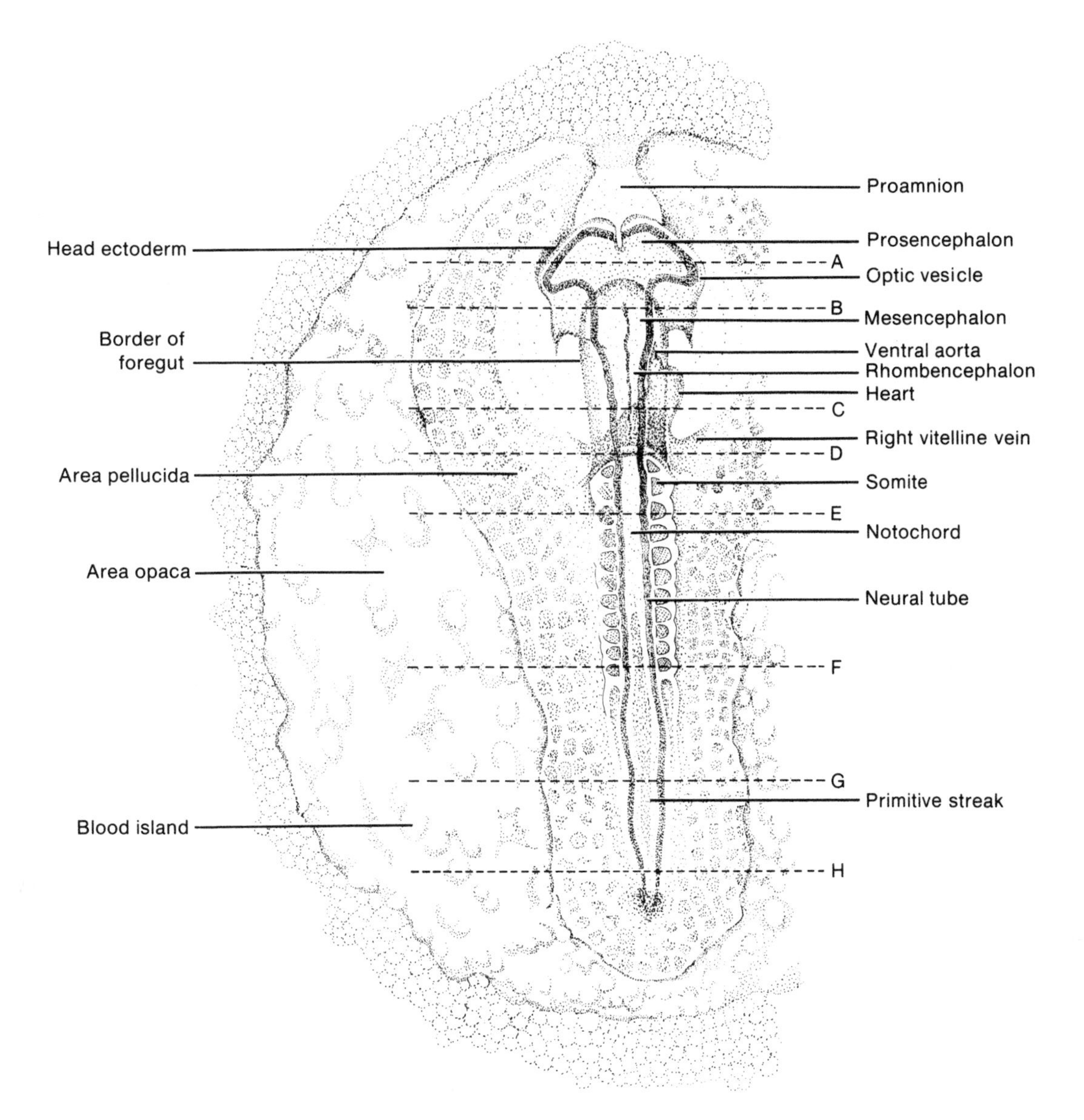

FIG. 21-7
Dorsal view of 33-hour chick embryo. (The letters A through H refer to the sections shown in Fig. 21-9.)

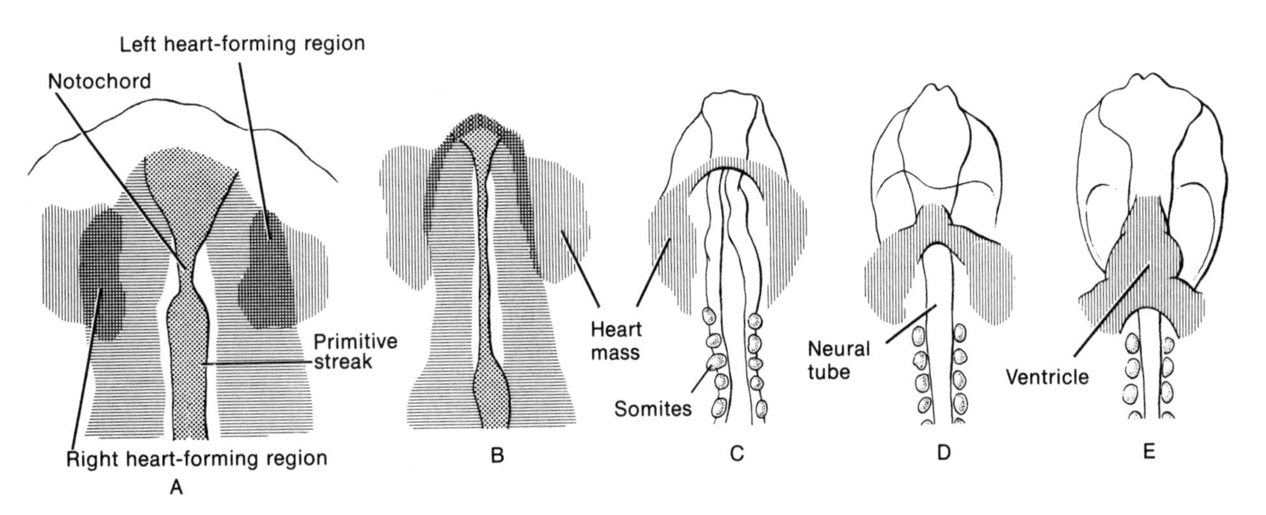

FIG. 21-8
Formation of the heart in the developing chick embryo (ventral view).

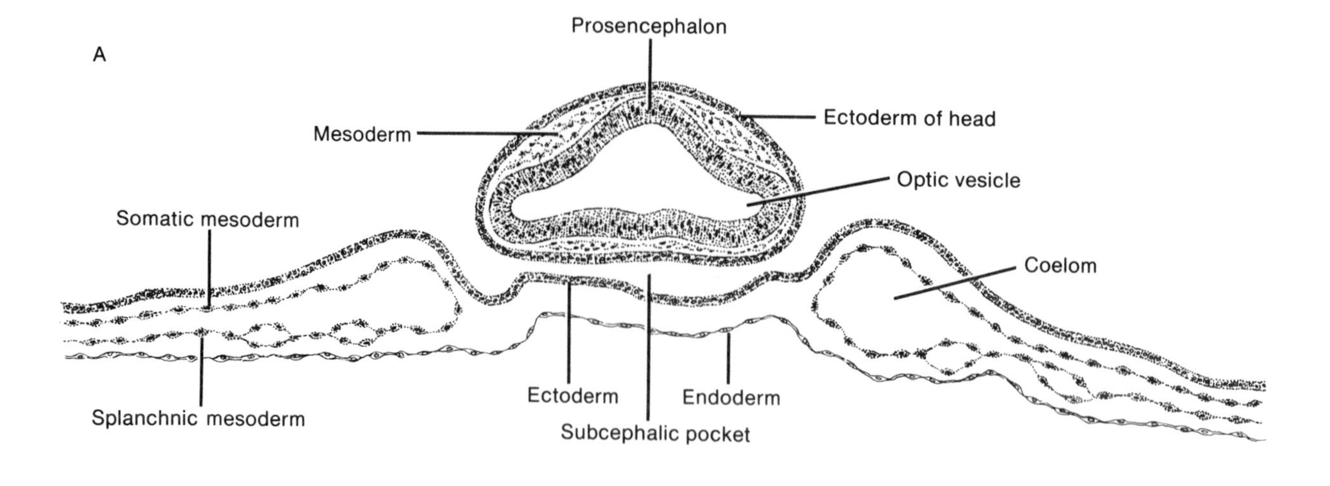

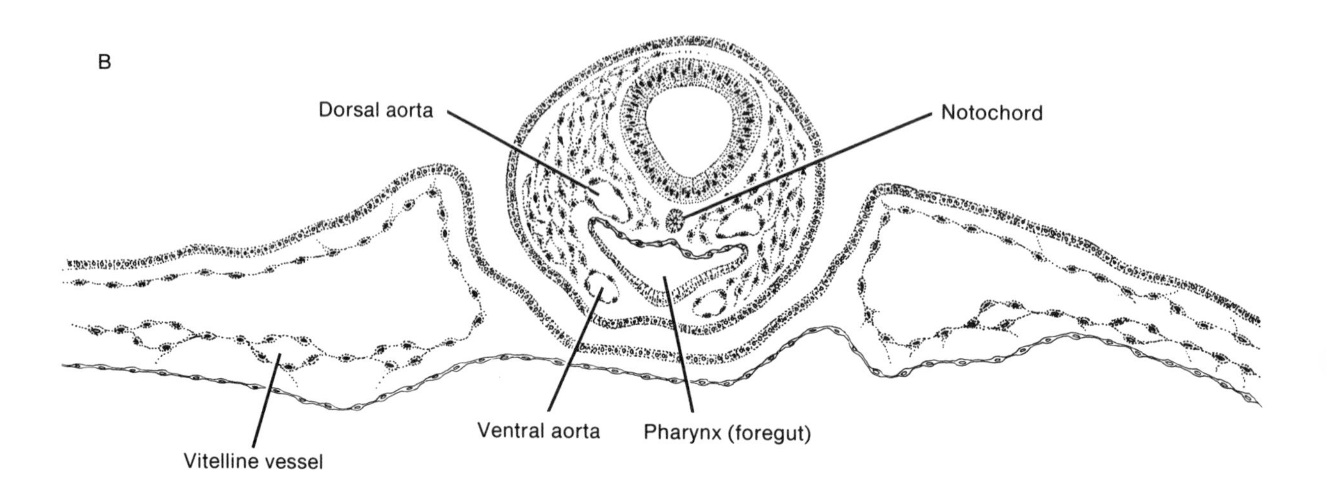

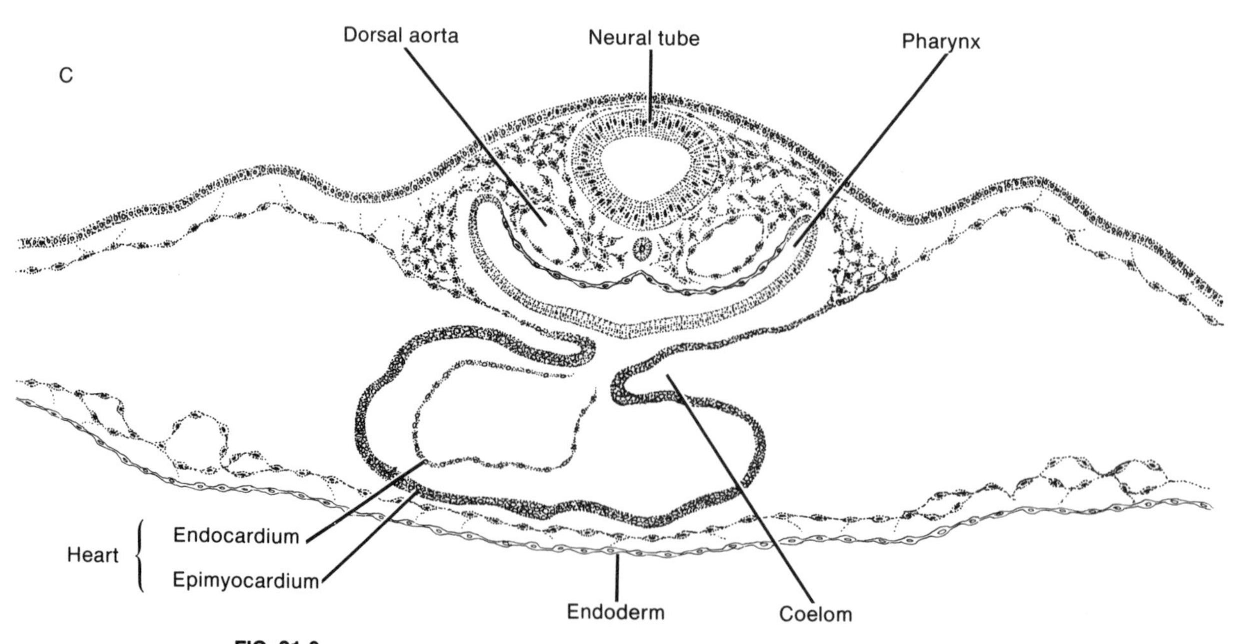

FIG. 21-9
Serial sections through selected regions of a 33-hour (12–15 somite) chick embryo. (After *General Zoology Laboratory Guide,* by J. E. Wodsedalek and C. F. Lytle. W. C. Brown, 1971.)

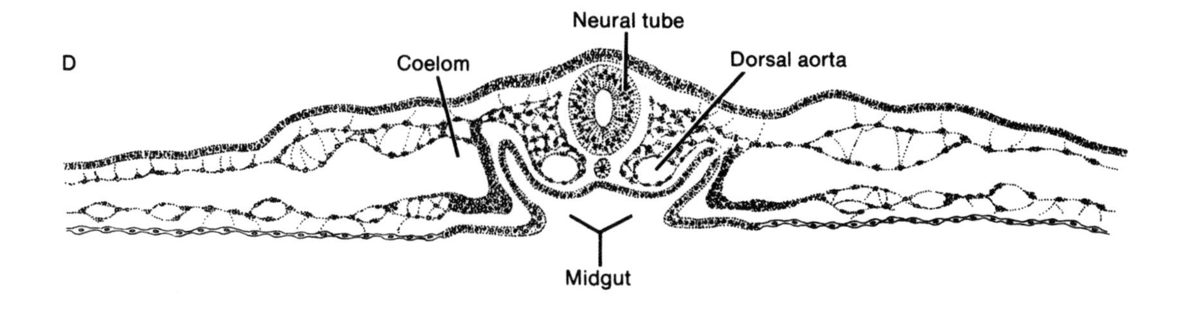
D
Neural tube
Coelom
Dorsal aorta
Midgut

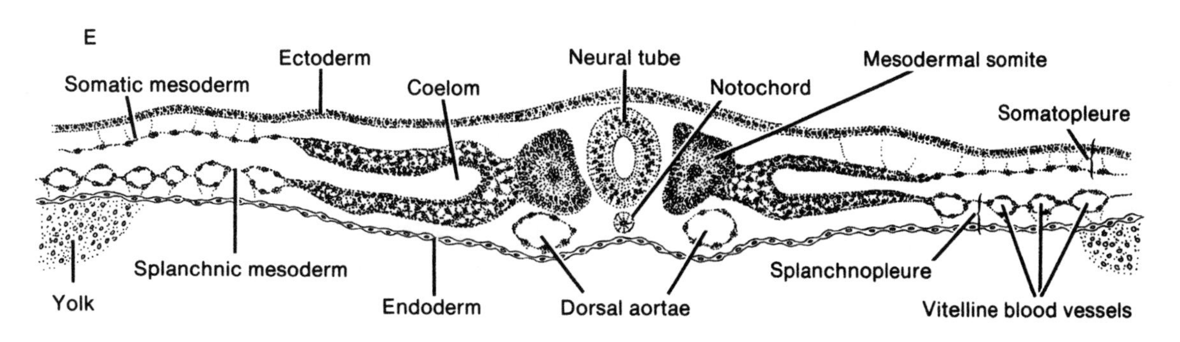
E
Ectoderm
Somatic mesoderm
Coelom
Neural tube
Notochord
Mesodermal somite
Somatopleure
Splanchnic mesoderm
Yolk
Endoderm
Dorsal aortae
Splanchnopleure
Vitelline blood vessels

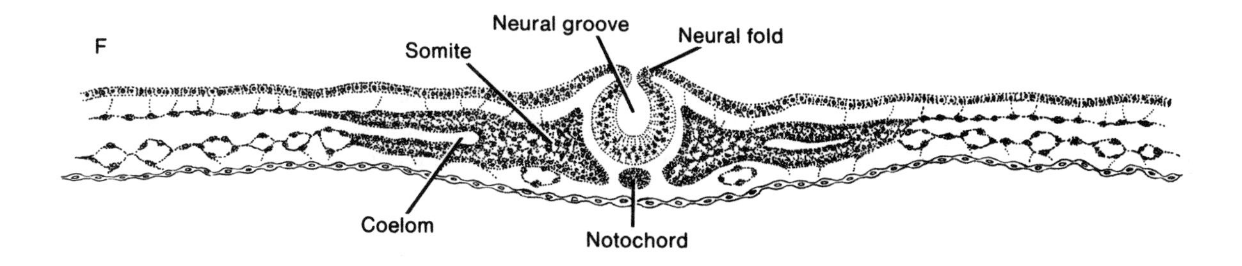
F
Neural groove
Neural fold
Somite
Coelom
Notochord

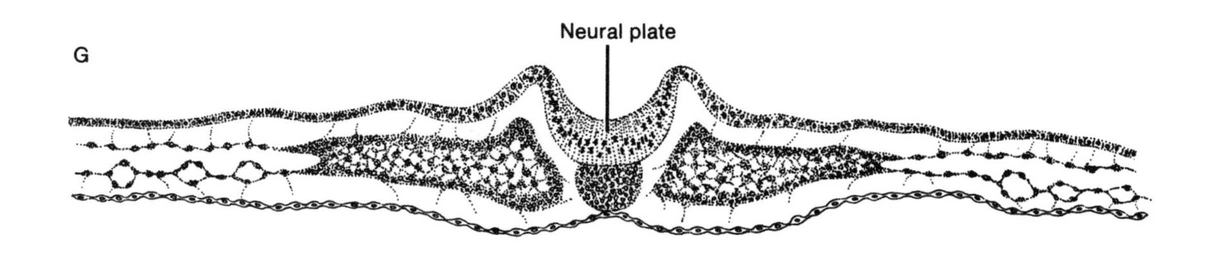
G
Neural plate

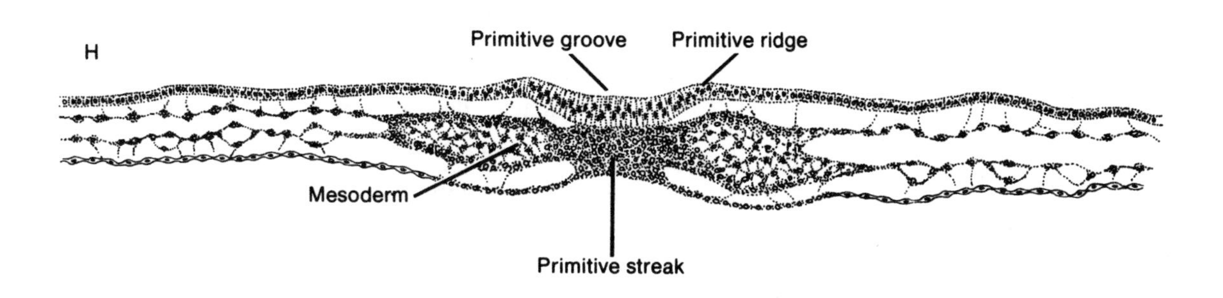
H
Primitive groove
Primitive ridge
Mesoderm
Primitive streak

the rest of the blastoderm by a definite space, called the **subcephalic pocket.** Note that the blastoderm immediately under the free head region consists of only ectoderm and endoderm.

Section B is through the head posterior to the optic cups. You can see the notochord, the dorsal and ventral aortas, and the pharynx (foregut).

Section C is through the region of the heart. Note the **epimyocardium,** the thick outer wall of the heart, and the **endocardium,** the thin inner lining. In this region, the embryo is no longer physically separated from the rest of the flat blastoderm.

Section D is through a region a short distance posterior to the heart. Observe the **midgut,** which is that part of the digestive tract just below the pharynx. Note the paired dorsal aortae, which will later fuse into a single dorsal aorta.

Section E is through a region that includes the third mesodermal somite. Note the absence of the digestive tract.

Sections F and G are through regions near the posterior end of the embryo and show the origin of the **neural tube,** which forms by a folding of the neural plate (Section G).

Section H is through the region of the primitive streak. Note the three germ layers that meet and fuse along the longitudinal axis of the embryo. A depression of the ectoderm forms the primitive groove bounded by two primitive ridges.

3. The 72-Hour Embryo

Remove the embryo from the yolk as shown in Fig. 21-6, and locate the various regions described below. At this stage, the anterior part of the embryo has turned to lie on its left side, whereas the posterior half remains dorsal side up (Fig. 21-10). Note the three pairs of **gill slits** and the blood vessels—the **aortic arches**—that pass through the gill arches. What is the significance of the appearance of these structures in the embryos of all higher vertebrates?

The central nervous system and the sense organs have now advanced considerably beyond their development in the 33-hour stage. The neural tube has developed into a five-part brain anteriorly and the spinal cord posteriorly. The optic vesicles have greatly enlarged and, by invaginating, have given rise to the double-walled **optic cups.** Concurrently, the surface ectoderm of the head immediately over the optic cups invaginates, thickens, and detaches to form the optic lens. Dorsal to the gill slits on each side is an invaginated thickening of the ectoderm forming the **auditory vesicle** ("ear").

The heart, which was tubular and single-chambered in the earlier stage, has become twisted and transformed into a two-chambered structure in this stage. The part of the heart next to the body is the **auricle** (atrium), and the part that dips ventrally is the **ventricle.** Note the **truncus arteriosus;** it arises from the ventricle and branches out into the aortic arches, which unite to form the **dorsal aortae.** Locate the **vitelline arteries,** a pair of large transverse blood vessels that leave the embryo about one-third of the distance from the posterior end and branch out into the **yolk sac,** and the **vitelline veins** that enter the auricle.

E. EXPERIMENTAL SURGERY ON CHICK EMBRYOS

Numerous studies employing surgical procedures have been carried out on young chick embryos. Some have been done to determine the effects of altering the normal structural integrity of various parts; others to determine the effect of the complete removal of structures on subsequent development. The following study will serve as an introduction to the fundamental techniques used in experimental developmental biology.

1. Surgical Removal (Extirpation) of Leg Limb Bud

a. Preliminary Procedures

Thoroughly wipe the work area with a disinfectant and assemble the following equipment and instruments:

several 72- to 84-hour eggs, because a fair number of embryos may die following this operation
dissecting microscope with spotlight illuminator
egg candler
styrofoam or cotton egg holder
sharp hacksaw blade
single-edge razor blade
clean Pasteur pipets
beaker containing 70% ethyl alcohol for sterilizing instruments
watchmaker's forceps
microsurgical needle
absorbent cotton

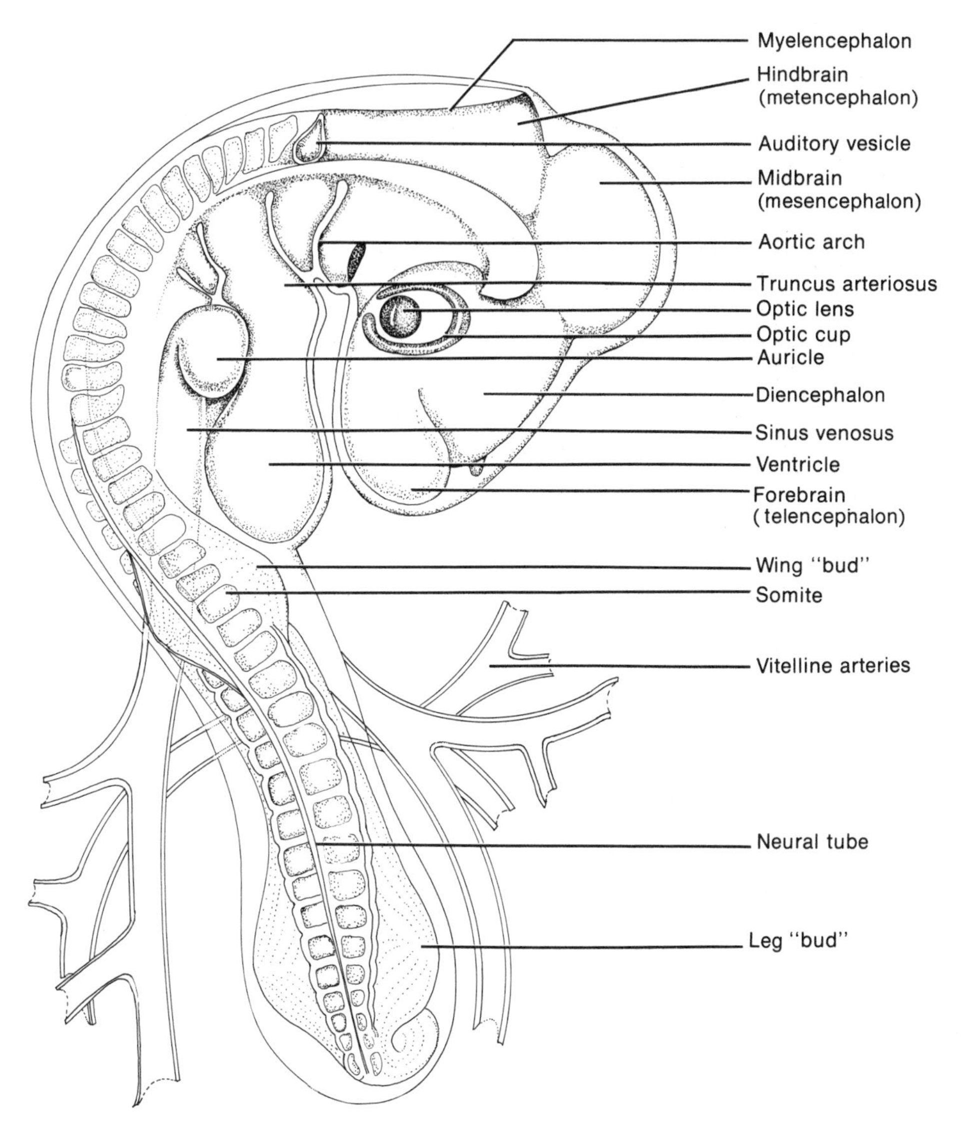

FIG. 21-10
Dorsal view of 72-hour chick embryo.

sterile Ringer's (physiological saline) solution

alcohol lamp

Curad tape (flesh color)

Note: Whenever you use an instrument, remove it from the beaker containing 70% ethanol and flame off the alcohol by lighting it with an alcohol lamp. This procedure completes the sterilization of the instrument. The exception to this is the microsurgical needle, which should not be kept in alcohol. To sterilize this needle, merely pass the tip through the alcohol flame.

1. Place an egg that has been incubated for 72–84 hours on the egg candler to determine that the egg was fertilized and contains an embryo. Your instructor will demonstrate the procedure for "candling" eggs.

2. Place an egg that has been candled in a styrofoam or cotton holder. Wipe the upper surface of the egg (nearer the blunt end) with cotton moistened with 70% alcohol. After the surface has dried use a hacksaw blade to make three cuts in the shell as shown in Fig. 21-11A. Place your finger tip against the edge of the blade to prevent it from sliding around while "sawing." Stop cutting when the blade "catches" rather than cuts smoothly.

3. Flame the razor blade and then lift up the shell flap as shown in Fig. 21-11B. If you have difficulty

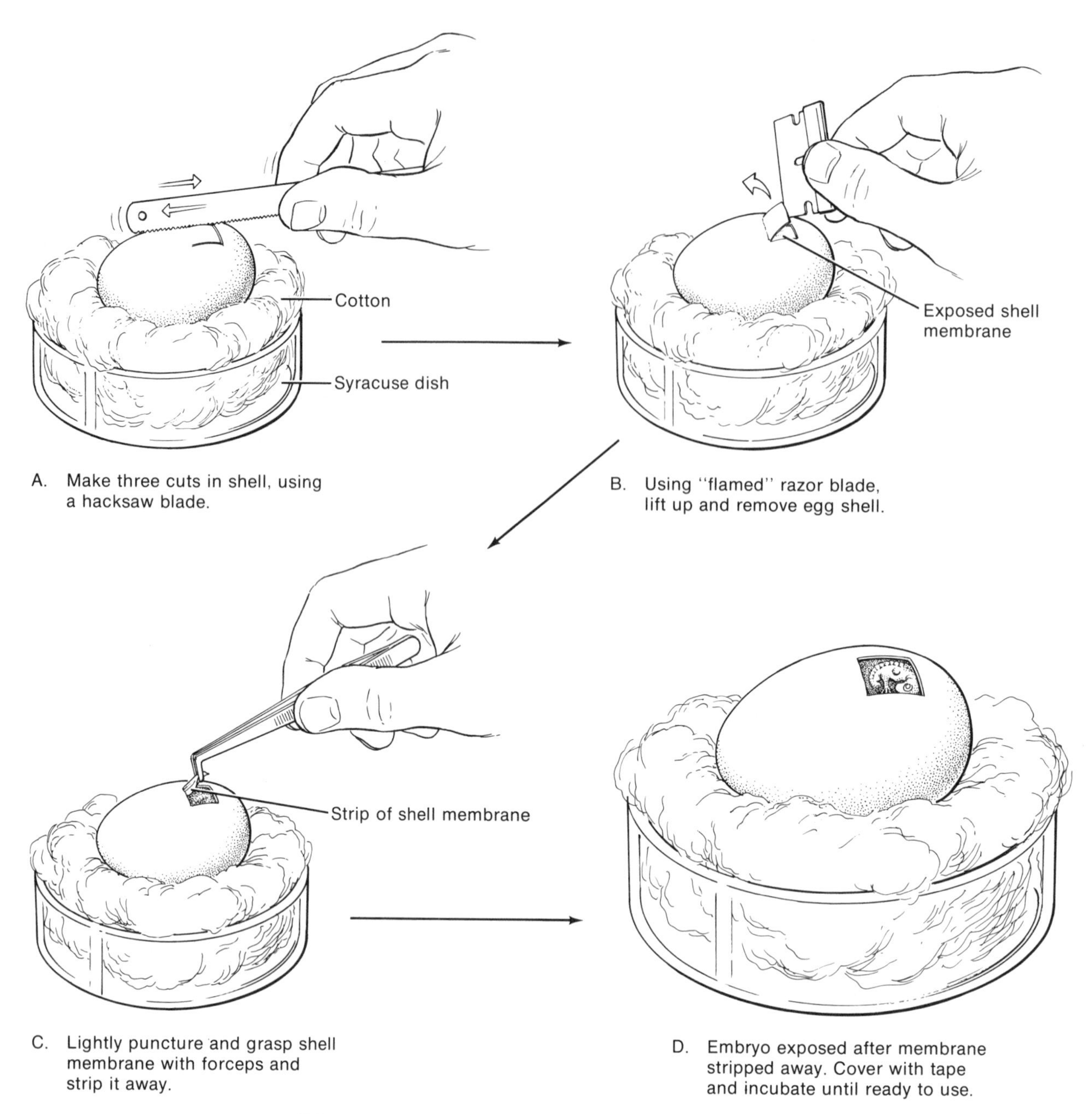

FIG. 21-11
Procedure for cutting window in egg shell and exposing embryo.

lifting the flap, make the cuts a little deeper with the hacksaw blade.

4. Remove the shell flap to expose the shell membrane (Fig. 21-11B). Pipet a small volume of sterile Ringer's solution onto the membrane to wash away the debris left from sawing. Gently roll the egg back and forth to break any adhesions of the shell membrane to the membranes covering the embryo. Then, with your forceps held at an oblique angle (to prevent damaging the embryo), lightly puncture and grasp the shell membrane stripping it away as shown in Fig. 21-11C. This procedure will result in a "window" exposing the embryo (Fig. 21-11D).

5. Adjust the light to provide maximum illumination of the embryo. In doing this you may notice a reflection from the vitelline membrane that surrounds the embryo. This membrane is easily ruptured and will pose no problems during the following surgical procedures, so do not be concerned if you do not see it.

6. After you have completed these preliminary procedures, cover the opening with Curad tape to prevent drying. Return the egg in its holder to the incubator until ready to use.

b. Removal of the Leg Limb Bud

Remove the Curad tape from the egg to expose the embryo. Using a dissecting microscope, locate the leg limb bud and, with a needle or watchmaker's forceps, cut away a substantial part of the leg bud (Fig. 21-12). It may be necessary to hold the limb bud steady using both instruments. Using a pencil, mark the date and procedure on the egg. Cover the window with tape and replace in the incubator. Record the results of your procedure using drawings and written observations. For this operation, eggs should not be opened until about 2 weeks after removal of the limb. Check for any deficiencies in growth or abnormal types of development.

After you have become proficient at this procedure you might want to determine the effects of such operations as

- removing various amounts of limb tissue
- removing a limb and replacing it upside down
- removing a limb and implanting it on a different part of the embryo
- placing carbon particles on different areas of the limb and, by observing the movement of the particles for a period of time, determining the fate of that particular area (i.e., does it develop into digits, bones [radio-ulna] of the lower arm, humerus of upper arm, etc.)

2. Disruption of the Neural Tube

Normal development of the spinal cord depends on the neural tube being intact and closed. In a normal 72-hour chick embryo the tube is closed. If it fails to close, a condition known as **spina bifida** results. The severity of this condition ranges from minor vertebral malformations to a severe deformation in which the unprotected spinal cord protrudes from the surface of the back. This is a relatively common birth defect in human beings.

In this exercise, you will attempt to induce spina bifida experimentally by surgically cutting open the posterior region of the neural tube. In a 72-hour chick embryo such an opening does not repair itself and normal structural relationships are permanently disrupted. Be especially careful during this operation because a major blood vessel, the dorsal aorta, lies a short distance under the neural tube. Puncturing this blood vessel will kill the embryo. Some damage to smaller vessels can be tolerated. However, if you notice large amounts of spurting blood, discard the egg.

Using a surgical needle with the tip bent at an angle, cut open a small section of the neural tube at the level of the leg buds (Fig. 21-13). To do this, try to enter the neural tube from the rear. This can be easily done by lifting the tip of the needle as it enters

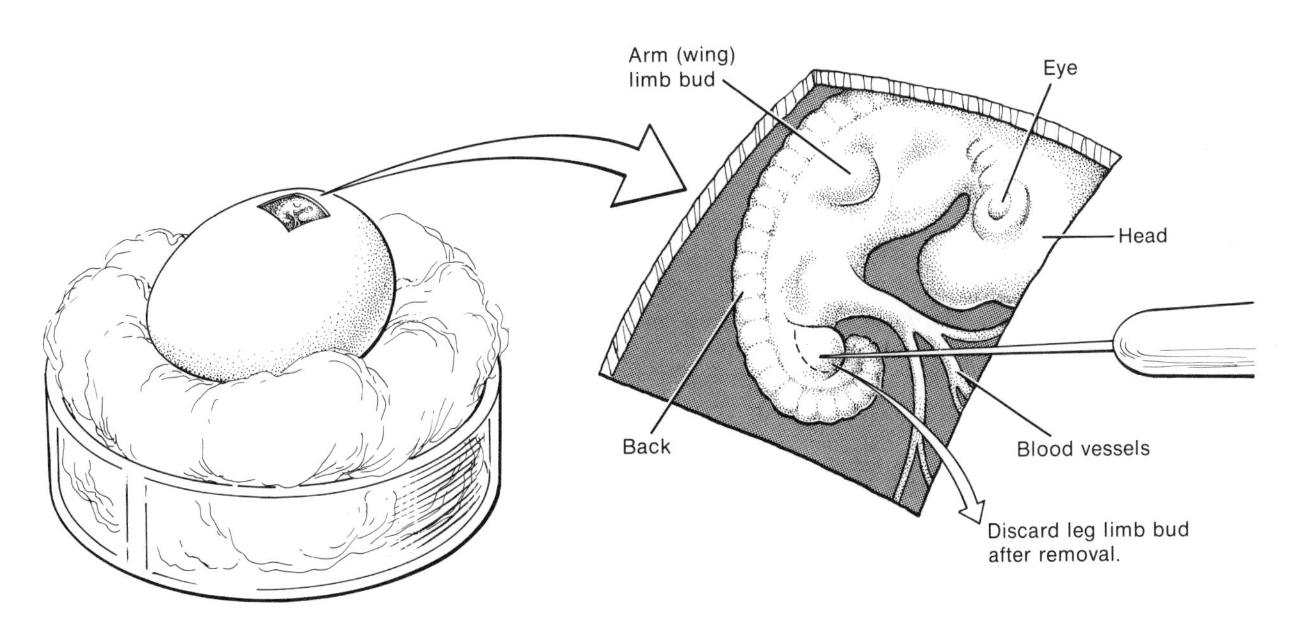

FIG. 21-12
Removal of right limb bud from 72-hour embryo.

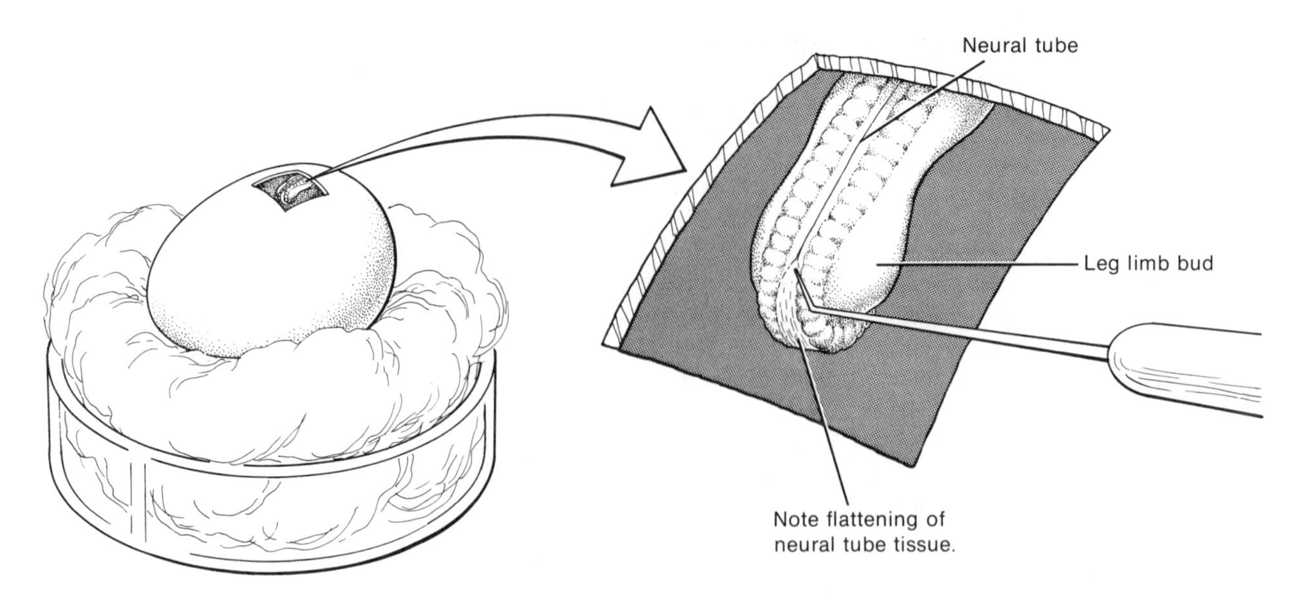

FIG. 21-13
Disruption of neural tube in 72-hour embryo.

the neural canal. If you are successful, the side walls of the tube will collapse and flatten out and the tube will no longer be "tubular" in appearance.

When you think you have observed the collapse of the neural tube, reseal the egg and place it back in the incubator in its holder. Further observations should not be made for about 2 weeks. At the end of this time, open the eggs and remove any feathers that may be present so that you can more readily observe the extent of the malformation. If time permits the embryo can be dissected to determine the full nature of the abnormality of the spinal cord and surrounding areas. Keep accurate records of your procedures and observations.

REFERENCES

Balinsky, B. I. 1981. *An Introduction to Embryology.* 5th ed. Saunders College/Holt, Rinehart and Winston.

Bradley, S. J. 1970. An Analysis of Self-Differentiation of Chick Limb Buds in Chorio-allantoic Grafts. *Journal of Anatomy* 107:479–490.

Browder, L. W. 1980. *Developmental Biology.* Saunders College/Holt, Rinehart and Winston.

Hamburger, V. 1960. *A Manual of Experimental Embryology.* University of Chicago Press.

Karp, G., and N. J. Berrill. 1981. *Development.* 2d ed. McGraw-Hill.

Lehman, H. E. 1977. *Chordate Development.* Hunter.

Patten, B. M. 1971. *Early Embryology of the Chick.* 5th ed. McGraw-Hill.

Rugh, R. 1962. *Experimental Embryology: Techniques and Procedures.* 3d ed. Burgess.

Slack, J. M. 1983. *From Egg to Embryo: Determinative Events in Early Development.* Cambridge University Press.

Torrey, T. W., and A. Feduccia. 1979. *Morphogenesis of the Vertebrates.* 4th ed. Wiley.

Trinkaus, J. P. 1984. *Cells into Organs.* 2d ed. Prentice-Hall.

Wolpert, L. 1978. Pattern Formation in Biological Development. *Scientific American* 239:154–164 (Offprint 1409). *Scientific American* Offprints are available from W. H. Freeman and Company, 41 Madison Avenue, New York 10010, and 20 Beaumont Street, Oxford OX1 2NQ, England. Please order by number.

EXERCISE 22

Plant Growth and Development

A higher plant proceeds through a continuous series of changes during its life cycle. The outward expression of this activity begins with the embryo in the seed and proceeds, sequentially, through germination, the appearance and subsequent enlargement of stems, leaves, and roots, to the production of flowers, fruit, and seeds. The location of these organs and their size and shape are the visible manifestations of the correlated activities of those cells produced by meristems. This complex and highly ordered series of events is called **development.**

Development may be divided into separate phases called growth, cell differentiation, and organogenesis. **Growth** may be defined as an irreversible increase in volume, which is usually—though not always—paralleled by an increase in weight. **Cell differentiation** includes all those events involved in the specialization of structure and function. **Organogenesis** is the development of the various plant organs.

A. COMPARATIVE STUDY OF SEEDS

You will begin the study of plant growth and development by investigating the structure and function of seeds.

Seeds develop from ovules and are formed following the sexual union of the sperm and egg, an event that occurs in the ovary of the flower. If you are not familiar with floral anatomy, refer to Fig. 22-1 and a flower model or living flowers. The ovary and, in some cases, other parts of the flower become the fruit. Within the seed is the embryo, which consists of three parts: the **cotyledons** (sometimes called *seed leaves*), the **epicotyl,** which is located above the point of attachment of the cotyledons, and the **hypocotyl,** found below the cotyledons. The mature seed also may have an **endosperm,** which functions as a nutritive tissue for the developing embryo.

Obtain previously soaked bean, pea, and corn seeds. (The so-called seed of corn is really a fruit.

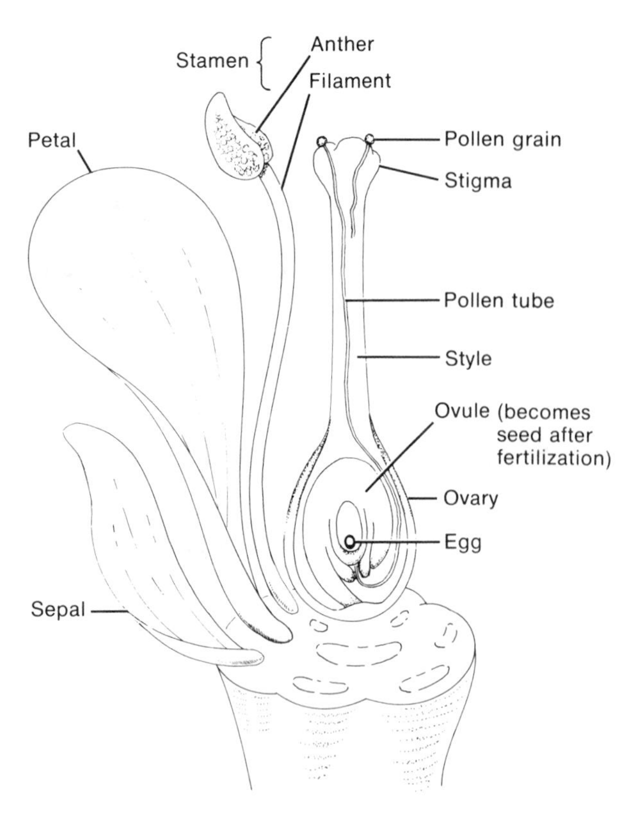

FIG. 22-1
Diagram of a flower. (Adapted from *Biology, Its Principles and Implications*, 3d ed., by Garrett Hardin and Carl Bajema. W. H. Freeman and Company. Copyright © 1978.)

The seed coat is intimately fused with a hard outer tissue called the **pericarp,** which is part of the fruit.) Carefully remove the outer seed coats from each seed. Then separate the other parts of each seed, comparing them with each other and with Fig. 22-2.

1. Cotyledons

Scrape the surface of the cotyledons of the bean and pea, and add a drop of iodine. On the basis of your observations, what appears to be the role of the cotyledons in the development of these seeds?

____

____

2. Epicotyl

This part of the embryo gives rise to the shoot system of the mature plant. What indication is there, if any, that this region will produce stem and leaf tissues?

____

3. Hypocotyl

This region will primarily give rise to all root tissues and, in some cases, to the lower part of the stem.

4. Endosperm

The endosperm (although found in many seeds) is present only in the corn "seed" of the group you are examining. Depending on whether the endosperm is composed of starches or sugars, corn is "starchy" or "sweet." Remove the endosperm from several corn "seeds," and cut it up into fine pieces. Then add it to 5 ml of Benedict's solution in a test tube, and heat in a hot water bath for several minutes. If a reducing sugar is present, the solution will change in color (from blue to green to orange to red to brown), depending on the amount of sugar present.

Cut through a kernel of corn, as shown in Fig. 22-2. Scrape the surface of the endosperm with a razor blade, and add a drop of iodine. The appearance of a blue color indicates the presence of starch. What commercial value does this starchy corn have?

____

____

B. GERMINATION

The seed is a resting stage in the development of a plant, serving to carry the plant over periods of unfavorable environmental conditions. When provided with optimal growing conditions, the seed, if living, will germinate and produce another plant. The production of viable seeds is required for maintaining the existence of any plant species in nature. It is also commercially important to seed growers that they be able to judge the viability of the seed from any given crop. For biologists who grow research plants from seed, it is important that they be assured that the percentage of germination will be high. In this part of the exercise, you will determine the degree of viability of a batch of seeds. One way of doing this is to germinate sample batches of seed under standardized conditions. This **germination test** has one major

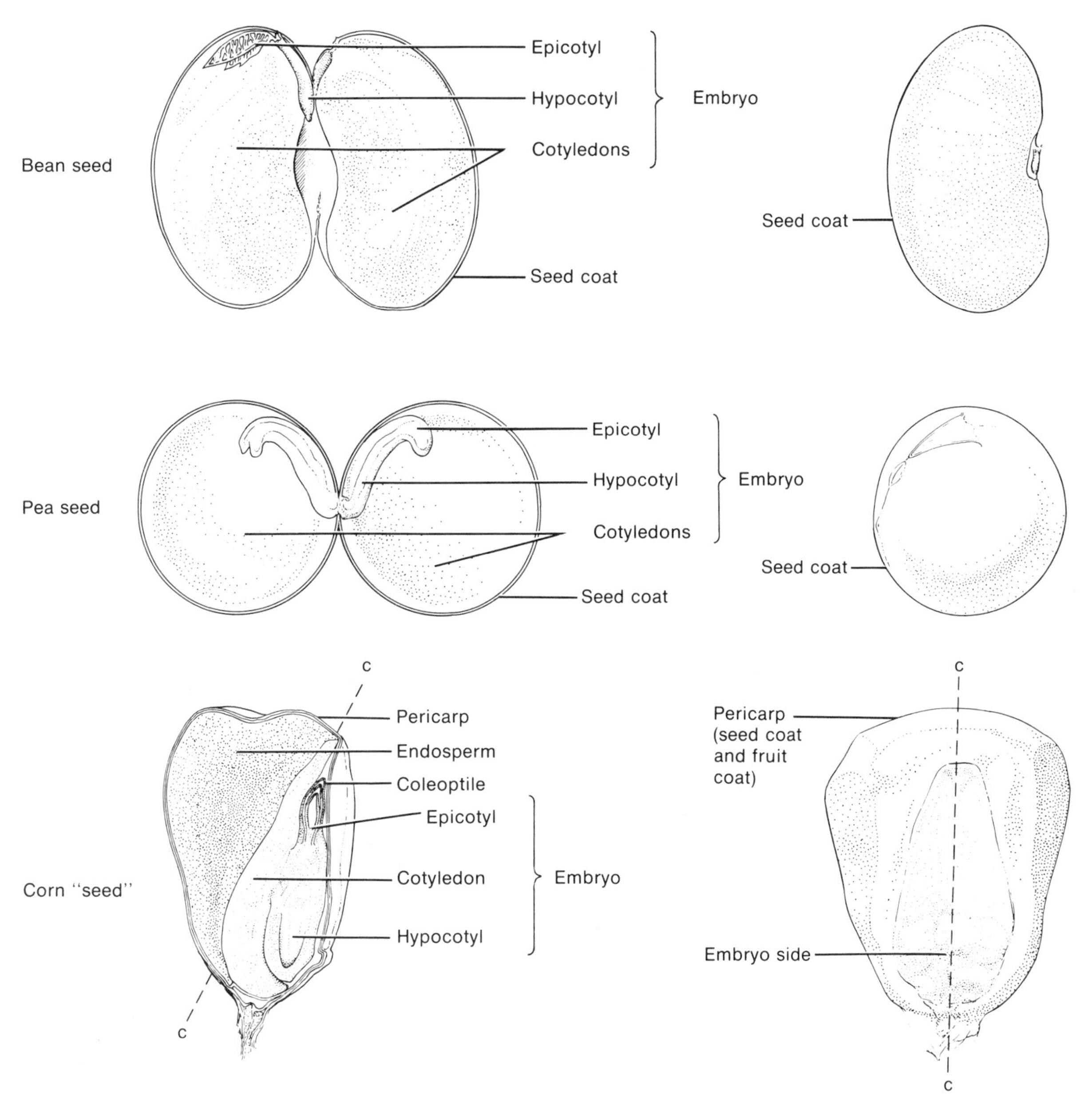

FIG. 22-2
Comparative anatomy of various seeds.

drawback—it takes from 7–10 days for many seeds to germinate. A quicker and easier method is to use the fact that embryos in viable seeds respire and are able to change colorless dyes, such as tetrazolium, into highly colored forms.

1. Germination Test

Obtain 100 bean seeds that have been soaked in water. Boil 50 of these seeds for 10 minutes to kill them. Plant 25 living seeds and 25 killed seeds about 15 mm deep in separate rows in the planting tray. The remaining boiled and unboiled seeds will be used for the tetrazolium test. Label each row as to section number, team number, date, number of seeds, and treatment (boiled or unboiled). After a week, determine the total number of seeds in each treatment that have germinated and are growing, and record these data in Table 22-1. Record your team results and (in parentheses) those of the combined class.

Examine the tray again after 2 weeks, and determine the percentage of germination. Record your data in Table 22-1. How do you account for the difference, if any, between the results obtained after 1 week and those after 2 weeks?

2. Tetrazolium Test

Take the boiled and unboiled seeds not used in the germination test, and cut them longitudinally through the embryo, as shown in Fig. 22-3. Place each unboiled seed in a petri dish, with the cut surface up. Pour in enough tetrazolium chloride solution to cover the seeds. Repeat this procedure with seeds that have been boiled. *Make sure each dish is labeled as to contents.* Set the seeds aside in a dark place for 30–45 minutes and then examine them. If the seed is viable, a red or pink color will be evident on the cut surface. Determine the percentage of viability, and record your data in Table 22-1. Record your results and (in parentheses) those of the combined class.

How does the percentage of viability obtained by the germination test compare with that obtained by the tetrazolium test?

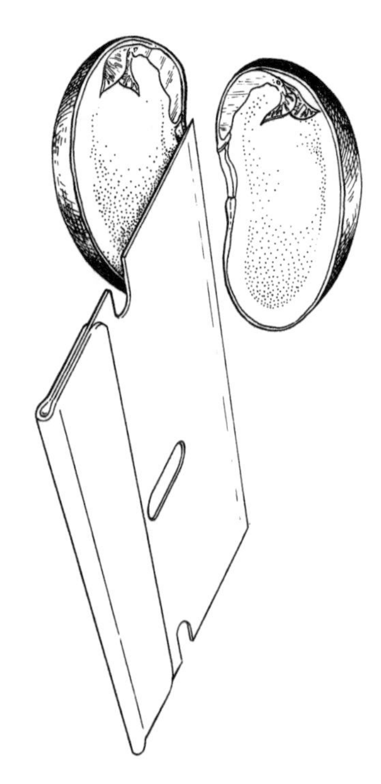

FIG. 22-3
Procedure for cutting bean seeds for tetrazolium test.

If results obtained by your team were different from those of the class, account for the difference.

TABLE 22-1
Germination and tetrazolium data.

Data	Germination test				Tetrazolium test	
	Unboiled seeds		Boiled seeds		Unboiled seeds	Boiled seeds
	Week 1	Week 2	Week 1	Week 2		
Total number of seeds	( )	( )	( )	( )	( )	( )
Number of seeds germinated	( )	( )	( )	( )	( )	( )
Germination (%)	( )	( )	( )	( )	( )	( )

3. Factors Affecting Germination

The seeds used in the first part of this exercise germinated in a short period of time when placed under conditions suitable for germination. Viable seeds of many plants, however, fail to germinate even when provided with ample water and oxygen and a suitable temperature. Such seeds are in an arrested state of development called **dormancy.** In this part of the exercise, you will study one factor involved in the dormancy of seeds.

Place moistened filter paper in the bottoms of three petri dishes. Obtain 40 dry honey locust seeds (other possible seeds are sweet clover, okra, and alfalfa) and 20 honey locust seeds that have been soaking for 2–3 hours in a 70% sulfuric acid solution. Carefully pour off the acid into a sink containing running water. With a rubber band, fasten cheesecloth over the beaker, and wash the seeds in running water for 5 minutes.

Place the 20 acid-treated seeds in one petri dish. Put 20 of the untreated seeds into a second petri dish. With a sharp razor blade, remove a small chip from the 20 untreated seeds, exposing the inner tissues. Place these seeds with their cut surface down in the third dish. Cover all dishes, tape them shut (to retard evaporation), and label each as to contents. Put the dishes in the dark at room temperature (or wrap them in aluminum foil and take them home for observation). Periodically examine the seeds over the next 3–5 days, and determine the percentage of germination for each group of seeds. Record your team data and (in parentheses) those of the class in Table 22-2. What is the "factor" that controls dormancy in these seeds?

____

____

Of what advantage is dormancy to seeds?

____

____

C. MEASUREMENT OF PLANT GROWTH

Growth may be defined as an irreversible increase in volume generally accompanied by an increase in weight. Growth, therefore, is quantitative and can be measured. In these experiments you will determine the locus of growth, and plot a growth curve for a specific plant organ.

1. Localization of Plant Growth

Prepare a moist chamber as shown in Fig. 22-4A. Cover a glass plate (9 × 10 cm) with moist toweling, and place it in the chamber. Cover the container to prevent drying.

From the germinating seeds provided by the instructor, select one having a fairly straight root about 1.5–2 cm long. Blot the root to remove any excess moisture, and then lay it against a millimeter ruler. Carefully mark 10 lines, each 1 mm apart, *starting from the tip of the root* (Fig. 22-4D). Make sure you wipe the excess ink from the thread, or the mark will become smudged, making accurate observations difficult. Repeat the marking operation until five roots have been marked. Avoid drying the roots during marking. Lay the seedlings on the paper-covered glass plate, and hold them lightly in place with a rubber band so that the whole length of each root is touching the moist paper (Fig. 22-4F). Place the plate and seedlings in the moist chamber, and then cover and keep in a dark place.

TABLE 22-2
Factors affecting the germination of seeds.

Data	Seed treatment		
	Untreated	Seed coat cut	Treated with H_2SO_4
Total number of seeds	( )	( )	( )
Number of seeds germinated	( )	( )	( )
Germination (%)	( )	( )	( )

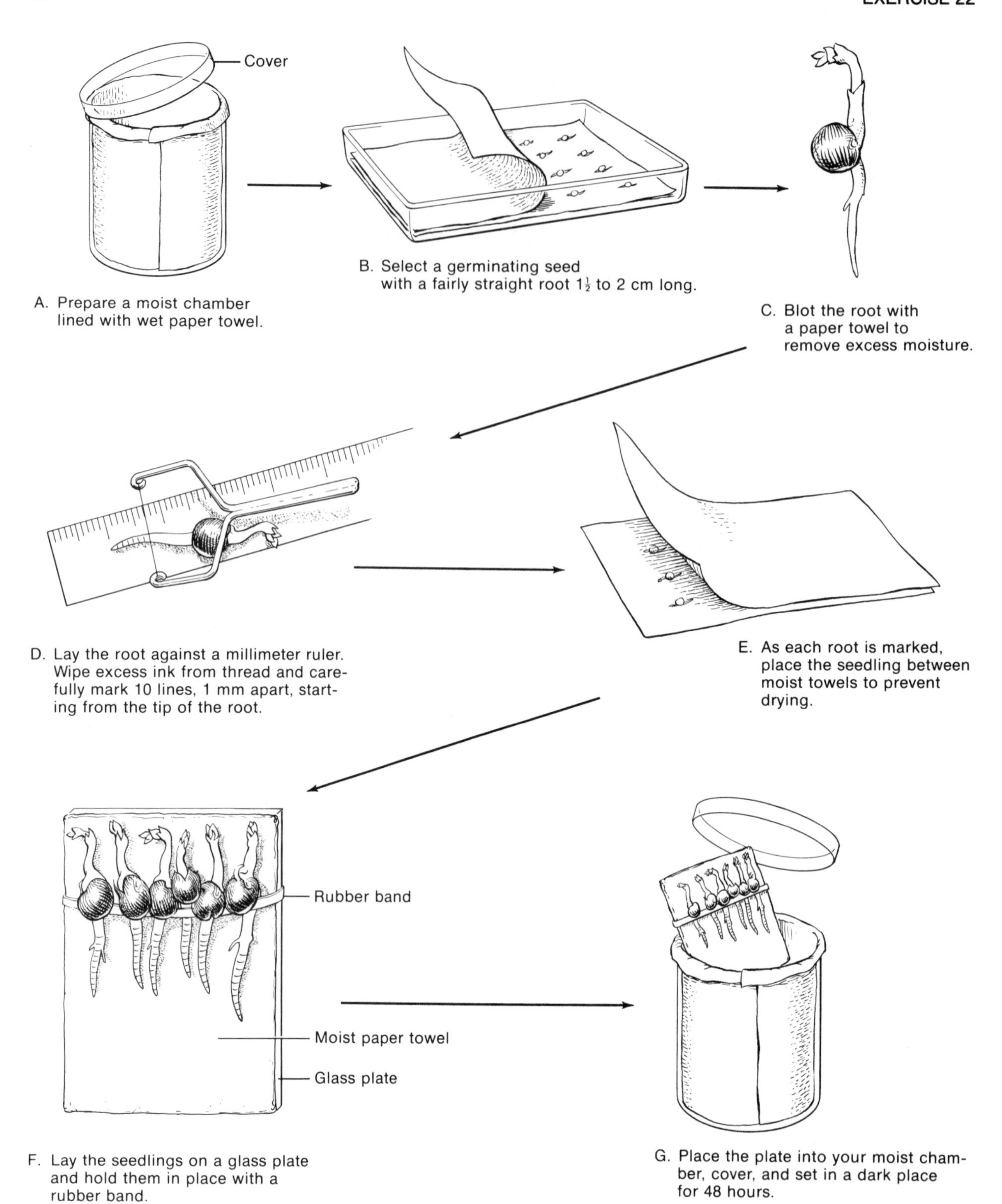

FIG. 22-4
Procedure for determining locus of growth in roots.

After 48 hours, measure the distance between the ink marks on each root. Average the lengths for each interval. Record your data in Table 22-3, and plot your data in Fig. 22-5.

Where does most of the growth occur?

If the ink lines were initially sharp and clear, how do you account for the smudging of the first, and possibly the second, line?

If you were to cut a longitudinal section through a young root and examine it microscopically, what would you expect to see in the first millimeter or so

TABLE 22-3
Data for locus of root growth.

Root tip no.	Interval									
	1	2	3	4	5	6	7	8	9	10
1										
2										
3										
4										
5										
Total										
Average										
Control										

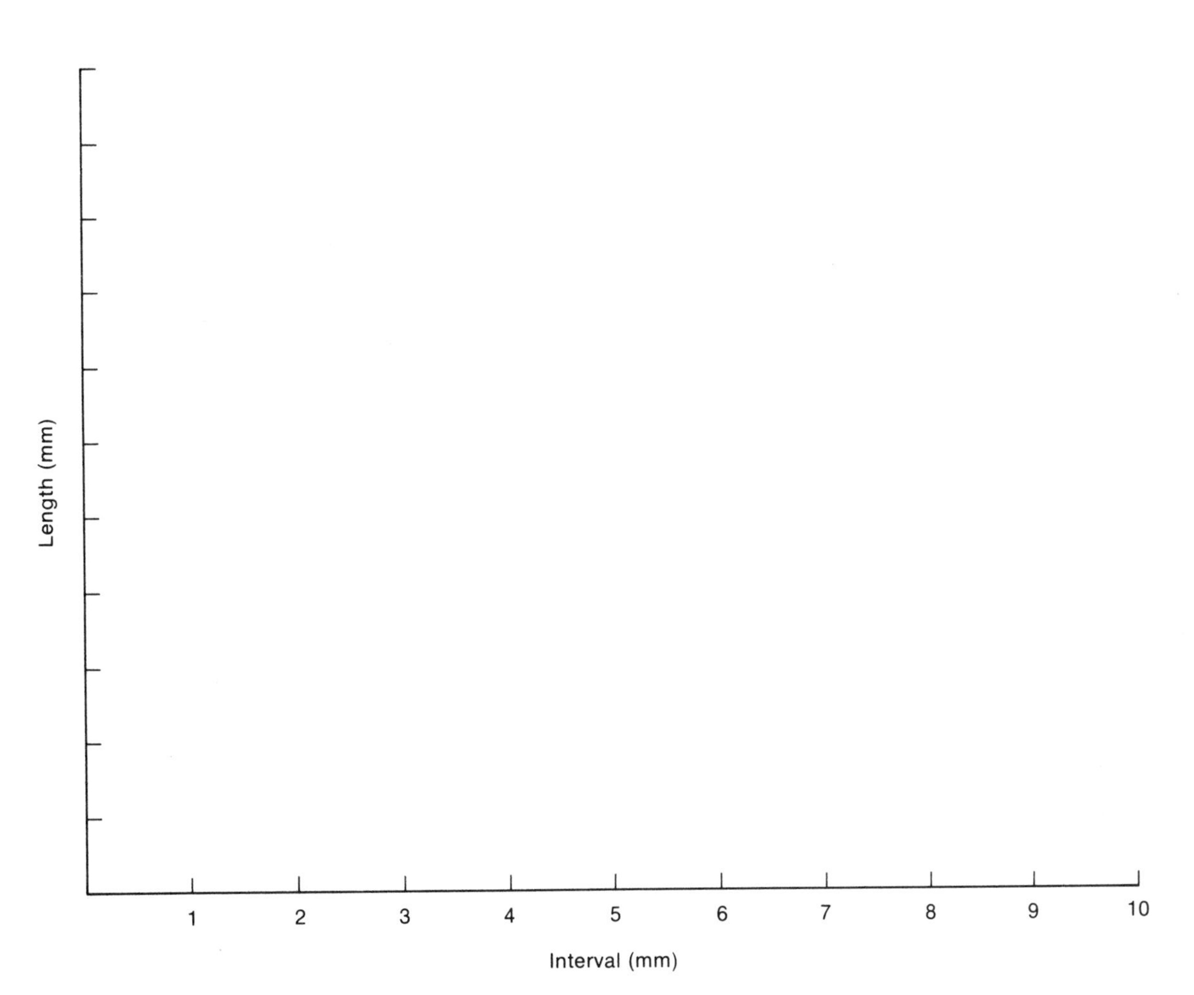

FIG. 22-5
Localization of root growth.

that would account for the results you have observed?

Examine a prepared slide of a longitudinal section through an onion root tip (compare with Fig. 22-6). Locate a cone-shaped mass of loosely arranged cells covering the tip of the root. This is a root cap, which serves to protect the meristematic region of the root. Closely examine the cells of the meristematic region. What occurred in this region when the tissue was living to account for the results obtained in the growth measurement experiment?

Locate the region of the root in which cell differentiation is occurring. Approximately how far back from the tip does this region occur? (The diameter of your low-power field, 10× objective, is approximately 2 mm.)

What is the consequence of cell division and the subsequent enlargement of cells at the root tip?

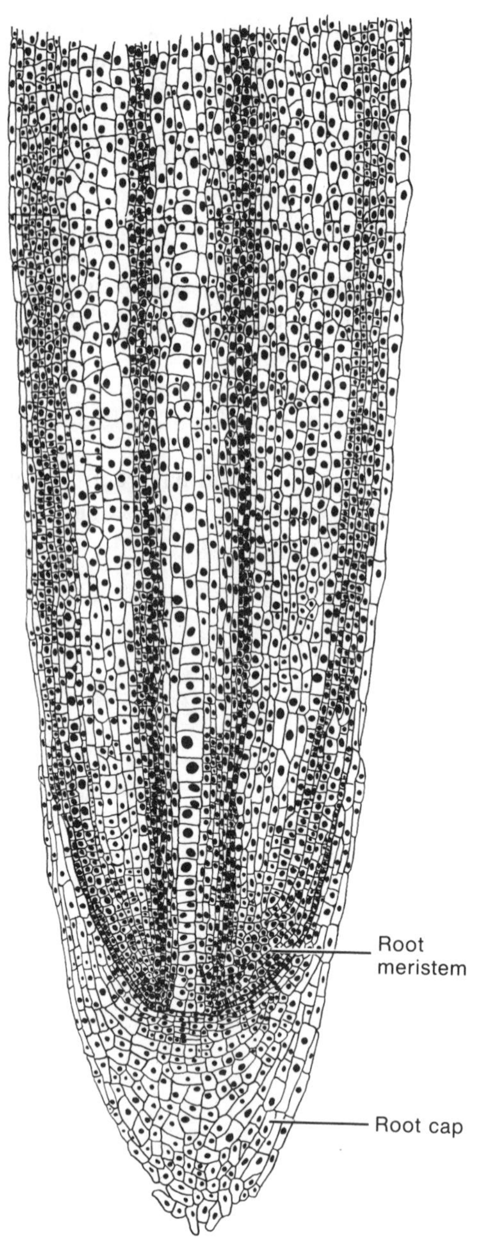

FIG. 22-6
Longitudinal section of onion root tip.

2. Growth Curve of Leaves

Select three bean seeds that have been soaking in water for several hours. Split them open and, with a millimeter ruler, measure the length of the embryonic foliage leaves (Fig. 22-7A). Determine the average length and record this figure under Date "0" in Table 22-4. This is the first in a series of measurements to be made of the growth of the first two foliage leaves.

Plant 25 of the soaked bean seeds about 7 mm deep in the container provided. Water thoroughly,

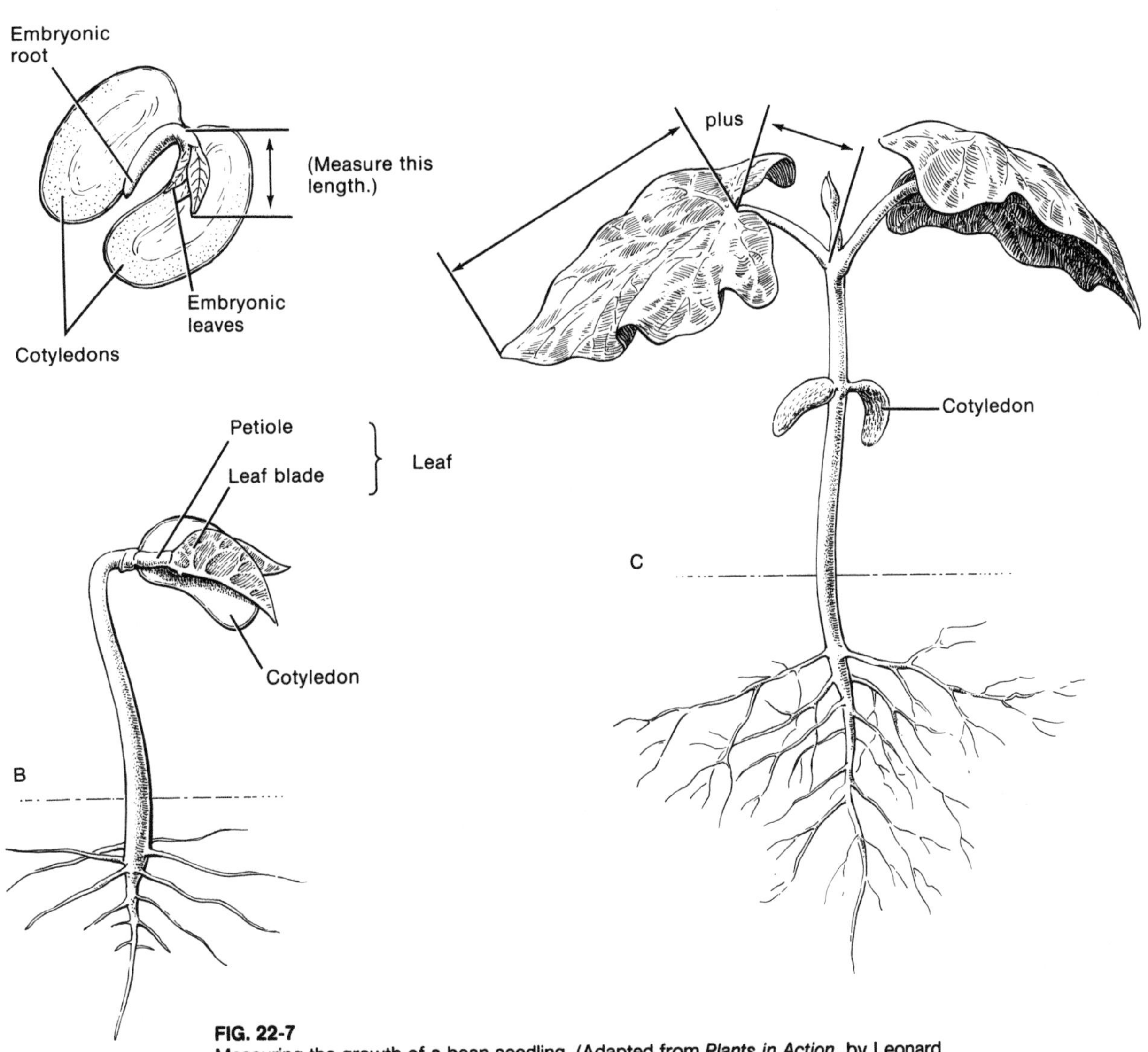

FIG. 22-7
Measuring the growth of a bean seedling. (Adapted from *Plants in Action*, by Leonard Machlis and John G. Torrey. W. H. Freeman and Company. Copyright © 1956.)

and place them in the greenhouse. In 2 or 3 days, dig up 3 of the seeds and measure the length of the leaves, including the petiole (Fig. 22-7B). Enter the average length in Table 22-4, and discard the young plants.

In succeeding laboratory periods, select three plants and measure the leaves as above (Fig. 22-7C). *Do not remove these plants from the container. Use the same three plants for all measurements beginning with the third measurement.*

Plot your data in Fig. 22-8. List several other organisms (or organs) that would show a similar growth curve.

____

____

____

____

TABLE 22-4
Growth curve data.

	Date of measurement						
	0						
Average length (mm)							

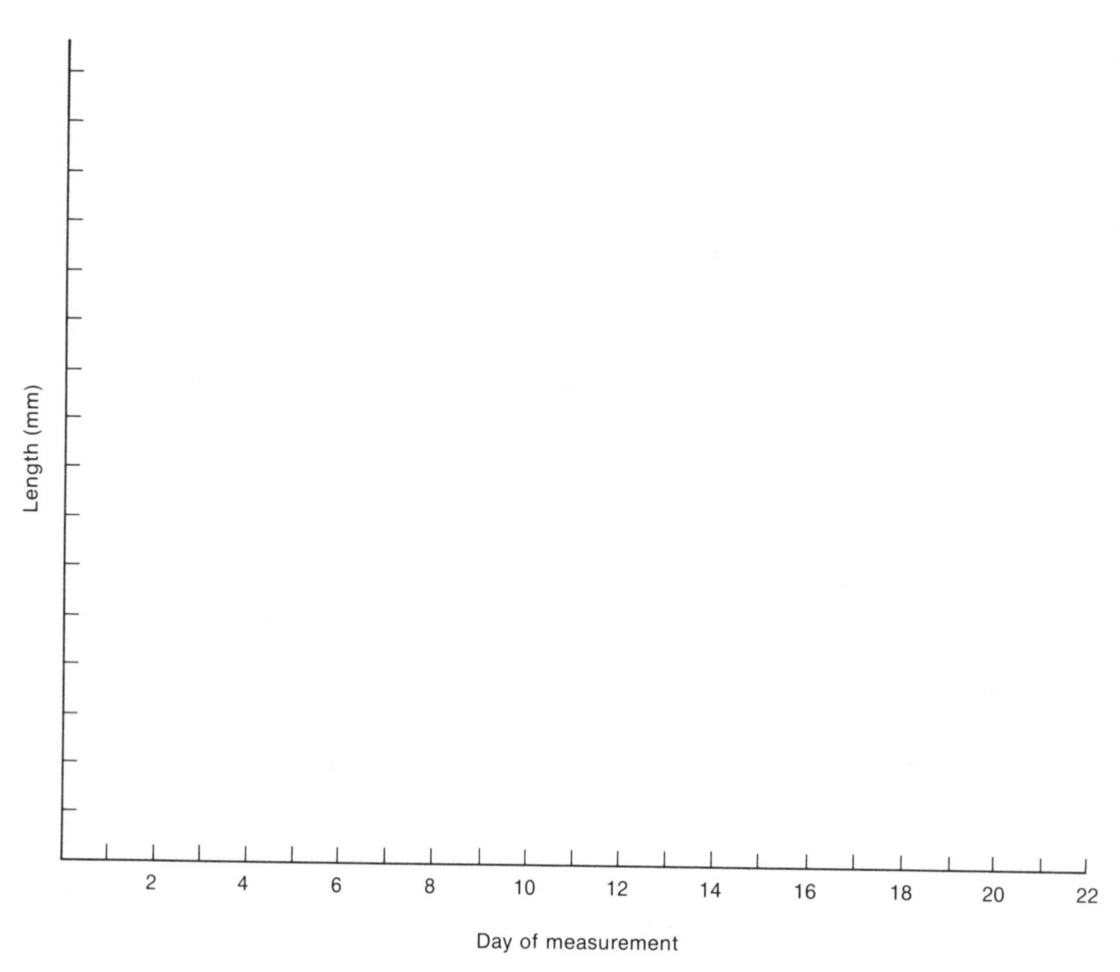

FIG. 22-8
Growth curve of bean plant leaves.

REFERENCES

Cutter, E. G. 1978. *Plant Anatomy: Cells and Tissues.* Part 1. 2d ed. Addison-Wesley.

Cutter, E. G. 1982. *Plant Anatomy: Organs.* Part 2. 2d ed. Addison-Wesley.

Foster, A. S., and E. M. Gifford, Jr. 1974. *Comparative Morphology of Vascular Plants.* W. H. Freeman and Company.

Galston, A. W., P. J. Davies, and R. L. Satton, 1980. *The Life of the Green Plant.* 3d ed. Prentice-Hall.

Greulach, V. 1983. *Plant Structure and Function.* 2d ed. Macmillan.

Jacobs, W. P. 1979. *Plant Hormones and Plant Development.* Cambridge University Press.

Leopold, A. C., and P. E. Friedemann. 1975. *Plant Growth and Development.* 2d ed. McGraw-Hill.

Raven, P. H., R. F. Evert, and H. Curtis. 1981. *Biology of Plants.* 3d ed. Worth.

Ray, P. M., T. A. Steeves, and S. A. Fultz. 1983. *Botany.* Saunders.

Wareing, P. F., and I. D. J. Phillips. 1981. *The Control of Growth and Differentiation in Plants.* 3d ed. Pergamon Press.

EXERCISE 23

Plant Development: Hormonal Regulation

One of the challenging problems facing biologists is understanding the basic mechanisms that regulate the development of plants and animals. The systematic differentiation of millions of cells into various tissues and organs, and the development of specific form and shape, requires precise coordination among the individual parts of the organism during each stage of growth.

The development of a seed into a mature plant is one aspect of this remarkable process. It consists of growth by cell division and cell elongation, differentiation of new organs such as roots, stems, leaves, and flowers, and a complex series of chemical changes. The final form of the plant is a blend of the plant's genetic "blueprint" and the modifying effects of the environment. When the seed begins to germinate, it absorbs large amounts of water, and cells at the meristems of the plant embryo begin to divide. For reasons not yet understood, the root almost always begins to develop before the shoot. At both root and shoot ends of the embryo, new cells are formed by the **meristematic** (dividing) areas of the growing points, followed by elongation and differentiation of these cells.

Ultimately, the control of growth and differentiation is in the DNA of the nucleus, which controls the production of hormones and other regulatory chemicals. Our knowledge of the mechanisms of regulation of the various patterns of growth and differentiation is expanding rapidly as new experimental evidence is reported.

Some of the growth-regulating substances in plants are auxins, gibberellic acids, cytokinins, and various growth inhibitors.

Auxins are produced in meristematic tissues. They influence cell elongation, inhibit the growth of lateral buds, promote the initiation of roots, and, in a few plants, regulate the differentiation of flower buds.

Gibberellic acids, synthesized in meristems and plant seeds, are highly complex substances that affect cell elongation, cell division, and flowering responses in some plants.

Cytokinins, which appear to be related to a component of RNA, promote cell division and, in the presence of auxin, induce the differentiation of roots and shoots.

Inhibitors of different types regulate such responses in plants as flowering, dormancy of buds and seeds, and rates of growth.

In the following parts of the exercise, you will study the effects of certain environmental and chemical factors on the patterns of growth and differentiation in plants.

A. GIBBERELLIC ACID

Gibberellic acids, first discovered by Japanese scientists in the 1920s, went largely unnoticed until the early 1950s, when English and American biologists became interested in these compounds. They were first isolated from a fungus, *Gibberella.*

In this part of the exercise, you will determine some of the more obvious effects of gibberellic acid on plant development.

Before initiating your study, make a hypothesis (or prediction) as to what you expect the results to be. One way of doing this is to use the "If-Then" predictive approach. For example,

If *gibberellic acid acts to inhibit growth,* then *plants treated with gibberellic acid should grow less than control plants that have not been treated with gibberellic acid.*

Make a prediction as to what you expect the results to be in this study.

____

____

____

____

____

____

1. Working on teams of three, obtain 40 bean seeds that have been soaking in water for several hours.

2. Plant 20 seeds (about 15 mm deep) in moist vermiculite in a tray. Label the tray "Gibberellic Acid" (Fig. 23-1A). Plant the remaining 20 seeds in a second tray labeled "Control."

3. When the plants are 7–8 cm tall (about 3 inches), select 10 plants in each try that are about the same size. Tag each individual plant with a number, along with the date. Cut the remaining plants at the ground level and discard the parts you have cut off (Fig. 23-1B).

4. Measure the height of each plant (in millimeters) from the cotyledons to the tip of the shoot apex. Can you think of other measurements you might take? If so, use these in place of the one suggested.

5. Record the date, height, and appearance of the plants in Table 23-1 under Day "0."

6. Apply a drop of gibberellic acid to the shoot apex of each plant in the tray labeled "Gibberellic Acid." What will you apply to the control plants?

____

7. Apply gibberellic acid to the plants weekly, and record the height of each plant and the general appearance of all plants in Table 23-1. As the plants continue to grow, it may be advisable to "stake" the plants using thin bamboo sticks and twine. Be careful not to crush the stems when tying them to the sticks.

8. At the conclusion of the experiment (as determined by your instructor), plot the data in Fig. 23-2. Analyze the data, and determine whether the growth response obtained with gibberellic acid is significantly different from that of the controls. (See Appendix F for a method of statistical analysis of the data.)

Given the results of your statistical analysis, do the data you obtained support your hypothesis? Explain.

____

____

____

____

____

B. PLANT GROWTH INHIBITORS

The idea that inhibitors were important in the regulation of plant growth first gained importance when

TABLE 23-1
Effect of gibberellic acid on plant growth.

Day	Date	Treated with gibberellic acid		Controls	
		Height of plants	Appearance of plants	Height of plants	Appearance of plants
0		1.___ 6.___ 2.___ 7.___ 3.___ 8.___ 4.___ 9.___ 5.___ 10.___ Avg.___		1.___ 6.___ 2.___ 7.___ 3.___ 8.___ 4.___ 9.___ 5.___ 10.___ Avg.___	
		1.___ 6.___ 2.___ 7.___ 3.___ 8.___ 4.___ 9.___ 5.___ 10.___ Avg.___		1.___ 6.___ 2.___ 7.___ 3.___ 8.___ 4.___ 9.___ 5.___ 10.___ Avg.___	
		1.___ 6.___ 2.___ 7.___ 3.___ 8.___ 4.___ 9.___ 5.___ 10.___ Avg.___		1.___ 6.___ 2.___ 7.___ 3.___ 8.___ 4.___ 9.___ 5.___ 10.___ Avg.___	
		1.___ 6.___ 2.___ 7.___ 3.___ 8.___ 4.___ 9.___ 5.___ 10.___ Avg.___		1.___ 6.___ 2.___ 7.___ 3.___ 8.___ 4.___ 9.___ 5.___ 10.___ Avg.___	
		1.___ 6.___ 2.___ 7.___ 3.___ 8.___ 4.___ 9.___ 5.___ 10.___ Avg.___		1.___ 6.___ 2.___ 7.___ 3.___ 8.___ 4.___ 9.___ 5.___ 10.___ Avg.___	
		1.___ 6.___ 2.___ 7.___ 3.___ 8.___ 4.___ 9.___ 5.___ 10.___ Avg.___		1.___ 6.___ 2.___ 7.___ 3.___ 8.___ 4.___ 9.___ 5.___ 10.___ Avg.___	
		1.___ 6.___ 2.___ 7.___ 3.___ 8.___ 4.___ 9.___ 5.___ 10.___ Avg.___		1.___ 6.___ 2.___ 7.___ 3.___ 8.___ 4.___ 9.___ 5.___ 10.___ Avg.___	

it was discovered that dormant buds of ash trees contained large amounts of inhibitors. The inhibitor concentration declined as dormancy ended. More recently, a substance called **dormin** has been isolated that is believed to cause plants to stop growing and enter into their dormant state. Dormin has also been called **abscissin** because it is responsible for causing leaves to fall off (absciss) plants late in the growing season. (Both dormin and abscissin have now been shown to be **abscisic acid.**) Thus, a single substance isolated from different plants not only controls leaf fall but also regulates dormancy and many other growth processes.

Before initiating this exercise, hypothesize on

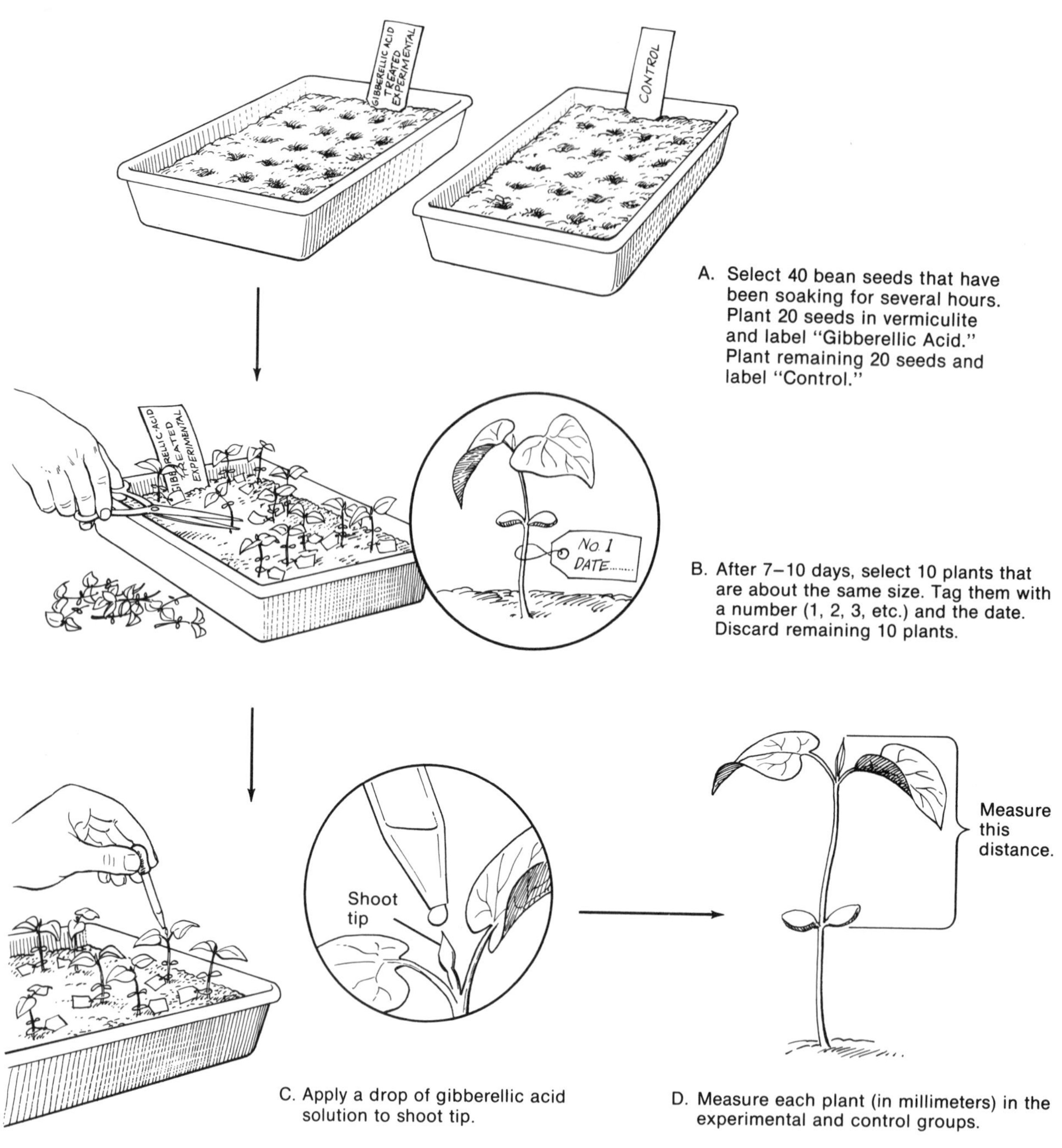

FIG. 23-1
Procedure for determining the effect of gibberellic acid on plant growth.

what you expect the results to be. Use the "If-Then" format given as an example in Part A.

____

____

____

____

____

____

Select three bean or sunflower plants that are in about the same stage of development (only the first two leaves should have expanded). Tag them (1)

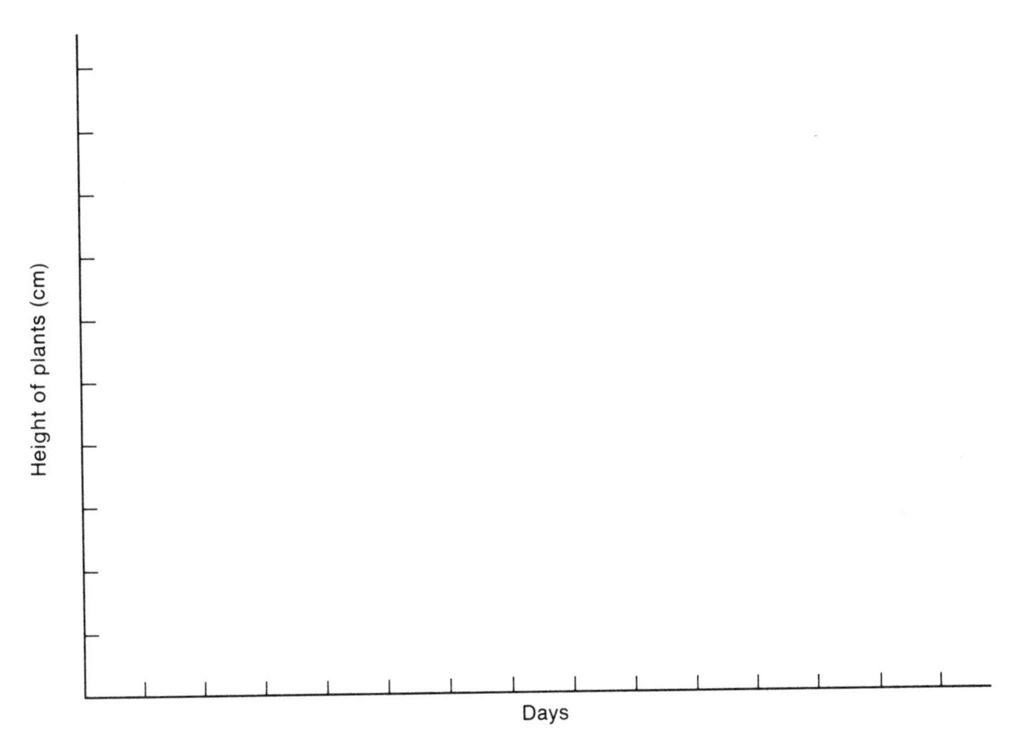

FIG. 23-2
Effect of gibberellic acid on plant growth.

TABLE 23-2
Effect of phosfon and gibberellic acid on plant growth.

Date	Phosfon			Gibberellic acid			Control		
	Average height (mm)	Color of leaves	Other	Average height (mm)	Color of leaves	Other	Average height (mm)	Color of leaves	Other
0									

"Phosfon," (2) "Gibberellic Acid," and (3) "Control," along with your name and the date (Fig. 23-3B). Record the average beginning height of the plants in Table 23-2. In Fig. 23-4, draw the plants as they appear before treatment.

Using a wooden match or similar applicator, apply a ring of phosfon-lanolin paste around the stem of the plant labeled "Phosfon" about 15 mm below the first pair of leaves (Fig. 23-3C). Then place the plant in bright sunlight. (*Note:* The phosphon and gibberellic acid is mixed with lanolin, an inert paste, which allows for even application of the growth substances to the treated part of the plant.)

Using a different applicator, apply gibberellic-acid lanolin paste to the pot labeled "Gibberellic Acid." What should be applied to the control plants and why?

____

____

Examine your plants and those of other students every 2–3 days for the next 3 weeks. Record the average height of the plants, and other information in Table 23-2.

At the conclusion of the study, plot your data in Fig. 23-5. In Fig. 23-4, draw the plants as they appear at the end of the study.

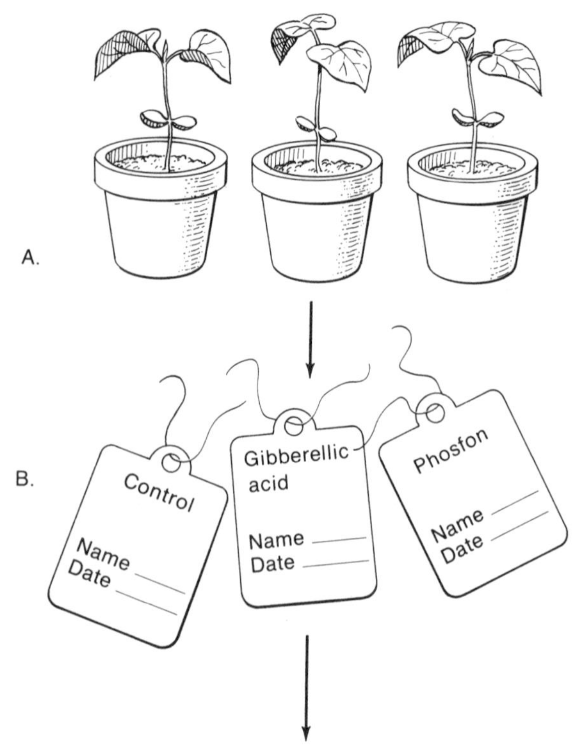

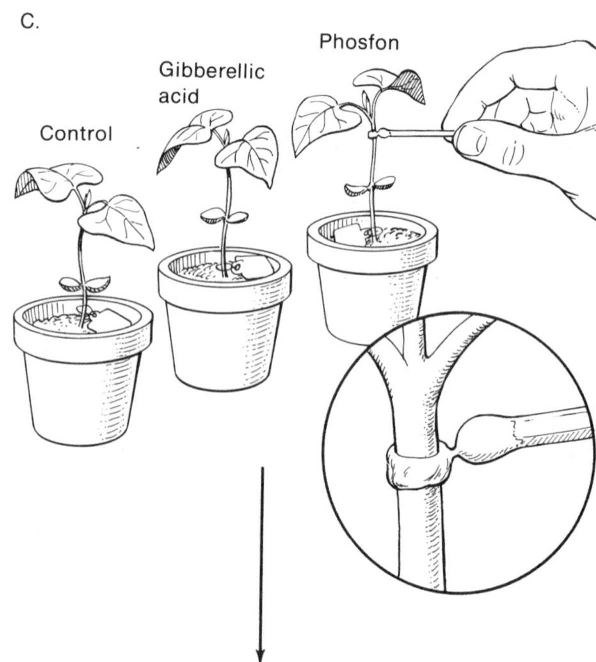

D. Examine the plants every 2 to 3 days for the next 3 weeks.

FIG. 23-3
Effect of growth inhibitors on plant development.

C. DETECTION OF PLANT GROWTH

Hormonal regulation of growth in animals has been known for a number of years. By 1930, it was commonly accepted that plant growth was also under the regulation of special hormonal substances, now called **auxins.** The principal naturally occurring auxin is called **indoleacetic acid,** or **IAA** (Fig. 23-6).

Before indoleacetic acid was discovered, the term *auxin* was used when referring to the natural growth hormone. The term now is used to describe any of a large number of compounds having physiological activity similar to indoleacetic acid. IAA has several effects on growth. A critical effect is the ability to induce cells to elongate. Other effects include fruit development, seasonal initiation of the activity of the vascular cambium, inhibition of lateral buds, and branch root development. Its effects appear to depend on the particular target tissue and the presence of other growth factors.

The isolation and identification of auxins was simplified by the use of a biological testing procedure called a **bioassay.** Numerous biologically active compounds are present in living organisms in amounts so minute that they cannot be detected by the usual chemical procedures. In a bioassay, the whole organism, or some part of it, is used to detect

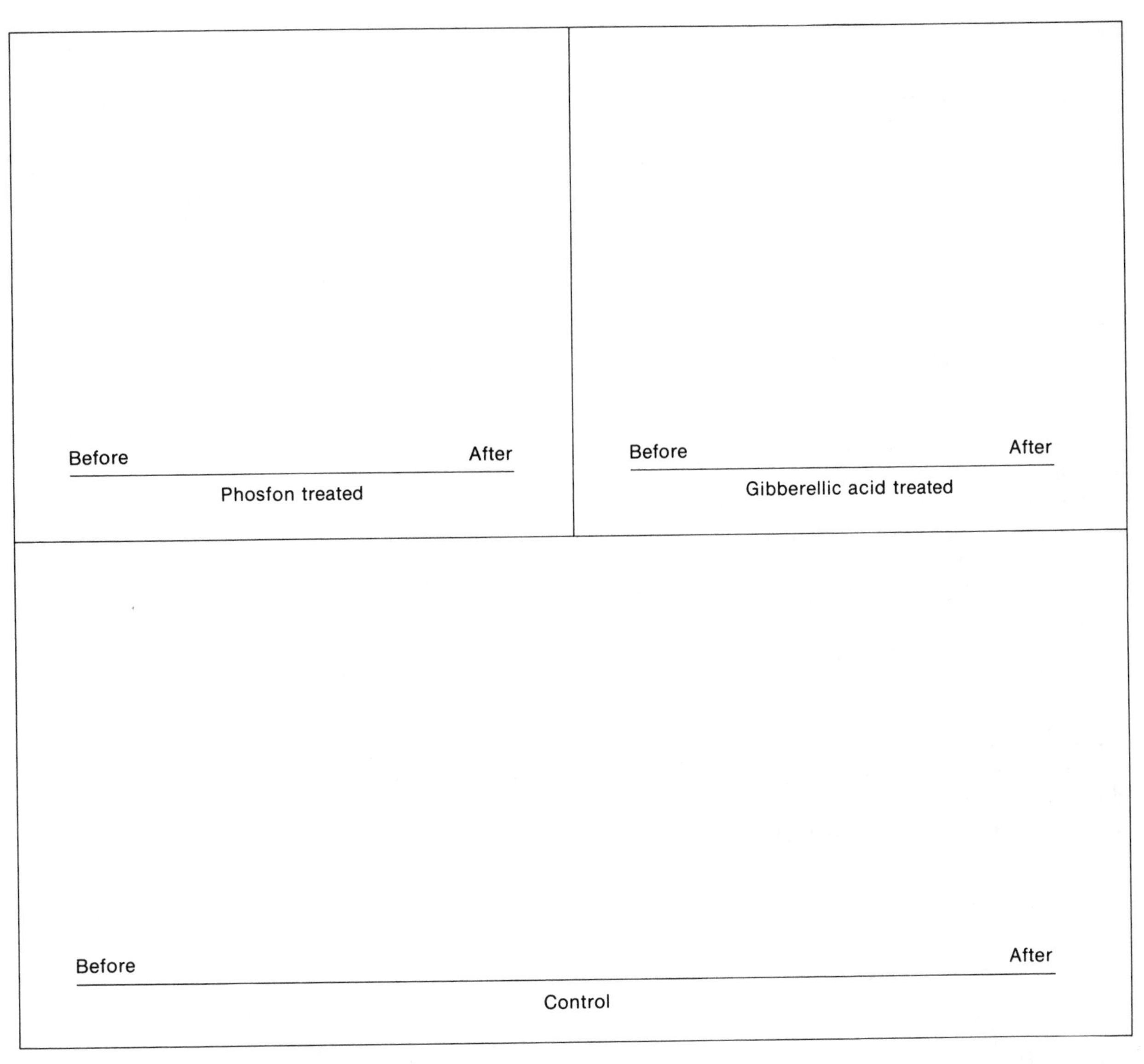

FIG. 23-4
Effect of phosfon and gibberellic acid on plant growth.

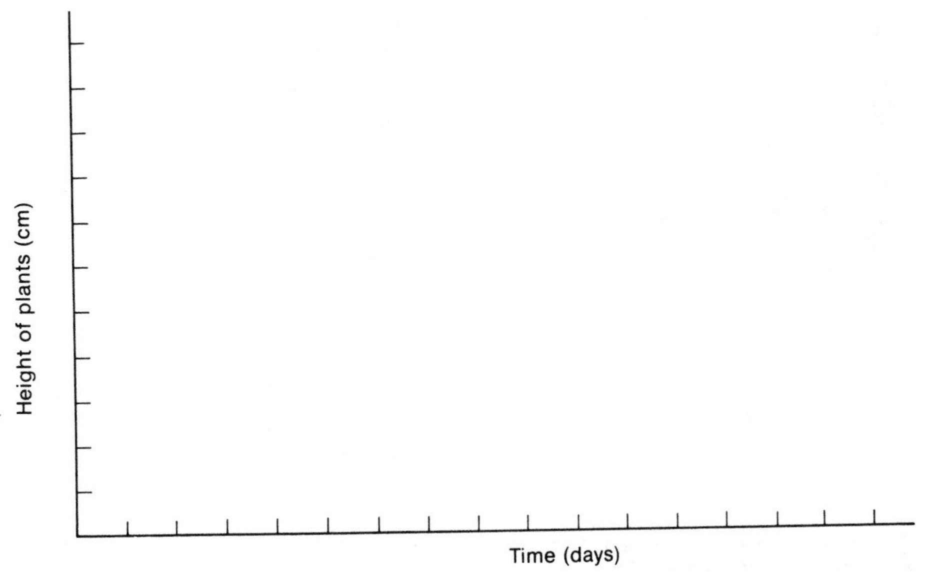

FIG. 23-5
Effect of phosfon and gibberellic acid on plant growth.

Indole nucleus

Acetic acid side chain

FIG. 23-6
Structure of indoleacetic acid.

the presence of, and to measure the amounts of, biologically active substances.

The most sensitive bioassay for auxins is the ***Avena*** **curvature test.** This test makes use of the coleoptile of the *Avena* (oat) seedling (Fig. 23-7). The **coleoptile** is a cellular sheath that surrounds the embryonic leaves of the developing seedling. If oat seeds are germinated in the dark, cells of the coleoptile continue to divide until the coleoptile is approximately 1 cm long. For the next 3–4 days, while the coleoptile reaches its maximum length of 5–6 cm, the coleoptile extends solely by cell elongation. This elongation is brought about by auxin produced in the tip of the coleoptile. When grown in the dark, the auxin is directed downward, causing the embryonic cells of the coleoptile to elongate. If the tip is removed, no cellular enlargement occurs. If the decapitated tip is replaced by an agar block containing auxin, growth of the coleoptile resumes. If the block of agar is displaced to one side, the cells on that side elongate more rapidly than the cells on the opposite side with the result that the coleoptile curves. This is the basis for the curvature test. It has been shown that the degree of curvature is, within limits, proportional to the concentration of auxin in the agar block. This test is a highly reliable and sensitive quantitative test for detecting substances that have auxin activity.

Although the *Avena* curvature test is very sensitive, it also requires rigidly controlled conditions of temperature and humidity and accurate measurements of curvature. For these reasons, it is difficult to conduct such a test in most introductory laboratory courses. A simpler bioassay uses oat coleoptiles and is less sensitive in terms of the concentration of auxin measurable. Called the ***Avena*** **straight-growth test,** it measures straight growth as reflected in an increase in length of coleoptile sections. To save time, the preliminary steps of soaking and planting the oat seeds will be carried out before the laboratory meeting.

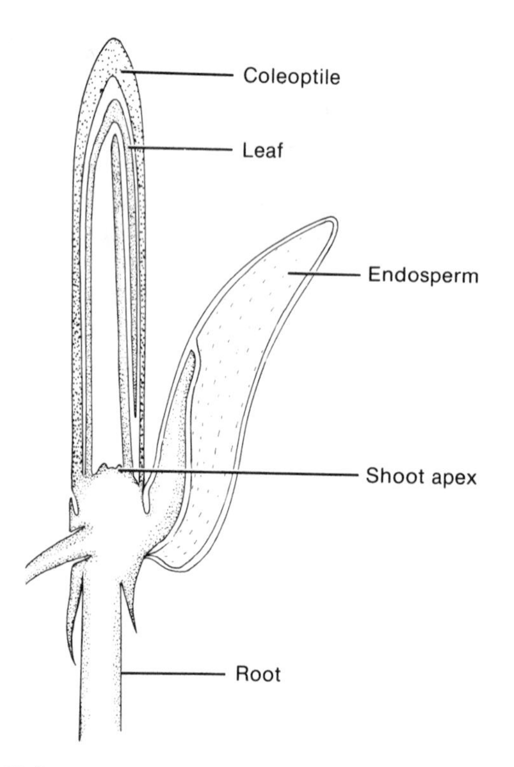

FIG. 23-7
Avena seedling. (From *Plants in Action,* by Leonard Machlis and John G. Torrey. W. H. Freeman and Company. Copyright © 1956.)

1. The *Avena* Coleoptile Straight-Growth Test

1. For each laboratory section, soak 600–700 Brighton hull-less oat seeds for 2 hours in a liter of distilled water. After soaking, rinse the seeds two or three times with distilled water. Plant the oats on moist germinating paper (or Kleenex) in a tray; cover and pack tightly with about 15 mm of moist vermiculite; cover and place in the dark at room temperature. Germinate for 70–72 hours, exposing the seeds to 1 hour of red light out of each 24.

2. Label six test tubes and fill them as indicated in Table 23-3. Your instructor will provide the IAA dilutions and incubating medium (a buffered nutrient solution containing sucrose).

3. Using the information given in the note at the bottom of Table 23-3, describe how you would prepare the various dilutions used in this experiment. (*Hint:* Notice that 1 ml of each dilution series is further diluted with 1 ml of incubating solution.)

____

____

____

4. Approximately 70 hours after planting, the coleoptiles will be 20–30 mm in length. Working in a room illuminated with red light (or use a room darkened as much as possible—but work quickly) select 60 coleoptiles measuring 25–30 mm in length. Cut off the root and seed and place the coleoptiles on moistened filter paper in a petri dish.

5. Place four or five coleoptiles at a time on a paraffin block and cut them with the special cutter, as shown in Fig. 23-8. This cutter will divide each coleoptile into three parts: a 3-mm tip portion, a 5-mm section, and a base of about 20 mm. Discard the 3-mm tips and the bases. Place 10 of the 5-mm sections in each of the six test tubes.

6. Stopper the test tubes with cotton or styrofoam plugs and place in the dark for 24 hours at 25°C. After 24 hours, remove the sections from the tubes and measure them with a millimeter ruler to the nearest 0.5 mm. Record the average length in Table 23-4.

7. Plot your data in Fig. 23-9 to obtain the standard curve for these concentrations. Does the curve show a proportional increase in growth with increasing IAA concentration? Explain.

____

____

____

____

____

____

8. Plot the unknown IAA concentration on the graph. On the basis of the position of the unknown, estimate the concentration of the IAA in the unknown solution.

____

9. What is the reason for having Tube 1?

____

____

____

TABLE 23-3
Protocol for the *Avena* coleoptile straight-growth test.

Tube contents	Tube no.					
	1	2	3	4	5	6
Incubation solution (ml)	1	1	1	1	1	1
Distilled H_2O (ml)	1	—	—	—	—	—
10^{-7} *M* IAA solution (ml)	—	1	—	—	—	—
10^{-6} *M* IAA solution (ml)	—	—	1	—	—	—
10^{-5} *M* IAA solution (ml)	—	—	—	1	—	—
10^{-4} *M* IAA solution (ml)	—	—	—	—	1	—
Unknown IAA solution (ml)	—	—	—	—	—	1

Note: The dilution series uses solutions of varying molar *(M)* concentration. A molar solution is made by dissolving the molecular weight of a compound, in grams, in a liter of solvent. The molecular weight of IAA is 175.2. Thus a 1.0 *M* solution of IAA contains 175.2 g/1000 ml of water. In a similar manner

$$
\begin{aligned}
1/10\ M &= 10^{-1}M = 17.52000000\ \text{g/liter} = 17520.00000\ \text{mg/liter}\\
1/100\ M &= 10^{-2}M = 1.75200000\ \text{g/liter} = 1752.00000\ \text{mg/liter}\\
1/1000\ M &= 10^{-3}M = 0.17520000\ \text{g/liter} = 175.20000\ \text{mg/liter}\\
1/10{,}000\ M &= 10^{-4}M = 0.01752000\ \text{g/liter} = 17.52000\ \text{mg/liter}\\
1/100.000\ M &= 10^{-5}M = 0.00175200\ \text{g/liter} = 1.75200\ \text{mg/liter}\\
1/1{,}000{,}000\ M &= 10^{-5}M = 0.00017520\ \text{g/liter} = 0.17520\ \text{mg/liter}\\
1/10{,}000{,}000\ M &= 10^{-7}M = 0.00001752\ \text{g/liter} = 0.01752\ \text{mg/liter}
\end{aligned}
$$

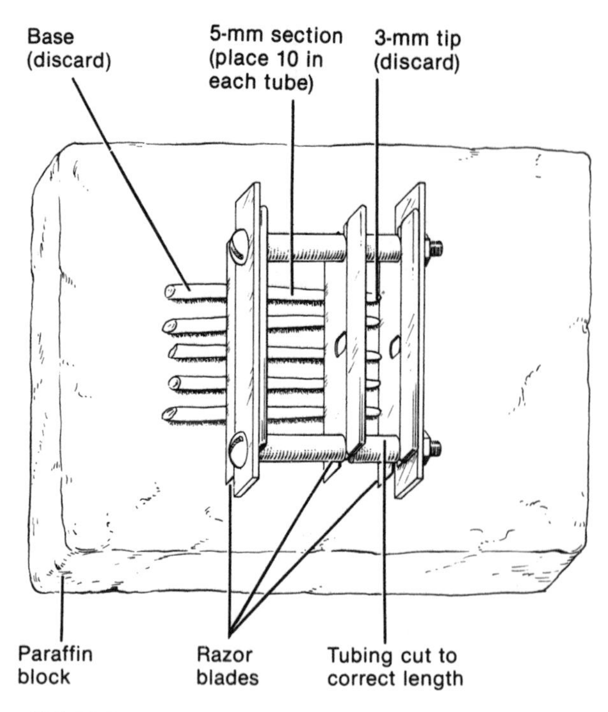

FIG. 23-8
Method of cutting coleoptiles for *Avena* straight-growth test.

2. Colormetric Test for IAA

There is a simpler chemical test for detecting one of the natural auxins (such as IAA), much less sensitive than either of the bioassays, but useful for **in vitro** (test tube) studies of auxin concentrations well above those normally found in plant tissue. For example, an enzyme found in plant tissues, called *indoleacetic acid oxidase,* breaks down IAA. This test would be useful for studying the action of such an enzyme.

The color test you will perform makes use of the fact that IAA, in the presence of iron chloride in Salkowski reagent, forms a red color complex that may be quantitatively measured with a colorimeter (see Appendix C for a discussion of spectrophotometry).

You will be provided with a stock solution of IAA at a concentration of 100 mg/liter. Make 10-ml samples of the following concentrations: 40, 20, 10, 1, and 0 mg/liter. This is done as follows. Collect 10 ml of the stock solution in a test tube. Pipet 4 ml of this solution into a test tube and dilute with distilled water to 10 ml. Label this tube "40 mg/liter." Repeat this procedure using 2 ml, 1 ml, and 0.1 ml of the stock solution, diluting each to 10 ml. Mark these tubes "20 mg/liter," "10 mg/liter," and "1 mg/liter," respectively. For the 0 mg/liter concentration, merely pipet 10 ml of distilled water into a tube. This dilution series will be used to establish a standard curve. In addition, you will be given a solution containing an unknown concentration of IAA.

To conduct the color test, add 2 ml of the solution being tested to a test tube containing 8 ml of Salkowski reagent. (*Note:* Because full color development is a function of time, stagger your tests to allow time for each colorimetric measurement.) Shake the mixture thoroughly (*and carefully,* as the reagent contains sulfuric acid!) and set aside for exactly 30 minutes to allow the color to develop. While waiting for the color reaction, warm up the colorimeter and adjust the wavelength to read 510 nm (see Appendix B for instructions on the use of the Bausch & Lomb Spectronic 20 colorimeter).

Standardize the colorimeter, using a solution consisting of 2 ml of distilled water and 8 ml of Salkowski reagent. Immediately transfer about 3 ml of this mixture to a colorimeter tube and place in the Spectronic 20. Adjust the instrument to read 100% transmittance. Why is this step required?

____

____

At the end of 30 minutes, record the percent transmittance for each of the IAA concentrations and the unknown, and record the data in Table 23-5. Plot your data in Fig. 23-10. What is the concentration of the unknown IAA solution?

____

TABLE 23-4
Data for *Avena* straight-growth test.

	IAA (moles)					
	O Incubating medium	10^{-7}	10^{-6}	10^{-5}	10^{-4}	Unknown
Average coleoptile length (mm)						

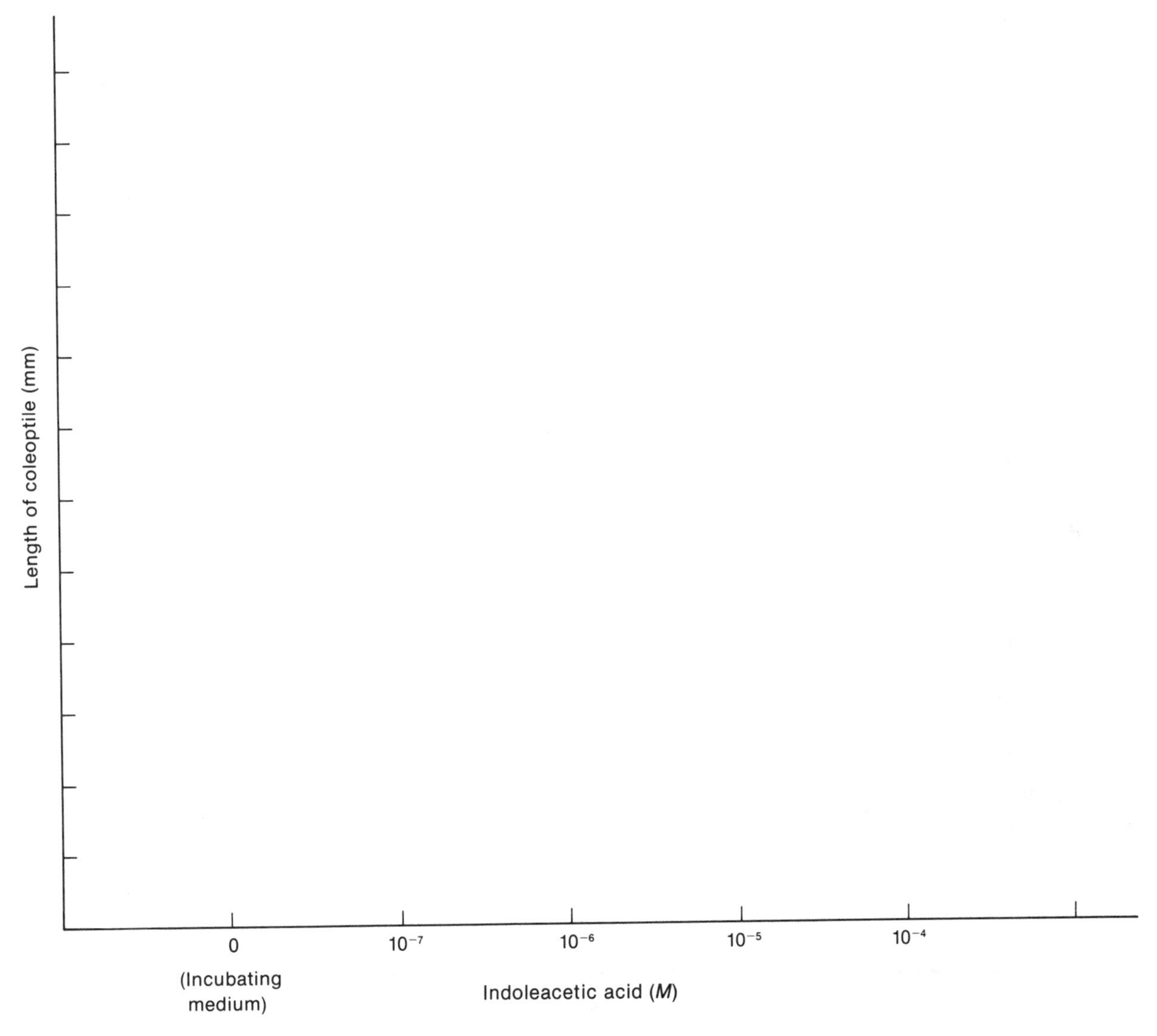

FIG. 23-9
Standard curve for *Avena* straight-growth test.

TABLE 23-5
Data for colorimetric determination of IAA

	Tube contents (mg/liter)					
	O Control	1	10	20	40	Unknown
Transmittance (%)						

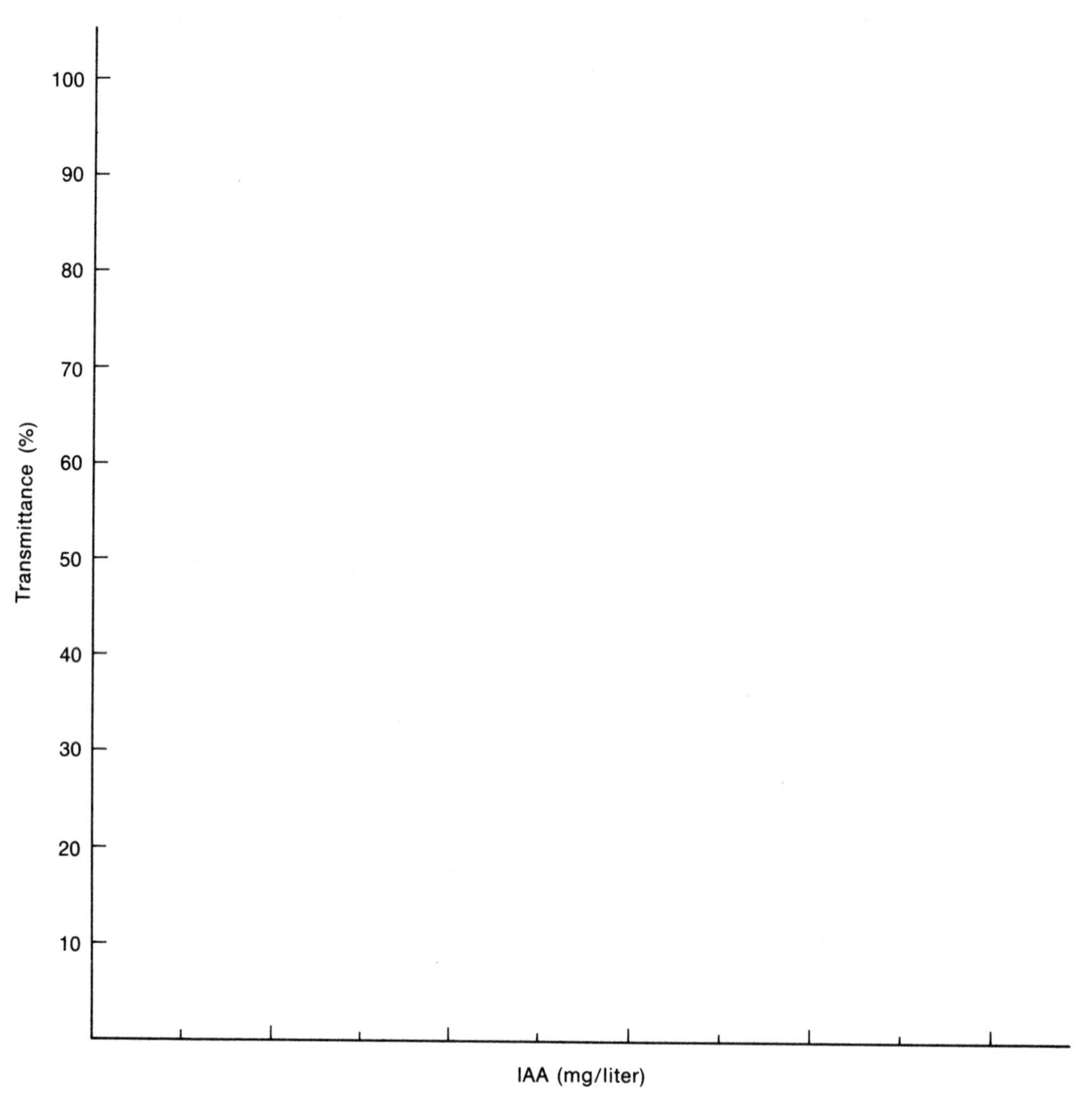

FIG. 23-10
Colorimetric determination of IAA.

Of the three methods discussed in this exercise, which would you use if you had to assay accurately for the total amount of auxin in 500 mg of plant tissue? Explain.

REFERENCES

Cutter, E. G. 1982. *Plant Anatomy: Organs.* Part 2. 2d ed. Addison-Wesley.

Galston, A. W., P. J. Davies, and R. L. Satter. 1980. *The Life of the Green Plant.* 3d ed. Prentice-Hall.

Jacobs, W. P. 1979. *Plant Hormones and Plant Development.* Cambridge University Press.

Leopold, A. C., and P. E. Kriedemann. 1975. *Plant Growth and Development.* 2d ed. McGraw-Hill.

Raven, P. H., R. F. Evert, and H. Curtis. 1981. *Biology of Plants.* 3d ed. Worth.

Steeves, T., and I. Sussex. 1972. *Patterns in Plant Development.* Prentice-Hall.
Torrey, J. C. 1967. *Development in Flowering Plants.* Macmillan.
Van Overbeek, J. 1968. The Control of Plant Growth. *Scientific American* 219:75–81 (Offprint 1111). *Scientific American* Offprints are available from W. H. Freeman and Company, 41 Madison Avenue, New York 10010, and 20 Beaumont Street, Oxford OX1 2NQ, England. Please order by number.
Wareing, P. F., and I. D. J. Phillips. 1981. *Growth and Differentiation in Plants.* 3d ed. Pergamon Press.

EXERCISE 24

Plant Development: Effect of Light

The morphogenetic effects of light on plants have been known since the latter part of the nineteenth century. As early as 1880, experiments conducted with carbon arc lamps showed that artificially prolonged days promoted plant growth. After the invention of the electric light bulb, a large amount of data began to accumulate on the morphogenetic effects of light on plants. These observations ranged from such anatomical effects as variations in fiber and vascular tissue development to establishing the presence of a pigment called **phytochrome,** which may regulate many of the responses of plants or plant parts to light.

In this exercise, you will study the effects of light on development and the effect of day length **(photoperiod)** on flowering and dormancy of plants.

A. PLANT GROWTH IN THE ABSENCE OF LIGHT

Obtain two pots or trays that contain vermiculite. Label one "Dark" and the other "Light." Include your name, the date, and section number on each label. Plant 10 bean seeds about 15 mm deep in each container. Water thoroughly, and then place one of the pots in the sunlight and the other in the dark (Fig. 24-1). (*Note:* Water plants daily throughout the period of the study.)

After about 4 or 5 days, examine the plants that have been growing in the dark. Compare them with those growing in the light. Record your observations in Table 24-1.

Reverse the conditions under which the plants have been growing (that is, place the "dark" plants in the light and the "light" plants in the dark). Examine the plants after 4 or 5 days. Describe the appearance of the plants.

____

____

Record your observations in Table 24-2.

The growth response observed in the plants grown

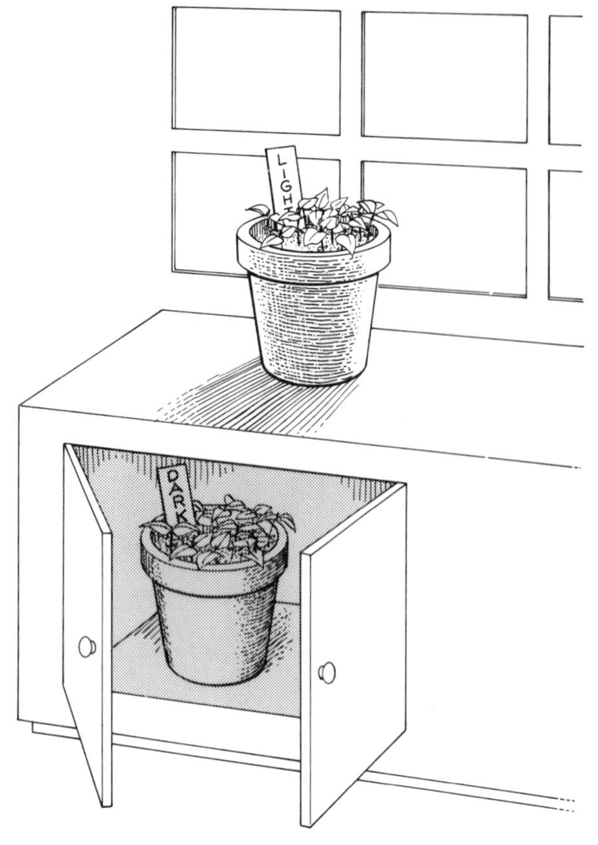

FIG. 24-1
Effect of darkness on plant growth.

in the dark is called **etiolation.** What advantage does such a response have for plant survival?

____

____

B. EFFECT OF DAY LENGTH ON PLANT GROWTH

Day length varies with the seasons. In the Northern Hemisphere, the day length increases during spring to a maximum on June 21, and thereafter it decreases to a minimum on December 21. This day length is called the photoperiod, and the response of plants to photoperiods of varying lengths is called **photoperiodism** or a **photoperiodic response.** In this part of the exercise, you will examine the effects of the photoperiod on two aspects of plant development.

1. Regulation of Flowering

There are basically two types of photoperiodic flowering responses. These are called **long-day** and **short-day responses.** Plants that exhibit these responses are called long-day (L-D) and short-day (S-D) plants. L-D plants flower when the photoperiod *exceeds* some critical value, which is generally in excess of 14 hours (Table 24-3). At day lengths less than the criti-

TABLE 24-1
Effect of darkness on plant growth.

Observations and measurement	Treatment	
	Plants grown in dark	Plants grown in light
Color of shoot		
Average number of leaves		
Average length of first pair of leaves		
Average length of shoot between cotyledons and first leaves		

TABLE 24-2
Effect of placing "light-grown" plants in the dark and "dark-grown" plants in the light.

Observations	Plants grown in dark → light	Plants grown in light → dark
Color of shoot		
Other observations		

cal value the plant usually grows only vegetatively. An S-D plant (really a long-night plant) will flower when the photoperiod is *less* than some critical value, usually about 10 hours, and grows vegetatively when this critical day length is exceeded. Other plants, not responsive to day length, are called **day-neutral** plants.

The photoperiodic control of flowering involves a complex mechanism about which few facts are known. These responses appear to be under the control of the light-sensitive proteinaceous pigment called **phytochrome**. This bluish green pigment has been extracted and isolated with a high degree of purity.

The effect of phytochrome on flowering appears to be related to its having two forms in the plant. One form, called P_{660}, absorbs red light (with maximum absorption at 660 nm) and is then converted into a second form, called P_{730}. The P_{730} form of the pigment strongly absorbs far-red radiation (maximally at 730 nm) and may then be reconverted to the P_{660} form. The photoperiod responses in various plants may be a result of differences in the rate of conversion of one form of phytochrome to another.

TABLE 24-3
Flowering responses of long-day, short-day, and day-neutral plants.

Plant	24-hour period		Flowers
	Light	Dark	
Long-day	————→		no
	———— (Critical day length)	————→	yes
Short-day	————→		yes
	———— (Critical day length)	————→	no
Day-neutral	————→		yes
	———— (Critical day length)	————→	yes

There is also some evidence for a second photomorphogenic receptor involving a system of pigments absorbing blue and far-red radiation. These appear to operate independently of the red, far-red phytochrome system.

In this study, you will be given two different species of plants and will attempt to determine whether they are L-D or S-D plants.

The L-D and S-D conditions are shown in Fig. 24-2.

1. Obtain two trays of plants provided by your instructor. Label one of the trays "L-D," along with your name, section number, and the date. Label the other tray "S-D" (Fig. 24-2A).

2. Place the trays under light conditions so that the L-D plants receive 16 hours of light and 8 hours of darkness (Fig. 24-2B). The S-D plants should receive 8 hours of light and 16 hours of darkness (Fig. 24-2C).

3. Examine the plants weekly, and watch for the appearance of flower buds. The detection of flowers is difficult in the early stages of formation, so close inspection may be necessary. However, avoid handling the plants excessively.

4. If, at the conclusion of the experiment (as determined by your instructor), no flowers are visible, remove the buds from each plant and place them in separate, labeled petri dishes containing moistened filter paper. Using a dissecting microscope, dissect the buds and determine if flower primordia are present or if the buds are still producing leaves. Your instructor will have floral buds and vegetative buds on demonstration to help you distinguish between the two.

5. Determine the photoperiod response of the plants, and confirm your results with your instructor. What economic use can be made of the photoperiodic response?

___

___

___

___

Many other aspects of plant growth are under photoperiodic control. Basing your answer on your reading, list several of these here.

___

___

___

___

2. Dormancy

In the broadest sense, and without reference to the factors involved, dormancy can be defined as a temporary suspension of visible growth and development. In this exercise, you will determine the effects of the photoperiod in regulating the initiation of the dormant, winter bud in a woody plant. If you are unfamiliar with the appearance of a winter bud, examine the demonstration of dormant buds set up by the instructor.

1. Working in teams, obtain two potted seedlings that are approximately the same height. Label each with your name, section number, the date, and treatment (L-D or S-D).

2. Place the plants on the L-D and S-D benches in the greenhouse. Measure the length of the main shoot on each plant and record this and any other observations that seem appropriate under Day 0 in Table 24-4.

3. Examine the plants during succeeding laboratory periods and record the date and stem length. What evidence can you see that the plant (under one treatment or the other) is forming a dormant bud? Examine the plants of other students, but do not handle them excessively. Make similar measurements and observations over the next several weeks, or until your instructor concludes the experiment.

4. At the conclusion of the experiment, plot the "stem length" data in Fig. 24-3. Suggest reasons for the differences in stem length, if any, between the plants given the L-D and S-D treatments.

___

___

___

Under which photoperiod has a dormant bud been formed?

___

___

A. Label "L-D" (long-day) or "S-D" (short-day), along with your name, date, and section number.

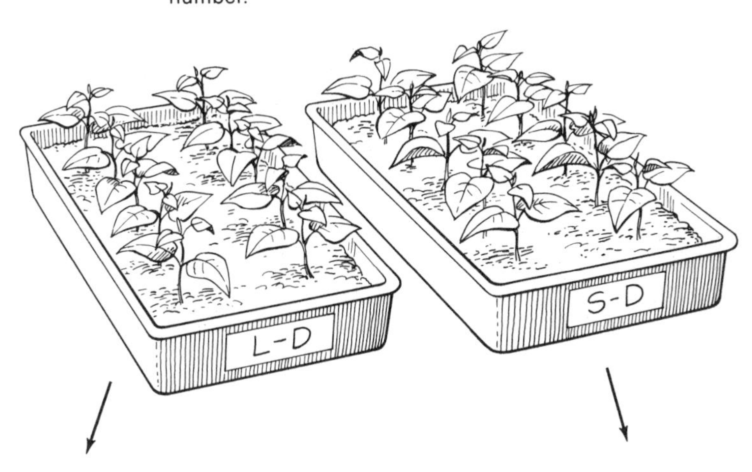

B. Place one group of plants under L-D conditions.

C. Place the second group under S-D conditions.

L-D (short-night) conditions

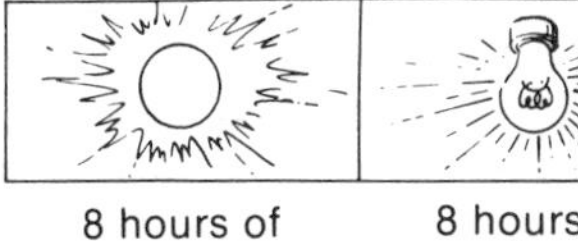

8 hours of sunlight | 8 hours of artificial light | 8 hours of darkness

S-D (long-night) conditions

8 hours of sunlight | 16 hours of darkness

D. Examine the plants weekly for the appearance of flowers. After 3 to 4 weeks (if no flowers are visible) dissect the buds to determine if they have flowers or leaves in them.

FIG. 24-2
Effect of photoperiod on flowering.

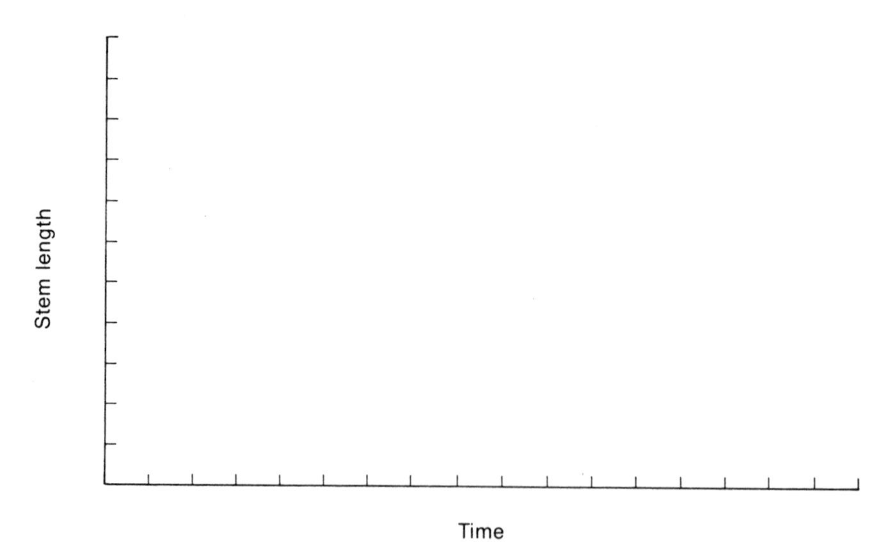

FIG. 24-3
Stem length data.

In view of your knowledge of growth regulators, suggest a mechanism by which the photoperiod is inducing a state of dormancy in these plants.

5. Remove all the leaves of the dormant plants, and return the plants to the same photoperiod that induced dormancy. Examine the plant weekly for 2–3 weeks. List your observations below, and discuss how the plant may receive the light stimulus.

TABLE 24-4
Observations of plant response to photoperiod.

Day	Date	L-D treatment		S-D treatment	
		Length of shoot (cm)	Other observations	Length of shoot (cm)	Other observations
0					

REFERENCES

Galston, A. W., and P. F. Davies. 1980. *Control Mechanisms in Plant Development.* 3d ed. Prentice-Hall.

Leopold, A. C., and P. E. Kriedemann. 1975. *Plant Growth and Development.* 2d ed. McGraw-Hill.

Mohr, H. 1972. *Lectures on Photomorphogenesis.* Springer-Verlag.

Ray, P. M., T. A. Steeves, and S. A. Fultz. 1983. *Botany.* Saunders.

Smith, H. 1975. *Phytochrome and Photomorphogenesis.* McGraw-Hill.

Vince-Prue, D. 1975. *Photoperiodism in Plants.* McGraw-Hill.

Wareing, P. F., and I. D. J. Phillips. 1981. *The Control of Growth and Differentiation in Plants.* 3d ed. Pergamon Press.

EXERCISE 25

Kingdom Monera: Divisions Schizophyta and Cyanobacteria

In this manual, we use the Whittaker five-kingdom system of classification shown in Fig. 25-1. The kingdom Monera (Greek *moneres,* "single, solitary") comprises the simplest known organisms and includes all of those that lack true nuclei in their cells: the **prokaryotes.** The organisms in the other four kingdoms possess nuclei and therefore are collectively referred to as the **eukaryotes.** The three "higher" kingdoms (Plantae, Fungi, Animalia) are all multicellular but differ in their modes of acquiring nutrients. Plants acquire their nutrients by photosynthesis, fungi by absorption, and animals by ingestion. The remaining kingdom, Protista, includes all the lower organisms that do not fit into one of the other four kingdoms. Thus it contains a diverse group of organisms, some of which are less closely related to each other than to certain members of the other four kingdoms.

Organisms in the kingdom Monera are unicellular but sometimes aggregate into filaments or other loose collections of cells. These prokaryotes lack membrane-bound organelles such as mitochondria, plastids, lysosomes, Golgi complexes, and endoplasmic reticulum. Their plasma membranes often have many folds and convolutions extending into the cytoplasm, which increase the surface area. They obtain nutrition predominantly by absorption, although some groups are photosynthetic or chemosynthetic. Reproduction is primarily asexual, by budding or fission, but conjugation, a mechanism of sexual reproduction, occurs in some species.

In this exercise, you will study examples of the following two of the three divisions of Monera:

Division Schizophyta (bacteria): unicellular, reproduction usually asexual by cell division, nutrition usually heterotrophic

Division Cyanobacteria (formerly called the blue-green algae): unicellular or colonial, have chlorophyll but no plastids, reproduction by fission, nutrition usually autotropic

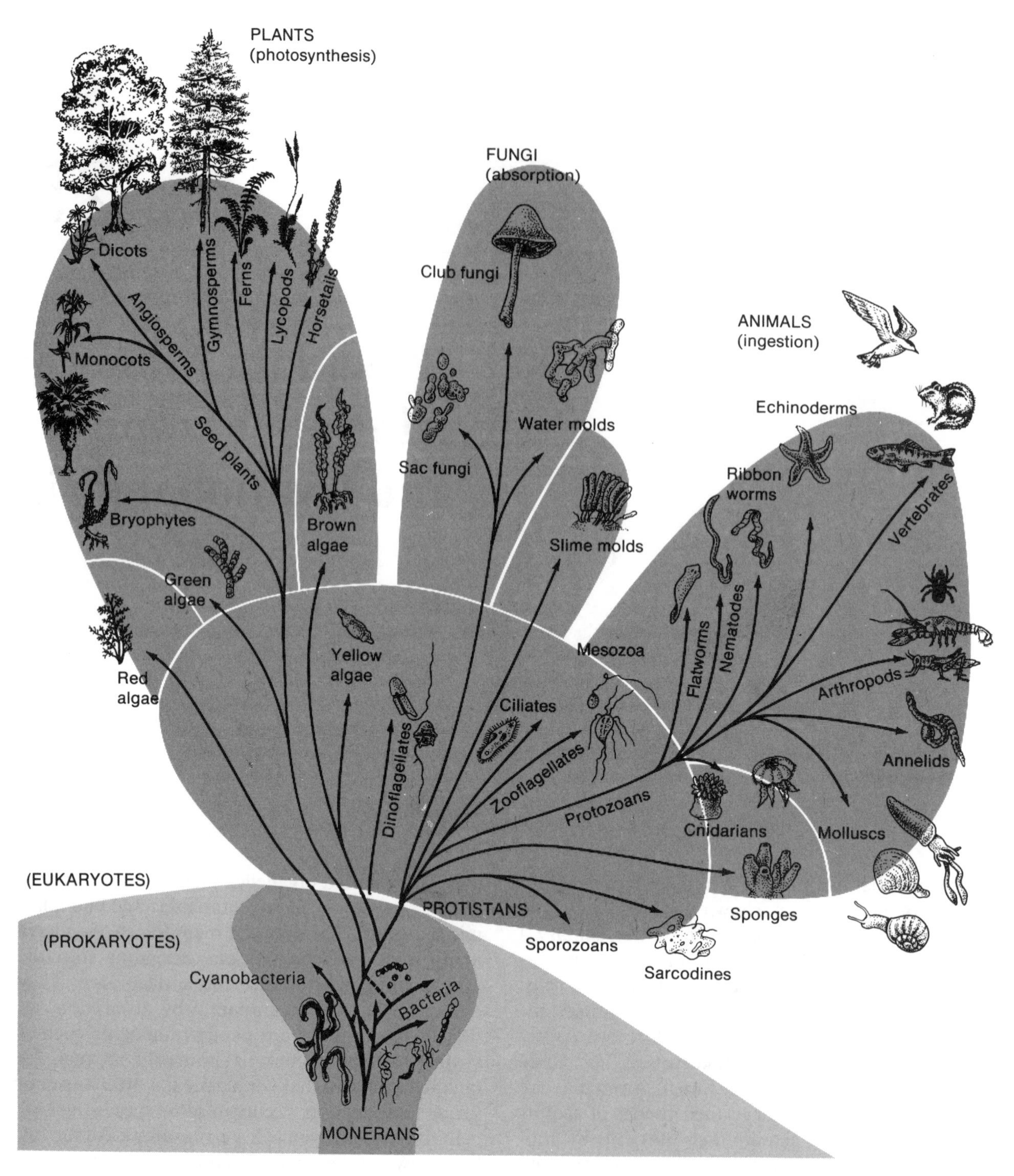

FIG. 25-1
The Whitaker five-kingdom system of classification. (After *Biology Today,* 3d ed., by David L. Kirk. CRM Books, a division of Random House, Inc. © 1980.)

A. DIVISION SCHIZOPHYTA (BACTERIA)

Bacteria, as a group, are the most ancient of all organisms, having been found in fossil records as far back as 3.5 billion years, which is far beyond any known fossil eukaryote. Despite their minute size and apparently simple structure, bacteria have adapted to a wide variety of environments. In fact, they are more widely distributed in nature than any other group or organisms and can survive in environments that support no other form of life. Some bacteria are **obligate anaerobes;** that is, they can survive and multiply only in the absence of free oxygen. Other bacteria, the **facultative anaerobes,** can survive without oxygen but grow more vigorously when it is available in their environment.

Bacteria are especially abundant in the soil. Some species play an important role in the nitrogen cycle —where nitrogen-fixing bacteria convert nitrogen gas in the atmosphere into nitrogenous salts that are used by plants for growth.

Bacteria are also responsible for the decay of organic matter, such as dead plants and animals. During the breakdown of this material, carbon dioxide is released and may then be reused in photosynthesis. Although most bacteria are **heterotrophs,** some are **autotrophs** and can make their own sugars, fats, amino acids, and so forth. One group of bacteria has the ability to obtain energy for its synthetic activities by converting light energy into chemical energy by a process similar to photosynthesis. Photosynthesis in bacteria, however, differs from that in eukaryotic organisms in that water (H_2O) is not used as the reductant for carbon dioxide. Rather, hydrogen sulfide (H_2S) may be used. Thus, in bacterial photosynthesis sulfur is evolved rather than oxygen.

A second group of autotrophic bacteria produces its own organic materials from carbon dioxide, ammonia, or nitrate by obtaining the energy for these syntheses from the oxidation of inorganic substances. For example, a bacterium present in the soil is capable of oxidizing ammonia to nitrate, thus generating useful energy. This organism is a typical **chemosynthetic** autotroph. Because these bacteria contain carbohydrates, fats, proteins, nucleic acids, and vitamins, they represent magnificent synthetic factories for making protoplasm.

Other bacteria are widely used in the commercial production of bakery goods, alcohol, vinegar, chemicals, enzymes, antibiotics, and many other products. These microorganisms are also responsible for food spoilage, food poisoning, and for many plant and animal diseases, including tuberculosis, scarlet fever, pneumonia, and diphtheria. Bacteria, therefore, directly or indirectly influence the survival of humanity.

The functional unit of these organisms is a single cell, the smallest of which is barely visible with the light microscope. For many years, bacteria were thought to reproduce primarily by an asexual process called *fission,* during which the cell pinches in two. It is now known that bacterial cells also exchange genetic material through sexual reproduction. Although sexual mechanisms have been found in only a few species, future research in this area may reveal that sexual reproduction is actually widespread among bacteria.

Before beginning this exercise the student should become familiar with the asepetic techniques described in Appendix I, which are essential when working with microorganisms.

1. Bacterial Staining

Since living bacteria are almost colorless (that is, they do not show enough contrast with the water in which they are suspended), they are frequently stained to contrast with their surroundings and, thus, be seen more readily.

Stains usually consist of a dye that is generally a salt consisting of negatively and positively charged ions. In dyes used for staining, one of the ions is usually colored and is termed the **chromophore.** For example, the simple stain methylene blue is the salt of methylene blue chloride, which dissociates as follows:

$$\text{Methylene blue chloride} \longrightarrow \text{methylene blue}^+ + \text{chloride}^-$$

The chromophore of this simple stain is the positively charged methylene blue ion. Thus methylene blue is called a **basic dye** or **stain** that reacts with negatively charged components of cells such as nucleic acids and some polysaccharides. Basic stains are sometimes referred to as **nuclear stains** since they stain only nuclei in those cells that have them. In contrast, stains in which the chromophore is the negative ion are called **acidic** stains because they react with positively charged constituents (i.e., many proteins) of cells, thus staining only cytoplasm. They are therefore referred to as *cytoplasmic stains.*

While there are numerous and varied techniques for staining bacteria, we will limit staining to those involving simple stains and differential stains and to one for determining the nature of reserve materials stored within the bacterial cell.

a. Preparation of Slides

Clean, grease-free slides are a must for obtaining good preparations. Two methods for cleaning slides follow; use either one.

1. *With Alcohol.* Wipe both surfaces of a slide with alcohol. Allow the alcohol to dry and then, using a forceps to hold the slide, pass the slide through the flame of an alcohol lamp or bunsen

burner to "flame off" any residual alcohol. Allow the hot slide to cool.

Note: When handling cleaned slides, always hold the slide by the edges. Grease or oil, even from your fingers, will cause water to form into drops, which will interfere with the even spreading of bacteria on the slide and, in some cases, will interfere with the adhering of bacteria to the slide in the "heat fixing" process.

2. *With Cleanser.* Moisten the tip of your finger and then rub it over a cake of cleanser, such as Bon Ami. Rub the paste formed on your finger over both surfaces of the slide. Allow the paste to air dry, then remove it using a clean paper towel. Remember, handle all cleaned slides by their edges to prevent the oil from your fingers getting onto the surfaces of the slides.

b. Fixing Bacteria to Slides

Prior to staining, the microorganism must be smeared on the slide and "fixed," that is, made to stick to the surface. If this isn't done, the cells will be washed away during the staining procedure. The general method for "fixing" bacteria shown in Fig. 25-2 uses bacteria removed from a broth (liquid) culture, but the same procedure can be used with agar cultures.

Obtain bacterial cultures from your instructor and fix them to the slides as shown in Fig. 25-2. Label your slides as to the bacteria being used.

c. Simple Staining Techniques

For this procedure you will use the simple stain methylene blue, which is a basic stain. Methylene blue stains rather slowly and takes 30–60 seconds to properly stain a microbial preparation.

1. Place the slides you fixed earlier onto a wire screen or other support in your staining rack.

2. Apply methylene blue stain a drop at a time until the entire smear is covered (Fig. 25-3A).

3. Stain for 60 seconds. (*Note:* If you notice the stain drying on the slide during the staining period, add additional stain drop by drop. *Do not allow the stain to dry!*)

4. After staining, gently and *briefly* wash the stain from the slide (Fig. 25-3B).

5. Remove excess water from the slide by touching one corner to a paper towel or other absorbent paper (Fig. 25-3C). Then blot the slide dry (Fig. 25-3D).

6. Examine your preparations with the oil immersion objective lens. At this point it will be evident if you have made your smear too thick (that is, transferred too many bacteria to the slide or, alternately, not spread them out enough). If you properly prepared the smear, you will observe numerous individual bacterial cells.

d. Negative Staining

Acidic dyes, such as sodium eosinate or nigrosin, are used in this procedure. These dyes have their chromophore in the negatively charged ion. Since most bacterial cells carry a net negative charge, the stain is not incorporated into the cells but forms a deposit around the cells. The cells thus appear as clear, colorless objects against a dark background. Under these conditions it is much easier to determine cell size and/or structure.

1. Transfer a loopful of the common bacterium *Escherichia coli (E. coli)* (or other bacteria provided by your instructor) to a clean slide. *Do not smear!*

2. Place a drop of nigrosin solution next to the bacterial suspension, and mix them together, *but do not smear them out* (Fig. 25-4A).

3. Following the procedure shown in Fig. 25-4C–D, spread the suspension smoothly over the slide, much as if you were preparing a blood smear.

4. Air dry the smear; *do not* heat "fix."

5. Examine the slide microscopically. If the smear was properly made, you should see a graduated thickness of the smear from one end of the slide to the other. It will probably be too thick at one end and you will not see any detail. At the other extreme, it will be too thin and there will not be any contrast. However, at some point, the smear will be sufficiently thin (or thick) to provide good contrast and individual cells will be outlined against the darker background provided by the stain (Fig. 25-4E).

e. The Gram's Stain

This staining procedure is one of the most widely used in microbiology. It is important as a **differential stain;** one that distinguishes between two different species of bacteria that are *morphologically* indistinguishable. Based upon their reaction, bacteria are classified into two large groups: those that *retain* the crystal violet stain throughout the entire procedure are characterized as **gram-positive.** Those that *lose* crystal violet after rinsing with alcohol, and are subsequently counterstained with safranin, are classified as **gram-negative.**

This procedure requires four different solutions:

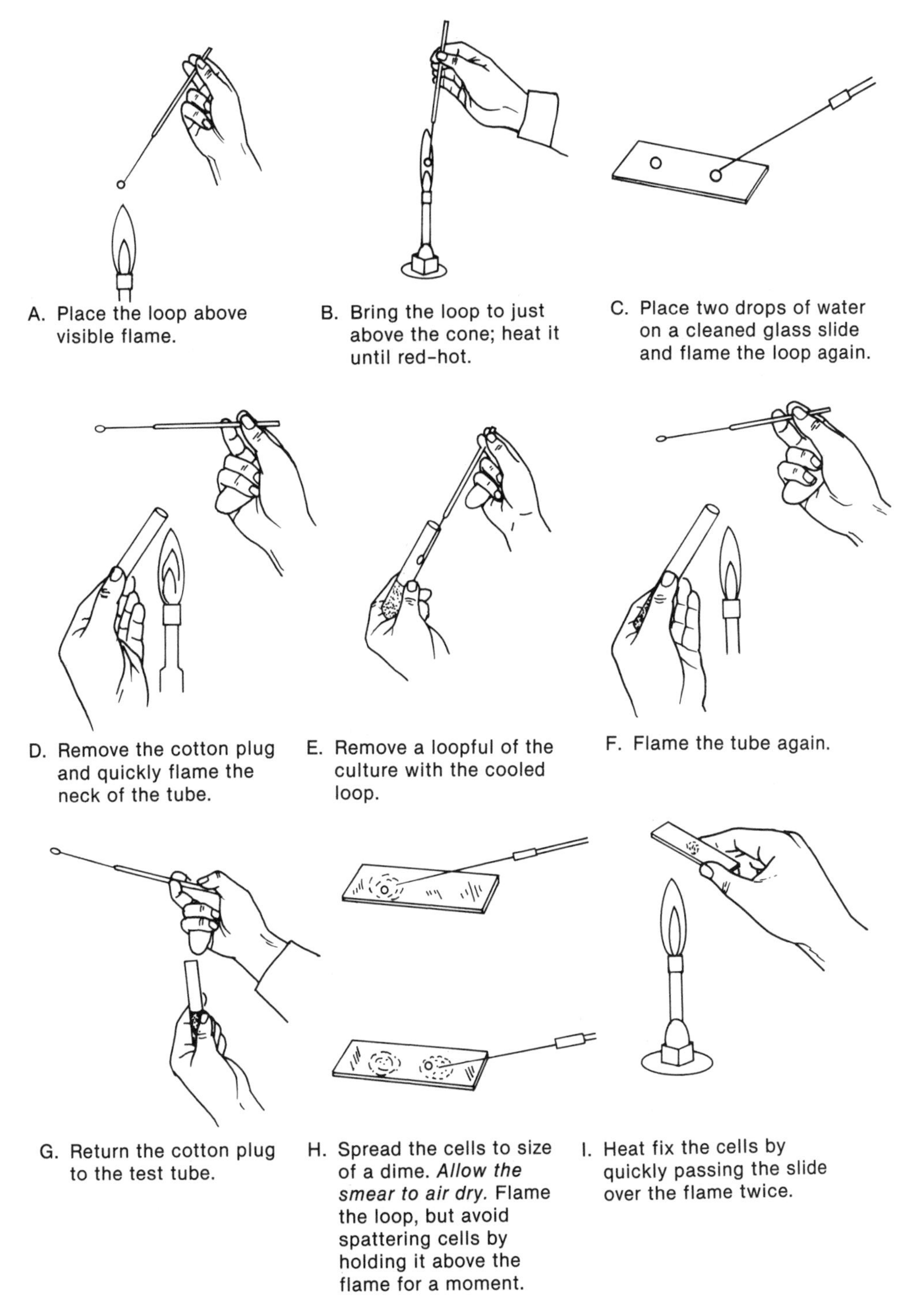

FIG. 25-2
Aseptic technique used in the preparation of bacterial smears.

Crystal violet, a basic dye

Iodine, which acts as a mordant—that is, a substance that increases the affinity of the cell for the dye

Alcohol (or acetone), which serves to remove the dye from the cell

Safranin, a basic dye of a different color that is used to stain the cell if the original stain has been removed with the decolorizing solution

1. Prepare smears of *E. coli* and *Bacillus subtilis* on clean microscope slides as diagrammed in Fig. 25-5A. Make certain that the smears are separate from each other.

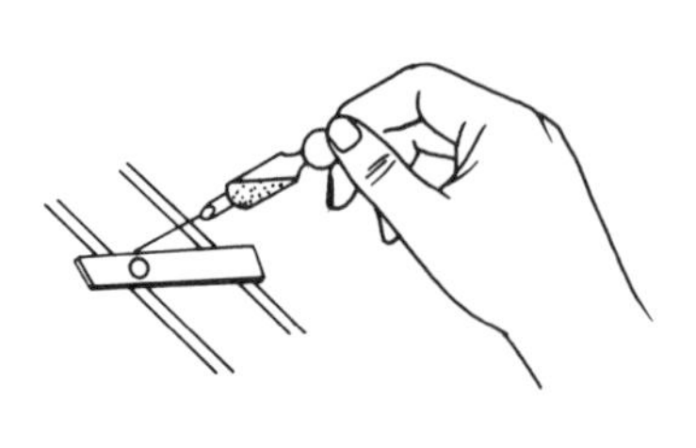

A. Add stain a drop at a time until the entire smear is covered.

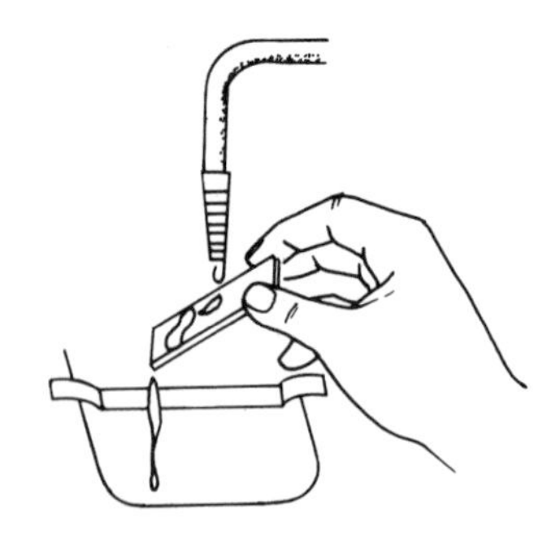

B. Briefly wash off the stain under gently running tap water.

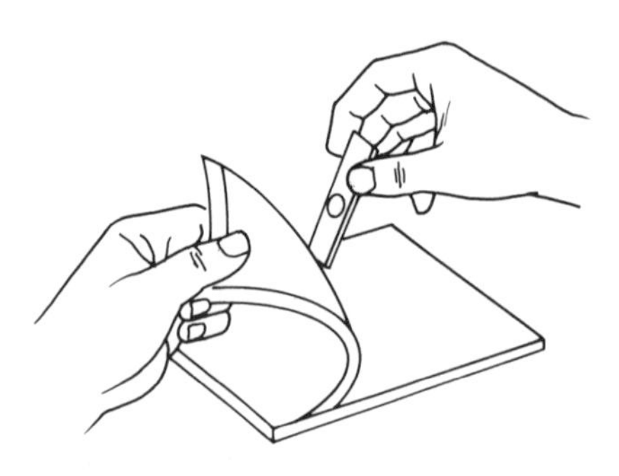

C. Remove excess water from the slide by touching one corner of the slide to the absorbent pad.

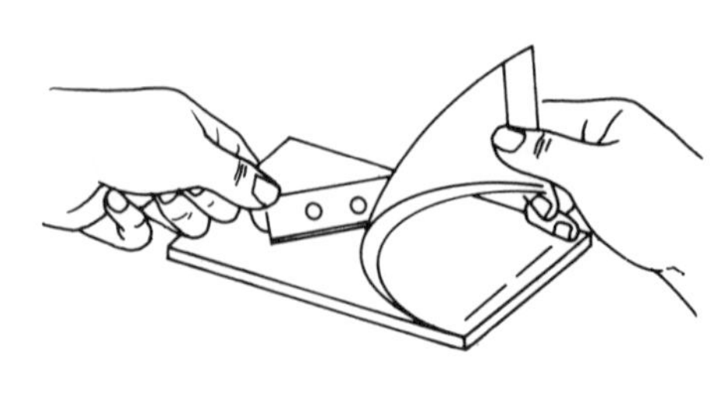

D. Place the slide between a clean absorbent paper pad and blot dry.

FIG. 25-3
Simple staining procedure.

2. Air dry and heat fix the smears.

3. Place the slide on your staining rack and flood it with crystal violet for 1–2 minutes. *Do not allow the stain to dry on the slide.*

4. Pour off the excess stain and *gently* wash the slide with tap water.

5. Flood the smears with Gram's iodine solution and leave on for 1 minute. Pour off the iodine, then add more iodine and allow to react for another minute.

6. Wash off the iodine solution with tap water and blot the slide, *gently,* between paper towelling.

7. Holding the slide at an angle, apply 95% alcohol a drop at a time to each smear until the violet color no longer appears in the wash (Fig. 25-5B). This typically takes 10–30 seconds. *It is important not to continue to rinse with alcohol when no more stain is seen in the rinse.*

8. Quickly rinse off the alcohol with tap water and then blot the slide dry.

9. Counterstain the smears with safranin for 30 seconds.

10. Gently wash the stain off with tap water. Drain off the excess water and allow the slide to air dry.

11. Examine the center smear with oil immersion. This will allow you to contrast gram-negative and gram-positive bacteria. Then examine each of the other smears. Is *E. coli* gram-negative or gram-positive?

Is *Bacillus subtilis* gram-negative or gram-positive?

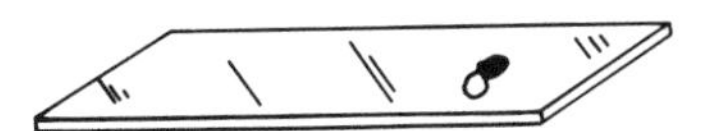

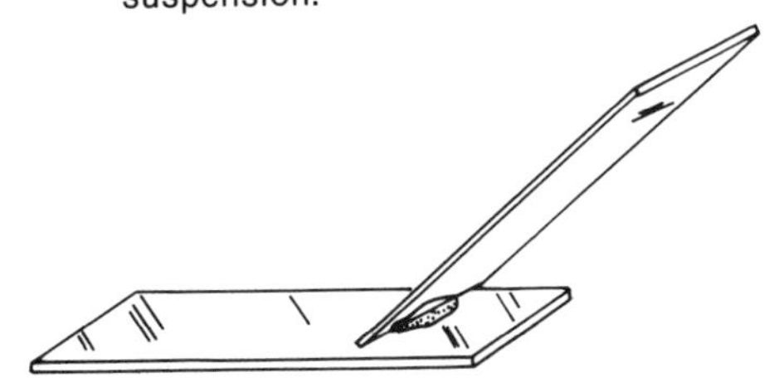

E. Appearance of your stain under the microscope.

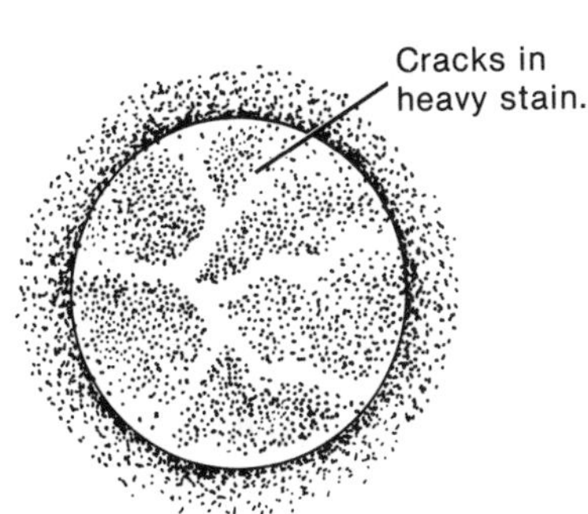

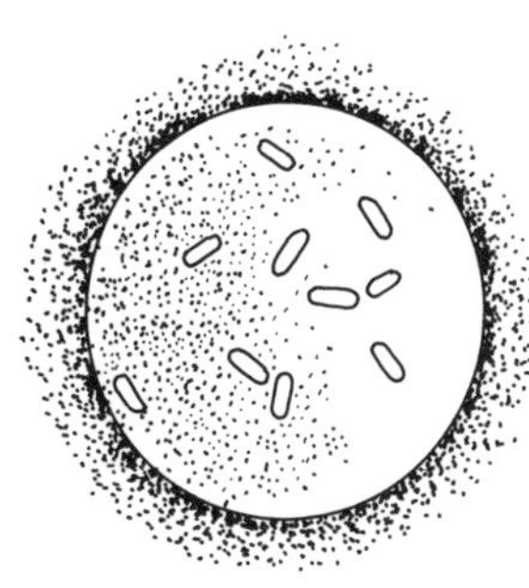

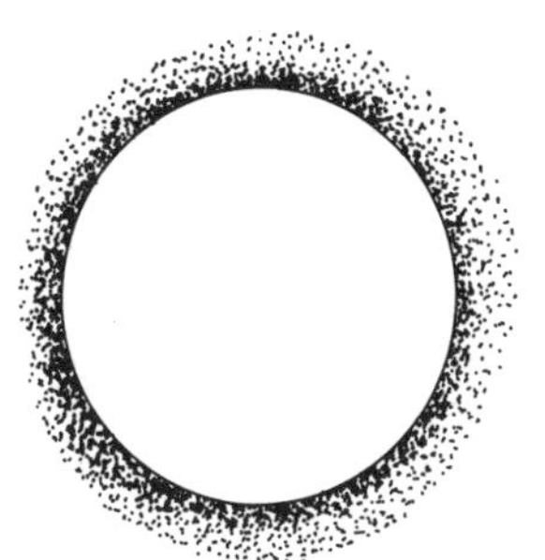

FIG. 25-4
Negative staining of bacteria.

12. If available examine other bacterial cultures provided by your instructor and determine which are gram-negative and which are gram-positive.

2. Control of Bacterial Growth

The control of bacterial infections is a major medical problem. Antibiotics provide an effective method of controlling infectious bacteria and other disease-causing organisms. The story of antibiotics began in 1929, when the British biologist Alexander Fleming found that a petri dish containing a bacterial culture had become contaminated by a mold called *Penicillium.* Noticing that the growth of the bacteria around the mold colony was inhibited, he concluded that the mold produced a diffusable chemical agent capable of inhibiting bacterial

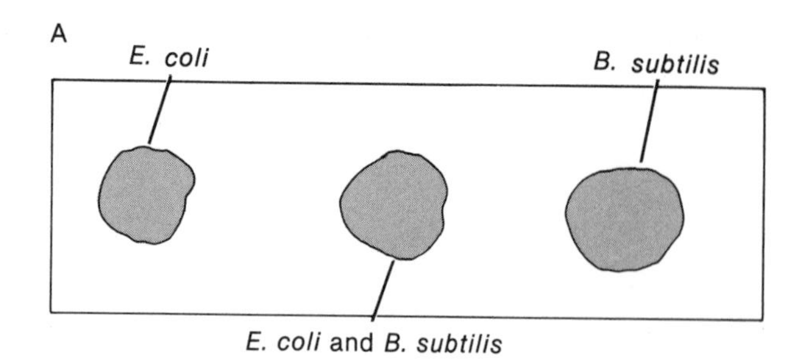

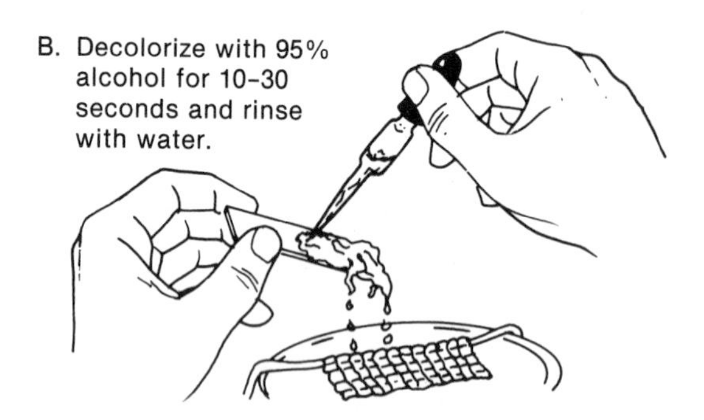

FIG. 25-5
Gram staining of bacteria.

growth. Fleming later isolated this antibacterial chemical, calling it *penicillin.* Such chemicals, which are produced by living organisms and have the ability to retard the growth of certain other organisms, are called **antibiotics**. Among the wide variety of such antibiotics are streptomycin, chloromycetin, terramycin, erythromycin, kanamycin, penicillin, and tetracycline. The use of such antibiotics has significantly lessened the occurrence and virulence of a number of infections.

In this part of the exercise, you will examine the effect of different antibiotics on the growth of one of the more common bacteria.

1. Obtain a petri dish containing nutrient agar—a mixture of chemicals that is optimal for the growth of the bacterium you are to study. Divide each plate into four sections labeled "1," "2," "3," and "Control" by marking the bottom of the dish with black pencil (Fig. 25-6A).

2. Lift the cover of the petri dish slightly, and add 10 drops of the bacterial suspension (provided by your instructor) to the plate (Fig. 25-6B, C).

The following three bacterial cultures, provided by your instructor, can be used in this study:

Escherichia coli
Bacillus subtilis
Serratia marcescens

Your instructor may provide alternate or additional cultures for this study.

Holding the plate at eye level, tilt the plate, allowing the suspension to run to the far edge (Fig. 25-6D). Then, holding the plate level, jerk it toward you. Repeat this several more times, rotating the plate partway between each jerk (Fig. 25-6). In this way, the bacterial suspension will cover the entire surface of the agar.

3. Incubate the plate for 24 hours at 37°C (or 48 hours at room temperature) before proceeding with the next step. Why might incubation for a period of time be desirable before adding the antibiotic discs?

____

____

____

____

4. After incubation, partially lift the cover. Then, using flamed forceps place three antibiotic discs in the numbered sections of the plate (Fig. 25-6G). The following antibiotic discs are commercially available and can be used in this study:

chloromycetin	novobicin
erythromycin	penicillin
kanamycin	streptomycin
neomycin	tetracycline

Use a different antibiotic in each section.

What should be placed in the control section of the plate?

____

Incubate the plate. Examine daily for the next 3 or 4 days. Record your observations in Fig. 25-7 and Table 25-1 using the following criteria:

"R" (for resistant) if there is no clear zone of inhibition around the colony and growth goes up to the disc

"HS" (highly sensitive) if there is a distinct zone of no growth around the disc (no matter what size the zone is)

"S" (sensitive) if there is an inhibited zone, but some colonies appear to have grown back in

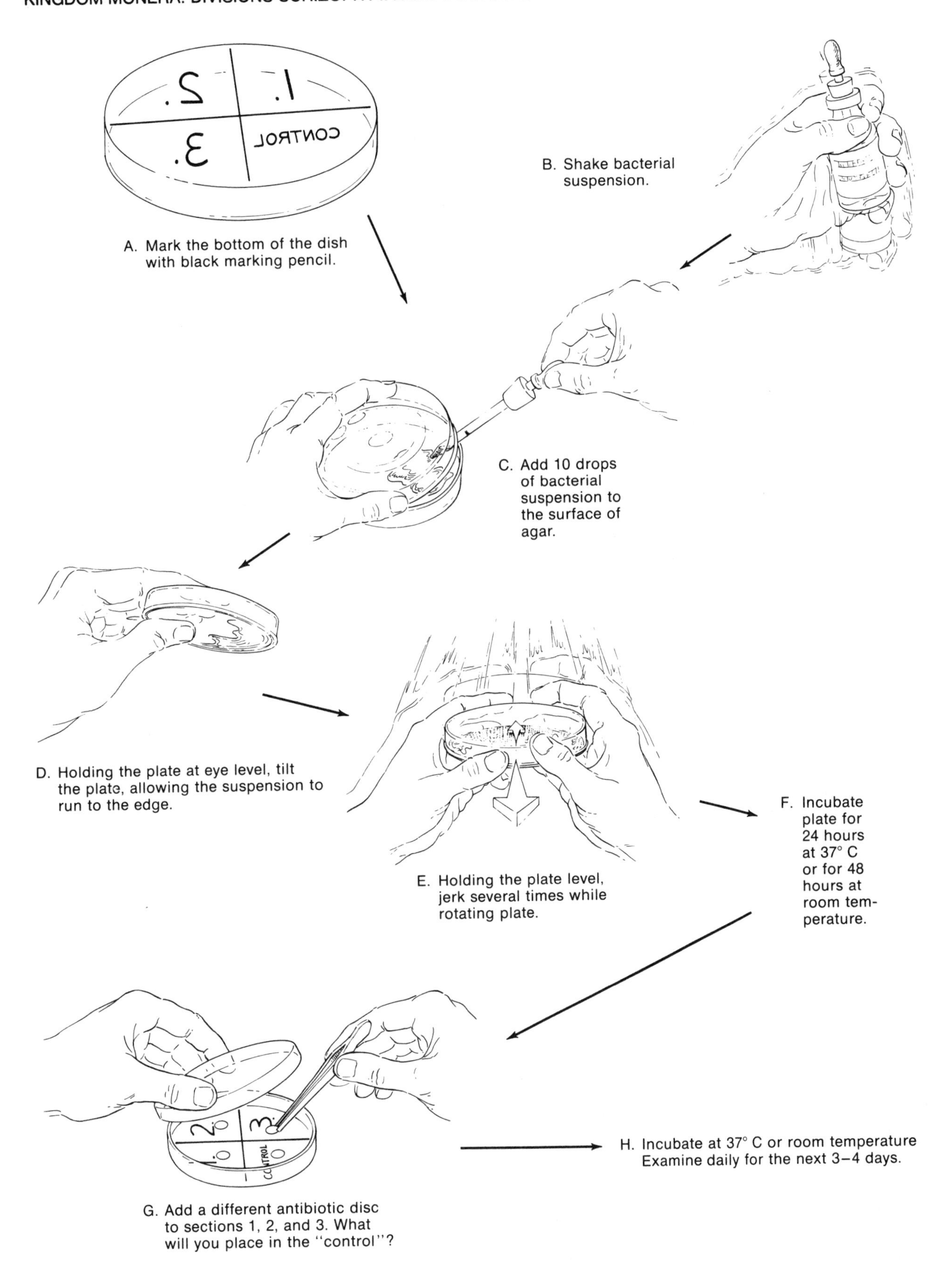

FIG. 25-6
Procedure for determining the effect of antibiotics on bacterial growth.

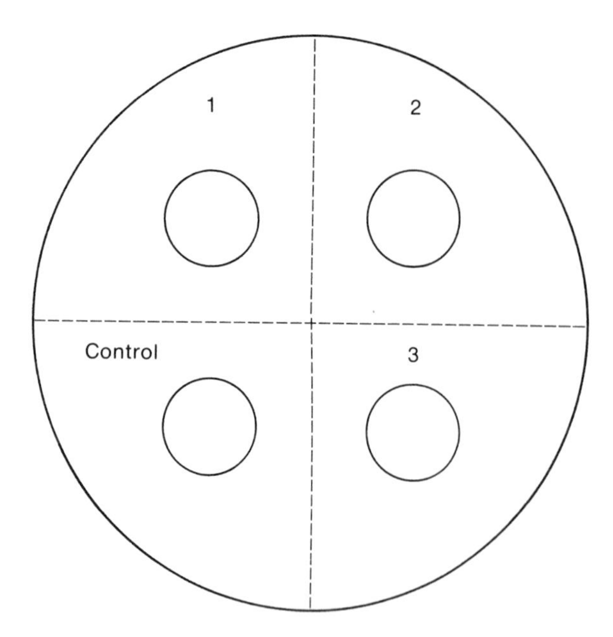

FIG. 25-7
Effect of antibiotics on bacterial growth.

Some bacteria have an enzyme called *penicillinase.* Would the growth of such bacteria be inhibited by penicillin? Explain.

____

____

____

____

TABLE 25-1
Bacterial sensitivity to antibiotics.

Antibiotic	Antibiotic sensitivity			
	Escherichia coli	*Bacillus subtilis*	*Serratia marcescens*	Other bacteria
Chloromycetin				
Erythromycin				
Kanamycin				
Neomycin				
Novobiocin				
Penicillin				
Streptomycin				
Tetracycline				

____

____

3. Bacteria in Milk

Milk is an excellent medium for the growth of bacteria. Therefore, great care must be used in its processing. Harmful bacteria in milk are killed by the process of **pasteurization**. In this exercise, you will examine milk from different sources and determine the quality of the milk in terms of its bacterial population.

1. Bring to class three to four samples of milk of varying age or from different sources (fresh milk from a dairy farm, milk in an unopened carton or bottle, milk in a carton or bottle that has been opened and in the refrigerator for 1–3 or more days, powdered milk, canned milk, raw cow's or goat's milk, if available, and any other milk sample you might like to test).

2. Fill separate test tubes one-third full with each of these milk samples (Fig. 25-8A). Label each tube, and then add 1 ml (20 drops) of methylene blue solution (Fig. 25-8B). Mix by shaking.

3. Plug each tube with sterile cotton, and incubate or place the tubes into a water bath at 37°C (Fig. 25-8C). Record the time in Table 25-2.

When bacteria are actively growing in milk, they consume oxygen. The reduction in oxygen levels can be detected by using methylene blue, which becomes colorless as the oxygen content of the milk dimin-

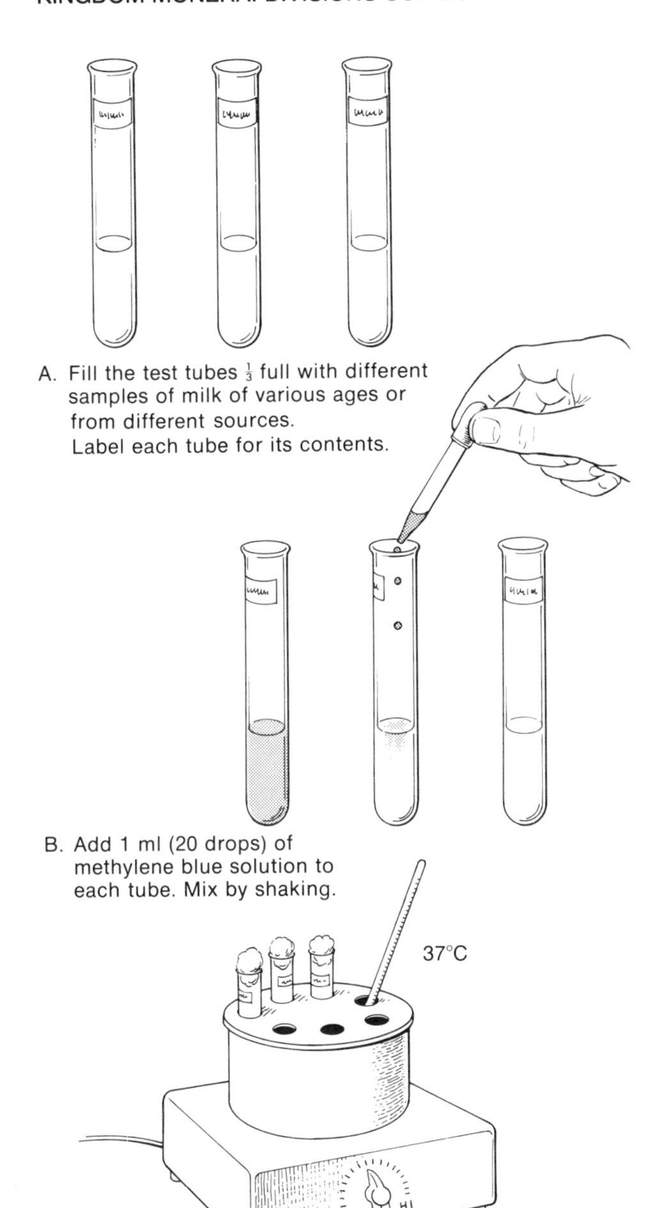

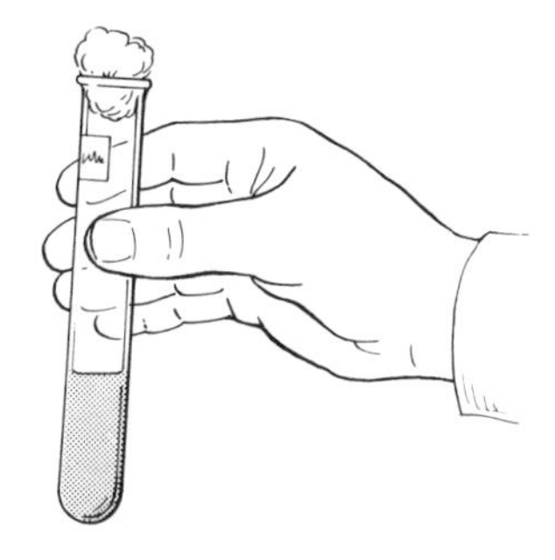

FIG. 25-8
Procedure for estimating bacterial contamination of milk.

ishes. If the number of bacteria present is high, the methylene blue-milk mixture will rapidly lose its color. If the bacterial population is low, more time will be required for decolorization.

For the purposes of this study, the quality of the milk with respect to the number of bacteria present can be rated by the time it takes to decolorize the methylene blue solution. Periodically examine the milk samples. In Table 25-2, record the time it takes to decolorize each sample. Rate each sample according to the description given in Table 25-3. Which sample rated the highest in bacterial contamination?

____

Which rated the lowest?

____

From your results, what appears to lead to the contamination of milk?

____

What appears to reduce contamination?

____

B. DIVISION CYANOBACTERIA

The group of organisms known as **cyanobacteria,** formerly called the *blue-green algae,* exhibits two types of body structure: unicellular and colonial. The unicellular condition is considered to be more primitive. In addition, cyanobacteria, although considered to be prokaryotic organisms on the basis of their cellular structure, use water in photosynthesis as the reducing agent for CO_2 and, therefore, evolve oxygen as a byproduct of photosynthesis. In this respect, they are biochemically similar to photosynthetic eukaryotic organisms, like the algae and plants. In this exercise, you will examine cyanobacteria to become familiar with their cellular morphology and range of complexity.

1. Unicellular Forms

Prepare a wet mount of *Chroococcus* and *Gloeocapsa,* two common cyanobacteria. To see the characteristic gelatinous sheath that surrounds the cells of these organisms, add a drop of India ink to the slide. The sheath will stand out against the dark

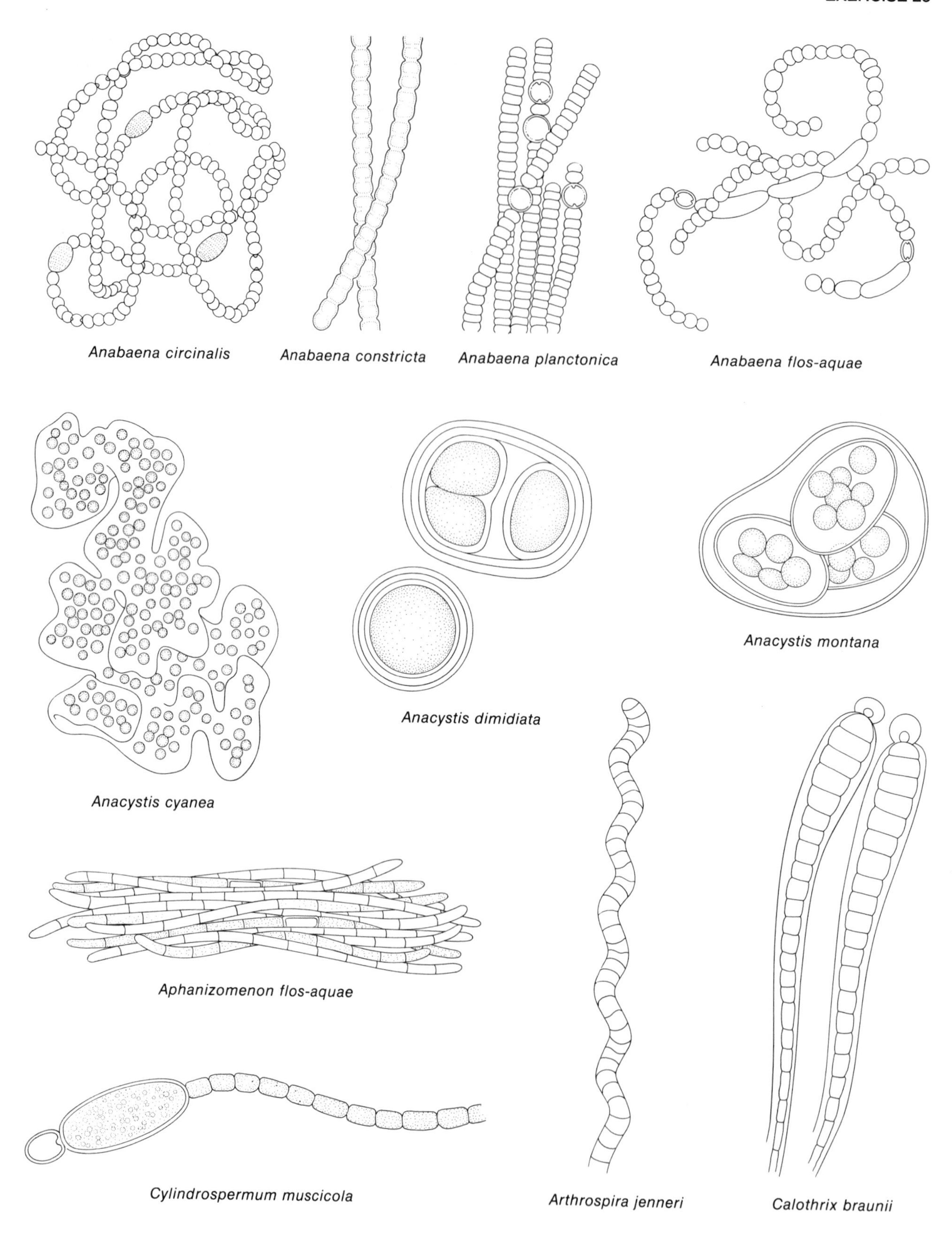

FIG. 25-9
Cyanobacteria that contaminate water supplies.

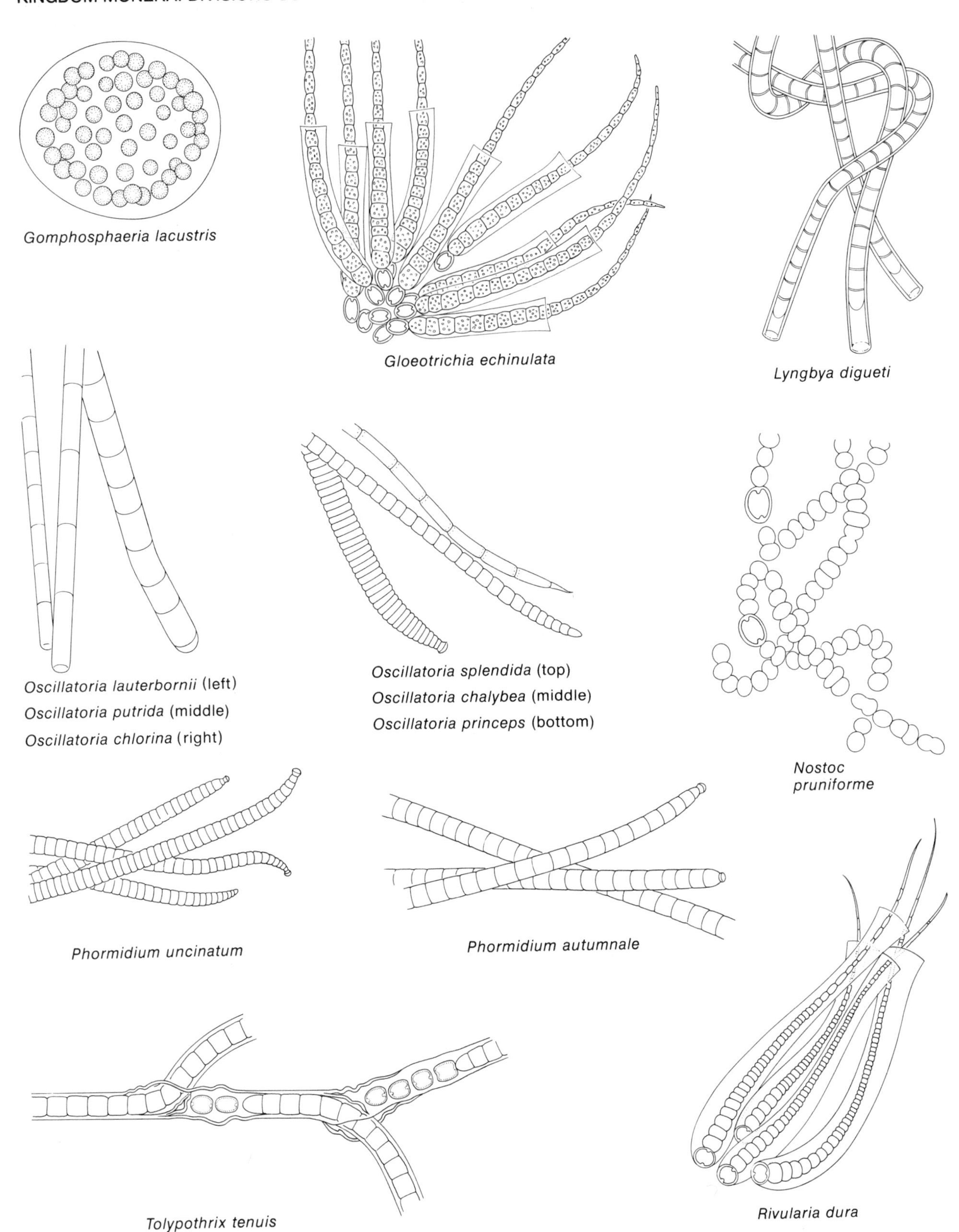

Gomphosphaeria lacustris
Gloeotrichia echinulata
Lyngbya digueti
Oscillatoria lauterbornii (left)
Oscillatoria putrida (middle)
Oscillatoria chlorina (right)
Oscillatoria splendida (top)
Oscillatoria chalybea (middle)
Oscillatoria princeps (bottom)
Nostoc pruniforme
Phormidium uncinatum
Phormidium autumnale
Tolypothrix tenuis
Rivularia dura

TABLE 25-2
Rating of milk samples of different ages or different sources.

Contents of tube	Time methylene blue added	Length of time to decolorize methylene blue	Rating
1			
2			
3			
4			

background. Where several cells are clustered together, do they share a common sheath?

Although these are unicellular forms, they frequently form small clusters of cells. Would you consider these clusters as representing multicellular organisms? Explain.

2. Colonial Forms

In colonial forms, the shape of the colony is largely determined by the plane of cell division and may result in filaments, plates, or spheres. Examine the following algae and classify each as to body form (e.g., filamentous) and plane of cell division (single, double, or irregular) that resulted in the form of the colony.

1. *Merismopedia:* Locate the gelatinous sheath. Frequently cells within the colony may be seen in cytokinesis.

2. *Oscillatoria:* What evidence is there of cellular differentiation with this colony?

Describe any motility you observe.

How does this organism reproduce?

3. *Rivularia:* How does this organism differ from *Oscillatoria?*

TABLE 25-3
Decolorization times.

Time to decolorize methylene blue	Rating
Less than 20 minutes	Highly contaminated
20 minutes to 2 hours	Poor
2–5½ hours	Fair
5½–8 hours	Good
More than 8 hours	Excellent

Locate the heterocyst. What is its function?

___

___

___

4. *Gloeotrichia:* How does this organism differ from *Rivularia?*

___

___

___

5. *Anabaena:* Crush the cells of the water fern *Azolla* to release this cyanobacterium. What function might this organism carry out in the cells of *Azolla?*

___

___

___

TABLE 25-4
Problems caused by cyanobacteria in water supplies.

Problem	Organism
Taste and odor	*Anabaena circinalis* *Anacystis cyanea* *Aphanizomenon flos-aque* *Cylindrospermum muscicola* *Gomphosphaeria lacustris*
Filter clogging	*Anabaena flos-aque* *Anacystis dimidiata* *Gloeotrichia eschinulata* *Oscillatoria princeps* *Oscillatoria chalybea* *Oscillatoria splendida* *Rivularia duro*
Reservoir wall growth	*Calothrix braunii* *Nostoc pruniforme* *Phormidium uncinatum* *Tolypothrix tenuis*
Polluted water	*Anabaena constricta* *Anacystis montana* *Arthrospira jenneri* *Lyngbya digireti* *Oscillatoria chlorina* *Oscillatoria putrida* *Oscillatoria lauferbornii* *Phormidium autumnale*

C. CYANOBACTERIA IN WATER SUPPLIES

The increasing human population and rapid development of industry and agriculture has resulted in a phenomenal increase in the use of water. This need for water has produced many difficult problems in procuring an adequate and suitable water supply.

As population and industrial demands increase, villages and cities are turning from groundwater sources to surface waters, such as lakes, streams, and reservoirs, for their water supplies. Groundwaters are essentially free from contaminating organisms. Surface waters, on the other hand, contain many organisms that in one way or another contribute to the unpalatability of the water supply. Such organisms affect the odor and taste of water, clog filters, grow in pipes, cooling towers, or on reservoir walls, form mats or blooms on the surface of the water, produce toxic materials, and so forth. Furthermore, the present methods of waste disposal are intensifying the problems of such nuisance organisms in water. Materials such as sewage and organic wastes from paper mills, fish-processing factories, slaughterhouses, and milk plants, to name a few, greatly increase the growth of algae and other organisms. Many of these cause problems when they become abundant.

Cyanobacteria are among the common nuisance inhabitants of surface waters. Collect surface water samples from different sources—for example, lakes, streams, reservoirs (including the walls), swimming pool filters, or water treatment plants. Table 25-4 characterizes some of the more common problems associated with an overabundance of these organisms. Use Fig. 25-9 to identify some of the cyanobacteria in your water samples.

REFERENCES

Bold, H. C., and C. L. Hundell. 1977. *The Plant Kingdom.* 4th ed. Prentice-Hall.

Brock, T. 1984. *Biology of Microorganisms.* 4th ed. Prentice-Hall.

Cairns, J. 1966. The Bacterial Chromosome. *Scientific American* 214(1):36–44 (Offprint 1030). *Scientific American* Offprints are available from W. H. Freeman and Company, 41 Madison Avenue, New York, 10010, and 20 Beaumont Street, Oxford OX1 2NQ, England. Please order by number.

Delevoryas, T. 1977. *Plant Classification.* 2d ed. Holt, Rinehart and Winston.

Pelczar, M. J., R. D. Reid, and E. C. S. Chan. 1983. *Microbiology.* 5th ed. McGraw-Hill.

Sharon, N. 1969. The Bacterial Cell Wall. *Scientific American* 220(5):92–98 (Offprint 1142).

Stanier, R. Y., and M. Doudoroff. 1976. *The Microbial World.* 4th ed. Prentice-Hall.

Walsby, A. E. 1977. The Gas Vacuoles of Blue-Green Algae. *Scientific American* 237(2):90–97 (Offprint 1367).

Whitaker, R. H. 1969. New Concepts of Kingdoms of Organisms. *Science* 163:150–160.

Wollman, E. L., and F. Jacob. 1956. Sexuality in Bacteria. *Scientific American* 195(1):109–118 (Offprint 50).

EXERCISE 26

Kingdom Protista I: Algae and Slime Molds

Protists are typically unicellular, but some members form simple colonies of similar cells. Although protists are frequently called *simple, lowly,* or *primitive,* they possess a tremendous degree of specialization at the cellular level. This has allowed them to adapt to a wide range of highly diverse habitats. These organisms are found in brackish, fresh, and marine waters and exhibit a variety of symbiotic and parasitic relationships.

This kingdom is divided into a number of divisions, but only the following six, which include the algae and slime molds, will be studied in this exercise:

Division Euglenophyta: plantlike, autotrophic organisms (i.e., those that synthesize all needed organic molecules from simple inorganic molecules).

Division Chrysophyta (diatoms and golden-brown algae): unicellular algae with plastids containing golden-yellow pigments.

Division Chlorophyta (green algae): unicellular and multicellular organisms having chlorophylls *a* and *b*; various carotenoids, and food stored as starch. Motile cells have two flagella.

Division Phaeophyta (brown algae): multicellular, marine organisms having chlorophylls *a* and *c* and the pigment fucoxanthin. Food stored as a carbohydrate called *laminarin.* Motile cells have two flagella, one anterior, the other trailing. Multicellular forms, unlike green algae, show considerable differentiation. Some have specialized conducting cells.

Division Rhodophyta (red algae): mostly marine plants having chlorophyll *a* and phycobilins. Food stored as carbohydrate called *floridean starch.* No motile cells in life cycle. Lack specialized conducting cells found in brown algae.

Division Gymnomycota (slime molds): heterotrophic (i.e., utilizing organic molecules synthesized by other organisms) amoeboid organisms that mostly lack a cell wall but form sporangia at some stage in their life cycles.

A. ALGAE

More than 20,000 species of algae are grouped into several divisions. All members of these divisions are (with few exceptions) photosynthetic. They have a relatively simple plant body, which may consist of a single cell, filaments of cells, plates of cells, or a structure somewhat comparable to that of some land plants. They do not have the complex organization of tissues found in the vascular plants.

Algae differ from each other in the type of flagella (if they consist of motile cells) and in several biochemical characteristics. While all contain chlorophyll, there is a wide variety of carotenoids distributed in the various groups. Indeed, the names given to the various divisions are derived from the various pigments that mask the green color imparted by the chlorophylls. Great diversity exists in the type of food storage products found in the different groups.

Reproduction is accomplished asexually by fragmentation of the parent body or through the production of spores capable of developing into other individuals. In addition, new individuals may arise sexually as a result of the union of two gametes. The zygote that is formed as a result of this union may develop directly into another alga or may produce spores.

1. Division Euglenophyta (Euglenoids)

The euglenoids seem to be protists that, in the long course of evolution, have acquired chloroplasts with biochemical properties similar to those of the green algae. The division takes its name from *Euglena,* a photosynthetic autotroph.

Euglena is a green, unicellular organism commonly found in the surface scum of standing or very slowly moving waters.

Place a drop of culture of living *Euglena* on a slide. Add a drop of 10% methyl cellulose, which will slow down the rapid movement of the organism. Add a coverslip and examine microscopically. Observe the method of locomotion of an actively moving specimen. The motility of the organism is due to the movement of a long **flagellum,** which pulls the organism through the water. Does the beating of the flagellum move from the base toward the tip, or in the opposite direction?

Euglena also exhibits a wormlike movement in which a series of contractions pass along the body. This type of movement is unique to *Euglena* and consequently has been termed **euglenoid movement.**

With the help of Fig. 26-1A, locate the following morphological features of *Euglena.* You may find it useful to supplement your observations by studying a commercially prepared slide of this organism. Using high power, identify the **pellicle,** a thin, elastic membrane that may be striated because of spiral thickenings; the **ectoplasm,** the peripherally located cytoplasm; the **endoplasm,** the denser, internal cytoplasm; the **chloroplasts,** organelles containing the green pigment chlorophyll; and the **nucleus,** which is centrally located. More readily seen in prepared slides is a dense staining body in the nucleus, called the **endosome,** and the **cytostome** (cell "mouth"), a funnel-shaped depression near the anterior end that leads to the **cytopharynx** (cell "gullet"), which is enlarged at the base to form the **reservoir.** Adjacent to the reservoir is a water-expulsion vesicle (**contractile vacuole**) that periodically collects excess cell water and discharges it into the reservoir and then out through the cytopharnyx. The **stigma,** a reddish-orange eye spot located near the anterior end of the animal and adjacent to the cytopharnyx, is a light-sensitive organelle. What function might it serve?

When you have finished making your observations of the living organism, add a drop of iodine solution to the edge of the coverslip and let it be drawn under by capillary action. The iodine will stain the flagellum so that it is more easily seen.

Nutrition of *Euglena* is normally **autotrophic** (Greek *autos,* "self"; *trophe,* "nourishment"); that is, the organism is able to synthesize its organic needs using inorganic molecules obtained from its surroundings and sunlight in a process called **photosynthesis.** Some species of *Euglena,* however, are able to grow and reproduce, even in the absence of any visible chloroplasts or chlorophyll. They exhibit **heterotrophic** (Greek *heteros,* "other"; *trophe,* "nourishment") nutrition. Suggest ways in which these organisms obtain their food requirements.

Reproduction in *Euglena* is by **binary fission** (Fig. 26-1B). The nucleus first divides by mitotic division. Then the anterior organelles (e.g., reservoir, flagella, and stigma) are duplicated. This is followed by a longitudinal splitting of the protist beginning at the anterior end. If available, examine demonstration slides showing this type of reproduction.

2. Division Chrysophyta (Diatoms and Golden-Brown Algae)

This division includes two major classes: the diatoms, with almost 10,000 species, and the golden-brown algae, with about 1500 species. The chrysophytes are characterized by plastids with golden-yellow pigments, rigid cell walls often impregnated with silicon, and food stored in the form of oils rather than starch. Fresh water containing large numbers of chrysophytes may have an unpleasant oily taste, as will the fish caught in such waters.

Diatoms are found throughout the world in the soil and in fresh and salt water. They are a major component of the plankton (small marine organisms found in large numbers in the upper levels of the water) where they serve as an important source of food for marine animals.

Add a drop of a culture of living diatoms to a slide, add a coverslip, and examine microscopically. If living diatoms are unavailable, examine a slide containing a mixture of various diatoms. The most striking feature of diatoms is their beautiful, ornamented cell walls, which are composed of pectin impregnated with silica. Observe the fine markings of the cell wall that gives each species of diatom its distinctive morphologic features. These markings are actually pores giving the protoplasm access to the external environment.

The walls of diatoms consists of two overlapping portions called *valves,* which fit together like a petri dish. The valves are either pennate or centric. Pennate diatoms are boat-shaped or rod-shaped and are bilaterally symmetrical in their pattern of markings. In the center of the valve of most pennate diatoms is

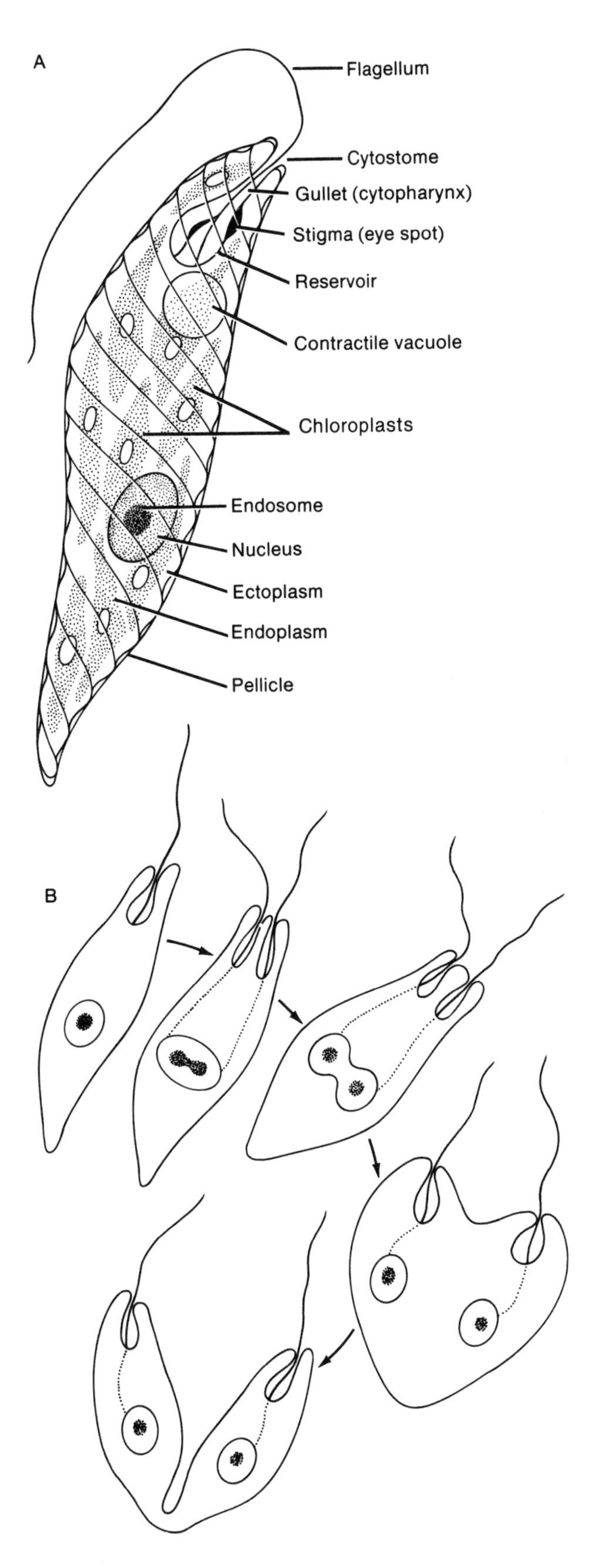

FIG. 26-1
Euglena: (A) morphology and (B) reproduction.

a groove or raphe. Centric diatoms are usually circular, oval, or elliptical in shape and have radial symmetry. Examine your slide preparation and identify pennate and centric diatoms.

Diatomaceous earth (diatomite) is found in vast deposits of silicon shells of diatoms deposited over millions of years in former ocean bottoms. It is used as a fine abrasive material in silver polish, some toothpastes, and for filtering and insulating materials.

3. Division Chlorophyta (Green Algae)

The green algae exhibit the greatest diversity, both in form and reproductive patterns. Of the approximately 7000 species, most are aquatic, although other habitats include the surface of snow, hot springs, trunks of trees, and (as symbionts) in lichens, protozoa, and hydra. Many of the green algae are microscopic, although some of the marine forms, called *kelps,* achieve lengths in excess of 25 meters.

a. The Volvocine Line of Evolution in Green Algae

Two general trends are evident in the evolution of this group: (1) there is an increase in complexity of the plant body from a single cell to a colonial type. Unicellular forms are considered the most primitive. Colonial forms probably had their origin in an ancestral unicellular alga and achieved their complex forms through cell divisions occurring in different planes. (2) There is evidence of increasing differences between the type of gametes and/or reproductive organs, with isogamy (sexual reproduction in which the gametes are morphologically alike) being the most primitive and oogamy (sexual reproduction in which one of the gametes, usually the larger, is not motile) being the most advanced.

In the volvocine line, a primitive, unicellular alga is thought to have given rise to increasingly complex colonial forms. In this study, you will examine *Chlamydomonas,* which could be considered the primitive form; *Gonium* and *Pandorina* as progressively complex forms; and *Volvox,* which could be considered the peak of the evolutionary line.

Observe these differences as you examine various representatives of this group.

1. **Chlamydomonas.** *Chlamydomonas* is an isogamous, single-celled, motile alga commonly found in damp soil, lakes, and ditches. It is typically egg-shaped and has a large cup-shaped chloroplast containing a proteinaceous body, the **pyrenoid,** which functions in starch formation. A **stigma,** or eyespot, is located inside the chloroplast. Suggest a function for this structure.

___

___

The nucleus is difficult to observe in living material.

Prepare a slide of living *Chlamydomonas,* and examine microscopically (Fig. 26-2). Add a drop of methyl cellulose to slow the movement of this alga. Locate the conspicuous chloroplast and the pyrenoid. The activity of threadlike structures, **flagella,** located at the anterior end of the cells, enables the organism to move through water. The flagella can be more easily observed by closing the iris diaphram to reduce the light.

At the start of **asexual reproduction,** the flagella are retracted and movement ceases. During this quiescent period, mitosis results in the formation of daughter protoplasts, which may divide a second and third time (Fig. 26-2). A cell wall is formed around each of the daughter protoplasts, resulting in the formation of temporary colonies that soon rupture and release individual daughter cells called **zoospores.** Locate nonmotile, vegetative colonies showing two, four, or eight daughter cells.

In **sexual reproduction,** haploid gametes are formed in the majority of the green algae. In *Chlamydomonas,* however, vegetative cells may function as gametes, one functioning as the male and another as the female gamete. The gametes are identical in size and appearance, although in some species the female gamete may be slightly larger. Gametes that are morphologically indistinguishable from each other are called **isogametes.**

Sexuality in *Chlamydomonas* can be demonstrated by using mating strains having separate sexes characterized as plus (+) and minus (−). Place a drop of each of the mating types next to each other on a slide. Do not mix. While examining them with the dissecting microscope, mix the drops together. Note the peculiar "clumping" phenomenon that precedes union of the gametes. Using a compound microscope, locate cells that have paired. At which end of the cell has union occurred?

___

___

As a result of sexual union, a diploid zygote ($2N$) is formed that secretes a thick, spiny wall and then enters a period of dormancy. Examine the preparation closely and observe zygotes. Under favorable

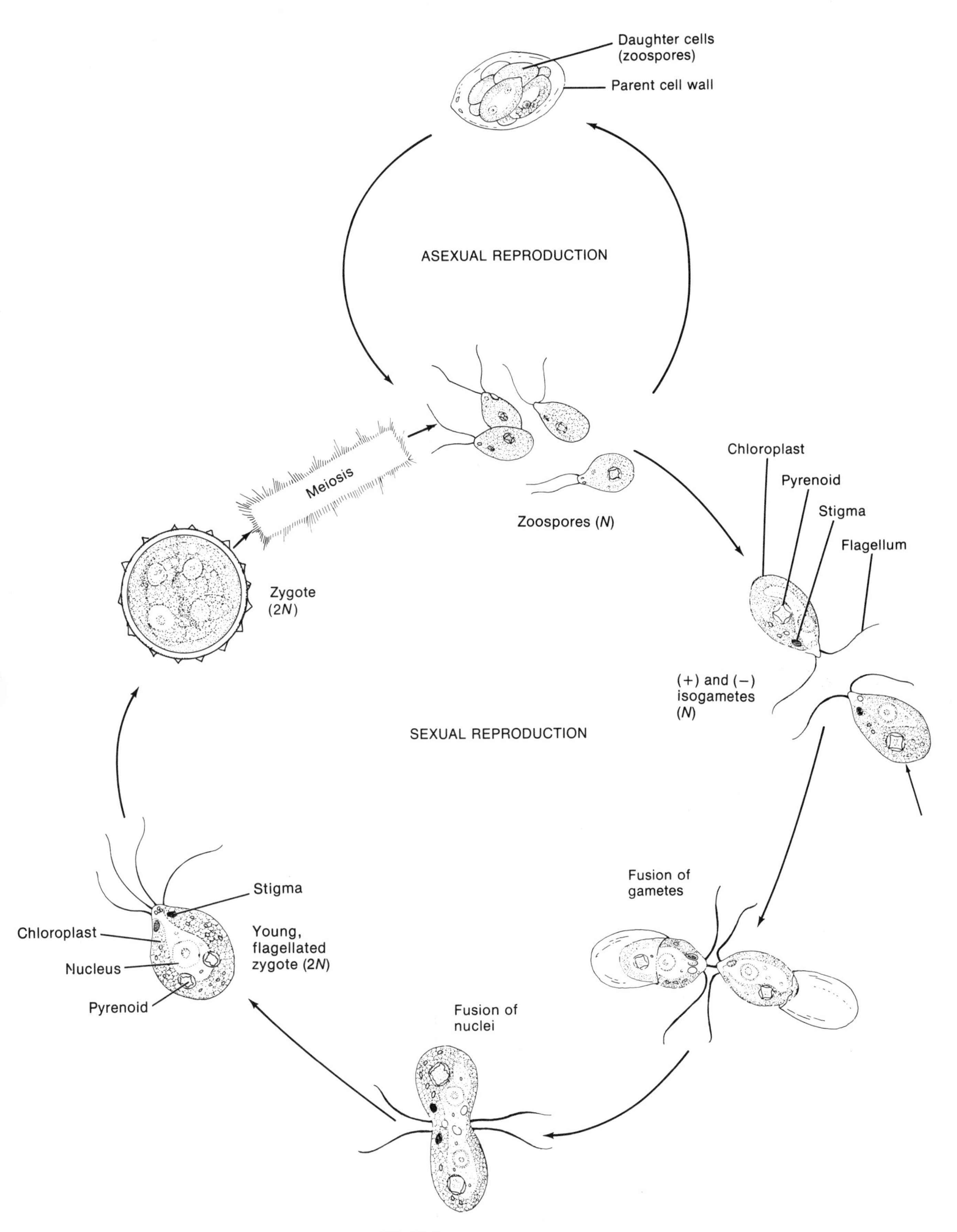

FIG. 26-2
Life cycle of *Chlamydomonas*.

conditions, the zygote nucleus undergoes meiotic division to form four haploid *(N)* nuclei. The protoplast then divides and forms four uninucleate zoospores, which soon escape from the old zygote wall, develop flagella, and swim away.

2. **Gonium.** *Gonium* is an isogamous, but colonial, organism made up of *Chlamydomonas*-like cells that are loosely arranged into a flat, platelike colony (Fig. 26-3A). The cells are held together by a gelantinous matrix.

Gonium represents a stage in evolution where the plant body has become larger by the assembly of a few cells exhibiting a relatively simple degree of coordination. The cells in this colony swim in unison so that the entire plate of cells moves as a unit.

Examine living colonies of *Gonium.* Is the number of cells in each colony constant?

If not, describe the variability in cell number found in different colonies.

---

3. **Pandorina.** Examine living colonies of *Pandorina,* another colonial type composed of *Chlamydomonas*-like cells (Fig. 26-3B). This organism is also isogamous. How does the colony of *Pandorina* differ from *Gonium* in shape?

In number of cells?

In *Pandorina,* as well as in *Gonium,* each cell of the colony, when mature, gives rise asexually to a new colony within the gelatinous envelope of the parent colony.

In sexual reproduction, isogametes fuse to form a zygote that undergoes meiosis to form zoospores. Each zoospore is capable of developing into a new colony.

A

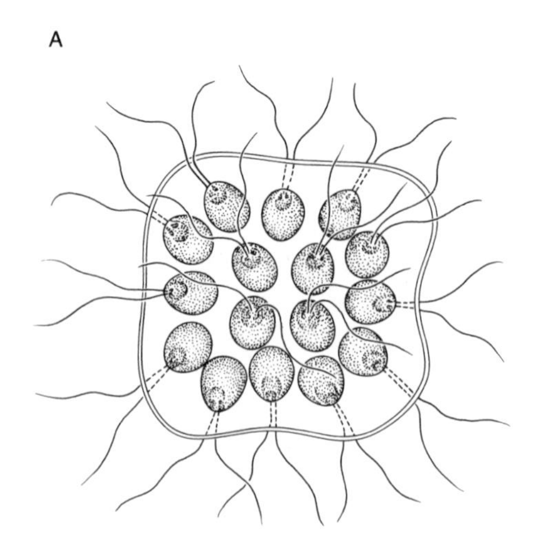

B

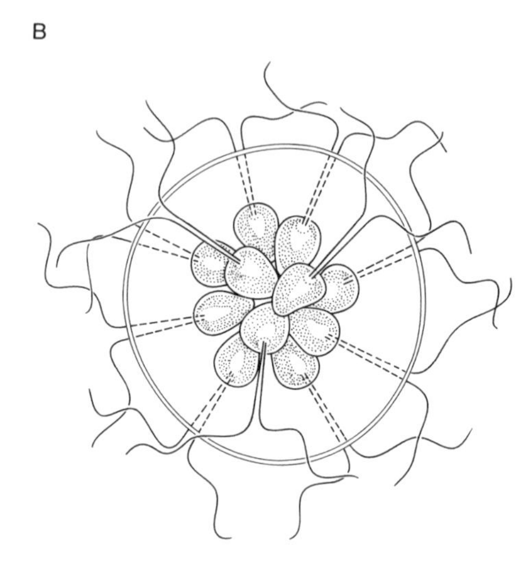

FIG. 26-3
Colonial green algae: (A) *Gonium* and (B) *Pandorina.*

4. **Volvox.** *Volvox* is a colonial alga in which the cells are bound together in a common matrix (Fig. 26-4A). Individually, the cells exhibit many of the features seen in *Chlamydomonas;* that is, they have a stigma, flagella, and large chloroplasts. However, in *Volvox* only specialized cells are able to function during reproduction.

Asexual reproduction is accomplished by the enlargement and subsequent division of certain of the cells within the colony. In early development, these cells form a flat plate, which soon rounds up into a sphere having a small pore at the posterior end. The colony continues to increase in size, and the sphere evaginates through the pore and turns itself inside out to form a daughter colony, which is released on disintegration of the parent colony.

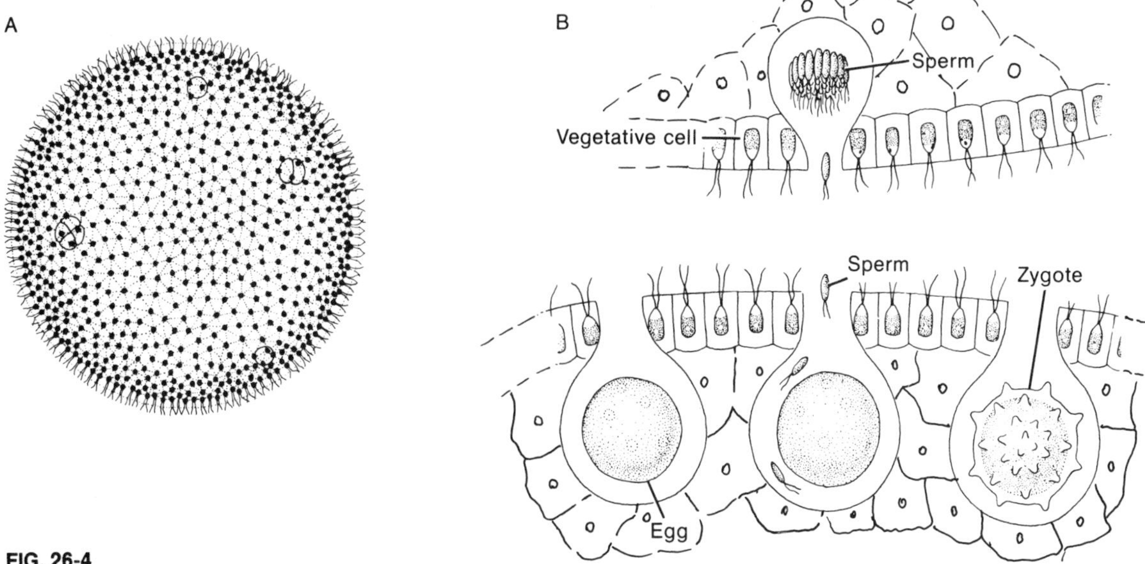

FIG. 26-4
Volvox: (A) adult colony and (B) sexual reproduction.

Examine a sample of living *Volvox,* with a dissecting microscope. Describe the motility of *Volvox.*

____

____

Describe the shape of the colony.

____

____

Examine living material and prepared slides of *Volvox* for presence of daughter colonies.

The green algae studied thus far all produced isogametes. In contrast, *Volvox* produces oogametes, gametes that are morphologically differentiated into sperm and eggs (Fig. 26-4B). These are developed from cells that become differentiated as the colony grows. In forming an ovum, a cell of the colony increases greatly in size, takes on a rounded form, and becomes filled with food materials, especially lipids. The male gametes are formed from other cells, which give rise to flat bundles of flagellated sperm. When mature, the eggs are fertilized by the sperm. They then develop heavy spiny walls and become zygotes. Germination of the zygote occurs in the spring and results in the formation of a new colony. Examine prepared slides of *Volvox,* and locate sperm, ova, and zygotes.

b. Filamentous Green Algae

1. **Spirogyra.** *Spirogyra* is a free-floating alga found in small freshwater pools in the spring. It is frequently referred to as "pond scum" (Fig. 26-5). Prepare a fresh mount and examine microscopically. Is any branching evident?

____

Examine a single cell under high power. What is the shape of the chloroplast?

____

____

Locate several small **pyrenoids** within the chloroplast. The nucleus, suspended in the center of the cell by cytoplasmic strands, is difficult to observe unless stained. Apply a drop of methylene blue to the edge of the coverglass. After a few minutes, reexamine and locate the nucleus, which will now appear as a bluish body in the central part of the cell.

In sexual reproduction, filaments of opposite mating types come to lie adjacent to one another and small projections appear in opposing cells of each filament (Fig. 26-5). The projections increase in length and eventually contact each other. At the point of junction, the cell walls dissolve, forming a **conjugation tube.** The protoplasts of the conjugating cells

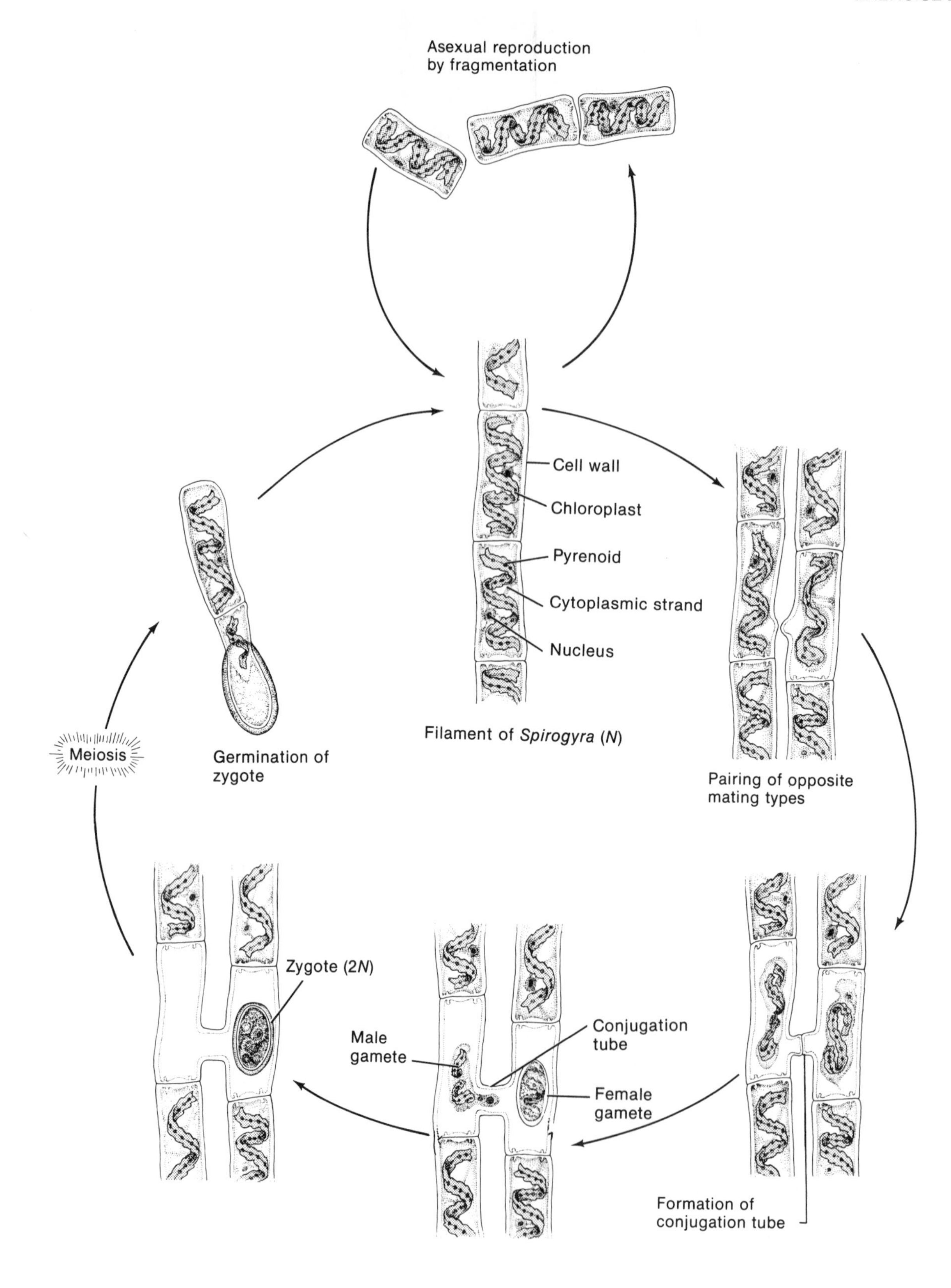

FIG. 26-5
Life cycle of *Spirogyra*.

become isogametes. One gamete functions as the male and migrates through the conjugation tube to unite with the nonmotile female gamete. Fusion of the gametes produces a zygote, which is released on disintegration of the filament. Before germination, the nucleus of the zygote undergoes meiosis, and three of the four haploid nuclei that are formed disintegrate. On germination of the zygote, a short protuberance is formed, which contains the fourth nucleus.

Mitosis, followed by cell division, results in a filament of cells similar to the parental filament. Examine your living material, and look for various stages in gametic union. If your preparation does not show any stages of conjugation or zygote formation, examine prepared slides showing this process.

No means of asexual reproduction is known to occur in *Spirogyra* other than fragmentation of the individual filaments.

2. **Ulothrix.** *Ulothrix,* like *Spirogyra,* consists of a simple, unbranched filament of cells. Unlike *Spirogyra,* it is not free-floating, but has a basal **holdfast** cell used to attach it to rocks or other objects in fresh and, in some cases, salt water (Fig. 26-6).

Examine a prepared slide or living specimens of *Ulothrix.* Note that all cells, with the exception of the holdfast, are alike. The holdfast cell may not be present. Why?

___

___

___

The chloroplast is C-shaped and may have one or more pyrenoids. Is an eye spot (stigma) present?

___

If not, would you expect this organism to have such a structure?

___

___

___

With the exception of the holdfast, any cell is capable of reproducing the plant by asexual or sexual means. In the asexual phase, the protoplast of the parent cell undergoes mitosis and produces four to eight daughter cells that, when released from the parent cell (the **zoosporangium**), become flagellated **zoospores** (Fig. 26-6). After swimming around for a short period of time, each spore loses its flagella, settles to the bottom, and, through a series of mitotic divisions, gives rise to a new filament.

In the sexual phase, the parent cell (called a **gametangium**) produces from 32–64 isogametes. These differ from zoospores in being smaller and having two rather than four flagella. Gametes from different filaments fuse to form a zygote that enters a dormant period. At this time it is called a **zygospore.** Under favorable conditions, this resting spore undergoes meiosis. The four haploid zoospores each then develop into a new filament of cells.

Is the plant body in *Ulothrix* haploid or diploid?

___

Explain.

___

___

Which type of reproduction (asexual or sexual) is primarily responsible for increasing the number of individuals in the population?

___

Explain.

___

___

___

Under what environmental conditions would you expect sexual reproduction to occur and why?

___

___

___

___

3. **Oedogonium.** *Oedogonium,* like *Spirogyra* and *Ulothrix,* has a simple, unbranched plant body consisting of a series of cells. It differs from the latter algae in that some of the cells have become specialized reproductive structures that produce distinctively different gametes.

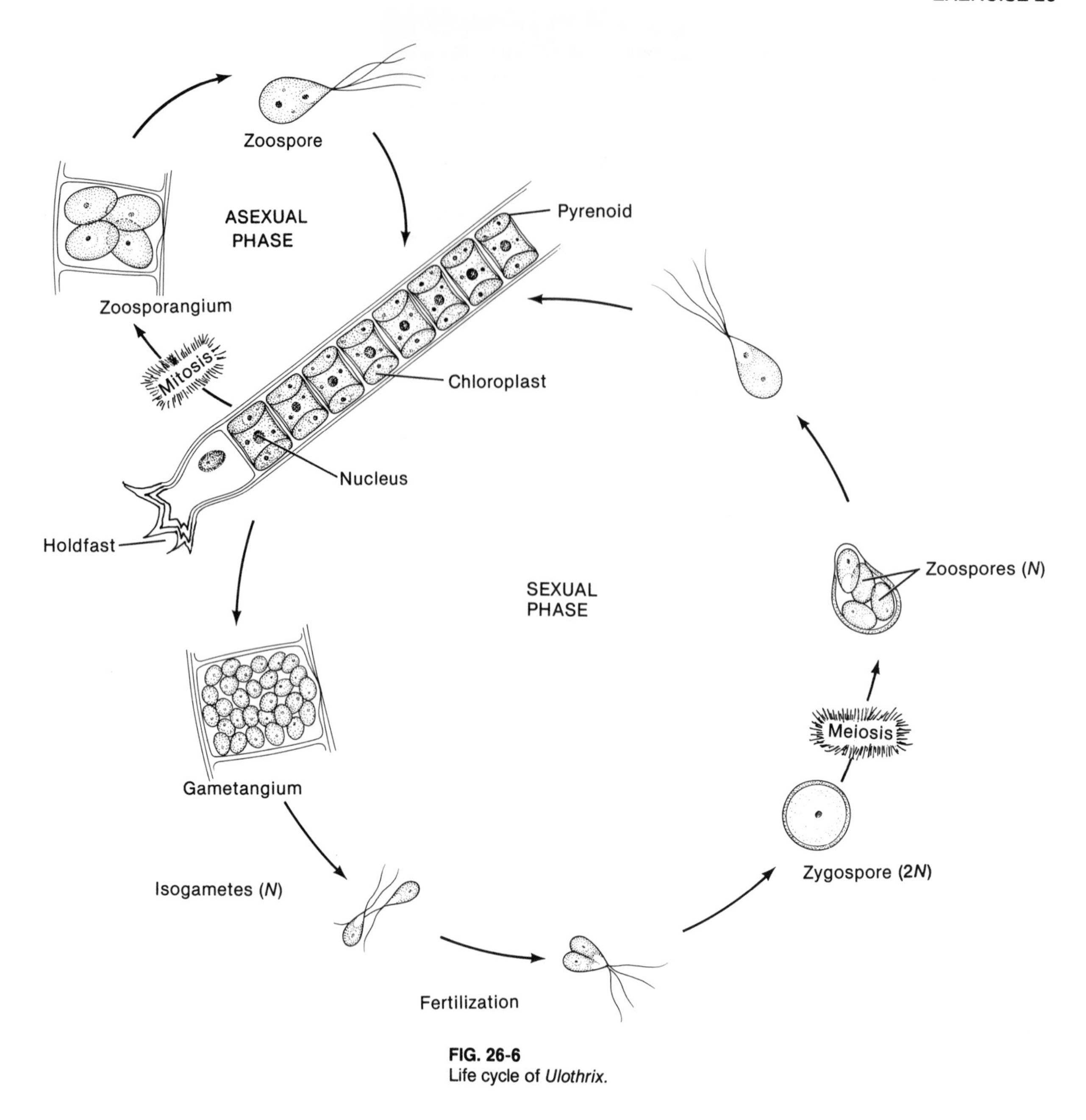

FIG. 26-6
Life cycle of *Ulothrix*.

Examine living specimens or prepared slides of *Oedogonium*.

A holdfast cell is present in some species.

How does the shape of the chloroplast differ from those found in *Spirogyra* and *Ulothrix?*

Where are the pyrenoids found?

Despite any differences in the shape of the cells found in the plant body of *Oedogonium*.

Asexual reproduction can take place in either of two ways. The plant body can simply fragment, with each fragment increasing in size through cell divi-

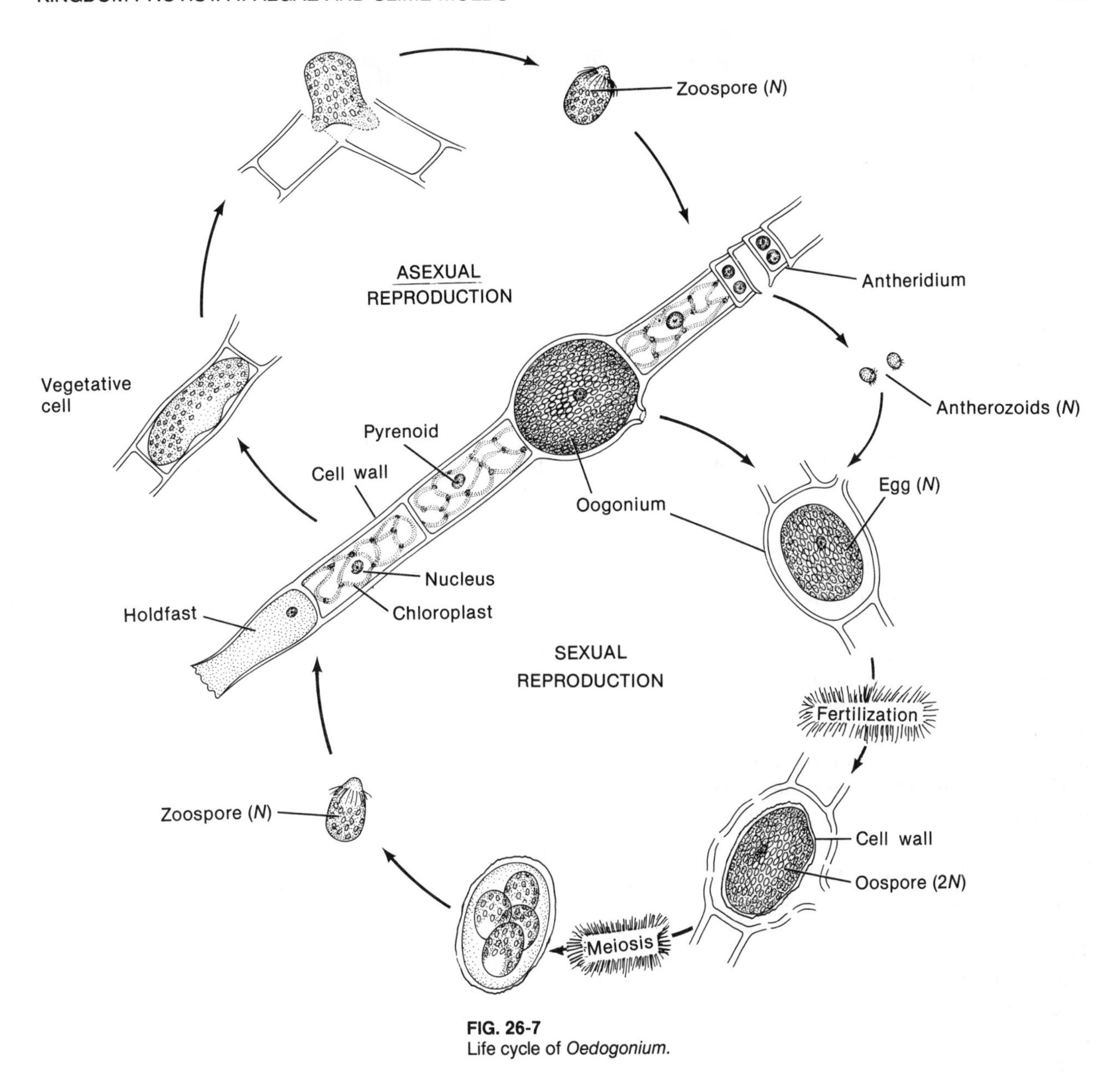

FIG. 26-7
Life cycle of *Oedogonium.*

sion, or the protoplast of any **vegetative cell** (i.e., nonreproductive cell or holdfast cell) can become a zoospore that, when released, actively swims for a while then produces a new filament of cells (Fig. 26-7).

To study the sexual phase, examine a prepared slide or living material showing **antheridia** are **oogonia,** the specialized reproductive cells. The antheridia are shortened, disc-shaped cells in contrast to the elongated vegetative cells of the filament. Each antheridium gives rise to two male gametes called **antherozoids** (sperm). They are small, egg-shaped cells with a ring of flagella at the pointed end.

The *oogonium* is a large, spherical cell in which the protoplast rounds up to form a single gamete—the egg. As the egg matures, a small pore, or transverse crack, appears in the oogonial wall. Antherozoids swimming near the oogonium are attracted to the egg and enter the oogonium through the pore or crack. Fertilization results in a **zygote** that is retained in the oogonium. The zygote soon develops a thick wall and becomes dormant. At this time, it is called an **oospore.** Disintegration of the oogonial wall releases the oospore, which remains dormant for several months. When environmental conditions are conducive to growth, the oospore undergoes meiosis producing four haploid zoospores each of which develop into new filamentous plants.

In what way is *Oedogonium* more advanced than *Spirogyra* or *Ulothrix?*

___

___

___

___

___

4. Division Phaeophyta (Brown Algae)

The brown algae are almost exclusively marine organisms ranging in size from microscopic to in excess of 100 meters in length. These algae comprise most of the conspicuous seaweeds of the temperate regions, where they dominate rocky shores. Economically, the brown algae provide us with several useful products, such as alginic acid, which is used as a stabilizer to give ice cream a smooth body and texture. It is also used in the manufacture of fire-resistant paints, cosmetics, and many pharmaceuticals.

In many parts of the world, the brown algae are an important food source. In oriental countries, *Laminaria* (Fig. 26-8A) is grown on ropes suspended between bamboo poles driven into the sea bottom along the coasts. A food product prepared from these algae, called "Kimbri," is a dietary staple in Japan.

Macrocystis, the giant Pacific kelp, is being cultivated on an experimental basis along the California coast to determine it's potential as a source of methane fuel.

Examine specimens of brown algae, noting the variation in size and complexity of the plant body. Many of the "kelps" (Fig. 26-8B) are externally differentiated into parts looking like "roots" (the **holdfast**), "stems" (the **stipe**), and "leaves" (the **blade**). Also observe that many of the specimens have air-filled bladders. How are such "floats" advantageous to the organism?

___

___

___

If available, examine *Sargassum,* a floating brown alga typically found in tropical waters (Fig. 26-8C). These plants occur in large numbers and may extend over thousands of acres of the ocean's surface. The Sargasso Sea takes its name from this alga.

5. Division Rhodophyta (Red Algae)

Like the brown algae, red algae are primarily marine organisms. Although occasionally found in cooler regions, they are most abundant in tropical and warm water. Unlike brown algae, the red algae always grow attached to solid objects and usually grow below the tide level. In warm waters, they are found at greater depths than any other group of algae, typically 100–200 meters below the surface. Water-soluble accessory pigments called phycobilins mask the color of chlorophyll *a* and give these algae their distinctive red color. These pigments are particularly well suited to the absorption of the wavelengths of light that penetrate into deep water.

The plant body of red algae is similar to that of the brown algae in that it is differentiated into a holdfast, stipe, and blade. In some, known as **corallines,** the plants are heavily impregnated with limestone and are considered to be just as important to the formation of coral reefs and atolls as the coral animals.

Red algae provide several useful products. For example, a gelatinous cell wall material provides agar, which is used extensively in laboratories for the culture of bacteria and fungi and some higher plants. Carrageenan, another colloidal substance, is used as an emulsifying agent in dairy products such as chocolate milk, where it prevents the chocolate from settling out.

The red alga *Porphyra* is used as food in several oriental dishes. In one, called *sushi,* various meats and rice are wrapped in sheets of this alga.

Examine various specimens of red algae on demonstration.

B. SLIME MOLDS (DIVISION GYMNOMYCOTA)

Slime molds are divided into two groups based on the form of the vegetative feeding phase of the life cycle. Because the vegetative phase of the cellular slime molds consists of masses of single amoeboid cells, they are thought to be more closely related to the amoebas than to any other group and, therefore, have been traditionally classified as protozoa . The acellular, or plasmodial slime molds, have a vegetative phase consisting of naked masses of protoplasm (plasmodia) of indefinite size and shape. Both types live predominantly on decaying plant material, primarily microorganisms (especially bacteria).

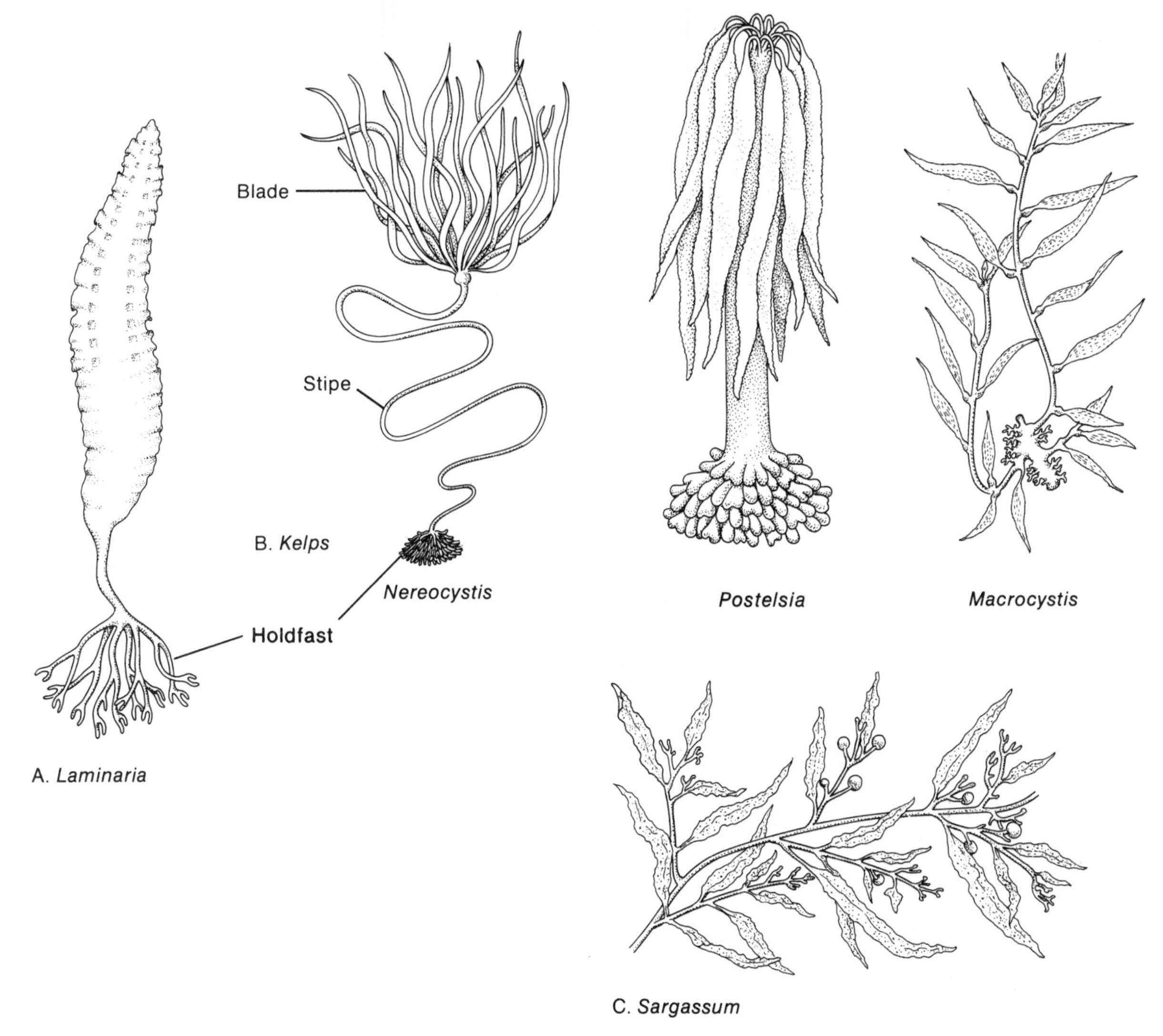

FIG. 26-8
Brown algae: (A) *Laminaria,* (B) kelps—*Nereocytis, Potelsia* and *Macrocystis,* (C) *Sargassum.*

In this part of the exercise, you will study the plasmodial slime molds by observing their growth and production of fruiting structures. Obtain a small piece of slime mold **sclerotia** (a dry, resting phase of the organism) and place it on the surface of the agar in a petri dish (Fig. 26-9).

Sprinkle a few oatmeal flakes over the sclerotium. Add two or three drops of water to moisten the oatmeal.

Replace the cover on the dish and set it aside in a dark place. After 24 hours, examine the plates for growth of the slime mold. Record your observation by drawing the growing organism in Fig. 26-10.

Once growth has begun, examine the slime mold with a hand lens, a dissecting microscope, or the low power of your compound microscope. Describe the unique pattern of cytoplasmic movement observed in the plasmodium of the mold.

____

____

Puncture a branch of the slime mold with a needle and watch it for a few minutes.

After the dish becomes covered with the plasmodium of the slime mold, partially remove the cover. Under these conditions, the slime mold will begin to dry out and in the process will initiate the formation of fruiting bodies. Examine the culture during the next few days and describe the shape of the fruiting body that is formed.

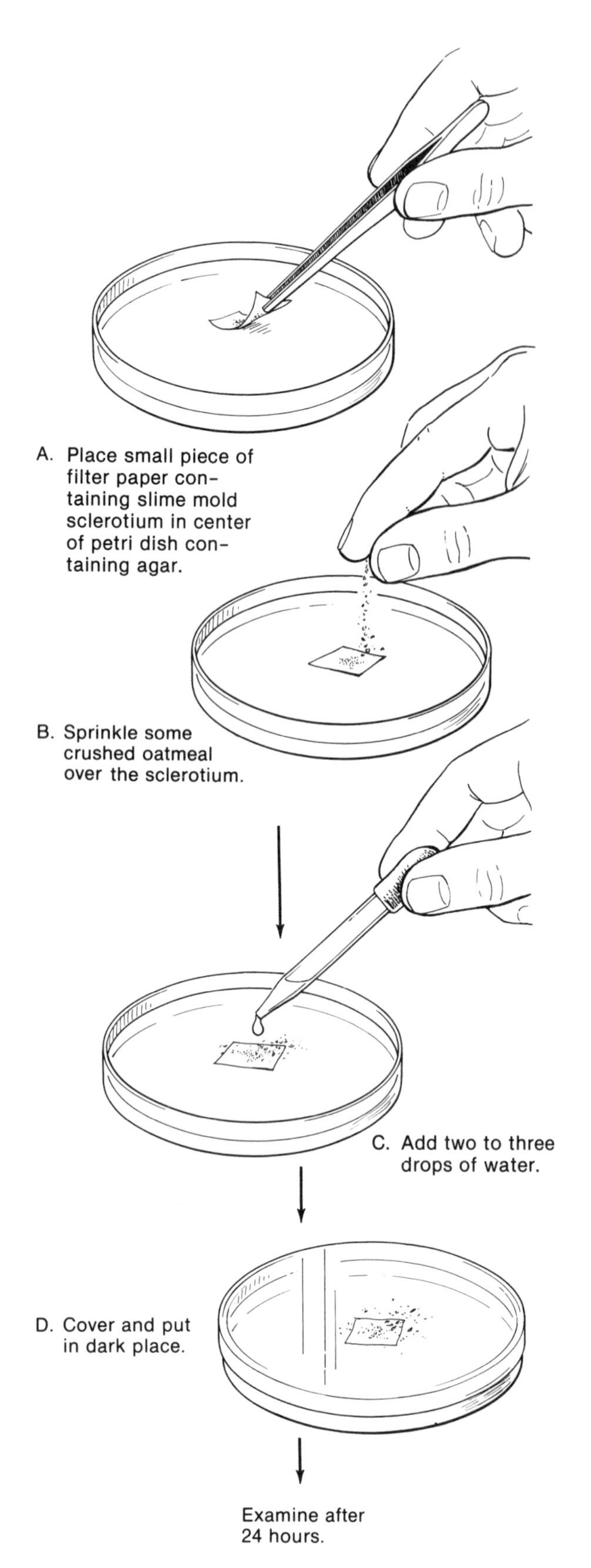

FIG. 26-9
Procedure for growing slime mold.

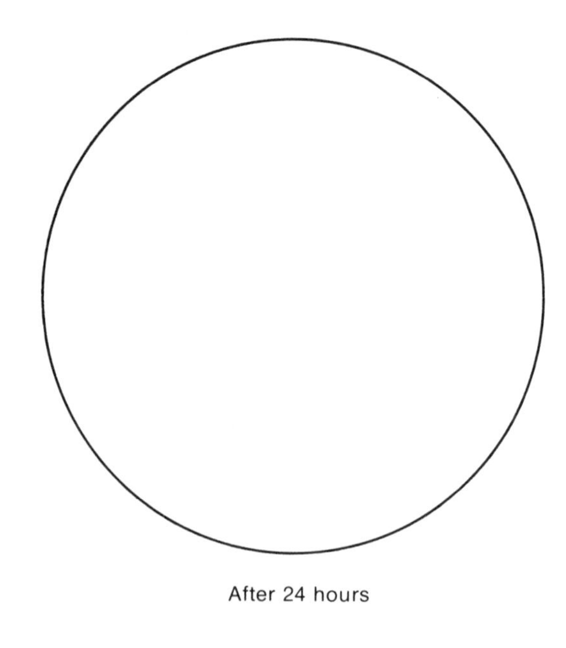

After 24 hours

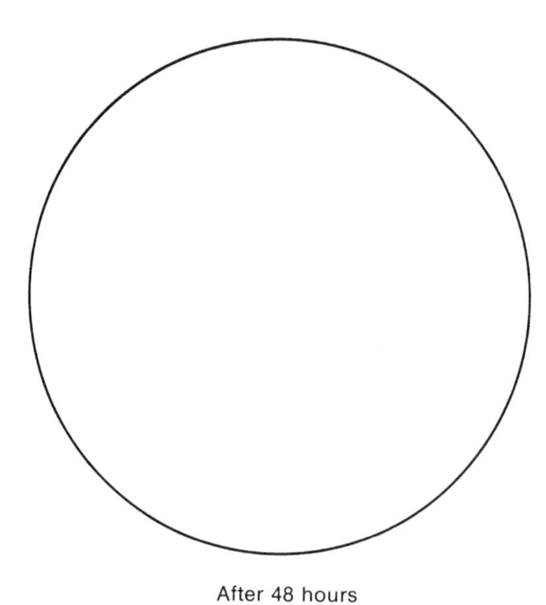

After 48 hours

FIG. 26-10
Growth of slime molds.

REFERENCES

Bold, H. C., and C. L. Hundell. *The Plant Kingdom.* 4th ed. Prentice-Hall.

Bold, H. C., and M. J. Wynne. 1985. *Introduction to the Algae: Structure and Function.* 2d ed. Prentice-Hall.

Chapman, A. R. O. 1979. *Biology of Seaweeds: Levels of Organization.* University Park Series

Delevoryas, T. 1977. *Plant Diversification.* 2d ed. Holt, Rinehart and Winston.

Klein, R. M. 1979. *The Green World.* Harper and Row.

Prescott, G. W. 1978. *How to Know the Freshwater Algae.* 3d ed. Brown.
Raven, P. H., R. F. Evert, and H. Curtis. 1981. *Biology of Plants.* 3d ed. Worth.
Ray, P. M., T. A. Steeves, and S. A. Fultz. 1983. *Botany.* Saunders.

EXERCISE 27

Kingdom Protista II: Protozoa

Protozoa are heterotrophic, one-celled organisms that exhibit a remarkable degree of subcellular organization. Instead of organs and tissues, they have functionally equivalent subcellular structures, called **organelles.**

Protozoans are found in a great variety of habitats. Most of them are free-living and inhabit fresh and marine waters. A number of protozoans also inhabit the bodies of other organisms in relationships described as **commensalistic** (intimate association of two or more organisms in which one benefits and the other is neither harmed nor benefited), **mutualistic** (intimate association of organisms in which both benefit), or **parasitic.** Of the five divisions of the Protista that are commonly referred to as *protozoa,* only the following four will be studied in this exercise.

Division Mastigophora (Flagellates): unicellular, mostly heterotrophic organisms that move by means of long, whiplike flagellae.

Division Ciliophora (Ciliates): unicellular, heterotrophic organisms that move by cilia. (Typically have macro- and micronuclei and unique reproductive patterns.

Division Sarcodina: heterotrophic organisms that lack cilia or flagella and move by irregular cytoplasmic extensions called *pseudopodia.*

Division Sporozoa: parasitic organisms that live for a part of their life cycle in cells of other organisms. Most have very complex reproductive cycles, and move by means of flagellae or pseudopodia for brief periods in their life cycle.

A. DIVISION MASTIGOPHORA (FLAGELLATES)

The members of this division move by means of **flagellae.** A few of the flagellates exist as free-living organisms in fresh or salt waters. However, most live as symbionts in the bodies of higher plants or animals.

1. Trichonympha

Trichonympha is a flagellate that inhabits the intestine of wood-eating termites (Fig. 27-1A). Although termites ingest bits of wood, they are unable to digest cellulose, wood's chief constituent. *Trichonympha* forms pseudopodia that engulf the wood fragments ingested by the termite. The cellulose walls of the wood is digested into soluble carbohydrates that can then be used by the termite. Neither the termite nor *Trichonympha* can live without the other. Examine slides of this organism. Note the large number of flagellae that cover the upper part of the organism and the small fragments of wood in the cytoplasm.

2. Trypanosoma

Trypanosoma (Fig. 27-1B) is a slender flagellate that inhabits the blood of vertebrates and is carried from one host to another by bloodsucking invertebrates. The trypanosomes that cause human sleeping sickness in Africa are transmitted by the bloodsucking tsetse fly. Examine stained slides of human blood showing these parasitic flagellates.

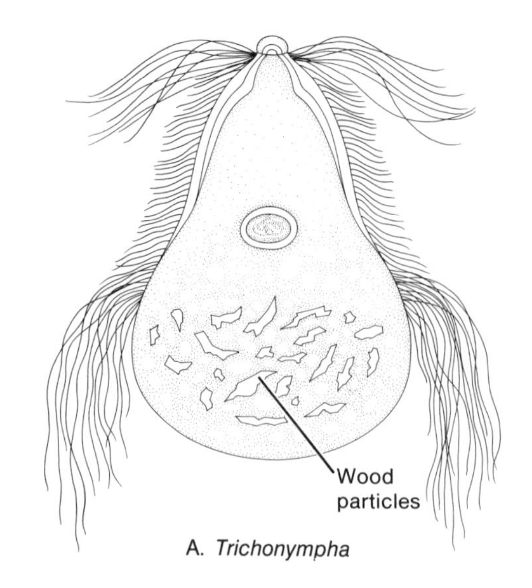

A. *Trichonympha*

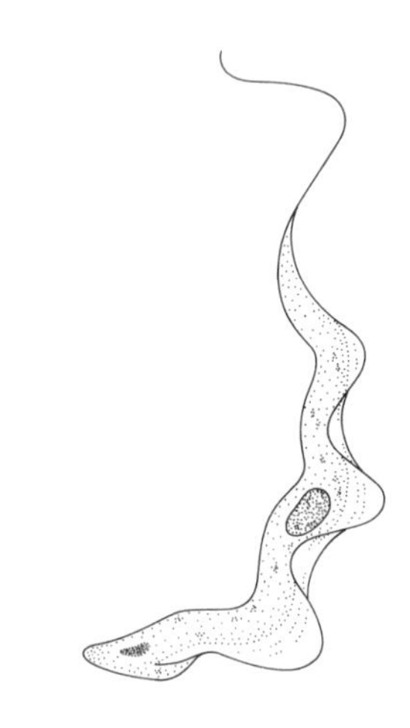

B. *Trypanosoma*

FIG. 27-1
Flagellates: (A) *Trichonympha* and (B) *Trypanosoma.*

B. DIVISION CILIOPHORA (CILIATES)

Compared with other protists, the ciliates are relatively large and complex. They are also distinguished from other protists by having **cilia,** two types of nuclei (**micro,** small; and **macro,** large), and a unique type of reproduction. Most of the ciliates are free-living (nonparasites) and are commonly found in fresh and salt waters. A few are parasitic to human beings. These organisms also play an important role in the aquatic food chain, serving as food for small multicellular animals, which, in turn, are eaten by larger animal forms.

1. Paramecium Caudatum

a. Morphology

Paramecium caudatum, one of several species of paramecia, is often selected as a typical ciliate to study because it is easy to obtain and, owing to its size, is easy to observe. In nature, this ciliate is frequently found in pond water that contains large amounts of decaying vegetation. Because it can be easily grown in large numbers under controlled laboratory conditions, it is used extensively in studies of nutrition, cancer, behavior, genetics, and ecology.

Place a small drop of *Paramecium* culture on a clean slide and add a small drop of methyl cellulose to slow down its movements. Add a coverslip and examine microscopically using the low-power objective. You may have to adjust the iris diaphragm to increase contrast.

Paramecium caudatum somewhat resembles a twisted slipper. With the aid of Fig. 27-2 and commercially prepared slides, locate the following structures on the living organism.

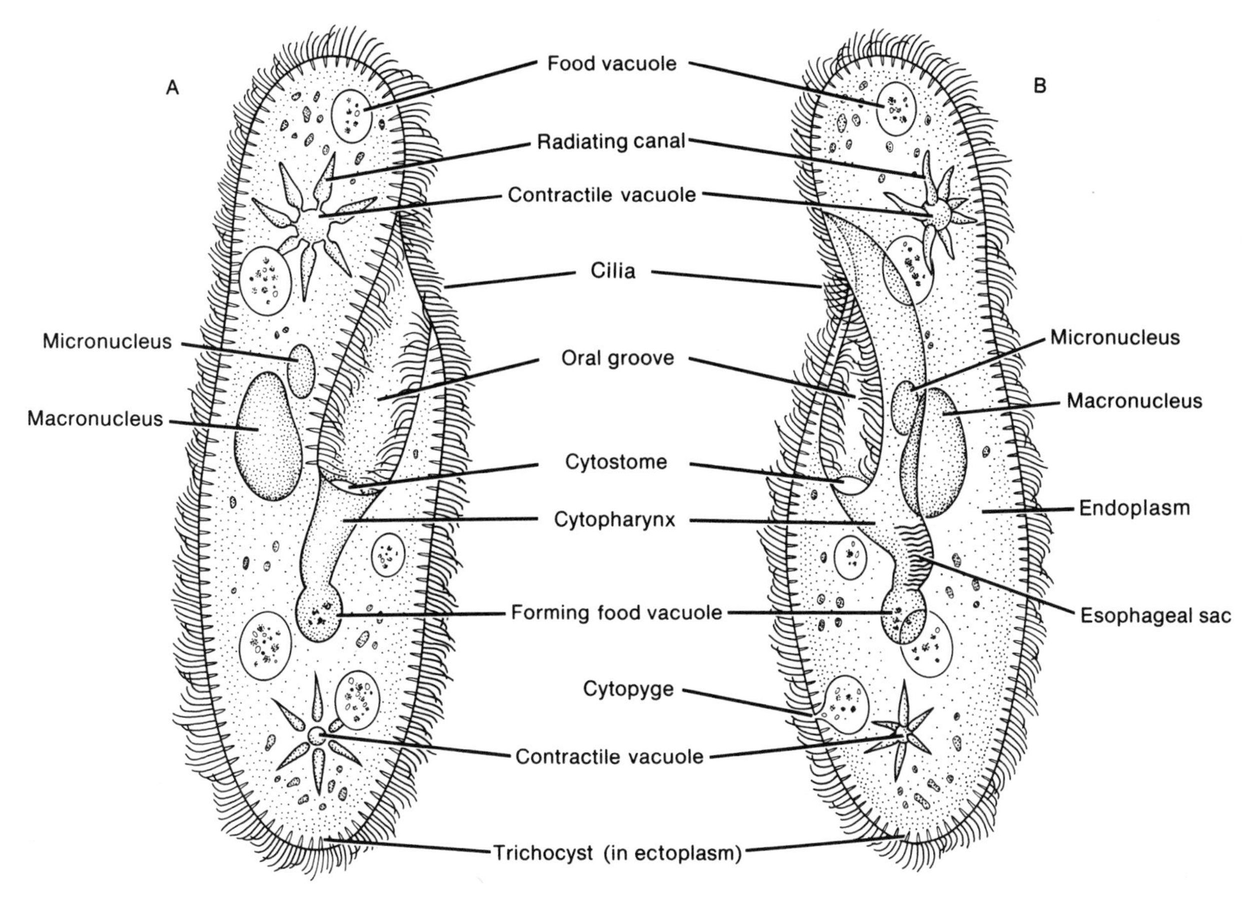

FIG. 27-2
Paramecium: (A) ventral view and (B) lateral view.

The **oral groove** begins at the anterior end, runs diagonally toward the posterior end, and leads to the **cytostome** (cell mouth) and then to the **cytopharynx.** Food is carried into the oral groove by beating **cilia,** which may be seen by using the high-power objective and adjusting the iris diaphragm.

A **contractile vacuole** is located at each end of the organism. Do these vacuoles move or are they stationary?

____

Do they contract at the same time, or alternately?

____

Locate the canals that radiate outward from each vacuole. Suggest a function for these structures.

____

____

____

What is the function of contractile vacuoles?

____

____

____

The **macronucleus** can be seen as a relatively large, clear area in the central part of the cell **(endoplasm).** A **micronucleus** is located adjacent to the macronucleus. Because these nuclei are difficult to observe in living material, add a drop of acetocarmine or methyl green stain to one side of the coverslip, and draw it under by applying a piece of paper toweling to the other side. (Supplement these observations using commercially prepared slides of *Paramecium* specifically stained to show the nuclei.) The macronucleus is associated with vegetative functions (i.e., feeding and digestion), whereas the micronucleus is reproductive.

Trichocysts, located in the peripheral layer of cytoplasm (**ectoplasm**) are carrot-shaped structures containing coiled, barbed threads. The function of trichocysts in *Paramecium* is not known. However, in other animals that have them, they serve to capture and hold smaller organisms on which the animals feed.

Add a small drop of iodine or acetic acid at the edge of the coverslip. Allow it to run underneath and observe the discharge of the trichocysts.

b. Nutrition and Feeding

The majority of protists, including *Paramecium,* are **holozoic;** that is, they ingest solid food such as other protists, bacteria, or organic **detritus** (decaying material) from the water. These food particles must be digested before they can be used for growth, repair, or reproduction.

Place a drop of *Paramecium* culture on a slide and add a *small* drop of congo red-stained yeast *(Note:* The resulting color should be pink, not red.) Add a coverslip, and locate a *Paramecium* under high magnification. Observe the vortex of water produced by the cilia in the vicinity of the oral groove that carries the stained yeast into the oral groove and then to the mouth and cytopharynx. The cytopharynx leads to an **esophageal sac** where a *food vacuole* (a membrane-bound sac containing water and suspended food particles) is formed. As soon as the food vacuole is formed, it is carried away by a rotary movement of the cytoplasm (cyclosis) and the formation of another vacuole begins. Food vacuoles travel in a defined route through the organism (Fig. 27-3). They first pass posteriorly, then anteriory, and finally in a posterior direction again to the region of the oral groove where the undigestible contents are eliminated through the **anal pore.**

Locate a food vacuole. You should observe that, when it is first formed, the vacuole is bright red-orange. Closely observe the vacuole as it moves through the animal. You will see that its contents changes color in the course of digestion from red-orange to blue-green to yellow and finally back to red-orange (Fig. 27-3). The reason for this is that congo red is an indicator dye that changes color with pH: it is blue-green under acidic conditions and red-orange under alkaline conditions. What does this indicate about the content of the food vacuoles?

____

____

____

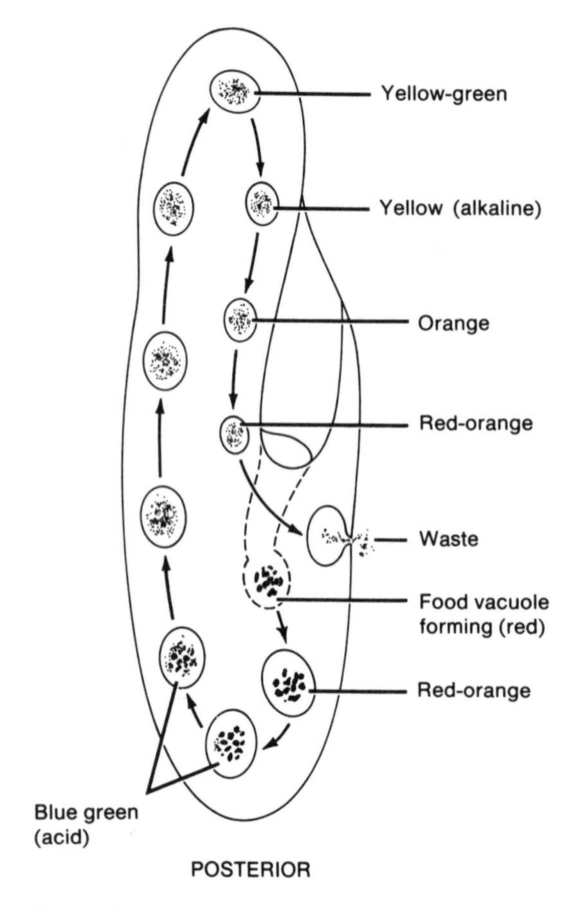

FIG. 27-3
Digestion in *Paramecium.* Arrows indicate pathway of food vacuoles.

What similarity is there between the acidity and alkalinity of the food vacuoles as they pass through this animal and that of the mouth, stomach, and intestines in human beings?

____

____

____

____

____

c. Reproduction

Asexual Reproduction. The most common type of reproduction in protists is **binary fission.** In this kind of asexual reproduction, the cell divides into two roughly equal parts, giving rise to daughter cells that are genetically identical. In flagellates, the plane of division was longitudinal; in ciliates, the cell divides transversely.

Examine demonstration slides of *Paramecium* showing the various stages in binary fission (Fig. 27-4).

Sexual Reproduction. Ordinarily, paramecia divide by binary fission. Occasionally, however, these organisms reproduce sexually by a process called **conjugation** in which micronuclei are exchanged. The process of conjugation is diagrammed in Fig. 27-5. Conjugation in living *Paramecium* can be observed by following the directions outlined below. For this study, mating strains of *Paramecium bursaria,* which is symbiotic with a green alga, will be used. You should supplement your observations by examining commercially prepared slides showing conjugation.

1. Into a depression or deep-well slide, add a *small* drop of one of the two mating strains.

2. While examining the slide with a dissecting microscope, add a drop of the *second* mating strain.

3. You should see, almost immediately the **agglutination** (or clumping) of opposite mating strains, which brings the cells together for the subsequent transfer of nuclear material.

4. Place the slide in a covered petri dish containing moist filter paper to prevent dessication of the culture.

5. Examine periodically. Conjugating paramecia may be seen up to 48 hours, after which few or no conjugants will be found.

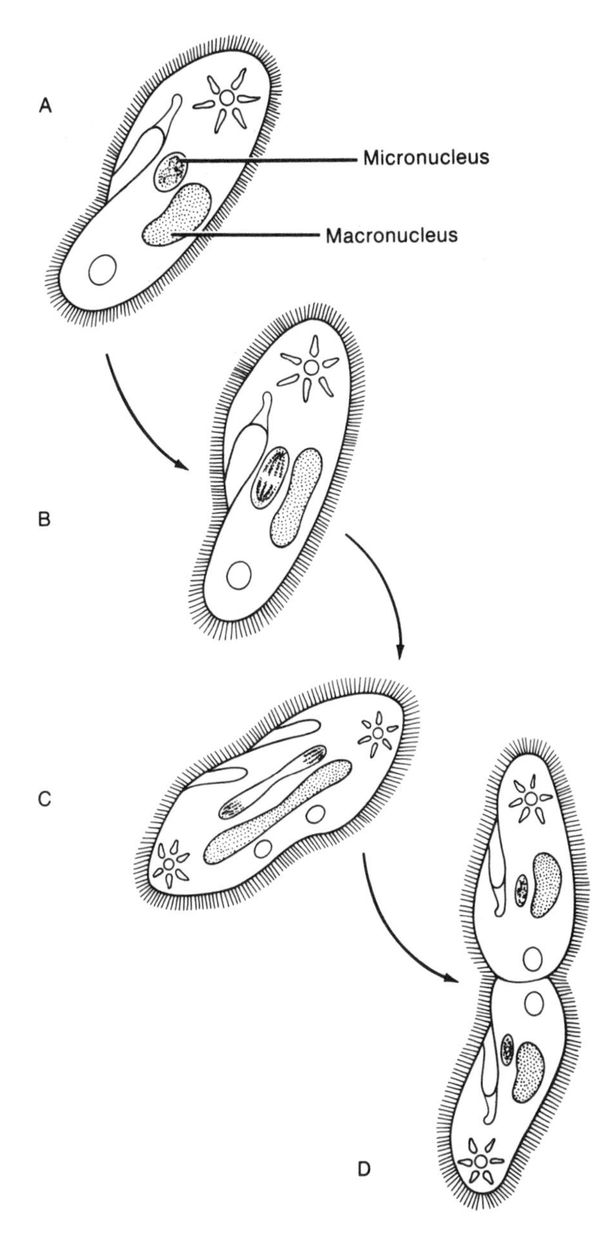

FIG. 27-4
Asexual reproduction (binary fission) in *Paramecium.*

2. Other Ciliates

Obtain samples of the following ciliates and examine microscopically.

a. Stentor

This organism is characteristically trumpet-shaped, blue in color when living, and has a macronucleus in the shape of a string of beads (Fig. 27-6A). Food is brought into the organism through activity of cilia that surround the mouth.

b. Vorticella

This freshwater ciliate somewhat resembles an inverted bell attached to a stalk that anchors it to submerged vegetation and stones (Fig. 27-6B). Locate an organism in which the stalk is straight and extended. While watching the animal, gently tap the slide and describe what occurs.

____

____

____

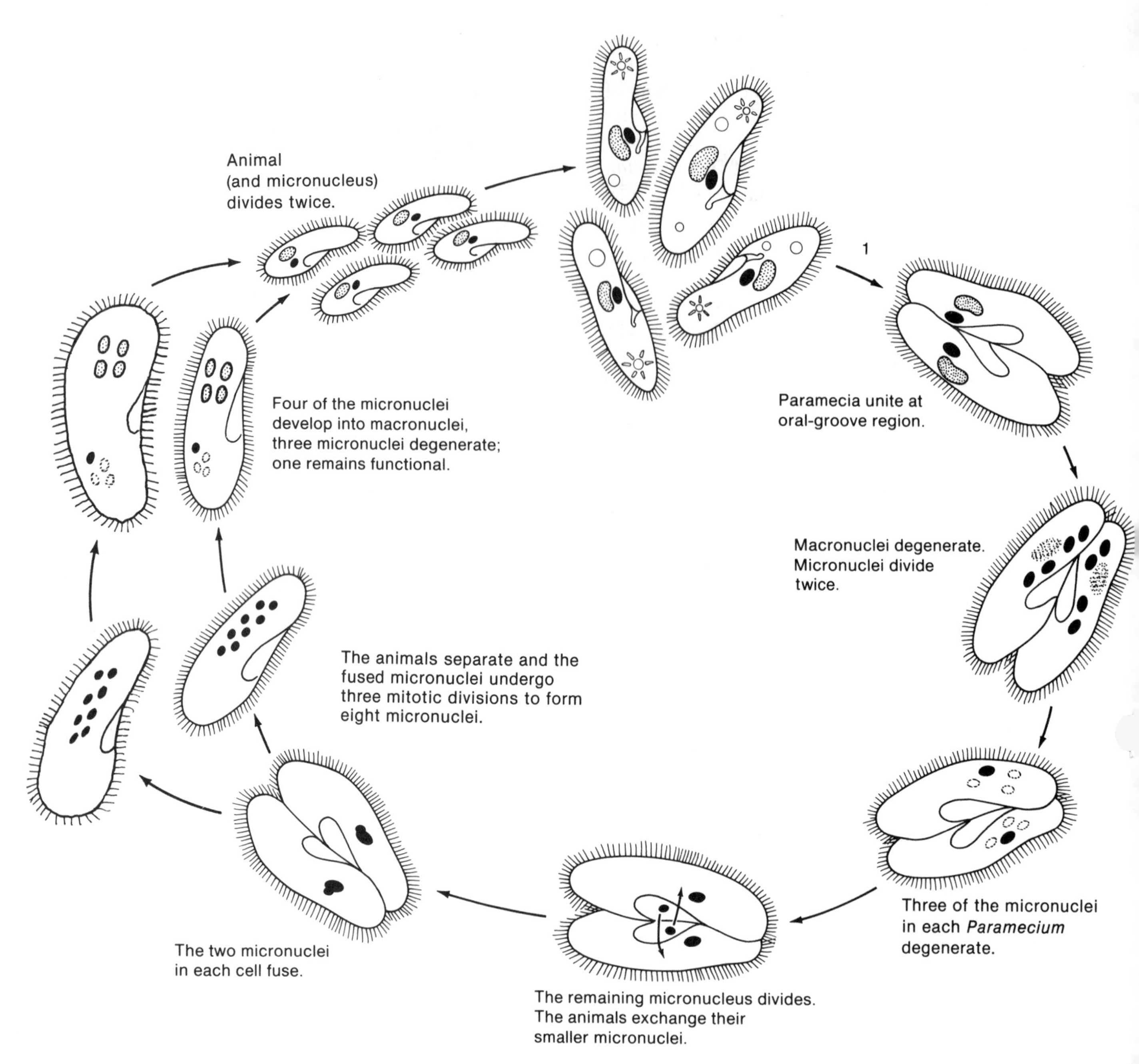

FIG. 27-5
Sexual reproduction (conjugation) in *Paramecium*.

c. Balantidium coli

This ciliate (Fig. 27-6C) is the only ciliate parasite found in human beings. It is found in the intestinal tract where it frequently burrows into the wall of the colon and causes ulcers. It is **pathogenic** (disease producing) and may cause symptoms similar to amoebic dysentery. The most common source of infection is the pig.

Examine prepared slides of *Balantidium coli* and note that the whole body is covered with cilia arranged in five rows. The macronucleus is slightly curved and is associated with a very small micronucleus. Food particles are swept into a cytostome by the currents set up by the beating cilia. There are two contractile vacuoles, and food vacuoles circulate in the cytoplasm. Like other ciliates, *Balantidium* divides by transverse fission.

d. Tetrahymena

This small ciliate (Fig. 27-6D), which is relatively

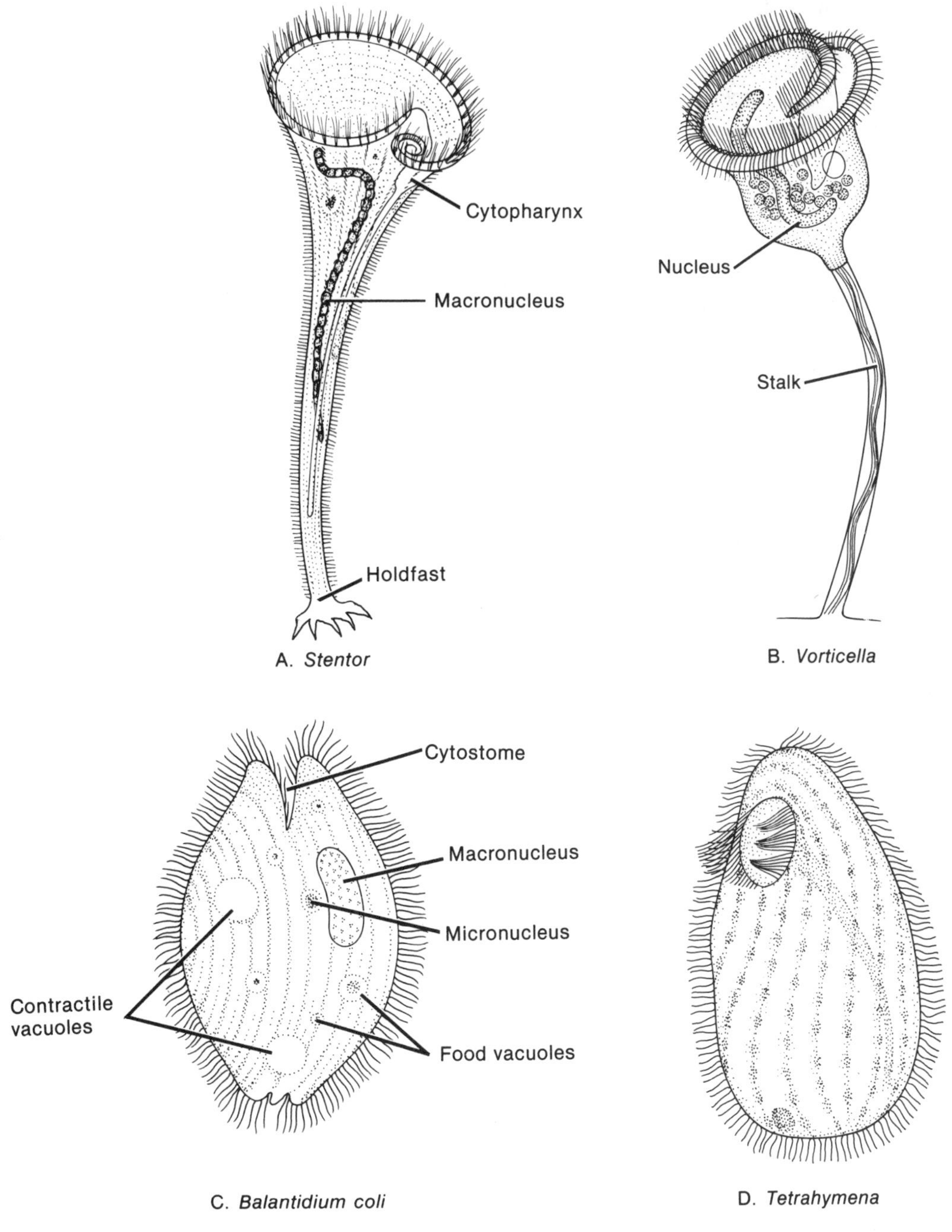

FIG. 27-6
Other ciliates.

easy to grow in pure culture, has been widely used in physiological and genetic research. Its mitotic division can be synchronized by appropriate heat shocks and, therefore, this protozoan is a useful tool in studying the details of mitosis.

Examine living specimens or prepared slides of *Tetrahymena.*

C. DIVISION SARCODINA

1. Amoeba

Amoebas (Greek *amoibe,* "change") are protists common to freshwater ponds and streams. Under the microscope, they look like gray, irregular masses that continually change shape due to the extension and withdrawal of fingerlike protuberances called **pseudopodia** (Fig. 27-7).

Using a clean dropper pipet, obtain a sample of amoebas from the bottom of the culture dish. Place a few drops of the culture in a depression slide (alternatively, place a few drops on a clean glass slide, and add a few pieces of debris from the culture and a few grains of sand or some small pieces of broken coverslips to prevent the organism from being crushed when you place a clean coverslip over the preparation). To visualize the three-dimensional form of an amoeba, examine the culture using a stereoscopic dissecting microscope. Adjust the spot lamp to give transmitted light first, and then reflected light. You may have to adjust the intensity of the light. These protists will be usually found on the bottom on the culture dish. Study your preparation under low power (10 ×). You will have to reduce the amount of light by closing the iris diaphragm, because the amoebas are nearly transparent and thus almost invisible under bright light. With the help of Fig. 27-7 and commercially prepared slides, locate the following: the **pseudopodia;** the **endoplasm,** the inner granular material that makes up most of the cytoplasm; the **ectoplasm,** a thin layer of cytoplasm surrounding the endoplasm; the **cell membrane (plasmalemma);** the **plasmagel,** the gellike outer layer of the endoplasm; the **plasmasol,** the fluid, central region of the endoplasm (streaming of the cytoplasm should be evident in this area); the **nucleus,** a somewhat transparent structure that is not fixed in position and sometimes has a wrinkled or folded appearance; and the **contractile vacuoles,** clear spherical vacuoles found in the endoplasm that collect water from the cell and discharge it to the outside. These vacuoles function to maintain water balance in the cell. Would you expect a marine (saltwater) form of amoeba to have contractile vacuoles? Explain.

____

____

Food vacuoles contain ingested food and enzymes for digestion. How are food vacuoles formed?

____

____

____

____

How are the products of digestion made available to the cell?

____

____

____

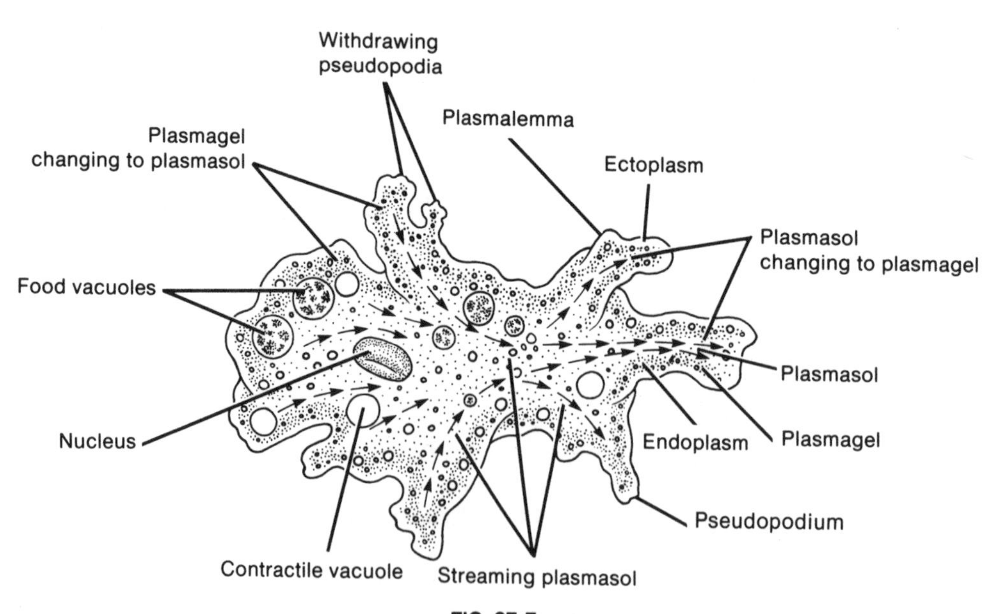

FIG. 27-7
Amoeba.

Reproduction in *Amoeba* is by binary fission in which the cytoplasm and nucleus divide to form two genetically identical daughter amoebas.

2. Other Sarcodina

Radiolarians (Fig. 27-8A) are amoebas that have silicious skeletons secreted by the cytoplasm, usually taking the form of an intricate latticework, through which extend stiff radiating spines. Examine prepared slides of radiolarian skeletons.

Foraminiferans (Fig. 27-8B) are a large group of marine amoebas that secrete slimy, snaillike shells made of calcium carbonate about themselves. Long, delicate feeding pseudopodia stream out of minute pores that perforate the surface of the shells. Examine prepared slides showing a variety of these foraminiferan shells.

Entamoeba histolytica (Fig. 27-8C) is the organism that causes amoebic dysentery. Examine prepared slides showing both the active (vegetative) stages and the cysts (resistant, infective stages of the organism). In its vegetative stage, the organism has a single nucleus; the cysts have four nuclei, as well as a heavy wall surrounding the entire cell.

Actinophrys and *Actinosphaerium* (Fig. 27-8D) are members of a group of sperical amoebas that have

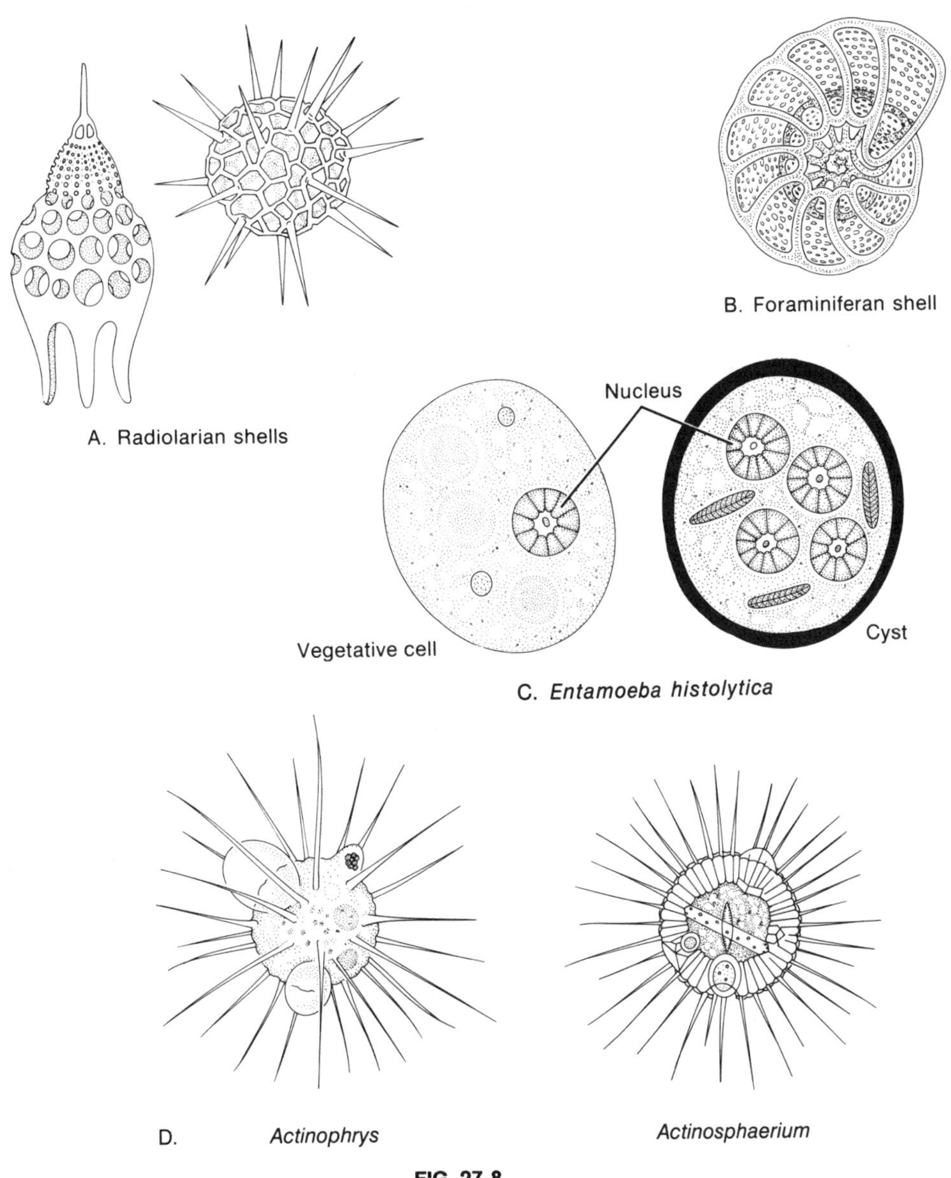

FIG. 27-8
Other Sarcodina.

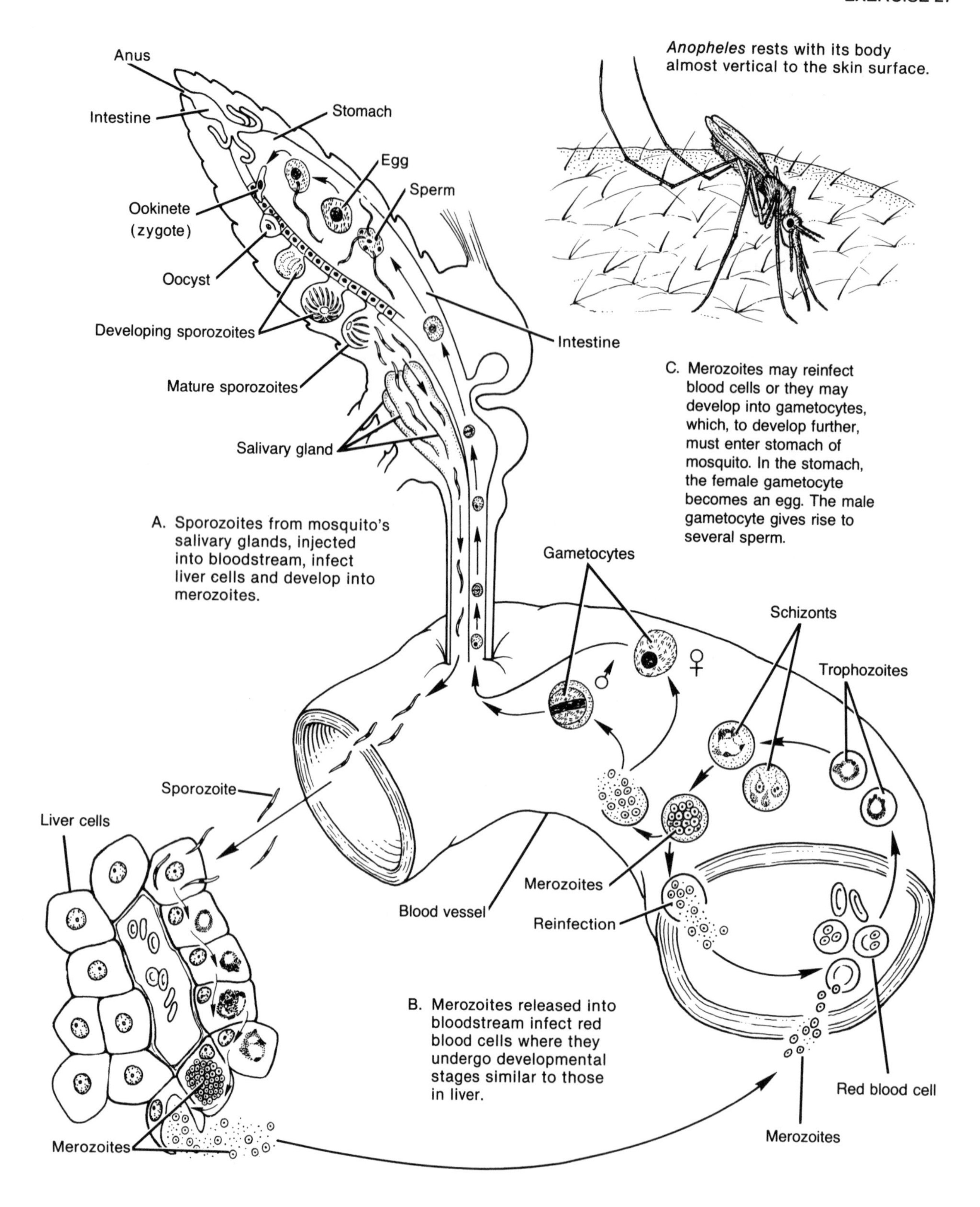

FIG. 27-9
Life cycle of *Plasmodium vivax*.

delicate pseudopodia that are supported by axial rods. Most species of this group live in fresh water. Examine slides of *Actinosphaerium.* Note in particular the delicate pseudopodia, which have suggested the name "sun animalcules" for these organisms.

D. DIVISION SPOROZOA

All members of this phylum are parasites and can be found infecting nearly all the major groups of the animal kingdom, ranging from simple invertebrates to human beings. In this exercise, you will become familiar with *Plasmodium vivax,* a xsporozoan vhat causes one form of mclaria in human beings.

Malaria is probably one of the most devastating diseases of humankind with respect to sickness, mortality (death), and economic loss. Although partly under control, it is not yet eliminated and indeed seems to be increasing in incidence once again.

There are four kinds of malaria. One, **benign tertian malaria,** caused by *Plasmodium vivax,* is characterized by attacks of fever every 48 hours. *Plasmodium ovale* also causes fever every 48 hours. **Quartan malaria,** which usually produces fever every 72 hours, is caused by *Plasmodium malariae.* **Pernicious malaria,** in which the fever is virtually continuous, is produced by *Plasmodium falciparum;* this form has the highest mortality rate.

Because all four kinds have fairly similar life cycles, *Plasmodium vivax* will be the representative sporozoan studied in this exercise. Refer to Fig. 27-9 and commercially prepared slides of the life cycle.

Malaria is transmitted through the bite of female mosquitoes; male mosquitoes cannot infect because they lack the mouth parts for piercing the skin and sucking blood. Many animals can be infected with malaria that is transmitted by a variety of different mosquitoes; however, the malarial parasite is transmitted to human beings only by mosquitoes of the genus *Anopheles.*

When the mosquito's mouth parts enter the skin, saliva, containing anticoagulants, enters the wound. If the mosquito is a carrier, **sporozoites,** the form of the parasite in its infective stage, enter the bloodstream (Fig. 28-9A), but do not invade the erythrocytes. Instead, they first penetrate certain cells of the liver, where they grow and multiply to form **merozoites.** On release from the liver cells, the merozoites enter the bloodstream and penetrate the erythrocytes, where they first become ring-shaped and then irregular in shape. At this stage, the parasite is called a **trophozoite** (Fig. 27-9B). The trophozoite undergoes maturation (the time required depending on the species) and then divides, by a type of fission called **schizogony,** to form another stage, called a **merozoite** (Fig. 27-9C). These merozoites are released from the red blood cells, where they reproduce asexually to form more merozoites. At regular intervals, depending on the species of *Plasmodium,* all infected red blood cells burst, releasing merozoites. Toxic substances that are released along with the merozoites account for the chills and fever that are typical of malarial attacks. Each merozoite in turn penetrates another erythrocyte and becomes a trophozoite. This cycle is repeated a number of times, with ever-increasing numbers of erythrocytes being affected.

Eventually, some merozoites, instead of becoming trophozoites, develop into male and female **gametocytes** (Fig. 27-9C). As long as the gametocytes remain in the human host, they are of no significance. However, if they are sucked up by an *Anopheles* mosquito, they will pass into the insect stomach and become active (Fig. 27-9C). One female gametocyte develops into one egg; one male gametocyte gives rise to several sperm by a process termed *exflagellation.* Union of the egg and sperm produces a zygote called an **ookinete,** which migrates through the stomach epithelium and becomes embedded in the wall of the stomach. The nucleus undergoes successive mitotic divisions and produces large numbers of sporozoites within a structure now called an **oocyst.** When the oocyst ruptures, the sporozoites enter the body cavity of the mosquito and then migrate to the salivary glands. The next time a mosquito bites a host, sporozoites are injected into the human body and the life cycle is repeated.

REFERENCES

Fingerman, M. 1981. *Animal Diversity.* Saunders.
Grell, K. G. 1973. *Protozoology.* 2d ed. Springer-Verlag.
Sherman, I. W., and V. G. Sherman. 1976. *The Invertebrates: Function and Form: A Laboratory Guide.* 2d ed. Macmillan.
Sleigh, M. A. 1973. *The Biology of the Protozoa.* Elsevier.
Villee, C. A., W. F. Walker, and R. D. Barnes. 1984. *General Zoology.* Saunders.

EXERCISE 28

Kingdom Fungi

The fungi, along with some bacteria, are the decomposers of the environment and, as such, they are as necessary to the biosphere as the food producers. Their metabolism releases carbon dioxide into the atmosphere and nitrogenous materials into the soil and surface waters where they are utilized by green plants and, ultimately, by animals.

Fungi are primarily terrestrial organisms. Some are unicellular; most are filamentous and may be organized into highly structured shapes of which mushrooms are an example. All fungi are **heterotrophic** and obtain their food as **saprobes** (i.e., they live on nonliving organic matter) or as **parasites** (i.e., they feed on living organic matter). Fungi do not ingest their food but absorb it. They secrete enzymes that break down the food outside the fungus. These partially digested molecules are then transported through the fungal membrane. All fungi have cell walls and the majority produce some type of spore.

There are seven major divisions in this kingdom.

Division Chytridiomycota, commonly called *chytrids:* parasitic or saprophytic organisms that live in fresh and salt water or soil. Cell walls contain chitin. Plant body **(thallus)** is coenocytic (multinucleate; nuclei not separated by cell walls) and when mature, gives rise to flagellated spores.

Division Oomycota (oomycetes): mostly saprophytic; a few are parasitic and pathogenic. All are oogamous (differentiation of gametes into distinct eggs and sperm) but gametes are nonmotile. Many are aquatic (called *water molds*) and some are terrestrial. Cell walls consist primarily of cellulose and occasionally contain chitin.

Division Zygomycota (zygomycetes): mostly saprophytic and terrestrial. Some are parasitic on plants and insects. Sexual reproduction results in thick-walled zygospore. Chitin is predominant in the cell walls.

Division Ascomycota (ascomycetes): largest division. Includes yeasts, mildews, and molds. Many are pathogenic. Hyphae divided by cross walls containing pores. Sexual reproduction results in the formation of a structure called an **ascus.**

Division Basidiomycota (basidiomycetes): terrestrial fungi whose most familiar members are the mushrooms. Sexual basidiospores are produced on a specialized hypha called a **basidium.**

Division Deuteromycota (fungi imperfecti): fungi that have characteristics of the ascomycetes but in which no method of sexual reproduction has been observed. Many are parasitic and pathogenic to plants and animals, including man, in which they cause infections of the skin and mucous membranes (ring worm and athletes' foot). Some fungi imperfecti are involved in the production of certain cheeses (roquefort and camembert) and antibiotics, including penicillin.

Division Mycophycophyta (lichens): lichens make up a large and diverse group of ascomycetes existing in a symbiotic relationship with green algae or cyanobacteria. Occasionally a basidiomycete or deuteromycete may be involved. The plant body (thallus) of lichens is quite distinctive in appearance, looking somewhat like that of liverworts and mosses. Three major growth forms are evident: crustose, foliose, and fruticose. Reproduction frequently involves the formation of ascospores by the fungal symbiont, or the production of soredia (small fragments consisting of at least one algal cell surrounded by fungal hyphae).

In this exercise you will study representative examples of the Zygomycetes, Ascomycetes, Basidiomycetes, and lichens.

A. DIVISION ZYGOMYCOTA (ZYGOMYCETES OR BREAD MOLDS)

One of the more common members of this division is *Rhizopus stolonifer,* the black bread mold that grows as cottonlike masses on bread, fruit, or other organic material. Infection begins when a spore germinates and forms masses of filamentous hyphae which differentiate as rhizoids (these anchor the fungus to its substrate, secrete digestive enzymes, and absorb partially digested organic materials) and sporangiophores (aerial hyphae that produce sporangia containing spores at their tips). As sporangia mature they become black in color, thus giving the fungus its name.

1. Asexual Reproduction

Examine a petri dish containing black bread mold. Here, the mold is not growing on bread, but on agar containing various organic compounds needed for growth. If you wish, bring some bread from home (but remember that most mass-produced breads contain mold inhibitors) and place a piece in a petri dish. Moisten it with several drops of water, and then place a small piece of the mold from the agar on the bread. Cover the dish and set it aside for 1–2 days.

Examine the mold with your dissecting microscope. To prevent numerous spores from being released into the air, *do not remove the cover*! Note the whitish mass of filaments growing over the surface of the agar (Fig. 28-1A). Each filament is called a **hypha** (plural **hyphae**). The total mass of hyphae is a **mycelium**. Some hyphae grow upward and form small, black, globelike structures called **sporangia** (Fig. 28-1C). Inside the sporangia are cells called **spores.** They are released when the sporangia open. What is the function of the spores?

____

____

Other specialized hyphae penetrate the agar. Turn the dish over, and focus downward through the agar to locate small, rootlike hyphae called **rhizoids** growing into the agar (Fig. 28-1C). Suggest a function for the rhizoids.

____

____

Remove a small piece of the mold, and place it in a drop of water on a slide (Fig. 28-1B). Add a coverslip and examine with your microscope. Locate sporangia, spores, hyphae, and rhizoids.

2. Sexual Reproduction

Sexual reproduction occurs when two different mating strains meet and fuse, attracted by hormones that diffuse from the hyphae. The two different strains are designated as (+) and (−) since they show no morphological differences from which to designate them as male or female.

Your instructor has inoculated an agar plate with a (+) strain and a minus (−) strain. The growth of each strain has brought them into contact. At the

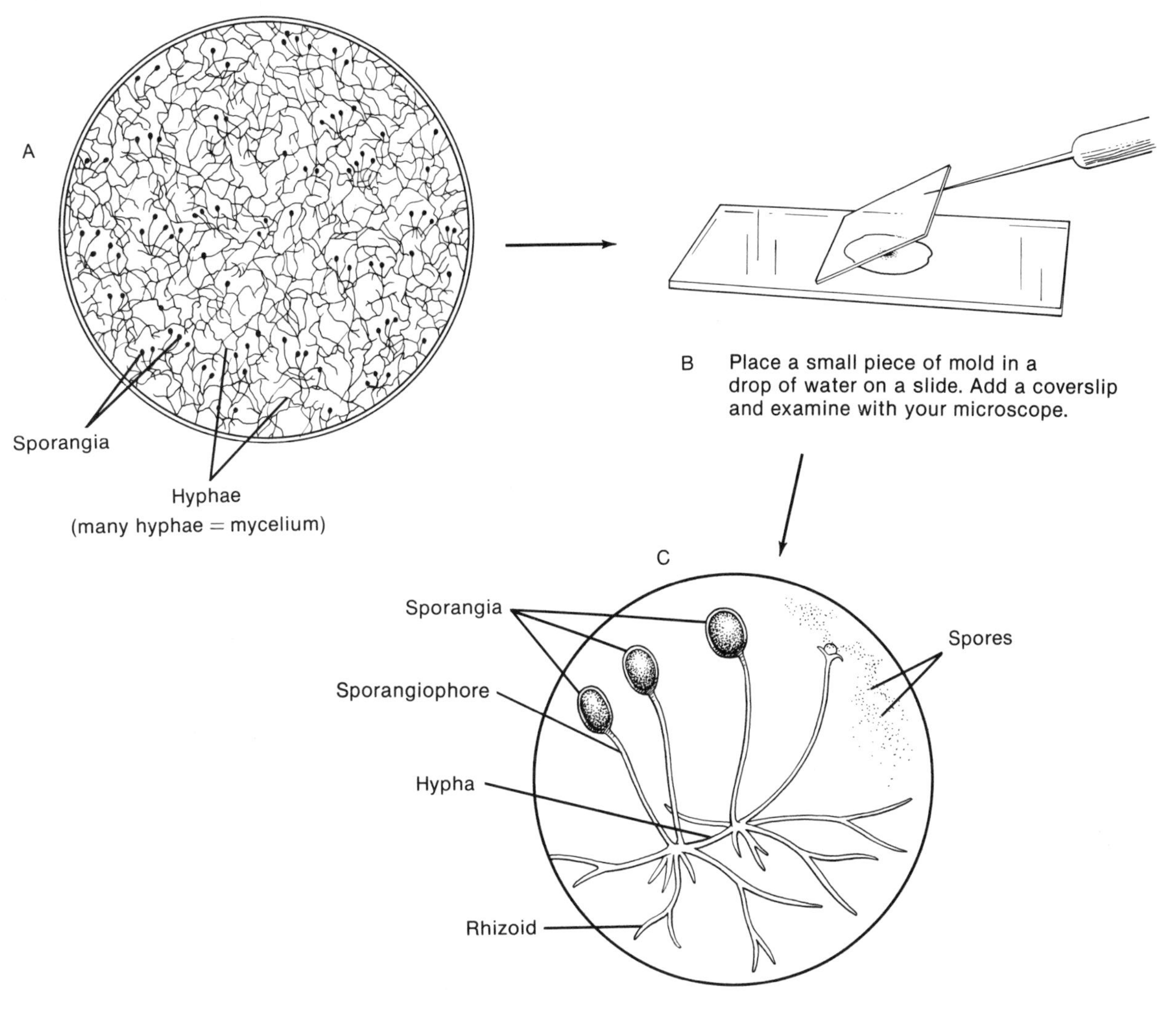

FIG. 28-1
Structure of the black bread mold *(Rhizopus stolonifer)*.

points of contact, gametes are formed. Fusion of the gametes (fertilization) has resulted in the formation of a line of black, thick-walled **zygospores** (Fig. 28-2B, C) across the culture. Keeping the dish closed, locate the zygospores with your dissecting microscope. Would you consider each strain to have a different sex? Explain.

____

B. DIVISION ASCOMYCOTA (SAC FUNGI)

The members of this group produce spores in a sac-like structure called an **ascus.** These spores, usually eight in number, are called **ascospores.**

1. Yeasts

Yeasts are among the simpler ascomycetes. Mount some living yeast cells on a slide and examine microscopically. Look for nuclei and small glistening food granules. Note that some of the yeast cells may exhibit small, rounded projections called **buds** (Fig. 28-3). What do these buds represent?

____

____

If none of the yeast cells are budding, examine a demonstration slide that shows this process.

Yeasts are classified as ascomycetes because at some period in their life cycle an ascus is formed.

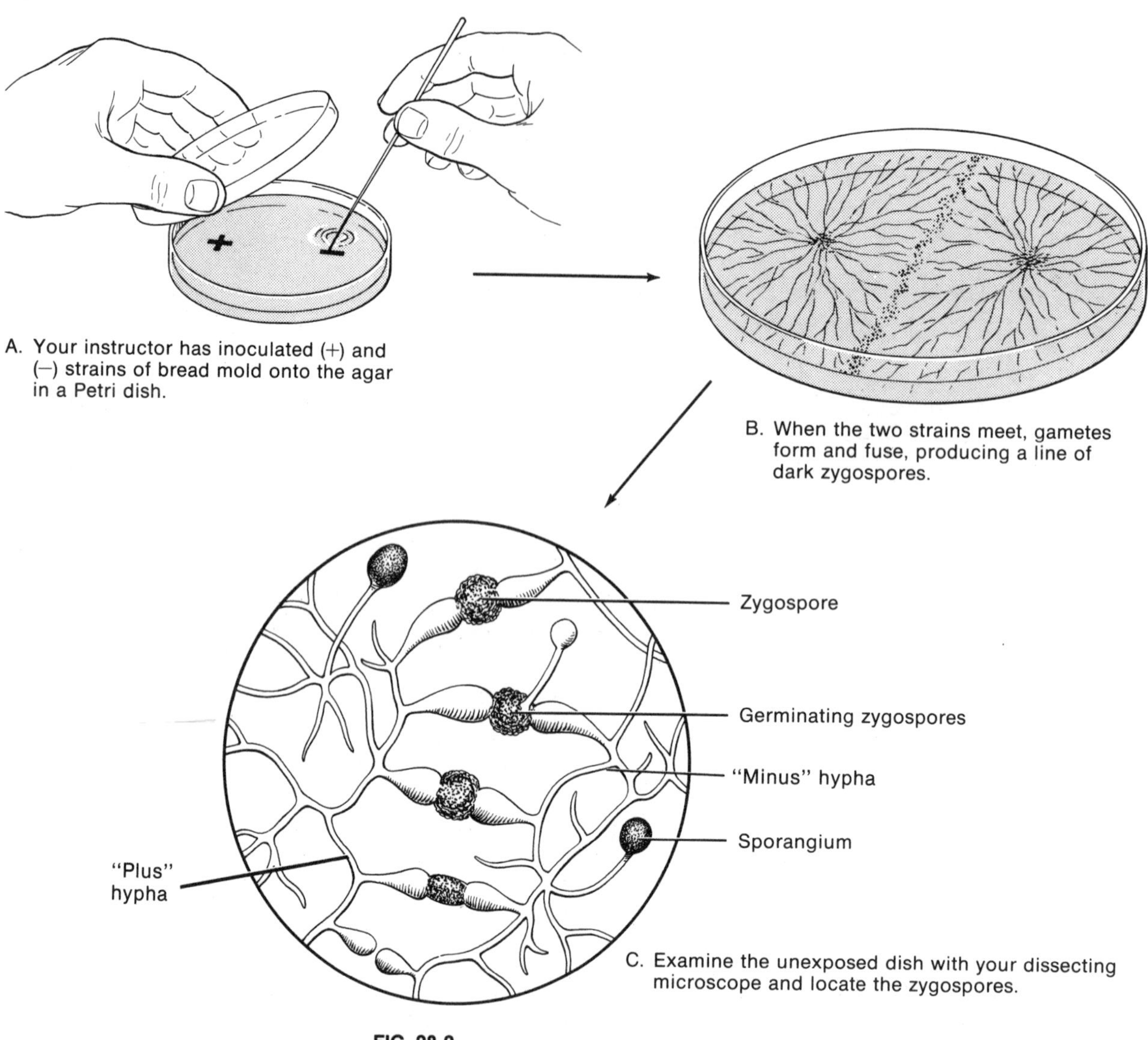

FIG. 28-2
Sexual reproduction in the black bread mold.

Examine a demonstration slide that shows the ascus of yeast and its contents. List several ways in which yeasts are economically important to human beings.

____

____

____

____

____

____

____

____

____

2. Powdery Mildew (Microsphaera)

Examine lilac plant leaves infected with this fungus. Why is this organism called *powdery mildew?*

The mycelium that grows over the surface of the leaf occasionally penetrates the epidermal cells (Fig. 28-4). These hyphae are called **haustoria** (singular haustorium). What is their function?

____

Examine demonstration slides that show haustorial

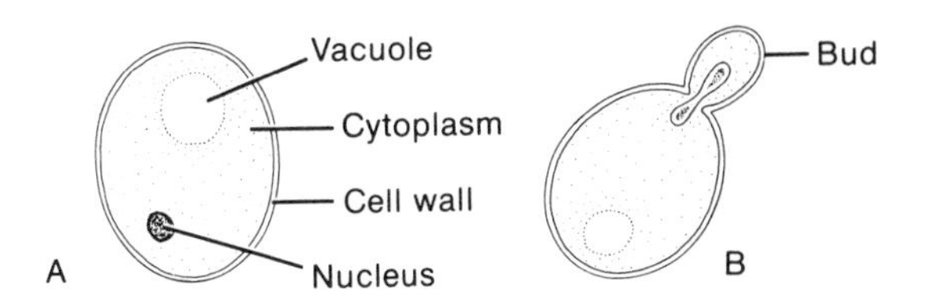

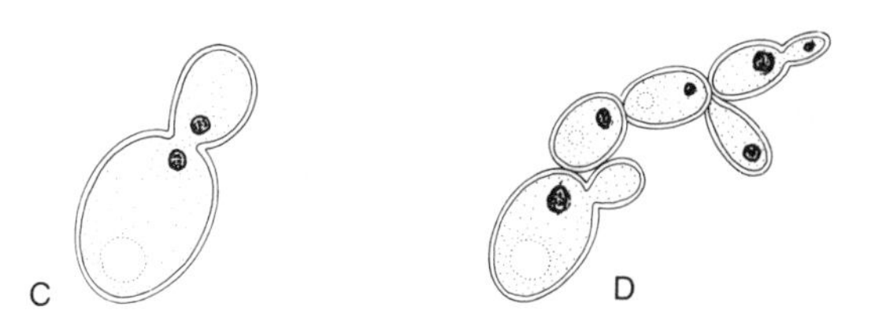

FIG. 28-3
Budding in yeast.

penetration of the epidermal cells of the host plant. What kind of nutrition does this fungus exhibit?

During late spring, large numbers of asexual spores **(conidia)** are produced and disseminated by the wind to spread the infection to the same or other lilac plants. Examine slides of powdery mildew, and locate the conidia.

Toward the end of summer, large numbers of spherical fruiting bodies are formed as a result of sexual reproduction. Examine an infected leaf with a dissecting microscope, and locate some of these fruiting bodies. Scrape the surface of the leaf, and mount the material in a drop of water on a slide. Examine microscopically. Locate a fruiting body, and note the elaborate appendages. How might these appendages be useful to the fungus?

While examining the fruiting body, gently apply pressure to the coverslip. Note the saclike asci that are extruded. Locate an ascus that contains ascospores. How many spores does each ascus contain?

What becomes of the ascospore after it is released?

C. DIVISION DEUTEROMYCOTA (FUNGI IMPERFECTI)

Much spoilage of food, leather, and cloth occurs as a result of the fungus *Penicillium*. Examine a living culture of *Penicillium*. Why is it sometimes referred to as a *blue-green mold?*

Examine prepared slides showing the special hyphal branches that produce asexual spores or conidia (Fig. 28-5). How does *Penicillium* differ from *Rhizopus* in the way in which spores are produced?

How is it similar to *Microsphaera?*

One species, *Penicillium notatum,* was found to produce a potent antibiotic, penicillin. It has proven to be one of the most effective antibiotics for combating bacterial infections.

D. DIVISION BASIDIOMYCOTA (BASIDIOMYCETES OR CLUB FUNGI)

The basidiomycetes represent another large and varied group of fungi that have both saprophytic and parasitic members. They are characterized by having a **basidium** (a club-shaped structure on which are usually produced four spores—**basidiospores**) formed at some point in the life cycle.

1. Mushrooms

Mushrooms are characterized by plates or gills on

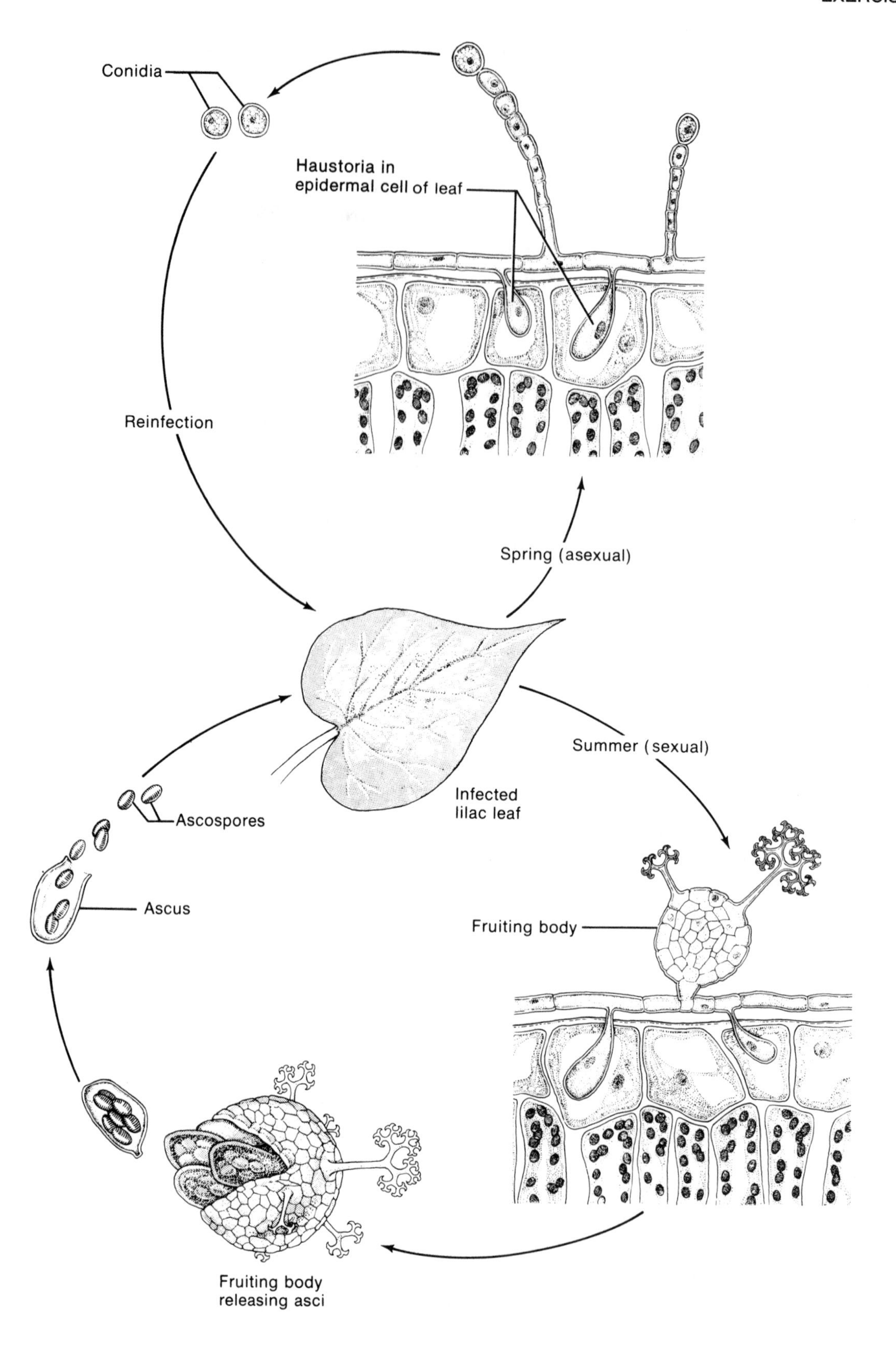

FIG. 28-4
Life cycle of a powdery mildew.

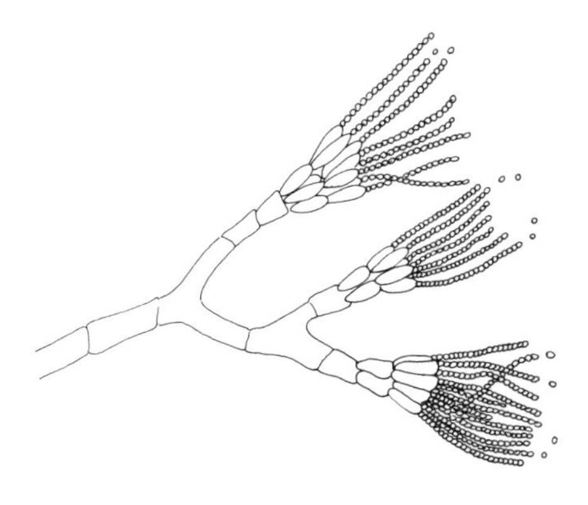

FIG. 28-5
Conidia of *Penicillium.*

the undersurface of a fruiting body. The basidia extend from the free surface of the gills (Fig. 28-6).

The mushroom mycelia grow beneath the surface of their substrate. When haploid mycelia of different strains unite, another mycelium is formed that is diploid. This mycelium may produce a fruiting body, commonly called a *mushroom,* which consists of a stalk and cap.

Examine the fruiting body of *Agaricus campestris,* the common commercial mushroom. Note the stalk and cap. Examine the undersurface of the cap and locate the gills. If the mushroom is young, the gills may be covered by a thin membrane that extends from the stalk to the outer margin of the cap.

Mount one of the gills in a drop of water on a slide and microscopically examine the edges of the gills. Locate basidia and basidiospores. If available, examine a prepared slide of *Coprinus,* showing a cross section through the cap. Identify gills, basidia, and basidiospores.

2. Bracket Fungi

These fungi are parasitic or saprophytic on various trees. The mycelium of the fungus may grow within the trunk several years before forming the character-

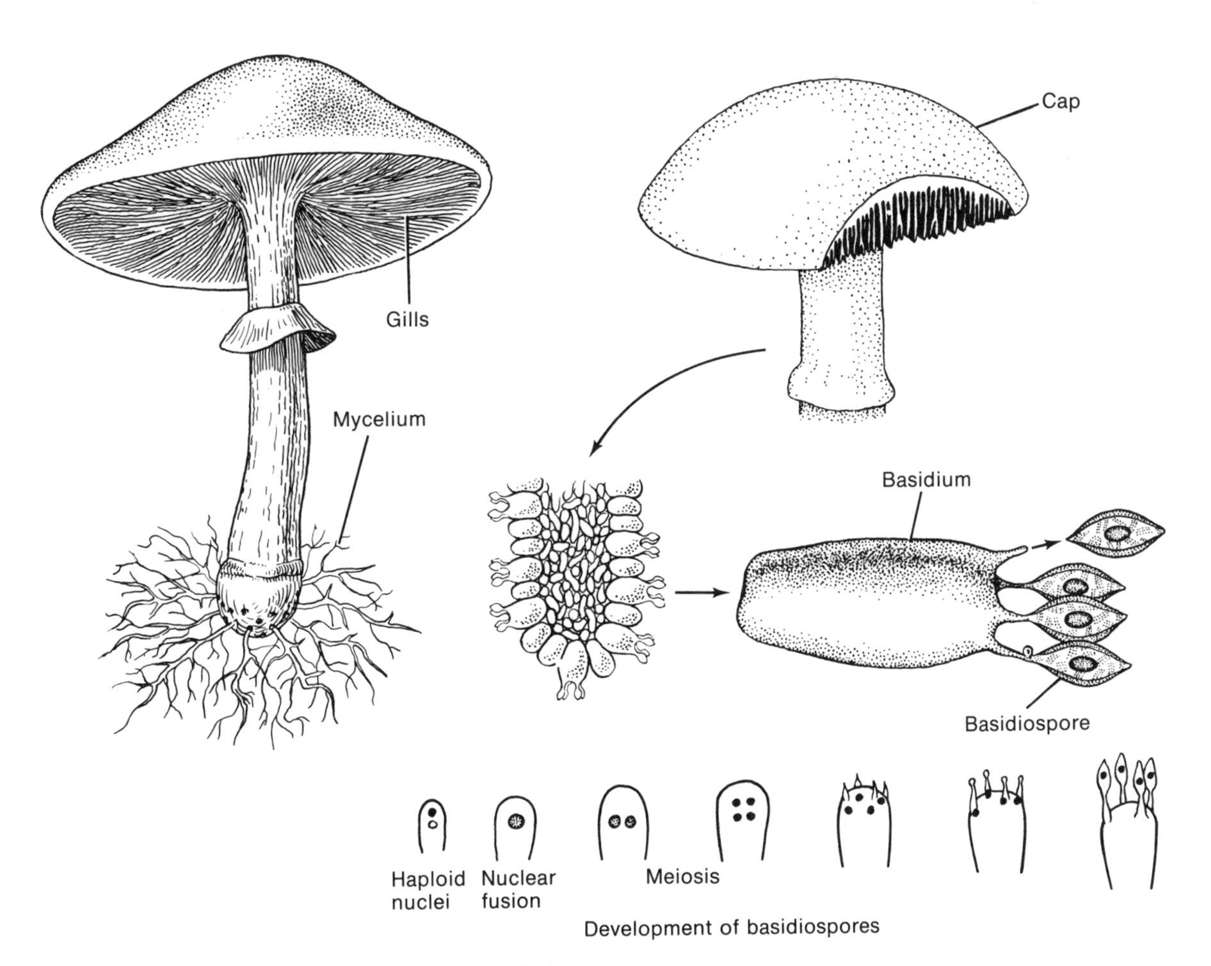

FIG. 28-6
Details in the life of a mushroom.

istic woody fruiting body on the outside of the tree. Examine several bracket fungi, and note the several growth layers that are evident. How could these growth layers be used to estimate the age of a bracket fungus?

____

____

____

____

Examine the undersurface of one of the fruiting bodies with a dissecting microscope. How does the undersurface of a bracket fungus differ from a mushroom?

____

____

Examine a prepared slide showing a cross section of the fruiting body of a bracket fungus. What do the circular openings in the cross section represent?

____

____

Locate basidia and basidiospores.

3. Rusts

The rusts are a group of parasitic basidiomycetes that infect almost every living species of seed plants as well as some of the ferns. All species of rust produce at least two distinctly different types of spores; some have three, four, or five different kinds.

The most widely known rust, and one that causes perhaps the greatest economic loss, is the stem rust of wheat, *Puccinia graminis.* Infection by this parasite affects the wheat plant in several ways. First, many of the host's cells are killed as the fungus uses the content of the cells for its own growth. Second, the fungus is robbing the host of food that otherwise would be used for growth in parts of the plant not infected by the fungus. Third, due to the killing of cells that may contain chloroplasts, photosynthetic processes are greatly reduced. Thus, as a result of infection by *Puccinia,* the wheat plant becomes pale green, stunted in overall growth, and ripens prematurely, with its small, shrunken, kernals having very limited food reserves.

a. Phases of the Rust in the Wheat Plant

1. **Urediospores.** Examine stems and leaves of wheat showing reddish patches (hence the name *rust*) on the surface. These masses (called **uredia;** singular, **uredium**) contain large numbers of reddish orange, binucleate **urediospores** (Fig. 28-7A). These are the first of several different types of spores produced and appear in late spring. They continue to be produced throughout the growing season until the plant is mature. Urediospores are transported by various means (but primarily by the wind) to other wheat plants where the spore germinates and sends out a hyphal branch that enters a stoma and penetrates the intercellular spaces of the leaf. The mycelium becomes highly branched. Haustoria penetrate and absorb food from the host cells. After a period of growth, the hyphal mass forms a layer of cells just beneath the epidermis, which then develops into the stalked urediospores, which are produced in such abundance that the epidermis ruptures.

With a razor blade, scrape a small amount of the uredia into a drop of water on a slide, and add a coverslip. Locate and examine some urediospores. Next examine a slide of a cross section through a uredium. Locate urediospores, mycelia, and haustoria.

2. **Teliospores.** As the wheat plant approaches maturity, the mycelium, which until this time was producing urediospores, now begins to produce **teliospores** (Fig. 28-7B) located in masses called **telia.** These spores have thicker walls than urediospores and generally do not germinate until the following spring.

Scrape a small amount of material from a wheat stem or leaf containing telia into a drop of water on a slide. Add a coverslip and examine microscopically. Locate and examine some teliospores. How many cells make up each teliospore?

____

Next examine a slide of a cross section through a telium. Locate immature teliospores, which when first formed are binucleate, and mature spores, in which the nuclei have fused and the spores are uninucleate.

b. Phases of the Rust in the Barberry (Alternate Host)

In the spring, the nuclei of cells making up the teliospore undergo meiosis so that each cell contains four haploid nuclei. Each cell then forms a short myce-

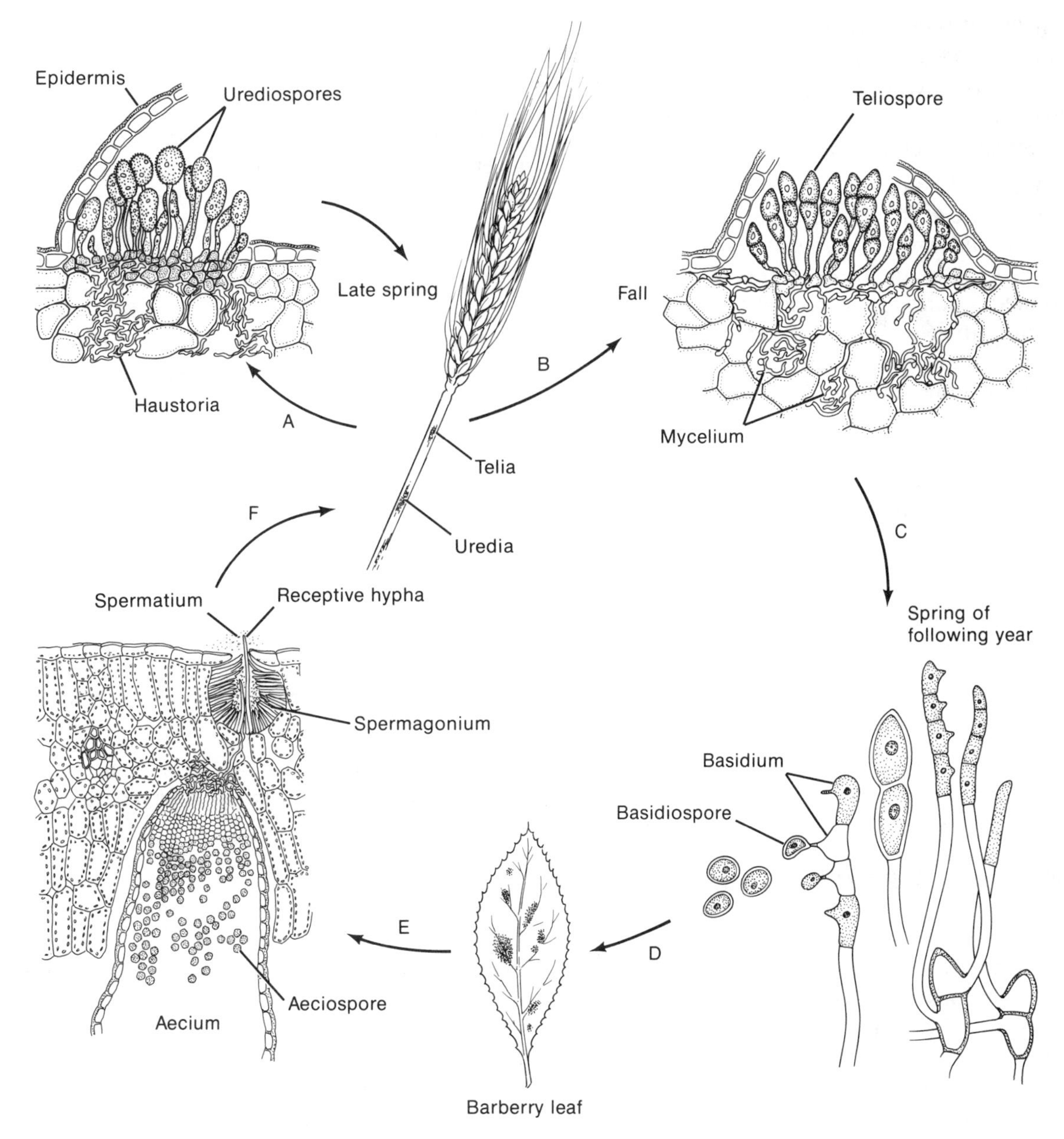

FIG. 28-7
Life cycle of wheat rust. (Adapted from *A Textbook of General Botany,* 5th ed, by G. M. Smith et al., Macmillan, 1956.)

lium called a **basidium.** The nuclei migrate into the basidium, which then is partitioned into four cells, each of which contains a haploid nucleus. Each cell of the basidium forms a small projection, into which will pass the nucleus. The swollen end of this projection becomes a **basidiospore.**

The wheat rust requires an alternate host to complete its life cycle. Basidiospores cannot reinfect wheat. They can only infect the alternate host—the common barberry (Fig. 28-7C, D).

When it lands on a fruit, branch, or leaf of the barberry, the basidiospore germinates and forms a hyphal branch that penetrates the tissue. In the tissue, the mycelium proliferates extensively. Ultimately, two types of structures are formed: **spermagonia** (singular, *spermagonium*) and **aecia** (singular, *aecium*)—see Fig. 28-7E. These are best observed by examining cross sections through them.

1. **Spermagonium.** The spermagonium is a flask-shaped structure just below the epidermis. It contains numerous hairlike hyphae at the ends of which are sporelike cells called **spermatia** (singular, *spermatium*).

Spermatia may be carried by insects to **receptive hyphae** protruding from other spermagonia where the nucleus from the spermatium enters the receptive hyphae. This initiates the development of another type of mycelium that ultimately produces cuplike structures called *aecia* that protrude from the lower surface of the barberry leaf (Fig. 28-7E).

2. **Acium.** Masses of **aeciospores** are formed within the aecia. If these spores are carried by air currents to the leaves or stems of wheat, they germinate and form an intercellular mycelium that then forms urediospores to complete the life cycle.

Examine infected leaves of barberry and locate spermagonia and aecia. Then examine slides showing cross sections through these structures. Locate spermatia, receptive hyphae, and aeciospores.

How do you control the spread of a fungus, such as wheat rust, whose infective spores can be carried by air currents to infect plants thousands of miles away?

___

___

___

___

E. DIVISION MYCOPHYCOPHYTA (LICHENS)

Lichens are called *pioneer plants* because they often grow on bare rock and are the first "plants" to cover burnt out regions. Lichens, however, are not plants. They are a symbiotic partnership between a fungus (usually an ascomycete) and a green alga or cyanobacterium. Because of their ability to obtain their nutrition from their photosynthetic partner, lichens have invaded some of the harshest environments in the world. They occur from arid desert regions to the arctic and grow on bare soil, tree trunks, rocks, and even underwater.

1. Vegetative Features

The plant body, or thallus, of lichens is quite distinctive in appearance, and indeed lichens are classified largely according to their form of growth. The following are descriptions of three main types of thalli.

1. *Crustose:* grows closely appressed to the substrate with only the upper surface visible

2. *Foliose:* more loosely attached to substrate with both upper and lower surfaces visible

3. *Fruticose:* attached at one point to substrate and grows erect or pendant

Examine various lichens both with the eye and with a dissecting microscope. Using the descriptions given above, determine which type of thallus each lichen has. Record your observations in Table 28-1.

2. Microscopic Anatomy

With a sharp razor blade, cut a thin cross section through a foliose or crustose lichen. Alternately, use commercially prepared slides. Examine microscopically, and, referring to Fig. 28-8, locate the following:

1. *Upper cortex:* a protective, dense aggregation of fungal hyphae.

2. *Algal layer:* a layer of algal cells and loosely interwoven masses of thin-walled fungal hyphae.

3. *Medulla:* a somewhat thick layer of loosely interwoven, colorless, hyphae. This layer comprises about two-thirds of the thallus and is believed to serve as a storage region.

4. *Lower cortex:* a thinner layer than the upper cortex with projections, called **rhizines,** that attach the lichen to its substrate.

3. Asexual Reproduction

Examine various specimens with a dissecting microscope and look for small, granular masses on the surface. These are **soredia,** which contain one or more algal cells surrounded by the fungal hyphae. Soredia are dispersed by wind or rain, and each is capable of growing into a new lichen thallus.

Scrape a few soredia into a drop of potassium hydroxide (or other wetting agent) and examine

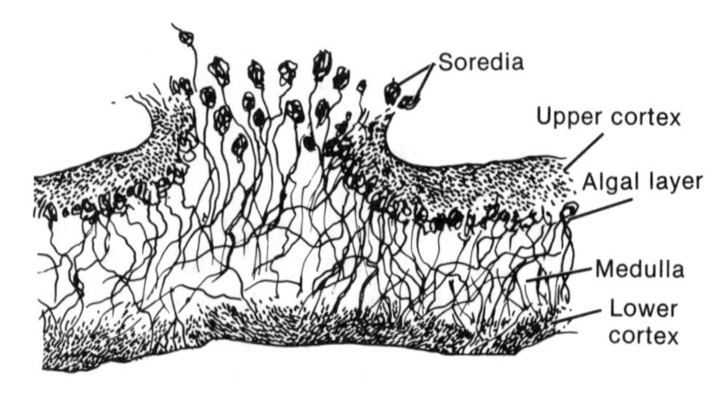

FIG. 28-8
Microscopic structure of lichen.

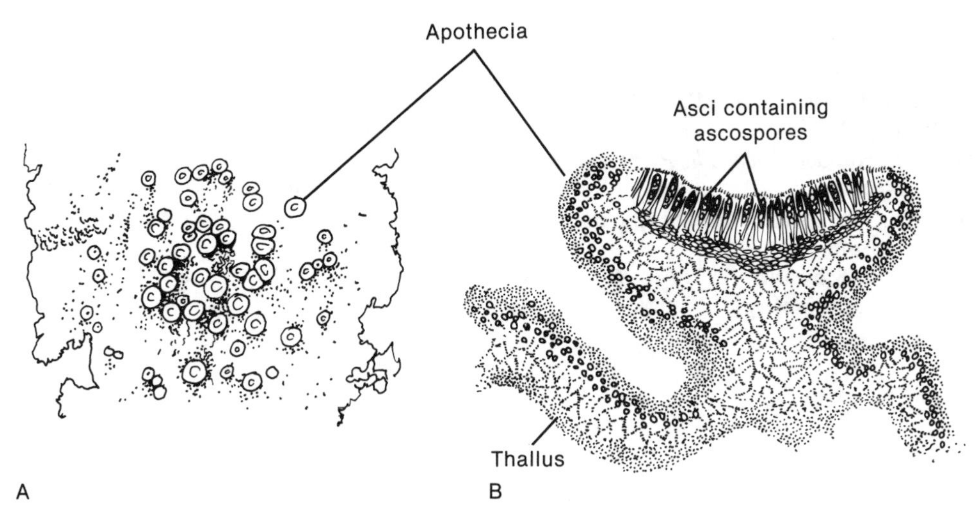

FIG. 28-9
(A) Apothecia on surface of lichen. (B) Cross section through apothecium showing asci containing ascospores.

with your compound microscope. Supplement your observations using commercially prepared slides through a lichen thallus (Fig. 28-8).

4. Sexual Reproduction

Select a lichen that has small, cup-shaped structures on its surface, called **apothecia** (Fig. 28-9A). Cut a thin cross section through an apothecium (or examine a prepared slide) and examine microscopically. Locate elongated asci (Fig. 28-9B). How many spores are found in each ascus?

____

How are they arranged?

____

Which of the lichen symbioants is reproducing sexually?

____

____

What is the reason for your answer?

____

____

____

____

____

TABLE 28-1
Classification of vegetative growth of lichen.

Name	Type of thallus

REFERENCES

Alexopoulos, C. J., and C. W. Mims., 1979. *Introductory Mycology.* 3d ed. Wiley.

Bold, H. C., and C. L. Hundell, 1984. *The Plant Kingdom.* 4th ed. Prentice-Hall.

Delevoryas, T. 1977. *Plant Diversification.* 2d ed. Holt, Rinehart and Winston.

Margolis, L., and K. V. Schwartz. 1982. *Five Kingdoms.* W. H. Freeman and Company.

Moose-Landecher, E. 1982. *Fundamentals of the Fungi.* 2d ed. Prentice-Hall.

Pritchard, H. N., and P. T. Bradt. 1984. *Biology of Nonvascular Plants.* Times Mirror/Mosby.

Richardson, D. H. S. 1975. *The Vanishing Lichens: Their History, Biology and Importance.* Hafner.

Webster, J. 1970. *Introduction to Fungi.* Cambridge University Press.

Ray, P. M., T. A. Steeves, and S. A. Fultz. 1983 *Botany.* Saunders.

EXERCISE 29

Kingdom Plantae: Division Bryophyta

Bryophytes are inconspicuous plants that grow in moist habitats. They are not fully adapted to the land because their sperm must swim through water to fertilize the egg. In addition, because they generally lack vascular tissue, they rely on the surrounding water to transport fluids and salts necessary for growth.

The three classes of bryophytes have the following characteristics:

Class Hepaticae (liverworts): Plant body (gametophyte) green and dorsoventrally flattened and bilaterally symmetrical, reproductive structures multicellular, water necessary for fertilization.

Class Musci (mosses): Plant body (gametophyte) erect and radially symmetrical; has stem and leaflike structures. Reproductive structures multicellular and require presence of water for fertilization. Lacks true vascular tissues.

Class Anthocerotae (hornworts): Plant body (gametophyte) is multilobed thallus growing closely appressed to rocks and soil. Gametophytes may contain nitrogen-fixing blue-green algae. Reproductive structures are multicellular. Spores produced continuously over several months.

In this exercise, several representatives of the liverworts and mosses will be studied, with special emphasis on form, habitat, and reproductive processes. In the algae and fungi the plant bodies are reproduced by fragmentation, by production of asexual structures (spores), or by the union of isogamous or anisogamous gamates. The reproductive structure is single-celled in each case. In the liverworts and mosses, the gametes are produced in multicellular sex organs—the **archegonia** (female) and **antheridia** (male).

A. CLASS HEPATICAE (LIVERWORTS)

1. The Plant Body

The liverworts were given their rather curious name because the lobing of the plant body in some species resembled the lobes of the liver. The term *wort* meant "plant," particularly a herbaceous (nonwoody) plant. Examine living gametophyte plants of *Marchantia.* Note the characteristic Y-shaped (**dichotomous**) branching of the thallus. Meristematic tissue found in the notch of the Y produces the peculiar growth pattern seen here. Is the plant body diploid or haploid with respect to the chromosome number in the cells?

Remove a small piece of the thallus, and examine the ventral (bottom) surface with a dissecting microscope. At the edges, locate slender, hairlike structures, the rhizoids. What is the function of the rhizoids?

Associated with the rhizoids are long, flattened structures, the scales. Locate these structures. Examine the dorsal surface. Observe the minute "pores" in the center of small, diamond-shaped areas. Because *Marchantia* is photosynthetic, what role might be ascribed to these openings?

What structures in the higher plants appear to be morphologically and functionally similar to these openings?

Examine prepared slides of a cross section of the thallus of *Marchantia* (Fig. 29-1). Locate the upper epidermis and the "pores" that were seen in the living material. Below the epidermis, find a series of small air chambers that are partitioned by branching filaments arising from the floor of the chambers. In living material, the filaments contain chloroplasts. Suggest a function for the air chambers.

The tissue underlying the air chambers is several cells thick. Many of the cells contain leucoplasts. A

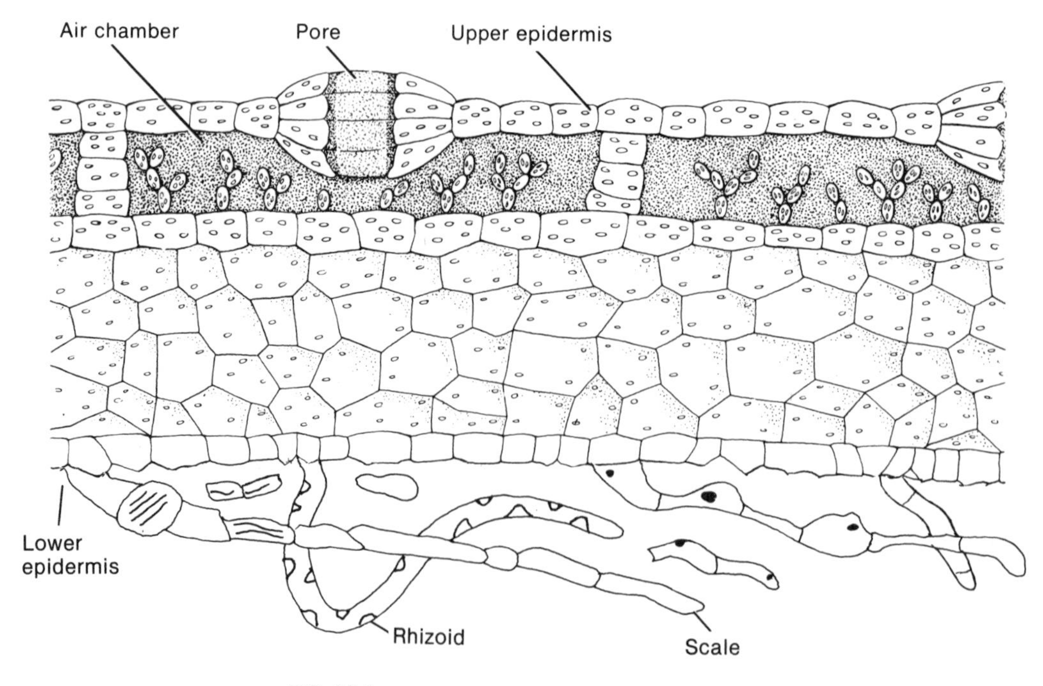

FIG. 29-1
Diagrammatic cross section of a *Marchantia* thallus.

few large cells contain mucilage, a gelantinous substance that tends to absorb water. What is the advantage of having this substance in the cells?

On the lower epidermis, locate the rhizoids and scales. How do scales differ morphologically from rhizoids?

2. Asexual Reproduction

The gametophyte of *Marchantia* reproduces asexually by means of small "plantlets" **(gemmae),** produced from cells located in the bottom of gemma cups. These are found on the dorsal surface of the thallus (Fig. 29-2). Examine a demonstration of the gemma cups using a dissecting microscope. Describe the appearance of any gemmae present in the cups.

If none are present, how could you account for their absence?

Are gemmae haploid or diploid?

3. Sexual Reproduction

The reproductive (sex) organs of *Marchantia* are borne on special upright branches (Fig. 29-2). Each branch is composed of a stalk and terminal disc. Many species of *Marchantia* are **dioecious;** that is, there are separate male and female gametophytes and therefore two kinds of reproductive branches. Examine a male gametophyte, which bears an **antheridial branch.** What is the shape of the terminal disc?

Compare the antheridial branch with the **archegonial branch** of the female gametophyte. Describe the female reproductive branch.

What is the chromosomal condition (haploid or diploid) of the antheridial and archegonial branches?

Examine slides showing cross sections of the antheridial branch. Locate the antheridia just below the upper surface of the disc (Fig. 29-2). Note that the antheridium is contained within a chamber that opens to the surface of the disc through a pore. Examine an antheridium closely, and note that it is attached to the base of the chamber by a stalk. The body of the antheridium consists of an outer layer of cells, the **jacket.** Within the jacket are found numerous small cells, which will develop into male gametes called *sperm.* If available, examine slides of *Marchantia* sperm.

Obtain a slide showing a cross section of an archegonial disc, on the ventral surface of which are located the archegonia. Locate an **archegonium** that shows the following parts: an elongated neck, an enlarged venter (containing the egg), and the stalk.

What type of division (mitosis or meiosis) is involved in the formation of the sperm and egg? Explain your answer.

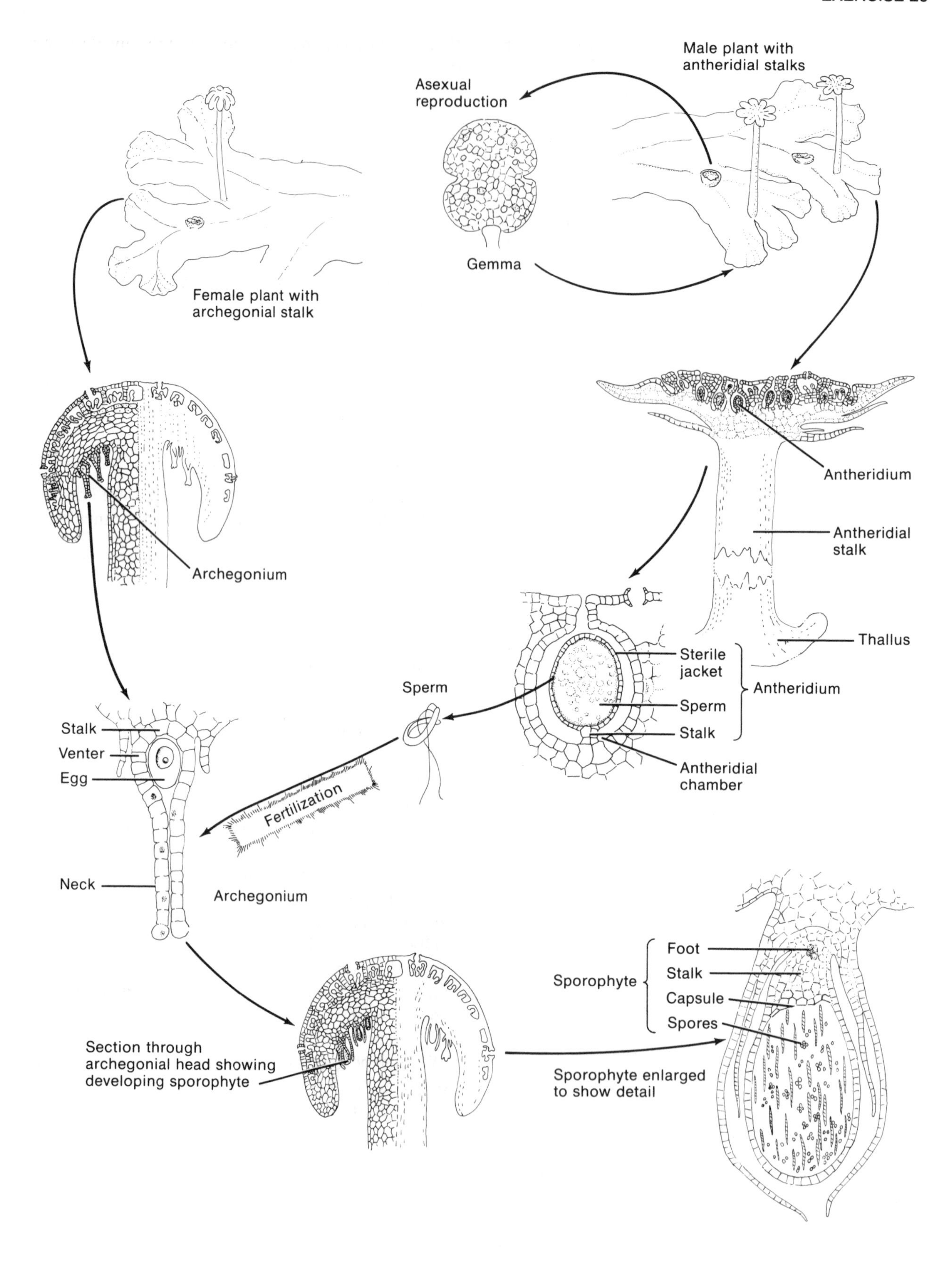

FIG. 29-2
Life cycle of *Marchantia*.

4. Fertilization

When the antheridia are mature, the sperm are released and fertilize the egg. The zygote (Fig. 29-2), which remains within the **venter,** undergoes cell division and develops into a multicellular embryo (young **sporophyte**). The embryo differentiates into three distinct parts: the foot, stalk, and capsule. What is the chromosomal condition of the sporophyte?

____

Examine a prepared slide of a developing sporophyte. Locate the foot embedded in gametophytic tissue. What are some functions of the foot?

____

____

Note the large, terminal capsule within which are found spores. By what type of division were the spores formed?

____

What is the chromosomal conditions of these spores?

____

____

Into what structure will the spores develop?

____

Among the spores, locate elongated cells with spirally thickened walls. These **elaters** help disperse the spores when the capsule breaks open. Note that the sporophyte matured within the archegonium. Before the release of the spores, the capsule is pushed free of the gametophyte by the elongation of the stalk.

Label the various structures in Fig. 29-2 with an N or $2N$ to indicate whether they are haploid or diploid.

B. CLASS MUSCI (MOSSES)

Mosses differ from the liverworts in that the gametophyte of the moss begins as a filamentous, branching structure—a **protonema**—and is differentiated into "stems" and "leaves." In addition, the capsule of the sporophyte contains a central, sterile region of cells —the **columella**—which greatly reduces the amount of **sporogenous** (spore-producing) tissue. The capsule contains a highly complex series of "teeth," which regulate the dispersal of the spores.

1. Sporophyte of *Polytrichum*

Examine living specimens of the common moss, *Polytrichum,* or, if this is not available, study preserved specimens that consist of both the gametophytic and sporophytic generations. The sporophyte is readily distinguishable, because it consists of a terminal capsule, (often covered by a hairy cap, the **calyptra**), a slender stalk (the **seta**), and a foot that is embedded in the top of the "leafy" gametophyte (Fig. 29-3). Carefully separate the sporophyte from the gametophyte. Remove the calyptra and save for later examination. Note that the capsule has a lid. Gently remove the lid, and examine the exposed surface of the capsule with a hand lens. Describe what you see.

____

____

Suggest a function for these structures.

____

____

Crush the capsule in a drop of water on a slide. Examine microscopically. If the sporophyte is mature, numerous spores will be present. These spores represent the first cells of the gametophyte generation. What is the chromosomal condition of these spores?

____

What type of division, meiosis or mitosis, is involved in the development of these spores?

____

2. Gametophyte

After release from the capsule, if environmental conditions are suitable, the spore soon germinates and grows into the filamentous structure of the pro-

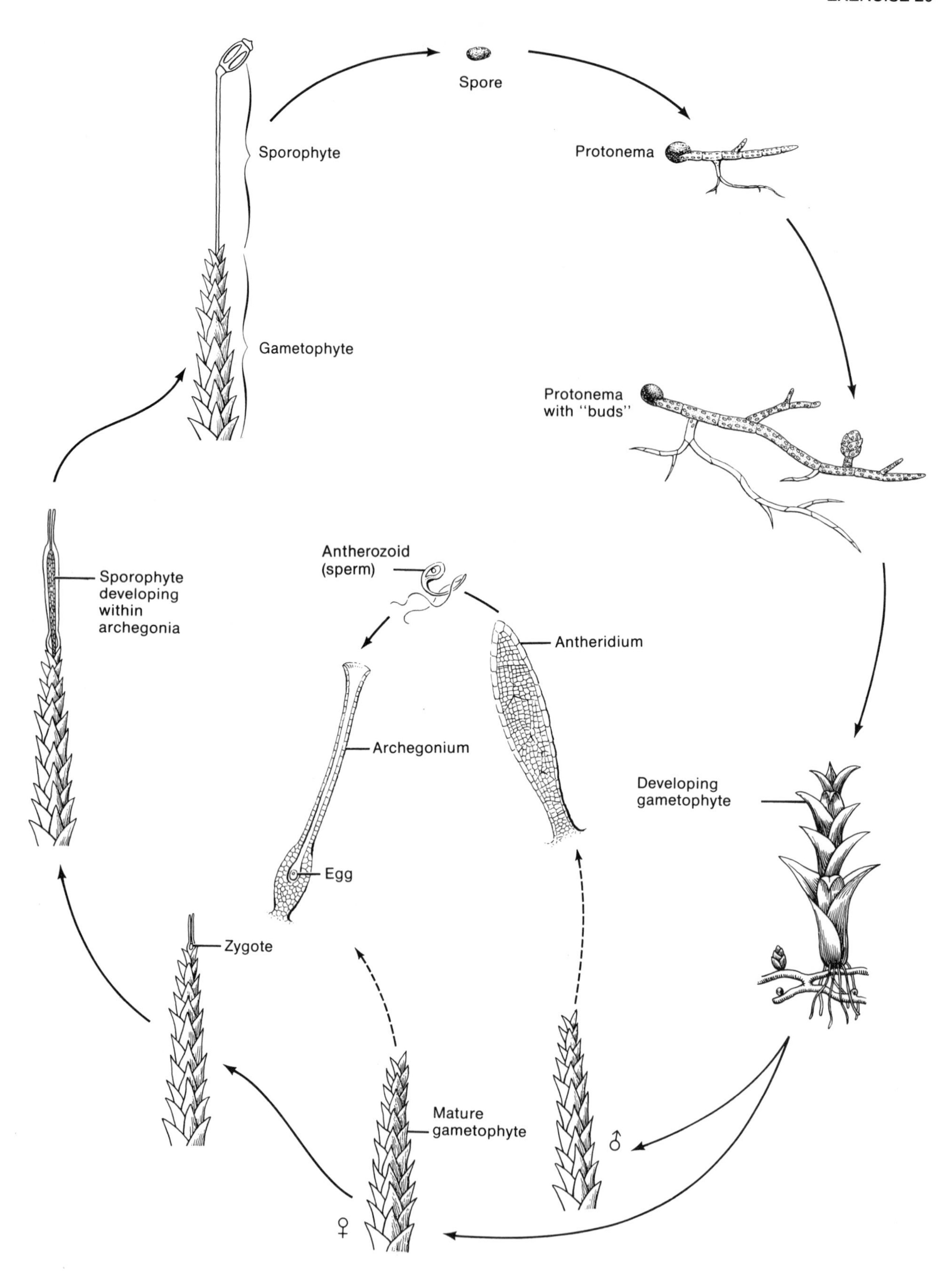

FIG. 29-3
Life cycle of *Polytrichum*.

tonema. Note the similarity of the protonema to filamentous algae. What evolutionary significance, if any, is indicated here?

Look for small, budlike structures along the length of the filament. These will later develop into mature, "leafy" gametophytes (Fig. 29-3).

Examine the leafy part of *Polytrichum* that remains after removal of the sporophyte. The gametophyte is photosynthetic. The mature gametophyte consists of "leaves," a "stem," and rhizoids. How could you determine if these are true stems and leaves?

Polytrichum is a dioecious moss, with the reproductive organs located at the tips of separate male and female plants. With forceps, carefully remove some of the leaves from the apex of a female plant. After removing as many leaves as possible, cut off the stem and mount in a drop of water on a slide. With your probe, gently tease the tip into small fragments. Examine microscopically. Locate the archegonia (Fig. 29-3). Note the canal that leads through the neck and terminates in the venter of the archegonium. Where is the egg located in the archegonium?

In a similar manner, mount a stem tip of a male plant. Locate the antheridia, which are the elongated, saclike structures consisting of an outer sterile jacket and an inner mass of cells destined to become **antherozoids** (sperm). If available, also examine prepared slides of moss antheridia and archegonia.

3. Fertilization

The nonmotile egg is retained within the venter of the archegonium. When the archegonium is mature, the cells lining the inside of the stalk disintegrate and a canal is opened leading to the female gamete. The male gamete swims through this canal and unites with the egg. What must be present for fertilization to occur?

The zygote undergoes a series of divisions and becomes a mass of undifferentiated cells, still held in the venter. The developing embryo differentiates into three parts: foot, seta, and capsule. The remains of the archegonium is retained as the calyptra and may be found enclosing the capsule. Is the calyptra sporophytic or gametophytic tissue?

Label the various structures in Fig. 29-3 with N or $2N$ to indicate whether they are haploid or diploid.

C. ADVANCES IN COMPLEXITY SHOWN BY THE BRYOPHYTES

Complete Table 29-1. Indicate the advances shown by the bryophytes over the algae with respect to morphology, habitat, and reproduction. The following suggested terms can be used:

Morphology: *unicellular, filamentous, spherical*

TABLE 29-1
Advances in complexity shown by the bryophytes over the algae.

Character	Algae	Liverworts	Mosses
Morphology			
Habitat			
Asexual reproduction			
Sexual reproduction			

colonies, prostrate thallus, erect thallus, radial symmetry, bilateral symmetry

Habitat: *fresh water, salt water, ditches, stagnant pools, damp woods, terrestrial*

Asexual reproduction: *fragmentation, cell division, zoospores, nonmotile spores, gemmae*

Sexual reproduction: *unicellular or multicellular reproductive structures, motile or nonmotile gametes, alternation of generations, isogametes or anisogametes, necessity of water for fertilization, dependency or nondependency of sporophyte, antheridia, archegonia*

REFERENCES

Bold, H. C., and C. L. Hundell. 1984. *The Plant Kingdom.* 4th ed. Prentice-Hall.

Delevoryas, T. 1977. *Plant Diversification.* 2d ed. Holt, Rinehart and Winston.

Margulis, L., and K. V. Schwartz. 1982. *Five Kingdoms.* W. H. Freeman and Company.

Raven, P. H., R. F. Evert, and H. Curtis. 1981. *Biology of Plants.* 3d ed. Worth.

Ray, P. M., T. A. Steeves, and S. A. Fultz. 1983. *Botany.* Saunders.

Watson, E. V. 1972. *Mosses.* Oxford University Press.

EXERCISE 30

Kingdom Plantae: The Vascular Plants

The members of this group represent a significant advancement in the plant kingdom. In these plants, we find for the first time a system of specialized tissues that carry water, dissolved minerals, and other organic products throughout the plant. They are called **vascular tissues** and the plants that have them are called **vascular plants,** or **tracheophytes.** These plants are able to transport great quantities of water and minerals over long distances in relatively short periods of time.

The vascular plants consist of several divisions and numerous classes. Only the following classes will be studied in this exercise:

Class Filicinae (ferns): gametophyte small, more or less free-living, and photosynthetic. Multicellular sex organs (gametangia). Mostly homosporous (i.e., having one type of spore) but may be heterosporous (i.e., having different types of spores; usually two).

Class Gymnospermae (conifers): seed plants having needlelike leaves. Ovules in cones and not enclosed. Sperm not flagellated.

Class Angiospermae (flowering plants): seed plants in which the flower is the reproductive structure containing the male (stamen) and female (pistil) parts. Pollination affected by wind and insects. Gametophytes highly reduced. Double fertilization (of egg and endosperm nuclei) is common feature. Seeds are typically enclosed in carpels. Mature seeds contained in fruits, which may be fleshy or hard.

In this exercise, you will become familiar with the structure and reproductive patterns of ferns, conifers (which include pines, spruces, and other evergreens), and the lily (a flowering plant).

A. CLASS FILICINAE (FERNS)

1. Sporophyte

Examine the mature sporophyte of *Polypodium.* Locate the horozontal stem, which may lie on, or just beneath, the surface of the soil. What name is given to this type of stem?

Note the large, deeply lobed leaves **(sporophylls)** of the plant (Fig. 30-1).On the undersurface of some of the leaves, locate small yellowish brown spots or **sori.** Each sorus contains many **sporangia.** What are produced in the sporangia?

Obtain part of a fern sporophyll that bears sori. Examine the sorus with a hand lens or dissecting microscope. Scrape the contents of a sorus into a drop of water on a slide. Examine under the low power of the microscope. Each sporangium is composed of a stalk and an enlarged capsule. Examine the capsule under high power. Note a ridge of cells **(annulus)** that extends around the capsule. These cells have thickened inner and radial walls. As the capsule matures, the annular cells lose water and the thin outer walls tend to shrink. This results in a considerable amount of tension on the lateral, thin-walled "cheek cells," which then rupture, releasing the spores.

Remove the coverslip from the slide, and blot the excess water with a piece of filter paper. Examine your preparation (without the coverslip), and observe the sporangia as the slide dries. Describe your observations.

2. Gametophyte

On germination, the fern spore develops into a short filament of cells, which differentiates into a platelike structure called a **prothallium.** The mature prothallium or gametophyte resembles a heart-shaped structure with a notch, containing the growing point, at one end. With a dissecting microscope, examine living fern gametophytes. Locate rhizoids on the ventral surface.

Among the rhizoids of mature gametophytes are found numerous antheridia (Fig. 30-1). Examine the gametophyte with a dissecting microscope. Locate archegonia on the ventral surface just posterior to the apical notch. Only the neck of the archegonium is visible. The venter is embedded within the gametophytic tissue. The sex organs are found on the ventral surface of the plant. Of what significance is this?

Examine prepared slides of a mature gametophyte, and locate the structures indicated above.

In the development of the new sporophyte, the zygote undergoes a series of divisions and develops into an embryo. The young embryo, still held in the venter of the archegonium, differentiates into four lobes. One lobe develops into the "foot." A second lobe grows down into the soil and becomes a primary root. A third lobe develops into the primary leaf, and the last lobe becomes the stem. Until the primary root and leaf become functional the young sporophyte is dependent on the gametophyte but as soon as the root begins to absorb water and the leaf photosyntheizes, the sporophyte becomes independent. Soon after this, secondary leaves and adventitious roots are formed and the primary root, leaf, and gametophyte die.

B. CLASS GYMNOSPERMAE (CONIFERS)

Conifers are characterized by producing a seed in the life cycle. In the conifers, the seeds are exposed (naked) on scalelike structures. Like the ferns, seed plants have a well-defined alternation of generations. In contrast to the ferns, the gametophyte is microscopic and completely dependent on the large free-living sporophyte, which has stems, leaves, and roots. Your study will be confined to only those conifers that have familiar representatives in the pines, spruces, cedars, firs, and other trees and shrubs.

1. Pine Sporophyte

Study specimens of pine branches, and note that two kinds of leaves are produced by pines. Locate needle-like, photosynthetic leaves borne on short spur shoots and inconspicuous scale leaves found at the bases of the spur shoots. How many needle leaves are borne on each spur shoot?

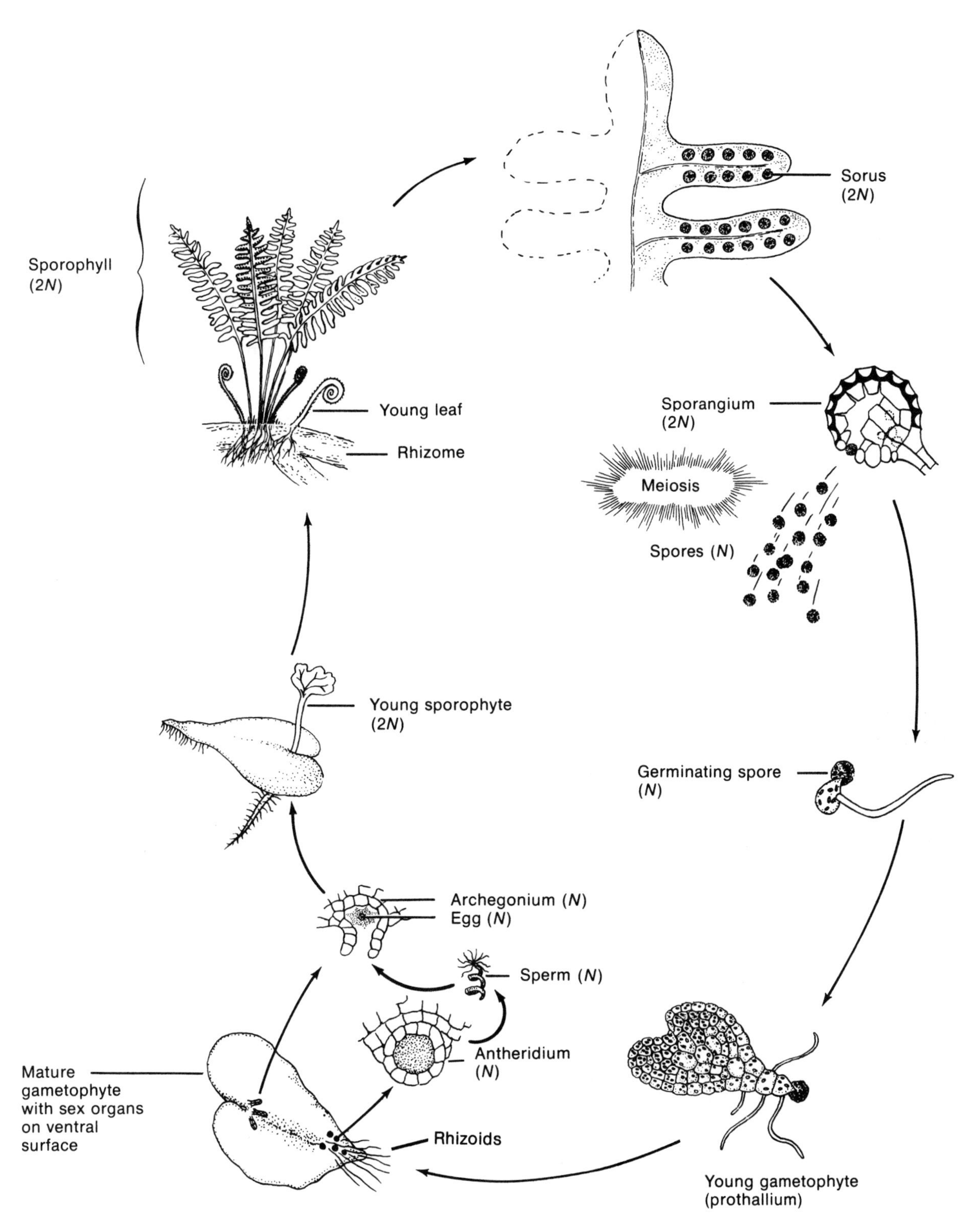

FIG. 30-1
Life cycle of a fern.

Examine branches of other species of pine. Describe any variation in the number of leaves in a cluster.

Would you consider this difference a means of separating one group of pine from another?

Why are pines called *evergreens*?

Examine cross sections of young and old pine stems (Fig. 30-2). Identify the epidermis, cortex, and vascular tissue, and note the numerous resin canals. What economically important products come from pines?

In the older stems, note the annual rings. How were these formed?

Examine cross sections of a pine needle leaf (Fig. 30-3). The leaves of pines, as those of numerous conifers, are uniquely suited for growth under dry conditions. The epidermis is covered with a thickened **cuticle** beneath which are one to several layers of thick-walled cells called the **hypodermis.** Stomates are sunken below the epidermis. In what way does the position of the stomates conserve water?

The leaf **mesophyll** consists of cells having peculiar ridges that project into the cells so that each cell looks like the piece of a puzzle. The mesophyll also

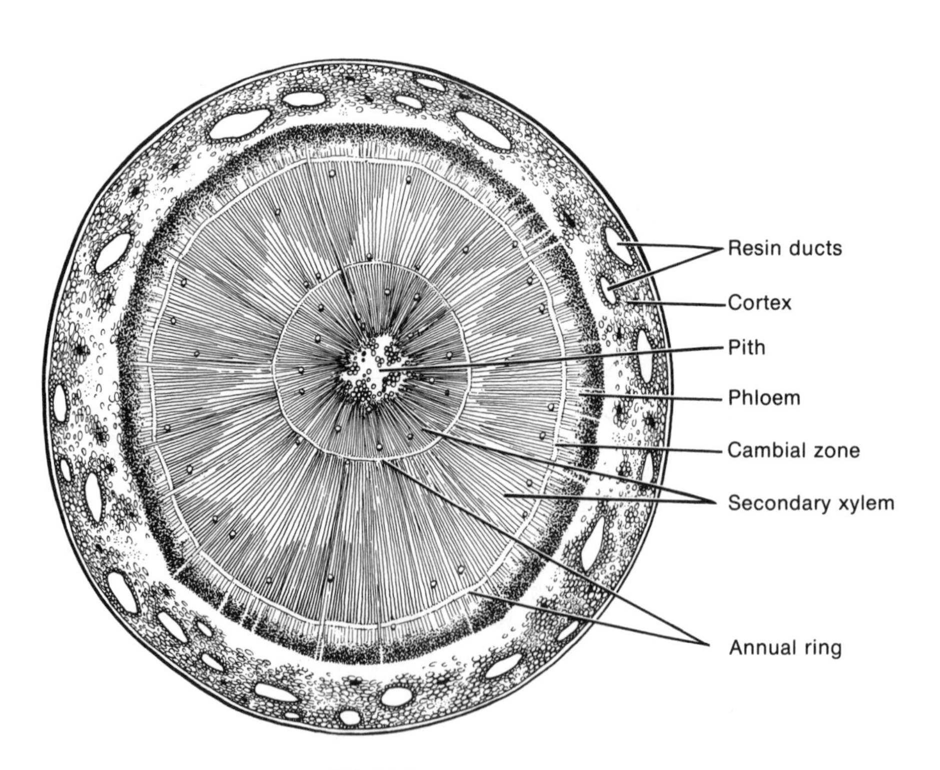

FIG. 30-2
Diagram of an older pine stem.

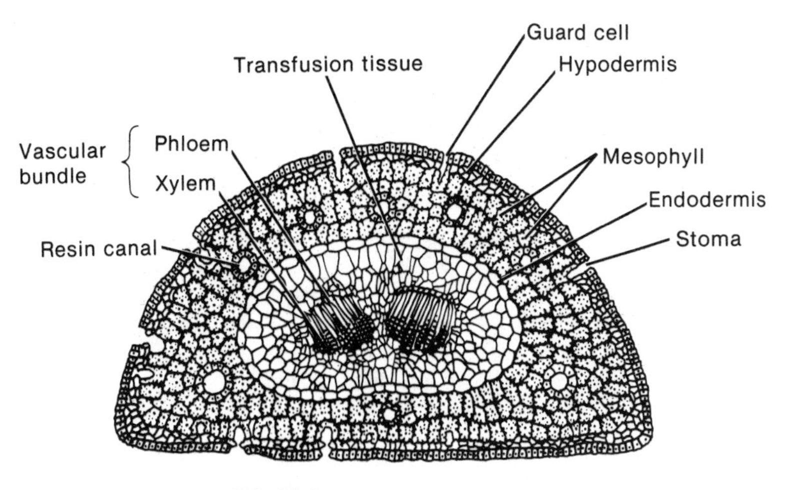

FIG. 30-3
Cross section of a pine needle leaf.

contains resin canals. Of what commercial importance are the substances found in these canals?

One or two vascular bundles (veins) are found in the center of the leaf. These are surrounded by a **transfusion tissue** that is thought to be involved in the conduction of materials between the mesophyll and vascular tissue. The transfusion tissue in turn is separated from the leaf mesophyll by the **endodermis.**

Examine cross sections of the leaves of other species of pines. Describe any variation in structure among the various needles.

Examine the slides of the various leaves with a dissecting microscope. What correlation exists between the number of leaves in a cluster and the shape of an individual needle leaf?

---

2. Pine Gametophyte

The pine tree produces two kinds of cones on the same tree. Each cone (**strobilus**) is an auxiliary shoot composed of a central axis that bears spirally arranged sporophylls. Examine branches bearing **staminate** or **pollen cones,** usually in clusters at the ends. Dissect one of the staminate cones and locate the central axis to which are attached scalelike structures called **microsporophylls** (Fig. 30-4). Remove a microsporophyll, and examine it with a dissecting microscope. Note the two saclike **microsporangia.** Break the microsporangia open, and examine the contents under the high power of your microscope. Depending on the stage of development, you will observe either young **microspores** (consisting of a single cell) or **microgametophytes** (consisting of two prominent cells and two degenerate cells). What is another name for microgametophyte of pine?

Note the lateral, winglike appendages. How might these "wings" be advantageous?

Examine prepared slides of pine pollen.

A second kind of cone found on the tree is the female **seed cone,** which is borne on short, lateral branches near the apex of young branches. They are partially hidden by the terminal bud. Examine branches bearing seed cones, and compare them to the staminate cones on the basis of size, shape, and location. Locate older seed cones that have opened and discharged their seeds. Dissect a seed cone and

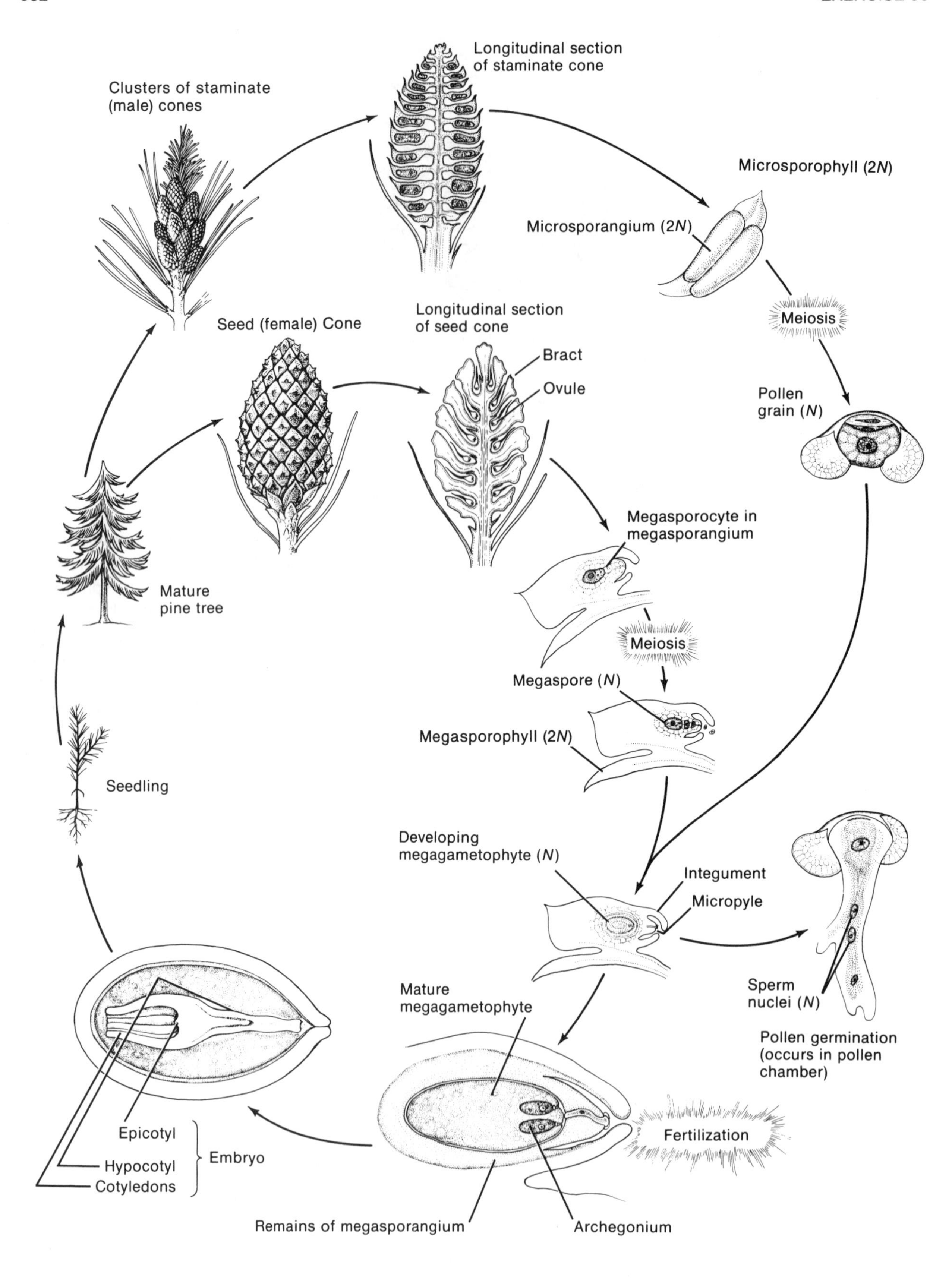

FIG. 30-4
Life cycle of a pine.

locate the overlapping, bractlike appendages (Fig. 30-4) attached to the central axis. On the upper surface of each bract, or **megasporophyll,** locate two small, white structures, the ovules. Note that the ovules are not attached directly to the bract but are borne on paper-thin scales, which in turn are attached to each bract. Why are gymnosperms classed as having "naked" seeds?

____

____

Examine longitudinal sections of a young pine ovule with a dissecting microscope (Fig. 30-4). Locate an outer layer of tissue that almost encloses the inner tissues except for a small opening at one end. The outer layer is the **integument;** the opening is the **micropyle.** Within the integument, locate the **megasporangium,** in which the **megagametophyte** is found. At thc micropylar end of the megagametophyte, locate archegonia. From what cell did the megagametophyte arise?

____

____

Note the pollen chamber just inside the micropyle.

3. Pollination and Fertilization

The pollen grains are released from the staminate cones toward the end of May. At about the same time, the young seed cones open and the pollen, by sifting down through the cone scales, comes to lie in the pollen chamber. The pollen grain germinates, produces a slender pollen tube containing the male gametes, and digests its way to the archegonia. Because the growth of the pollen tube and the development of the megagametophyte is quite slow, fertilization may not occur for as long as 13 months following pollination.

Remove the seed coats from soaked pine seeds, and cut through the gametophytic tissue to locate the embryo. Locate the epicotyl and hypocotyl. At the tip of the hyptocotyl, locate a coiled suspensor, which, by its growth, pushed the developing embryo deeply into the gametophytic tissue that contained large quantities of stored food. If available, examine stages in the germination of a pine seed. Note the peculiar manner in which the embryo comes out of the seed.

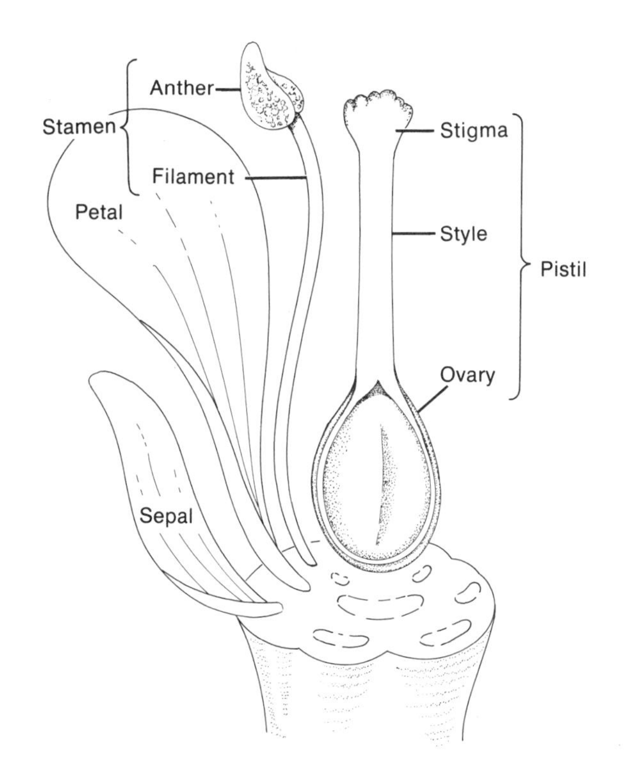

FIG. 30-5
Diagram of a flower.

C. CLASS ANGIOSPERMAE (FLOWERING PLANTS)

Sexual reproduction requires the union of gametes. The result of this union (fertilization) is a single cell, the zygote, which by a series of morphogenetic changes, develops into a new individual. In the flowering plants, the gametes are produced in modified branches that we call *flowers.* Examine a flower of a lily and locate the structures described below and as shown in Fig. 30-5, of a typical flower.

1. **Sepals** are modified leaves and are the outermost structures of the flower; they are typically green, although they may be other colors and in some cases may be absent. Collectively, the sepals constitute the **calyx.**

2. **Petals** are internal to the sepals, and may be white or highly colored. Collectively, the petals make up the **corolla.**

3. **Stamens** are structures internal to the petals. Each stamen consists of a stalk (the filament) and a terminal capsule (the anther), which when mature contains the pollen grains. The pollen grains produce the male gametes.

4. **Pistils**—one or more—are found at the center of the flower. Pistils are made up of one or more

carpels—leaflike structures bearing ovules. It is postulated that in evolution the carpels have rolled inward, enclosing the ovules. The pistil consists of three parts: an enlarged basal region (the ovary), a slender stalk (the style), and a somewhat flattened tip (the stigma). The ovary contains seedlike structures called ovules, which, when mature, contain the female gametes. In the process of pollination, the pollen grains are transported to the stigma, where they germinate and send a long tube down to the ovules. How does pollination in the flowering plants differ from that in conifers?

The pollen tube penetrates the ovule and releases two male gametes, one of which unites with the egg to form the zygote.

The lily plant is called a **sporophyte,** a stage in the life cycle of the plant during which spores are produced within specialized structures called **sporangia.** These are formed on or within modified leaves called **sporophylls.** What flower parts described above can be considered sporophylls?

Certain cells—**spore mother cells** or **sporocytes**—undergo meiotic division within the sporangium to form these spores. The spore undergoes morphogenetic changes and develops into microscopic plants (gametophytes) that produce the gametes. In the flowering plants, the gametophytic plants is parasitic on the sporophyte. Because the male gamete is smaller than the female gamete, it is called a **microgamete** to distinguish it from the larger female **megagamete.** Similarly, the male gametophyte is called the *microgametophyte* as opposed to the female *megagametophyte.* This same terminology applies to all male and female structures. This scheme is summarized in Fig. 30-6.

In studying the stamens (microsporophylls), recall that the stamen consists of an anther and filament. The anther contains the microsporangium in which are found the microsporocytes. Examine slides of a young lily anther (Fig. 30-7). Locate the microsporocytes. What is the chromosomal condition of these cells (i.e., N or $2N$)?

What indication is there that some of the microsporocytes have divided?

What kind of division has occurred?

The cells formed as a result of the division of the microsporocytes are called *microspores.* These will develop into mature microgametophytes or pollen grains. Remove an anther from your flower, crush it in a drop of water on a slide, and examine microscopically. For a more detailed examination of the pollen grain, obtain slides showing a section through an older lily anther. Locate the numerous pollen grains. The pollen grains are shed from the anther and are transferred by various means to the stigma of the same or a different flower. In this process of pollination, a pollen tube is formed that is chemotropically oriented to grow down through the style to the ovules. During the growth of the pollen tube, gametes are formed as a result of the division of one of the nuclei in the pollen tube. Examine cultures of germinating pollen and observe the pollen tubes. If possible, locate the two male gametes. In Fig. 30-7 trace the stages in the development of the microgametophyte.

In the ovule, a centrally located mass of tissue enlarges to form the megasporangium or **nucellus** (Fig. 30-7). Concomitant with the development of the megasporangium, adjacent cells grow up and around the nucellus to form the integuments, which later develop into the outer coverings of the seed. The integuments do not fuse but leave an opening (the **micropyle**) at the apex of the nucellus, through which the pollen tube enters the ovule. As the nucellus grows, the nucleus of the centrally located megasporocyte undergoes meiosis to form four megaspore nuclei.

One of the megaspore nuclei migrates to the micropyle end; the other three migrate to the opposite (chalazel) end. This 3 + 1 arrangement repre-

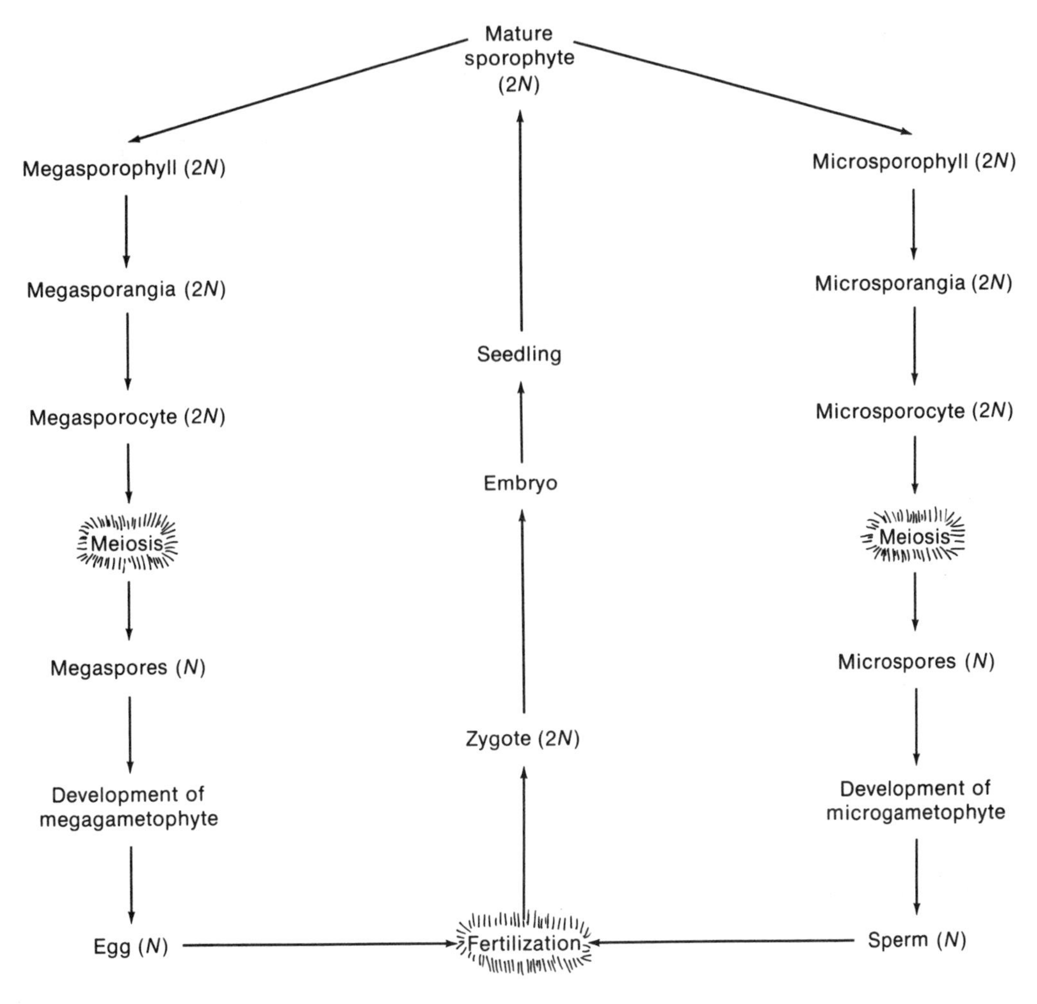

FIG. 30-6
Generalized scheme of the reproductive cycle of higher plants.

sents the first 4-nucleate stage in the development of the embryo sac.

The three haploid nuclei then fuse to form a $3N$ nucleus. This stage is followed by a mitotic division of the $3N$ and N nuclei to form two triploid nuclei and two haploid nuclei. This is the second 4-nucleate stage. A final division occurs resulting in an 8-nucleate embryo sac consisting of four triploid and four haploid nuclei. At this point, three of the triploid nuclei migrate to the chalazel end of the embryo sac where cell walls are formed around them. These cells, called **antipodals,** are probably nutritive in nature.

Three haploid nuclei migrate to the micropylar end of the embryo sac. The middle one becomes the egg. The role of the two lateral ones (called **synergids**) is not fully understood although recent work has shown that the pollen tube enters the synergid first and then the egg. Thus, the synergids might function in guiding the pollen tube. The two remaining nuclei migrate to the center of the embyro sac and form a fusion nucleus. What is the chromosome content of this nucleus?

____

Examine slides that show the developmental stages indicated above. As you study these slides, refer to Fig. 30-7.

Fertilization is complete when the pollen tube, having reached the micropyle of the ovule, digests its way through the nucellus, reaches the embryo sac, and releases both male gametes. One gamete unites with the egg to form the zygote, which subsequently develops into an embryo. The second gamete unites with the fusion nucleus to form the endosperm nucleus. What is the chromosome condition of the endosperm nucleus?

____

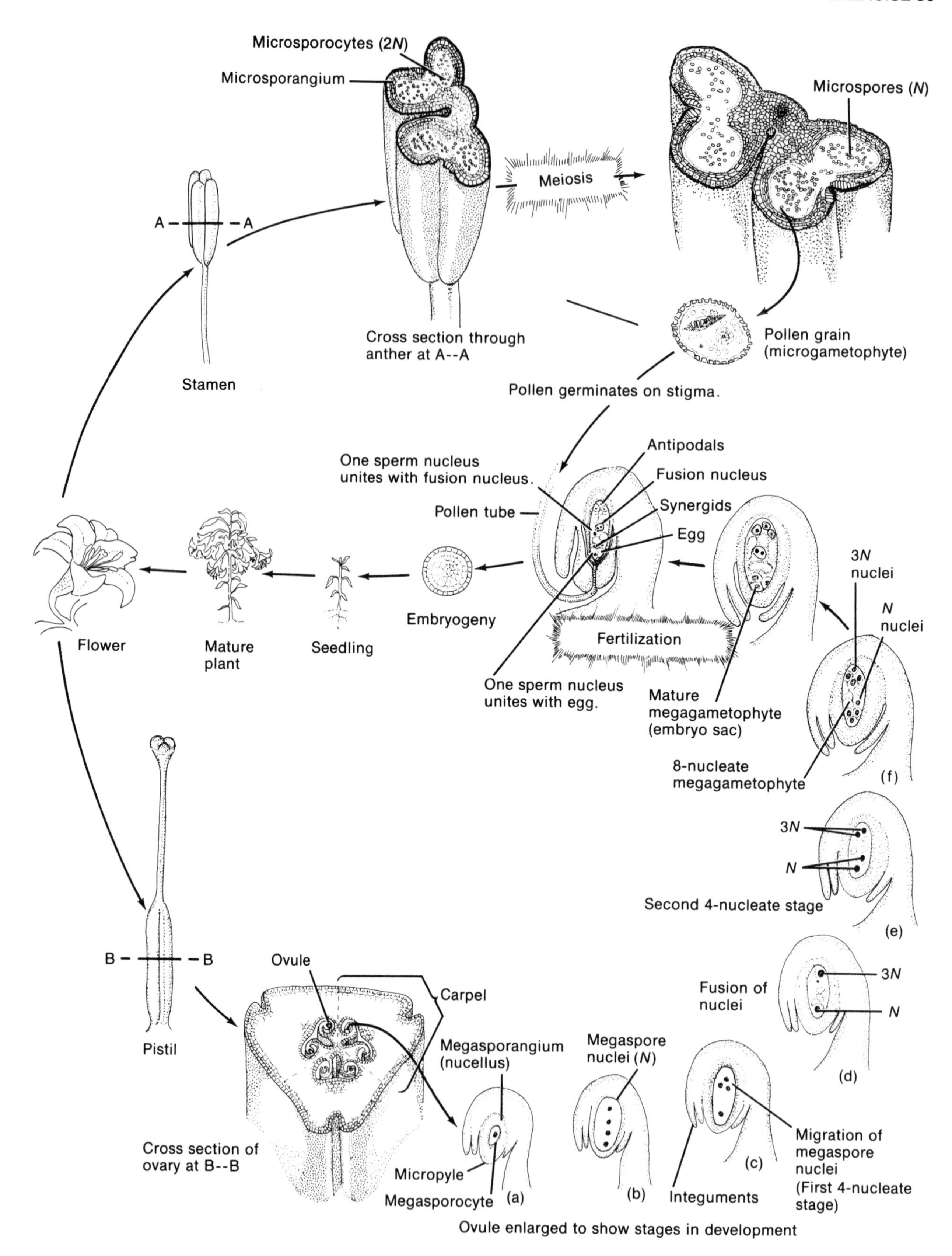

FIG. 30-7
Life cycle of a lily.

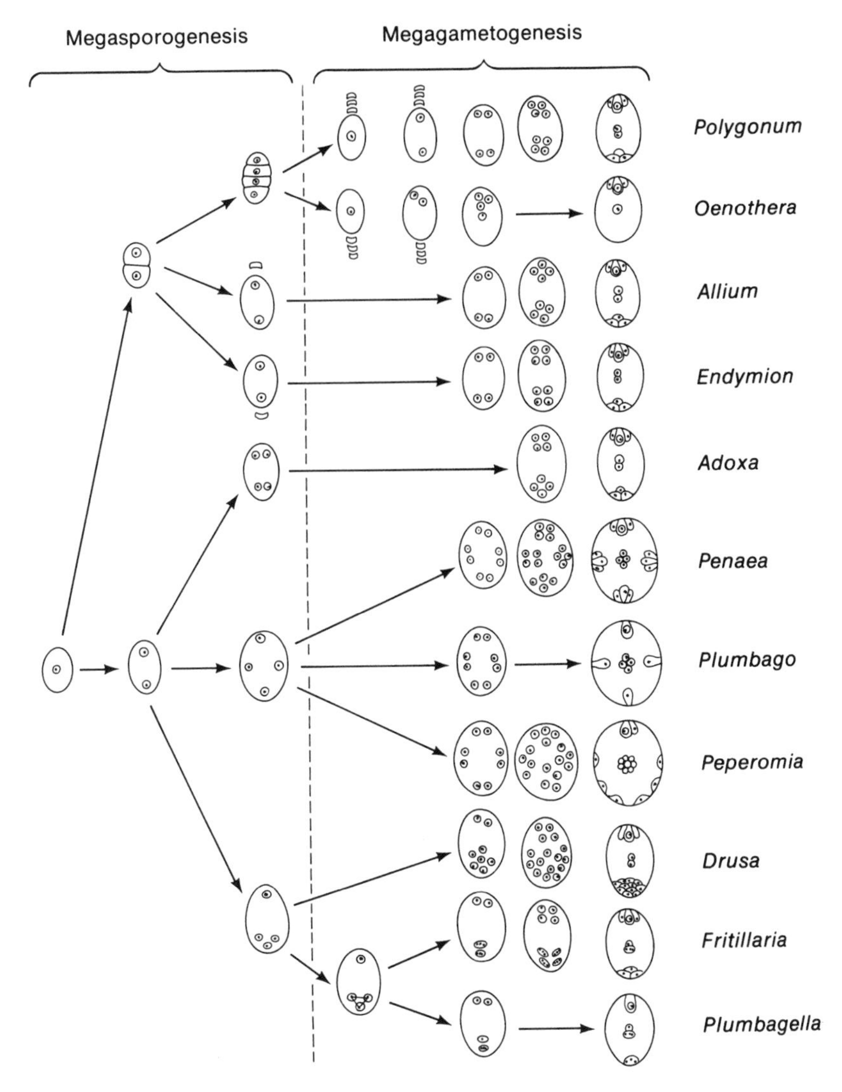

FIG. 30-8
Comparison of megagametophyte development in angiosperms. (Adapted from *Comparative Morphology of Vascular Plants,* by Adriance S. Foster and Ernest M. Gifford, 2d ed. W. H. Freeman and Company. Copyright © 1974. Redrawn from Johri in *Recent Advances in the Embryology of Angiosperms,* edited by P. Maheshwari, University of Delhi, 1963.)

Subsequent division of the endosperm results in a mass of nutritive tissue that is used as a food source for the developing embryo. This double fertilization is a phenomenon peculiar to flowering plants and has no counterpart in the lower plants or in the animal kingdom.

D. TYPICAL EMBRYO SAC DEVELOPMENT

The events in the development of the lily embryo sac (as described in Part C) are really only one of several different types of development that take place in the formation of the megagametophyte in angiosperms (Fig. 30-8). Many are much more complex than that which occurs in the lily. Of the various examples shown in Fig. 30-8, which plant shows a pattern of embryo sac development comparable to the lily?

____

(*Note:* The number of "dots" in each nucleus indicates the chromosome complement. For example, two "dots" is diploid, three "dots" is triploid, and so forth.)

Although you should be aware of the complexities that occur, we can simplify this developmental event by describing a composite picture of embryo sac development. This is shown in Fig. 30-9.

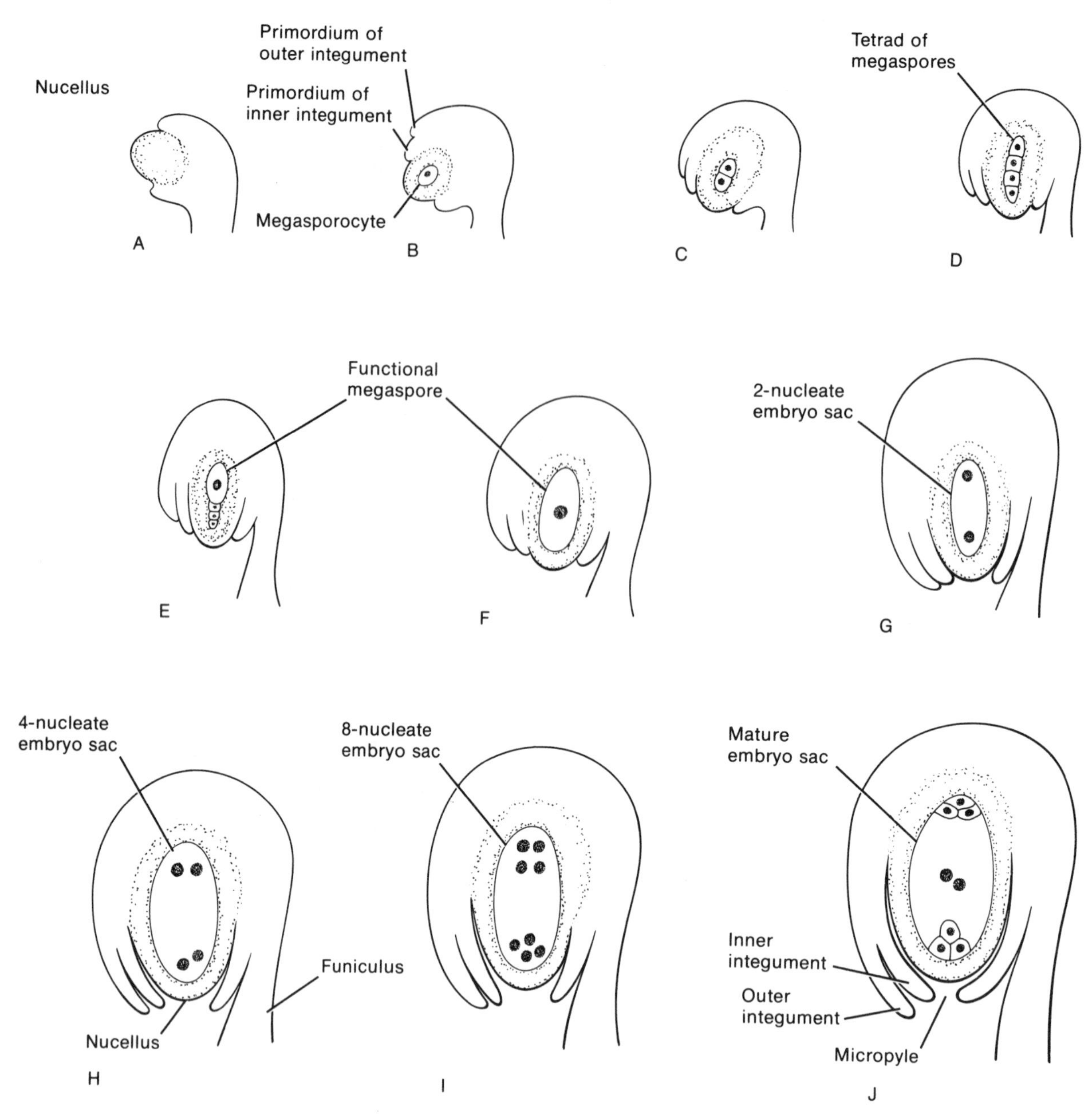

FIG. 30-9
Megagametophyte development in a typical angiosperm. (Adapted from *Comparative Morphology of Vascular Plants*, by Adriance S. Foster and Ernest M. Gifford, 2d ed. W. H. Freeman and Company. Copyright © 1974. Redrawn from *A Textbook of General Botany*, 4th ed., by R. M. Holman and W. W. Robbins. Wiley, 1951.)

E. ADVANCES IN COMPLEXITY SHOWN BY VASCULAR PLANTS

Complete Table 30-1. Indicate the increases in complexity shown by the tracheophytes over the bryophytes with respect to morphology and reproduction. The following are suggested terms to use.

Morphology: *true leaves present or absent, true stems and roots present or absent, vascular tissue present or absent, erect or prostrate plant body*

Asexual reproduction: *spores, fragmentation, gemmae, death and decay of older parts, and so on*

Sexual reproduction: *unicellular or multicellular reproductive organs, motile or nonmotile gametes, alternation of generations, isogametes, anisogametes, water necessary or unnecessary for*

TABLE 30-1
Advances in complexity shown by the vascular plants.

Character	Liverworts and mosses	Vascular plants
Morphology of sporophyte		
Morphology of gametophyte		
Asexual reproduction		
Sexual reproduction		

fertilization, antheridia, archegonia, dependency or interdependence of sporophyte, pollination

REFERENCES

Bold H. C. 1980. *Morphology of Plants and Fungi.* 4th ed. Harper & Row.

Bold, H. C., and C. L. Hundell. 1984 *The Plant Kingdom.* 4th ed. Prentice-Hall.

Cutler, D. F. 1978. *Applied Plant Anatomy.* Longman.

Delevoryas, T. 1977. *Plant Diversification.* 2d ed. Holt, Rinehart and Winston.

Greulich, V. 1983. *Plant Structure and Function.* 2d ed. Macmillan

Klein, R. M. 1979. *The Green World.* Harper & Row.

Raven, P. H., R. F. Evert, and H. Curtis. 1981. *Biology of Plants.* 3d ed. Worth.

Ray, P. M., T. A. Steeves, and S. A. Fultz. 1983. *Botany.* Saunders.

EXERCISE 31

Kingdom Animalia: Phyla Porifera, Cnidaria, and Ctenophora

According to Whittaker's classification scheme, the kingdom Animalia includes multicellular organisms having eukaryotic cells that lack cell walls, plastids, and photosynthetic pigments. Most members of this kingdom take in nutrients by ingestion, with digestion taking place in an internal cavity. Some take in nutrients by absorption and lack an internal digestive cavity.

The higher forms in the animal kingdom have evolved highly sophisticated levels of organization and tissue differentiation, including sensory neuromotor systems and modes of motility based on contractile fibers.

Reproduction for most of the organisms in this kingdom is sexual, and except for some of the lowest phyla haploid cells occur only in the gametes.

In this exercise, you will study representatives of the phyla Porifera (sponges), Ctenophora (comb jellies), and Cnidaria (coelenterates), the first of the multicellular or "tissue" animals.

A. PHYLUM PORIFERA (SPONGES)

The phylum Porifera (Latin *porus,* "pore"; *ferre,* "to bear") consists of the sponges, which are characterized by bodies with radial symmetry and an internal skeleton consisting of minute spicules or fibers made up of **spongin.** The body surface of sponges has many pores, which are connected to canals and chambers lined by flagellated cells.

Based largely on the structural organization of the internal skeleton, this phylum is divided into three major classes:

Class Calcarea (calcareous sponges): Skeletons are composed of calcium carbonate spicules.

Class Hexactinellida (glass sponges): Skeletons are composed of siliceous material.

Class Demospongiae (natural sponges): Skeletons are composed of organic fibers (spongin) or siliceous spicules or both.

1. Class Calcarea

Scypha (formerly called *Sycon* or *Grantia*) is a small, slender sponge that rarely exceeds an inch in height and is found in clusters adhering to various objects in shallow marine waters.

Using a dissecting microscope, examine specimens of *Scypha* and note their slender, vase-shaped appearance. Observe the basal end by which this organism attaches itself to various objects and the large single excurrent opening, the **osculum,** at the free end. Minute pores **(ostia)** along the body wall allow water to enter the single tubular central canal, or **spongocoel.** Locate bristlelike structures along the body surfaces. These are the spicules, which give rigidity to the body. The spicules are chiefly of four kinds: (1) long, straight spicules surrounding the osculum that protrude beyond the body surface, producing a bristly appearance; (2) short, straight spicules surrounding the ostia; (3) T-shaped spicules lining the spongocoel; and (4) three-branched spicules embedded in the body wall. Examine slides showing isolated spicules.

Study a prepared stained cross section of *Scypha* under both low and high magnification (Fig. 31-1). Note the thick body wall and the numerous short radial canals. Water enters the sponge through the ostia, which lead into the **incurrent canals.** These canals extend almost to the wall of the spongocoel, or the digestive cavity. The spongocoel connects to the outside through an opening called an **osculum.** The incurrent canals are connected to the **excurrent (radial) canals by** a series of small pores called **prosopyles.** The excurrent canals, in turn, open into the spongocoel. The exterior surface of this sponge is covered by a thin **dermal epithelium;** the spongocoel is lined with **gastral epithelium.** The radial canals are lined with small, flagellated collar cells, or **choanocytes.** These cells resemble some of the flagellated protozoans—indicating a probable evolution

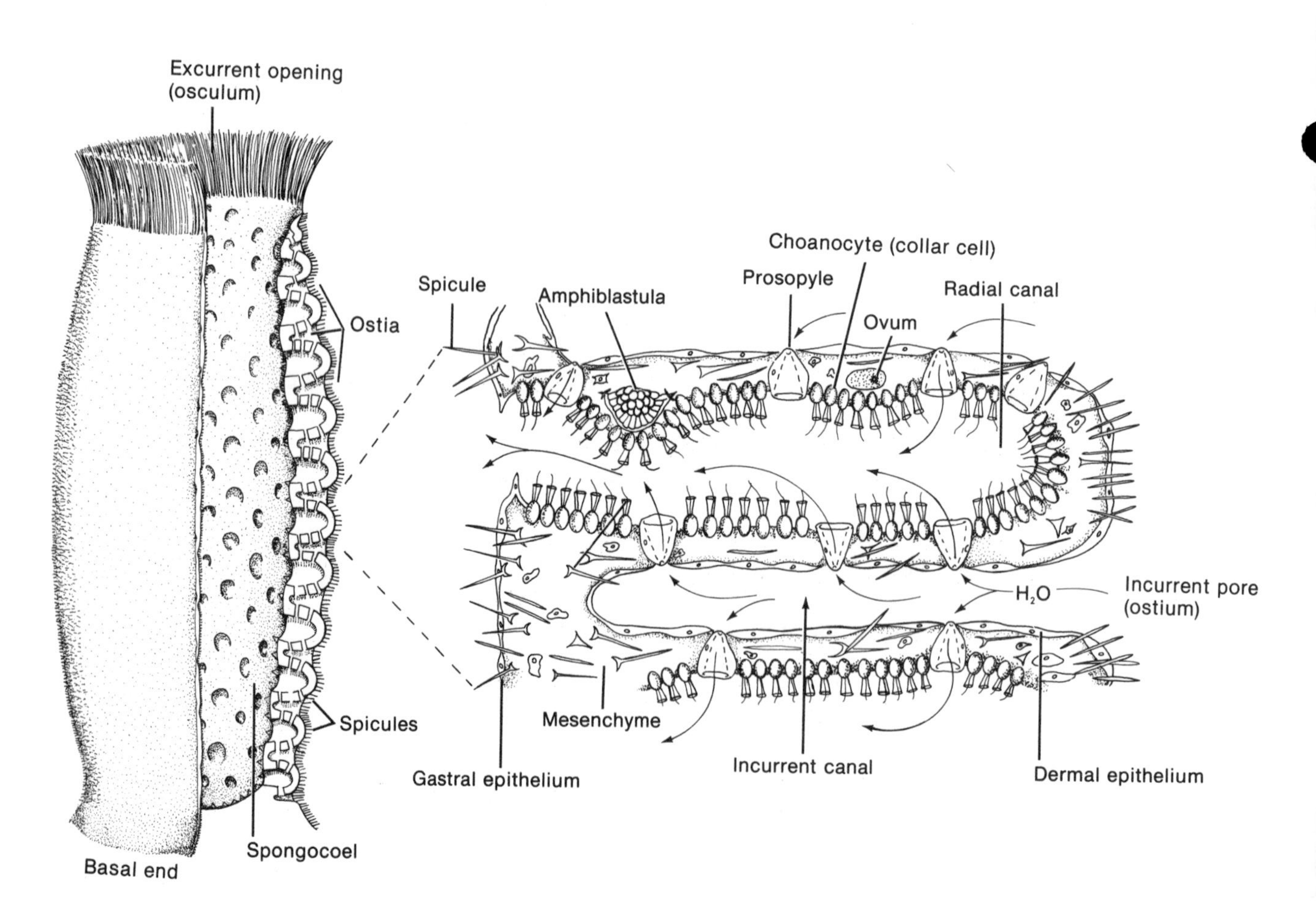

FIG. 31-1
Canal system of *Scypha*.

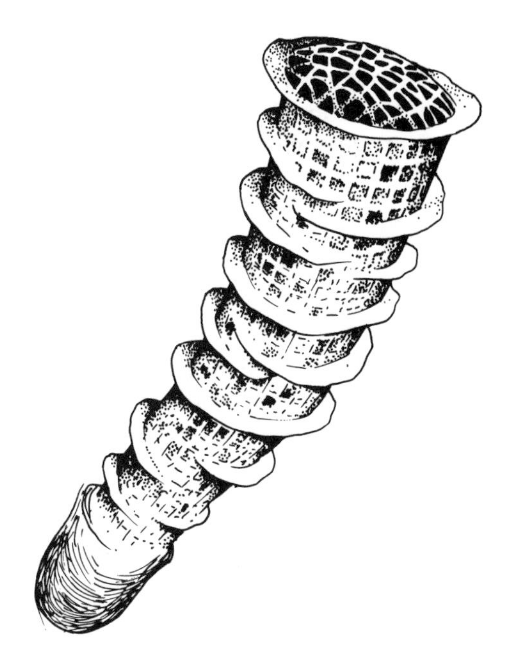

FIG. 31-2
Skeleton of *Euplectella* (Venus's flower basket).

of the sponges from protozoans. Between the dermal and gastral epithelium is a gelatinous **mesenchyme.**

Water, taken into the body through the ostia, is passed over the collar cells, which remove tiny bits of food, and is then extruded through the osculum. The food consists of plankton (microscopic plants and animals) and bits of organic matter that are digested in the food vacuoles of the choanocytes. Indigestible material is extruded into the canal system for extrusion through the osculum by way of the spongocoel.

Sponges can reproduce by both asexual and sexual means. Parts of a sponge lost by injury can be replaced asexually by regeneration. Many kinds of sponges reproduce by simple budding; the buds either separate from the original sponge as growth proceeds or remain attached to form clusters. Because *Scypha* is **monoecious (hermaphroditic),** sexual reproduction is accomplished by male (sperm) and female (ova) gametes produced by the choanocytes. Fertilization is internal. When the embryo reaches the oval shaped **amphiblastula** stage, it escapes through the osculum, swims about for a short time, then attaches to a solid object to begin growth as a young sponge.

2. Class Hexactinellida

Hexactinellids are beautiful vase- or funnel-shaped sponges whose skeletons are composed of six-rayed spicules made of siliceous materials; they are therefore referred to as *glass sponges.* They are deep-water forms of sponges, and usually only the skeleton is recovered.

Examine specimens of the sponge *Euplectella* (Venus's flower blasket) (Fig. 31-2). What you are observing is merely the skeleton of the living sponge, because all of the protoplasm has dried and decayed away. Note that the skeleton is composed of a delicate network of siliceous spicules fused at their tips.

3. Class Demospongiae

This class includes the familiar bath sponges and freshwater sponges (Fig. 31-3). Examine the cleaned and dried skeletons of a bath sponge (*Spongia*); it is the sponge we commonly use because of its water-absorbing capacity. The skeleton can hold water because of capillary forces in the fine spaces of the

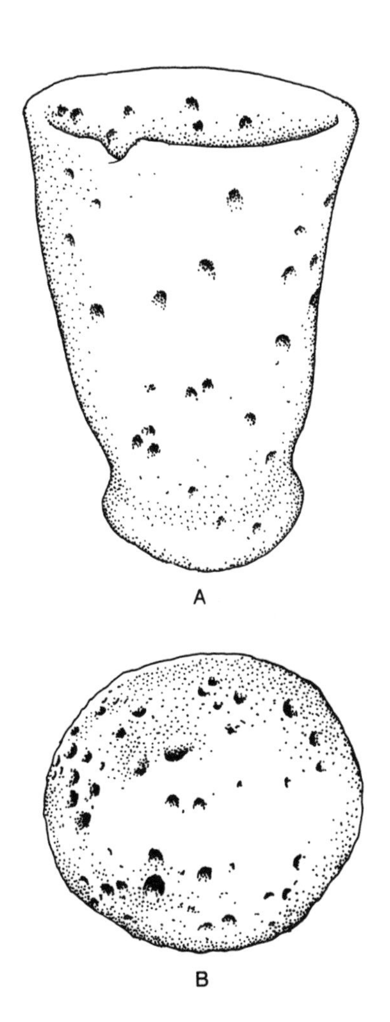

FIG. 31-3
Some common sponges: (A) *Gelliodes*, the elephant's ear sponge, and (B) *Spongia*, the common bath sponge.

irregular spongin network. Synthetic sponges have largely replaced the natural variety in most areas of the world.

B. PHYLUM CNIDARIA (COELENTERATES)

The phylum Cnidaria consists of a diverse group of radially symmetrical organisms characterized by a digestive (**gastrovascular**) body cavity, some muscle fibers, and many minute stinging capsules called **nematocysts.** The members of this phylum may be one of two forms: a polyp or medusa. **A polyp** is a tubular organism closed at one end with a mouth surrounded by tentacles at the other end. **A medusa** is an umbrella-shaped, jellylike, free-swimming organism with a mouth at the end of a central projection called a **manubrium.** The phylum is divided into three classes:

Class Hydrozoa (hydras): In which the polyp stage is predominant and gives rise in many species to small free-swimming medusae.

Class Scyphozoa (true jellyfish): Includes the larger jellyfish, in which the medusa stage, consisting largely of gelatinous **mesoglea** or "jelly," is predominant. The polyp stage is minute or lacking. (Scyphozoans can be roughly distinguished from the medusae of the hydrozoans by their size, which ranges from 25 mm to 2 m in diameter.)

Class Anthozoa (corals and sea anemones): In which there is only a polyp stage.

1. Class Hydrozoa

a. Hydra

The common freshwater hydrozoan, *Hydra,* is an excellent organism for the study of the polyp form of coelenterates. It is also of fundamental interest because it illustrates a definite, though primitive, level of organization into tissues. Commonly studied species are the brown (*Hydra oligactis*) and green hydras (*Hydra viridissimus*); the latter are green because of the minute algae living symbiotically within their bodies.

Using a hand lens or a dissecting microscope, examine living hydras in a Syracuse dish or deep-well slide containing pond water. These slender organisms vary from 10–25 mm in length. At the upper end of the animal, observe several elongate, actively moving tentacles, which are used to capture food (Fig. 31-4). At the center of this circle of tentacles is the mouth. Tap the edge of your slide or dish, and observe the extreme contractility of this organism. Note that the animal expands after contracting. Compare the rate of elongation with that of contraction.

____

____

So that you can observe the hydras ingesting food, your instructor will add to your slide or dish several brine shrimp or *Daphnia* that have been thoroughly rinsed with tap water. Why are the brine shrimp washed before being fed to the hydra?

____

____

____

Using a dissecting microscope, observe your specimen and describe the feeding process, beginning with the first contact of the hydra's tentacles with its food.

____

____

____

It is not unusual for a single hydra to capture and ingest 10 or more *Daphnia* in the course of a 2-hour feeding period.

While observing the feeding process, you may have noticed that as the tentacles of the hydra came in contact with its prey, the shrimp or *Daphnia* jerked for several minutes before becoming still. The food organism reacted in this way because it had been stung by numerous stinging cells released by the tentacles of the hydra. Close examination of the tentacles of your specimen will reveal the presence of wartlike structures called **nematocysts.** Until it is discharged, each nematocyst is enclosed in a **cnidoblast** cell, with the coiled thread inside it (Fig. 31-4C). Observe the trigger (**cnidocil**) located on the edge of the undischarged nematocyst. When it is discharged, the coiled thread is turned inside out as it is released from the cnidoblast. Note the barbs and long thread in the discharged nematocyst. The release of these nematocysts can be observed by placing a hydra in a drop of water on a clean slide and then adding a drop of dilute acetic acid at the edge of the water.

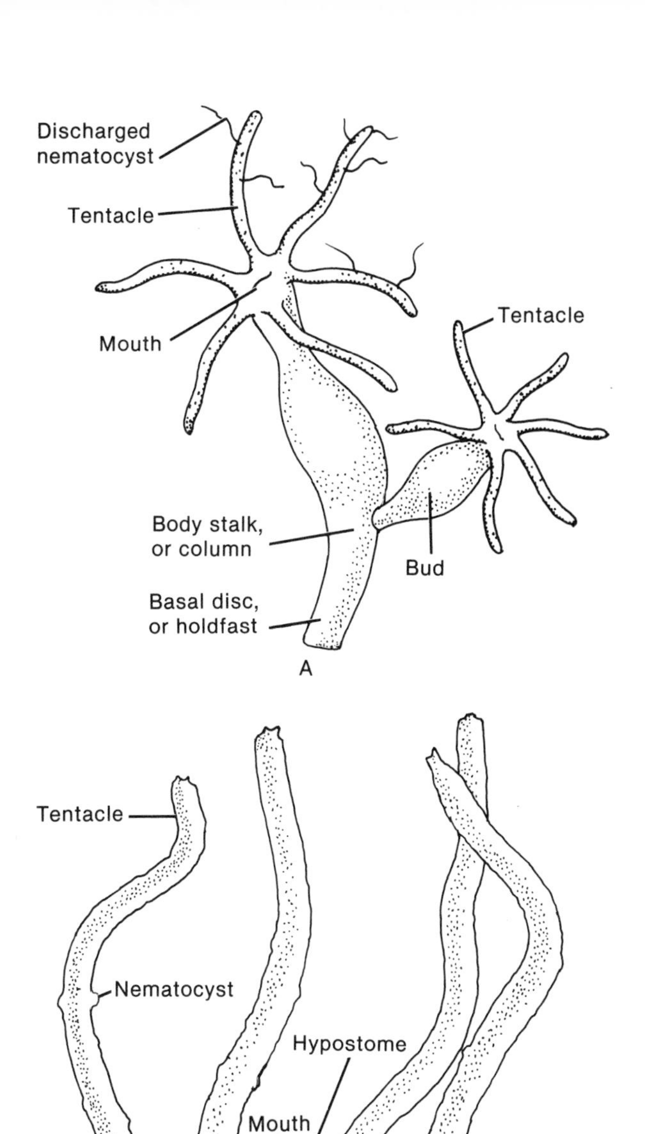

Observe the tentacles as the acid diffuses toward them. Note the discharge of the nematocysts. The structures of the various types of nematocysts can be observed by adding a drop of dilute methylene blue or other suitable stain at the edge of the water.

Hydras reproduce asexually by simple budding and sexually by the production of sperm and ova. Budding consists of a simple outgrowth from the body wall (its gastrovascular cavity continuous with that of the parent) separating upon maturity from the parent. Examine specimens of budding hydras.

The gametes are produced in organs called **spermaries (testes)** and **ovaries,** which are protubernances along the body wall. Both sex organs may be present in a single organism. Examine prepared slides showing the location and structure of these organs.

To study the histology of *Hydra,* examine stained cross sections. Note that the body wall surrounding the **gastrovascular cavity** consists of two distinct layers of cells: an outer layer of **ectoderm** and an inner layer of **endoderm** (Fig. 31-4C). These two

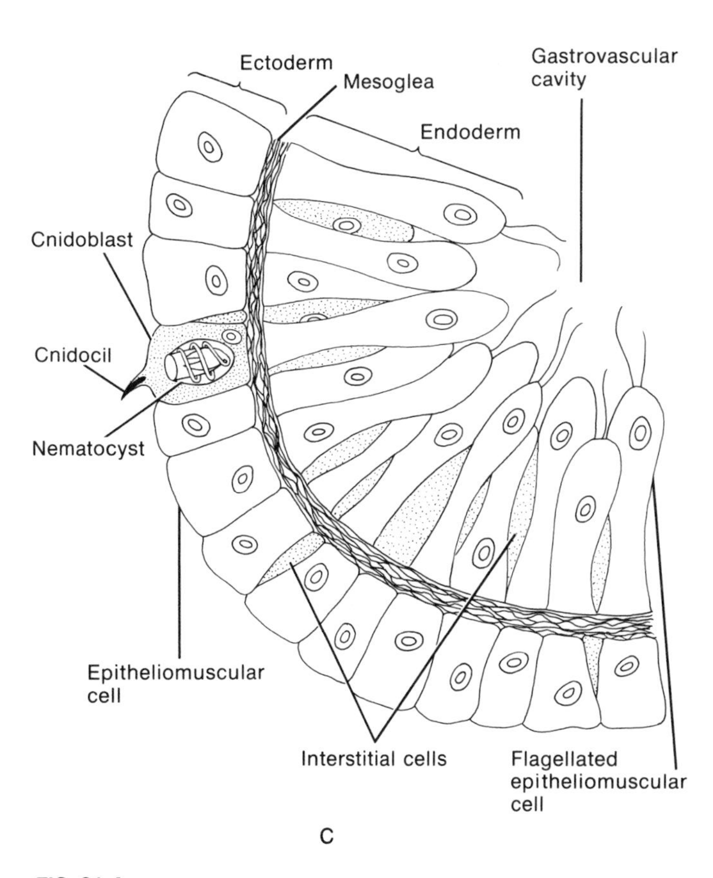

FIG. 31-4
Hydra: (A) overall view, (B) longitudinal section, and (C) section of body wall.

layers are held together by a thin, noncellular layer —the **mesoglea.** In the ectoderm are two principal cell types: the larger **epitheliomuscular cells** containing contractile fibrils at their base and the smaller **interstitial cells** located between them. What might you expect the function of the epitheliomuscular cells to be?

___

___

The ectoderm of the tentacles also contains nematocysts.

Many of the large endodermal cells contain food vacuoles in which intracellular digestion takes place. These food vacuoles are formed as a result of the engulfing of food particles by pseudopodia that are extended into the gastrovascular cavity. Many of the digestive cells also contain flagellae, whose beating activity directs food toward the pseudopodia. Extracellular digestion in hydra is by the secretion of enzymes directly into the gastrovascular cavity, followed by absorption of the digested food material.

b. Obella

Unlike *Hydra,* most members of the hydrozoans are colonial marine organisms. *Obelia,* a typical coelenterate, is found attached to seaweeds (particularly the kelps), rocks, shells, or pilings in the shallow waters off seacoasts. Reproduction in this animal is **metagenetic** (alternation of sexual and asexual generations), a common characteristic of plants. The **polyp** stage is the asexual generation, and the **medusa** stage is the sexual one.

Place a small piece of *Obelia* in a Syracuse dish and examine under a dissecting microscope. The plantlike colony consists of numerous branches, which terminate in two kinds of polyps. The feeding, or **nutritive,** polyps **(hydranths)** possess tentacles and resemble *Hydra.* The **reproductive polyps (gonangia)** are club-shaped and lack tentacles (Fig. 31-5). The branches of the colony are covered by a transparent, noncellular sheath—the **perisarc.** This covering expands around the hydranths and gonangia to form vase-shaped protective structures called **hydrothecae** and **gonothecae,** respectively. The inner cellular core of the colony is called the **coenosarc.**

c. Gonionemus

Because the medusae of *Obelia* are too small to study in detail, the large medusa stage of *Gonionemus,* a common jellyfish, will be used to illustrate the typical structure of a coelenterate medusa.

Using a hand lens or dissecting microscope, examine a specimen in a Syracuse dish (Fig. 31-6). Note that the medusa is umbrella-shaped. Turn the organism over, and note that the margin of the "umbrella" extending inward is a muscular ring of tissue called the **velum.** What might a function of the velum be?

___

___

The medusa usually swims with its convex surface upward so that the tentacles are facing the surface of the water. Observe the number of tentacles around the bell. Extending into the cavity of the medusa is the **manubrium,** which contains a mouth at its tip surrounded by four **oral lobes.** Extending from the manubrium along the inner surfaces are four **radial canals** that join with the circular canal at the margin and connect with the cavities of the tentacles. What is the function of these canals?

___

___

Observe the numerous **nematocyst batteries** and the **suctorial pads** on the tentacles.

At the base of each tentacle are round, pigmented structures that are believed to be **photoreceptors.** Between each tentacle is a **statocyst,** a small capsule containing a tiny, solid body, which serves as a balancing organ. The reproductive organs **(gonads)** open into the radial canals. *Gonionemus,* like most coelenterates, is **dioecious;** that is, it has the male and female organs in separate organisms. The eggs and sperm are released into the sea, where fertilization takes place. The zygote develops into a **planula larva,** which swims about and then settles and attaches to a submerged object, where it transforms into a microscopic polyp. Refer to Fig. 31-5 for this pattern of sexual reproduction as it occurs in *Obelia.*

d. Other Hydrozoans

Examine demonstration specimens of other hydrozoans to see the great diversity of structure among the individuals forming a colony. In this respect, pay particular attention to *Physalia,* the Portugese man-of-war, which shows a highly specialized form of **polymorphism** (many forms). In addition to a modified medusa that forms the gas-filled float, there are several kinds of tentacles that serve to paralyze and capture food for the colony.

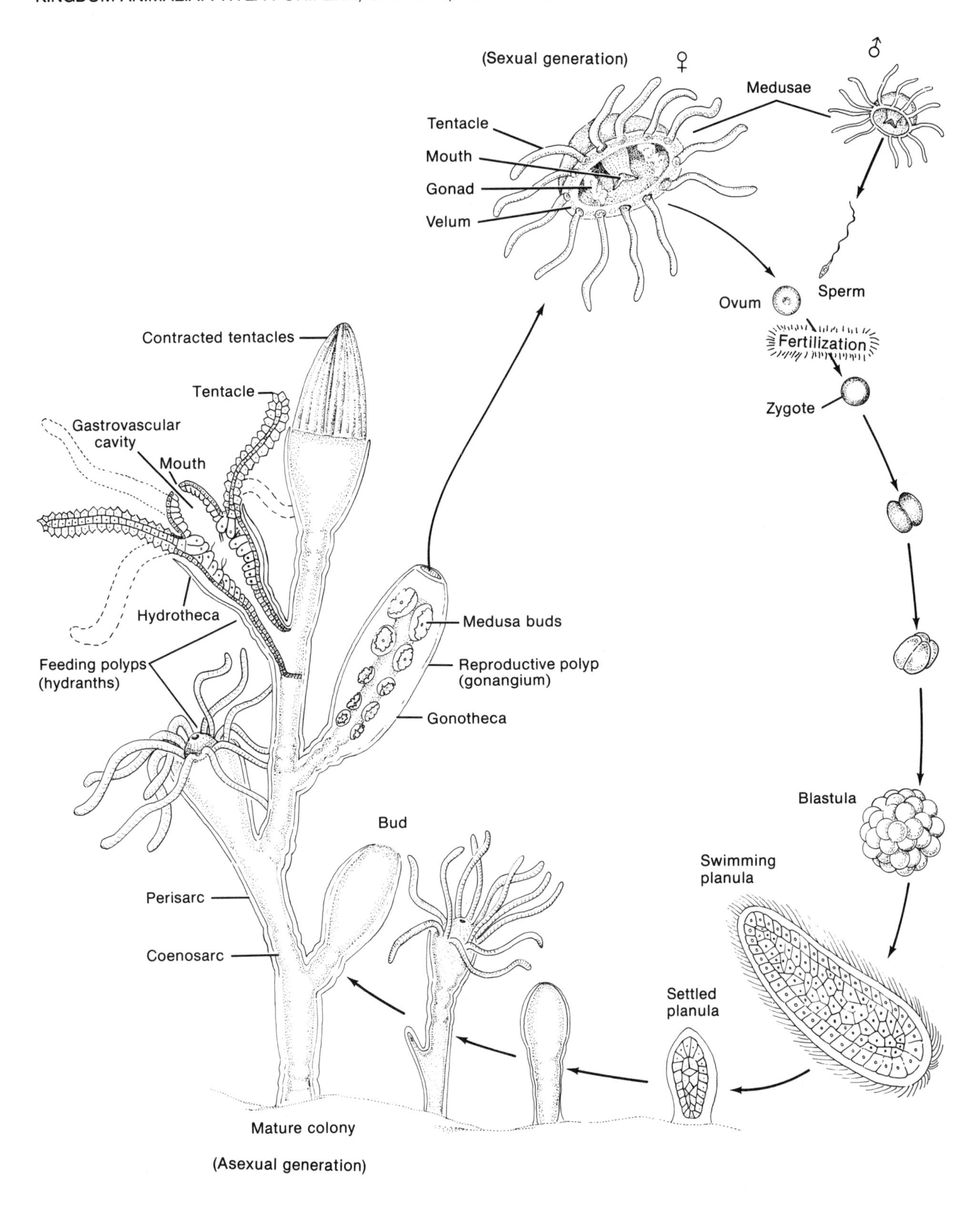

FIG. 31-5
Obelia life cycle.

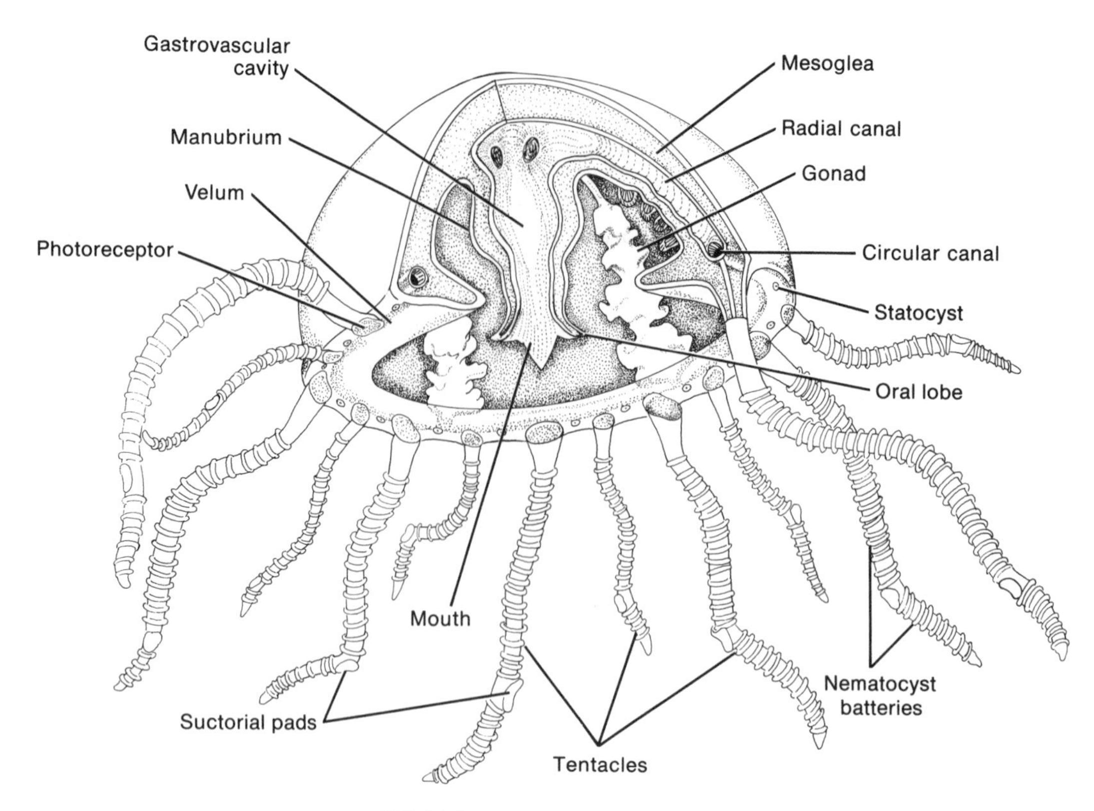

FIG. 31-6
The internal anatomy of *Gonionemus.*

2. Class Scyphozoa

In scyphozoans (Greek *skyphos,* "cup", *zoon,* "animal"), the medusa stage is quite large, ranging from 25 mm to 2 meters in diameter. The polyp stage is minute or lacking.

Aurelia, one of the commonest of the scyphozoan jellyfish, exists throughout the world. *Aurelia* is often seen in large groups, drifting along or swimming slowly by rhythmic contractions of the shallow, almost saucer-shaped bell. Great numbers are sometimes cast on shore during storms.

Examine a specimen of *Aurelia.* Identify as many of the following structural characteristics of this jellyfish as you can. The body is shallowly convex above and concave below and is fringed by a row of closely spaced marginal tentacles (Fig. 31-7). The tentacles are interrupted at eight equally spaced intervals by indentations, each of which contains a **sense organ.** The sense organs consist of a pigmented eyespot, sensitive to light; a hollow statocyst, containing minute calcareous particles whose movements set up stimuli that direct the swimming movements; and two sense pits, lined with cells that are thought to be sensitive to food or other chemicals in the water. Many circular muscle fibers are present in the margin. What is the function of these muscle fibers?

___

___

The mouth is located in the center of the oral (concave) surface at the end of a very short **manubrium.** The manubrium lies between four tapering **oral arms,** each of which has a ciliated groove. Stinging capsules in the lobes paralyze and entangle small animals, which are then swept up the grooves, through the **mouth,** and into a **digestive cavity** that extends into four pouches, in which there are tentaclelike projections. These projections are covered with nematocysts that paralyze prey that arrives in the pouches still alive and struggling. Many **radial canals** extend through the **mesoglea** from the pouches to a *ring canal* in the bell margin. Flagella lining the entire gastrovascular cavity maintain a steady current of water, which brings a constant supply of food and oxygen to the animal's internal parts and removes waste from them.

The four horseshoe-shaped, colored bodies by which *Aurelia* is recognized are the **gonads** (testes or

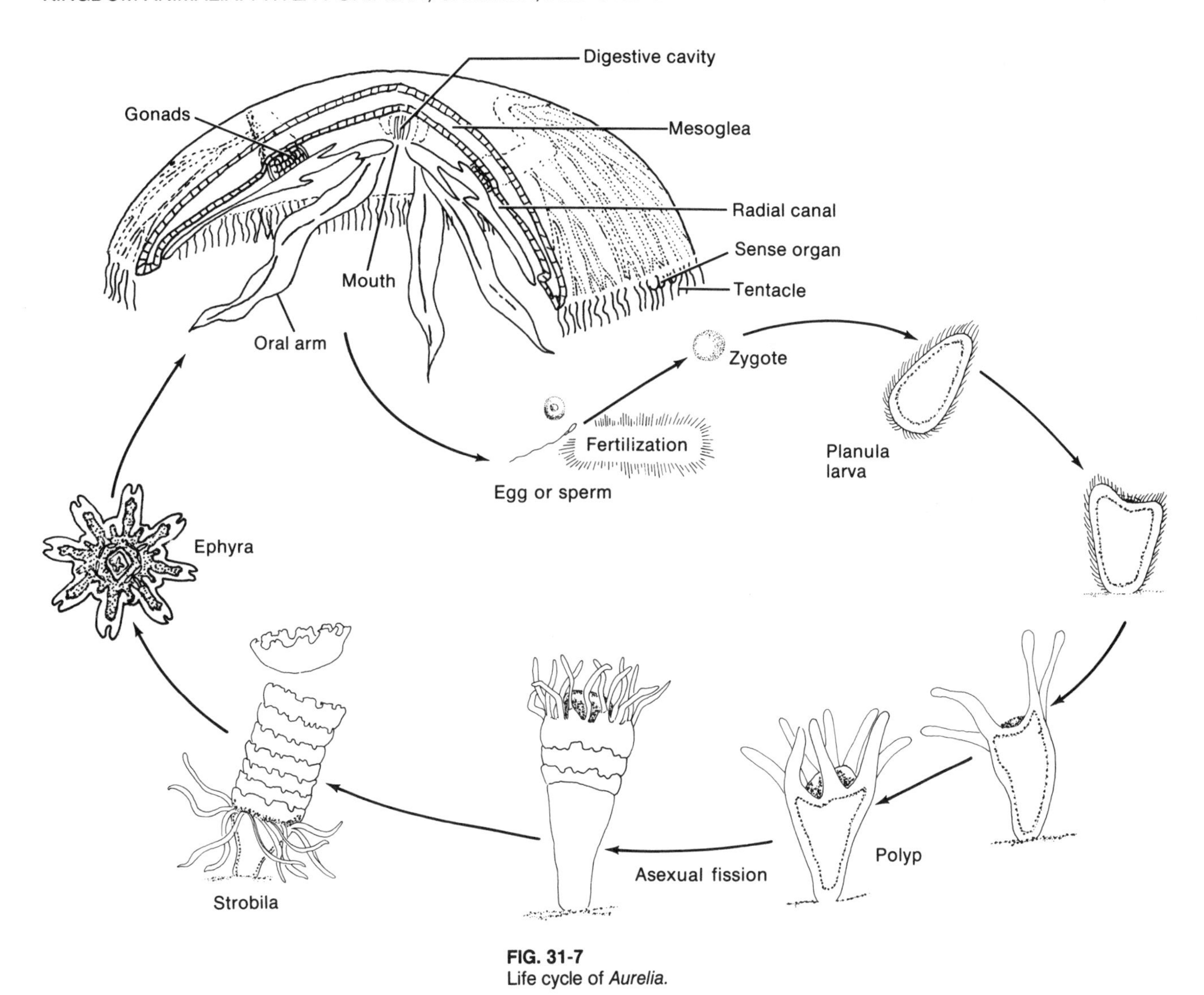

FIG. 31-7
Life cycle of *Aurelia.*

ovaries); they are located on the floor of the large central part of the gastrovascularcavity. In a male medusa, the sperm cells are discharged into the gastrovascular cavity then to the outside through the mouth. The eggs of the female are discharged into the cavity, where they are fertilized by sperm cells entering the mouth along with incoming food.

The zygotes emerge and lodge on the oral lobes, where each develops into a ciliated **planula larva.** This larva soon escapes, swims about for a while, then settles and attaches to the sea bottom. Losing the cilia, it becomes a minute trumpet-shaped polyp (**scyphistoma**) with a basal disc, mouth, and tentacles. This polyp feeds and stores food and many survive in this stage for many months, meanwhile budding off other small polyps like itself.

Usually in the fall and winter, the polyp develops a series of horizontal constrictions **(strobilation)** that resembles a pile of minute "saucers" **(ephyra)** with fluted borders. One by one, the "saucers" pinch off from the parent and swim away as little medusas that will develop into adult jellyfish.

Examine demonstration specimens and slides showing the stages in the life cycle of *Aurelia.*

3. Class Anthozoa

Anthozoa (Greek *anthos,* "flower"; *zoon,* "animal") are marine polyps of flowerlike form that have no medusa stage. They are distinguished from hydrozoan polyps by a gastrovascular cavity that is divided by a series of vertical partitions and surface ectoderm that turns in at the mouth to line the gullet. Externally, there is no difficulty in distinguishing the large sea anemones or the stony corals of this class from most of the small, fragile hydrozoan polyps. Besides the familiar sea anemones and stony corals, this class includes the soft, horny, and black corals; the colonial sea pens and sea pansies; and others. They are abundant in warm, shallow waters, but also inhabit polar seas. Other species inhabit areas ranging from the tide lines to depths of 5000 meters.

a. Sea Anemones

Of the polyp type of coelenterates, sea anemones are among the most highly specialized. They have a well-developed nerve net, mesenchyme cells between ectoderm and endoderm, and several sets of specialized muscles.

Examine the common sea anemone *Metridium.* Note the stout cylindrical body, expanded at its upper end into an **oral disc** around a slitlike **mouth** surrounded by several rows of tentacles. When undisturbed and covered by water, the body and tentacles are widely extended. If irritated or exposed by a receding tide, the oral disc may be completely turned inward and the body tightly contracted. Cilia on the tentacles and oral disc beat to keep these surfaces free of debris. The basal end forms a smooth, muscular, slimy **pedal disc** on which the anemone can slide about very slowly and by which it holds onto rocks so tenaciously that it is likely to be torn if someone tries to pry it loose. They commonly attach themselves to the shells of crabs or other shelled animals in the ocean. In this way the anemones are widely dispersed. Many anemones are exquisitely colored. Examine a preserved or a living specimen, and identify the external and internal features described in the following discussion.

From the mouth, a muscular **gullet** hangs down into the **gastrovascular cavity** (Fig. 31-8). The gullet is not cylindrical but is flattened and lined with cilia that beat downward, thus drawing a current of water into the gastrovascular cavity and steadily supplying the internal parts with oxygen. At the same time, other cilia lining the gullet beat upward, creating an outgoing current of water that takes with it carbon dioxide and other wastes. When small animals touch the tentacles, the cilia of the gullet reverse their beat, and the food is swept down the gullet into the gastrovascular cavity. The gullet is connected with the

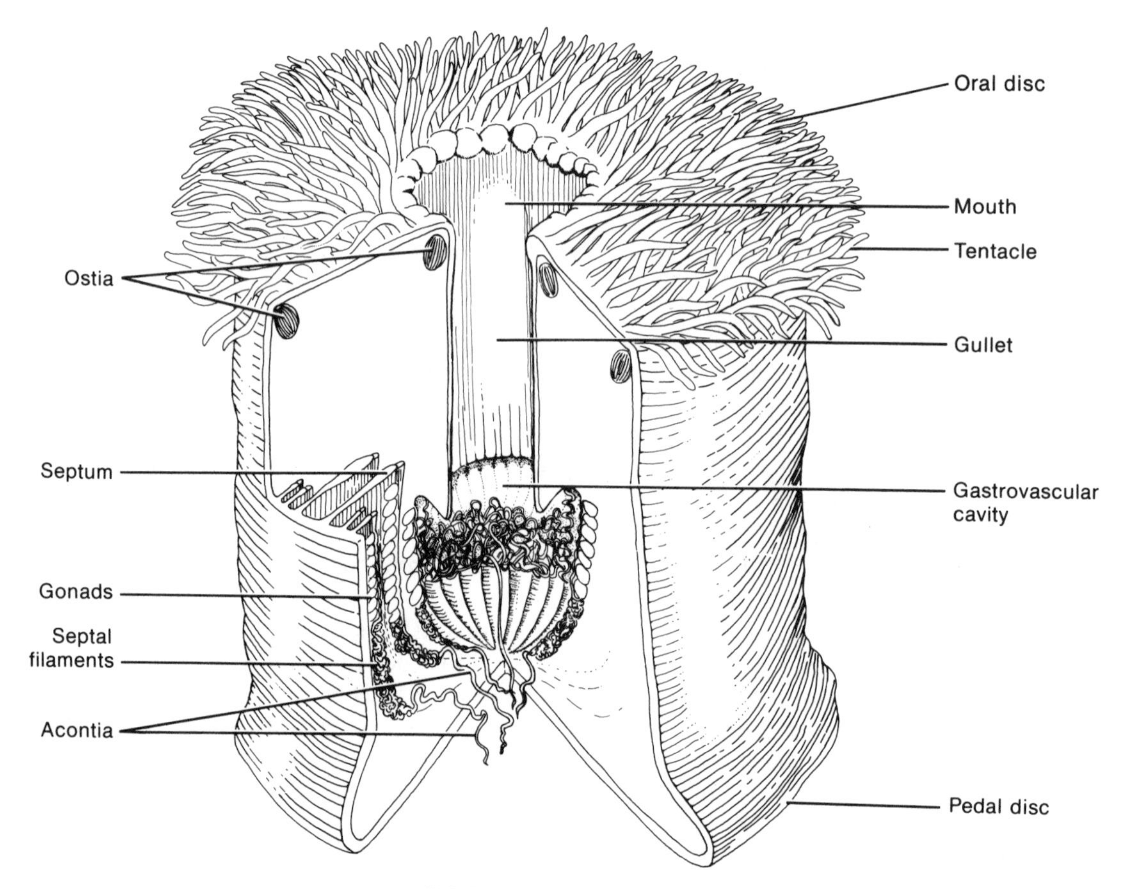

FIG. 31-8
Internal anatomy of *Metridium.*

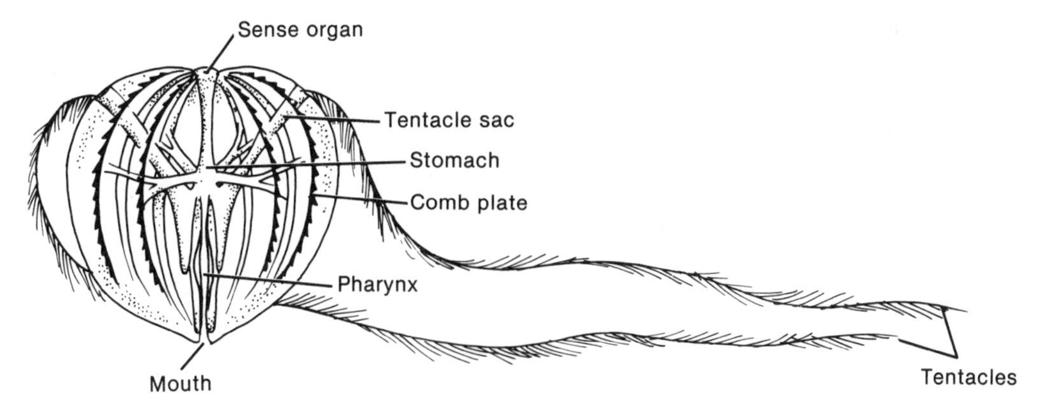

FIG. 31-9
Pleurobrachia, a ctenophore.

body wall by a series of vertical partitions (**septa**). What is the function of these septa?

____

____

In the septa, beneath the oral disc, are openings, or **ostia,** through which water can pass between the internal compartments. The free, inner margin of each septum is a thick, convoluted **septal filament** that becomes a threadlike **acontium** toward the bottom. Both parts bear nematocysts for paralyzing prey and gland cells for the secretion of enzymes needed in digestion.

Anemones sometimes reproduce asexually by separating into two longitudinal halves; each half then regenerates into a new adult. Sexual reproduction is accomplished by the small, rounded **gonads** located at the edges of the septa. The sexes are separate. Eggs and sperm leave the gonads through the mouth, and fertilization takes place in the water. The fertilized egg develops into a ciliated planula larva, which finally settles down in a rocky crevice and grows into an adult anemone.

b. Corals

Although small and fragile polyps, corals can combine to form huge colonies consisting of millions of individuals. Each member of a colony secretes a protective skeleton of limestone containing a pocket into which the polyp partly withdraws when attacked. New individuals build their skeletons on the skeletons of dead ones, and thus, over many years, huge undersea ledges known as coral reefs are built up by the action of these minute animals. Three main types of coral reefs are recognized: a **fringing reef** grows in shallow water, bordering the coast or separated from it by only a narrow stretch of water that can be waded when the tide is out. A **barrier reef** also is parallel to the coast but is separated from it by a channel (lagoon) deep enough to accommodate large ships. The best-known of these is the Great Barrier Reef along the northwest coast of Australia, which is about 2000 kilometers long and from a few to 150 kilometers off shore. An **atoll** is a ring-shaped island surrounding a central lagoon. Atolls are hundreds or thousands of kilometers from the nearest land and slope off into the depths of the ocean.

Examine samples of coral on demonstration in the laboratory, and note the variety of different forms. Locate and examine the small pockets that contained the polyps.

C. PHYLUM CTENOPHORA (COMB JELLIES)

The ctenophores are called *comb jellies* after the eight longitudinal bands of fused cilia that look like combs (Fig. 31-9). The synchronized beating of these cilia propel the animals, mouth-end forward, through the sea. Examine preserved specimens of ctenophores and note the jellylike consistency of their body walls, which is due to the gelatinous mesogloea. Observe the two long tentacles that contain specialized cells, which secrete a sticky substance used to catch their prey.

Reproduction in the ctenophores is only sexual, each organism having male and female reproductive organs. The fertilized eggs develop into free-swimming larvae that gradually develop into adults.

REFERENCES

Barnes, R. D. 1980. *Invertebrate Zoology.* 4th ed. Saunders.

Buchsbaum, R. 1975. *Animals Without Backbones.* 2d ed. rev. University of Chicago Press.

Fingerman, M. 1981. *Animal Diversity.* Saunders.

Lane, C. E. The Portugese Man-of-War. 1960. *Scientific American* 202(3):156–168.

Meglitsch, P. A. 1972. *Invertebrate Zoology.* 2d ed. Oxford University Press.

Storer, T. I., R. L. Usinger, R. C. Stebbins, and J. W. Nybakken. 1979. *General Zoology.* 6th ed. McGraw-Hill.

Villee, C. A., W. F. Walker, and R. D. Barnes. 1984. *General Zoology.* 6th ed. Saunders.

EXERCISE 32

Kingdom Animalia: Phylum Platyhelminthes

The members of this phylum are characterized by a dorsoventrally flattened body and are, therefore, commonly called *flatworms.* The space between the ectoderm-derived body wall and the endoderm-derived internal organs is filled with loose parenchymal tissue of mesenchymal origin. Since they have a solid body and no coelom, they are called **coelomates.** Because these organisms possess a third tissue layer, the mesoderm, they are considered to be **triploblastic.** They have advanced considerably beyond the coelenterates in that they possess well-developed nervous, excretory, muscular, digestive, and reproductive systems. The phylum includes the following three major classes:

Class Turbellaria: free-living flatworms, most of which inhabit fresh or salt water or moist places on land

Class Trematoda: flukes, which are internal or external parasites

Class Cestoda: tapeworms, the adults of which are intestinal parasites of vertebrates

A. CLASS TURBELLARIA (FREE-LIVING FLATWORMS)

Flatworms are best exemplified by a group of small freshwater organisms collectively called the *planarians.* The genus *Dugesia* is commonly found in slow streams or ponds of cool, fresh water, either adhering to or crawling over sticks, stones, leaves, and other debris.

Examine a living planarian in a Syracuse watch glass with a hand lens or dissecting microscope. Observe its general shape and size (Fig. 32-1A). From its movements, determine its anterior, posterior, dorsal, and ventral sides. What type of symmetry is evident?

What is its coloration?

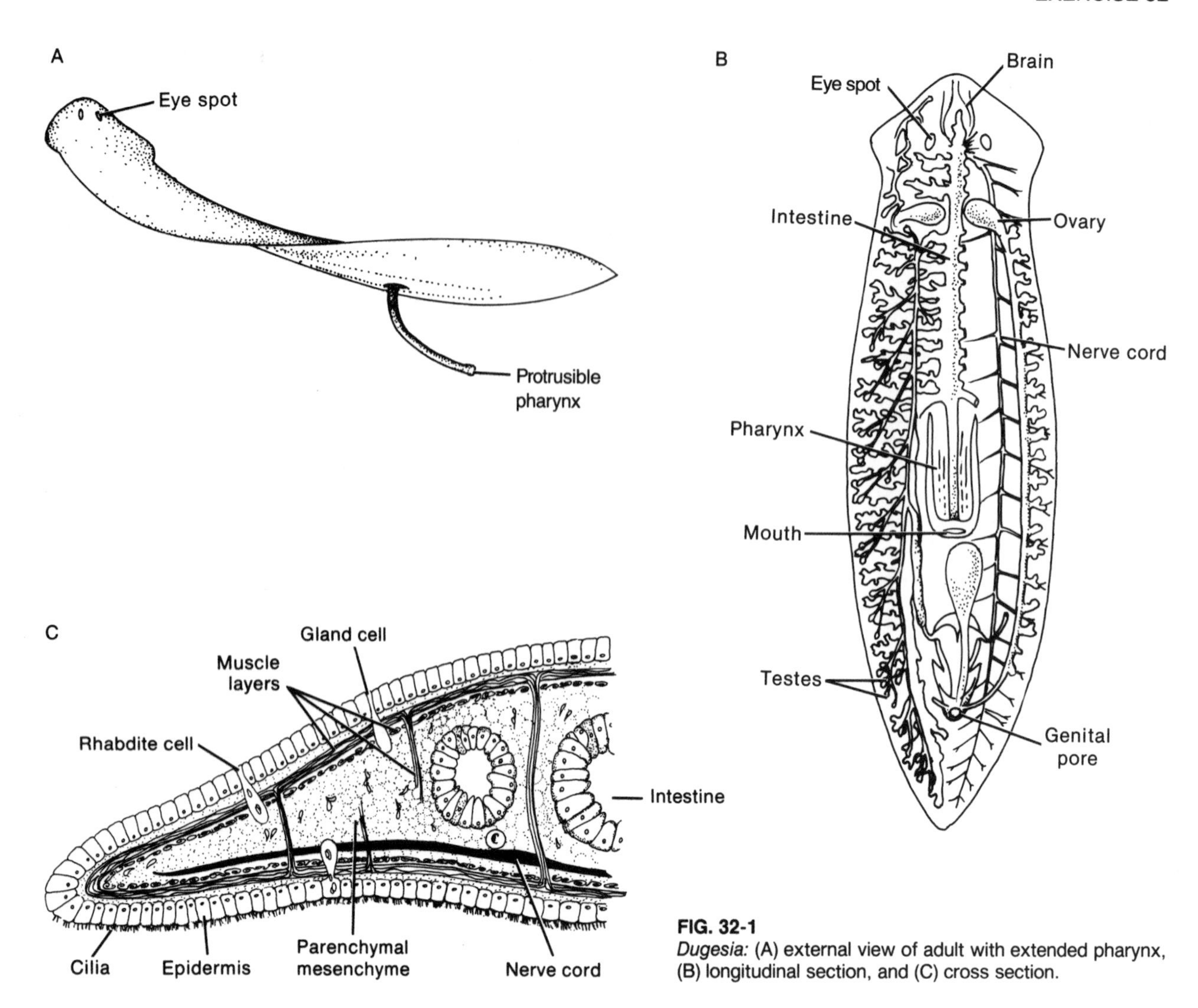

FIG. 32-1
Dugesia: (A) external view of adult with extended pharynx, (B) longitudinal section, and (C) cross section.

What are its reactions to such stimuli as jarring, contact with a toothpick, or being turned on its dorsal surface?

Observe the protrusible **pharynx** on the ventral surface. What is the advantage of such a structure?

Note the large eye spots on the anterior dorsal surface.

Using a dissecting microscope, examine a prepared slide of a whole mount of a planarian cleared to show the internal anatomy. Observe the branched digestive tract leading from the pharynx. What is the function of the numerous lateral projections from each of the three main branches of the digestive tract?

Excretory tubes and nerve elements are difficult to observe because of the thickness of the animal. Fig. 32-1B is a longitudinal section of a flatworm showing the location of the internal organs.

Examine cross sections of a planarian cut through the pharynx and through regions anterior and posterior to the pharynx. Observe the parts of the digestive tract in each of these regions (Fig. 32-1). There is no body cavity; the space between the surface epidermis and the endodermal linings of the digestive tract is filled with a mass of cells of mesenchymal origin **(parenchyma).** Observe cilia on some of the ectodermal surface cells. Where are they most prominent?

The epidermis contains rhabdite cells. It is thought that the rhabdites are secreted into the water where they swell and form a protective gelantinous sheath around the animal.

What is the arrangement of the muscle layers?

How does such an arrangement of muscles give the planarian flexibility?

Locate the two nerve cords just inside the ventral epidermis, one on either side.

Planarians have been used extensively in studies of **regeneration.** Their regenerative powers are readily demonstrated. Add several drops of the anesthetic tricaine methanesulfonate or magnesium sulfate to a Syracuse dish, and allow 15 minutes for the planarian to become completely anesthetized. Carefully cut the planarian with a sharp razor blade wiped clean of any oil. Diagram your cut in Fig. 32–2. (Your instructor will indicate different cuts to be made.) Add fresh pond water and place your dish, covered to prevent evaporation, in a cool, dark spot in the laboratory. Do not allow the water to evaporate and remove any degenerating parts, which will appear grey. Observe the planarian throughout the next 2 weeks, and make sketches of the regeneration.

B. CLASS TREMATODA (FLUKES)

Flukes are characterized as having a thick cuticle and no cilia. The mouth is usually surrounded by a sucker. Rarely, one or more suckers may also be present on the ventral surface. All members of this class are parasitic; most of their hosts being vertebrates.

1. Opisthorchis (Clonorchis) sinensis (Chinese Liver Fluke)

Although not normally found in the Western Hemisphere, *Opisthorchis sinensis,* which is prevalent in certain parts of Asia, is commonly used to illustrate the typical morphology and life cycle of flukes. The Chinese liver fluke lives in human beings and has a life cycle that requires two intermediate hosts. It is a common parasite in China, Japan, and Korea, where it has infested millions of human beings.

The adult fluke, about 15 mm long, lives in the bile ducts of the human liver where it remains firmly attached by means of a pair of suckers. It is hermaphroditic and capable of self-fertilization, though cross-fertilization is much more common. What advantage is gained from possessing the capability to self-fertilize?

The fertilized eggs that are released into the bile ducts are eventually evacuated in the feces of the host. If the feces get into water, the eggs might then be eaten by snails (Fig. 32-3). Within the digestive system of the snail, the egg opens and a larval form, called a **miracidium,** emerges and makes its way through the tissues of the snail, where it becomes asexually transformed into other larval forms. It has been estimated that a single miracidium can ultimately give rise to 250,000 infective larvae called **cercariae.** Why is it essential that one miracidium give rise to such a large number of cercariae in the life cycle of the fluke?

The cercariae escape from the snail and swim about until they come in contact with a second, intermediate, host, a Chinese golden carp. They burrow through the skin of the fish, lose their tails, and encyst to form **metacercariae.** When raw or partly cooked fish is eaten by human beings, the cyst walls are digested in the stomach and metacercariae are released. They then migrate to the bile ducts and mature into adult flukes to complete the life cycle. In light of what you know about the life cycle of the

FIG. 32-2
Regeneration in *Dugesia.*

fluke, what would be the most effective methods of controlling the infestation of human beings by this organism?

____

____

____

____

____

Examine a prepared slide of the Chinese liver fluke, and note the flattened, leaflike shape of the worm. Observe an anterior **oral sucker,** in the center of which is the mouth, and a **ventral sucker** about one-third of the way back (Fig. 32-3). Observe the bilobed intestine, and note that it has no opening other than the mouth. A pair of irregularly branched testes occupies the posterior third of the body. Tiny **sperm ducts** convey sperm cells to the **genital pore** located just anterior to the ventral sucker. The long, coiled uterus lies behind the ventral sucker. The **ovary,** a single body lying near the middle of the animal is connected to a lighter-staining **seminal receptacle** used for the storage of sperm. After copulation, the sperm cells of each animal are stored in the seminal receptacle of the other. Also connected to the ovary by means of two delicate tubules are the **yolk glands,** which consist of many small, rounded bodies characteristically found in the lateral midparts of the body. These glands supply the eggs with yolk as they develop. At the posterior end of the fluke is the **excretory pore,** through which nitrogenous wastes are excreted.

2. Fasciola hepatica (Sheep Liver Fluke)

Although generally similar to *Opisthorchis, Fasciola hepatica* is considerably larger, ranging from 15–50 mm, and has a more complex reproductive system. It commonly infests the livers of sheep and cattle,

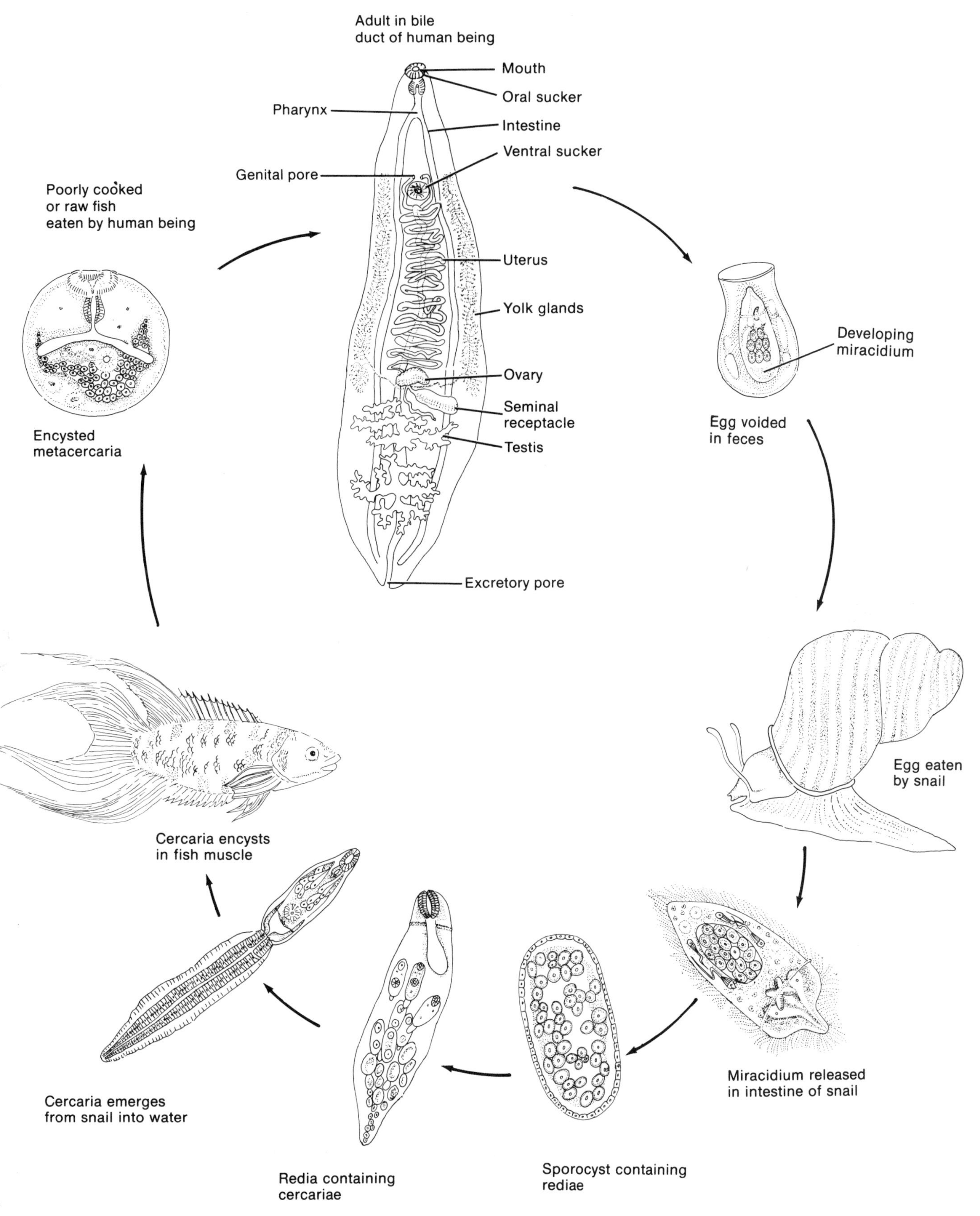

FIG. 32-3
Life cycle of *Opisthorchis sinensis*.

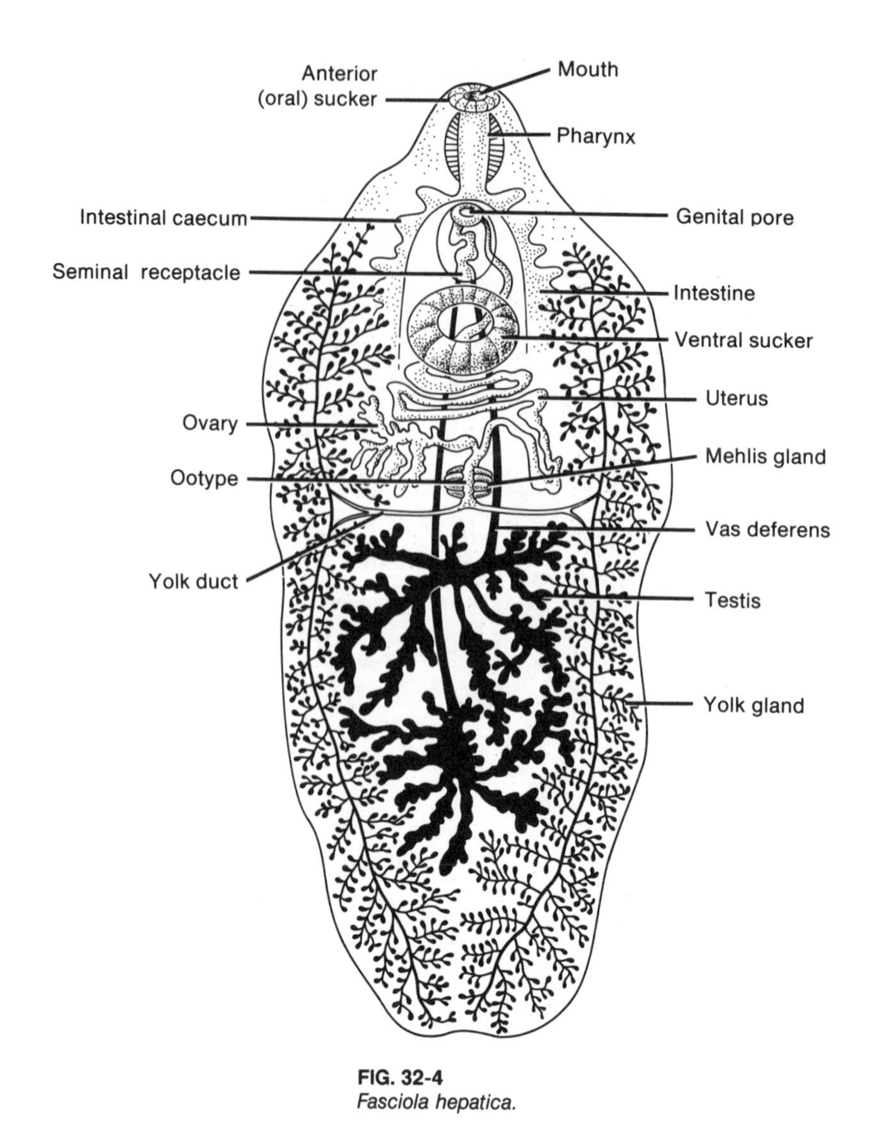

FIG. 32-4
Fasciola hepatica.

but it may also infest horses, rabbits, camels, pigs, goats, and even human beings. Human infestations of this fluke have been common in the Orient, where human feces are used to fertilize ponds and fish from these ponds are eaten raw.

Examine a demonstration slide of this fluke, and note its flat, leaflike body, which is rounded anteriorly and bluntly pointed posteriorly (Fig. 32-4). Locate the **anterior sucker** surrounding the mouth and the **ventral sucker** a short distance back from it. The digestive system consists of the mouth, a short muscular pharynx, and a branched intestine that extends along both sides of the fluke. Each of the branches has many smaller lateral branches, called **caeca.** What is the function of these caeca?

____

____

The female reproductive system, found mainly in the anterior half of the body, consists of a branched ovary connected to a median **ootype** by an oviduct. The ootype is surrounded by a **Mehlis gland.** Along each side of the body are many yolk glands that are joined to two yolk ducts with a common entry into the ootype. The male reproductive system has two highly coiled testes, each of which is connected by a vas deferens to a seminal receptacle.

3. Fasciolopsis buski (Giant Intestinal Fluke)

This organism, measuring approximately 75 mm in length, is probably the largest of the flukes found in human beings. It is common in Southeast Asia, where it is estimated to have infested 10 million people. Except for an unbranched intestine, the arrange-

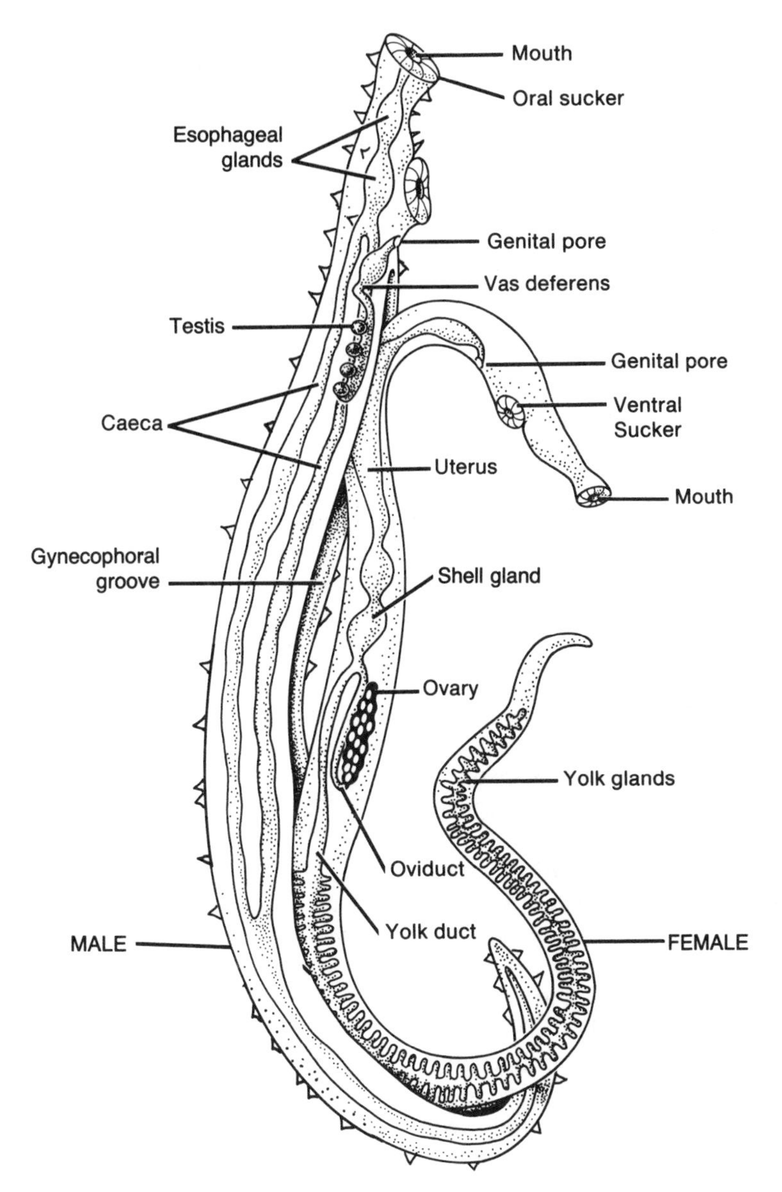

FIG. 32-5
Longitudinal section of *Schistosma* male and female that are copulating.

ment of organs in *Fasciolopsis buski* generally resembles that of the Chinese liver fluke. Examine demonstration slides of this fluke.

From your study of the life cycles and morphologies of flukes, explain how they are adapted to their parasitic life.

____

____

____

____

____

4. Schistosoma (Blood Fluke)

The schistosomes are trematodes that inhabit the intestinal veins of human and other vertebrates and are therefore commonly called **blood flukes.** Examine demonstration slides of the adults of one of the species of *Schistosoma,* and observe that they are long slender flukes in which the sexes are separate (Fig. 32-5).

The life cycle of *Schistosoma* is very similar to that of *Opisthorchis* and *Fasciola;* the snail is the intermediate host for the various larval stages. However, rather than being ingested with food, the cercariae invade their human hosts primarily by burrowing through the skin, particularly that of the feet, or by being taken in with drinking water. Cercariae of certain schistosomes that do not normally infest human beings can burrow into the skin and produce a "swimmer's itch."

C. CLASS CESTODA (TAPEWORMS)

Tapeworms are usually long, flat, ribbonlike organisms. The extent to which the tapeworm has adapted to a parasitic existence is manifest in its degenerate digestive and nervous systems and its highly developed reproductive system. The flatness of these worms enhances cellular diffusion so that there is no need for respiratory and circulatory systems.

1. Dipylidium caninum (Dog Tapeworm)

In this exercise, the common cat or dog tapeworm, *Dipylidium caninum,* will be studied as a typical tapeworm.

Fleas are the hosts for the larval stage of this species, and children are sometimes infected by it if they accidentally swallow fleas while playing with dogs or cats.

Examine a prepared slide showing characteristic segments of an adult tapeworm (Fig. 32-6). Note the head, or **scolex,** with its four suckers and several rows of hooks used for attachment to the intestine of its host. Behind the scolex is a short neck, followed by a **strobilus** consisting of a series of segments, called **proglottids.** The mature proglottids possess a high degree of internal organization. They contain both male and female reproductive organs in duplicate. The smaller ducts leading from the lateral **genital pores** are the oviducts, which lead into the diffuse ovaries. The testes are distributed throughout the proglottid and the **vas deferens,** by which the sperm leave the proglottid, is the larger duct leading to the genital pore. Lateral and transverse **excretory canals** are present in the mature proglottids, as are thin, longitudinal nerve cords.

The proglottids near the posterior end of the tapeworm become filled with many ovarian capsules, each of which contains a number of eggs. The reproductive organs observed in the mature proglottids have atrophied in these **gravid** segments, and the uterus has enlarged, filling them entirely. The gravid proglottids break loose from the tapeworm and are released in the feces of the host. The embryos released from these segments are eaten by the larvae of the dog or cat flea and develop into **cysticercoids,** which are infective. When swallowed by a dog or cat, such an infected flea will be digested by the host and the cysticercoid released. It attaches to the intestinal wall and develops into an adult tapeworm.

2. Taenia saginata (Beef Tapeworm)

The most common intestinal tapeworm to infest human beings is the beef tapeworm, which ordinarily grows to a length of 5–7 meters. In addition to cattle, the buffalo, giraffe, and llama have been recorded as natural hosts for the cysticercus stage. Meat inspection has reduced the occurrence of this parasite to about 1% of the cattle in the United States. In countries in which beef is prepared by broiling large pieces over open fires, the incidence of infection is quite high. Why?

____

____

____

Examine preserved specimens of this tapeworm, as well as slides showing scolices and mature and gravid proglottids.

3. Taenia solium (Pork Tapeworm)

The pork tapeworm that infests human beings closely resembles the beef tapeworm and has a similar life cycle, except that the larval stages, called **bladderworms** (cysticerci), develop in pigs. A mature pork tapeworm can be 2–8 meters long. People are infested by eating raw or partly cooked pork. Meat inspection has made it uncommon in the United States.

4. Dibothriocephalus latus (Fish Tapeworm)

The broad, fish tapeworm, which can grow to a length of 18 meters, is the largest tapeworm to live in human beings. This tapeworm requires two intermediate hosts to complete its life cycle. The eggs, when released into water, develop into larvae that are eaten by small crustaceans, called *copepods.* These in turn are eaten by fish, in which the larval stage **(plerocercoid)** develops. People acquire the parasite when they eat raw or partly cooked fish.

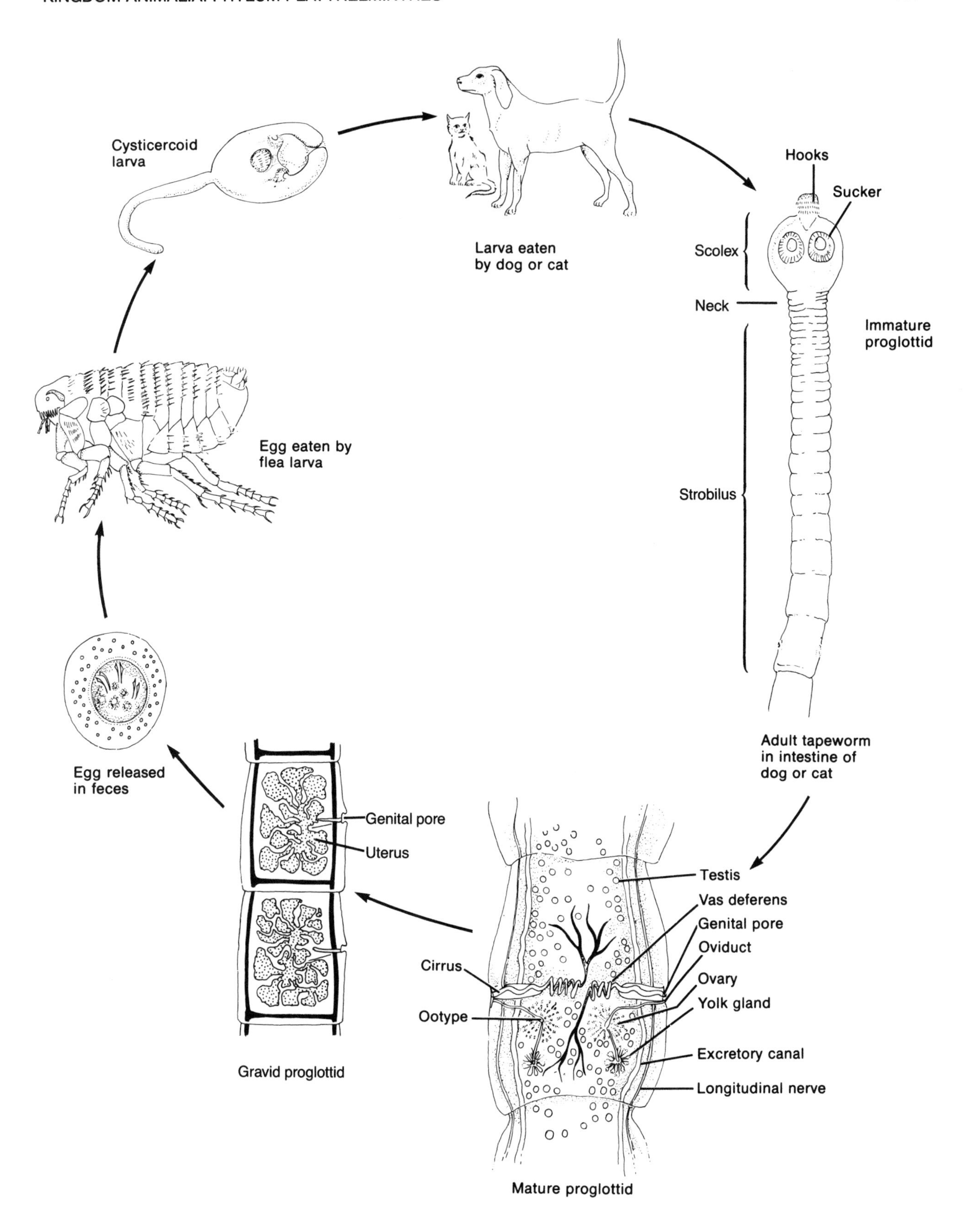

FIG. 32-6
Life cycle of *Dipylidium caninum.*

This tapeworm has recently become well established in the western Great Lakes region of North America, where it commonly infests pike and pickerel.

Examine prepared slides of the fish tapeworm, and note that the scolex has sucking grooves, rather than the suckers and hooks found in the other tapeworms you have studied. Also note that the proglottids are broader than they are long and that the reproductive organs are located in the center of a proglottid. Do you consider tapeworms to be more, or less, adapted to parasitism than flukes? Give the reasons for your answer.

____

____

____

____

____

____

REFERENCES

Barnes, R. D. 1980. *Invertebrate Zoology.* 4th ed. Saunders.

Buchsbaum, R. 1975. *Animals Without Backbones.* 2d ed. rev. University of Chicago Press.

Fingerman, M. 1981. *Animal Diversity.* Saunders.

Meglitsch, P. A. 1972. *Invertebrate Zoology.* 2d ed. Oxford University Press.

Noble, E. R., and G. A. Noble. 1982. *Parasitology.* 5th ed. Lea and Febiger.

Schmidt, G. D., and L. S. Roberts. 1981. *Foundations of Parasitology.* 2d ed. Mosby.

Storer, T. I., R. L. Usinger, R. C. Stebbins, and J. W. Nybakken. 1979. *General Zoology.* 6th ed. McGraw-Hill.

Villee, C. A., W. F. Walker, and R. D. Barnes. 1984. *General Zoology.* 6th ed. Saunders.

EXERCISE 33

Kingdom Animalia: Phyla Nematoda and Rotifera; Parasites of the Frog

The Nematoda and six related minor phyla are all characterized by having a pseudocoelom, which is a body cavity that is not entirely lined with mesoderm and is thus referred to as **pseudocoelomates.** The more advanced coelomate phyla have a body cavity that is entirely lined with mesoderm. The pseudocoelomate phyla have no circulatory system: food molecules and dissolved gases are moved from one part of the body to another by the fluid that fills their pseudocoeloms.

The pseudocoelomates include a heterogeneous group of organisms that includes rotifers, gastrotrichs, kinoryncha, nematomorphs (horsehair worms), acanthocephala, entoprocts, and one of the most successful groups in the animal kingdom—the nematodes, or roundworms. In addition to a pseudocoelom, all of the organisms are characterized by having a complete digestive tract with a separate musculature and a nonliving body covering, called a *cuticle.* The reproductive organs are found in the fluid-filled pseudocoelom.

The pseudocoelomates include the following seven phyla:

Phylum Nematoda: roundworms, including both free-living and parasitic worms having elongate cylindrical bodies tapered at both ends

Phylum Rotifera: microscopic, free-living animals found primarily in freshwater habitats and characterized by an anterior locomotor and feeding organ and a specialized internal grinding organ

Phylum Gastrotricha: microscopic, free-living aquatic animals in which the cuticle is covered by scales or spines

Phylum Kinorhyncha: microscopic marine organisms in which the cuticle is superficially segmented, with lateral spines along the body

Phylum Nematomorpha: horsehair worms that are long and slender, having a cylindrical body rounded at both ends

Phylum Acanthocephala: spiny-headed worms with a protrusible spiny proboscis; parasites in the digestive tract of vertebrates

Phylum Entoprocta: small, free-living marine animals that have a circle of ciliated tentacles on a headlike region, called a *calyx,* which is attached to a stalk anchored to the substratum

In this exercise, you will only study representatives of the rotifers and nematodes and parasites of the frog.

A. PHYLUM NEMATODA (ROUNDWORMS)

Among multicellular animals, roundworms are probably second only to insects in number. Most of the 500,000 species of nematodes are free-living. Some are found in fresh water, some in salt water, and some in mud, field, or garden soils. Other nematodes are parasites of roots, stems, leaves, and even the seeds of many horticultural and agricultural plants, causing inestimable damage to crops. Many thousands of species of roundworms are parasites of invertebrate and vertebrate animals. Among the vertebrates, it is doubtful that any species is not parasitized by one or more of the roundworms. Although more than 50 species are known to infest human beings, only about a dozen are parasitic in human beings. Most of the species that are either free-living or parasitic in plants or invertebrates are barely visible to the eye, transparent, and have simple life cycles. The species that are parasitic in vertebrates, on the other hand, can be several meters in length and have more complicated life cycles.

1. Ascaris lumbricoides (Pig Roundworm)

Ascaris is a roundworm commonly parasitic in the intestine of hogs. The sexes are separate, the male worm being shorter and more slender than the female and readily distinguished by a sharply curved posterior end. In addition, the male has a pair of hairlike structures **(spicules)** extending from the anal opening, which are used during copulation. Both male and female worms are several centimeters long. Examine preserved specimens of males and females. Using a hand lens or dissecting microscope, observe the terminal **mouth** surrounded by three lobelike **lips.** Locate the **excretory pore** just behind the mouth. Note the external chitinous cuticle.

Using a scalpel, carefully slit a female worm along the whole length of the body. Expose the internal organs by pinning the body wall back in a dissecting pan (Fig. 33-1). Cover the worm with water. Locate the long, flat digestive tract extending the length of the worm. The Y-shaped female reproductive system occupies most of the body cavity. Identify the single **vagina,** which divides into a pair of large, straight, and uncoiled **uterine tubes.** The uteri narrow down into the **oviducts,** which in turn continue into the long, thin, extremely coiled **ovaries.**

The male reproductive tract consists of a single, long, very coiled tube, which is divided into a long slender **testis, sperm duct,** an enlarged **seminal vesicle,** and a small **ejaculatory duct** opening into the terminal end of the digestive tract.

Examine a prepared slide of a cross section of the female (Fig. 33-2). The body wall consists of an outer, noncellular **cuticle** and an inner **hypodermis.** What is the function of the cuticle?

____

____

____

____

The hypodermis has four extensions into the body cavity. Two of these extensions contain the **dorsal** and **ventral nerve cords,** and the other two contain the **excretory tubes.** The greater part of the body wall is composed of muscle tissue. The digestive tract is centrally located. Observe the thin-walled intestine, and note that the intestinal wall consists of a single layer of large gastrodermal cells. The body cavity of *Ascaris* is a **pseudocoelom.** Locate the two large cross sections of the uteri, and note the shelled ova.

Sections through the oviducts and ovaries can be distinguished by their smaller sizes and by the absence of shells around the egg cells.

Look for meiotic stages in the oviduct (e.g., primary and secondary oocytes, metaphase I and II).

2. Trichinella spiralis

Infection by this parasitic roundworm, which causes **trichinosis** in human beings, results from eating raw or partly cooked pork containing the encysted larvae of this parasite. In the human intestine, the larvae are released from their cysts and develop into adult worms. The mature female worm gives birth to larval worms, which then migrate through the body and **encyst** in the muscles of the diaphragm, ribs,

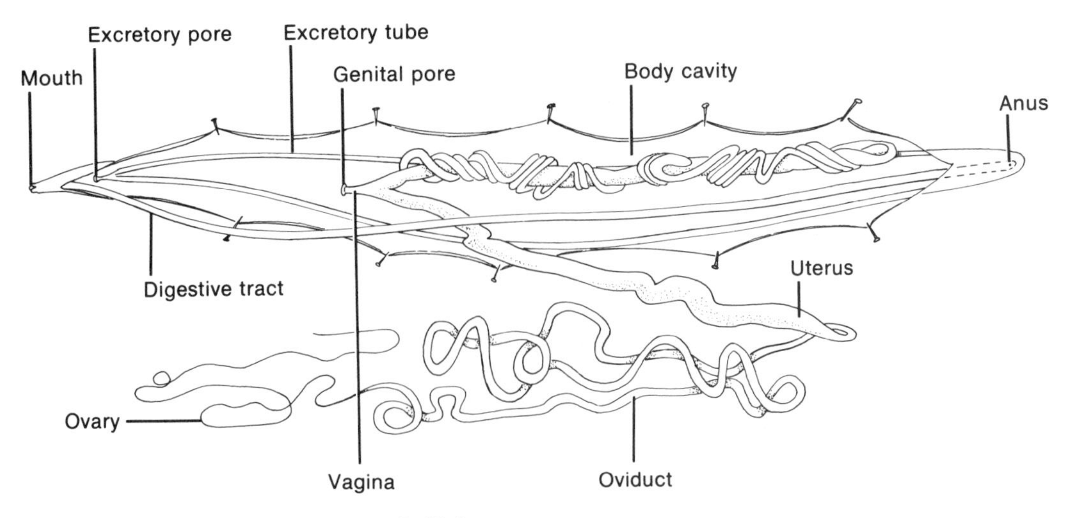

FIG. 33-1
Internal anatomy of female *Ascaris.*

and tongue. The greatest damage is done during this migration phase. Why?

____

____

____

The incidence of human infection in the United States is astonishingly high—approximately 17% of the population is infected.

Examine prepared slides of adult worms and encysted larvae.

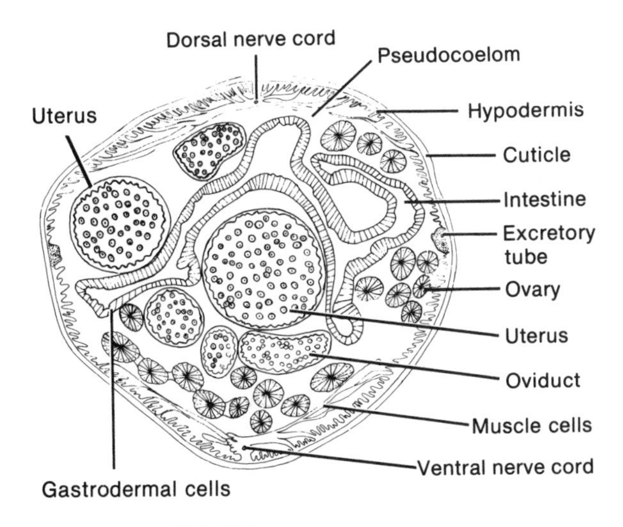

FIG. 33-2
Cross section of female *Ascaris.*

3. Necator americanus (Hookworm)

No group of nematodes causes more injury to human beings or greater economic loss through attacks on domestic animals than the hookworms. These tiny parasites have highly developed mouth cavities that contain plates by which the worm holds onto the intestine while it sucks blood and tissue. Larvae develop from eggs on moist soil in the warm climates. Infestation of human beings is by penetration of the skin by the larvae and is very often a result of going barefoot in areas where the parasite is prevalent.

Examine prepared slides of hookworms. Locate the **buccal capsule** at the anterior end of this worm, and note the cutting plates inside the capsule. At the posterior end, locate the **bursa,** an umbrellalike expansion of the cuticle, which is used in copulation.

4. Enterobius vermicularis (Pinworm)

This roundworm is one of the most common parasites of children, and its incidence in Canada and the United States is estimated to be between 30–60%. The parasite enters the body as an ovum through either the mouth or the anus: ova are easily carried under fingernails, get into the bedclothes of an infected person, or may be on toilet seats. Although pinworms cause no serious or fatal disease, infections in children lead to severe itchiness in the anal region, insomnia, restlessness, loss of appetite, and irritability.

Examine slides of adult pinworms. These small, spindle-shaped worms are characterized by a lateral expansion of the cuticle at their anterior ends. The posterior end of the male is strongly curved.

5. Free-Living Nematodes

Free-living nematodes can be found in almost any loose soil containing organic matter. Using a dissecting microscope or the low power of a compound microscope, examine a small pile of loose soil that your instructor placed in a petri dish containing agar several days earlier. Scrape several of the worms onto a slide, add a drop of water, and examine under the low and high magnifications of your microscope. With the exception of the nervous system, these nematodes are transparent, and the various organ systems can be observed. If the movements of the worms are too rapid, add a drop of dilute hydrochloric acid to your slide. Describe and account for the type of movement exhibited by these worms.

____

____

____

A very common free-living nematode is the vinegar eel *(Anguillula aceti,* formerly *Turbatrix aceti),* which is frequently found on the bottom of a vinegar barrel where it feeds on the bacteria and yeasts that have settled there. Mount a drop of vinegar-eel culture on a slide and examine microscopically. Because the vinegar eel is quite transparent, its internal organs can be easily seen and, being viviparous (i.e., giving birth to living young), all stages of development can be studied. For better observation, add a drop of 0.2% neutral red stain to the drop of vinegar and add a coverslip. If the animals are moving too vigorously for observation, slow them down by withdrawing solution with a small piece of filter paper or paper toweling placed at the edge of the coverslip.

Use Fig. 33-3 to identify the sexes and study the internal anatomy of this free-living nematode.

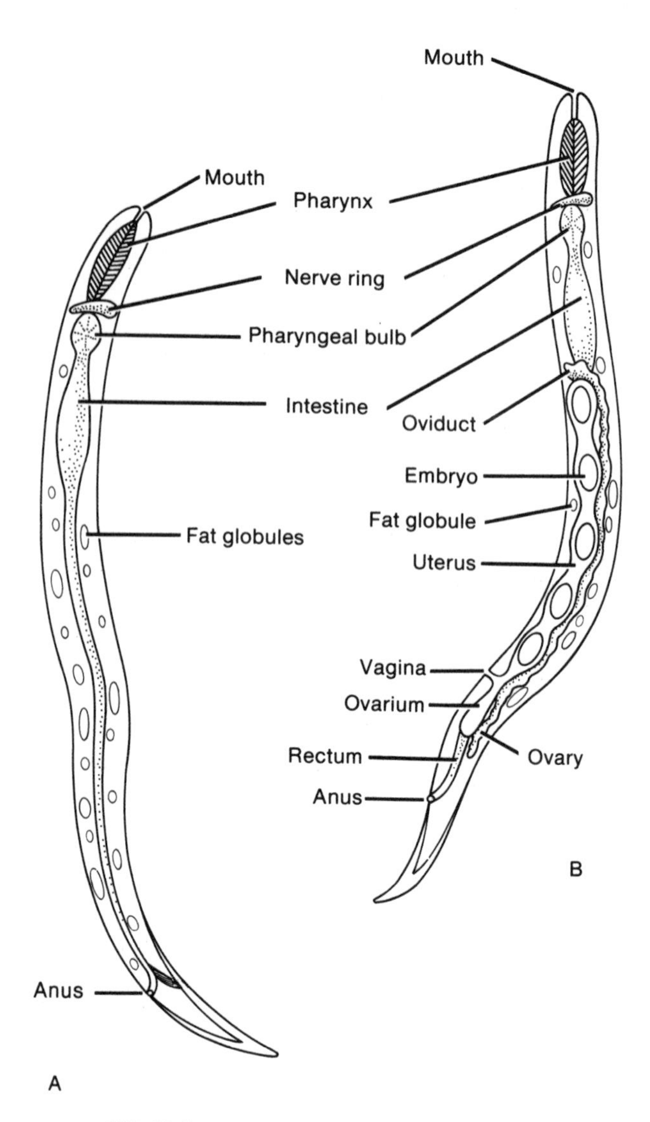

FIG. 33-3
Anguillula aceti: (A) male and (B) female adults.

B. PHYLUM ROTIFERA

Rotifers are microscopic organisms that are common and abundant in freshwater lakes, ponds, quiet streams, and wayside ditches; they are even found in street gutters and eave troughs of buildings. They constitute an important component of the plankton that is a source of food for many fish species.

On a clean slide, prepare a wet mount of a culture of rotifers that contains one or more of the genera shown in Fig. 33-4. Observe the elongate cylindrical body, which is divided into three general regions: the head, trunk, and foot. Although the cuticle is divided into several segments, the internal organs are not; therefore, this segmentation is not true, as it is in the annelids. You may observe, in a rotifer that is moving about, a telescoping of these segments as the animal expands and contracts. Observe the long, tapering **foot** and its two **spurs** at the posterior end of the body. The anterior end of the body is expanded into a crown of cilia, called a **corona,** which is used primarily for locomotion and for gathering food (Fig. 33-5). The coronal cilia create a current of water that draws food particles toward the mouth. Because the beating motion of these cilia gives the

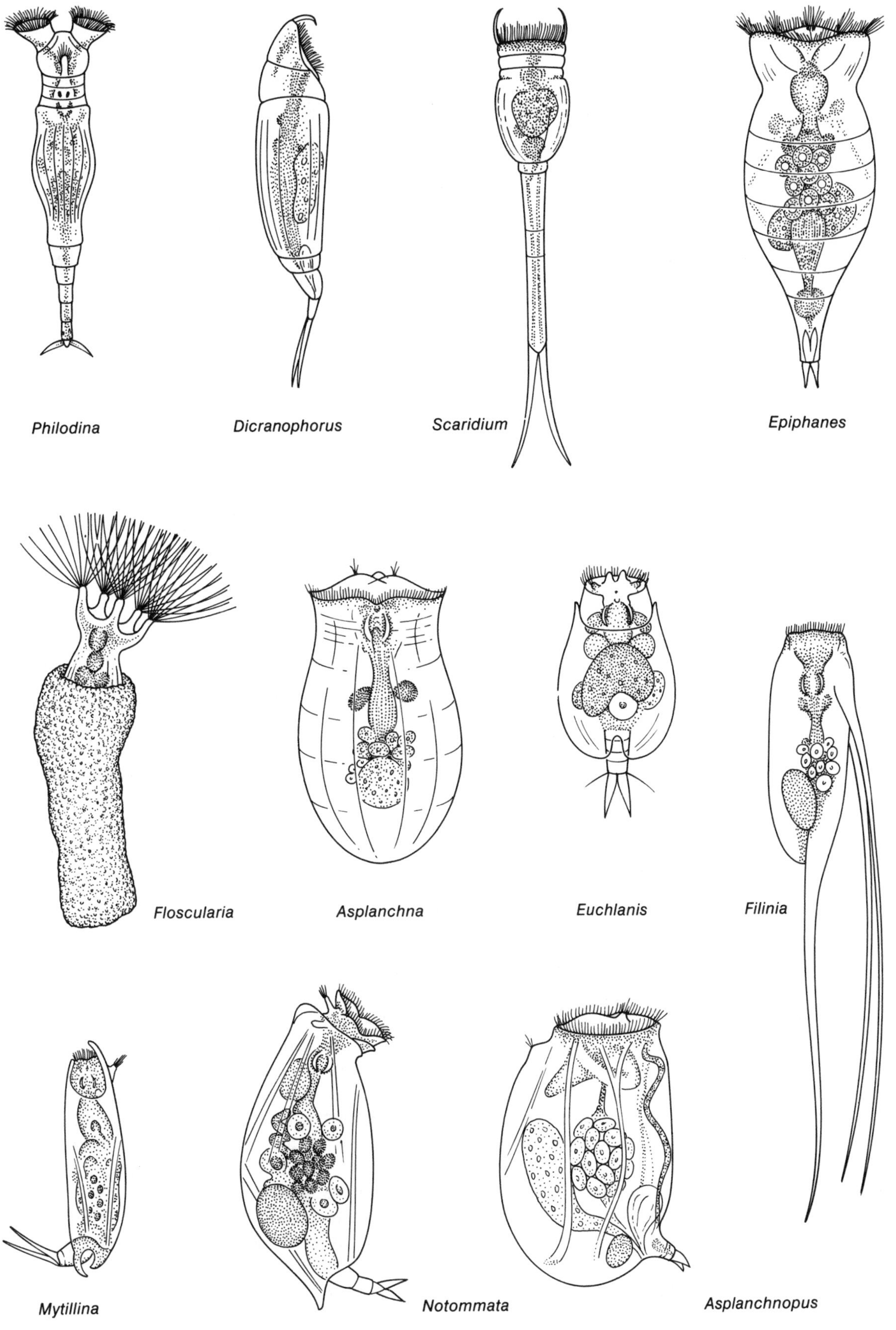

FIG. 33-4
Representative rotifers.

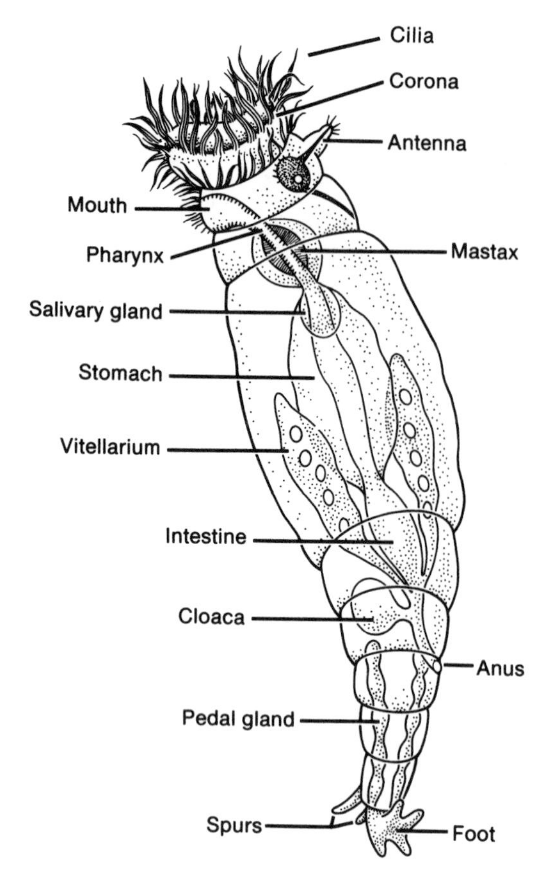

FIG. 33-5
Anatomy of a representative rotifer.

impression of a turning wheel, the organisms are sometimes called *wheel animalcules.* The mouth leads into a muscular pharynx that has a grinding apparatus, called a **mastax.** This structure can be readily observed in living rotifers because of its constant movement. Posterior to the pharynx is the narrow **esophagus,** surrounded by the large **salivary glands** and saclike **stomach.** The digestive tract is lined with cilia that move food into the **intestine** and waste products to the **cloaca,** where they are then eliminated through the **anus,** located near the base of the foot.

The female reproductive system of rotifers consists of a single ovary and a yolk-producing **vitellarium.** The eggs pass from the oviduct into the cloaca and then to the outside. Males are smaller than the females and have a single testis and a penis, which is inserted into the female oviduct during copulation.

C. PARASITES OF THE FROG

Frogs are the hosts of a number of animal parasites and are therefore excellent organisms for the study of parasite-host relationships. Pith a frog supplied by your instructor, and then cut through the ventral body wall to expose the internal organs. For the pithing procedure, see Appendix E. Refer to Fig. 33-6 in identifying some of the common frog parasites.

1. Parasites of the Blood

Place a drop of blood from one of the larger blood vessels or the kidney, place a drop on a clean slide, and draw it out in a thin film, using the edge of another slide. Add a coverslip, and examine microscopically for blood parasites. The most easily recognized forms are the extracellular flagellate trypanosomes, which are about twice the size of a red blood cell. Less readily observed are the intercellular parasites of the red blood cells. Examine the red blood cells under high power, and locate clear oval or spindle-shaped parasites in the cytoplasm. Two genera are known to parasitize the frog: *Lankesterella* and *Cytamoeba.*

2. Parasites of the Lung

Carefully remove the lungs by cutting them at their point of attachment to the larynx. Place them in a Syracuse dish containing room-temperature frog Ringer's solution, and examine for dark bodies. Cut open the lungs, and observe the movements of these parasites. Among the species of trematodes commonly found in the lung is a relatively large one called *Haematoloechus.* Mount the parasites on a slide, add a drop of Ringer's solution, and examine for details of internal anatomy.

3. Parasites of the Digestive Tract

With a scissors, slit open the digestive tract from the rectum to the esophagus. Examine each region for the presence of roundworms and flatworms. Several species of parasites are commonly found in the frog digestive tract. Carefully examine the contents of the intestine for small trematodes. *Diplodiscus,* a relatively large trematode, lives in the rectum and is recognized by its large ventral sucker.

Take a small part of the contents of the rectum, smear it thinly on a slide, add a coverslip, and identify the parasites that are present. Protozoans common to the rectum of the frog are *Opalina,* which is a large wedge-shaped, multinucleate ciliate, and *Nyctotherus,* a bean-shaped ciliate with a well-developed cytostome and cytopharynx. Other protozoan

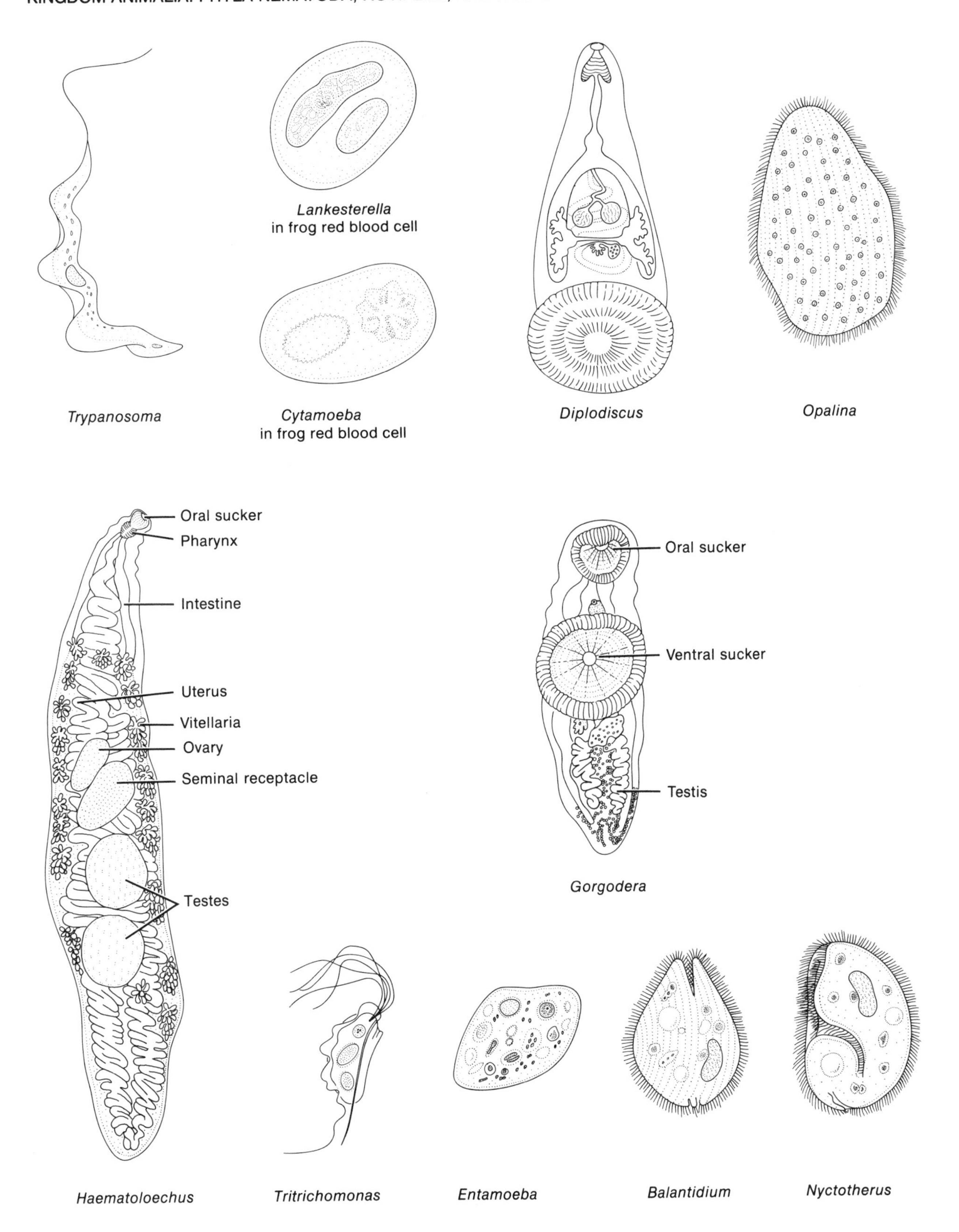

FIG. 33-6
Common parasites of the frog.

parasites that may be found in the digestive tract are *Tritrichomonas* and *Entamoeba.*

4. Parasites of the Bladder

Cut open the urinary bladder, and examine for the presence of parasites. At least six species of trematodes have been reported to exist in the frog bladder. One of the more common trematodes is *Gorgodera.*

5. Other Parasites

Carefully examine all other regions of the frog body for parasites (e.g., lining of body cavity, mesenteries, mouth, and surfaces of internal organs). Identify and list the kinds of parasites found (e.g., roundworms, flukes, tapeworms, or protozoa).

REFERENCES

Barnes, R. D. 1980. *Invertebrate Zoology.* 4th ed. Saunders.

Croll, N. A., and B. E. Matthews. 1977. *Biology of Nematodes.* Wiley.

Meglitsch, P. A. 1972. *Invertebrate Zoology.* 2d ed. Oxford University Press.

Noble, E. R., and G. A. Noble. 1982. *Parasitology.* 5th ed. Lea and Febiger.

Rainis, K. G. 1976. Enteric Protozoa in Frogs. *Carolina Tips* 39(13):49–50.

Schmidt, G. D., and L. S. Roberts. 1981. *Foundations of Parasitology.* 2d ed. Mosby.

Storer, T. I., R. S. Usinger, R. C. Stebbins, and J. W. Nybakken. 1979. *General Zoology.* 6th ed. McGraw-Hill.

Villee, C. A., W. F. Walker, and R. D. Barnes. 1984. *General Zoology.* Saunders.

EXERCISE 34

Kingdom Animalia: Phylum Mollusca

Molluscs (Latin *molluscus,* "soft"), as the origin of the name suggests, are soft-bodied animals having an internal or external shell. Included in the phylum are snails, oysters, slugs, clams, octopuses, and squids. Most molluscs are **bilaterally symmetrical** and have well-developed respiratory, excretory, circulatory, and digestive systems. Some may have a calcareous shell surrounding the body mass.

Molluscs are similar to annelids in their development. Both have **trochophore** larvae. Molluscs differ from annelids, however, in the absence of **segmentation.** Further, the coelom, so prominent in the annelids, is greatly reduced in the molluscs and is generally restricted to an area surrounding the heart.

Most molluscs are slow moving, but the bodies of several species have been highly modified for rapid locomotion. Although primarily marine organisms, some molluscs are found in fresh water (clams and snails) and on land (snails and slugs).

The molluscs are characterized by having three main body regions: a **head-foot,** which is the sensory and locomotive part of the body; a **visceral mass** containing the excretory, digestive, and circulatory organs; and the **mantle,** which secretes the **shell.** The **gills,** which function in respiration, are located inside the mantle.

Representatives of the following five major classes of molluscs will be studied in this exercise:

Class Bivalvia: characterized by containment of the visceral mass in a shell having right and left valves that are hinged together and a hatchet-shaped foot that extends out between the valves during locomotion (mussels, clams, scallops, and oysters)

Class Polyplacophora: characterized by an elliptica body covered by a shell of eight plates, although some species lack these plates, called *chitons*

Class Scaphopoda: characterized by a visceral mass enclosed in an elongated, tapered, and toothlike shell; the foot is cone-shaped; and the gills absent (tooth or tusk shells)

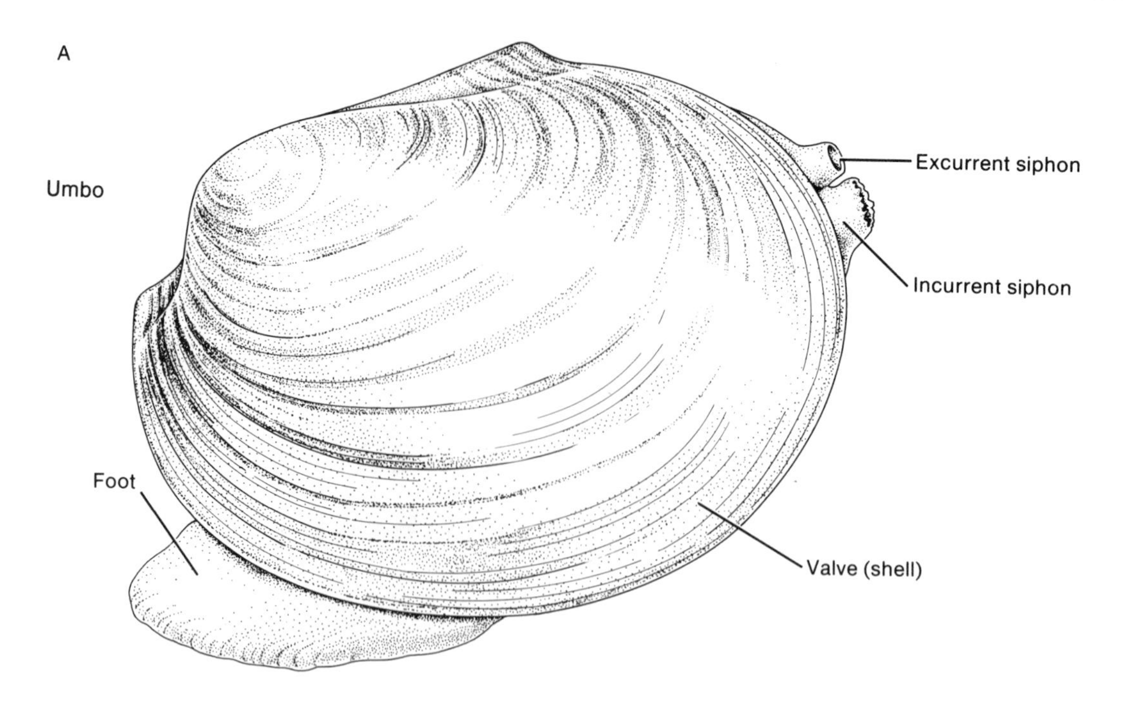

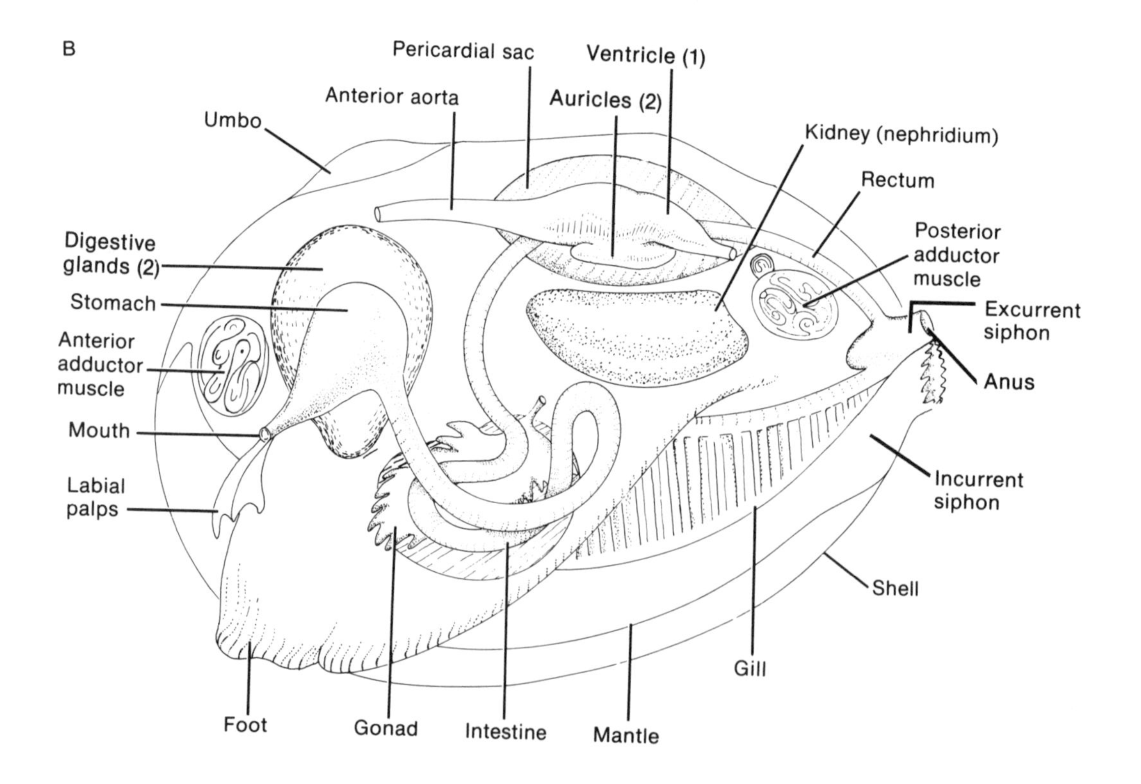

FIG. 34-1
Morphology of the freshwater clam *Anodonta:* (A) external and (B) internal anatomy.

Class Gastropoda: characterized by a viscera contained in a spirally coiled shell (although the shell in some is greatly reduced or absent); a distinct head with one or two pairs of tentacles; and a large, flat foot (snails and slugs)

Class Cephalopoda: characterized by either an internal or an external shell; a large and prominent head with conspicuous, complex eyes; and a mouth surrounded by 8–10 or more tentacles (squid and octopus)

A. CLASS BIVALVIA (MUSSELS, CLAMS, SCALLOPS, OYSTERS)

Members of this class, characterized by having a shell consisting of two **valves,** include common freshwater clams and mussels and marine varieties of clams and oysters. Freshwater clams, which will be studied in this section, are abundantly distributed and live on the bottoms of lakes, rivers, ponds, and streams where they feed on **plankton**—microscopic plant and animal life.

1. General Structure

Study the shell of the freshwater clam, *Anodonta,* and note that it consists of two valves hinged together along the dorsal side (Fig. 34-1A). On the anterior part of each valve is a protruding region, the **umbo.** The concentric lines that extend outward from the umbo are lines of growth. The valves are held together by two large **adductor muscles** located at opposite ends of the shell (Fig. 34-1B). Cut these muscles by inserting a knife or scalpel between the two valves of the shell and cutting toward the "hinge" where the valves are joined together. Open the valves, and observe that they are lined by a glistening **mantle.** The outer epithelial layer of the mantle secretes the shell, which is made up of three layers (Fig. 34-2). The thin outer layer, the **periostracum,**

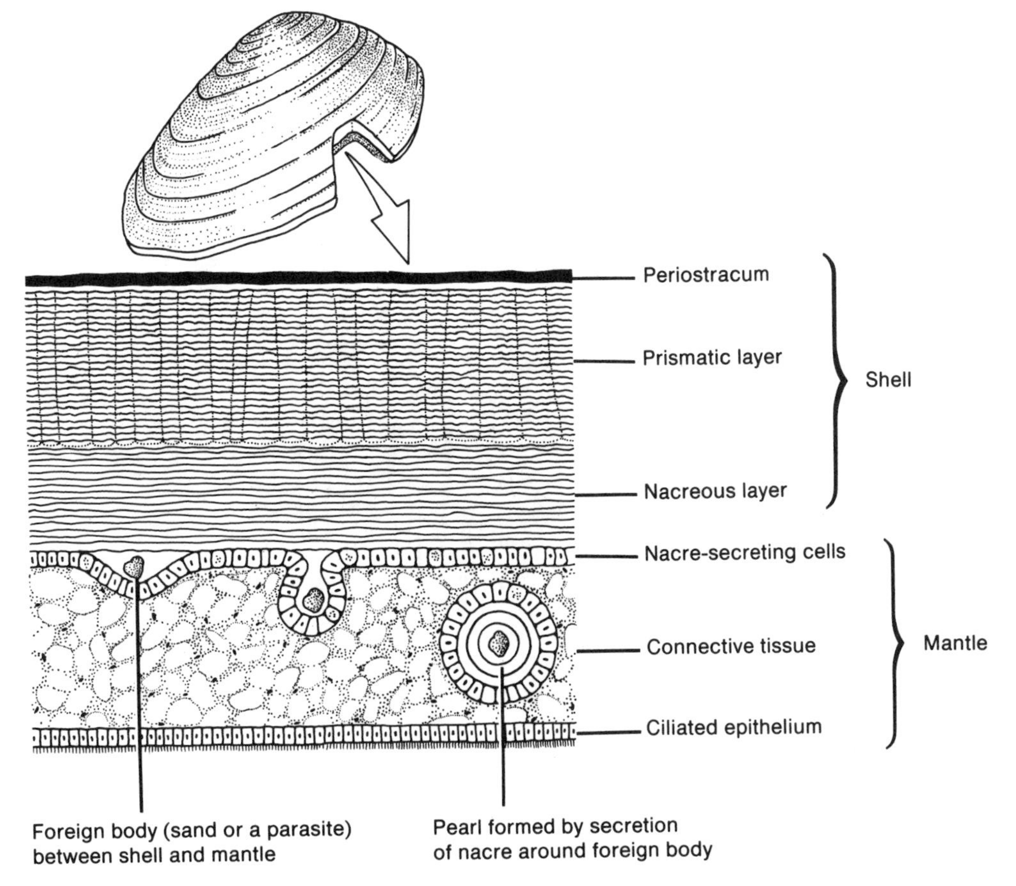

FIG. 34-2
Diagram of structure of shell and mantle.

functions to protect the underlying parts of the shell from acids in the water. It also gives the shell its color. The middle **prismatic layer** is composed of calcium carbonate. The inner **nacreous layer,** called *mother-of-pearl,* is made up of numerous layers of calcium carbonate and has an iridescent sheen.

At the posterior margin of the shell, the mantle is thicker, coming together to form two openings called the **siphons.** Water enters the clam through the ventral **incurrent siphon,** circulates through the mantle cavity and over the gills, and then leaves through the dorsal **excurrent siphon.**

Remove one of the valves and its mantle, thus exposing the mantle cavity and the internal organs. Observe the large muscular **foot** extending down from the visceral mass. Locate the **gills,** which hang down into the mantle cavity. How many gills are there? What is their function?

____

____

Locate two pair of flaplike **labial palps** on the anterior edge of the visceral mass, near the anterior muscle. These palps surround the mouth. Dorsal to the gills is the **pericardial sac,** which encloses the heart. Carefully cut open this sac, and locate the heart. Note that the heart consists of three chambers: two lateral **auricles** and one **ventricle** (Fig. 34-3). Carefully cut away the gills, and locate the kidney **(nephridium),** a dark-colored organ lying near the gills and just below the pericardial cavity. What is the function of this organ?

____

____

Most of the digestive system is located within the visceral mass (Fig. 34-1B). *Carefully* cut the visceral mass lengthwise into left and right halves. The **mouth,** located between the palps, leads by way of a short **esophagus** into an expanded **stomach,** which is flanked on either side by two large **digestive glands.**

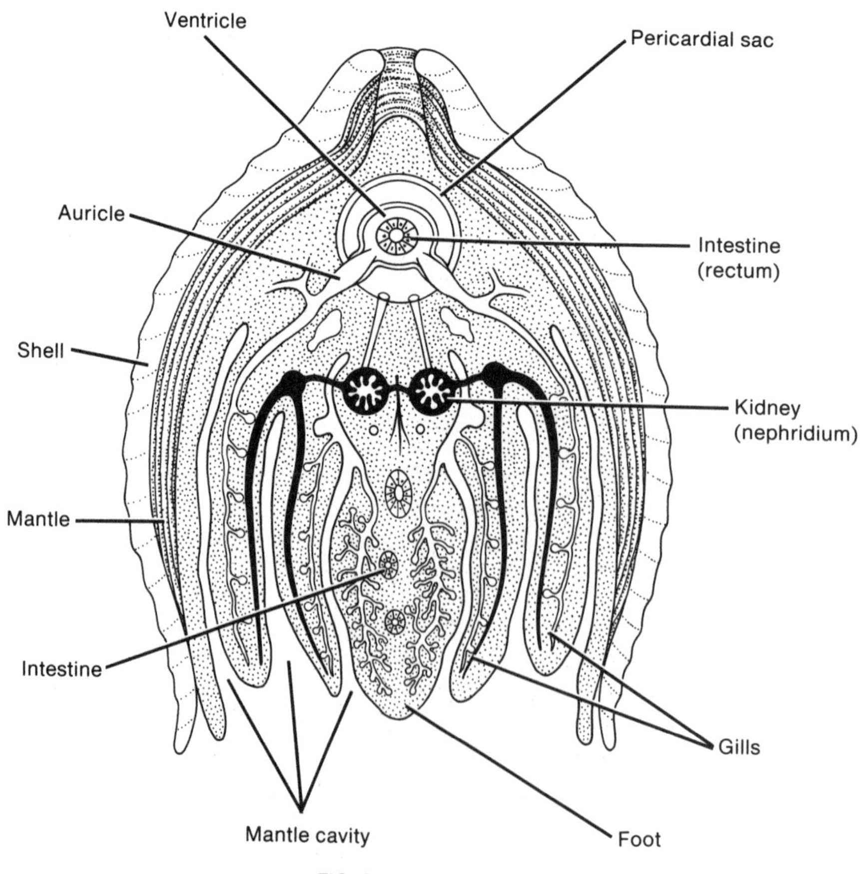

FIG. 34-3
Cross section of a clam.

Remove these glands carefully, and note that the stomach leads into an **intestine** that winds through the visceral mass and then passes through the pericardial sac as the **rectum.** The rectum empties into the **excurrent siphon** through the **anus.** Locate as many parts of the digestive tract as possible in your dissection.

2. Reproduction

Most mussels and clams are male or female; a few species are hermaphroditic. The reproductive cycle in these organisms is quite interesting in that the juvenile stage is parasitic on fish (Fig. 34-4). The eggs are released into the cavity of the gills where fertilization takes place. Each zygote then develops into a larva, called a **glochidium.** The larvae stay within the gill through the winter and are released into the water the following spring. If they come into contact with a fish, a contact stimulus causes them to close their valves and thus become attached to the gills or the fins of the fish host. What is the advantage of their being attached to the gills or fins instead of the body wall?

____

____

____

____

____

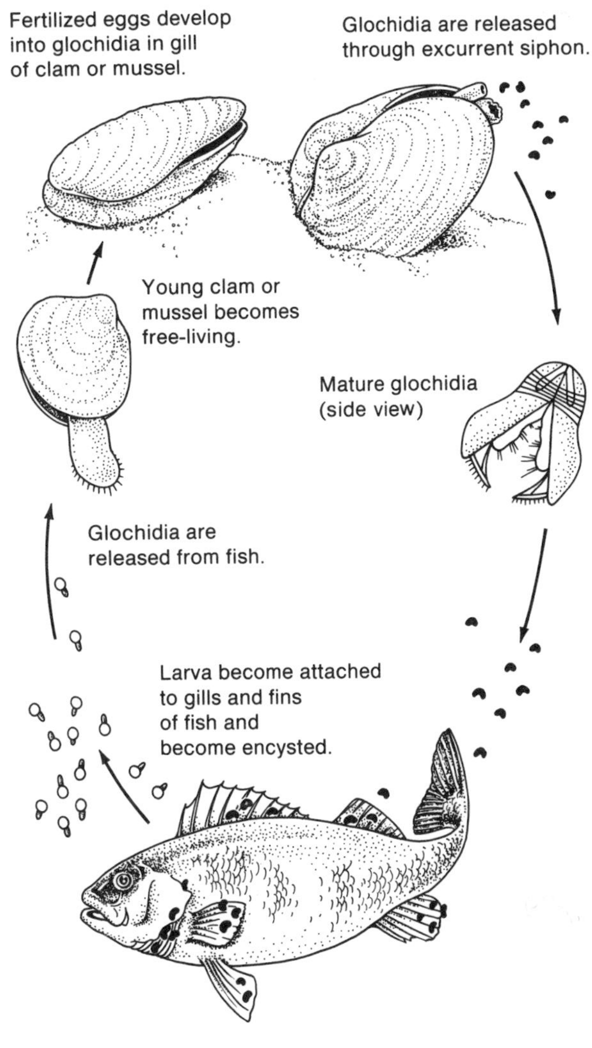

FIG. 34-4
Life cycle of a clam or mussel.

The tissue of the fish reacts by growing around the larva.

After several weeks, the parasitic larval form is released and begins a free-living existence. Examine slides or preserved specimens of glochidia. Note the toothlike appendages on the valves, which function to attach the larva to the fish.

B. CLASS POLYPLACOPHORA (CHITONS)

These molluscs, sometimes called *coat-of-mail shells,* are primitive marine organisms. They have a characteristic shell composed of eight plates (Fig. 34-5A). A large, broad, flat muscular foot occupies the greater part of the ventral surface. They are usually found on rocky seashores or in water less than 25 fathoms deep. They attach tightly to rocks through the suction produced by the foot. When pulled off, they tend to roll up, much like an armadillo, with their soft parts covered by the hard shell. Examine the chitons on demonstration.

C. CLASS SCAPHOPODA (TOOTH, OR TUSK, SHELLS)

The visceral mass of these marine molluscs is enclosed in a long, tubular, toothlike shell that is open at both ends (Fig. 34-5B). The foot is typically cone-shaped and is used for burrowing into mud or sand head first, leaving the narrow end of the shell exposed above the surface. Examine the tooth shell on demonstration.

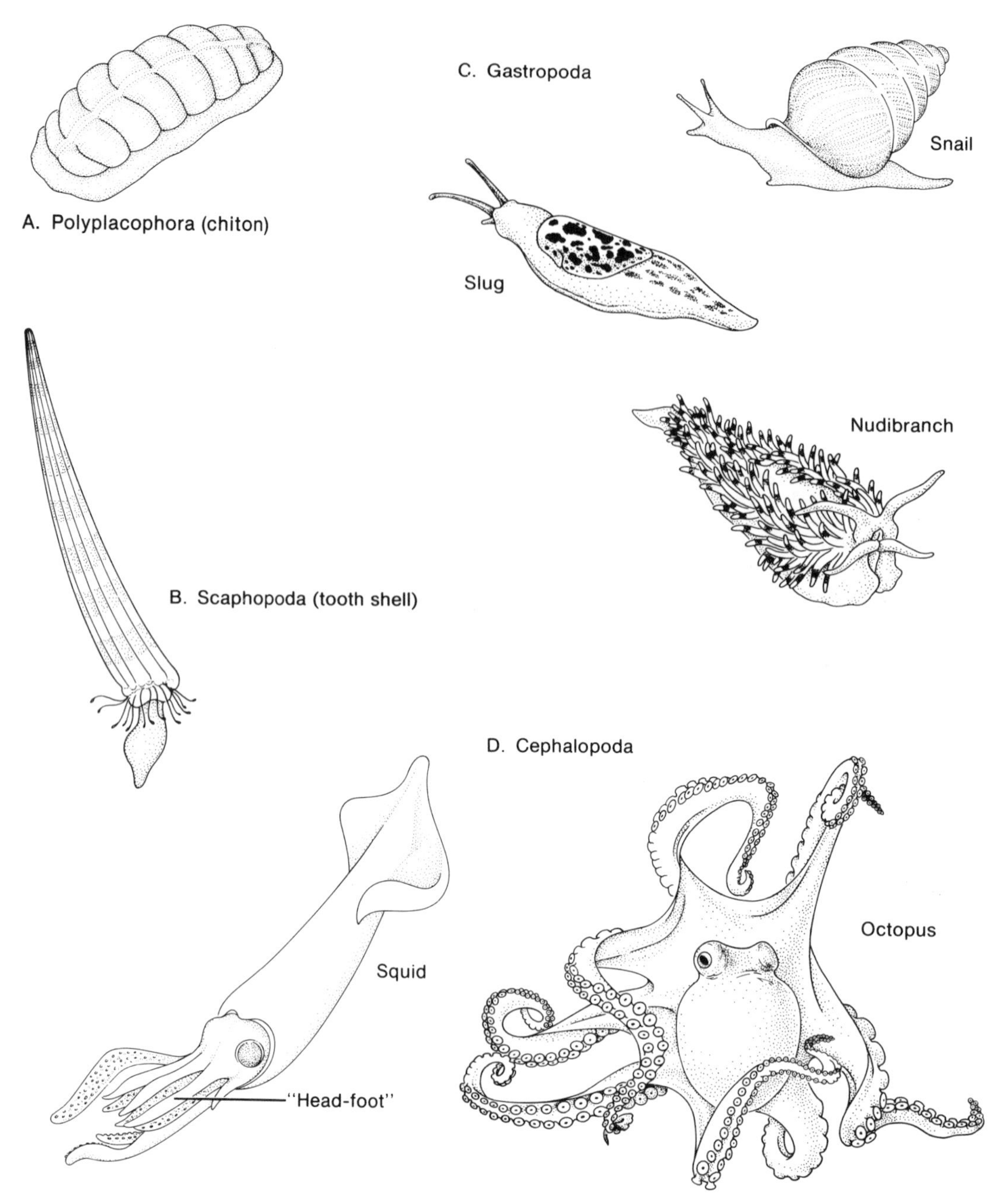

FIG. 34-5
Representative classes of other molluscs.

D. CLASS GASTROPODA (SNAILS AND SLUGS)

This is by far the largest and most diverse class of molluscs (Fig. 34-5C). Most gastropods are marine, but some live in fresh water and others have become adapted to land. For all forms, the visceral mass is enclosed in a coiled shell during the early stages of development. In most species of gastropods, adults retain the coiled shell, but in slugs, not only has the shell been lost but some species have become land dwellers.

Snails have an unusual mode of locomotion: a **slime gland,** located in the forward part of the foot, secretes a thin film of mucus over which the snail then moves by means of wavelike contractions of the muscles of the foot.

Slugs, because they lack a protective shell, are restricted to moist areas to prevent desiccation. Consequently, they are not very active in the daytime, feeding at night instead.

Sea slugs **(nudibranchs)** live among the seaweeds. Some species so resemble the plants that they are difficult to see. Others have **warning coloration** and are protected by stinging cells.

Examine demonstrations of various gastropods.

E. CLASS CEPHALOPODA (SQUIDS AND OCTOPUSES)

The cephalopods are considered to be the most advanced and highly developed class of molluscs (Fig. 34-5D). In contrast with other molluscs, the cephalopods are active, free-swimming animals.

All cephalopods are marine organisms characterized by the modification of a foot to form tentacles and a head with prominent, highly developed eyes. The eye is remarkably similar to the vertebrate eye in that it has an eyelid, iris, pupil, lens, cornea, and retina. In some species in the class Cephalopoda, the shell is external; in others, it is internal. Some species reach sizes of several meters in length. The giant squid of the North Atlantic is the largest living invertebrate.

Examine a squid. What morphological characteristics can you see on this organism that are adaptations to a predatory existence?

____

____

____

____

Because the squid relies on its ability to swim rapidly for protection, it has no need of a cumbersome external shell. Consequently, the shell is a vestigial structure, consisting of a horny plate buried in the visceral mass. Interestingly, the shell, called a **cuttle bone,** is sold in pet stores. It is attached to a bird's cage to be used by the bird to "sharpen" its beak.

Next examine an octopus. How are the squid and octopus similar?

____

____

____

____

How are they different?

____

____

____

____

REFERENCES

Barnes, R. D. 1980. *Invertebrate Zoology.* 4th ed. Saunders.

Buchsbaum, R. 1975. *Animals Without Backbones.* 2d ed. rev. University of Chicago Press.

Meglitsch, P. A. 1972. *Invertebrate Zoology.* 2d ed. Oxford University Press.

Purchon, R. D. 1977. *The Biology of the Mollusca.* 2d ed. Pergamon.

Solem, A. 1979. *The Shall Makers.* Wiley.

Villee, C. A., W. F. Walker, and R. D. Barnes. 1984. *General Zoology.* 6th ed. Saunders.

EXERCISE 35

Kingdom Animalia: Phylum Annelida

Organisms in the phylum Annelida (Latin *anellus,* "ring") are referred to as segmented worms to distinguish them from nonsegmented flatworms and roundworms. The body is composed of a series of similar parts called **somites.**

Annelids exhibit **cephalization;** that is, the nervous system is concentrated at the anterior end of the organism in structures called **cerebral ganglia.** A ventral nerve cord arises from these ganglia and passes posteriorly. The excretory and circulatory systems are well developed. The digestive tract is straight and tubular and, in contrast with that of the roundworms, is supplied with its own musculature allowing it to function independently of the muscular activity associated with the body wall. Annelids, like roundworms, have a fluid-filled body cavity separating the digestive tract from the body wall. However, in annelids this cavity is a **true coelom** that is formed through splitting of the mesoderm in the course of development.

The habitats of members of this phylum are widespread: they are found in marine or fresh waters, or on the land. The phylum includes the following three classes, all of which will be studied in this exercise:

Class Polychaeta: having conspicuous segmentation and lateral projections called **parapodia,** a prominent head that has tentacles, and separate sexes in most species; most are marine organisms (*Neanthes:* clam worm)

Class Oigochaeta: having conspicuous segmentation but lacking a well-defined head; members are hermaphroditic and are usually found in fresh water or moist soils (earthworms)

Class Hirudinea: having dorsoventrally flattened bodies with a large posterior sucker and inconspicuous segmentation; members are hermaphroditic and are found in fresh and marine waters (leeches)

A. CLASS POLYCHAETA

The polychaetes consist primarily of free-living marine organisms that are abundant in depths ranging from the low-tide line to about 50 meters. Many are found in the intertidal zone and a few have been found in depths of more than 4500 meters.

Because of their abundance (thousands have been found in a square meter), these animals are important in the marine food chain. They are eaten by hydroids, flatworms, starfish, fish, and even other marine annelids.

Polychaetes have a particularly interesting external form to study. Even with the numerous and varied modifications that have enabled them to adapt to a wide variety of habitats, these organisms have retained a fundamental organization that is common to all polychaetes. This basic body structure can be observed in *Neanthes virens,* the marine clam worm that lives in burrows in the sand at tide level (Fig. 35-1A). It remains in its burrow throughout the day, coming out at night to feed.

Examine a specimen of *Neanthes.* Note that the long, slender body is somewhat compressed, being rounded dorsally and flattened ventrally. It is composed of about 200 segments, called **somites.** The head is generally well developed and contains the mouth, which is retracted except when the organism is feeding, at which time the pharynx is everted and the jaws used in capturing small animals are exposed (Fig. 35-1B, C). Over the mouth region is the highly modified segment called the **prostomium,** which has a pair of **antennae,** pairs of **simple eyes,** and laterally placed conical **palps.** The **peristomium** bears four tentacles. What function might these tentacles serve?

____

____

____

Posterior to the head region are a variable number of similar segments, each of which bears a pair of fleshy outgrowths called **parapodia.** Each parapodium contains several terminal, bristlelike structures called **setae.** Suggest a function for the setae.

____

____

____

B. CLASS OLIGOCHAETA

The common earthworm, *Lumbricus terrestris,* will be studied in detail to illustrate the main characteristics of the oligochaetes. The bodies of most earthworms are divided externally and internally into well-defined segments that are separated from each other by membranous partitions. Except for the tail and head regions, all segments are essentially alike. The earthworm hunts food at night and thus has been called a "night crawler." It extends its body from the surface opening of a small tunnel, which it makes by "eating" its way through the soil. The rear end of the worm's body remains near the opening while the head end forages for decaying leaves and animal debris.

It has been estimated that an acre of good soil contains more than 50,000 earthworms. By their continual foraging and tunneling, these worms turn over from 18–20 tons of soil per acre and bring more than 25 mm of rich soil to the surface every 4–5 years. Thus, indirectly, the earthworm enriches farmland, assisting farmers in the production of greater quantities of food for a rapidly expanding population.

1. External Anatomy

Obtain a specimen of *Lumbricus* and, using Fig. 35-2 as a guide, study its external anatomy. A hand lens or dissecting microscope will be helpful in identifying the smaller features.

At the anterior end (**prostomium**) is a small fleshy projection over the mouth. It is not considered to be a segment of the worm. At the posterior end is the **anus,** the opening at the end of the digestive tract through which solid wastes are expelled.

About one-third of the way back from the mouth region is a thick cylindrical collar—the **clitellum.** This structure, which functions in reproduction, will be considered later.

Place the worm so that the ventral side is uppermost. The ventral surface of a living worm is a lighter color than the dorsal (upper) surface. With your finger, lightly stroke the ventral surface in an anterior direction. The bristles you feel are the setae and are used by the worm in movement. How many pairs of setae are there in each segment of the worm?

____

____

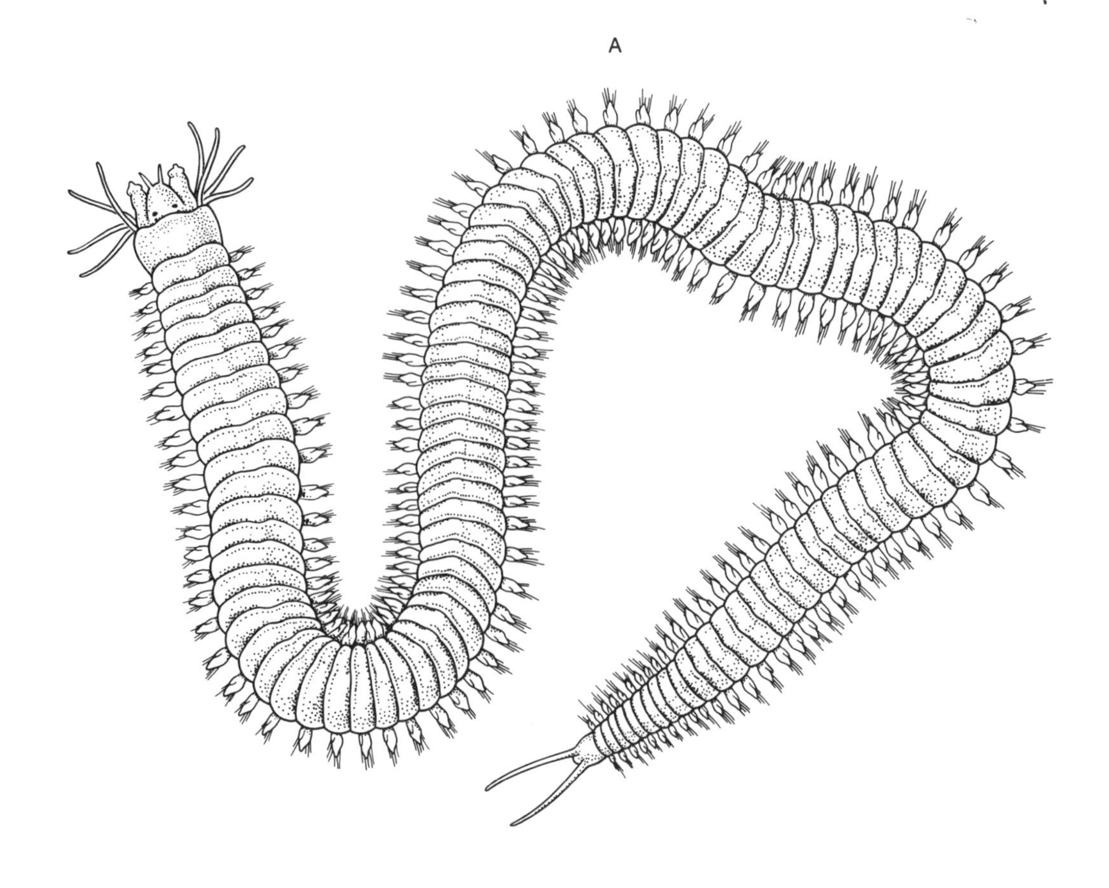

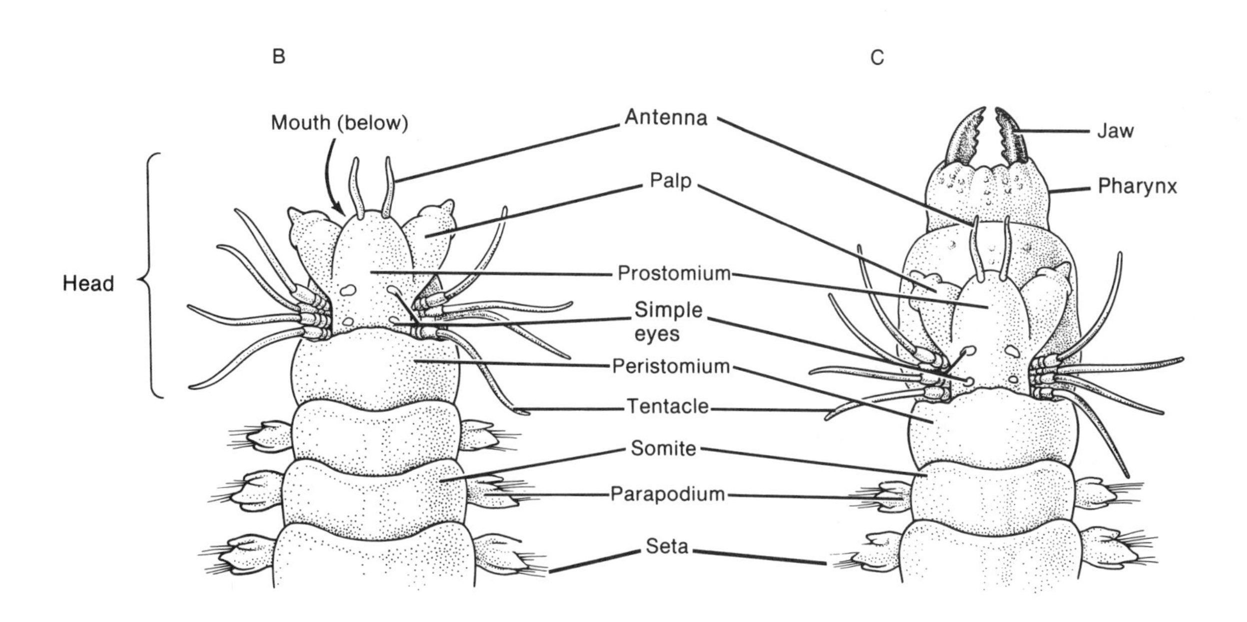

FIG. 35-1
(A) External anatomy of *Neanthes virens.* (B) Dorsal view of head with pharynx retracted. (C) Dorsal view of head with pharynx everted.

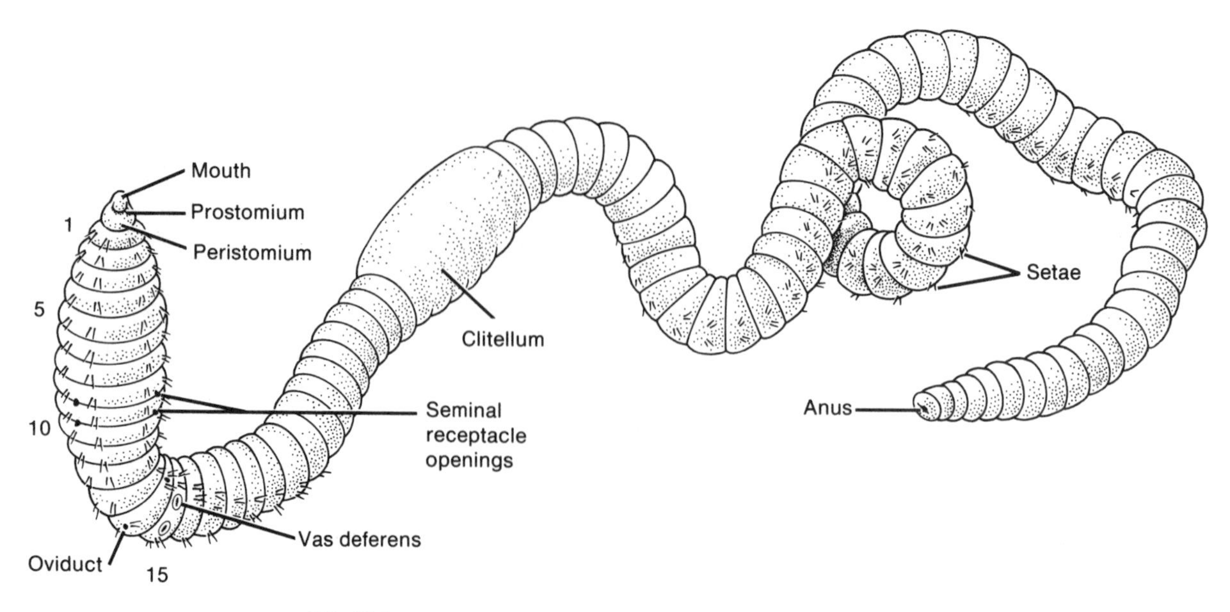

FIG. 35-2
External anatomy of the common earthworm, *Lumbricus terrestris.*

Which segments do not have setae?

Each segment (except the first three and the last one) contains pores, called **nephridiopores.** These small openings connect with the **metanephridia,** which are the primitive kidneys of the earthworm. Liquid wastes, which collect in the body cavity, are excreted through these nephridial openings. A pair of rather large openings of the **vas deferens** (male) are located on each side of Segment 15. Sperm are released from the worm through these openings. Eggs are released through openings of the **oviducts** (female) located in Segment 14.

2. Internal Anatomy

Place the earthworm on the dissecting tray, *dorsal side up,* and pin into position. To expose the internal organs, dissect the worm as outlined in Fig. 35-3. With the help of Figs. 35-4 and 35-5, locate the structures described next.

a. Digestive System

The **mouth** is located at the anterior end. The opening is just below the overlapping prostomium.

The mouth leads to a slightly expanded and muscular **pharynx.** Food taken in by the animal is passed on by muscular contractions in the pharynx through the **esophagus,** which is covered by three pairs of whitish **seminal vesicles,** to the **crop** where it is temporarily stored.

The crop opens into a thick-walled, highly muscular **gizzard** where, with the aid of small soil particles taken in during feeding, food is ground up. It is then passed into the **intestine,** where it is digested and absorbed.

Solid waste products of digestion are passed to the exterior through the **anus.**

b. Circulatory System

An intestine feature of the circulatory system is that it is a "closed" system in which the blood circulates within a series of blood vessels. The blood is red because it contains hemoglobin, the same pigment that gives the red color to human blood. The hemoglobin, however, is not contained in cells but is dissolved in the plasma.

The major vessels of the circulatory system are a **dorsal longitudinal vessel** lying on top of the digestive tract and a **ventral blood vessel** lying below it. These two vessels are connected to each other by five pairs of vessels passing around the esophagus. They are larger than the other blood vessels and constitute the **hearts.** Although the hearts contain valves and contractile tissue and serve as pumps to circulate blood through the animal, they are thought to play a minor role in circulation. The dorsal and ventral blood vessels are the major circulatory pumps. Associated with the hearts and surrounding the eso-

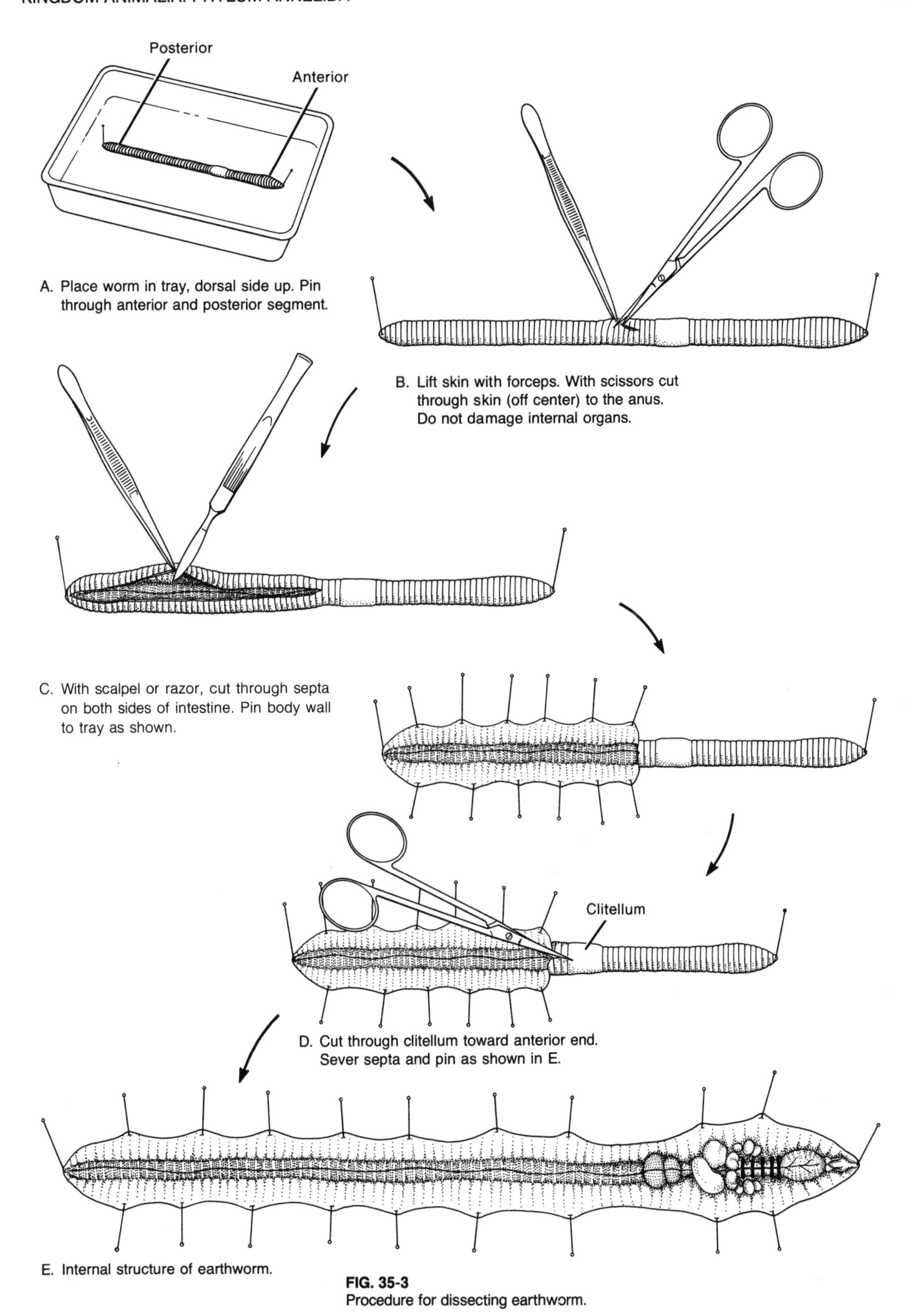

FIG. 35-3
Procedure for dissecting earthworm.

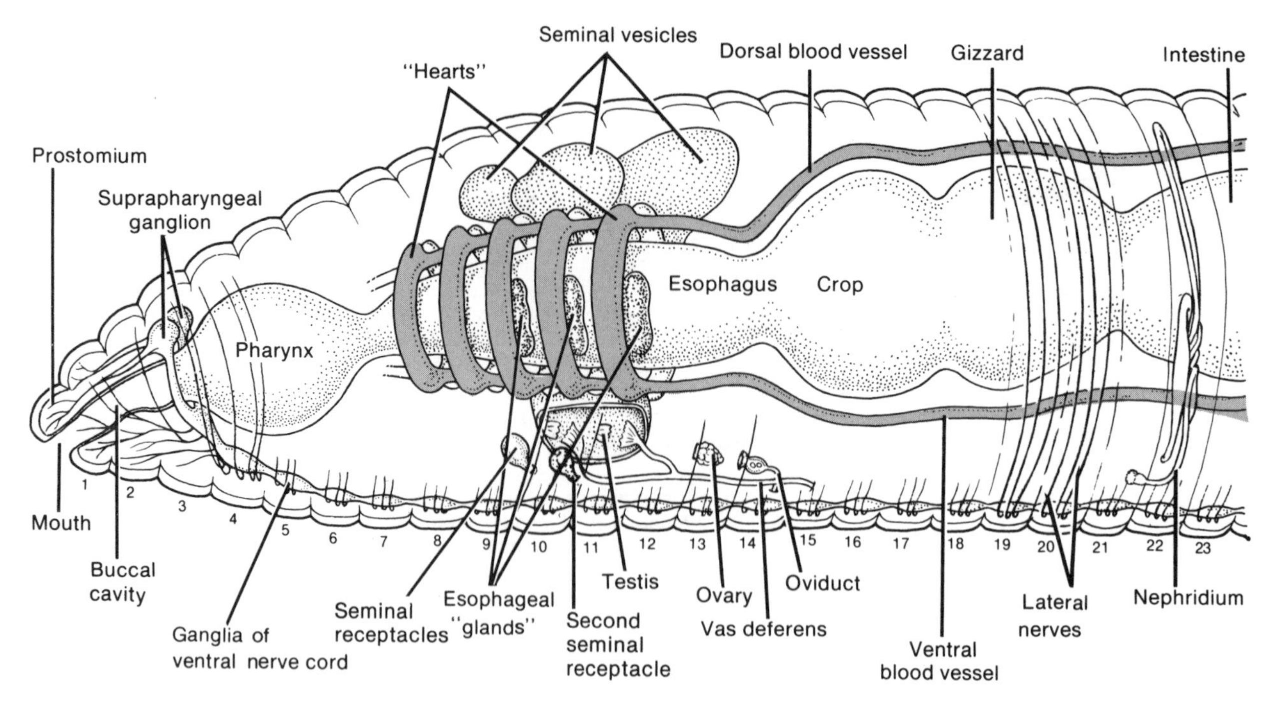

FIG. 35-4
Internal structure of *Lumbricus:* lateral view.

phagus are two pairs of **esophogeal** (calciferous) **glands,** which are considered to be excretory organs that function in controlling the amount of calcium in the blood.

Most of the other blood vessels are difficult to observe unless the material has been sectioned for microscopic study.

c. Reproductive System

To obtain a clear view of the reproductive organs, cut through the intestine near the clitellum. Carefully lift the intestine and tease it free as far forward as the posterior end of the pharynx. Then cut it out.

As previously stated, earthworms are **hermaphroditic,** having complete sets of male and female reproductive organs. The male system consists of a pair of trilobed **seminal vesicles** located between Segments 9 and 13 (Fig. 35-5). Within these vesicles, sperm are produced in the **testes.** Tear off one of the lobes, smear it in a small drop of water on a slide, and add a coverslip. By using high magnification, you should be able to observe sperm. The vesicles connect with the **vas deferens** (sperm duct), which exits to the outside in Segment 15.

To observe the female reproductive organs, it may be necessary to remove the seminal vesicles. The female system consists of (1) a pair of **ovaries** in Segment 13; (2) a pair of **oviducts,** opening into Segment 13 by way of a ciliated **egg funnel** and an opening to the outside through the ventral surface of Segment 14; (3) an **egg sac** (a small pouch branching from the egg funnel); and (4) two pairs of **seminal receptacles** in Segments 9 and 10.

Although earthworms are hermaphroditic, they do not undergo self-fertilization. Rather, sperm are transferred between worms during **copulation:** two worms come together along their ventral sides and become temporarily joined together by the secretion of a "slime tube"; sperm are discharged from the seminal vesicles of each worm, pass long seminal grooves on the ventral body surfaces, and enter the **seminal receptacles** of the other worm.

When eggs leave the ovaries, the glandular clitellum secretes a tube of mucus that slides over the anterior segments and picks up eggs from the oviducts in Segment 14 and sperm from Segments 9 and 10. The tube of mucus finally slips over the anterior end of the worm to form the egg cocoon, from which the young eventually hatch.

d. Nervous System

The nervous system in the worm is difficult to study. Its major component is the **ventral nerve cord,** which runs the length of the worm on the inner ventral surface. At its anterior end, the cord divides and passes around the front part of the pharynx where it

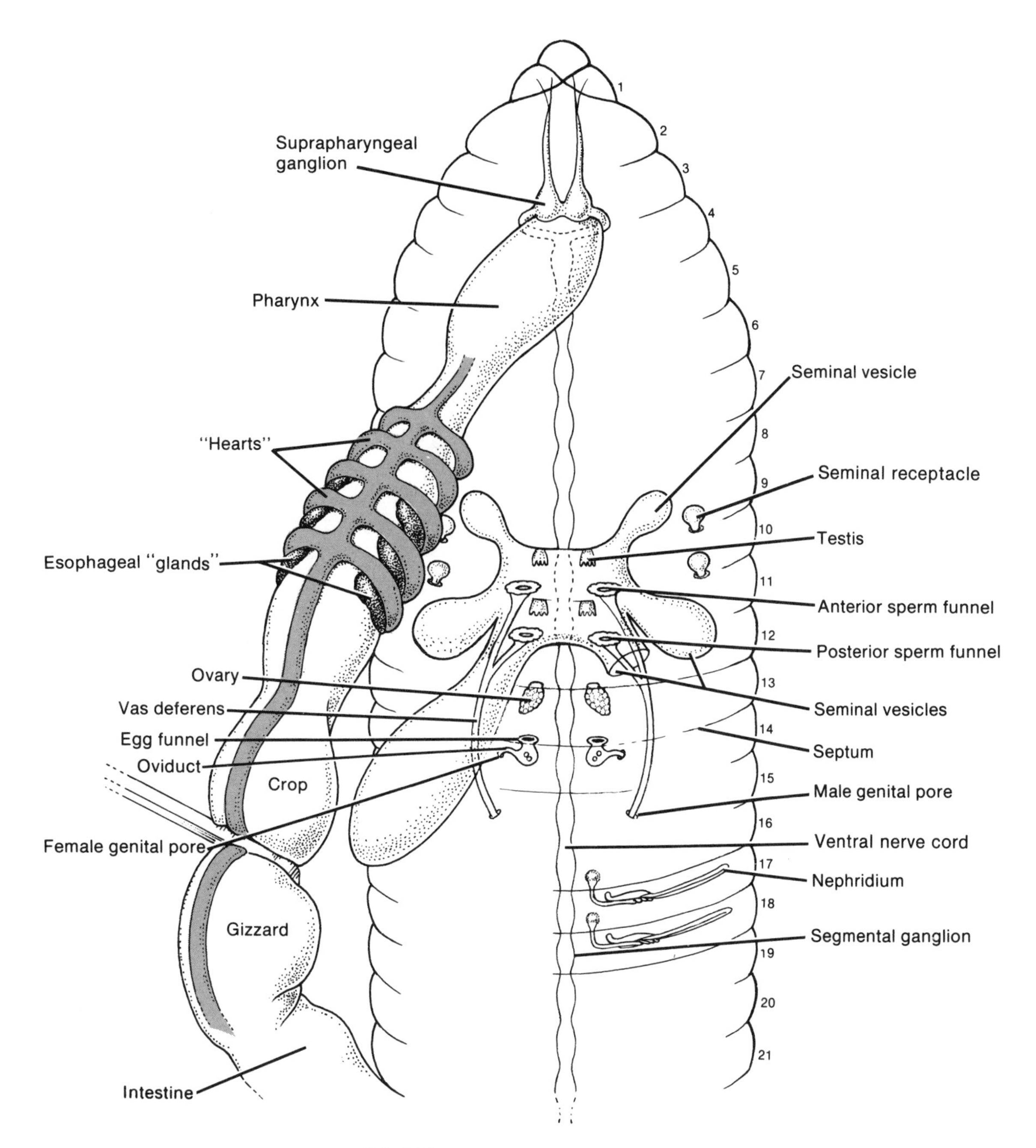

FIG. 35-5
Internal structure of *Lumbricus:* dorsal view.

enlarges to form two swellings, the **suprapharyngeal ganglia.** These might be considered primitive "brains."

Along the length of the cord, lateral nerves extend from segmental ganglia to the muscles of the body wall.

e. Excretory System

In *Lumbricus,* every body segment, except the first three or four and the last, contains an excretory organ called a **nephridium** (Fig. 35-6). Each nephridium opens into the coelomic cavity just anterior to the segment to which it is located through a ciliated funnel (**nephrostome**). The nephrostome continues as a finely coiled tubule, surrounded by capillaries, that leads to a **bladder.** The bladder discharges to the outside through a **nephridiopore,** located near the ventral surface of the body wall.

The cilia of the nephrostome create currents that draw wastes from the coelom into the tubule. Various materials are exchanged and reabsorbed by the

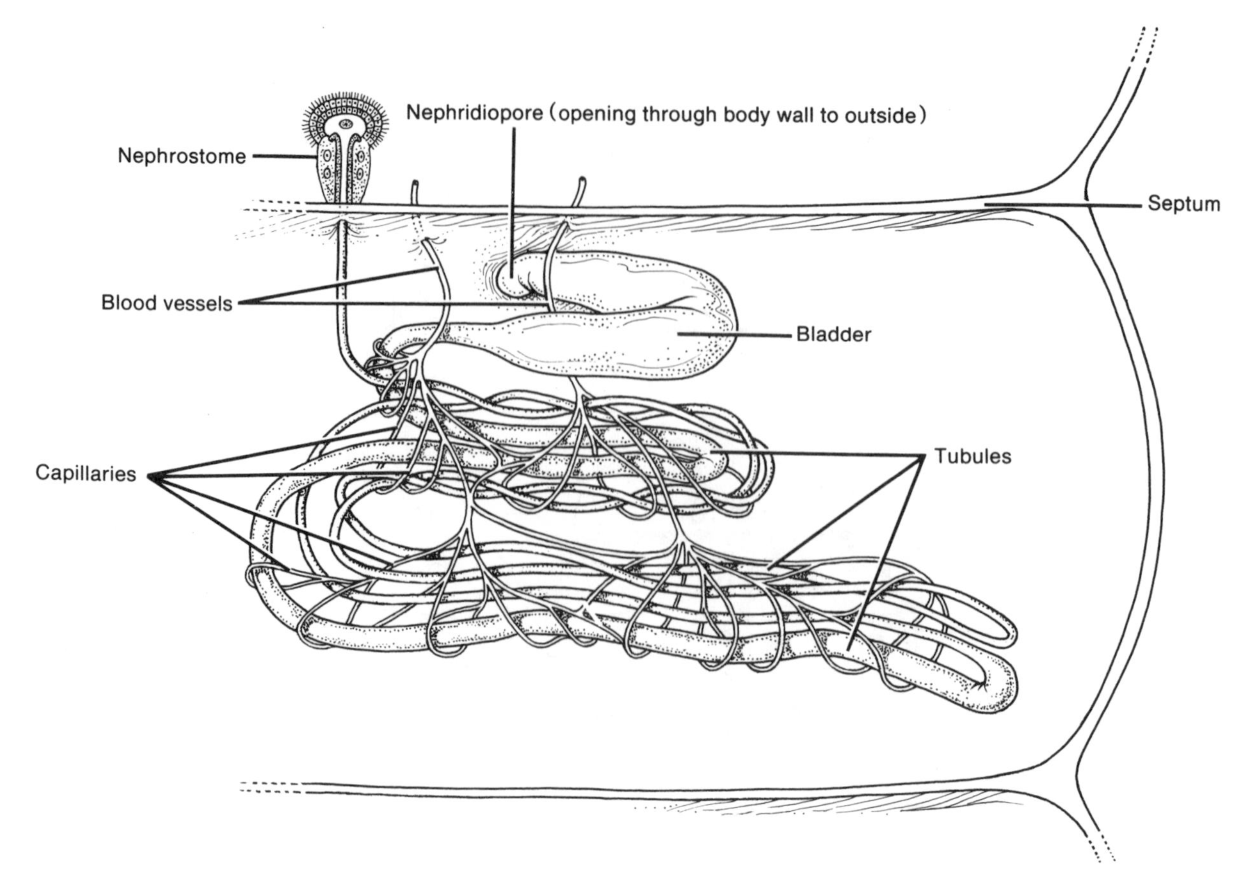

FIG. 35-6
Structure of a nephridium in *Lumbricus.*

blood in the capillaries surrounding the convoluted tubules, and thus a salt and water balance in the body is maintained.

3. Microscopic Anatomy

Examine a cross section of *Lumbricus* for details of its internal anatomy (Fig. 35-7).

The body surface is covered by a noncellular **cuticle,** which is secreted by the **epidermis** directly beneath. The epidermal layer consists of a layer of columnar epithelial cells and mucus-secreting cells, sensory cells, and photoreceptor cells that function in orienting the animal to its environment.

Under the epidermis is a layer of **circular muscles.** What effect would contraction of this muscle layer have on locomotion of the worm?

____

____

____

Under the circular-muscle layer is a thick band of **longitudinal muscles** that are arranged in a featherlike pattern. What effect would contraction of this muscle layer have on the body?

____

____

____

The innermost layer of the body wall is covered with thin, flattened cells that form the **peritoneum.** The space between the body wall and the digestive cavity is the **coelom.** Locate the **dorsal** and **ventral blood vessels** and the **ventral nerve cord** in the coelomic cavity. You may be able to observe nephridia, located in the coelomic cavity on either side of the intestine, in your cross section, as well as setae.

The wall of the intestine is composed of three layers. The innermost layer is the **mucosa,** a layer of

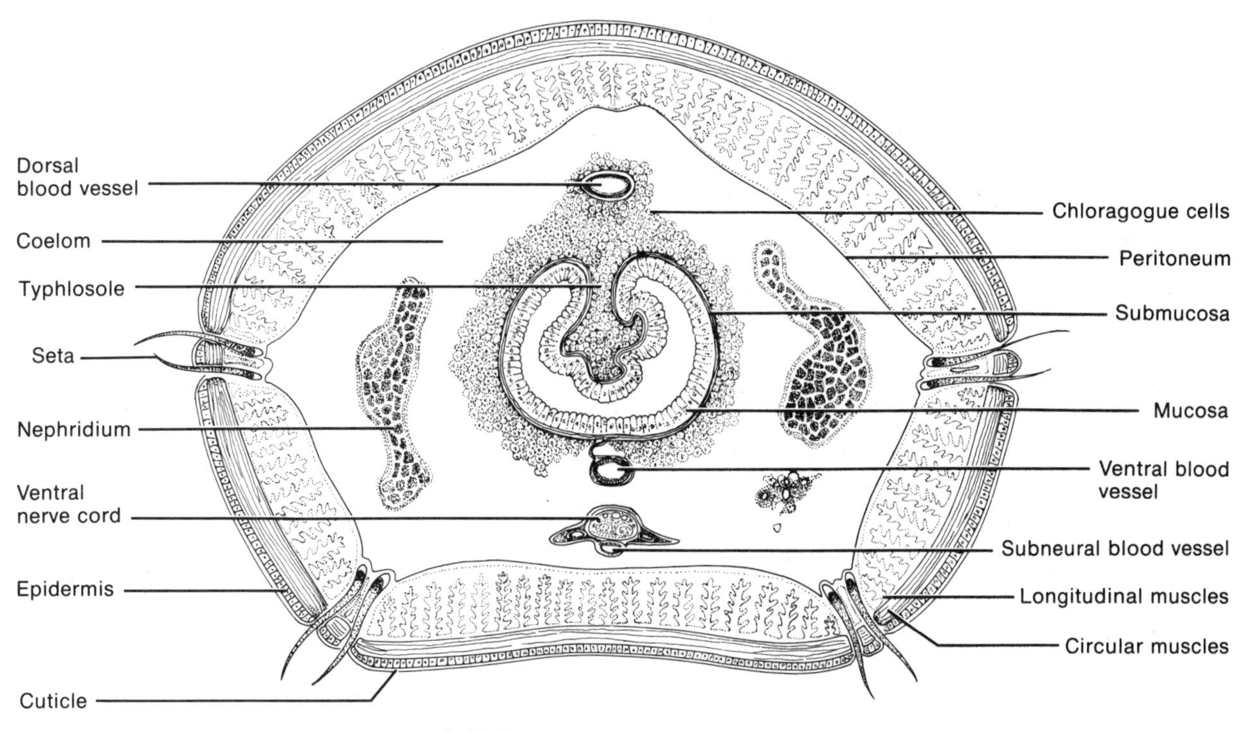

FIG. 35-7
Cross section through segment of earthworm.

narrow, ciliated, columnar cells. The intermediate layer is the **submucosa,** which contains circular and longitudinal muscle fibers and numerous small blood vessels. The outermost layer is made up of slender columnar epithelial cells containing numerous granules and **chloragogue cells,** which aid in the elimination of wastes from the blood. What is the function of these muscle layers and blood vessels in the intestine?

____

____

____

C. CLASS HIRUDINEA

Leeches are highly specialized annelids, most of which live in freshwater habitats although a few are marine organisms and some have become adapted to a terrestrial existence in tropical countries.

Contrary to common belief, leeches are not true parasites. Rather, they can be considered predators, feeding on the blood of various invertebrates and vertebrates. It is true that some species of leeches live permanently on a single animal and might be termed **ectoparasites.** However, many others do not.

Examine various leeches on display in the laboratory. Note that they are dorsoventrally flattened and tapered at both ends (Fig. 35-8). Most species are 20–60 mm long. The largest is said to be as long as a half meter when it is crawling. Note the **anterior** and **posterior suckers,** which are used for attachment. The anterior sucker contains the mouth, which has jaws covered with chitinous teeth for biting. Blood that is sucked up is stored in an enormous crop, allowing the animal to ingest three times its weight in blood. This permits the leech to go as long as 9 months between feedings.

In most leeches, respiration takes place directly through the moist body surface, although some leeches have gills for this purpose. Waste products are removed from the coelomic fluid and blood by nephridia.

Leeches are hermaphroditic, but reproduction is by cross-fertilization. Copulation and the formation of a cocoon, an adaption of terrestrial life, that is similar to that of earthworms.

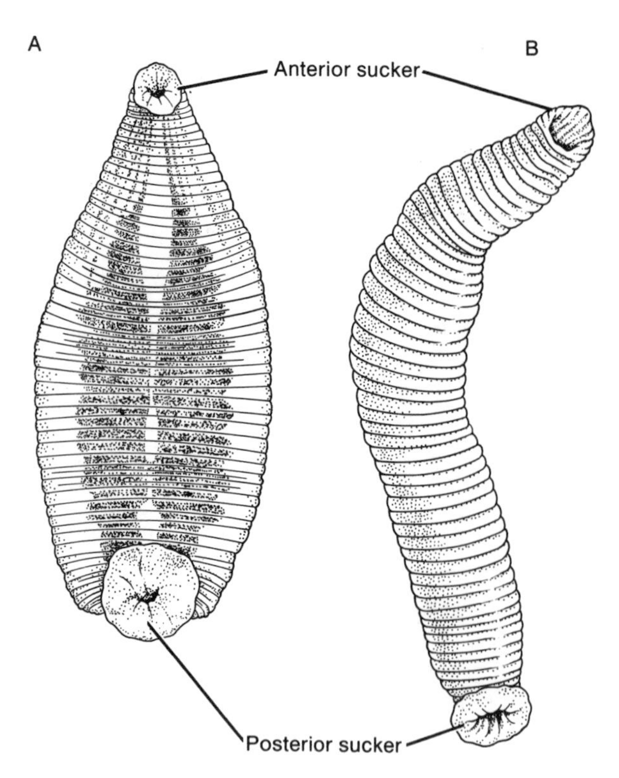

FIG. 35-8
Leeches: (A) *Placobdella,* commonly found on turtles, and (B) *Hirudo medicinalis,* a leech once used for bloodletting.

REFERENCES

Barnes, R. D. 1980. *Invertebrate Zoology.* 4th ed. Saunders.

Buchsbaum, R. 1975. *Animals Without Backbones.* 2d ed. rev. University of Chicago Press.

Dales, R. P. 1967. *Annelids.* 2d ed. Hutchinson.

Edwards, C. A., and J. R. Lofty. 1977. *Biology of Earthworms.* 2d ed. Chapman and Hall.

Villee, C. A., W. F. Walker, and R. D. Barnes. 1984. *General Zoology.* 6th ed. Saunders.

EXERCISE 36

Kingdom Animalia: Phyla Onycophora and Arthropoda

The phylum Onycophora (Greek *onychus,* "claw"; *phorus,* "bearing") includes a group of terrestrial organisms that have a distinct head, a somewhat cylindrical, unsegmented body, and a number of short, unjointed legs. They are primarily found in very moist land habitats, beneath stones, leaves, or logs in forests. In this exercise, you will study *Peripatus,* a representative genus of this phylum.

The phylum Arthropoda (Greek *arthros,* "joint"; *podos,* "foot") is the largest of all animal phyla. Of the million or so known species of animals, more than three quarters are arthropods. The arthropods are considered to have attained the greatest "biological success": they comprise the largest number not only of species, but also of organisms; they occupy the greatest variety of habitats, consume the largest amounts and kinds of food, and are capable of defending themselves against their enemies.

The arthropods are characterized by their rigid, chitinous body covering, called an **exoskeleton.** The body is segmented externally and internally to varying degrees, depending on the species, and the appendages are jointed and highly modified for a large variety of functions.

The classification of the Arthropoda is quite complex because of its large number of species and the tremendous diversity in form among its members. It is divided into three subphyla, which are subdivided into classes. Some of the major classes are included in the following:

Subphylum Trilobita: primitive marine arthropods that are now extinct were characterized by having a distinct head, a segmented body and appendages

Subphylum Chelicerata: relatively simple, chiefly terrestrial arthropods with the first two body parts (head and thorax) combined in a single cephalothorax and lacking antennae and mandibles but having six pairs of jointed appendages

Class Xiphosura (horseshoe crabs): aquatic arthropods having compound lateral eyes, a cephalothorax, and an abdomen with five or six pairs of appendages

Class Pychnogonida (sea spiders): small marine organisms having a short, thin body and a mouth on a long proboscis

Class Arachnida: terrestrial arthropods having simple eyes, no gills, and six pairs of jointed appendages

Subphylum Mandibulata: the largest group of arthropods, which are characterized by having three distinct body parts and three or more pairs of legs, one or two pairs of appendages modified to serve as antennae, and one pair adapted to function as mandibles (jaws)

Class Crustacea: the most common aquatic arthropods, which are characterized by respiring mainly by gills and having two pairs of antennae, one pair of jaws, and two pairs of maxillae

Class Insecta: mainly terrestrial arthropods having a distinct head, thorax, and abdomen and one pair of antennae, three pairs of legs, and two pairs of wings

Class Chilopoda: terrestrial arthropods having long, flattened bodies with many somites (segments), each somite having a pair of legs

Class Diplopoda: terrestrial arthropods characterized by having a long, usually cylindrical body with many somites, each somite having two pairs of legs

A. PHYLUM ONYCOPHORA

The members of this phylum possess characteristic features that are common to both the annelids and arthropods and thus come closer than any others to being the "missing link" between any two phyla.

Examine a preserved specimen or plastic mount of *Peripatus,* which resembles the annelids in the structure of the eyes, the nephridia, the ciliated reproductive ducts, and the simple gut. It resembles the arthropods in having a **hemocoel** (body cavity that functions as part of an open circulatory system) and a dorsal heart, and in the general structure of the reproductive organs. Note the absence of external segmentation, although there is a pair of legs for each internal segment of the body (Fig. 36-1). The legs terminate in claws, which superficially resemble those of arthropods but differ in not being jointed. The head has three segments, each of which bears three pairs of appendages: two short **antennae,** two blunt **oral papillae,** and two small horny **jaws.** Like the annelids, *Peripatus* possesses an internal system of excretory **nephridia** with external openings (**nephridiopores**) at the base of each appendage. The **anus** opens at the blunt posterior end and is preceded by a single genital opening, the **gonopore.**

B. PHYLUM ARTHROPODA

1. Subphylum Chelicerata

The body of the chelicerates is usually divided into two regions: a cephalothorax and abdomen. The first pair of appendages on the cephalothorax are modified as pincerlike or fanglike appendages called **chelicerae.** There are usually five other appendages on the cephalothorax which in some groups are all walking legs, while in others the first pair are modified as feeding appendages, called **pedipalps.**

a. Class Xiphosura

This class includes many fossil forms and living species of horeshoe crabs. Examine a preserved specimen of *Limulus polyphemus,* a horseshoe crab that is common in the shallow marine waters off the Atlantic coast from Nova Scotia to Yucatan. The dark brown outer covering, or **carapace,** is shaped like a horseshoe, thus giving it its common name of horseshoe crab (Fig. 36-2). Posterior to the carapace is an **abdominal shield** with short, lateral movable spines, to which is attached a tail spine (**telson**). The telson is used to right the body when it is overturned and to push it forward when burrowing into the ocean floor.

Turn your specimen over, and observe the seven pairs of jointed appendages. The first pair consists of two **chelicerae,** which function to some extent in locomotion but are used mainly in capturing and macerating food organisms. The next four pairs are leglike and end in small pincers, or **chelae.** The appendages in the next pair, also leglike, end in leaflike tips used to sweep away mud that clings to the body when burrowing. The seventh pair consists of very small, modified appendages, called **chilaria,** attached to the pregenital somite. Respiratory organs consisting of deep pleats that resemble the leaves of a book ("**book gills**") are on the hind surface of appendages 2 through 6. Gaseous exchange between the blood and the water that is constantly moving past these organs takes place across them as the book gills wave back and forth.

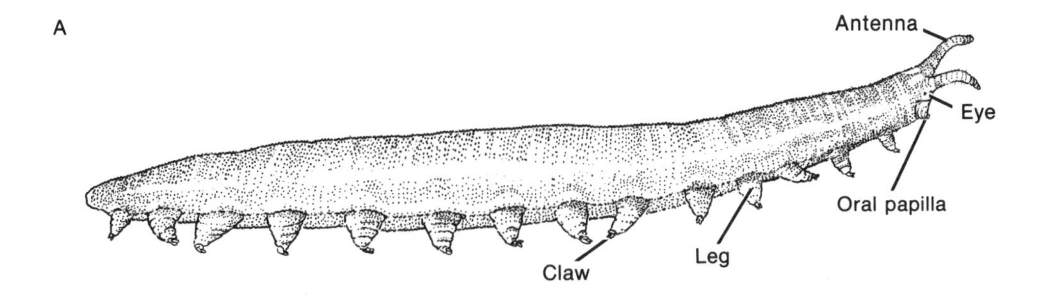

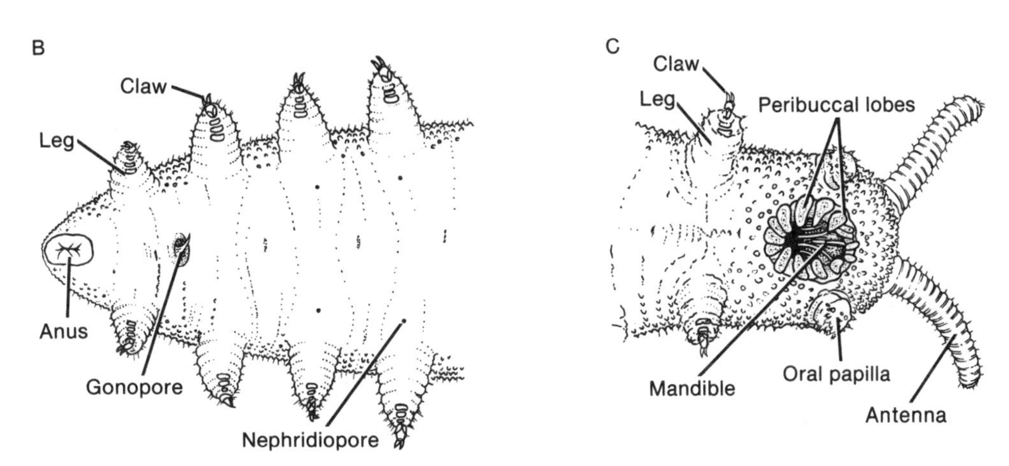

FIG. 36-1
Peripatus: (A) lateral view of whole organism, (B) posterior ventral surface, and (C) anterior ventral surface.

b. Class Arachnida

The arachnids are a diverse group of arthropods that include spiders, scorpions, daddy long legs, ticks, mites and their relatives. They are characterized by having body segments fused into two main regions: a cephalothorax bearing four pairs of walking legs and an abdomen. Arachnids further differ from the other arthropods by the absence of compound eyes, antennae, and true jaws.

Examine preserved specimens or plastic mounts of spiders, scorpions, ticks, and mites (Fig. 36-3).

2. Subphylum Mandibulata

The mandibulate arthropods differ from the chelicerates in having **mandibles** instead of chelicerae as the first pair of appendages. The mandibles in the different classes of this subphylum are modified for biting and chewing or piercing and sucking.

a. Class Crustacea

This class of arthropods includes fairy shrimps, water fleas, barnacles, crayfish, and crabs. Most crustaceans are marine, but many others inhabit inland waters, and a few, such as sow bugs, live in moist places on land. Many crustaceans are microscopic whereas others, such as giant crabs, may be as long as 5 meters.

Crustceans have three important arthropod features. The body is segmented and covered by a chitinous exoskeleton, and the appendages are jointed. However, the segmentation of the crustaceans is more specialized than that of the annelids. In the latter, the segments are nearly identical, whereas in the crustaceans and most arthropods there is considerable regional specialization, including a well-developed head.

The common crayfish (*Cambarus* or *Procambarus* sp.) is a cannibalistic scavenger that lives on the muddy bottoms of freshwater lakes, streams, and

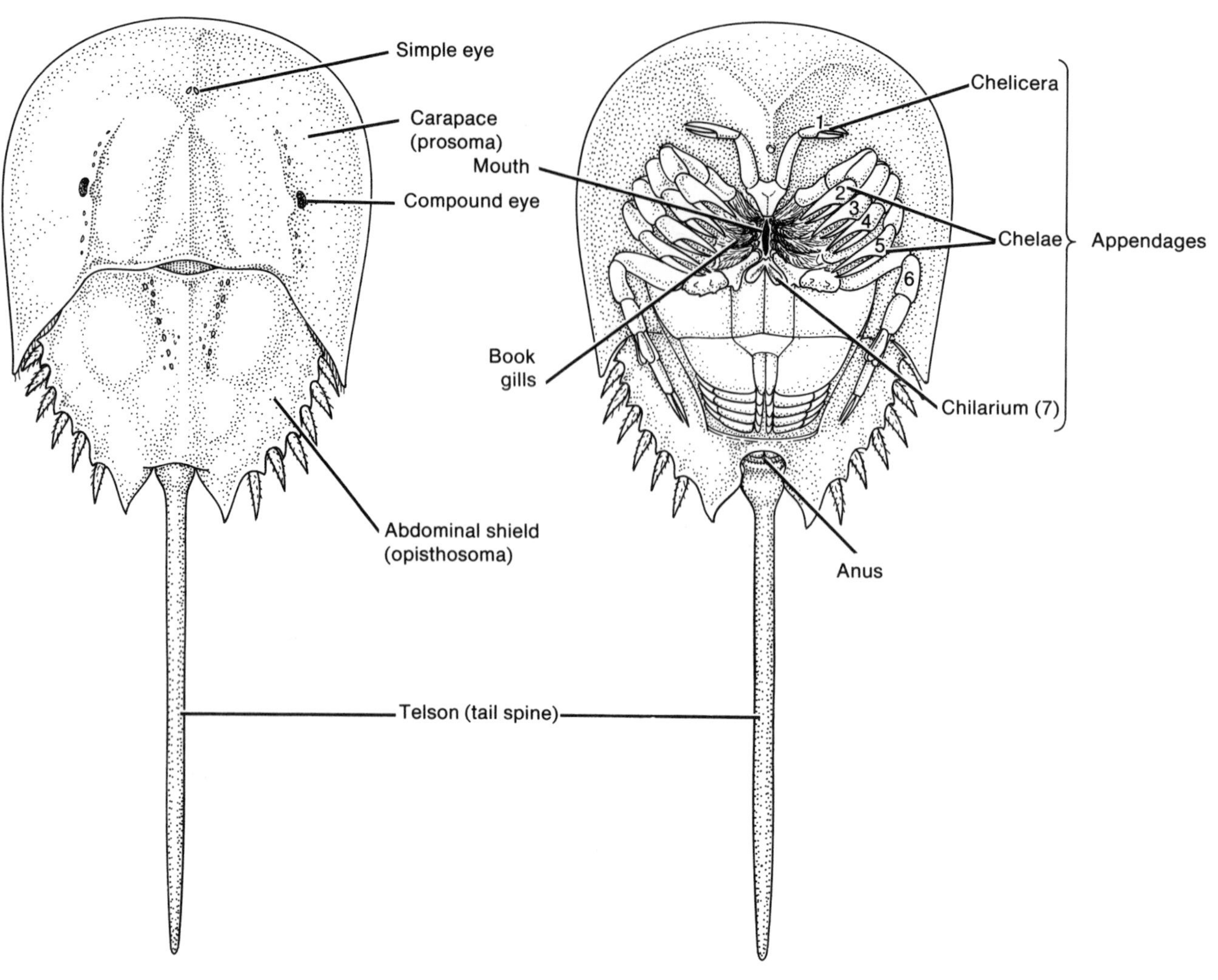

FIG. 36-2
Limulus: (A) dorsal view and (B) ventral view.

ponds. It emerges at night to feed on dead and decaying matter, insect larvae, worms, and other crayfishes. It can be as much as 15 cm long, and its appendages are differentiated to serve specific purposes.

If available, observe a living specimen of a freshwater crayfish in a shallow pan of water or in an aquarium. Study its manner of walking and swimming. In which direction does a crayfish normally crawl?

___

Hold a crayfish firmly against the bottom of the container, and then introduce a drop of India ink at the posterior end of the abdomen. From your observation, describe the direction of water flow and the mechanism used by the crayfish to set up such water currents.

___

___

___

Feed a crayfish a live insect (cockroach or cricket), and note the coordinated activity of the mouth.

1. **External Anatomy.** Examine a preserved specimen of the crayfish. The body is divided into an anterior **cephalothorax** (fused head and thorax) and a posterior **abdomen** (Fig. 36-4). The chitinous exoskelton effectively protects the crayfish from preda-

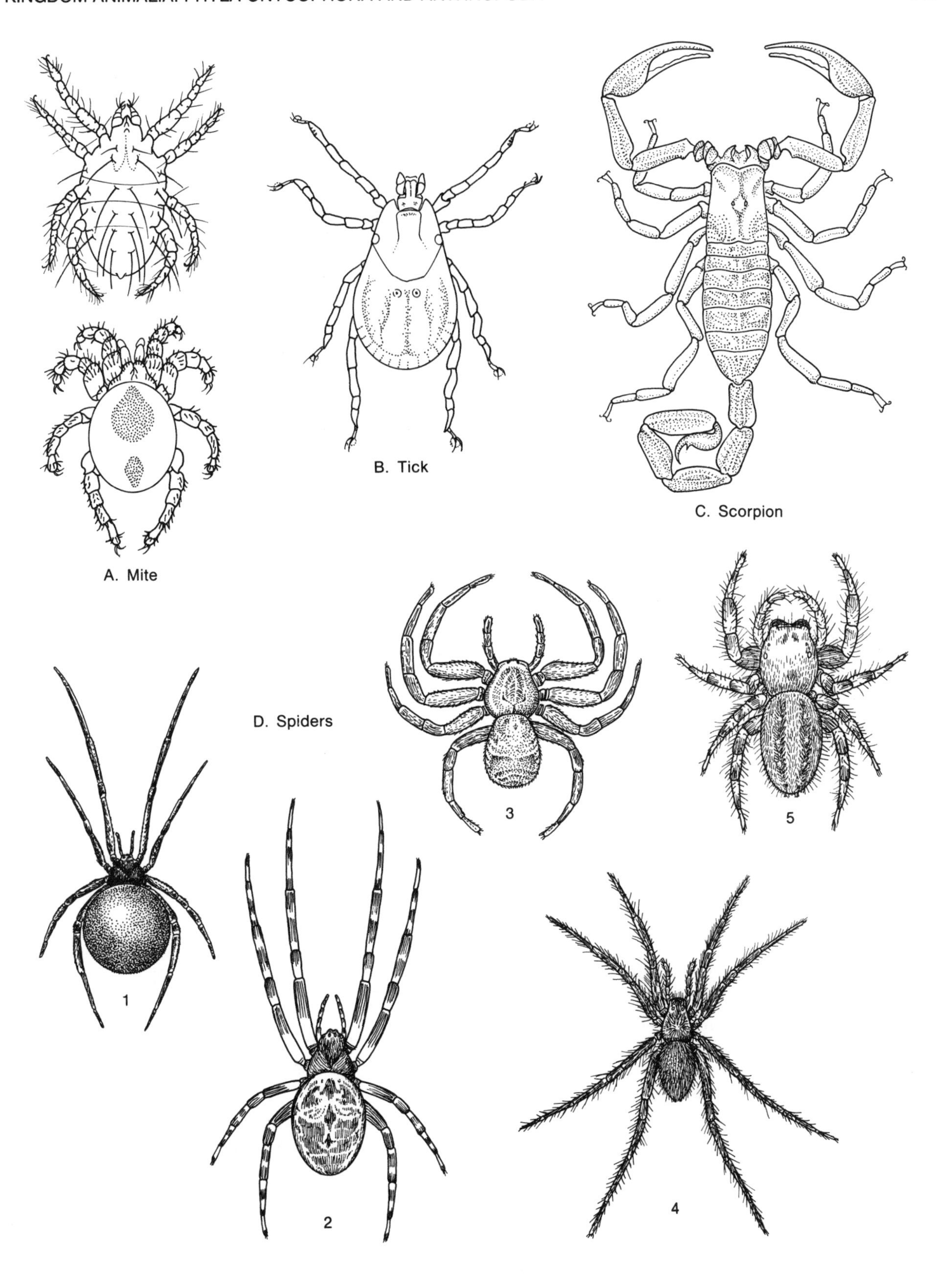

FIG. 36-3
Representative arachnids: (A) mite, (B) tick, (C) scorpion, and (D) spiders. The spiders shown are (1) black widow, (2) common garden, (3) crab, (4) jumping, and (5) wolf.

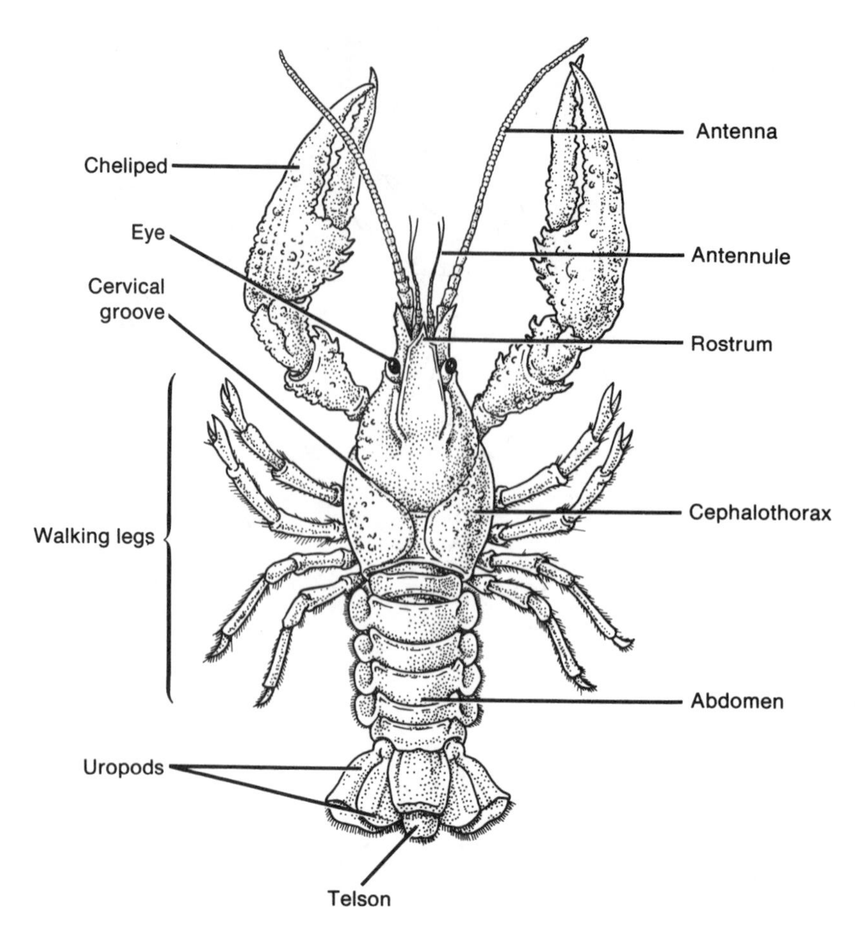

FIG. 36-4
Dorsal view of a crayfish *(Cambarus* or *Procambarus).*

tors. In order to grow, however, the crayfish periodically **molts** (i.e., sheds its exoskeleton). During the period in which a new skeleton is being formed, the crayfish is defenseless; it hides to escape its enemies.

The **carapace** is a saddlelike covering over the cephalothorax. A transverse groove separates the fused head from the thoracic region. Laterally, the carapace covers the gills. The **rostrum** is an anterior, pointed extension of the head. The eyes are located on either side of the rostrum.

The **abdomen** consists of several segments and is terminated by the telson, an extension of the last abdominal segment. To escape its enemies, the crayfish spreads the telson and the wide **uropods** like a fan, rapidly drawing the telson forward under its body. This motion causes the crayfish to dart backward into the muddy bottom, which becomes cloudy because of the flipping movements. Each abdominal segment contains a dorsal plate, a ventral plate, and two lateral plates. Examine the appendages on the lower side of the abdomen. These appendages are the **swimmerets.** What is the function of these appendages?

____

____

____

____

Note that all other appendages are modifications of this three-part plan. Although male and female crayfish have an equal number of swimmerets, in male crayfish those adjoining the thorax have been modified: they are elongate and can be brought together to form a troughlike channel, used for the transfer of sperm from the male to the seminal receptacles of the female. In female crayfish, the eggs, which look like clusters of grapes, are attached to the abdominal swimmerets and are aerated by gentle, waving movements of the swimmerets.

Examine the five pairs of walking legs and count the number of segments in each leg (Fig. 36-5). Note that there are no pincers on the fifth (posterior) pair of legs. Locate the male genital pores, which open into the base of the legs of this pair. The openings of the oviducts in the female are located at the base of each third walking leg. Carefully cut the membrane at the base of each leg, and remove the right walking legs, starting with the most posterior one and working forward. Arrange these appendages in order on a piece of paper. How many of these legs have feathery gills attached to them?

____

What is the advantage of the featherlike structures in these gills?

____

____

____

____

The legs of the first pair, the **chelipeds,** are much larger than the rest. Cut the exoskeleton from the dorsal surface of the claw, and study the action of the muscles in opening and closing the pincer.

The appendages of the crayfish and other crustaceans are **homologous** organs; that is, they are all fundamentally similar in structure and arise from a similar embryonic rudiment. The structural adaptions of the appendages in different regions are correlated with their functions (Fig. 36-4). When corresponding structures in different segments of the same animal are homologous, it is called **serial homology.** On the other hand, **analogous** organs are similar in function but not necessarily in structure. Cite examples of analogous organs.

____

____

____

____

2. **Internal Anatomy.** Using a sharp scalpel and scissors, carefully dissect away the dorsal half of the exoskeleton to expose the internal organs (Fig. 36-6). In the middorsal region, locate the diamond-shaped heart perforated by three pairs of openings called **ostia.** Identify the following blood vessels: **ophthalmic artery** extending anteriorly from the heart to the eyes; **hepatic arteries,** a pair of short arteries going to the digestive gland; **antennary arteries** going to the antennae and green glands; **dorsal abdominal artery,** a single artery leading posteriorly from the heart to the dorsal part of the abdomen; **sternal artery,** a single vessel leading from the heart to the ventral part of the crayfish where it splits and becomes the **ventral thoracic artery** anteriorly and the **ventral abdominal artery** posteriorly. Because there are no veins in the crayfish, the blood flows from the arteries into open spaces or **sinuses** instead of capillaries as it does in the earthworm. The blood, after being oxygenated in the sinuses of the gills, returns through efferent channels to the heart. This type of circulatory system is known as an **open system** to distinguish it from the **closed system** found in the annelids and many higher forms, including human beings.

Remove the heart, and locate the gonads. In the female crayfish, the ovaries are a pair of tubular structures bilaterally arranged in front of the heart and continuing behind it as a single mass. In the male, the testes are highly coiled white tubes.

Locate the mouth. This leads to a short, tubular esophagus, which in turn leads to a large, saclike stomach. The stomach is made up of two parts; the larger **cardiac stomach,** in which food is stored, and the small, posterior **pyloric stomach.** Most of the digestion takes place in the pyloric stomach as a result of the grinding action of the gastric mill and the enzymes that are secreted into the region by the digestive glands. The remaining part of the digestive system is the intestine, a small, straight tube leading to the anus. Anterior to the stomach and just behind each antenna are the excretory structures, commonly called the **green glands.** Find their external openings at the bases of the antennae.

The nervous system is similar to that of annelids, except for the fusion of several originally separated nervous elements. Remove the main organs from the thoracic and abdominal regions of your crayfish and locate the **ganglion,** or brains, in front of the esophagus. Expose the brain by careful dissection and find the nerves passing to the eyes, antennae, and antennules. The brain is connected to the **ventral nerve cord** by a pair of nerves that pass around the esophagus. Follow the ventral nerve cord as it passes through the thorax and abdomen, and count the number of ganglia. Each ganglion gives off pairs of nerves to the appendages and internal organs of the segment in which it lies.

LOCATION	APPENDAGES OF CEPHALOTHORAX	FUNCTION
Front of mouth	Antennules	Senses other organisms and helps to balance crayfish
Front of mouth	Antenna	Senses other organisms
Mouth	Mandible or jaw	Crushes food
Behind mandibles	First maxilla	Moves food to mouth
Behind mandibles	Second maxilla	Bails water in gill chamber
At anterior and ventral part of thorax region	First maxilliped	Holds food, touches, and tastes
At anterior and ventral part of thorax region	Second maxilliped	Holds food, touches, and tastes
At anterior and ventral part of thorax region	Third maxilliped	Holds food, touches, and tastes
Posterior to maxillipeds at ventral part of thorax	Cheliped	Grasps food
Posterior to maxillipeds at ventral part of thorax	Walking leg (Gill)	For locomotion

APPENDAGES OF THE ABDOMEN

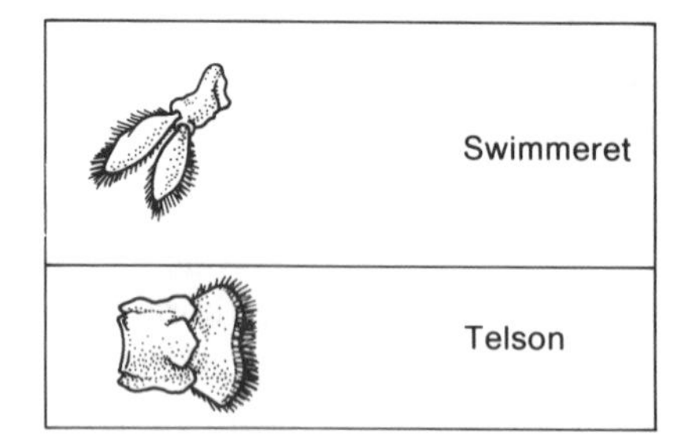

LOCATION	APPENDAGES OF THE ABDOMEN	FUNCTION
Ventral side of abdomen	Swimmeret	First swimmeret in male transfers sperm to female, which uses 2nd, 3rd, 4th, and 5th swimmerets to hold eggs and young
Posterior end	Telson	For swimming

FIG. 36-5
Location, structure, and function of crayfish appendages.

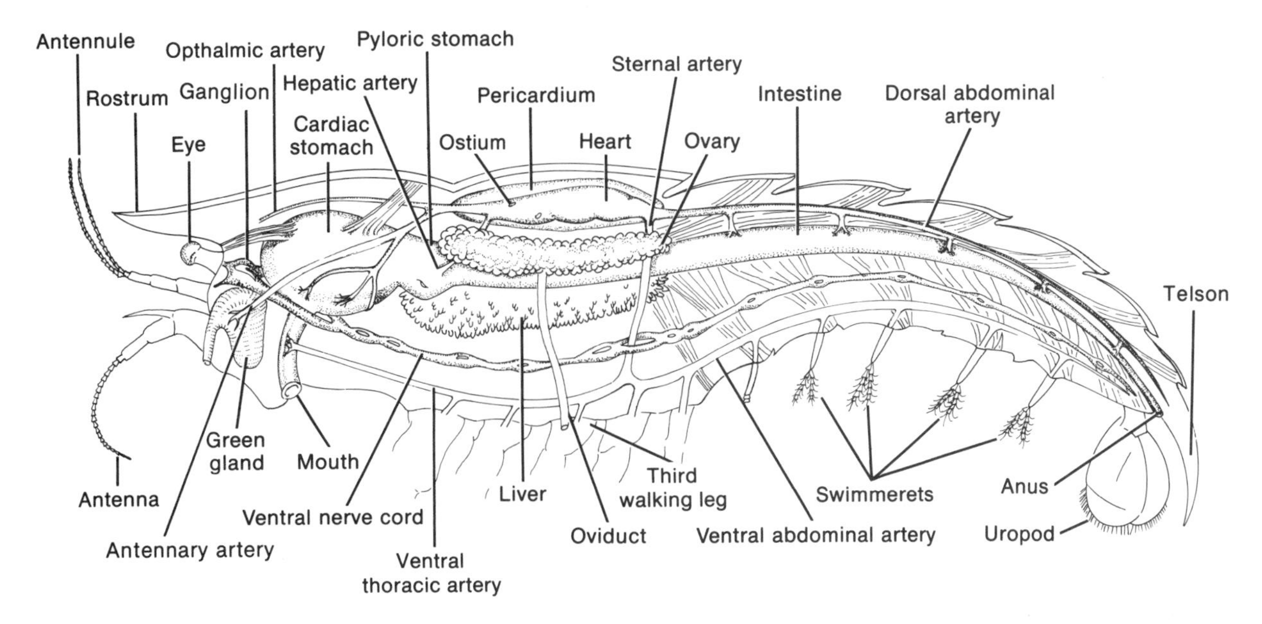

FIG. 36-6
Internal anatomy of the crayfish.

b. Class Insecta

It has been estimated that there may be as many as 5 million insect species, of which fewer than a million have been identified and classified (there are, for example, more than 300,000 species of beetles alone). Insects range in size from those no larger than a dust particle, and a species of hairy, winged beetle that can crawl through the eye of a needle, to the Atlas moth of India, which has a 30-cm wingspan. It has been estimated that the insect population of the world is at least 1×10^{18} and, taking the average weight of insects as 2.5 mg (i.e., less than 0.001 ounce), the weight of the earth's insect population exceeds that of its human inhabitants by a factor of 12.

Clearly, insects are the most successful and abundant of all land animals. They are the principal invertebrates that can live in dry environments and the only ones able to fly. Some insects can survive in temperatures as low as $-30°F$ ($-35°C$); others in temperatures as high as 120°F (49°C). These abilities are due to the chitinous body coating (**exoskeleton**) that protects the internal organs against injury and loss of moisture and to the system of tracheal tubes that enables insects to breathe air. The ability to fly permits them to readily find food. Under favorable conditions, insects can multiply rapidly. The insect's small size frees it from the need to compete with larger animals for a place in the environment.

Although insects are exceedingly diverse organisms, the basic external and internal organization can be effectively illustrated by a study of the large black lubber grasshopper, *Romalea microptera.* The following description, although primarily pertaining to the short-winged lubber grasshopper, will serve for any common species.

1. **External anatomy.** The body of the grasshopper is divided into a head consisting of six fused segments (**somites**); a thorax of three somites, to which are attached the legs and wings; and a long, segmented abdomen that terminates with the reproductive organs (Fig. 36-7). The outer **exoskeleton** consists largely of chitin, which is secreted by the epidermis. In order to grow, the grasshopper periodically sheds this exoskeleton (**molts**); adults do not molt. Pigment in and under the cuticle gives the grasshopper coloration resembling that of its surroundings, affording protection from predators.

The **head** (Fig. 36-8) has one pair of slender, jointed **antennae,** two **compound eyes,** and three simple eyes or **ocelli.** The mouth parts are of the chewing type and include a broad upper lip or **labrum;** a tonguelike **hypopharynx;** two heavy blackish lateral jaws or **mandibles,** each with teeth along the inner lateral margins for chewing food; a pair of **maxillae** of several parts, including palps (sensory appendages) at the side; and a broad lower lip or **labium,** with two short palps.

The **thorax** consists of three parts: a large anterior **prothorax,** the **mesothorax,** and the posterior **metathorax.** Each part bears a pair of jointed legs. Iden-

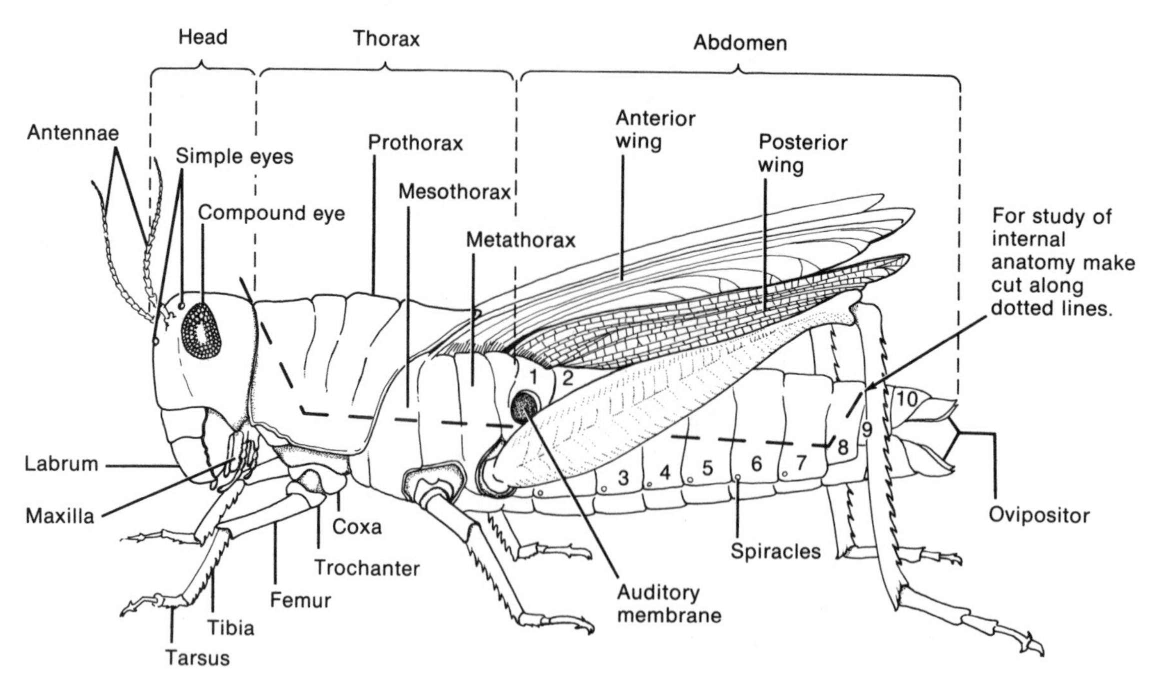

FIG. 36-7
External anatomy of a female grasshopper.

tify the leg segments indicated in Fig. 36-7. The meso- and metathorax each bear a pair of wings. The anterior wings of the grasshopper are thick and shield the larger pair of flight wings. Both pairs of wings are derived from the cuticle and have thick parts (veins) that strengthen them. Stretch out the wings, and examine the anterior protective wings and the flight wings.

The slender **abdomen** consists of 11 somites, the posterior ones being modified for reproduction. The male has a blunt terminal segment, whereas the female has four sharp conical prongs, the **ovipositors,** which are used in egg laying (Fig. 36-7). Along the lower sides of the thorax and abdomen are 10 pairs of **spiracles,** the small openings of elastic air tubes, or **tracheae,** that branch to all parts of the body and constitute the respiratory (**tracheal**) system of the grasshopper. This system of air tubes brings atmospheric oxygen directly to the cells of the body. The spiracles open and close to regulate the flow of air. The three most anterior pairs of spiracles are inhalatory, piping air directly to all body tissues. The other spiracles are exhalatory.

2. **Internal Anatomy.** It is difficult to preserve the internal organs of the grasshopper because the preservative often fails to penetrate the exoskeleton. Careful dissection is necessary to study the internal anatomy.

After removing the wings, start at the posterior end, and make two lateral cuts toward the head with a pair of scissors or fine scalpel as indicated in Fig. 36-7. Remove the dorsal wall. Locate the muscles on the inside of the body wall, and note their arrangement. What is their function?

____

____

A space between the body wall and digestive tract, the **hemocoel,** is filled with colorless blood.

Study the digestive tract and identify its parts (Fig. 36-9). Beginning at the anterior end, find the **mouth,** which is located between the mandibles and leads to a short **esophagus** followed by the **crop.** Next is the **stomach,** to which are attached six double finger-shaped digestive glands (**gastric caeca**); these glands produce enzymes that are secreted into the stomach to aid digestion. The digestive tract continues as the **intestine,** which consists of a tapered anterior part, a slender middle part, and an enlarged **rectum** that opens to the outside at the **anus.** During feeding, food held by the forelegs, labium, and labrum is lubricated by secretions from the salivary glands and chewed by the mandibles and maxillae. Chewed food is stored in the crop. Because most of the diges-

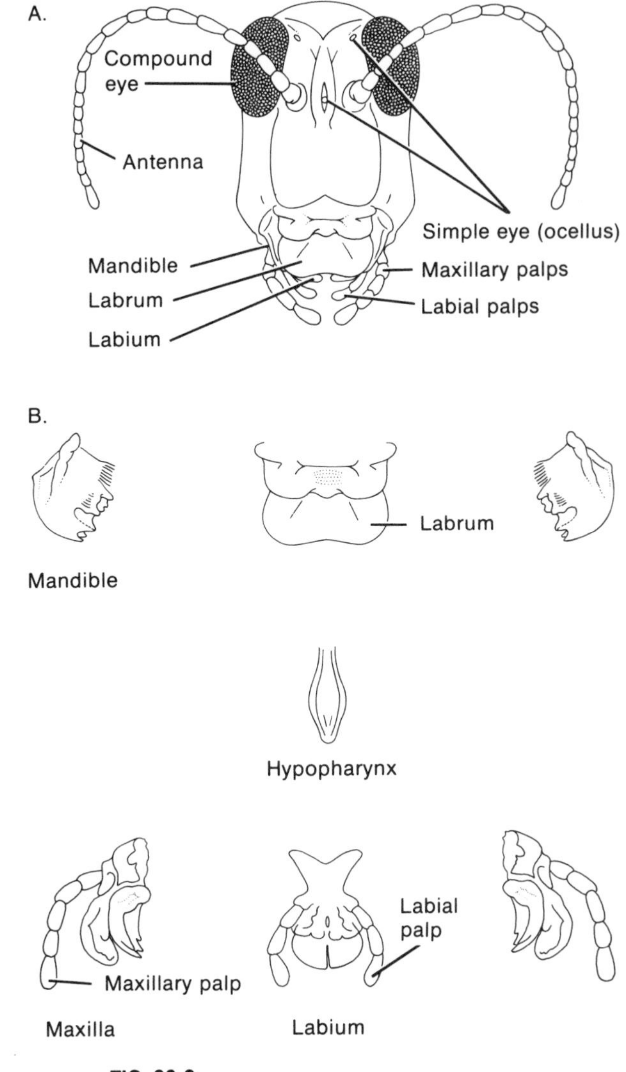

FIG. 36-8
Grasshopper (A) head and (B) mouth parts.

tive tract, except for the stomach and crop, is lined with chiton, which is impervious, digestion and absorption take place mainly in the stomach. Excess water is absorbed from any undigested food in the rectum.

The excretory system is made up of numerous tiny tubules—the excretory, or **Malpighian, tubules**—which empty their products into the anterior end of the intestine. These tubules remove urea and salts from the blood.

The sexes are separate, and their reproductive organs are in the terminal abdominal segments. In the male, each of the two **testes** is composed of a series of slender tubules, or follicles, and is located above the intestine; each testis is joined to a longitudinal **vas deferens** (Fig. 36-10). The vas deferens are joined to a single ejaculatory duct, to which **accessory glands** are attached. In the female, each **ovary** is composed of several tapering egg tubes (**ovarioles**), which produce the ova. Each ovary is joined to an **oviduct** leading to the **vagina,** to which a pair of accessory glands and a single **spermatheca** are attached. The latter organ is used to store sperm received at copulation.

The insect circulatory system can be studied by examining the wing veins of a living adult grasshopper or cricket. This can be done by preparing a plasticene or wax cell large enough to hold the insect on a microscope slide. Pin the animal down with two strips of paper, one across the thorax and the other across the body beneath the wings. Slip a piece of tinfoil or glazed white paper beneath the anterior

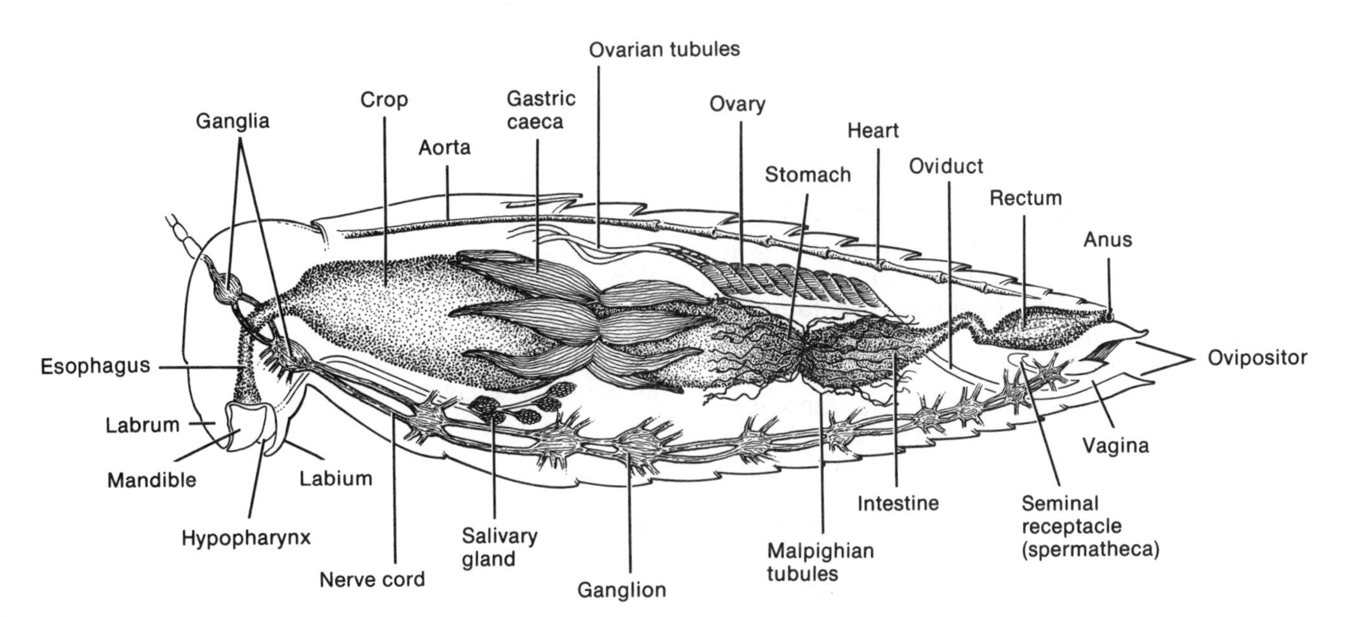

FIG. 36-9
Internal anatomy of a female grasshopper.

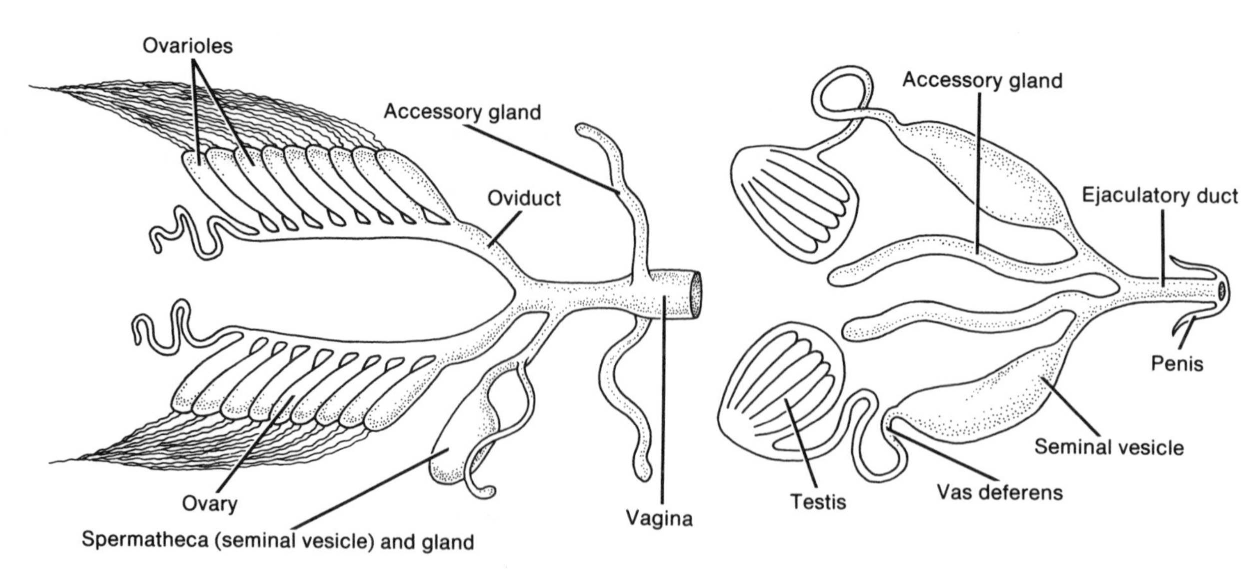

FIG. 36-10
Reproductive systems of the grasshopper.

wing, and examine under a dissecting microscope. Note the moving blood cells and the direction of flow of the hemolymph in various parts of the wing. In order to observe the beating heart, pin the insect's wings out to the side and focus light through the body with the concave mirror. What is the direction of blood flow in the dorsal tubular heart?

---

c. Class Chilopoda (Centipedes)

Centipedes live mainly in warm climates. They usually hide by day in a place that affords protection —in the soil or under stones, bark, or rotten logs. At night, they are very active, fast-running animals that prey on insects, earthworms, spiders, and other small animals, which are paralyzed by the centipede's poisonous claws. The smaller centipedes of the northern United States are harmless to human beings, but the larger ones of the South and the tropics inflict a painful bite.

Examine preserved or plastic-mounted specimens of centipedes. Observe that centipedes are elongate, flattened, wormlike animals with 15 or more pairs of legs (Fig. 36-11A). Each body segment bears a single pair of legs with the last 2 pairs being directed backwards and often different in form from the other pairs. The head bears a pair each of **mandibles** and long, jointed **antennae** and 2 pairs of **maxillae** (Fig. 36-11B). The first somite bears a pair of jointed **maxillipeds** that function as poisonous claws. The genital openings are located at the posterior end of the body, usually on the next to the last segment.

d. Class Diplopoda (Millipedes)

Millipedes, like centipedes, are found in humid, dark places—under leaves, in moss, under stones or boards, in rotting wood, or in the soil. Many species have "stink glands," which give off a foul-smelling fluid that can be so strong as to kill insects that are placed in a jar with the millipede. Millipedes, unlike some centipedes, do not bite human beings. Most millipedes are scavengers and feed on decaying plant material, but they also eat animal material.

Examine preserved or plastic-mounted specimens of millipedes. Observe that they are elongate, wormlike animals with 30 or more pairs of jointed legs (Fig. 36-11C). Most body segments have 2 pairs of legs. The head bears two clumps of many simple eyes and a pair each of mandibles and antennae (Fig. 36-11D). The thorax is short, of four single somites, all but the first having a pair of legs.

REFERENCES

Borror, D. J., and D. M. DeLong. 1976. *An Introduction to the Study of Insects.* 5th ed. Holt, Rinehart and Winston.

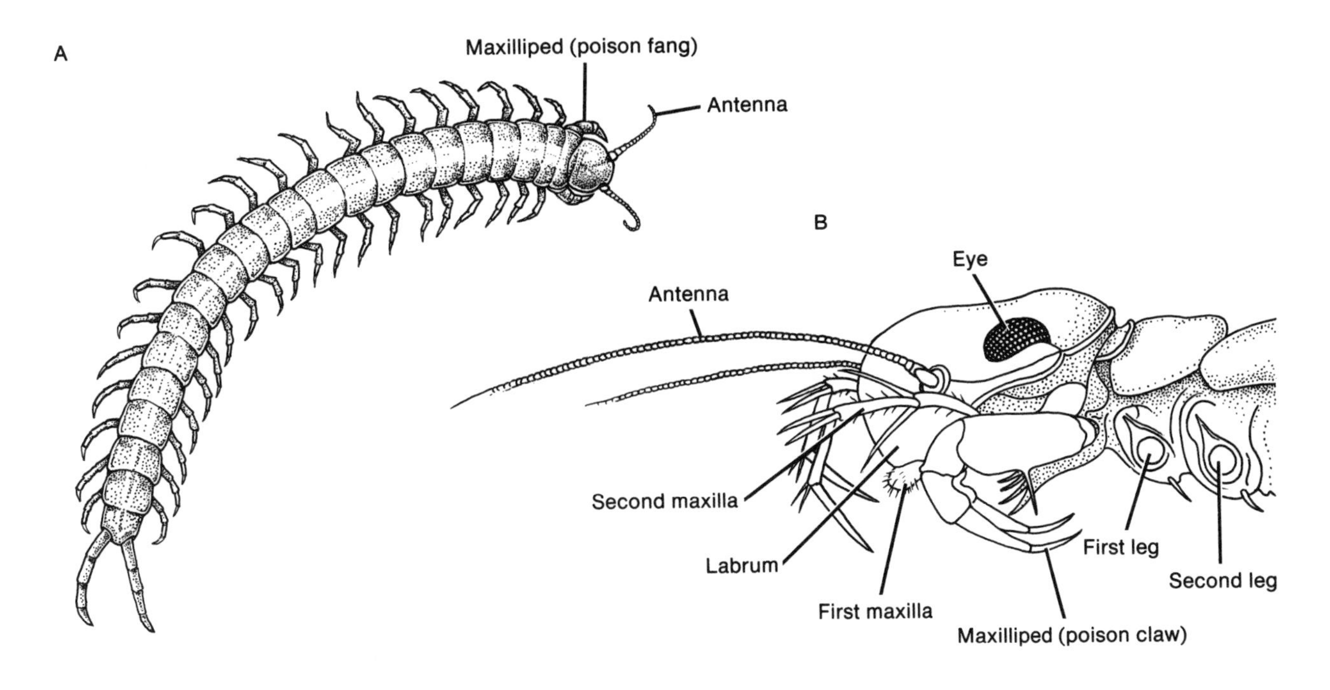

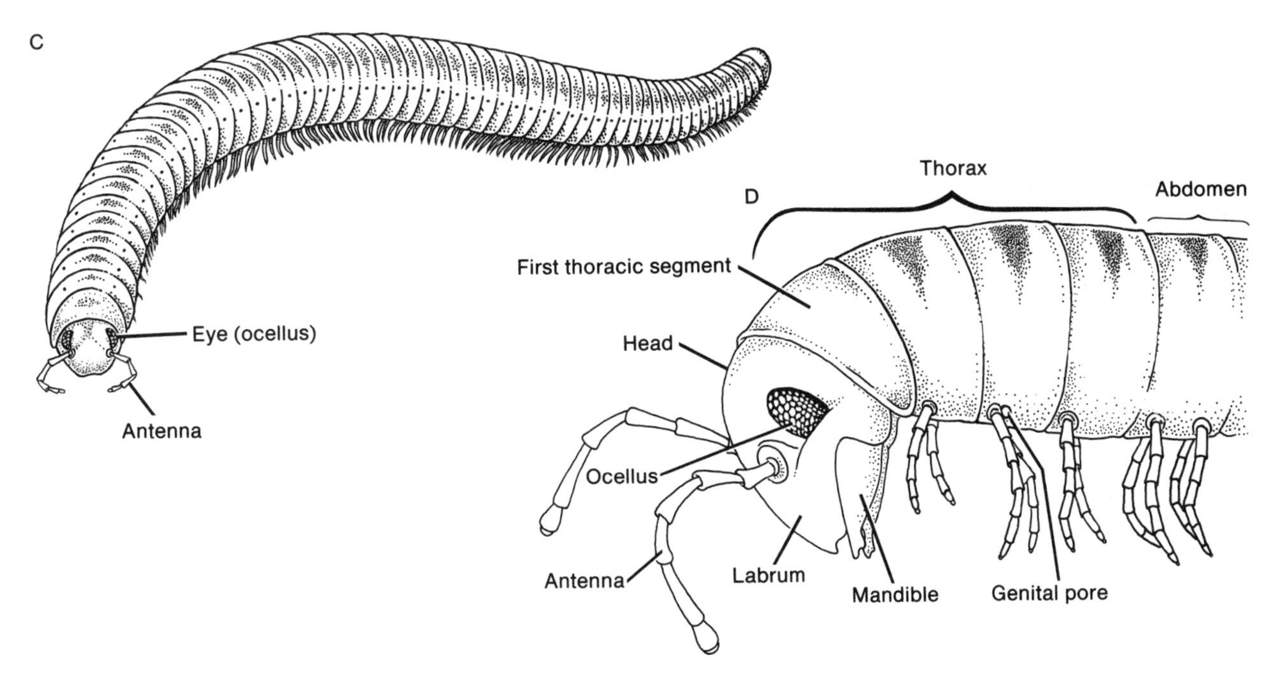

FIG. 36-11
(A) A centipede *(Scolopendra)*, (B) lateral view of head of *Scutigera*, a centipede, (C) a millipede *(Spirobolus)*, and (D) lateral view of head of *Spirobolus*.

Buchsbaum, R. 1975. *Animals Without Backbones.* 2d ed. rev. University of Chicago Press.

Meglitsch, P. A. 1972. *Invertebrate Zoology.* 2d ed. Oxford University Press.

Romoser, W. S. 1981. *The Science of Entomology.* 2d ed. Macmillan.

Storer, T. I., R. L. Usinger, R. C. Stebbins, and J. W. Nybakken. 1979. *General Zoology.* 6th ed. McGraw-Hill.

Villee, C. A., W. F. Walker, and R. D. Barnes. 1984. *General Zoology,* 6th ed. Saunders.

EXERCISE 37

Kingdom Animalia: Phylum Echinodermata

The phylum Echinodermata (Greek *echinos,* "hedgehog"; *derma* "skin") consists of mostly marine, bottom-dwelling animals commonly known as starfish, see urchins, sand dollars, sea cucumbers, and sea lilies. The phylum is so named because of the presence of spiny plates **(calcareous ossicles),** which form a dermal skeleton. Echinoderms are typically radially symmetrical (although the larva is bilaterally symmetrical), and they have true coeloms arising as outpocketings from embryonic mesoderm of the gut. It is on the basis of this last characteristic, and because the bipinnaria larva more closely resembles the chordate larva, that the echinoderms are said to be more closely related to the chordates.

A unique feature of the echinoderms is a derivative of the coelom known as the water vascular system, a system of tubes that are filled with a watery fluid. Although a circulatory system is present, it is greatly reduced. Thus, the coelomic fluid acts as the principal medium for the transport of food and respiratory gases. There are no excretory organs and, thus, echinoderms have little capacity for ionic exchange, which explains why this group of organisms has never invaded fresh waters.

The phylum includes many classes of extinct echinoderms but only the following five include living species:

Class Asteroidea (starfish or sea stars): having a star-shaped body with 5–25 arms, that is covered by a flexible, spiny skeleton

Class Crinoidea (feather stars and sea lilies): having a flowerlike body with many slender, branched arms

Class Ophiuroidea (brittle stars): having a body with a central disc and five distinct, slender, jointed arm

Class Echinoidea (sea urchins and sand dollars): having a cylindrical or disc-shaped body in a shell of fused plates that bear movable spines

Class Holothuroidea (sea cucumbers): having a soft, wormlike body with no arms or spines

A. CLASS ASTEROIDEA (STARFISH OR SEA STAR)

The simplest and perhaps the most familiar of all echinoderms is the starfish. The common starfish, *Asterias,* found along the Atlantic coast of North America, is a typical example. Starfish crawl on the shallow bottom or in tide pools among the rocks and sand of the seashore and coral reefs. They have been serious predators of oysters. At one time oyster fishermen caught starfish, cut them up, and threw them back into the ocean. Then it was discovered that each of the pieces could regenerate and grow into another starfish. Today, "sea mops" made of cloth are dragged over the oyster beds to entrap the starfish. They are then exposed to the sun to dry.

1. External Anatomy

Examine a preserved specimen, and note that the body is composed of a central **disc** from which radiate 5 **arms** or **rays** (Fig. 37-1A). Some specimens may have fewer arms, but this is usually because they have been broken off in handling. Some starfish have more than 5 arms; in rare cases, specimens with as many as 25 have been found. The oral, or ventral, surface of each arm contains grooves ex-

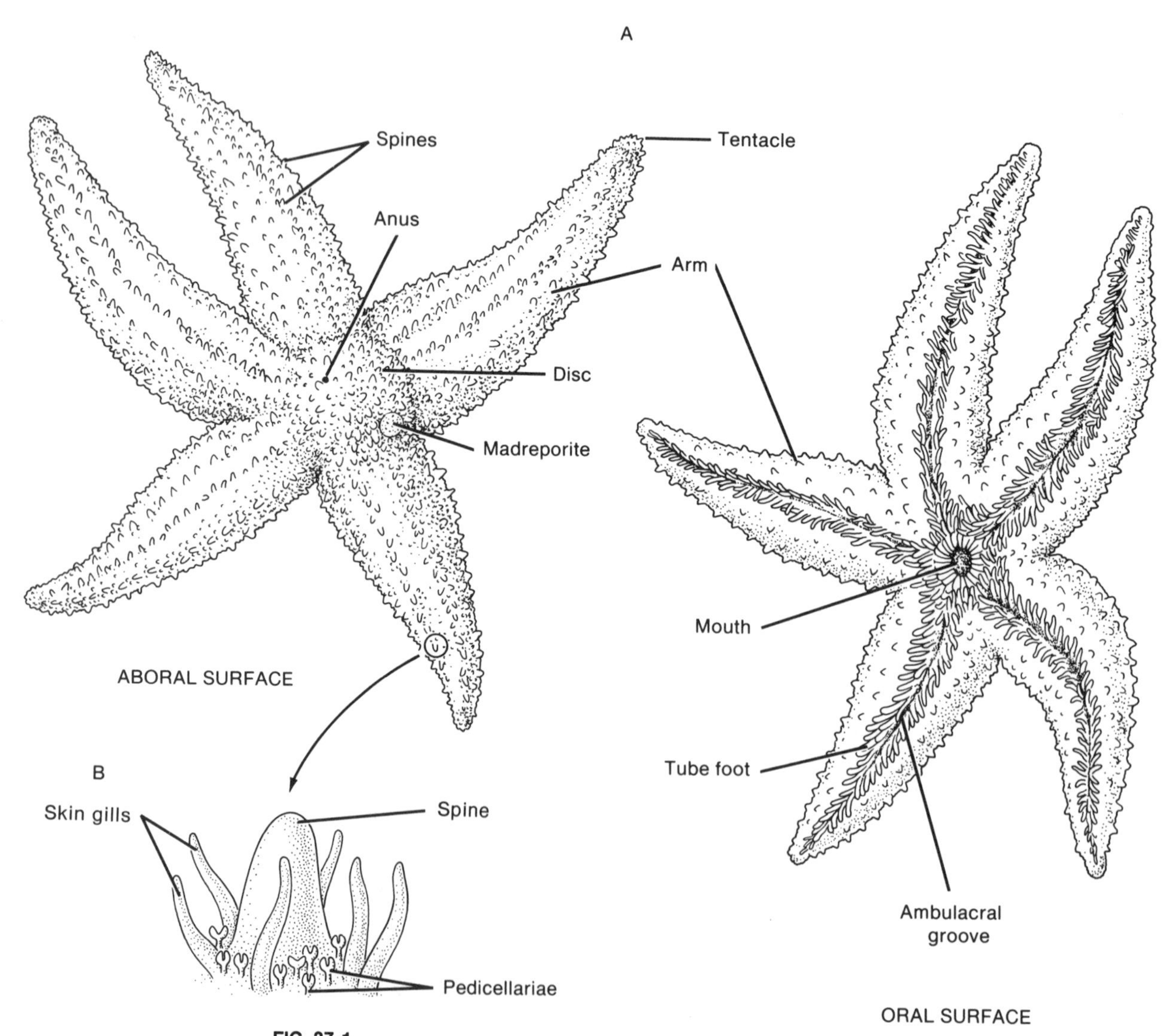

FIG. 37-1
Asterias: (A) external features of the oral and aboral surfaces and (B) spines, pedicellariae, and skin gills.

tending outward from the centrally located mouth. The spiny surface is the aboral, or dorsal, surface. Locate the **madreporite,** a small, porous, buttonlike structure, at one side of the aboral surface. The spines of the aboral surface are part of small calcareous plates (ossicles) that lie buried beneath the integument. These plates form the endoskeleton. Around the spines are many minute pincerlike **pedicellariae** and numerous tiny **skin gills,** which function in respiration (Fig. 37-1B). Each pedicillaria has two jaws, moved by muscles, that open and shut when touched; they keep the body surface clean of debris and may also help to capture food.

The groove in the oral surface of each arm is called the **ambulacral groove.** Along the sides of the ambulacral groove are a series of flexible **spines** that can lie across the groove for protection. Located in these grooves are rows of small, fingerlike projections called **tube feet,** which are organs of locomotion. If you separate the tube feet, you may be able to see the thick, white **radial nerve cord** that runs down the center of each arm. At the tip of each arm, locate the small, light-sensitive **eye spot.** These light-sensitive tips are thrust upward during locomotion.

2. Internal Anatomy

Cut off the tip of one of the arms, and then make longitudinal cuts on one side of the arm to the central disc. Carefully remove the aboral surface to expose the internal organs (Fig. 37-2A). Note that most of the coelom in each arm is taken up by two highly branched digestive glands, the **hepatic caeca.** Examine the glands with a hand lens or dissecting microscope, and note the numerous lobes that secrete digestive enzymes. The two main ducts of these glands join at the base of the arm to form the **pyloric duct,** which enters the centrally located, saclike stomach. A ventral **mouth** and a short **esophagus** lead directly into the stomach, which consists of a lower **cardiac stomach,** which is lobed, and an upper **pyloric stomach.** During feeding, the cardiac stomach is everted through the mouth. The food is partly digested and passed into the pyloric stomach, which empties into the anus located in the center of the aboral disc. Two small **rectal caeca** can usually be found near the anus. These caeca are thought to function as temporary storage areas for waste products.

Cut the pyloric duct where it enters the stomach, and remove the hepatic caeca. If the starfish was caught during the breeding season, the arms will be filled with the **gonads** (reproductive organs). At other times the gonads are normally very small.

The male and female gonads look alike. To determine the sex of the starfish, the gonads must be examined microscopically. To do this, remove a small piece of the reproductive organ, and mince it in a drop of water on a slide. Add a coverslip, and examine under the low and the high power of the microscope. The testes of the male will have flagellated sperm. The ovaries of the female produce spherical eggs that are considerably larger than the sperm. Eggs and sperm are discharged into the water, where fertilization takes place. The fertilized eggs develop into bilaterally symmetrical, ciliated larvae. A similar larval stage is formed during the development of the hemichordates. This larva may pass through several distinct stages before it develops into an adult. The similarity of the larval stage between the Echinodermata and Hemichordata (primitive chordates), as well as similarities in their early development, suggest the probability that both groups arose from a common ancestor at some remote time.

The **water vascular system** of the starfish consists of a series of seawater-filled ducts that function in locomotion and feeding (Fig. 37-2B). To study this system, carefully remove the reproductive organs and the remaining parts of the digestive system: the stomach and anus. Be careful not to damage the sieve plate. Water enters this system through a sievelike **madreporite,** which is connected to a circular **ring canal** by the **stone canal.** The water is then distributed to the **radial canals** that are in each of the rays. The margin of the ring canal bears **Tiedemann's bodies,** which apparently produce the free amoebocytes found in the water vascular system. Lining the ridge through which each radial canal passes is a double row of bulblike structures called **ampullae.** These structures are connected to the **tube feet,** which project from the **ambulacral groove** on the undersurface of each ray. Water from the radial canal collects in the ampullae. Contraction of the ampullae causes the tube feet to elongate as water is forced into them. Expansion of the ampullae results in shortening of the tube feet. Thus, through the use of small suction discs at the end of each tube foot and the alternate expansion and contraction of the ampullae, the starfish is able to move.

B. CLASS CRINOIDEA (FEATHER STARS AND SEA LILIES)

Examine preserved or plastic-mounted specimens of feather stars and sea lilies, the oldest and most primitive of the living classes of echinoderms. These flowerlike echinoderms live from just below the tideline to depths of more than 3600 meters. The

A

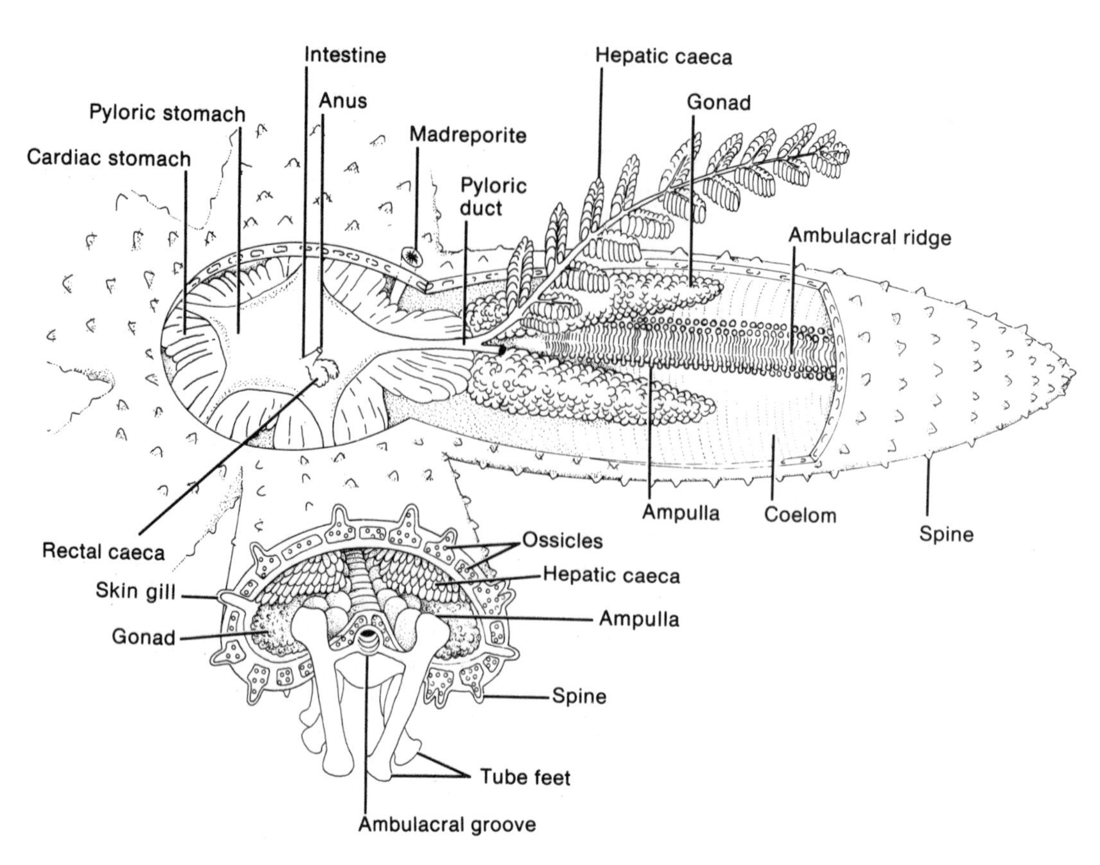

B

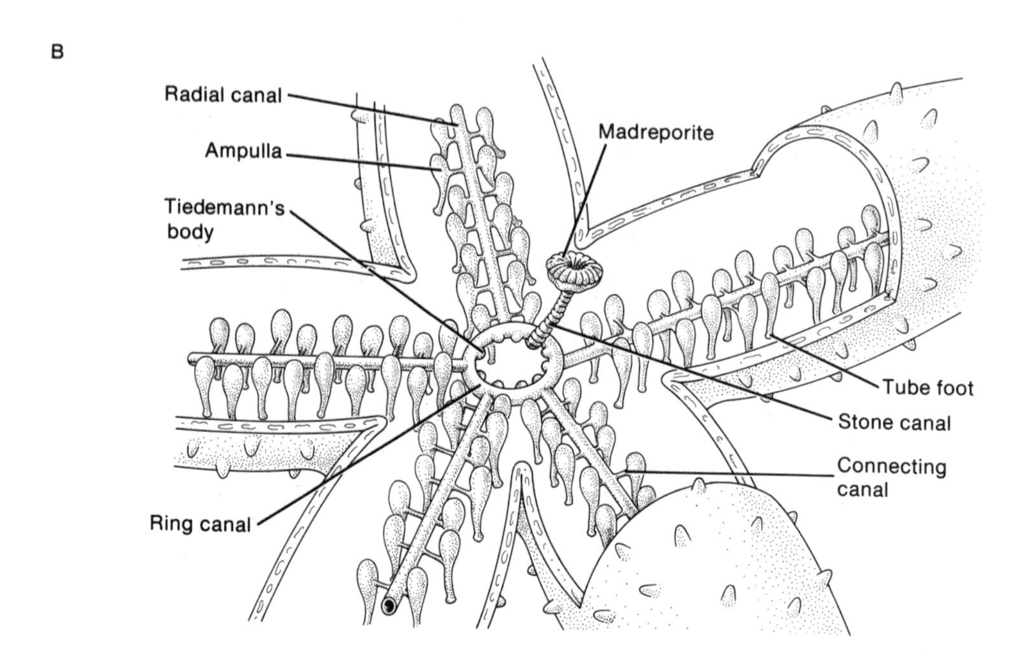

FIG. 37-2
Starfish: (A) internal organs of the starfish, and (B) the water vascular system.

body of a feather star *(Antedon)* is a small cup-shaped **calyx** of calcareous plates, to which are attached five flexible arms that branch to form 10 or more narrow appendages, each bearing many slender lateral **pinnules,** arranged like barbs on a feather, thus giving rise to its common name (Fig. 37-3A). The sea lily has a long, jointed **stalk** that extends from the lower surface of the calyx and attaches it to the sea bottom by rootlike outgrowth called **cirri** (Fig. 37-3B). Both mouth and anus are on the oral surface of the calyx. Each arm has an ambulacral groove, lined with cilia and containing tentaclelike tube feet.

C. CLASS OPHIUROIDEA (BRITTLE STARS)

Examine preserved or plastic-mounted specimens of brittle stars. These echiniderms have five arms like the sea stars, but the arms are longer and more slender and flexible. This echinoderm has a small,

FIG. 37-3
Crinoidea: (A) feather star *(Antedon)* and (B) sea lily *(Metacrinus)*.

rounded disc with five distinct arms that are long, slender, jointed, and very fragile (Fig. 37-4). The skeleton of the arms consists of two parts: an outer, superficial endoskeleton and a deeper, internal, articulated series of vertebral ossicles. This arrangement permits the solidly armored arm to move quite freely and thus enables this animal to crawl rapidly and swim. The arms break or can be cast off easily, such parts being readily regenerated.

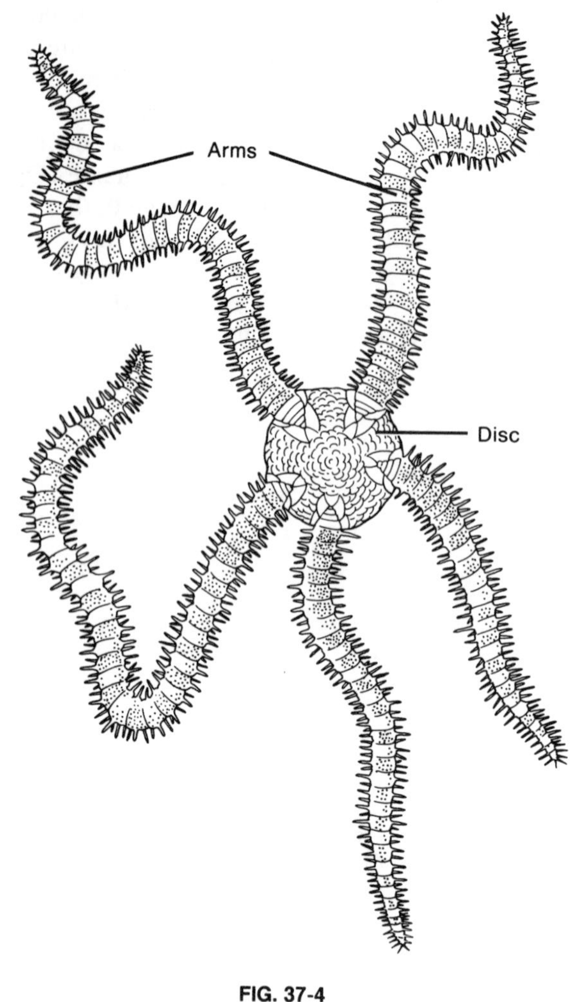

FIG. 37-4
Brittle star.

D. CLASS ECHINOIDEA (SEA URCHINS AND SAND DOLLARS)

Members of this class have globose, oval, or disc-shaped bodies lacking free arms or rays but covered with slender movable spines and five bands of tube feet.

1. Sea Urchin

Examine a preserved specimen of the sea urchin *Arbacia* or *Strongylocentrotus*. Study the surface of this animal and observe the sharp, movable spines that are anchored to the solid shell, or test (Fig. 37-5A). On the shell are the rounded tubercles over which the spines articulate. Among the spines are pedicellariae on long, flexible stalks. Some echinoids have several kinds of pedicellariae; a few bear poison-producing glands. Locate the long, slender tube feet, and note that they are restricted to five regions of the shell known as the **ambulacra.**

Turn the sea urchin so that the oral side faces you. (Fig. 37-5B). In the center of the oral surface is the mouth, which bears five protrusible teeth; these teeth are the principal chewing organs, making up the jaw apparatus known as **Aristotle's lantern.** A soft, membranous area called the **peristome** surrounds the mouth; it has specialized oral tube feet, which are believed to be chemoreceptive in function. At the edge of the peristome are five pairs of **gills.**

2. Sand Dollar

Examine the concave aboral surface of the sand dollar *Echinarachnius* (Fig. 37-6). Observe the arrangement of the **ambulcra** on the surfaces. In the center of the aboral surface, you will find the madreporite, at the end of which are five genital pores **(gonopores).** Turn your specimen over, and locate the mouth in the center of the disc and the anus at the edge.

E. CLASS HOLOTHUROIDEA (SEA CUCUMBERS)

Examine preserved specimens of the sea cucumber *Cucumaria* or *Thyone* (Fig. 37-7). Note that the body surface has no spines. The endoskeleton is reduced to microscopic spicules, thus giving the body wall a tough, leathery, quality. The mouth is located at the center of a conspicuous crown of **tentacles,** which are modified tube feet. In *Cucumaria,* you will also observe lengthwise zones of tube feet that are tactile and respiratory in function. In *Thyone,* the tube feet are distributed over the whole body.

The body wall of the sea cucumber is composed of a cuticle over a nonciliated epidermis, a dermis, a layer of circular muscles, and five double bands of longitudinal muscles. The action of these muscles enables the sea cucumber to extend or contract its body and to move by wormlike movements.

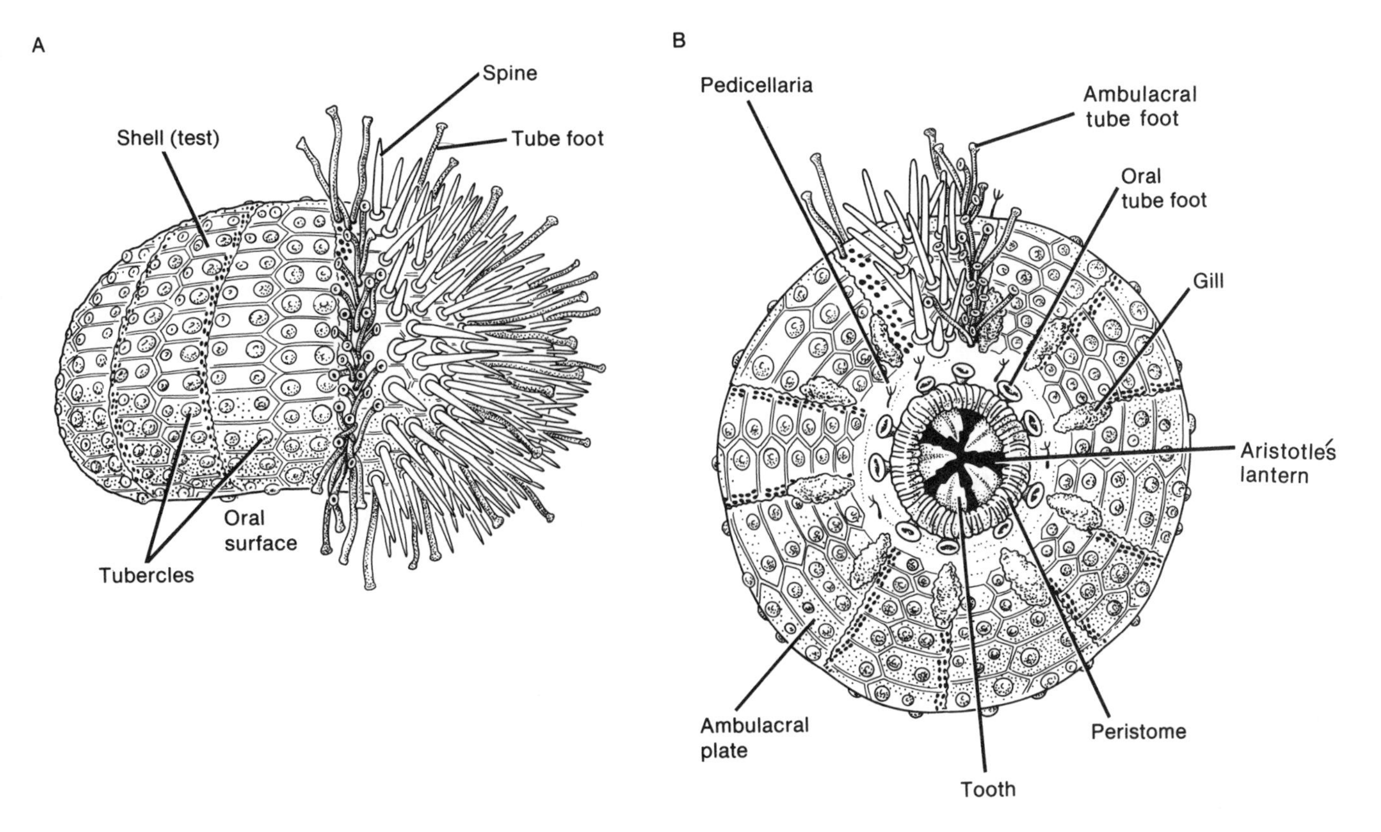

FIG. 37-5
Sea urchin: (A) lateral view and (B) oral surface. Spines and tube feet have been removed on the left to show structure of the test.

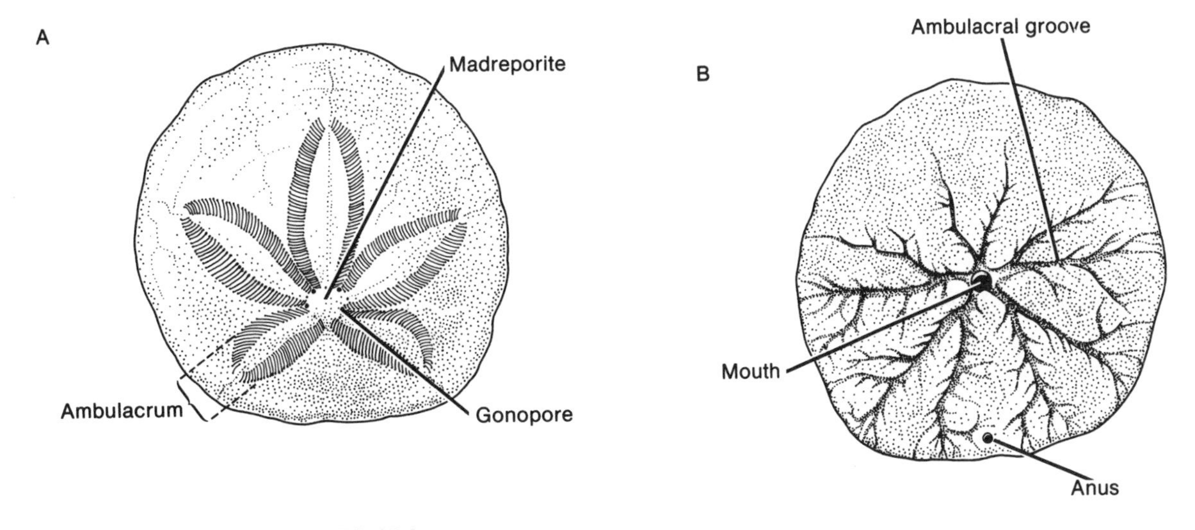

FIG. 37-6
Echinarachnius (sand dollar): (A) aboral and (B) oral surfaces.

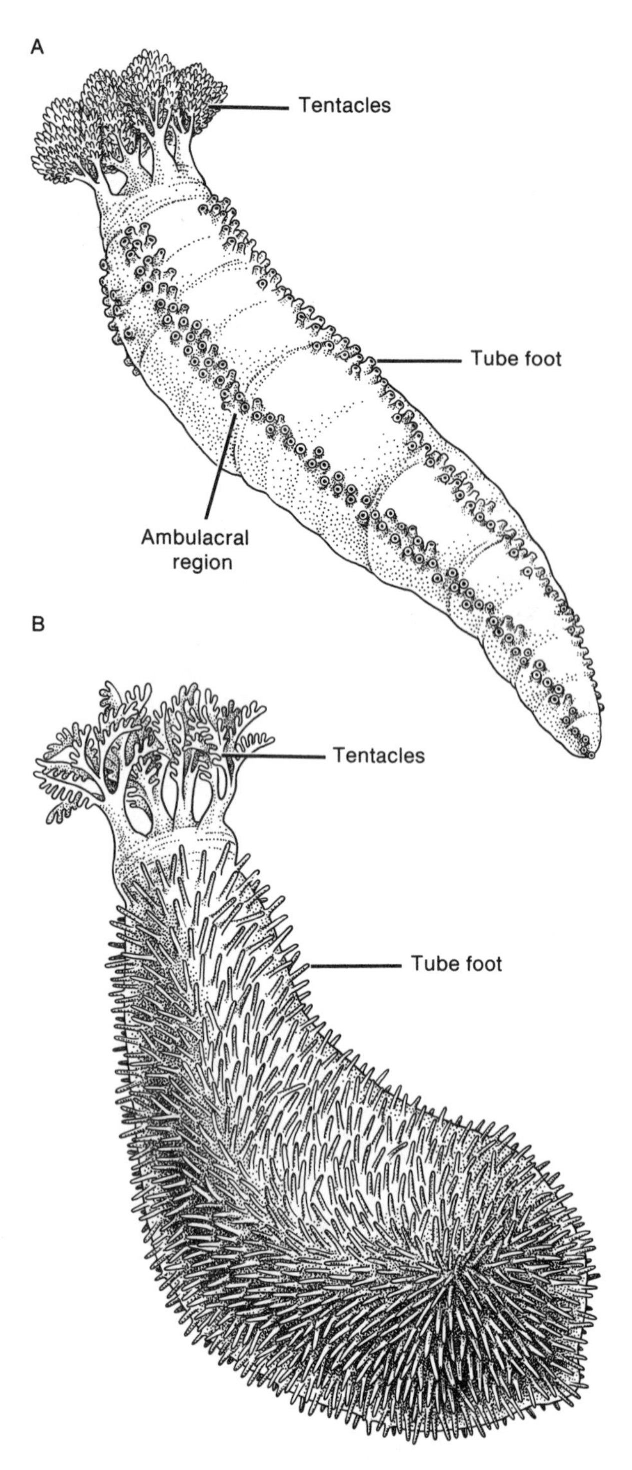

FIG. 37-7
Sea cucumbers: (A) *Cucumaria* and (B) *Thyone*.

REFERENCES

Barnes, R. D. 1980. *Invertebrate Zoology.* 4th ed. Saunders.

Buchsbaum, R. 1975. *Animals Without Backbones.* 2d ed. rev. University of Chicago Press.

Daly, H. V., J. T. Doyen, and P. R. Ehrlich. 1978. *Introduction to Insect Biology and Diversity.* McGraw-Hill.

Hyman, L. H. 1955. *The Invertebrates.* Vol. 4: *Echinoderms.* McGraw-Hill.

Meglitsch, P. A. 1972. *Invertebrate Zoology.* 2d ed. Oxford University Press.

Storer, T. I., R. L. Usinger, R. C. Stebbins, and J. W. Nybakken. 1979. *General Zoology.* 6th ed. McGraw-Hill.

Villee, C. A., W. F. Walker, and R. D. Barnes. 1984. *General Zoology.* 6th ed. Saunders.

EXERCISE 38

Kingdom Animalia: Phyla Hemichordata and Chordata

Representatives of the phylum Hemichordata (Greek *hemi,* "half"; *chorda,* "string") are a group of small, soft-bodied animals there are entirely marine. They are often found in U-shaped burrows on sandy or muddy sea bottoms. The hemichordates, because they possess both echinoderm and chordatelike characteristics, presumably represent an evolutionary link between the echinoderms and chordates. In this exercise, the acorn worm *Saccoglossus* will be studied as a representative of this phylum.

The phylum Chordata (Greek *chorda,* "string") is the largest and ecologically most significant phylum. It is divided into three subphyla, the Urochordata (tunicates), Cephalochordata (lancelets), and Vertebrata, all of which share three important characteristics: a notochord pharyngeal gill slits (or pouches) and a dorsal hollow nerve cord at some stage in their development.

Representatives of the following phyla and classes of vertebrates will be studied in this exercise:

Phylum Hemichordata (acorn worms): wormlike animals characterized by having bilateral symmetry, a well-developed enterocoelom, and gill slits, as well as a primitive dorsal nervous system and an internal skeleton in the form of a notochord

Phylum Chordata: chordates possessing, at some stage in their life cycles, well-developed gills or gill slits, a dorsal hollow tubular nerve cord, and a notochord

Subphylum Urochordata (tunicates): having a larval stage in which chordate characteristics are present, the neural tube and notochord being lost in the sedentary adult, although it does possess a primitive circulatory system

Subphylum Cephalochordata (lancelets): having a well-developed coelom, a circulatory system without a discrete heart, and a fusiform body that has prominent muscle segments (myotomes)

Subphylum Vertebrata (vertebrates): having a cranium (skull), visceral arches, and a spinal column of segmented vertebrae that are cartilaginous in lower forms and bony in higher forms; notochord extends from cranium to base of tail; enlarged brain; head region with specialized sense organs

Class Agnatha (cyclostomes): having a long, slender, and cylindrical body with median fins, a mouth located ventrally, 5–16 pairs of gill arches and a persistent notochord, but lacking true jaws and scales (hagfish and lampreys)

Class Chondrichthyes: having a cartilaginous skeleton with notochord, tough skin that is covered with scales, median and paired lateral fins, a ventrally located mouth with both upper and lower jaws, and pectoral and pelvic girdles (sharks, skates, and rays)

Class Osteichthyes: having a skeleton that is somewhat bony, a terminal mouth, gills covered by an operculum, median and paired fins, and (usually) skin that is covered with scales, although some are scaleless (perch, carp, and trout)

Class Amphibia: having moist, glandular skin that lacks scales; two pairs of limbs, but no fins; a bony skeleton; a terminal mouth with upper and lower jaws and a tongue that is often protrusible; being aquatic in the larval stage but usually terrestrial as an adult (frogs, toads, salamanders, and newts)

Class Reptilia: having a body that is dry and covered with scales; two pairs of limbs (absent in snakes), with digits adapted to running, crawling, climbing, or swimming; and a bony skeleton (lizards, snakes, turtles, crocodiles, and alligators)

Class Aves: warm-blooded animals having a body covered with feathers, forelimbs modified as wings, a bony but light skeleton, and a beak (includes all birds)

Class Mammalia: having mammary glands that secrete milk for nourishing the young, hair in varying quantities, and young that are born alive (human beings, dogs, cows, and mice)

A. PHYLUM HEMICHORDATA

Hemichordates are common marine animals of broad distribution. They are small, soft-bodied organisms that live singly or in colonies on sandy or muddy sea bottoms or in open ocean waters. The most common representatives are the acorn worms *Balanoglossus* and *Saccoglossus.*

Examine a preserved or plastic-mounted specimen of *Saccoglossus.* Note the softness of the body and its wormlike form (Fig. 38-1). The epidermis is ciliated and well supplied with mucus-secreting cells, which are important in burrowing and feeding. The body is divided into three regions: an anterior **proboscis,** a short **collar,** and an elongate **trunk.** The **mouth** opening is ventral and located at the base of the proboscis. The trunk is divisible into three regions: an anterior **branchial region** containing gill slits; a **genital region,** which, in other species, may be enlarged into large genital ridges; and, posteriorly, an **abdominal region** containing the intestine and lateral pouches of the hepatic caeca. The **anus** is terminal.

During development, a ciliated larva stage is formed that looks very much like the larvae of echinoderms. The similarity of the larvae in these

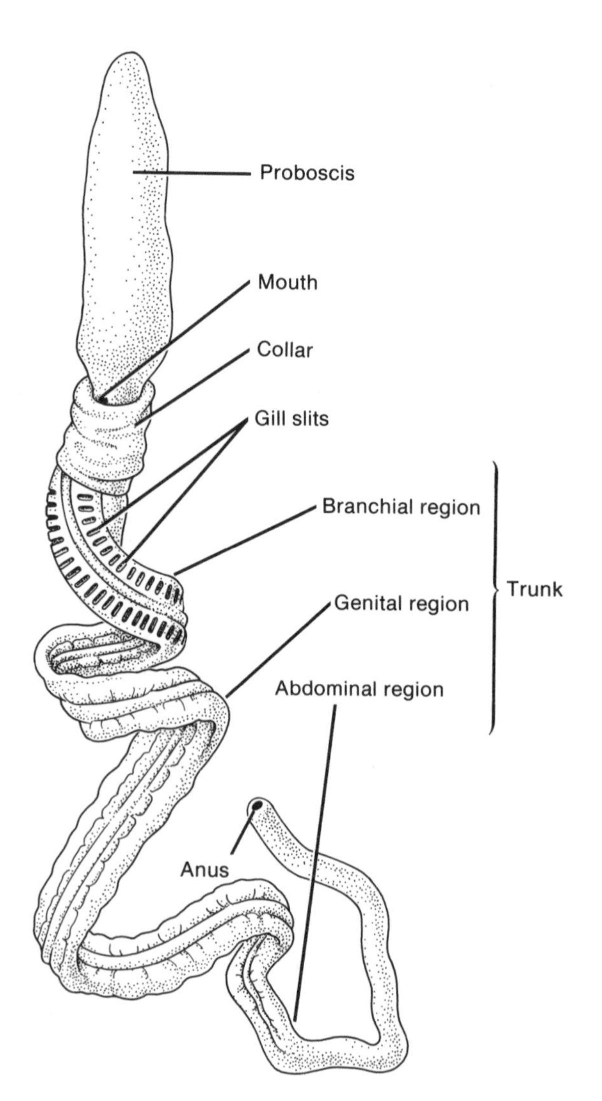

FIG. 38-1
External anatomy of the acorn worm *Saccoglossus.*

two phyla, as well as the similarities in their early development, suggest that the two groups probably arose from a common ancestor at some remote time.

B. PHYLUM CHORDATA

Chordates are defined by the presence of three major features: one is the dorsal, hollow **nerve cord,** which in mammals becomes the brain and spinal nerve cord. Another major characteristic is the **notochord,** a cartilaginous rod that develops dorsal to the primitive gut in the early embryo. In the lower chordates, the notochord persists throughout life; whereas, in the vertebrates, it is surrounded and later persists as the soft center of the intervertebral discs. The third chordate characteristic is the presence, during some stage in the life cycle, of gill slits in the pharynx or throat.

1. Subphylum Urochordata

The urochordates are a group of exclusively marine and sessile organisms that are covered by a firm tunic, from which they get the common name of tunicates.

Examine a preserved or plastic-mounted specimen of the tunicate *Ciona* or *Molgula* (Fig. 38-2A). The adult animal is saclike in appearance and is usually attached by its base to rocks, seaweeds, shells, wharf pilings, or ship hulls. At its free end are two openings: a mouth, or incurrent siphon, and an excurrent siphon. The body wall consists of a firm tunic, beneath which lies the mantle, which secretes the tunic and contains the muscles by means of which the body shape can be altered. The body is divided into an atrial cavity, containing a large branchial sac that has many gill slits, and a visceral cavity, containing the major organs (gonads, stomach, and heart).

Examine a prepared slide of the larval stage of *Molgula.* This transparent, free-swimming larva somewhat resembles an amphibian tadpole and thus is commonly called a **tadpole larva.** The larva has all three chordate characteristics, only one of which (gill slits) are retained in the adult tunicate. Its tail contains a supporting notochord, a dorsal hollow tubular nerve cord, serial pairs of lateral, segmental muscles, and **gill slits** in the pharynx (Fig. 38-2B). It swims about like a tadpole, settles down on some object, and then transforms into a sedentary adult. During this transformation, two of the chordate features (i.e., the notochord and nerve cord) disappear.

2. Subphylum Cephalochordata

Amphioxus *(Branchiostoma),* the most commonly studied member of the cephalochordates, is a small, fishlike animal found in shallow marine waters in many parts of the world. Although the animal can swim with lateral, undulating movements of its body, it spends most of the time buried in the sandy bottom with its anterior end projected.

In addition to gill slits, the cephalochordates have a notochord and a nerve cord that persist in the adult and extend the length of the body.

a. External Anatomy

Examine a preserved or plastic-mounted specimen of amphioxus. The animal is pointed at both ends and compressed laterally and, therefore, is commonly called a **lancelet** (Fig. 38-3A). Notice the absence of a distinct head, and the conspicuous chevron-shaped **muscle bands (myotomes)** of the body. A median **dorsal fin** extends almost the entire length of the body and a median **ventral fin** covers the posterior third of the animal. At the anterior end of the body is the funnel-shaped **buccal cavity** surrounded by a circle of oral tentacles **(cirri).** Approximately two-thirds of the way back from the anterior end, on the ventral side, locate the **atriopore,** through which water leaves after having been pumped through the pharynx.

b. Internal Anatomy

Study the internal anatomy of amphioxus by examining preserved or stained specimens under the dissecting microscope (Fig. 38-3A). Identify the **nerve cord,** which extends nearly the entire length of the animal's body. Above the nerve cord are short rods of connective tissue, called **fin rays,** that strengthen the dorsal fin. Locate the **notochord,** a cartilagelike rod that lies ventral to the nerve cord and extends the length of the body. The notochord is a longitudinal elastic rod of cells and serves as an internal skeleton. Ventral to the notochord is the digestive tract. The mouth, located in the buccal cavity, leads into the relatively large **pharynx.** Observe the **gill slits** in the wall of the pharynx. The pharynx is lined with cilia that beat inward to produce a steady current of water that enters the mouth, passes over the gill slits (leaving behind suspended food particles), and is then eliminated through the atriopore. Posteriorly, the pharynx joins the intestine. Close to the point at which it does is a ventral outgrowth, the **"liver,"** that secretes digestive enzymes into the intestine.

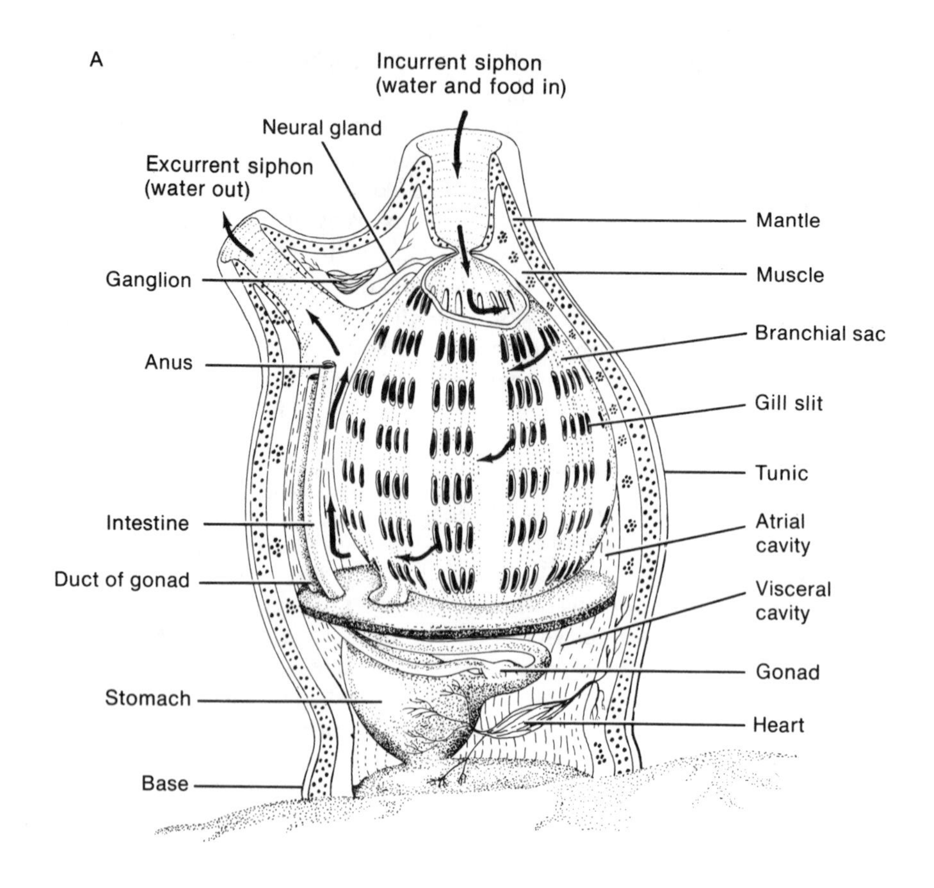

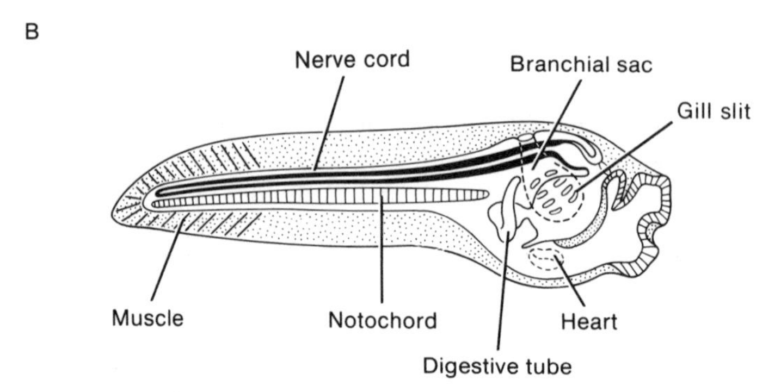

FIG. 38-2
Molgula: (A) external anatomy of an adult and (B) tadpole larva.

Examine a cross section of amphioxus taken from the region of the pharynx, and locate the following structures shown in Fig. 38-3B; the **dorsal fin** with its dorsal fin ray; **metapleural folds;** the skin consisting of a one-celled layer of **epidermis** and a thicker **dermis;** muscle bands **(myotomes);** the **nerve cord** and its **central canal;** the **notochord; dorsal aortae,** a pair of small blood vessels ventral to the notochord; the **pharynx,** which contains a middorsal furrow, the **hyperbranchial groove** lined with cells, and a midventral groove, the **endostyle,** having both gland cells and cilia; the **"liver,"** a tube on the right side of the pharynx; the **ventral aorta,** a small blood vessel just ventral to the endostyle; the **atrial cavity;** and the **coelom.**

On either side of the pharynx, locate the **gonads,** paired bodies containing the sex cells. Where would you expect fertilization to take place?

____

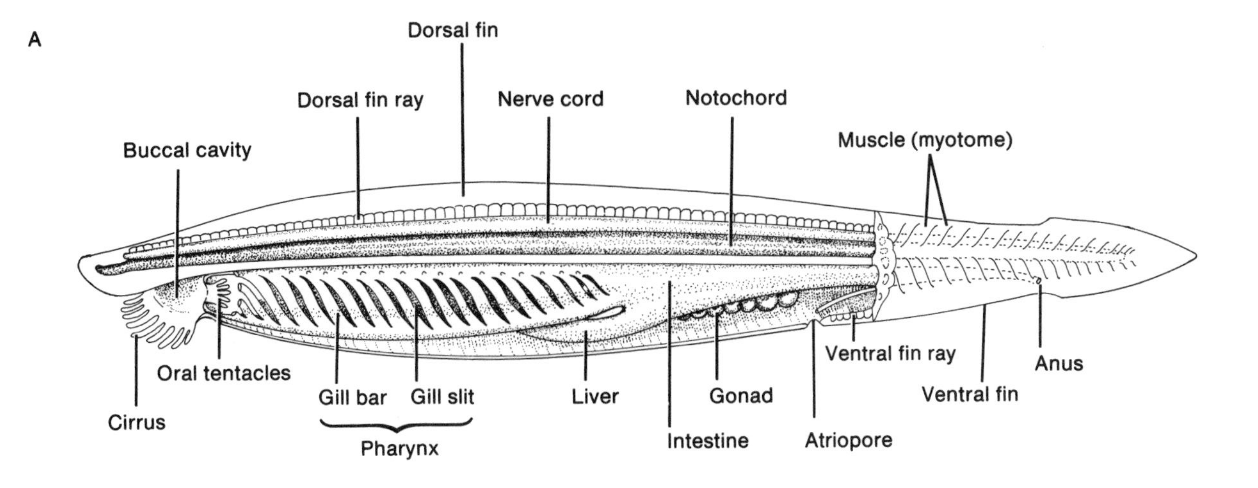

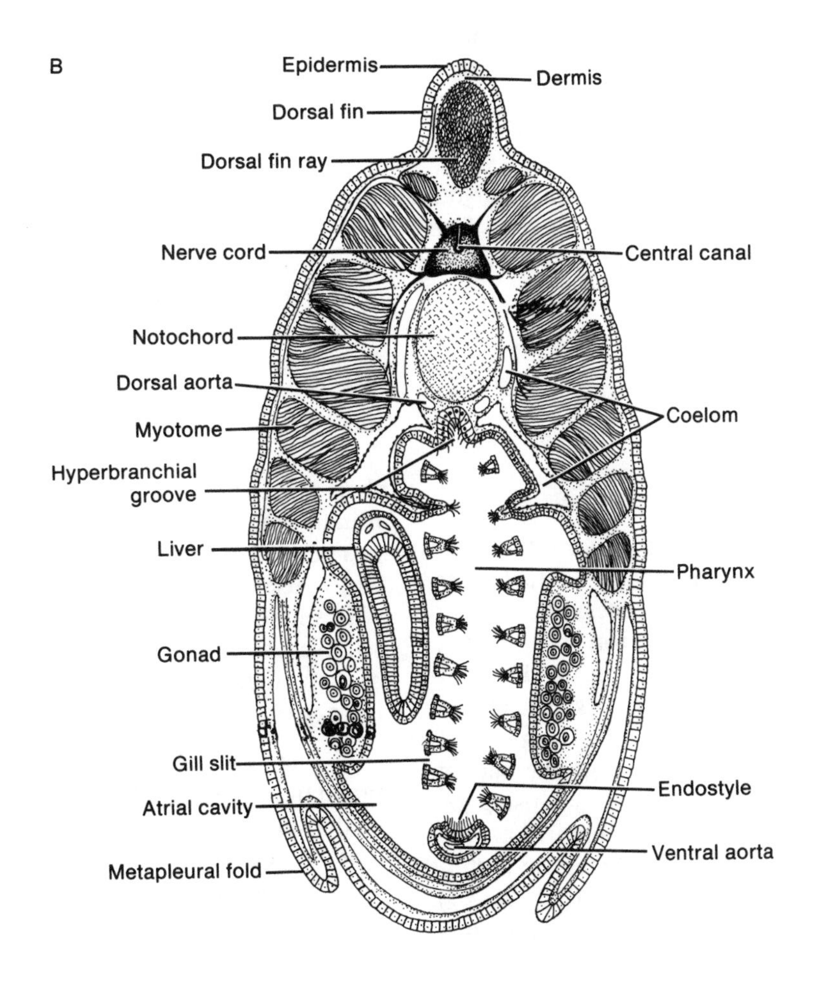

FIG. 38-3
Amphioxus *(Branchiostoma):* (A) lateral view of internal anatomy and (B) cross section.

The sex of your specimen can be determined by examining the cells of the gonads. The testes are made up of a large number of small, dense cells. The ovaries contain fewer, larger cells with vesicular nuclei. What is the sex of your specimen?

___

3. Subphylum Vertebrata

The Vertebrata is by far the largest and most important of the chordate subphyla. The vertebrates are characterized by an endoskeleton that includes a vertebral column, or backbone, composed of a series of vertebrae. The vertebrae develop around the notochord, which in most vertebrates is present only in the embryo.

a. Class Agnatha

The Agnatha include the hagfishes and lampreys (Fig. 38-4). Organisms in this class are also called *cyclostomes* ("round mouths"). All hagfishes are marine, whereas lampreys inhabit both marine and fresh water.

Examine the sea lamprey on demonstration. Note the shape of its body, which is similar to that of an eel, and the median fins. How many dorsal fins are there?

___

How many ventral fins?

___

At the anterior end of the body, locate the seven external gill slits, which open to the gills. Just anterior to the gill slits are the **eyes.** Between the eyes, on the dorsal surface, is a single **nasal opening** that functions as an olfactory (smell) organ. The ventrally located mouth lies within a suctorial disc called the **buccal funnel** (Fig. 38-4B). Lampreys attach themselves to their prey through the suction generated by the buccal funnel. Then, using the pointed horny teeth inside the buccal funnel and the rasplike tongue, the lamprey penetrates the flesh of the animal and feeds on the blood (Fig. 38-4D). Some years ago, the sea lamprey all but eliminated the commercial fish (whitefish, lake trout, lake perch) of the Great Lakes. Successful eradication measures, including chemical control and electric barriers to block migration, have virtually eliminated the lamprey, and the population of fish that are used commercially has steadily increased.

Examine the **ammocoete larva** of the lamprey. To which organism already studied does it bear a striking resemblance?

___

In terms of evolutionary relationships, what is the significance of this similarity?

___

___

___

b. Class Chondrichthyes

Chondrichthyes (Greek *chondros,* "cartilage"; *ichthys,* "fish") include sharks, skates, and rays. All are predators, and most live in the ocean, although a few are found in tropical rivers and lakes, and in fresh and brackish waters.

This class is characterized by skeletons that lack bones but are instead made of a softer, more flexible material, called **cartilage.** The skin is covered by small, pointed scales made of plates of dentine covered by enamel, which structurally resemble vertebrate teeth.

Sharks are active swimmers and are usually found in the open oceans. Their typical diet includes fish, squid, and small crustaceans. Some of the larger sharks capture sea lions and seals. Rays and skates have a principal diet of small invertebrates.

In many countries, including the United States, sharks and rays are used for food, although it is usually not identified as such. In Asia, the shark's fins are boiled to yield a gelatinous substance that is widely used as a flavoring for soups.

Examine the dogfish shark *(Squalus acanthias),* a relatively small species frequenting the eastern shore of the United States (Fig. 38-5A). Locate the following external features:

1. The **head,** which is bluntly pointed, contains a broad mouth having several rows of sharp teeth. The teeth, unlike those of the bony fishes and higher vertebrates, are not attached to the jaw but are embedded in the flesh. Teeth are continuously being formed and move forward to replace those that are lost.

2. The two **nostrils** contain olfactory organs, enabling the shark to smell materials dissolved in the water.

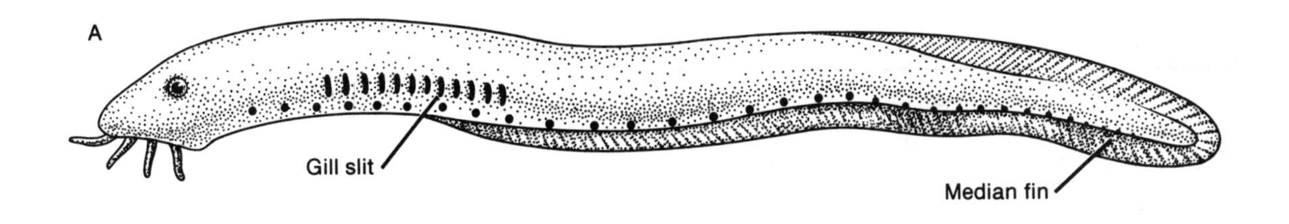

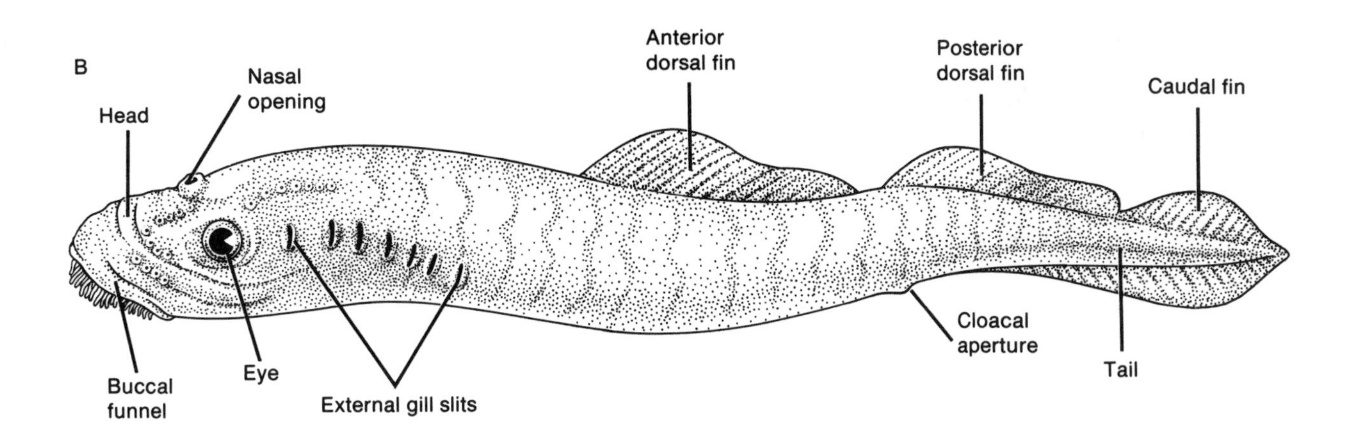

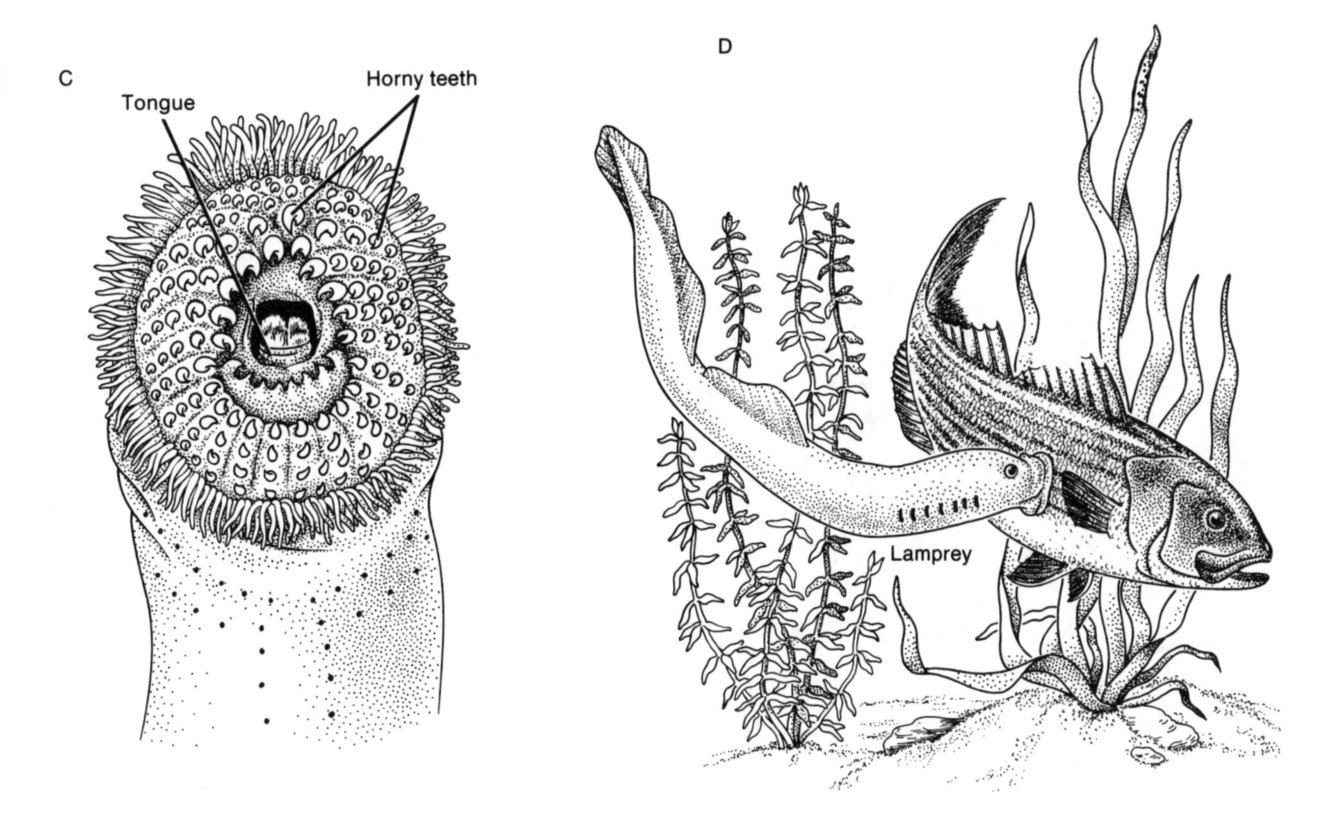

FIG. 38-4
Agnatha: (A) hagfish, (B) sea lampry, (C) buccal funnel of lamprey, and (D) lamprey attached to fish. (Parts B–D after *Atlas and Dissection Guide for Comparative Anatomy,* 2d ed., by Saul Wischnitzer. W. H. Freeman and Company. Copyright © 1972.)

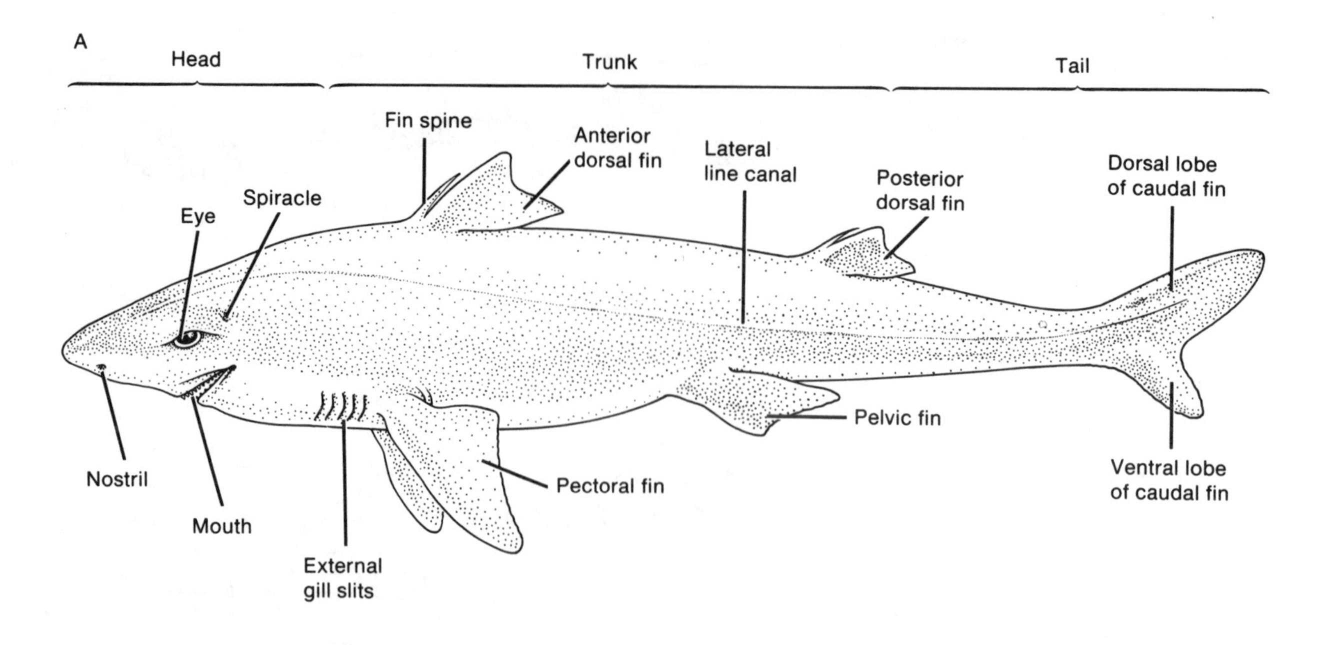

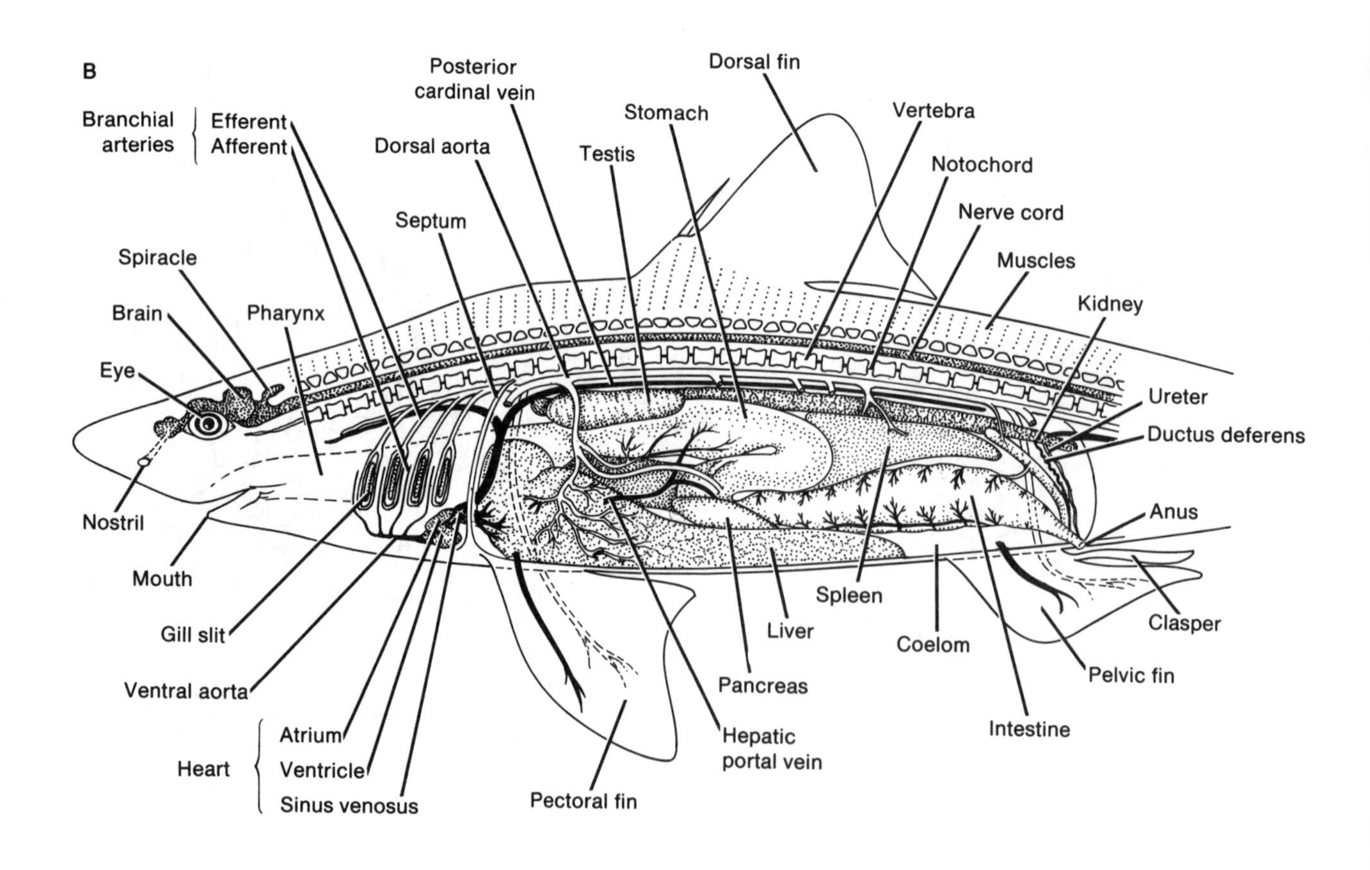

FIG. 38-5
Dogfish shark: (A) external anatomy and (B) internal anatomy. (After *Atlas and Dissection Guide for Comparative Anatomy,* 2d ed., by Saul Wischnitzer. W. H. Freeman and Company. Copyright © 1972.)

3. The **eyes** are without lids and are adapted for use in dim light.

4. Anterior to each **pectoral fin** are six gill sits. Five of them look like slits, but the sixth, which is located just behind the eye, is modified as the **spiracle.**

The **anus** is located between the **pelvic fins.** In the male, the pelvic fins possess **claspers,** which are brought close together during mating and inserted into the cloaca of the female. Seminal fluid containing sperm thus flows down the channel formed and enters the female.

6. There are separate medianly located **dorsal fins,** each having a characteristic spine just anterior to the fin. The **caudal (tail) fin** is bilobed and is used to propel the animal through the water.

7. The **lateral line,** a fine groove along each side of the body, contains a canal having numerous openings to the surface. In the canal are sensory hair cells that connect to the tenth cranial nerve. These sensory cells respond to low-frequency pressure stimuli in the water, serving in a way as a response to touch at a great distance.

8. The body itself is covered evenly by diagonal rows of **placoid** scales. Each scale is covered by enamel and has an inner dentine layer.

The internal anatomy of the shark is shown in Fig. 38-5B. A peculiar feature of the digestive system is the **spiral valve,** a spirally arranged partition covered with a mucous membrane that functions to increase the area for absorption.

c. Class Osteichthyes

The Osteicthyes (Greek *osteon,* "bone"; *ichthys,* "fish"), or bony fish, are characterized by skeletons made of bone, a skin covered with dermal scales and gills covered by an operculum. The members of this class range from the common yellow perch to the more unusual lungfish (Fig. 38-6).

This class includes most of the animals we think of as fish. Fish are a highly successful and adaptable group that inhabit virtually all types of water, including fresh, brackish, and salt. Although most fish are streamlined in shape, usually having spindle-shaped bodies, they have developed a wide variety of forms to assure success in given habitats. As a group, fish are vitally important as a source of food throughout the world. Many species are also caught for sport.

The common yellow perch *(Perca flavescens)* is typical of the bony fish and, thus, a good example to study (Fig. 38-7).

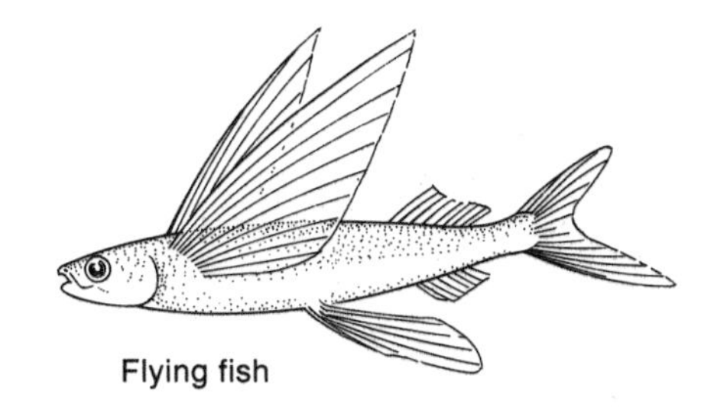

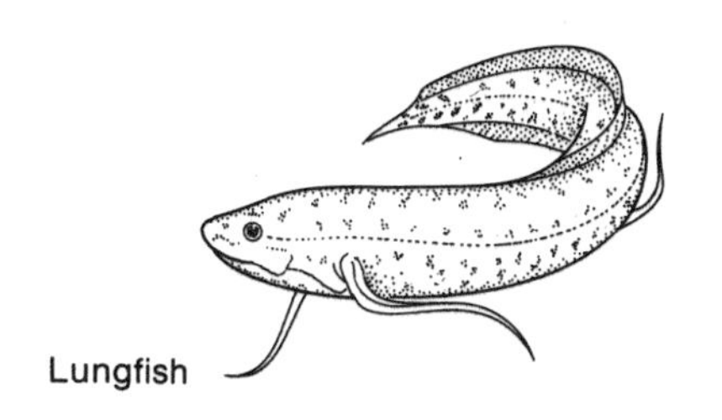

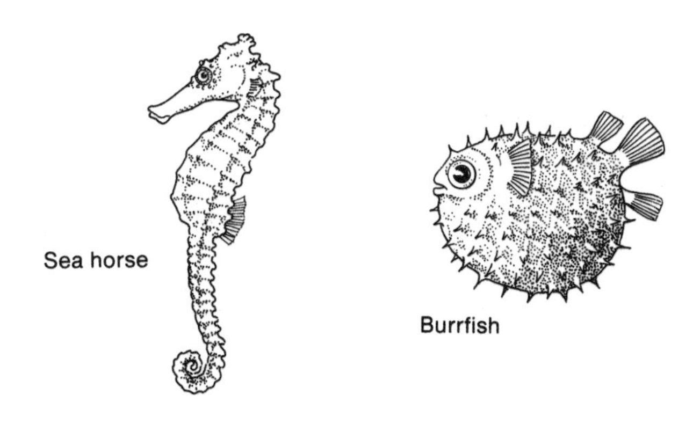

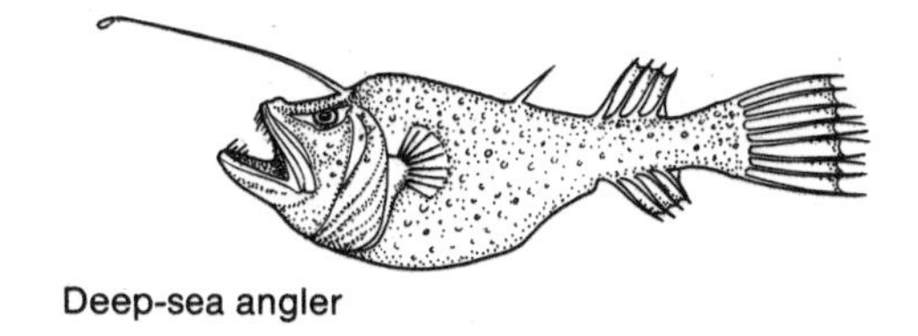

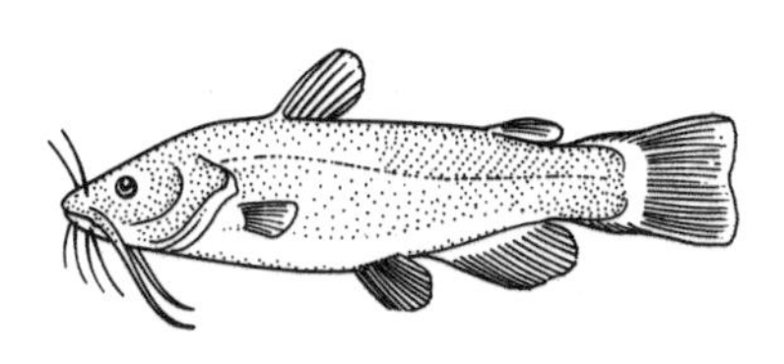

FIG. 38-6
Representative bony fish.

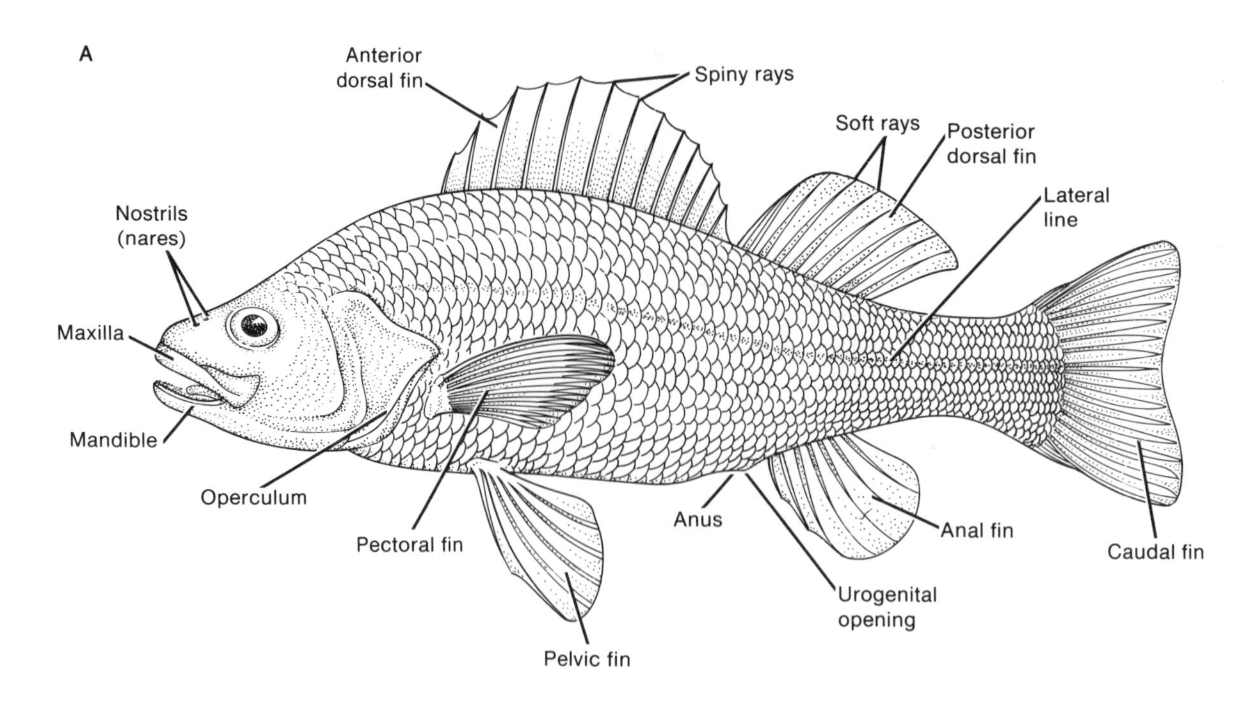

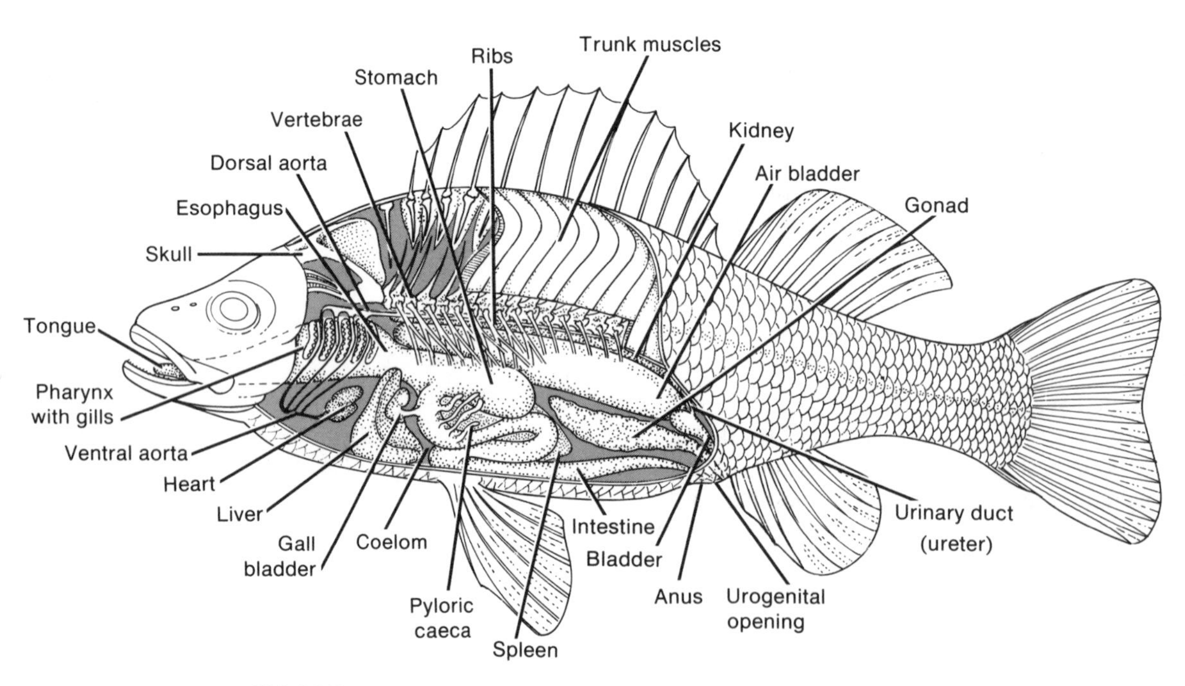

FIG. 38-7
Yellow perch *(Perca flavescens):* (A) external anatomy and (B) internal anatomy.

1. **External Anatomy.** The main body regions of the perch are the head, trunk, and tail **(caudal fin).** The **head** extends from the tip of the snout to the hind edge of the operculum, the **trunk** from this point to the anus, and the **tail** is the remainder (Fig. 38-7A). Examine the head, and observe the large mouth, which has distinct upper and lower jaws, the upper jaw being the **maxilla** and the lower jaw the **mandible.** Dorsally on the snout are two double external **nares (nostrils);** they are the openings to the **olfactory sacs,** which are highly sensitive to dissolved chemicals in the water. There is no external

ear, but each internal ear contains semicircular canals that are balancing organs that enable the fish to maintain the proper position in the water. The lateral eyes are without lids. Behind each eye is a bony **operculum,** which is a protective covering for the four comblike respiratory gills.

Cut away the operculum from the left side, and examine the gills. Cut away one gill arch, place it in water, and examine with a dissecting microscope. The spaces between the gills are called **gill slits,** or **gill clefts.** Attached to the lower edge of the operculum is the **branchiostegal membrane,** supported by rays of cartilage. This membrane serves as a one way valve allowing water to pass out of the opercular opening but preventing its return.

A lateral line extends along each side of the whole body of the perch. The lateral line is essentially a system of water-filled canals that communicate by means of pores with the water in which the fish swims. The system is believed to register vibratory currents made both by moving objects in the water and by the swimming movements of the fish itself. The system therefore functions as a sensitive "listening" device detecting and discriminating between different kinds of turbulence.

Two separate dorsal fins are on the back of the perch; a **caudal fin** is on the end of the tail; and an **anal fin** is on the ventral side of the tail. Just anterior to the anal fin are the anus and the **urogenital opening.** The lateral, paired fins are the **pectoral fins** behind the opercula, and the ventral, **pelvic fins** are just below them. The fins are membranous extensions of the integument supported by fin rays. All the fins except the anterior dorsal fins are flexible, being supported by soft rays. Each anterior dorsal fin has 13–15 solid calcified spines, and there are one or two similar rigid spines at the anterior edge of the other fins. The fins are used in swimming, steering, and maintaining equilibrium.

The trunk and tail are covered by thin, rounded scales in lengthwise and diagonal rows. Remove a scale, mount it in a drop of water on a slide, and examine under the low power of a microscope. Note the annual growth rings. As a fish grows, the size of each scale rather than the number of scales increases. As a scale grows, concentric rings are formed. Rings formed in the fall and winter are closer together than those formed in the spring and summer. By counting the number of regions of closely spaced concentric rings, you may determine the number of winters your specimen lived. What is the approximate age of your fish?

____

The entire body of the fish is covered by a soft mucus-producing epidermis that facilitates easy movement of the fish in water and protects against entry of disease organisms.

2. **Internal Anatomy.** Hold the fish with the ventral side up and the head pointing away from you. Insert the point of your scissors through the body wall in front of the anus, and cut up the midline of the body to the space between the opercula. Now lay the fish on its right side (with the head on your left) in the dissecting pan. Continue to cut up around the back edge of the gill chamber to the top of the body cavity. Make another incision from the starting point of the ventral incision close to the anus, and cut upward to the top of the body cavity. Be careful not to disturb the internal organs. With a scalpel, remove the whole lateral body wall by cutting along the top of the body cavity. This procedure will expose the body organs in their normal positions.

a. *Digestive System.* Locate the reddish brown **liver** in the anterior end of the body cavity (Fig. 38-7B). Raise the lobes of the liver and find the **gall bladder** attached to the lower side. Cut the liver free from its attachment to the body, and remove it. This will expose the short **esophagus** and **stomach.** Locate the **pylorus** where the stomach and intestine join. Three tubular **pyloric caeca,** secretory or absorptive in function, attach to the intestine. Examine the loops of the intestine, and trace it to the anal opening. Long masses of fat lie along the intestinal loops. Observe the **spleen,** attached to an internal mesentery near the stomach. It is a dark red gland that has no duct and no functional connection with the digestive system. Cut the esophagus at its anterior end and carefully remove the entire alimentary canal, from the mouth to the anus.

b. *Reproductive System.* Having removed the alimentary canal, you should be able to see the gonads and the urinary bladder.

In the female, the **ovary** lies between the intestine and air bladder. It is an epithelial sac filled with eggs. The posterior end of the ovary is tapered, and the eggs pass to the outside through the urogenital opening just behind the anus. The ovaries are paired in early stages, as they are in other vertebrates, but during development they fuse into a single organ.

In the male, the **testes** are a pair of white, elongated bodies lying just below the air bladder to which they are joined by a thin sheet of tissue, the **mesentery.** They fuse together toward the posterior end, and the sperm are passed to the outside through the urogenital opening.

c. *Air Bladder (Swim Bladder).* Locate the **air bladder** along the top of the body cavity. The bladder is filled with gases (oxygen, nitrogen, and carbon dioxide) and acts as a hydrostatic organ to adjust the specific gravity of the fish at different depths of water. By secretion or absorption of gases through blood vessels in the bladder wall, a fish makes the adjustment slowly as it moves from one depth to another.

d. *Excretory System.* Remove the air bladder from the body. Lying in the roof of the bladder, near the middle, is a large blood vessel, the **dorsal aorta.** Parallel to it is a pair of long, narrow, dark-colored kidneys. The **urinary ducts (ureters)** run along the edges of each kidney, join at their posterior ends, and empty into the **urinary bladder.** Fluid nitrogenous wastes removed from the blood are emptied by means of this excretory system through the urogenital opening.

e. *Circulatory System.* Using a point of a scissors, make a horizontal cut through the anterior wall of the body cavity in front of the liver. This will expose the **peritoneal cavity,** in which lies the two-chambered **heart.** The heart consists of a single, median, light colored **ventricle** and a larger thin-walled **atrium (auricle).** Behind the atrium is the **sinus venosus.** The large vessel carrying blood from the ventricle is the **ventral aorta.** It is greatly enlarged just anterior to the ventricle, forming the **conus arteriosus.** Trace the ventral aorta forward toward the gill region, and identify some of the branchial arteries. Rhythmic contractions of the ventricle force the blood through the conus arteriosus and short ventral aorta into four pairs of **afferent branchial arteries,** which distribute blood to capillaries in the gill filaments for oxygenation. In the gill filaments, the blood is collected into correspondingly paired **efferent brachial arteries** leading to the dorsal aorta, which has branches to all parts of the head and body.

f. *Nervous System.* Hold the fish with the dorsal side up and the head pointing away from you. Using your scalpel, cut the skin from

FIG. 38-8
Representative amphibians.

the skull, and scrape the skull carefully to wear away the bone. When the bone is very thin, remove the pieces with your forceps to expose the brain. Locate the **olfactory lobes** in front, the larger lobes of the **cerebrum** behind them, and the very thin **optic lobes** posterior to the cerebrum. The **cerebellum** is posterior to the optic lobes, and the **medulla** is the enlargement where the **spinal** cord joins the brain.

d. Class Amphibia

The amphibians include toads, frogs, mud puppies, and salamanders (Fig. 38-8). As the name of this class implies (Greek *amphi,* "dual"; *bios* "life"), these animals live both on the land and in the water. Indeed, their position in the evolutionary scale is between fish and reptiles in that they were the first chordates able to live on land. They exhibit several modifications that have allowed them to adapt to a terrestrial existence. These modifications include legs, lungs, nostrils, and sense organs that function both in water and air.

Most amphibians lay their eggs in the water and have an aquatic larval stage called a **tadpole.** Some amphibians are often confused with reptiles, particularly the salamanders and newts, which look like lizards. Reptiles, however, have scales, whereas salamanders and newts have smooth, slimy skin.

Amphibians are very beneficial to human beings. For example, frogs are extensively used for laboratory dissections, pharmacological experiments, and even fish bait. Extracts of frog skin are used to make the glue used in book bindings. More than 4 million frog legs are eaten each year. In Asia, toad skin is used for medicinal purposes; certain glands contain digitalislike secretions that increase blood pressure. In France and in tropical countries, frogs and toads have been introduced to keep insects under control.

The common leopard frog, *Rana pipiens,* was studied in detail in Exercises 7 and 8. Examine other representatives of this class, noting particularly the external adaptations that have allowed them to make the transition from the water to the land.

e. Class Reptilia

Reptiles include some of the most interesting and diverse chordates, such as the turtles, snakes, lizards, crocodiles, and alligators (Fig. 38-9). They are poikilothermic (i.e., they have variable body temperature), most are covered with scales or bony plates, and they respire by means of lungs. They were the first vertebrates to evolve that adapted to living in dry places, the dry skin and scales retarding moisture

FIG. 38-9
Representative reptiles.

loss from the body and allowing the animal to occupy rough land surfaces. The name of the class indicates its mode of locomotion (Latin *reptus,* from *repere,* "to creep").

Examine available representatives of the various modern reptiles, and note their great diversification in form. Note particularly the ways in which reptiles are morphologically more advanced than amphibians—ways that have better adapted them for life on land such as a dry, scaly skin to prevent desiccation and appendages (when present) suited for rapid locomotion on land or in the water.

Modern reptiles comprise only a fraction of the known orders that flourished during the Mesozoic era—the age of reptiles. During that time, reptiles were the dominant vertebrates and occupied most of the habitats available from dry uplands and deserts to marshes, swamps, and oceans.

f. Class Aves

Birds are probably the best known and most easily recognized of all vertebrates because they have feathers and are capable of flying (Fig. 38-10). No other animals have feathers, although bats (which are mammals) can fly.

Birds are considered to have a reptilian ancestry. Indeed, their early embryonic development parallels that of reptile. Further, a basic reptilian characteristic, scales, persists on the legs of birds. Fossil remains of some of the earliest birds indicate that they had reptilelike teeth.

Examine the demonstration of various birds. These animals have a number of striking adaptive features, allowing them to occupy a wide variety of habitats. Note the following morphological modifications.

1. **Coloration** is varied and striking. Although some birds are entirely one color, the feathers of most are marked with spots, stripes, or bars. This **protective coloration** allows a bird to blend with the environment and thus to be less visible.

2. The **bill** is a multipurpose structure. It functions both as a mouth and as hands, being used to preen the feathers, obtain and arrange nesting materials, and even in defense. The form of the bill tells something about the food habits of the bird. For example, seed-eating birds have conical bills, those that probe into cracks and crevices for insects have slender ones, flesh-eating species have sharp ones, and birds such as ducks that sieve food from the water have wide bills with serrated margins.

3. The **feet** are variously modified for swimming, climbing, perching, wading, and grasping.

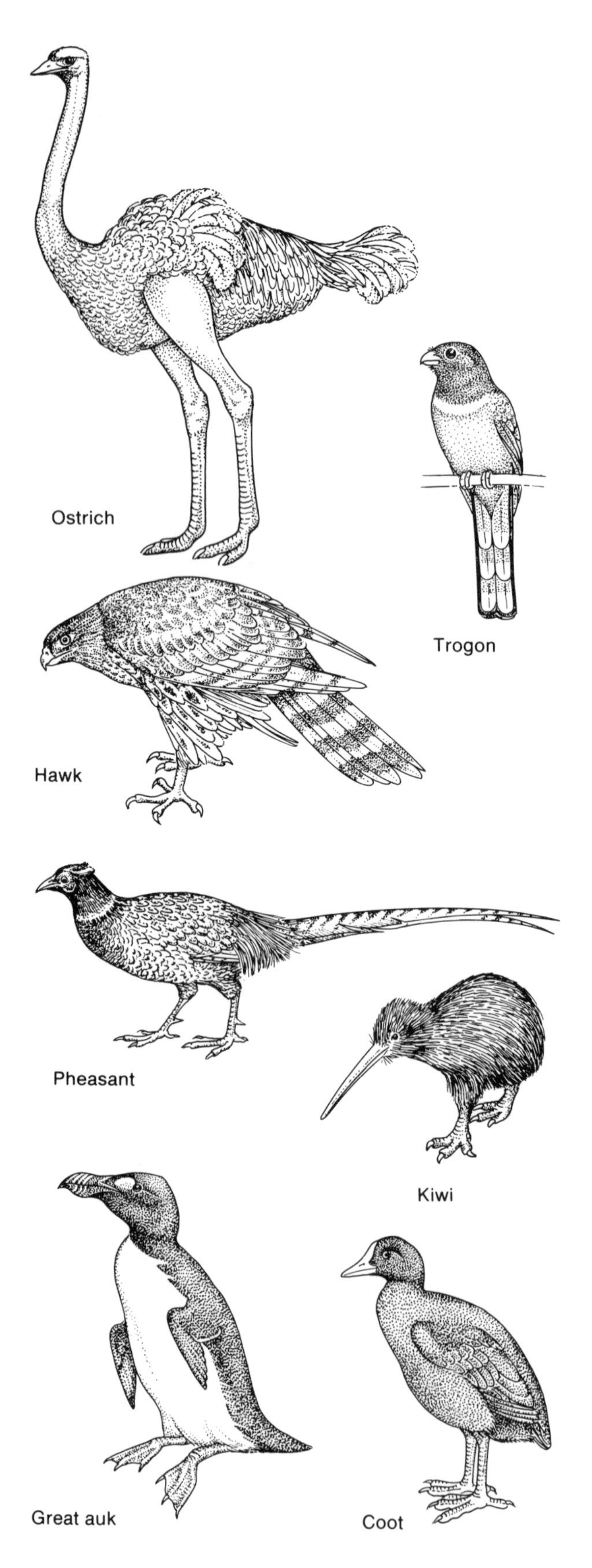

FIG. 38-10
Representative birds.

4. The **wings** are shaped like air foils and thus supply the "lift" that keeps the bird in the air. The wings of some birds (penguins) are further modified for "flying" under water. In a few species, the wings have so degenerated that the bird is no longer capable of flight, as is true of the ostrich and kiwi.

g. Class Mammalia

Mammals, often called the "highest" animals, include rodents, monkeys, bats, horses, whales, cows, deer, and human beings (Fig. 38-11). All are covered to varying degrees with hair and are warm-blooded. They are called *mammals* because the young are nourished by milk from the mammary glands of the females. Mammals have an interesting evolution with respect to the development and care of the young. Primitive mammals, such as the duckbill platypus, resemble reptiles in that they lay eggs. In more advanced mammals, such as the marsupials (opossum and kangaroo), the young have limited development within the uterus; they are born in a very immature condition and transferred to a pouch where they are suckled until they are more mature. The young of the most advanced mammals are retained in the uterus until they are in an advanced stage.

Mammals live in all kinds of habitats ranging from the tropics to the poles and from the oceans to the driest deserts. They have been able to move into these niches and survive, because of the wide diversity in their morphological, physiological, and behavioral features.

What, for example, is the advantage of retaining the embryo in the uterus during its development?

____

____

____

____

What advantage is there in nourishing the young with milk from the mammary gland rather than with the food in the yolk of a reptilian egg?

____

____

____

____

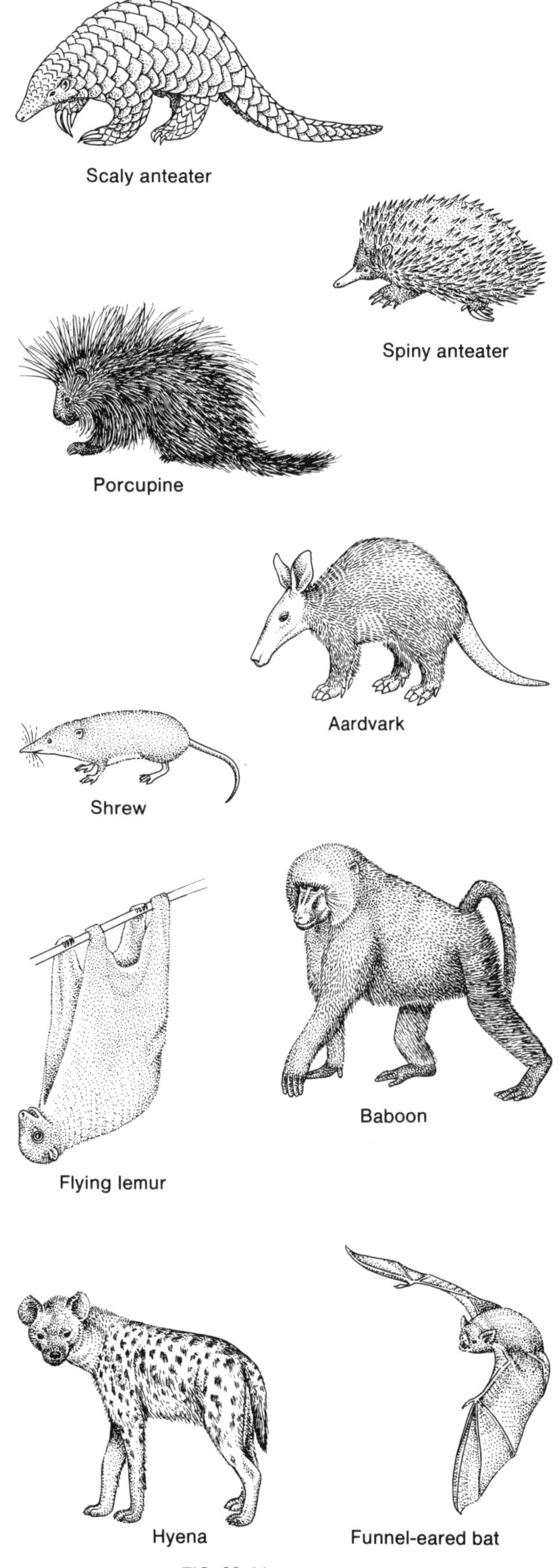

FIG. 38-11
Representative mammals.

What advantage is there in having an opposing thumb?

____

____

____

____

REFERENCES

Applegate, V. C., and J. W. Moffett. 1965. The Sea Lamprey. *Scientific American* 192(4):36–41.

Gilbert, P. W. 1962. The Behavior of Sharks. *Scientific American* 207(1):60–68 (Offprint 127). *Scientific American* Offprints are available from W. H. Freeman and Company, 41 Madison Avenue, New York, 10010, and 20 Beaumont Street, Oxford OX1 2NQ, England. Please order by number.

Romer, A. S., and T. S. Parsons. 1977. The Vertebrate Body. 5th ed. Saunders.

Storer, T. I., R. L. Usinger, R. C. Stebbins, and J. W. Nybakken. 1979. *General Zoology.* 6th ed. McGraw-Hill.

Vaughn, T. A. 1978. *Mammalogy.* 2d ed. Saunders.

Villee, C. A., W. F. Walker, and R. D. Baines. 1984. *General Zoology.* 6th ed. Saunders.

Young, J. Z., and M. J. Hobbs. 1975. *The Life of Mammals.* 2d ed. Oxford University Press.

Appendixes

APPENDIX A

Anatomical Terminology

Beginners to the study of any science are often confused by its jargon, and students of anatomy are no exception. However, the specialized terminology of anatomy is necessary to achieve precision. For example, when referring to a point on the body, what do we mean when we say *above, behind, over, on top of,* or *below?* Each of these words may be interpreted differently by different persons. To eliminate any ambiguity anatomists have developed a set of well-defined terms that are used universally to locate and identify body structures and features.

This appendix presents some of the more important terms used in gross anatomy, the study of body structures visible to the eye.

A. ORIENTATION AND DIRECTION

The following terms describe direction and the position of body parts with respect to each other. In studying these terms refer to Fig. A-1. Note that some terms have a different connotation when referring to a four-legged animal than when referring to a human.

Dorsal/Ventral (backside/bellyside): These terms are generally used with four-legged animals. *Dorsum* is Latin for "back." Thus, dorsal refers to the backside of the body or other structure. The backside of the arm, for example, is its dorsal surface. Ventral is derived from the Latin term *venter,* which means belly. Ventral thus refers to the bellyside of the animal. In humans, dorsal and ventral are synonomous with the terms *posterior* and *anterior.* In four-legged animals, dorsal and ventral are used interchangeably with superior and inferior.

Anterior/Posterior (front/back): These terms, used in reference to humans, describe surfaces that are facing forward or backward. The abdomen, the chest, and the face are on the anterior surface. Posterior surfaces, the back and buttocks, are on the back side of the body. These terms can also be used to describe the position

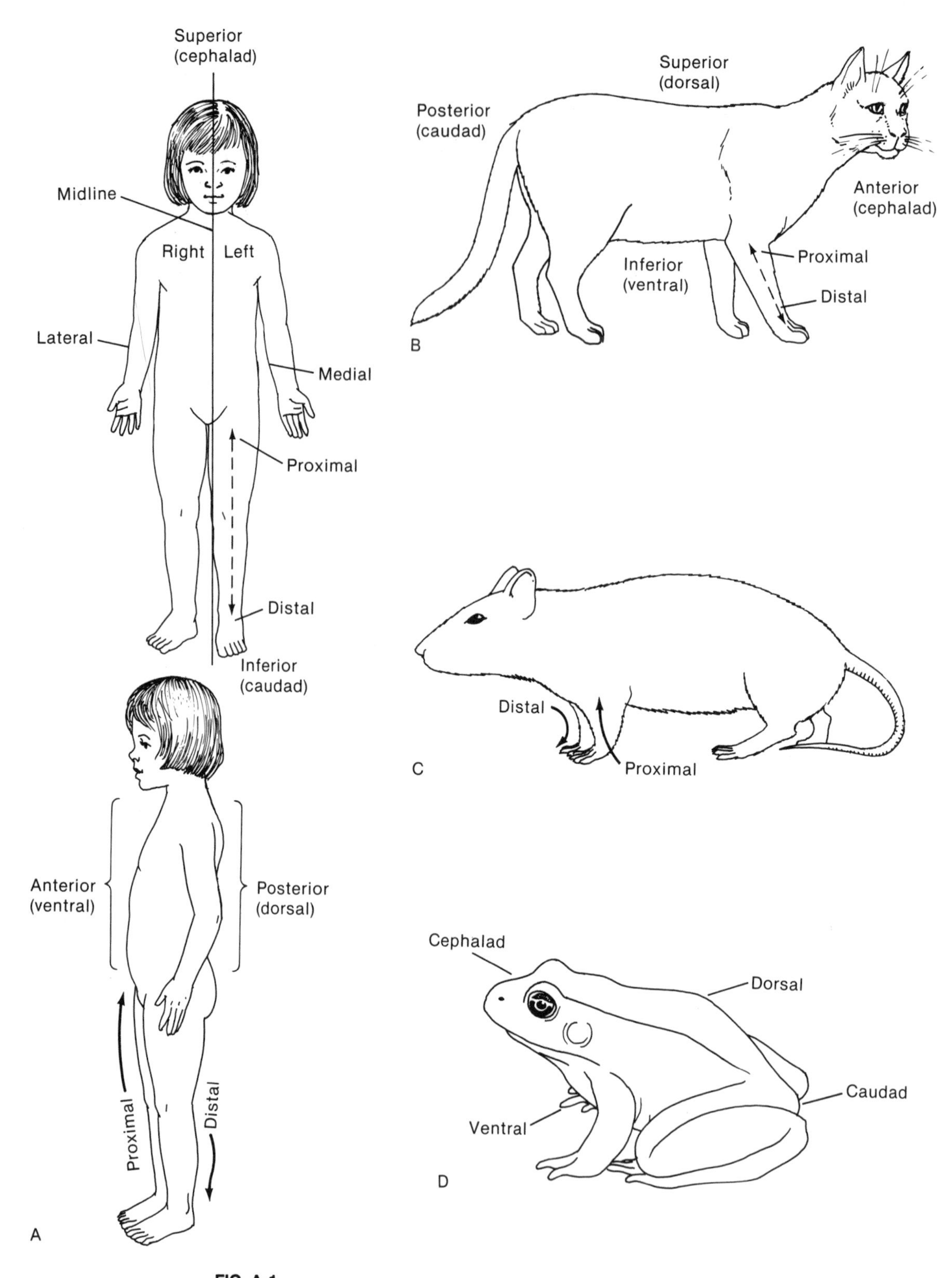

FIG. A-1
Terms used for purposes of anatomical orientation and direction in the (A) human, (B) cat, (C) mouse, and (D) frog.

of one structure in relation to another. For example, a structure can be anterior or posterior to another one; e.g., the molar teeth are posterior to the bicuspids.

Superior/Inferior (above/below): The Latin word *super* means above. Thus a structure located above another is considered to be superior; the eyes are superior to the nose, for example. *Inferus* is a Latin term for low or below. A structure is inferior to another structure if it is underneath or below that structure; the abdomen is inferior to the chest.

Cephalad/Caudad (toward the head/toward the tail): In four-legged animals these terms are interchangeable with anterior and posterior. In humans, they may be used alternately with superior and inferior.

Proximal/Distal (nearer or toward the body or attached end/further or away from the body or attached end): These terms are used to locate various parts of the limbs. For example, the fingers are distal to the wrist; the knee is proximal to the ankle; the elbow is distal to the shoulder.

Medial/Lateral (toward the midline, or median plane, of the body/away from the midline of the body): The midline of the body is an imaginary line on the plane that divides the body into left and right halves (Fig. A-1). Medial (from the Latin *medius,* meaning middle) refers to surfaces or structures closest to the midline. The inner surface of an arm or leg is its medial surface because it is closest to the body's midline. The term *lateral* is the opposite of medial. It is derived from the Latin term *lateralis,* meaning side. The outside surface of an arm is its lateral surface.

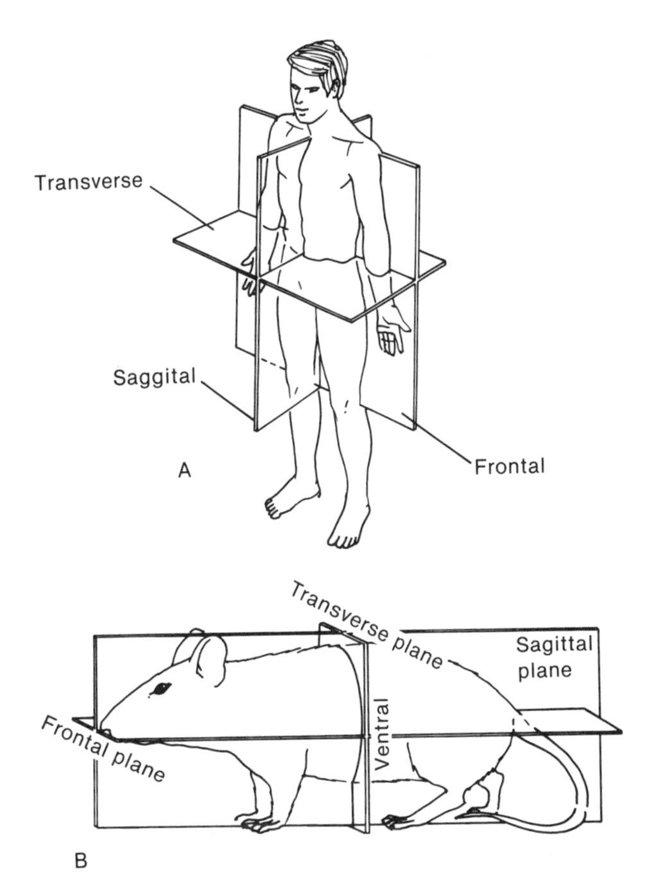

FIG. A-2
Planes and sections used in anatomical terminology shown in (A) a human and (B) a mouse.

B. PLANES AND SECTIONS

In order to observe internal structures and their positions relative to each other, it is necessary to make a cut, or **section,** through the body. When a section of the body is made, it is through an imaginary line called a **plane.** Three different sections have been classified by anatomists (Fig. A-2).

Longitudinal (Saggital) Section: A cut that is made parallel to the long axis of the body, and divides the body into right and left halves, is a longitudinal, or **saggital,** section. If the cut divides the body into equal left and right halves it is referred to as the midsaggital, or **median,** section. Cuts made parallel to the median (midsaggital) plane are called **parasaggital** sections.

Frontal (Coronal) Section: This section is cut along a longitudinal plane that divides the body into anterior (ventral) and posterior (dorsal) regions.

Transverse (Cross) Section: Any cut of the body that is made perpendicular to the longitudinal or frontal plane is a transverse, or cross, section. It divides the body into superior (cephalad) and inferior (caudad) parts.

Note: Longitudinal and transverse sections are frequently made of various organs and tissues for gross and microscopic observation. As shown in Fig. A-3, there is quite a difference between an organ or tissue cut in a transverse section and one cut in a longitudinal section.

C. BODY CAVITIES

The vertebrate body has two major cavities (Fig. A-4). These are

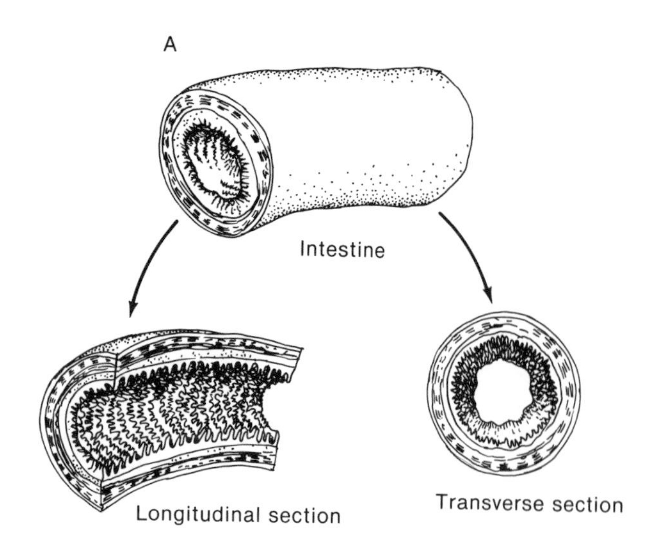

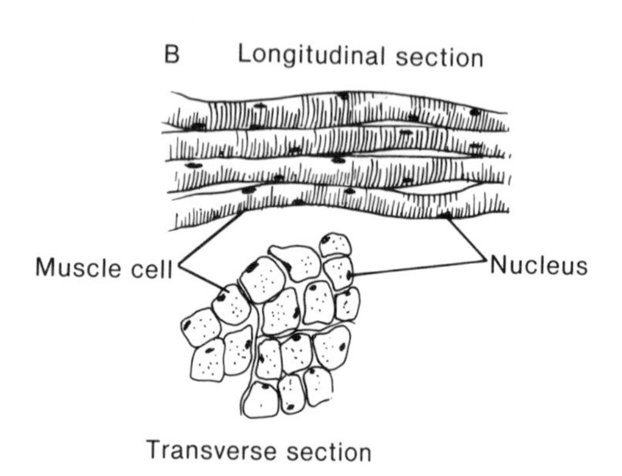

FIG. A-3
Longitudinal and transverse sections made of (A) an organ, the human intestine, and (B) a tissue, human muscle.

1. The **dorsal body cavity,** which consists of the **cranial** cavity and the **vertebral** or **spinal canal.** The cranial cavity is surrounded by the bones of the skull and contains the brain. The vertebral canal, surrounded by the vertebrae, contains the spinal cord, part of the spinal nerves, and the spinal fluid.

2. The **ventral body cavity,** or **coelom.** This cavity is located in the anterior (or ventral) part of the body and contains the major organs **(viscera)** of the body. The coelom is subdivided into two regions by the muscular diaphragm. The upper region is called the **thoracic** (or chest) **cavity.** This is further partitioned into two **pleural** cavities, containing the lungs, and the **pericardial cavity,** in which the heart is located.

The lower region, the **abdominopelvic cavity,** consists of two parts. The **abdominal cavity,** which is immediately inferior (caudad) to the diaphragm, contains the stomach, liver, spleen, pancreas, gallbladder, kidneys, ureters, small intestine, and most of the large intestine. Caudad to the abdominal cavity is the **pelvic cavity** containing the urinary bladder, sigmoid colon, rectum, and the reproductive organs.

D. GLOSSARY OF COMMON ANATOMICAL TERMS

The following glossary consists of the vocabulary commonly used in descriptions of vertebrate anatomy.

acetabulum (Latin *acetabulum,* "vinegar cup"; from *acetum,* "vinegar"): The cupshaped socket in the pelvic girdle that receives the head of the femur.

adrenal gland: An endocrine gland located cranial to the kidney (mammals) or on its ventral surface (frogs). Its hormones help the body adjust to stress and help regulate sexual development and the metabolism of salts, minerals, carbohydrates, and proteins. It is often called the suprarenal gland.

allantois (Greek *allas,* gen. *allantos,* "sausage"): An extraembryonic membrane in reptiles, birds, and mammals that develops as an outgrowth of the urinary bladder. It accumulates waste products and functions as a gas exchange organ in reptiles and birds; in mammals, it is part of the fetal placenta.

alveoli (Latin *alveolus,* "a small cavity"): The small, thin-walled, vascular sacs at the ends of the mammalian respiratory tree. Gas exchange occurs in these structures.

anus (Latin *anus,* "anus"): The terminal opening of the digestive tract in vertebrates.

aorta (Greek *aorte,* "to lift up"): The major artery that carries blood from the heart to the various parts of the body. It is also called the *dorsal aorta* to distinguish it from the ventral aorta of fish, which carries blood to the gills.

aortic valve: The valve, consisting of three semilunar-shaped folds at the base of the aorta, that prevents blood from flowing back into the left ventricle.

arrector pili (Latin *arrectus,* "upright"; *pilus,* "hair"): Small muscles within the skin at the base of hair follicles that raise the hairs to form "goose flesh."

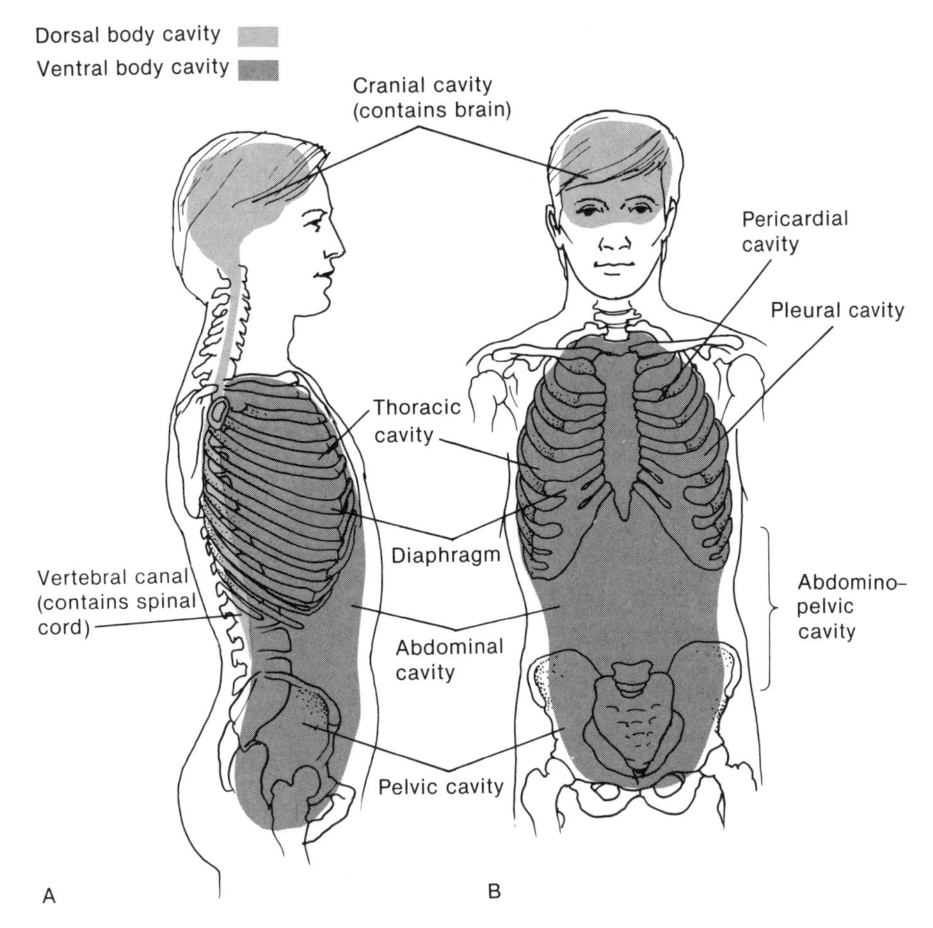

FIG. A-4
(A) Lateral and (B) frontal views of the human showing the location of the dorsal and ventral body cavities.

atlas (Greek mythology *Atlas):* The first cervical vertebra, which supports the skull at the top of the vertebral column.

atrioventricular valve (Latin *atrium,* "hall"; *ventriculus,* "little belly"): The valves between the atria and ventricles of the heart, which prevent the backflow of blood. In mammals, the right valve has three flaps (thus, it is also called the tricuspid valve); the left one has two flaps (and is sometimes called the bicuspid valve or mitral valve).

atrium (Latin *atrium,* "hall"): The two chambers of the heart that receive blood from the body (right atrium) and the lungs (left atrium).

axis (Latin *axis,* "hub, axle"): The second cervical vertebra. Rotation of the head occurs between the atlas and the axis.

Bowman's capsule: The thin-walled, expanded proximal end of a kidney tubule that surrounds the glomerulus.

bronchus (Greek *bronkhos,* "windpipe"): A major branch of the trachea that leads to a lung.

caecum or **cecum** (Latin *caecus,* "blind"): The first part of the large intestine, forming a dilated pouch into which opens the ileum, colon, and appendix. In many herbivores, including the rat, it is very long and often contains bacteria that digest cellulose.

canaliculi (Latin *canaliculus,* "little canal"): Microscopic canals in the bone matrix that allow the cell processes of bone cells to communicate with each other.

central canal: The cavity in the middle of the spinal cord. It connects with the fourth ventricle of the medulla oblongata and is filled with cerebrospinal fluid.

cervical (Latin *cervix,* "neck"): Pertaining to the neck, e.g., cervical vertebrae.

cervix: The neck of an organ, such as the cervix of the uterus.

chromatophore (Greek *chroma,* "color"; *phoros,* "to bear"): A cell in the skin of vertebrates that contains pigment granules.

cloaca (Latin *cloaca,* "sewer"): The exit chamber from the digestive system in lower vertebrates, which may also serve as the exit for the reproductive and urinary systems.

coelom (Greek *koilos,* "hollow"): A body cavity in vertebrates and many invertebrates that is completely lined by a simple, squamous epithelium of mesodermal origin.

collagen (Greek *kolla,* "glue"; Latin *genere,* "to beget"): A fibrous protein forming most of the intercellular material in cartilage, tendons, and other connective tissues.

conus arteriosus: A chamber of the heart of lower vertebrates into which the ventricle empties.

coronary vessels (Latin *corona,* "garland, crown"): Blood vessels that encircle the heart and supply and drain the cardiac muscles.

corpus luteum (Latin *corpus,* "body"; *luteus,* "yellow"): A yellowish endocrine gland within the ovary that develops from the follicle, after ovulation. It secretes estrogens and progesterone, which maintain the uterus during pregnancy.

cutaneous (Latin *cutis,* "skin"): Pertaining to the skin, e.g., cutaneous nerve.

dendrite (Greek *dendron,* "tree"): A filamentous process, usually branched, that carries nerve impulses toward the nerve cell body of a neuron.

diaphragm (Greek *dia,* "across"; *phragma,* "partition"): The complex of muscles and tendons that forms the partition between the thoracic and abdominal cavities in mammals.

ductus arteriosus: A fetal blood vessel connecting the pulmonary artery directly to the descending aorta, which permits much of the blood to bypass the lungs. It normally closes and atrophies after birth.

epidermis (Greek *epi-,* "on or over"; *derma,* "skin"): The outermost layers of cells in both plants and animals.

epithelium (Greek *epi-; thele,* "nipple"): A tissue composed of cells that cover all body surfaces and line all cavities including the lumen of blood vessels and ducts. Secretory cells of glands originate from epithelial layers during embryonic development.

foramen (Latin *foramen,* "opening"): A small opening in the skull or other organs.

foramen magnum (Latin *magnus,* "great, large"): The opening in the base of the skull through which the spinal cord passes.

foramen ovale (Latin *ovalis,* "oval"): An opening in the septum between the atria of a fetal mammalian heart. It permits much of the blood in the right atrium to enter the left atrium and thus bypass the lung. It closes at birth and becomes the fossa ovalis in the adult.

fossa ovalis (Latin *fossa,* "trench"): An oval-shaped depression in the median wall of the right atrium of an adult mammal. It is a vestige of the foramen ovale found in the fetus.

glomerulus (Latin *glomeris,* "ball"): A ball-like cluster of capillaries found within the Bowman's capsule at the head of a nephron in vertebrate kidneys.

hepatic (Greek *hepar,* "liver"): Pertaining to the liver.

hepatic duct: The duct that carries bile from the liver. Hepatic ducts usually join the cystic duct to form the common bile duct.

hypophysis (Greek *hypo-,* "under"; *physis,* "growth"): Often called the pituitary gland, it is attached to the underside of the hypothalamus. It produces a variety of hormones regulating growth, metabolism, sexual activity, and water balance.

hypothalamus (Greek *hypo; thalamos,* "inner chamber"): The small region of the brain lying just below the thalamus. It is an important center for the control of visceral activity and the regulation of the hypophysis.

insertion (Latin *in,* "into"; *serere,* "to join"): The place where a muscle attaches to a bone or other structure. When the muscle contracts the bone or structure is caused to move a greater distance at this end of the muscle.

intestine (Latin *intestinus,* "internal"): The primary digestive and absorptive parts of the digestive tract. Located between the stomach and the cloaca or anus, it is divided into small and large intestines.

islets of Langerhans: Patches of endocrine tissue in the pancreas that secrete hormones (insulin and glucagon) essential for the regulation of blood glucose levels.

jejunoileum (Latin *jejunus,* "empty"; *ileum,* "groin, flank"): That part of the small intestine beyond the jejunum.

kidney (Middle English, *kidenei,* "kidney"): The organ that removes nitrogenous waste products of metabolism from the blood and produces urine.

lacunae (Latin *lacuna;* plural *lacunae,* "pool"): Small cavities in the bone matrix that contain the bone cells (osteocytes).

ligament (Latin *ligamentum,* "bone"): A band of fibrous connective tissue connecting bones or cartilage, which serves to support and strengthen joints.

liver (Old English *lifer,* "liver"): The large gland located in the upper part of the abdominal cavity that secretes bile and metabolizes carbohydrates, proteins, and fats. It also synthesizes many plasma proteins, degrades toxins, and removes damaged red blood cells.

lymph node: Small, oval structures associated with the lymphatic vessels of higher vertebrates. Lymphocytes are produced here, foreign particles phagocytosed, and some immune responses initiated.

mediastinum (Latin *mediastinus,* "median"): The mass of tissues and organs separating the lungs from other organs within the thoracic cavity. It contains the aorta, esophagus, pericardial cavity and heart, thymus, and vena cava.

melanin (Greek *melas,* "black"): The black or brown pigment in the skin. It is contained within the melanocytes of frogs and other lower vertebrates.

mesenteries (Latin *mesos,* "middle"; Greek *enteron,* "gut"): The double layers of mesoderm that suspend the digestive tract and other internal organs within the coelom.

mucosa (Latin *mucus,* "mucus"): The lining of the digestive and respiratory tracts containing cells that secrete mucus.

muscle fiber (Latin *musculus,* "little mouse," because of the mouselike shape of some muscles): The elongated muscle cell.

myofibrils (Greek *myos,* "mouse"; *fibrilla,* "small fiber"): Microscopic fine longitudinal fibrils within a muscle fiber that act as the contractile elements. In striated muscle they bear cross striations.

myofilaments (Latin *filum,* "thread"): Ultramicroscopic filaments of actin and myosin that are components of the myofibrils.

naris (Latin *naris,* "external nostril"): The external nostril.

nasal (Latin *nasus,* "nose"): Pertaining to the nose, e.g., nasal bone, nasal cavity.

nephron (Greek *nephros,* "kidney"): The functional unit of a vertebrate kidney which includes the Bowman's capsule, glomerulus, and proximal and distal tubules.

nictitating membrane (Latin *nictare,* "to wink"): A membrane in the median corner of the eye of many terrestrial vertebrates, which slides across the surface of the eyeball. In human beings it consists of only a vestigial semilunar fold.

ocular (Latin *oculus,* "eye"): Pertaining to the eye, e.g., extrinsic ocular muscles.

omentum (Latin *omentum,* "membrane"): One of the two mesenteries that attach to the stomach. The greater omentum is a saclike fold between the body wall and the stomach; the lesser omentum extends from the liver to the stomach and duodenum.

optic (Greek *optikos,* "sight"): Pertaining to the eye.

oral cavity (Latin *os,* gen. *oris,* "mouth"): The mouth cavity; also known as the buccal (cheek) cavity.

origin of a muscle (Latin *origin,* "beginning"): That attachment of a muscle to a bone or other structure that moves the lesser distance when the muscle contracts.

ostium (Latin *ostium,* "opening"): The opening into a tubular organ (e.g., ostium of the uterine tube) or between two distinct cavities of the body.

ovarian follicle (Latin *ovum,* "egg"; *folliculus,* "little bag"): A group of cells within the ovary that envelop the developing egg. It is also an endocrine gland that produces estrogen.

ovulation (Latin *ovulum,* "little egg"): The release of egg(s) from the ovarian follicle(s) and ovary into the coelom from which they enter the oviduct or uterine tube.

pancreas (Greek *pan,* "all"; *kreas,* "flesh"): An elongated gland attached to the duodenum that secretes enzymes and precursors of enzymes that act on all categories of food. It also contains the islets of Langerhans that produce insulin and glucagon.

parathyroid gland (Greek *para,* "beside"): One of several endocrine glands embedded on the surface of the thyroid gland. It secretes a hormone essential for calcium and phosphorous metabolism.

peritoneum (Greek *peri-,* "around"; *tonus,* "stretched over"): A membrane that lines the body cavity and forms the external covering of the visceral organs.

pharynx (Greek *pharynx,* "pharynx"): That part of the digestive tract that lies between the mouth cavity and the esophagus; the throat.

pituitary gland: See hypophysis.

placenta (Greek *plakoeis,* "a flat object"): A mammalian organ connecting a mother and her fetus through which food, gases, and waste products are exchanged.

pleura (Greek *pleura,* "side, rib"): The epithelial membranes that cover the lungs (visceral pleura) and line the pleural cavities (parietal pleura).

polar body (Latin *polaris,* "pole"): A small cell located near the animal pole of a developing egg cell. Polar bodies result from an unequal division of the cytoplasm during the first and second meiotic divisions.

portal vein (Latin *portare,* "to carry"): A vein that carries blood from one organ to another rather than to the heart, e.g., the hepatic portal vein.

pulmonary (Latin *pulmo,* "lung"): Pertaining to the lung.

pylorus (Greek *pylorus,* "gate"; *ourus,* "guard"): The distal opening of the stomach surrounded by a strong band of tissue that closes the opening between the stomach and the duodenum.

rectum (Latin *rectus,* "straight"): The caudal part of the large intestine.

renal (Latin *ren,* "kidney"): Pertaining to the kidney, e.g., renal artery.

saliva (Latin *saliva,* "saliva"): The mucous secretions of several large glands that discharge into the mouth cavity. In mammals, it contains salivary amylase (ptyalin), which initiates the chemical breakdown of starch.

sarcolemma (Greek *sarkos,* "flesh"; *lemma,* "peel"): The thin covering of a muscle fiber.

sarcoplasm (Greek *sarkos; plasma,* "form"): The cytoplasm of a muscle fiber.

serosa (Latin *serum,* "watery fluid"): The epithelial and connective tissue membranes that line body cavities and cover visceral organs (e.g., peritoneum, pericardium, and pleura).

somatic (Greek *soma,* "body"): Pertaining to the body wall rather than the internal organs, e.g., somatic muscles.

sternum (Greek *sternon,* "chest"): The bone on the midventral surface of the chest; the breastbone. Costal cartilages attach it to the ribs in mammals.

submucoa (Latin *sub-,* "under"; *mucus,* "mucus"): A layer of vascular connective tissue in the wall of the digestive or respiratory tract that lies beneath the mucosa.

tendon (Latin *tendere,* "to stretch"): A band of connective tissue that attaches muscles to bones or to other muscles.

tendon of Achilles (Achilles, the hero of Homer's *Iliad,* was said to be invulnerable except for this tendon): The tendon that extends from the large muscle mass on the caudal surface of the leg to the calcaneus (heel) bone of the foot.

thalamus (Greek *thalamos,* "inner chamber"): One of two masses of gray matter located at the sides of the third ventricle of the brain. It is an important center for the sensory impulses traveling to the cerebrum.

thymus (Greek *thymos,* "thymus"): A lymphoid organ located in the ventral part of the thoracic cavity. It is well developed in the fetus, where it participates in the maturation of bone marrow stem cells into immunologically competent T lymphocytes.

thyroid gland (Greek *thyreos,* "shield"): The bilobed endocrine gland located near the cranial end of the trachea and over the thyroid cartilage of the larynx in human beings (hence its name). It is the source of the hormone thyroxine, which regulates the general level of metabolism.

truncus arteriosus: One of the two arterial trunks in a frog that lead from the front of the heart to arterial arches supplying (1) the lungs and skin, (2) the head, and (3) the main part of the body.

umbilical (Latin *umbilicus,* "navel"): Pertaining to the navel, e.g., umbilical cord, umbilical artery.

urostyle (Greek *oura,* "tail"; *stylos,* "pillar"): The spikelike caudal part of the frog's vertebral column. It consists of two fused caudal vertebrae and serves as the origin for certain muscles used in jumping.

vena cava (Latin *vena,* "vein"; *cavea,* "hollow"): One or more large veins in vertebrates that return blood to the heart from the body. It enters the sinus venosus (frog) or the right atrium (mammals) of the heart.

ventricle (Latin *ventriculus,* "stomach"): (1) A

muscular chamber of the heart that receives blood from an atrium and pumps blood out of the heart, either to the lungs or the body tissues. Frogs have a single ventricle. (2) One of the large chambers within the brain.

villi (Latin *villus,* "shaggy hair"): Microscopic projections of the mucosa of the small intestine that increase its surface area.

white matter: The whitish material of the brain and spinal cord. It is composed of the myelinated processes of neurons.

zygomatic arch (Greek *zygon,* "yolk"): The bony arch beneath the orbit (eye) of the mammalian skull, which joins the facial and cranial regions of the skull; the cheekbone.

REFERENCE

Nomina Anatomica. 1983. 5th ed. Williams and Wilkins.

APPENDIX B

Use of the Bausch & Lomb Spectronic 20 Colorimeter

The Bausch & Lomb Spectronic 20 colorimeter is an extremely versatile instrument that is useful for the spectrophotometric or colorimetric determinations of solutions.

The optical system is shown in Fig. B-1. White light is focused by a lens (1) onto an entrance slit (2), where it is collected by a second lens (3) and refocused on the exit slit (4) after being reflected and dispersed by a diffraction grating (5). Rotation of this grating by a cam (6) enables one to select various wavelengths of light in a range from 375–625 nm. The addition of a filter (9) can extend the usable wavelength to 950 nm. After the light passes the exit slit, it goes through the sample being measured (7) and is picked up by a phototube (8). A dial indicates the amount of light absorbed by the sample.

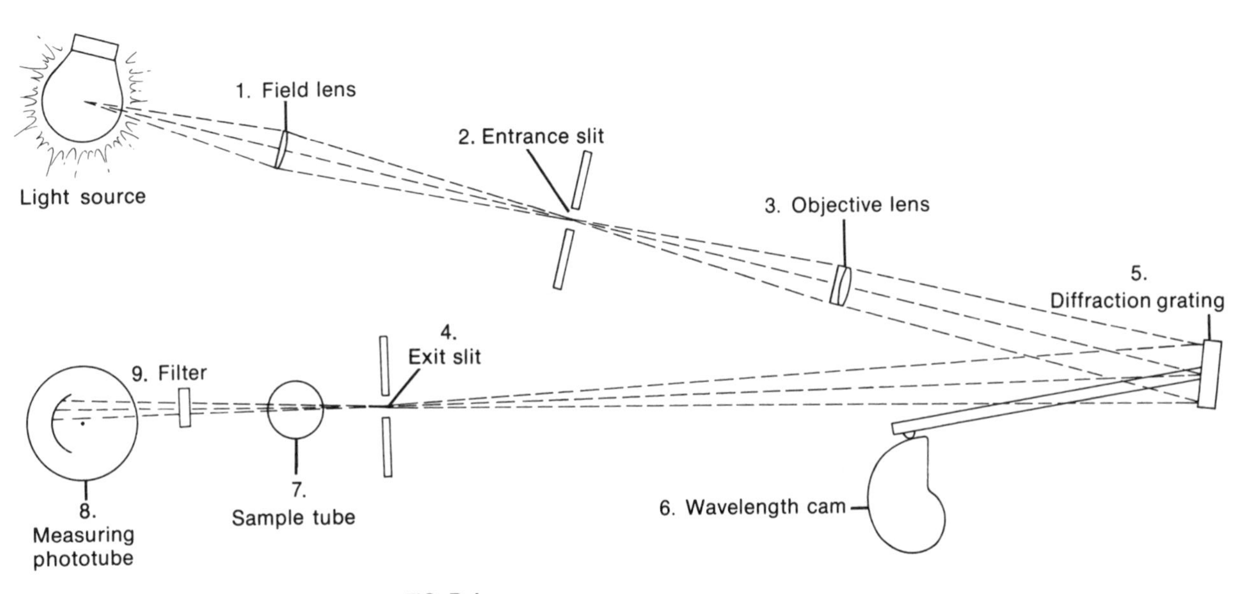

FIG. B-1
Optical system of the Spectronic 20 colorimeter.

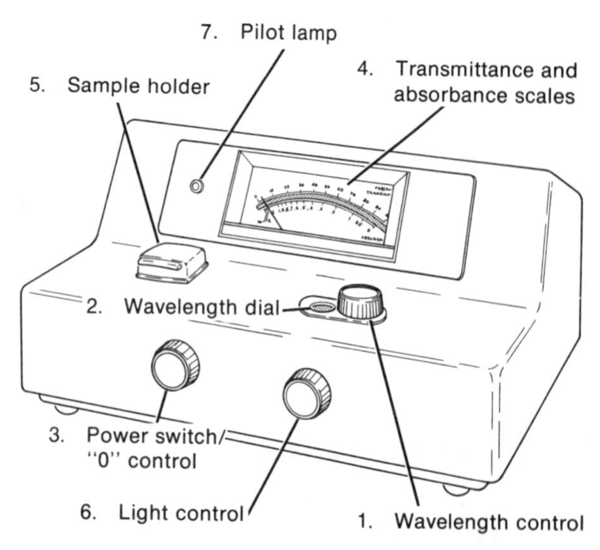

FIG. B-2
Controls on the Spectronic 20 colorimeter.

A. COLORIMETRY

Directions for colorimetric use are as follows:

1. Rotate the wavelength control (1) shown in Fig. B-2 until the desired wavelength is shown on the wavelength dial (2). The wavelength for a given substance can be found by referring to the literature or by determining it experimentally.

2. Turn the instrument on by rotating the "0" control (3) in a clockwise direction. Allow 5 minutes for the instrument to warm up.

3. Adjust the "0" control with the cover of the sample holder closed (5) until the needle is at 0 on the transmittance scale (4).

4. Place a colorimeter tube containing water or another solvent in the sample holder, and close the cover.

5. Rotate the light control (6) so that the needle is at 100 on the transmittance scale (0.0 absorbance). This control regulates the amount of light passing through the second slit to the phototube.

6. The unknown samples may then be placed in the tube holder, and the percent transmittance or the absorbance can be read. The needle should always return to 0 when the tube is removed. Check the 0% and 100% transmittance occasionally with the solvent tube in the sample holder to make certain the unit is calibrated.

Note: Always check the wavelength scale to be certain that the desired wavelength is being used.

The colorimetric measurements made with this apparatus employ standard matched tubes. They are selected so that variation in light transmitted through the tubes due to slight differences in diameter and wall thickness is minimal. You will be issued a set of such matched tubes. *They are to be used only for colorimetry.* The matched tubes must be handled carefully so as not to etch or scratch the surfaces exposed to the light beam. Obviously, the tubes will no longer be "matched" if scratched or etched, because such defects will cause the absorption and scattering of light.

B. SPECTROPHOTOMETRY

The method of operation for spectrophotometry is essentially the same as for colorimetry. The main difference is that the wavelength is reset for each reading, and thus a blank, or solvent, control must be readjusted at each new wavelength setting.

This procedure can be used when no information is available to determine the proper operating wavelength for an unknown substance. To do this, plot an absorption curve (absorbance versus different wavelengths) of the unknown substance (Fig. B-3). An operating wavelength may then be chosen according to the following:

1. Choose the wavelength at which the substance maximally absorbs the light (the minimum transmittance), because the greatest sensitivity will be obtained at this wavelength.

2. Do not choose wavelengths on the slope, because a small error in wavelength will cause a large error in reading.

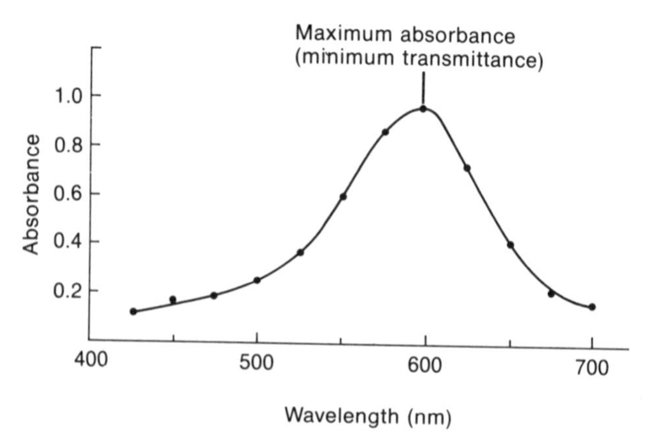

FIG. B-3
Absorption spectrum of an unknown substance.

REFERENCES

Karp, G. 1984. *Cell Biology.* 2d ed. McGraw-Hill.
Schleif, R. F., and P. C. Wensink. 1982. *Practical Methods in Molecular Biology.* Springer-Verlag.

APPENDIX C

Spectrophotometry

Many kinds of molecules interact with or absorb specific types of radiant energy in a predictable fashion. For example, when white light illuminates an object, the color that the eye perceives is determined by the absorption by the object of one or more of the colors from the source of the white light. The remaining wavelength(s) are reflected (or transmitted) as a specific color. Thus an object that appears red absorbs the blue or green colors of light (or both), but not the red.

The perception of color, as just described, is qualitative. It indicates what is happening but says nothing about the extent to which the event is taking place. The eye is not a quantitative instrument. However, there are instruments, called *spectrophotometers,* that electronically quantify the amount and kinds of light that are absorbed by molecules in solution. In its simplest form, a spectrophotometer has a source of white light (for visible spectrophotometry) that is focused on a prism or diffraction grating to separate the white light into its individual bands of radiant energy (Fig. C-1). Each wavelength (color) is then selectively focused through a narrow slit. The width of this slit is important to the precision of the measurement; the narrower the slit, the more closely absorption is related to a specific wavelength of light. Conversely, the broader the slit, the more light of different wavelengths passes through, which results in a reduction in the precision of the measurement. This monochromatic (single wavelength) beam of light, called the **incident beam (I_0),** then passes through the sample being measured. The sample, usually dissolved in a suitable solvent, is contained in an optically selected **cuvette,** which is standardized to have a light path 1 cm across. However, for special purposes, variations are available that are larger or smaller than the standard 1-cm size.

After passing through the sample, the selected wavelength of light (now referred to as the **transmitted beam, I**) strikes a photoelectric tube. If the substance in the cuvette has absorbed any of the incident light, the transmitted light will then be reduced in total energy content. If the substance in the

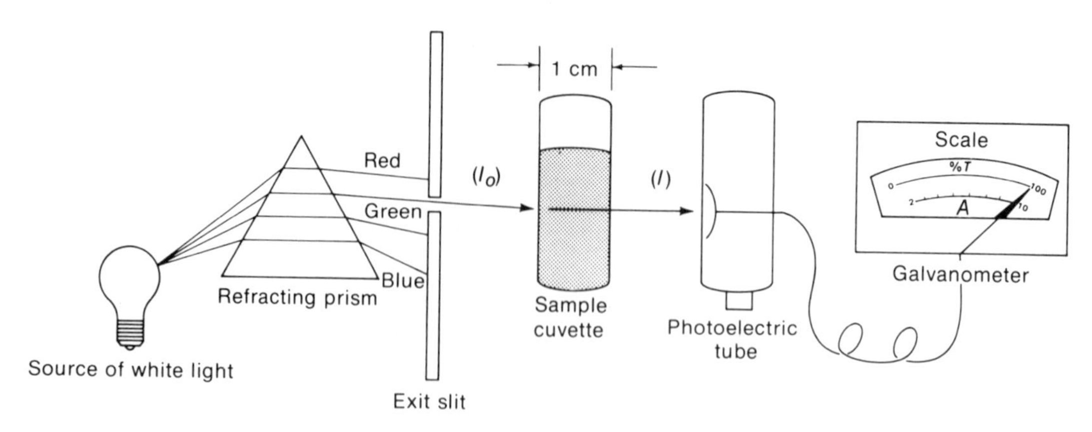

FIG. C-1
Typical photoelectric spectrophotometer.

sample container does not absorb any of the incident beam, the radiant energy of the transmitted beam will then be about the same amount as that of the incident beam. When the transmitted beam strikes the photoelectric tube, it generates an electric current proportional to the intensity of the light energy striking it. By connecting the photoelectric tube to a device that measures electric current (a galvanometer), a means of directly measuring the intensity of the transmitted beam is achieved. In the Bausch & Lomb Spectronic 20 spectrophotometer, the galvanometer has two scales: one indicates the **% transmittance (% *T*)**, and the other, a logarithmic scale with unequal divisions graduated from 0.0 to 2.0, indicates the **absorbance *(A)***. The term **optical density (OD)** may be used instead of absorbance; however, the latter term is more commonly accepted.

Because most biological molecules are dissolved in a solvent before measurement, a source of error can be the absorption of light by the solvent. To assure that the spectrophotometric measurement will reflect only the light absorption of the molecules being studied, a mechanism for "subtracting" the absorbance of the solvent is necessary. To achieve this, a "blank" (the solvent) is first inserted into the instrument, and the scale is set to read 100% transmittance (or 0.0 absorbance) for the solvent. The "sample," containing the solute *plus* the solvent, is then inserted into the instrument. Any reading on the scale that is less than 100% *T* (or greater than 0.0 *A*) is considered to be *due to absorbance by the solute only.* Other instruments are available that continuously give the desired ratio between sample and blank, both visually and on a strip-chart recorder. This kind of spectrophotometer "reads" the cuvettes for the *blank* and the *sample* simultaneously. The reading that appears on the absorbance scale is therefore a ratio between the *sample* and the *blank*.

Spectrophotometers are not limited to detecting absorption of only visible light. Some also have a source of ultraviolet light (usually a hydrogen or mercury lamp), which has wavelengths that range from about 180–400 nm. Ultraviolet wavelengths ranging from 180–350 nm are particularly useful in studying such biological molecules as amino acids, proteins, and nucleic acids, because each of these compounds have characteristic absorbances at different ultraviolet wavelengths. Other useful spectrophotometers have sources and suitable detectors of infrared radiation (780–25,000 nm). Numerous biological molecules can be effectively studied because of their infrared absorption spectra.

A. UNITS OF MEASUREMENT

The following terminology is commonly used in spectrophotometry.

1. Transmittance *(T)*: the ratio of the transmitted light *(I)* of the sample to the incident light (I_0) on the sample.

$$T = \frac{I}{I_0}$$

This value is multiplied by 100 to derive the % *T*. For example,

$$T = \frac{75}{100} = 0.75$$

and

$$\%\ T = \frac{75}{100} \times 100 = 75\%$$

2. Absorbance *(A)*: logarithm to the base 10 of the reciprocal of the transmittance:

$$A = \log_{10}\frac{1}{T}$$

For example,

1. Suppose a % T of 50 was recorded (equivalent to $T = 0.50$).
2. Then $A = \log_{10}\frac{1}{0.50} = \log_{10}2.0$.
3. Thus $\log_{10}2.0 = 0.301$ (A equivalent to a % T of 50).

Similarly, a % T of 25 = 0.602 A; a % T of 75 = 0.125 A; and so forth.

The absorbance scale is normally present along with the transmittance scale on spectrophotometers. The chief usefulness of absorbance lies in the fact that it is a logarithmic rather than arithmetic function, allowing the use of the Lambert-Beer law, which states that for a given concentration range the concentration of solute molecules is directly proportional to absorbance. The Lambert-Beer law can be expressed as

$$\log_{10}\frac{I_0}{I} = A$$

in which I_0 is the intensity of the incident light; I is the intensity of the transmitted light.

The usefulness of absorbance can be seen in the graphs shown in Fig. C-2, one graph showing the percent transmittance plotted against concentration and the other showing absorbance plotted against concentration. Using the Lambert-Beer relationship, it is necessary to plot only three or four points to obtain the straight-line relationship shown in Fig. C-2B. However, certain conditions must prevail for the Lambert-Beer relationship to hold:

Monochromatic light is used.

A_{max} is used (i.e., the wavelength maximally absorbed by the substance being analyzed).

The quantitative relationship between absorbance and concentration can be established.

The first condition can be met by using a prism or diffraction grating or other device that can disperse visible light into its spectra.

The second condition can be met by determining the absorption spectrum of the compound. This is done by plotting the absorbance of the substance at a number of different wavelengths. The wavelength at which absorbance is greatest is called the A_{max} (or λ_{max}) and is the most satisfactory wavelength to use because, on the slope, absorbance changes rapidly with slight wavelength deviations, whereas at the maximum absorbance, changes in wavelength alter absorbance less. Fig. C-3 shows an absorption spec-

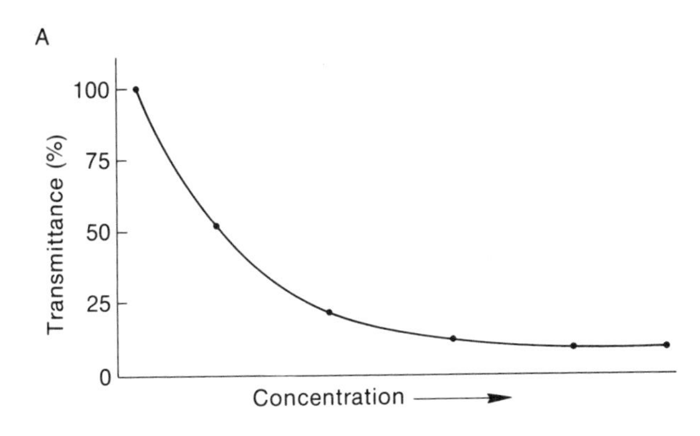

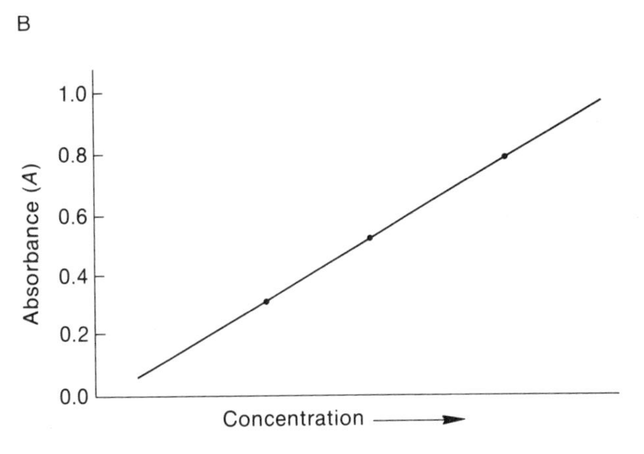

FIG. C-2
(A) Percent transmittance versus concentration. (B) Absorbance versus concentration.

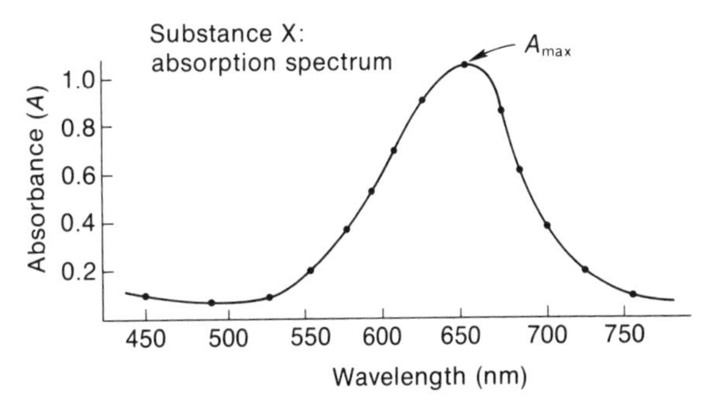

FIG. C-3
Determination of A_{max}.

trum of a hypothetical substance having an A_{max} of approximately 650 nm. Some compounds, however, can have several peaks both in the visible spectrum and in the ultraviolet range. An example of this is shown in Fig. C-4 for riboflavin.

To establish the quantitative relationship between absorbance and concentration of the colored substance, it is necessary to prepare a series of standards of the substance analyzed in graded known concentrations ("color standards"). Because absorbance is *directly* proportional to concentration, a plot of absorbance versus concentration of the standard yields a straight line. Such a plot is called a **concentration curve** or **standard curve** (Fig. C-5). After several points have been plotted, the intervening points can be extrapolated by connecting the known points with a straight line. It is not necessary to use dotted lines to indicate extrapolation on graphs; a dotted line is used in the illustration to indicate the parts of the line for which points were not determined but were presumed. When the Lambert-Beer law is followed, this is an acceptable and time-saving assumption; otherwise, points would need to be plotted throughout the entire line. In general, your graph should extend from a minimum of about 0.025 A to a maximum of about 1.0 A, or 94–10% T, this being the range of "readable" and reproducible values of absorbance. However, recall that the Lambert-Beer law operates at only certain concentrations. This is apparent in Fig. C-5 in which, at concentrations greater than 1.0 mg/ml, the curve slopes, indicating the loss of the concentration-absorbance relationship.

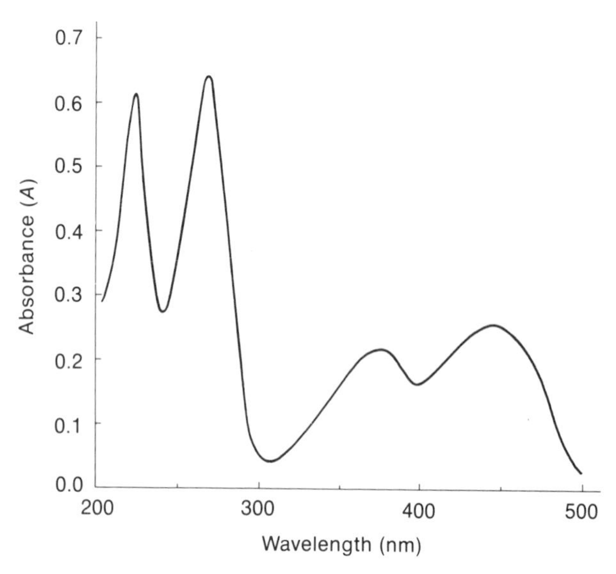

FIG. C-4
Absorption spectrum of riboflavin. (From *Experimental Biochemistry,* 2d ed., edited by John M. Clark, Jr., and Robert L. Switzer. W. H. Freeman and Company. Copyright © 1977.)

After a concentration curve for a given substance has been established, it is relatively easy to determine the quantity of that substance in a solution of unknown concentration by determining the absorbance of the unknown and locating it on the y axis, or ordinate (Fig. C-6). A straight line is then drawn par-

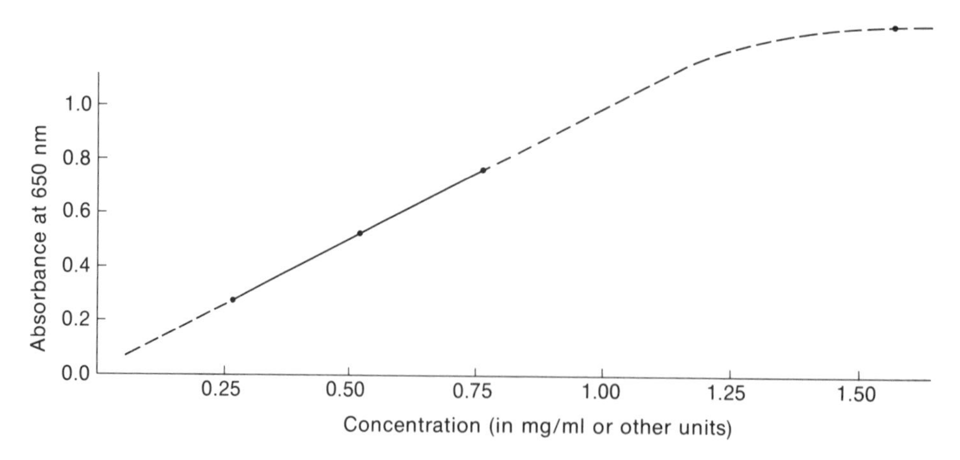

FIG. C-5
Preparation of concentration (standard) curve.

allel to the x axis, or abscissa, until it intersects with the experimental curve. A perpendicular is then dropped to the x axis, the value at the point of intersection indicating the concentration of the unknown solution. In this example, the unknown absorbance is 0.32, which indicates a concentration of about 0.37 mg. Concentrations are commonly expressed either as micrograms per milliliter (μg/ml) or as milligrams per milliliter (mg/ml).

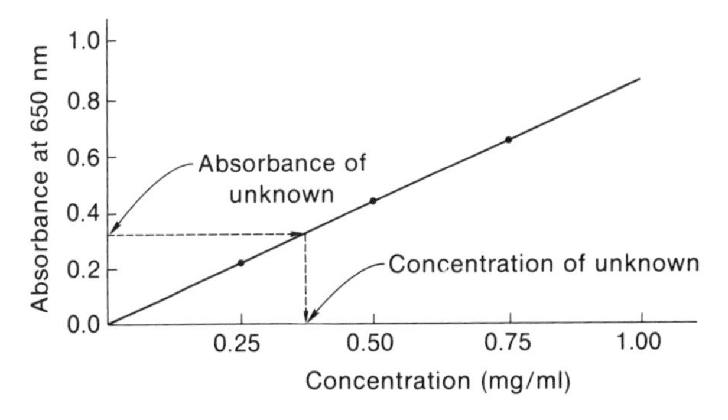

FIG. C-6
Determining concentration of "unknown."

If the absorbance value of the unknown is such that the line drawn parallel to the x axis intersects the experimental curve where it is curved (Fig. C-5), then you cannot accurately determine the concentration. In this event, dilute the unknown by some factor until the absorbance readings intersect with the straight-line part of the graph where concentration is proportional to absorbance. You can then determine the unknown concentration and multiply the value by the dilution factor.

REFERENCES

Karp, G. 1984. *Cell Biology.* 2d ed. McGraw-Hill.

Schleif, R. F., and P. C. Wensink. 1982. *Practical Methods in Molecular Biology.* Springer-Verlag.

Straughton, B. P., and S. Walker, eds. 1976. *Spectroscopy.* Chapman and Hall.

APPENDIX D

Chromatography

Chromatographic techniques are among the most useful methods available for the separation of complex mixtures of solutes based on their selective adsorption as they are passed over such adsorbents as charcoal, starch, cellulose powders, ion-exchange resins, and filter paper. The most widely used chromatographic techniques are column chromatography on ion-exchange resins, paper-partition chromatography on filter paper, and thin-layer methods using an adsorbent bound to a supporting material such as glass. These techniques are frequently employed to isolate proteins, enzymes, lipids, hormones, plant growth substances, pigments, and other naturally occurring organic materials.

A. PAPER CHROMATOGRAPHY

Paper chromatography has revolutionized the art of detecting and identifying small amounts of organic and inorganic substances. It permits the separation of mixtures on a very small scale, which no other simple method affords. The technique has found widespread application in many areas of biology and has been used to separate and identify such substances as amino acids, carbohydrates, fatty acids, antibiotics, and many other naturally occurring substances.

To employ this technique, a small spot of the mixture to be chromatographed is placed near one end of a length of filter paper. This end of the paper is then immersed in a solvent system that is usually composed of two or more miscible substances. In **descending paper chromatography,** the solvent is contained in a trough near the top of the chromatographic chamber and is allowed to irrigate the paper in a downward flow by means of capillary action and gravity (Fig. D-1). In **ascending chromatography,** the solvent is placed at the bottom of the chamber and is allowed to rise upward the length of the paper by capillary action.

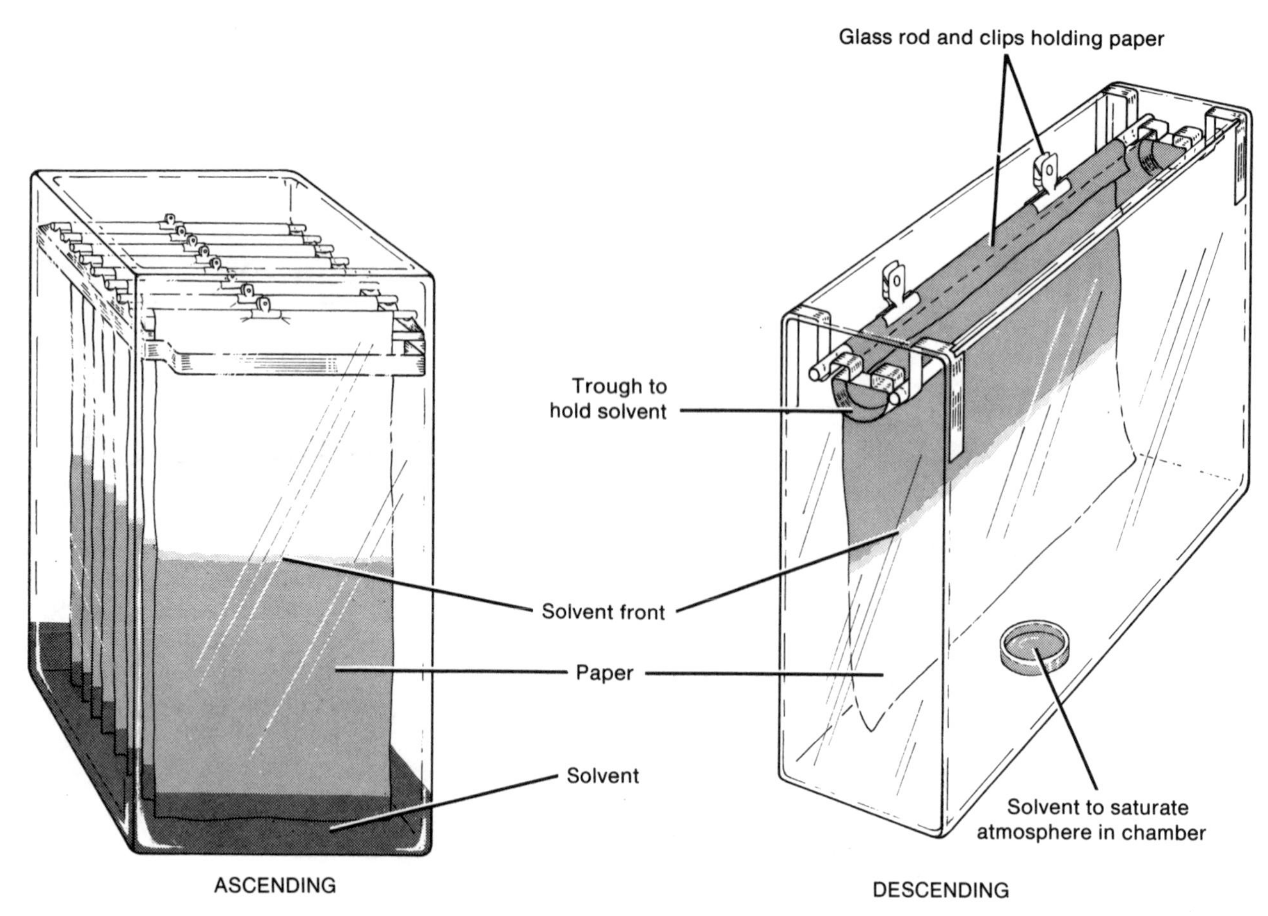

FIG. D-1
Apparatus for ascending and descending chromatography.

Although ascending chromatographic separation is typically used because the apparatus is quite simple, descending chromatography is faster (the solvent flow is faster because of gravity) and allows for the collection of the solvent (and solute) as it runs off the paper. This permits additional analysis of solute components using other procedures.

Locating or visualizing substances that have to be separated can be done in a variety of ways. Some substances have color and thus are easily seen on the chromatogram. For those compounds that are colorless, various procedures are employed. For example, amino acids become pink or purple spots when sprayed with ninhydrin. Reducing sugars become gray-black spots when treated with analine phthalate, and many organic substances become brown spots on a yellow background when exposed to iodine vapors.

So long as conditions such as temperature and purity of solvent are carefully regulated, it is possible not only to separate components of a mixture but to identify them. Under standardized conditions, identical substances move at characteristic rates in similar solvent systems. Thus one can calculate R_f values by comparing the distance the solvent travels with the distance the substance travels in accord with the following formula:

$$R_f = \frac{\text{distance of center of "spot" from origin}}{\text{distance of solvent front from origin}}$$

Tables of R_f values for various substances in different solvents permit preliminary identification of a substance by computing its R_f value. However, for more precise identification it is advisable to chromatograph known substances on the same sheet so that R_f values of the known and "suspected" substances can be determined under identical conditions.

B. THIN-LAYER CHROMATOGRAPHY

In thin-layer chromatography, an adsorbent is applied to a supporting material, frequently glass or aluminum, in a very thin layer. A binding agent is generally used to adhere the adsorbent to the supporting material.

In one of the more common procedures, a slurry of silica gel adsorbent, with calcium sulfate as the binder, is spread as a thin layer on a 20-cm^2 glass plate. The plates, dried to remove excess moisture, are then handled in much the same way as is paper in ascending chromatography. After separation, the spots can be scraped from the glass for detailed analysis or eluted and rechromatographed.

Separation is accomplished very quickly on thin-layer plates. Using paper chromatographic methods, the separation of some mixtures may take 24 hours. The same separation may take less than an hour when using thin-layer procedures.

REFERENCES

Heftman, E., ed. 1983. *Chromatography, Fundamentals and Applications of Chromatographic and Electrophoretic Methods.* Elsevier.

Karp, G. 1984. *Cell Biology.* 2d ed. McGraw-Hill.

Schleif, R. F., and P. C. Wensink. 1982. *Practical Methods in Molecular Biology.* Springer-Verlag.

Stahl, E., ed. 1969. *Thin-Layer Chromatography: A Laboratory Handbook.* 2d ed. Springer-Verlag.

APPENDIX E

Use of Live Animals in the Laboratory

A. GENERAL PROCEDURES

If this is the first time living vertebrate animals are being used in your class, it is important that you understand the purpose in bringing these animals into the laboratory. First, there must be a genuine reason for using live animals. Second, live animals must always be treated in a humane way: never cause them unnecessary irritation or injury. Accordingly, if any organ or tissue damage may result from legitimate experimentation, the animal is first put under an anesthetic, or the nervous system is treated to make the organism insensible to pain. Avoid injuring the animal's tissues or making it bleed; such damage makes the animal less capable of "normal" reactions.

When handling organs or tissues that have been exposed or removed, use a glass hook or a small camel-hair brush moistened with Ringer's saline solution. Never handle living tissues or organs with your fingers. When you need to lay an excised organ or part down, never place it on the table, or on dry paper; place it in a watch glass or other dish and keep it moistened. Apply Ringer's solution as often as necessary while working, to keep the tissue from drying out.

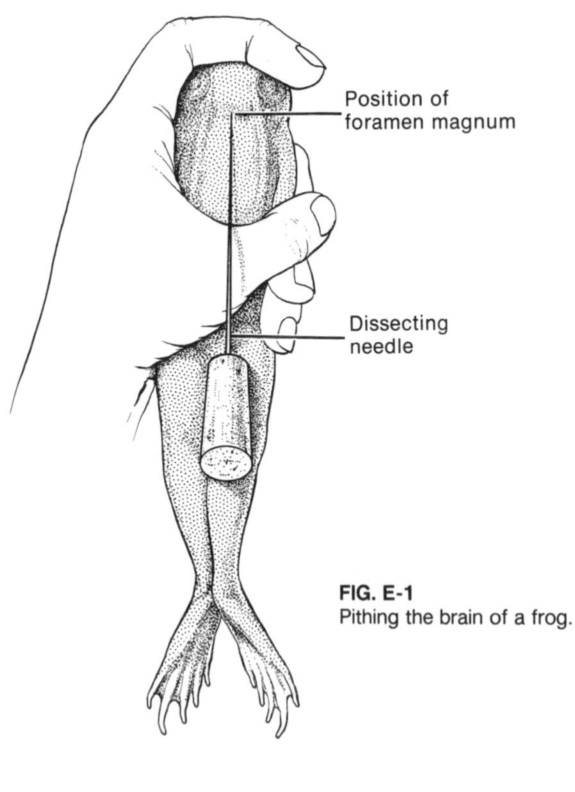

FIG. E-1
Pithing the brain of a frog.

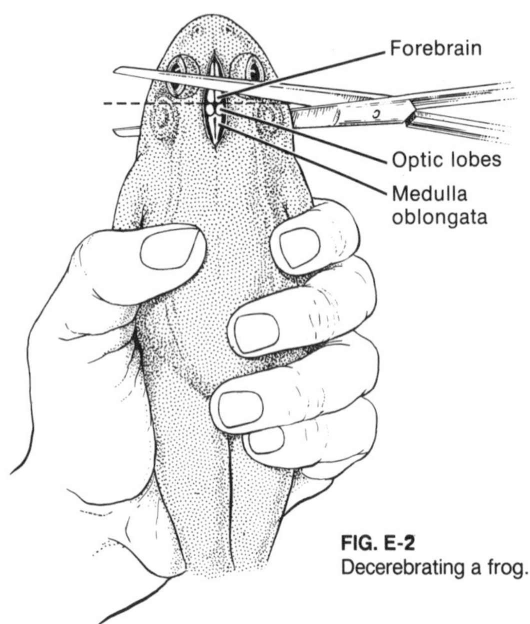

FIG. E-2
Decerebrating a frog.

B. PITHING PROCEDURE

A **spinal frog** (one whose entire brain has been destroyed) is prepared by a procedure called **pithing.** Grasp the animal as indicated in Fig. E-1, using the thumb and fingers to secure the limbs. With the index finger, depress the snout so that the head is at a sharp angle to the body. Run the tip of the dissecting needle down the midline of the head. At a point 2–3 mm behind the posterior border of the eardrums, the needle should dip. This marks the location of the **foramen magnum,** a large opening in the skull through which the spinal cord emerges from the **cranium.** Place the point of a dissecting needle in this groove, and with a sharp movement force it through the skin and foramen magnum into the brain. Twist and turn the needle to destroy the brain. Halt any bleeding, and treat the pithed frog in the same way as the decerebrate animal.

A. DECEREBRATION PROCEDURE

A **decerebrate frog** is one in which the forebrain is no longer functional. Select a large, active frog, and grasp it firmly in the left hand, pinioning the forelimbs and hindlimbs. Position the blades of a pair of *sharp* scissors as indicated in Fig. E-2, just behind the posterior margins of the eyes. Cut quickly and cleanly to avoid excessive nerve shock. Stop any blood flow with cotton or gauze pads. Lay the animal ventral side down on a damp paper towel. Be sure to moisten the skin from time to time, as the bulk of the respiration of such animals takes place through the skin.

APPENDIX F

Elementary Statistical Analysis

The inherent variability of organisms, coupled with the degree of error normally encountered in most measuring systems, demands the use of some form of data evaluation before valid conclusions can be made or inferences drawn. It is particularly important that students of the biological sciences become familiar with statistical methods of handling data. The following section briefly outlines some of the basic methods used in this form of analysis.

The procedures fall into two groups. The first is concerned with those statistics that define the nature and distribution of the data. The second outlines three procedures that may be used to compare two or more sets of data. This introduction to statistical methods is not to be regarded as complete, nor is it expected to substitute for formal training in this area. Wherever possible, the student should supplement the information given by use of the references listed at the end of this appendix.

A. DEFINING THE DATA

In order to define a sample of data, one must have some knowledge of the central tendencies and degree of dispersion of the data. The statistics usually used for this purpose are the arithmetic mean, the standard deviation, and the confidence interval of the mean.

1. Arithmetic Mean

The mean ($\overline{X}$) is computed by summing (Σ) the individual sample measurement (x_i) and dividing by the total number of measurements (N):

$$\overline{X} = \frac{\Sigma x_i}{N}$$

2. Standard Deviation

The mean of a group of data gives little information concerning the distribution of the data about the mean. Obviously, different numerical values can give the same mean value. For example,

Set 1: 32, 32, 36, 40, 40 $\overline{X} = 36$

Set 2: 2, 18, 24, 36, 100 $\overline{X} = 36$

The mean for both sets of figures are identical. Yet the variation from the mean in Set 2 is so great as to make the average meaningless. The standard deviation is a measure of data dispersion about the mean. The range covered by the mean plus or minus ($\pm$) one standard deviation includes about 68% of the data on the basis of which the mean was calculated. The range covered by the mean $\pm$ two standard deviations will include approximately 95% of the data. The calculation of the standard deviation is summarized below:

1. Compute the arithmetical mean $(\overline{X})$ by summing (Σ) the individual measurements (x_i) and dividing by the total number of measurements (N): $\overline{X} = \Sigma x_i/N$.

2. Calculate the deviation from the mean for each measurement: $(x_i - \overline{X})$.

3. Square each of the individual deviations from the mean: $(x_i - \overline{X})^2$. This allows one to deal with positive values.

4. Determine the sum of the squared deviations: $\Sigma(x_i - \overline{X})^2$.

5. Calculate the standard deviation(s) of the sample using the formula

$$s = \sqrt{\frac{\Sigma(x_i - \overline{X})^2}{N-1}}$$

A sample set of standard deviation measurement is shown in Table F-1, where

Number of individual measurements $(N) = 10$
Arithmetical means $(\overline{X}) = 220/10 = 22$
Degrees of freedom $(N - 1) = 9$

$$s = \sqrt{\frac{\Sigma(x_i - \overline{X})^2}{N-1}} = \sqrt{\frac{42}{9}} = \sqrt{4.66} = 2.16$$

3. Confidence Interval for a Sample Mean

The confidence interval (C) for a sample mean equals the standard deviation of the sample (s) divided by the square root of the sample number (N) and multiplied by a factor (t) that is determined by the probability level desired and the value of the sample number:

$$C = \pm t\left(\frac{s}{\sqrt{N}}\right)$$

It is highly improbable that a sample mean, based on a relatively small series of data, will correspond exactly to the true mean calculated from an infinitely large sample of the population. It is necessary, therefore, to define a range within which the true mean might be expected to lie. To define this range —the standard error of the mean (S.E.$\overline{X}$) must be known. The standard error equals the standard deviation divided by the square root of the number in the sample:

$$\text{S.E.}\overline{X} = \frac{s}{\sqrt{N}}$$

The confidence interval is then calculated by multiplying the S.E.$\overline{X}$ by t, whose value depends on the number in the sample N, and the level of probability selected. Normally, a probability or significance level of 0.05 is accepted in biological studies. This implies that in only 5% of the samples taken separately from a given population would the parameters defined by the sample *fail* to have significance.

The t values may be obtained from the t table (Table F-2). These values are listed in columns for 0.10, 0.05, and 0.01 probability levels. Note that it is necessary to have a value for the number of degrees of freedom (D.F.) of the sample. In this case, the D.F. for the sample equals the number (N) in the sample minus one $(N - 1)$.

B. COMPARISON OF DATA

1. Standard Error of the Difference of Means

The standard error of the difference of means is computed by using the formula

$$\text{S.E.}\overline{X}_1 - \overline{X}_2 = \sqrt{\frac{s_1}{N_1} + \frac{s_2}{N_2}}$$

where s_1 and s_2 represent the standard deviations of two different groups, N_1 and N_2 represent the number of individuals in each group (preferably at least 20), and $\overline{X}_1$ and $\overline{X}_2$ are the respective means. Using this formula, if the difference between the two means is *larger than two times* the standard error of the difference, it can be concluded that the difference between the groups was not due to chance

TABLE F-1
Sample calculation of the standard deviation.

Observation number	Individual measurement (x_i) of stem length (mm)	Deviation from mean ($x_i - \overline{X}$)	$(x_i - \overline{X})^2$
x_1	20	−2	4
x_2	24	+2	4
x_3	22	0	0
x_4	19	−3	9
x_5	26	+4	16
x_6	22	0	0
x_7	24	+2	4
x_8	20	−2	4
x_9	22	0	0
x_{10}	21	−1	1
	TOTAL = 220 = Σx_i	0	42 = $\Sigma(x_i - \overline{X})^2$

TABLE F-2
Significance limits of student's *t* distribution.

(*n*) degrees of freedom	Confidence levels		
	0.10	0.05	0.01
1	6.314	12.706	63.657
2	2.920	4.303	9.925
3	2.353	3.182	5.841
4	2.132	2.776	4.604
5	2.015	2.571	4.032
6	1.943	2.447	3.707
7	1.895	2.365	3.499
8	1.860	2.306	3.355
9	1.833	2.262	3.250
10	1.812	2.228	3.169
11	1.796	2.201	3.106
12	1.782	2.179	3.055
13	1.771	2.160	3.012
14	1.761	2.145	2.977
15	1.753	2.131	2.947
16	1.746	2.120	2.921
17	1.740	2.110	2.898
18	1.734	2.101	2.878
19	1.729	2.093	2.861
20	1.725	2.086	2.845
21	1.721	2.080	2.831
22	1.717	2.074	2.819
23	1.714	2.069	2.807
24	1.711	2.064	2.797
25	1.708	2.060	2.787
26	1.706	2.056	2.779
27	1.703	2.052	2.771
28	1.701	2.048	2.763
29	1.699	2.045	2.756
30	1.697	2.042	2.750
40	1.684	2.021	2.704
60	1.671	2.000	2.660
120	1.658	1.980	2.617
∞	1.645	1.960	2.576

Source: Adapted from Table III of Fisher and Yates, *Statistical Tables for Biological, Agricultural, and Medical Research,* Oliver & Boyd, Edinburgh.

alone, but was due to the treatment given. It can be further concluded that similar plants or animals under similar treatment could be expected to respond in a similar manner.

2. Student's *t* Test

The student's t test is used to determine whether, within a selected degree of probability, two groups of data represent samples taken from the same or different populations of data. In other words, it is used to determine if two groups of data are significantly different. This test uses both the means and standard deviations of the two samples. It is calculated as

$$t = \frac{(\overline{X}_1 - \overline{X}_2)\left(\sqrt{\dfrac{N_1 N_2}{N_1 + N_2}}\right)}{\sqrt{\dfrac{(N_1 - 1)(s_1{}^2) + (N_2 - 1)(s_2{}^2)}{N_1 + N_2 - 2}}}$$

where s_1 and s_2 represent standard deviations of two different groups, N_1 and N_2 represent the number of individuals in each group, and $\overline{X}_1$ and $\overline{X}_2$ are the respective means.

The calculated t value is then compared to the value in the t table (Table F-2) at the probability level chosen (usually 0.05) and at the combined degrees of freedom of the two samples ($N_1 + N_2 - 2$). If the value for t is *less* than that found in the table, then the two groups of data are not considered significantly different at the chosen level of probability. If the t value *exceeds* that in the table, then the two groups of data may be considered significantly different.

3. Wilcoxon Test

This test, although mechanically easier to carry out than the student's t test, should only be used on small amounts of data—such as in a preliminary study—to help determine if a particular experiment is worth pursuing.

Basically, this test employs a ranking system. A simple example, given below, is a comparison of the pulse rate *(PR)* between normal males and females to test for sexual differences.

The hypothesis to be tested (H_0—the null hypothesis) is

$$PR\,♀ = PR\,♂$$

that is, there is no difference in heart rate (in terms of pulse beat) between males and females.

a. Raw Data

Sex	*Pulse Rate*					
♂	74	77	78	75	72	71
♀	80	83	73	84	82	79

b. Calculation of Means

$$\Sigma\,♂ = 447;\ \overline{X} = 74.5$$

$$\Sigma\,♀ = 481;\ \overline{X} = 81.7$$

c. Ranking of Data

1. Rank the data in an order of increasing magnitude. Underline all values from the group having the smaller mean. In this example, it is the males

2. Also write the rank number under the data, beginning with 1 for the smallest value and proceeding to the largest value. Again, underline the rank number corresponding to the male values.

$$\underline{71}\ \ \underline{72}\ \ 73\ \ \underline{74}\ \ \underline{75}\ \ \underline{77}\ \ \underline{78}\ \ 79\ \ 80\ \ 82\ \ 83\ \ 84$$

$$\underline{1}\ \ \underline{2}\ \ 3\ \ \underline{4}\ \ \underline{5}\ \ \underline{6}\ \ \underline{7}\ \ 8\ \ 9\ \ 10\ \ 11\ \ 12$$

3. Next, reorder the rank numbers by sex, and add them up. If there is no real difference in pulse rate between sexes, then the sum of rank numbers should be equal.

$$♂\ \ \underline{1} + \underline{2} + \underline{4} + \underline{5} + \underline{6} + \underline{7} = 25$$

$$♀\ \ 3 + 8 + 9 + 10 + 11 + 12 = 53$$

d. Computing *U*

In this case, the sums of the rank numbers are different. But is there a significant difference between males and females? To determine this, the statistical value U must be determined:

$$U = W_1 - \tfrac{1}{2}\,n_1\,(n_1 = 1)$$

where W_1 = the total of the rank numbers belonging to the group with the smallest mean (i.e., the males: $W_1 = 25$).

n_1 = the number of individuals (or measurements) in the group having the smallest mean (i.e., the males: $n_1 = 6$)

Thus

$$\begin{aligned} U &= 25 - \tfrac{1}{2} \times 6(6 + 1) \\ &= 25 - 21 \\ &= 4 \end{aligned}$$

e. Determining Significance

1. Refer to Table F-3 (Wilcoxon table for unpaired data). Note that it has three columns labeled n_1, n_2, $C_{n_1 n_2}$, and Values of U.

TABLE F-3
Wilcoxon distribution (with no pairing). The numbers given in this table are the number of cases for which the sum of the ranks of the sample of size n_1, is less than or equal to W_1.

n_1	n_2	$C_{n_1n_2}$	Values of U, where $U = W_1 - \frac{1}{2}n_1(n_1+1)$																				
			0	1	2	3	4	5	6	7	8	9	10	11	12	13	14	15	16	17	18	19	20
3	3	20	1	2	4	7	10	13	16	18	19	20											
3	4	35	1	2	4	7	11	15	20	24	28	31	33	34	35								
4	4	70	1	2	4	7	12	17	24	31	39	46	53	58	63	66	68	69	70				
3	5	56	1	2	4	7	11	16	22	28	34	40	45	49	52	54	55	56					
4	5	126	1	2	4	7	12	18	26	35	46	57	69	80	91	100	108	114	119	122	124	125	126
5	5	252	1	2	4	7	12	19	28	39	53	69	87	106	126	146	165	183	199	213	224	233	240
3	6	84	1	2	4	7	11	16	23	30	38	46	54	61	68	73	77	80	82	83	84		
4	6	210	1	2	4	7	12	18	27	37	50	64	80	96	114	130	146	160	173	183	192	198	203
5	6	462	1	2	4	7	12	19	29	41	57	76	99	124	153	183	215	247	279	309	338	363	386
6	6	924	1	2	4	7	12	19	30	43	61	83	111	143	182	224	272	323	378	433	491	546	601
3	7	120	1	2	4	7	11	16	23	31	40	50	60	70	80	89	97	104	109	113	116	118	119
4	7	330	1	2	4	7	12	18	27	38	52	68	87	107	130	153	177	200	223	243	262	278	292
5	7	792	1	2	4	7	12	19	29	42	59	80	106	136	171	210	253	299	347	396	445	493	539
6	7	1716	1	2	4	7	12	19	30	44	63	87	118	155	201	253	314	382	458	539	627	717	811
7	7	3432	1	2	4	7	12	19	30	45	65	91	125	167	220	283	358	445	545	657	782	918	1064
3	8	165	1	2	4	7	11	16	23	31	41	52	64	76	89	101	113	124	134	142	149	154	158
4	8	495	1	2	4	7	12	18	27	38	53	70	91	114	141	169	200	231	264	295	326	354	381
5	8	1287	1	2	4	7	12	19	29	42	60	82	110	143	183	228	280	337	400	466	536	607	680
6	8	3003	1	2	4	7	12	19	30	44	64	89	122	162	213	272	343	424	518	621	737	860	994
7	8	6435	1	2	4	7	12	19	30	45	66	93	129	174	232	302	388	489	609	746	904	1080	1277
8	8	12870	1	2	4	7	12	19	30	45	67	95	133	181	244	321	418	534	675	839	1033	1254	1509

Source: Reproduced from Table H of Hodges and Lehmann, *Basic Concepts of Probability and Statistics,* published by Holden-Day, San Francisco, by permission of the authors and publisher.

2. The values of n_1 and n_2 in the data are 6 and 6. Refer to this combination in the first column of the table. The corresponding figure in the $C_{n_1n_2}$ column is 924.

3. Run along the U columns to $U = 4$, and then down the $U = 4$ column to the row corresponding to $C_{n_1n_2} = 924$. The value at these intersecting points is 12. This value represents the number of possible rank totals that would be less than or equal to ($\leqslant$) 25.

4. Suppose we consider that if an event occurs less than 5 times out of 100 ($\frac{5}{100}$ or $\frac{50}{1000}$, $P = 0.05$) it occurred, not as a result of chance, but as a result of the treatment, or in this case, was due to the difference in sex.

The value of 12 in the U column represents a probability of 12 times out of 1000 ($\frac{1.2}{100}$ or $P = 0.012$). This value is considerably lower than the $\frac{5}{100}$ that we will accept for considering that the difference in pulse rate is based on the difference in sex. Therefore, we would reject the hypothesis that

$$PR\,♂ = PR\,♀$$

and would say that the difference in pulse rates is significant and results from the difference in sex.

C. CHI-SQUARE ANALYSIS

The statistical test most frequently used to determine whether data obtained experimentally provide a good fit, or approximation, to the expected or theoretical data, is relatively simple to carry out. Basically, this test can be used to determine if any deviations from the *expected values* are due to chance alone or to factors or circumstances other than chance.

The formula for chi square (X^2) is

$$X^2 = \Sigma\left[\frac{(O-E)^2}{E}\right]$$

where O = the **observed** number of individuals
E = the **expected** number of individuals
and Σ = the sum of all values of $(O - E)^2/E$ for the various categories of phenotypes

The following example shows how this type of analysis can be applied to genetics data. It can be used in many other analyses as long as numerical data are used and not percentages or ratios.

TABLE F-4
Summary of the calculations of chi squares for the hypothetical cross given.

Phenotype	Genotype	*O*	*E*	*(O − E)*	$(O - E)^2$	$(O - E)^2/E$
Tall	*T_*	84	82.5	1.5	2.25	0.027
Dwarf	*tt*	26	27.5	−1.5	2.25	0.082
Total		110	110.0	0		0.109

TABLE F-5
Distribution of X^2.

n	Probability *(P)*													
	.99	.98	.95	.90	.80	.70	.50	.30	.20	.10	.05	.02	.01	.001
1	.00016	.00063	.00393	.0158	.0642	.148	.455	1.074	1.642	2.706	3.841	5.412	6.635	10.827
2	.0201	.0404	.103	.211	.446	.713	1.386	2.408	3.219	4.605	5.991	7.824	9.210	13.815
3	.115	.185	.352	.584	1.005	1.424	2.366	3.665	4.642	6.251	7.815	9.837	11.345	16.268
4	.297	.429	.711	1.064	1.649	2.195	3.357	4.878	5.989	7.779	9.488	11.668	13.277	18.465
5	.554	.752	1.145	1.610	2.343	3.000	4.351	6.064	7.289	9.236	11.070	13.388	15.086	20.517
6	.872	1.134	1.635	2.204	3.070	3.828	5.348	7.231	8.558	10.645	12.592	15.033	16.812	22.457
7	1.239	1.564	2.167	2.833	3.822	4.671	6.346	8.383	9.803	12.017	14.067	16.622	18.475	24.322
8	1.646	2.032	2.733	3.490	4.594	5.527	7.344	9.524	11.030	13.362	15.507	18.168	20.090	26.125
9	2.088	2.532	3.325	4.168	5.380	6.393	8.343	10.656	12.242	14.684	16.919	19.679	21.666	27.877
10	2.558	3.059	3.940	4.865	6.179	7.267	9.342	11.781	13.442	15.987	18.307	21.161	23.209	29.588

Source: Table IV of Fisher and Yates, *Statistical Tables for Biological, Agricultural and Medical Research,* published by Longman Group, Ltd., London (previously published by Oliver & Boyd, Edinburgh) and by permission of the authors and publishers.

1. Example of Chi-Square Analysis

In a cross of tall maize (corn) plants to dwarf plants, the F_1 generation consisted entirely of tall plants. The F_2 generation consisted of 84 tall and 26 dwarf plants. The question we want to answer is whether this F_2 data fits the expected 3:1 monohybrid ratio. Using the data given in Table F-4, chi square was calculated.

$$X^2 = \Sigma\left[\frac{(O-E)^2}{E}\right]$$

$$= \Sigma\ [0.027 + 0.082]$$

$$X^2 = 0.109$$

What does this chi-square value of 0.109 mean? Consider that if the observed values were exactly equal to the expected values (i.e., $O = E$) then we would have a perfect fit, and X^2 would equal 0. Thus if you obtain a small value of X^2, this would indicate a close agreement of the observed and expected ratios, whereas a large value for X^2 would indicate marked deviation from the expected ratios. However, deviations from the expected values are always bound to occur due to chance alone. The question is "Are the observed deviations within the limits expected by chance?"

Statisticians have generally agreed, for these types of studies, on the arbitrary limits of 1 chance in 20 (probability = 0.05 = 5%) for making the distinction between acceptance or rejection of the data as fitting the expected ratio.

The chi-square value for a two-term ratio (i.e., 3:1) that corresponds to a 0.05 or 1 in 20 probability is 3.841. Therefore, you would expect to obtain this value, *due to chance deviations only,* in only 5% of similar trials if the hypothesis is true. When X^2 for this two-term ratio is larger than 3.841, then the probability that the variation is due to chance alone is less than 5% or 1 in 20. You would therefore *reject* the hypothesis that the observed and expected ratios are in close agreement.

In our example, X^2 was 0.109, which is considerably less than 3.841. Thus we can say that the variation between the observed and expected values *was* due to chance alone and accept the data as fitting the 3:1 ratio.

Where did we obtain the value 3.841? Mathematicians have developed a variety of statistical tables. Table F-5 is an example of a table of chi-square values. The table is set up so that probability *(P)* values extend across the top, and "degrees of freedom" (*n*) values are down the left margin. The number of degrees of freedom in tests of genetic ratios is *generally always one less than the number of classes* in the ratio being analyzed. Thus in tests of such ratios as 1:1 or 3:1 there is one degree of freedom, a test of a 1:2:1 ratio would have two degrees of freedom, and a test of a 1:2:1:2:4:2:1:2:1 would have eight degrees of freedom. The general idea of degrees of freedom can be exemplified by the situation encountered by a small boy when he is putting on his shoes. He has two shoes, but only one degree of freedom. Once one shoe is filled by a foot, right or wrong, the other shoe is automatically committed to being right or wrong too. Similarly, in a two-place table, one value can be filled arbitrarily, but the other is then fixed by the fact that the total must add up to the precise number of observations made in the experiment, and the deviations in the two classes must compensate for each other. When there are four classes, any three are usually free, but the fourth is fixed. Thus, when there are four classes, there are usually three degrees of freedom.

In our example, we have two classes in the ratio (i.e., 3:1) and therefore have one degree of freedom when we interpret the chi-square table. Look at the one-degree-of-freedom row under the .05 probability column, and you will find the value 3.841. This number represents the **maximum value** for chi square that you should be willing to accept and yet consider the deviations observed as due to chance alone. If you were willing to accept a *P* value representing 1 chance in 10, what value of chi square would you accept as maximum?

In our example, chi square was calculated to be 0.109. Looking across the one-degree-of-freedom line in Table F-5, we find that this value falls between the .70 ($X^2 = .148$) and the .80 ($X^2 = .0642$) columns. This says that the probability that the deviations we obtained from the expected values could be attributed to chance alone is 70–80%. That is, if we were to repeat the study 100 times, we would obtain deviations as large as those observed about 70% of the time (i.e., 7 out of every 10 experiments). We can thus reasonably regard this deviation as simply a sampling, or chance, error.

REFERENCES

Bailey, N., ed. 1981. *Statistical Methods in Biology.* Halsted Press.

Milton, J., and J. Susan. 1983. *Statistical Methods in the Biological and Health Sciences.* McGraw-Hill.

Sokal, R. R., and F. J. Rohlf. 1981. *Biometry.* 2nd ed. W. H. Freeman and Company.

APPENDIX G

Logarithms

If $a^x = y$, then the logarithmic relation is defined as $x = \log_a y$. Thus the logarithm of a number to a given base is defined as the exponent of the power to which the base must be raised to give the number; that is, x is the logarithm of y, and y is the antilogarithm of x.

The fundamental theorems of logarithms are

$$\log (M \times N) = \log M + \log N$$

$$\log (M/N) = \log M - \log N$$

$$\log M^n = n \log M$$

$$\log \sqrt[n]{M} = \log (M/n)$$

In Table G-1 the logarithms are located in the body of the table, and the numbers from 1.0 to 9.9 are given in the left-hand column and the top row. To locate the logarithm of 4.7, for example, read down the left-hand column to 4 and across the columns to the .7 column to find 0.672 (the zero and decimal point are omitted for convenience). To find an antilogarithm, the reverse procedure is employed; for example, 0.532 is located in the 3 row down and in the .4 column across, and therefore its antilogarithm is 3.4.

Numbers less than 1.0 or greater than 10.0 are conveniently expressed as the product of a number between 1.0 and 10.0 and a power of 10. Thus, 4700 $= 4.7 \times 10^3$, and since the logarithm of a product is equal to the sum of the individual logarithms,

$$\log 4700 = \log 4.7 + \log 10^3 = 0.672 + 3 = 3.672$$

Similarly, if a number is less than 1.0,

$$\begin{aligned} \log 0.000047 &= \log 4.7 + \log 10^{-5} \\ &= 0.672 - 5 \\ &= -4.328 \end{aligned}$$

To find an antilogarithm, or the number represented by a given logarithm, the reverse procedure is employed, and the logarithm is converted into a sum consisting of a positive term between 0.0 and 1.0 and an integer. The antilogarithm of this sum will equal the product of the individual antilogarithms. Thus,

TABLE G-1
Three-place logarithms.

	0	.1	.2	.3	.4	.5	.6	.7	.8	.9
0	—	000	301	477	602	699	778	845	903	954
1	000	041	079	114	146	176	204	230	255	279
2	301	322	342	362	380	400	415	431	447	462
3	477	491	505	519	532	544	556	568	580	591
4	602	613	623	634	644	653	663	672	681	690
5	699	708	716	724	732	740	748	756	763	771
6	778	785	792	799	806	813	820	826	833	839
7	845	851	857	863	869	875	880	886	892	898
8	903	909	914	919	924	929	935	940	945	949
9	954	959	964	969	973	978	982	987	991	996

$$\begin{aligned} \text{antilog } 2.532 &= \text{antilog } 0.532 \times \text{antilog } 2 \\ &= 3.4 \times 10^2 \\ &= 340 \end{aligned}$$

$$\begin{aligned} \text{antilog } -7.468 &= \text{antilog } 0.532 - 8 \\ &= 3.4 \times 10^8 \\ &= 0.000000034 \end{aligned}$$

Finally, the following relationships should be remembered:

$$\log 1 = \log 10^0 = 0$$

$$\log 10 = \log 10^1 = 1$$

$$\log 100 = \log 10^2 = 2$$

$$\log 1000 = \log 10^3 = 3$$

$$\log 10000 = \log 10^4 = 4$$

$$\log 0.1 = \log 10^{-1} = -1$$

$$\log 0.01 = \log 10^{-2} = -2$$

$$\log 0.001 = \log 10^{-3} = -3$$

$$\log 0.0001 = \log 10^{-4} = -4$$

Although this is a mathematical definition of logarithms, and the one that must be employed in carrying out problems, a qualitative description may be useful in placing this function in perspective. At first glance, the logarithm may appear to be a fairly arbitrary and artificial way of expressing any measurement as compared with the simpler (arithmetic) expression. However, it should be noted that we are very often more concerned—and this is true particularly in biology—with *relative* measurements or relations rather than with absolute magnitudes. The logarithmic expression has exactly this property, that it gives equal weight to equal relative (rather than absolute) changes. Thus, a twofold change represents 0.3 logarithmic units anywhere on the scale, although this may represent a milligram, a kilogram, or a ton on an absolute scale. Further, measurements are usually made with the same *relative* accuracy—to 1%, or to 1 part in 10,000—rather than to a given *absolute* accuracy. In a logarithmic plot this given relative accuracy will be expressed by the same distance; for example, 10% accuracy will represent 0.04 log units (= log of 1.10).

That this term is an equally "natural" means of expressing many relations is further indicated by the fact that many relations or physical "laws" are actually expressed in logarithmic form; that is, the activity or action is proportional to the logarithm of some concentration or ratio rather than to its absolute amount. Outstanding examples are the relations between pH and salt concentrations in buffer solution, between oxidation-reduction potential and the concentrations of the reduced and oxidized substance, and between electromotive force and the concentration of substances in concentration cells.

Logarithms are also particularly useful in graphing relations that extend over a very wide range of values, since they have the property of giving equal relative weight to all parts of the scale. This is valuable in "spreading out" the values that would otherwise be concentrated at the lower end of the scale.

APPENDIX H

Radioisotopes

The nuclei of atoms are composed of neutrons and protons. In any single element, the nuclei contain the same number of protons. Some nuclei, however, may contain a *different number of neutrons* and thus have a different atomic mass. Atoms having nuclei with the same number of protons but different numbers of neutrons are called **isotopes.** Most isotopes are stable. However, others have certain combinations of protons and neutrons that make them unstable. Such isotopes tend to reach a stable state and in the process emit a distinctive type of radiation. During the disintegration of such isotopes, called **radioisotopes,** three types of radiation may be emitted:

Alpha (α) rays are particles consisting of two neutrons and two protons each. Thus, alpha rays are basically charged helium atoms.

Beta (β) rays are electrons emitted by some nuclei. The energy levels of these electrons may vary and produce hard beta radiation (highly charged electrons) or soft beta radiation (low-charged electrons). Phosphorus 32 is an example of a hard beta emitter; carbon 14 is a soft beta emitter.

Gamma (γ) rays are electromagnetic radiation and closely resemble X rays. Cobalt 60 is an example of a gamma emitter.

Because of the constant loss of particles, radioisotopes are said to "decay." The amount of radioactive material that remains will therefore be reduced with time. The rate of decay is known for many radioisotopes and is expressed as the **half-life** of the isotope. The half-life is the length of time it takes for a radioactive substance to lose half of its radioactivity. For example, the half-life of uranium 238 is 4,500,000,000 years; phosphorus 32 is 8.0 days.

Because radiation is damaging to human tissues, any radioactive materials used in the laboratory must be handled with extreme caution. Radioisotopes should be handled with the same care as harmful bacteria or strong acids. Remember, your "senses" cannot detect radiation. You *must* use proper techniques and follow certain rules.

SAFETY PRECAUTIONS IN THE USE OF RADIOISOTOPES

1. *Do not* eat, drink, smoke, chew gum, or apply cosmetics in the laboratory.

2. Wear an apron or laboratory coat.

3. To guard against contamination, place some disposable material (such as aluminum) over the table top on which you are working.

4. *Never* pipet any radioactive material by mouth. Use the special pipets provided by the instructor.

5. Discard paper, and other materials used, in the container designated for this purpose.

6. Never pour radioactive liquids into the standard drain. Pour them into the "waste" jar provided by the instructor.

7. Wash hands thoroughly before leaving the laboratory and after any suspected contact with radioactive material.

8. If you should spill radioactive material, notify your instructor immediately.

9. Clean up immediately after the experiment. Wash all contaminated equipment, and set it aside to be checked by the instructor for radioactivity.

REFERENCES

Chapman, J. M., and G. Ayrey. 1981. *The Use of Radioactive Isotopes in the Life Sciences.* Allen Unwin.
Chase, G. D., and J. L. Rabinowitz. 1967. *Principles of Radioisotope Methodology.* 3d ed. Burgess.
Wolf, G. 1964. *Isotopes in Biology.* Academic Press.
Wolfe, R. R. 1984. *Radioactive Tracers in Biochemistry.* A. R. Liss.

APPENDIX I

Aseptic Techniques

The term *aseptic technique* refers to those procedures that enable a microbiologist to keep the organisms used in an experiment separate from the millions of other microorganisms in the environment. With the working materials sterilized, aseptic techniques become a matter of simple technical procedures designed to prevent contamination. These procedures consist of ways of transferring organisms from test tube to test tube, from test tube to flask or petri dish, and from petri dish to petri dish or flask.

A. PROCEDURES FOR TRANSFERRING BACTERIA

When a nutrient medium is being prepared in a test tube or flask, the mouth of the container should be plugged before sterilization with long-fiber cotton, preformed plastic plugs, or plastic or metal caps. To inoculate the tube or flask with the desired organism, use the following procedure (refer to Fig. I-1). Use an inoculating loop for liquid cultures and a loop or needle for cultures growing on solid media. Hold the test tube containing the organisms to be transferred and the tube to which they are to be transferred as shown in Fig. I-1A. Then sterilize the inoculating needle, holding it like a pencil, by heating the wire part of it until it glows red in the flame of an alcohol lamp or Bunsen burner. Remove the cotton plugs or other closures by grasping them between the fingers and flame the mouths of the vessels briefly to eliminate loose cotton fibers and dust. Cool the loop or needle by touching it to the inside of the stock culture vessel (or to the agar). Pick up the organisms to be transferred by touching the wire loop or needle lightly to the growth, or by dipping it into the broth. Then transfer the culture to the new growth vessel by dipping the loop or needle gently into the broth or by drawing it across the surface of the agar. Pass the mouths of the containers rapidly through the flame again, reinsert the cotton plugs to prevent the entrance of other microbes, and kill the organisms remaining on the loop (or needle) by heating the wire to redness. **(CAUTION: To avoid**

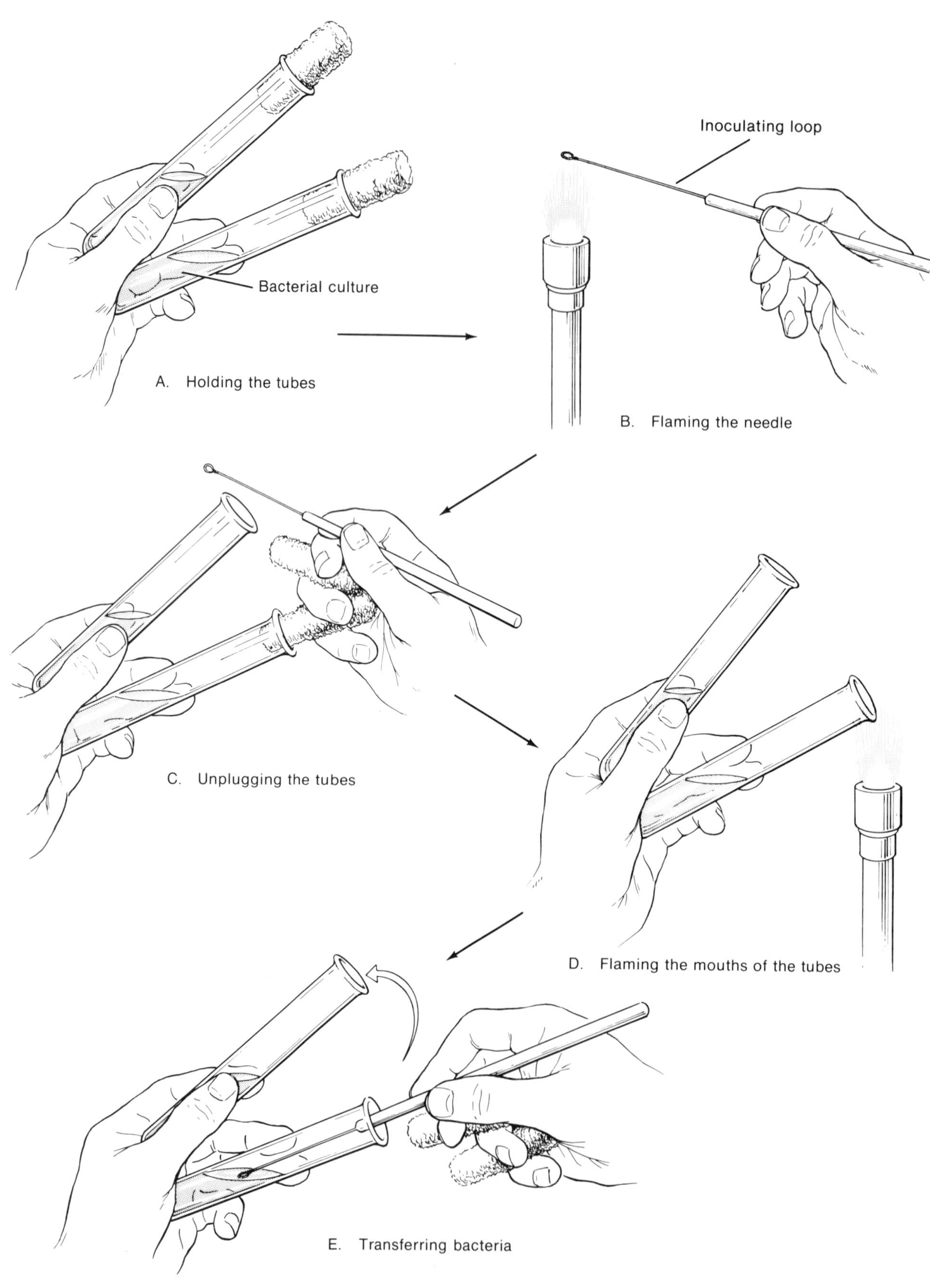

FIG. I-1
Procedure for inoculating bacteria.

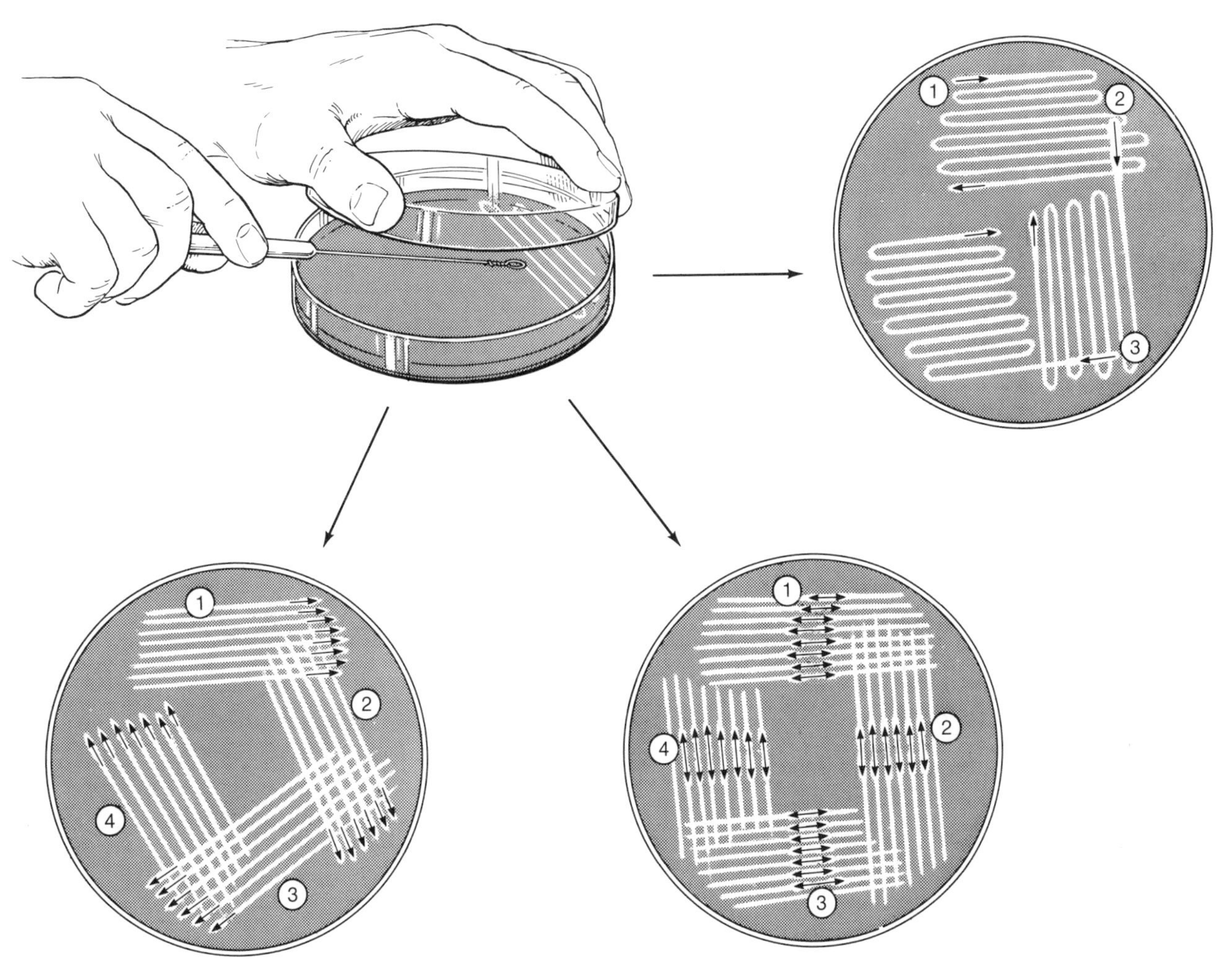

FIG. I-2
Procedure for plating bacteria.

potentially dangerous spatter, heat the wire gradually from the holder toward the tip.)

To transfer bacteria from one petri dish to another, use the following procedure (Fig I-2).

1. Sterilize the transfer needle.

2. Slightly lift the lid of the culture-containing dish.

3. Touch the hot needle or loop to the agar to cool it and then touch the bacterial growth on the agar.

4. Close the lid of the culture dish and slightly open the lid of the petri dish to which you are transferring a culture.

5. Streak the needle gently across the surface of the agar, using one of the "streaking" patterns shown in Fig. I-2. The objective of streaking is to isolate individual colonies at the completion of the third and fourth streak on the plate.

6. Close the lid and sterilize the needle.

B. DILUTION TECHNIQUES AND CALCULATIONS

Dilution techniques are among the more useful procedures used in the laboratory. They provide simple and accurate methods for (1) changing the concentration of a solution, (2) indirectly "weighing" a solute whose weight is well below the usual limits of analytical balances, or (3) determining the quantity of bacteria in a culture. For example, suppose you want to prepare a solution having a solute concentration of 0.001 mg/ml. Because 0.001 mg is $\frac{1}{100}$ of 0.1 mg, dissolving 0.1 mg in 100 ml would give the required concentration. By extrapolation it is easy to see how any amount of solute, regardless how small, could indirectly be "weighed" in this manner.

To understand dilution you should be familiar with the following basic terminology and information.

1. 1:10; 1:20; 1:100 means 1 part in a total of 10; 1 part in a total of 20; 1 part in a total of 100, respectively. Therefore,

1:10 means 1 + 9 or 0.1 + 0.9

1:20 means 1 + 19 or 0.1 + 1.9

1:100 means 1 + 99 or 0.1 + 9.9

2. $1{:}10 = \frac{1}{10} = 1{:}1 \times 10^1$; $1{:}100 = \frac{1}{100} = 1{:}1 \times 10^2$. This is a fraction and obeys the laws of simple algebra and arithmetic.

3. Any unit can be used provided that the same units are used throughout.

1 ml : 10 ml

1 gal : 10 gal

0.1 ml + 0.9 ml

1 g + 9 g

4. In dilution the amount present in the original suspension is reduced by the dilution factor or fraction. Thus if 100 particles per milliliter are present in the original suspension, after a 1:10 dilution there will be

$$\frac{100 \text{ particles}}{\text{ml}} \times \frac{1}{10} = \frac{10 \text{ particles}}{\text{ml}}$$

5. There are three methods of preparing a 1:10 dilution.

a. In the weight-to-weight method (w:w) 1.0 g of solute is dissolved in 9.0 g of solvent, giving 10 total parts by weight, one part of which is solute.

b. In the weight-to-volume (w:v) method, enough solvent is added to 1.0 g of solute to make a total volume of 10 ml. If a graduated cylinder is used, the dry solute is first placed in the container. The graduated cylinder is then filled to exactly 10.0 ml. In this method one part (by weight) is dispersed in 10 total parts (by volume).

Most biological solutions used in the laboratory are very dilute. Therefore, the accuracy of most work would not be affected if a previously weighed solute were dissolved in the desired volume of solvent, because most dilute solutions do not appreciably change in volume after adding small quantities of solute. Of course the weight of the solution will change. Thus, if 1 g of NaCl is dissolved in 10 ml of water, the solution will *weigh* 11.0 g but the volume will remain essentially unchanged at 10 ml. If the amount of solute added is large, this will not hold true and the volume will increase measurably.

c. If the solute is a liquid, a 1:10 solution is prepared as volume to volume (v:v). This would be the case if it was desired to prepare a 1:10 dilution of pure ethyl alcohol. For example, adding 1.0 ml of ethanol to 9.0 ml of water results in a 10-part solution of which alcohol is one part. Alternatively, because alcohol is a little lighter than water, 1.0 g of ethanol could be added to 9.0 g of water to obtain almost the same solution on a weight-to-weight basis. In either case, the method of preparing a solution should be clearly indicated to avoid confusion.

6. *Percent Concentration.* Percent means *parts per hundred.* Because 1 part in 10 is the same as 10 parts in 100, each of the solutions described in the preceding discussion (5) could be considered "10%" solutions. However, each is slightly different in actual concentration, so clarity is assured only if solutions are properly labeled. For example, 10% (w:w); 10% (w:v); or 10% (v:v). The percent-by-weight-to-volume method is frequently used.

7. The concentration of a solution is given as the amount of solute per volume. For example

particles/ml

g/liter

mg/ml

8. Some calculations

a. To determine the total dilution of the following three consecutive dilutions of an initial solution

1:10 (1 + 9)
1:10 (1 + 9)
1:5 (1 + 4)

simply multiply each factor

$$\frac{1}{10} \times \frac{1}{10} \times \frac{1}{5}$$

Thus the total dilution $= 1{:}500 = 5 \times 10^{-2} = 1{:}5 \times 10^2$.

b. To obtain a dilution of 1:2000, factor the denominator (2000)

$$10 \times 10 \times 10 \times 2 = 2000$$

or

$$10 \times 10 \times 20 = 2000$$

Then combine the following dilutions:

$$1{:}10,\ 1{:}10,\ 1{:}10,\ 1{:}2 = 1{:}2000$$

or

$$1{:}10,\ 1{:}10,\ 1{:}20 = 1{:}2000$$

Note: Dilutions can be made of any multiple of the factors of the denominator of the total dilution fraction. This denominator is called the "dilution factor." The d.f. equals the reciprocal of dilution: $\frac{1}{10}$ = d.f. of 10.

c. To calculate the original amount after determining the amount in dilution, multiply by the dilution factor. For example, suppose that, after plating out 1 ml of a 1:2000 dilution of an original culture containing *x* cells per milliliter, you find that 300 colonies develop. How many bacteria were originally in the culture?

$$\frac{300 \text{ bacteria}}{1 \text{ ml plated}} \times 2000 = 600{,}000$$

$$= 6.0 \times 10^5 \text{ bacteria/ml}$$

in original suspension.

d. Suppose that you find 300 colonies on a plate. The volume you plated was 0.2 ml (0.2 ml = $\frac{1}{5}$ ml). The dilution was 1:5000. Determine the amount of bacteria in the original suspension.

$$300 \text{ bacteria} \times 5000 \times 5$$

$$= 7{,}500{,}000 \text{ bacteria/ml}$$

$$= 7.5 \times 10^6 \text{ bacteria/ml}$$

in original suspension, or

$$\frac{300 \text{ bacteria}}{0.2 \text{ ml}} \times 5000$$

$$= 7.5 \times 10^6 \text{ bacteria/ml}$$

REFERENCES

Atlas, R. M., and A. E. Brown. 1984. *Experimental Microbiology.* Macmillan.

Benson, H. J. 1985. *Microbiological Applications.* 4th ed. Wm. C. Brown.

Bradshaw, L. J. 1979. *Laboratory Microbiology.* 3d ed. Saunders.

Collins, C. H., and P. M. Lyne. 1984. *Microbiological Methods.* 5th ed. Butterworths.

Kelley, S. G., and F. J. Post. 1982. *Basic Microbiology Techniques.* 2d ed. Star Publishing.

APPENDIX J

Mendel's Laws of Inheritance

Gregor Mendel was an Austrian monk trained in the science of mathematics. He began his studies of heredity in 1857 using the common garden pea. He chose plants because they were readily available, easy to grow, and grew rapidly. Furthermore, different varieties had distinctly different traits that "bred" true and reappeared year after year. Lastly, the pistil and stamens in the flower are entirely enclosed by the petals. Thus the flower typically self-pollinates (and thus self-fertilizes); accidental cross-pollination could not occur and confuse the experimental results. In Mendel's own words (a truism that holds even today), "The value and utility of any experiment are determined by the fitness of the material to the purpose for which it is used."

Mendel chose two variants for each of seven different traits (form of the seed, color of the seed, position of the flower, flower color, form of the pod, color of the pod, and stem length). For each trait, Mendel noted the apparent loss of one alternate characteristic in the next generation. For example, when he crossed short plants with tall plants, only tall plants were found in the next generation (also called the Filial 1 or F_1 generation). When these F_1 progeny were allowed to self-pollinate, both parental types reappeared in the succeeding, or F_2, generation. Furthermore, Mendel noted a consistent 3:1 ratio of tall plants to short plants. This 3:1 ratio is frequently called the **monohybrid** ratio; *hybrid* refers to the heterozygosity of the particular trait in the offspring or parents. Thus, in the above example (and for all others he tested), the trait was not lost in the F_1 generation, but was "masked" in some fashion such that it did not express itself. The trait that was expressed is **dominant** and that which is hidden, or not expressed, is **recessive.**

Mendel's research suggested to him that those "factors" (now called **genes**) that determined various traits were transmitted from generation to generation in a predictable manner. From his findings, Mendel suggested that each gene for a given trait could exist in two alternative forms (now called **alleles**) that were responsible for the phenotype expressed. Thus, plants or animals that expressed only

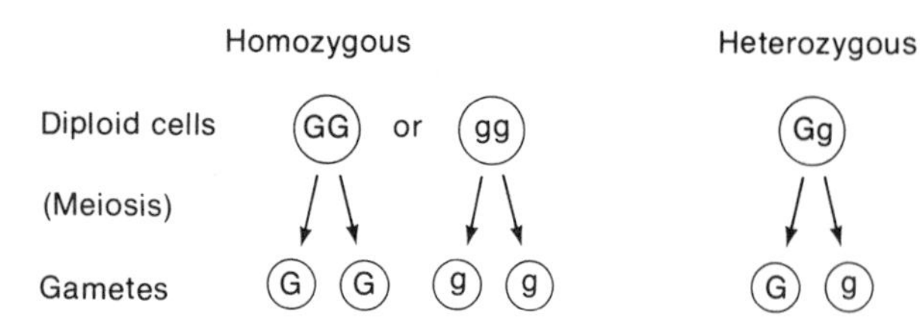

FIG. J-1
Types of gametes produced in homozygous and heterozygous crosses.

one phenotype (either dominant or recessive) through several generations were considered to be **homozygous** (or pure breeding) for that trait. Organisms that produced two phenotypes were **heterozygous** (or hybrid).

Today we know that the gametes are the physical link between two generations and that homozygous individuals produce only one type of gamete, whereas heterozygous organisms produce two types of gametes and in equal numbers.

From his experiments involving the transmission of single characteristics, Mendel formulated his first law of inheritance—the **law of segregation.** This law says that, in sexually reproducing organisms, there are factors (genes) that come together at fertilization and segregate at meiosis. The alleles involved do this without losing their identity; that is, they show no changes in the offspring, which thus rules out any consideration involving a "blending" theory of inheritance.

Mendel's second law, the **law of independent assortment,** evolved from his experiments involving crosses between individuals differing in two gene pairs. For example, if you cross a homozygous plant with smooth, yellow seeds with one having wrinkled, green seeds, the F_1 generation results in plants having smooth, yellow seeds. In this cross, what is the genotype for each of the parents for each of the traits considered?

In the space below, show the cross, gametes, and genotype and phenotype for the F_1 plants.

When the F_1 plants were self-fertilized, they gave rise to an F_2 generation divided into four phenotypic classes as follows:

$\frac{9}{16}$ of the population were smooth and yellow.

$\frac{3}{16}$ of the population were wrinkled and yellow.

$\frac{3}{16}$ of the population were smooth and green.

$\frac{1}{16}$ of the population were wrinkled and yellow.

This ratio is referred to as a **dihybrid ratio,** where *dihybrid* is defined as being heterozygous in terms of the two pairs of factors, or alleles, being studied. In the space below, diagram the cross just described. Show the phenotype and genotype of the F_1 plants, the gametes produced by the F_1 plants, the phenotype and genotype of the F_2 plants and the dihybrid ratio.

Interestingly, one can still demonstrate the validity of Mendel's first law of segregation if you consider the data for only one pair of genes at a time. For example, if you combine all the smooth seeds ($\frac{9}{16} + \frac{3}{16} = \frac{12}{16}$) and all the wrinkled seeds ($\frac{3}{16} + \frac{1}{16} = \frac{4}{16}$), you obtain a $\frac{12}{16}$:$\frac{4}{16}$ or 3:1 monohybrid ratio. Thus you again demonstrate that gene pairs act independently of one another and are not changed in their transmission from one generation to another. This indeed was the basis for Mendel's second law of inheritance.

Although Mendel's career as a "geneticist" was cut short by a beetle that devastated his pea crops and by being appointed head of the monastery, he left us with several important concepts, including his two laws of inheritance and the fact that results of various crosses are mathematically predictable.

REFERENCES

Avers, C. 1984. *Genetics.* 2d ed. W. Grant.

Gardner, E. J., and D. P. Snustad. 1984. *Principles of Genetics.* 7th ed. Wiley.

Lerner, I. M., and W. J. Libby. 1976. *Heredity, Evolution, and Society.* 2d ed. W. H. Freeman and Company.

Stern, C., and E. R. Sherwood. 1966. *The Origin of Genetics.* W. H. Freeman and Company.

Stine, G. J. 1973. *Laboratory Exercises in Genetics.* Macmillan.

APPENDIX K

Culture of *Drosophila*

A. LIFE CYCLE OF *DROSOPHILA*

There are four distinct stages in the life cycle of the fruit fly (Fig. K-1): egg, larva, pupa, and adult. At 25°C, a fresh culture of *Drosophila* will produce new adults in 9 or 10 days: about 5 days in the egg and larval stages and 4 days in the pupal stage. The adult flies may live for several weeks. *Drosophila* cultures should not be exposed to high temperatures (e.g., 30°C), because such exposure results in sterilization or death of the flies, nor to low temperatures (e.g., 10°C), because it prolongs the life cycle (perhaps 57 days) and reduces viability.

1. Egg

The adult *Drosophila* female starts to deposit eggs on the second day after emergence from the pupa. Each egg is about 0.5 mm long and is ovoid in shape and white in color. Extending from its anterior end are two thin stalks that expand into flattened, spoonlike terminal parts. Embryonic development of the egg takes about one day at 25°C.

2. Larva

The larva is white, segmented, and wormlike. It has black mouth parts (jaw hooks) in a narrowed head region, but no eyes and so is completely blind. The larva also lacks appendages and must literally eat and push its way through its environment. It breathes by trachea and has a pair of conspicuous spiracles (air pores) both at the anterior end and at the posterior end of the body.

The larval stage in the *Drosophila* life cycle is one of rapid eating and growing. It consists of three subdivisions called **instars.** The first and second instars terminate in molts. Each molt consists of a complete shedding of the skin and mouth parts of the larva and is the mechanism by which the animal grows. The third instar terminates in pupation. Just before pupation, the animal ceases to feed, crawls to some

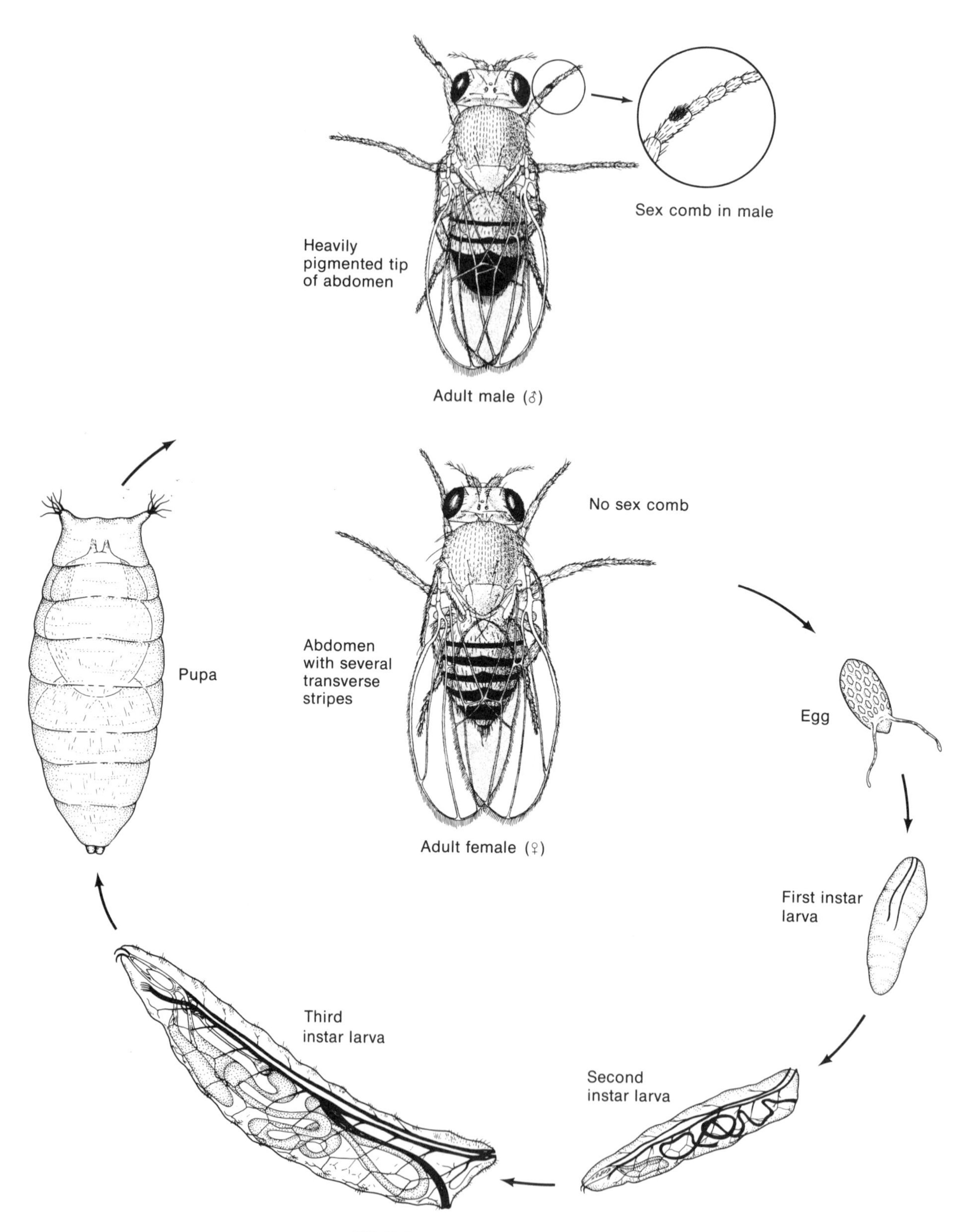

FIG. K-1
Life cycle of *Drosophila melanogaster*.

relatively dry surface, and everts its anterior spiracles. The larval stage takes about 4 days at 25°C for completion, at which time the third instar is about 4.5 mm long.

3. Pupa

The pupal stage is considered a reorganizational stage of the fly's life cycle in which most larval structures are destroyed and adult structures are developed from embryonic tissues called **anlagen** (imaginal discs). These embryonic tissues have been lying dormant in the animal since their differentiation in the egg stage. The animal pupates within the last larval skin, which is at first soft and white but slowly hardens and darkens in color. The transformation that takes place in the pupa results in the development of the body form and structures of the adult **(imago).** The pupal stage takes about 4 days at 25°C to be complete, at which time the adult fly emerges from the puparium (pupal case).

4. Adult

The adult stage is considered the reproductive stage of the life cycle. The fly emerges from the puparium by forcing its way through the anterior end. At first the adult fly is greatly elongated, with its wings unexpanded. Within an hour, however, the wings expand and the body gradually attains the more rounded form typical of an adult. The adult is also relatively light in color when it first emerges, but within the first few hours it darkens to its characteristic adult colors.

Adult *Drosophila* mate about 6 hours after emerging from the puparium. The sperm are then stored in the spermathecae and ventral receptacles of the female and are released gradually into the oviduct as eggs are produced and passed through the oviduct into the vagina. As stated earlier, the female begins to deposit eggs about 2 days after it has emerged. It may deposit as many as 50 to 75 eggs per day for the first few days. Thereafter, its egg production decreases with time. The average lifespan of adult flies is 37 days at 25°C.

B. CULTURE MEDIUM

A number of media have been developed for the culture of *Drosophila.* However, the easiest to use are the "instant media" available from many biological supply houses. A concentrated medium requires only the addition of water to be immediately usable. No cooking is necessary.

Pour the medium into chemically clean bottles. These bottles may be of any size, but 4-ounce (120 ml), wide-mouthed bottles or half-pint milk bottles are satisfactory. (It is best to transfer the mixture to a beaker for pouring. None of the medium should come into contact with the neck of the bottle.) Fill the bottles with medium to a depth of about 1 inch (2.5 cm). Then place a piece of *nonabsorbent paper such as brown wrapping paper* in each bottle so that it extends down into the medium (a double piece of paper is more satisfactory). The paper should be about an inch wide and should extend upward to a point about one-half inch (12 mm) below the neck of the bottle. This paper provides a dry place on which the larvae can pupate. Stopper the bottles with cotton plugs (such as those used in bacteriology) or with disposable foam plugs, which are more convenient. Sterilize the bottles in an autoclave at 20 pounds pressure for 20 minutes. At the same time, it is wise to sterilize other materials that you may wish to use in handling flies. Before placing the flies in the cooled bottles of medium, shake a small amount of dry yeast on the medium. The yeast will grow and serve as food for the developing fly larvae.

C. ETHERIZATION OF FLIES

It is necessary to etherize flies in order to keep them inactive while they are being examined and when they are being transferred into culture bottles for matings. The etherizing bottle can consist of any small container that has an opening the same size as that of the culture bottle and a tight-fitting cork. Into the bottom of the cork, tack a pad of cotton. Just before employing the technique outlined below (Fig. K-2), douse the cotton with a few dropperfuls of ether.

1. Shake the flies down into the bottom of the culture bottle by tapping the bottle on a rubber pad on a desk.

2. Remove the cotton plug from the culture bottle and, in its place, *quickly* insert the empty etherizing bottle. Make certain you have blown any residual fumes from the etherizing bottle before you invert it over the culture bottle.

3. Reverse the position of the two bottles so that the etherizer is now on the bottom.

4. Hold the two bottles firmly together and shake the flies from the culture bottle into the etherizer. This is best done by tapping the sides of the culture

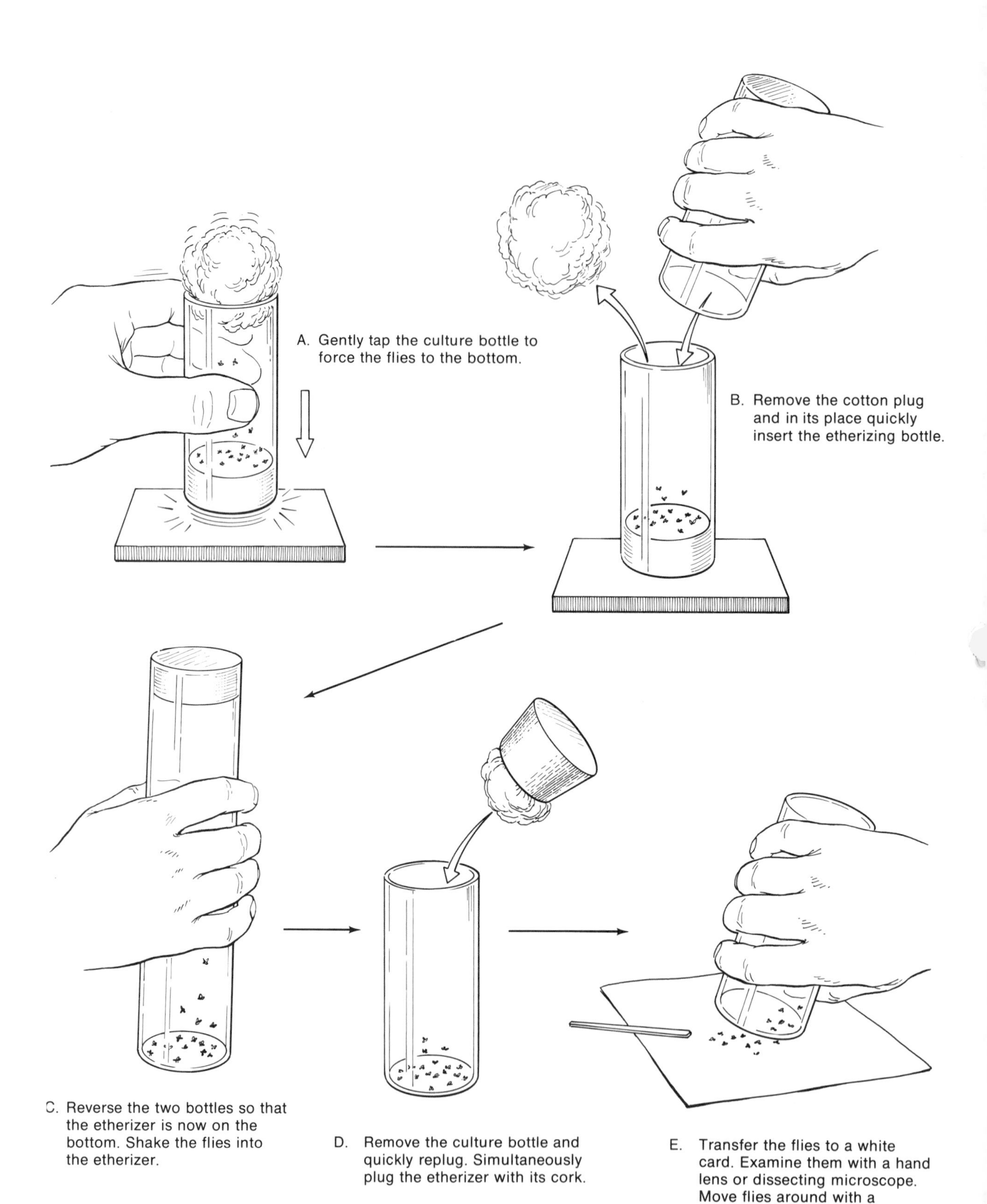

FIG. K-2
Procedure for etherizing flies.

bottle horizontally so that the flies are dislodged and fall into the etherizer. Make certain you do not shake the food loose in the culture bottle.

5. After you have shaken the flies into the etherizer, remove the culture bottle and quickly replug it. Simultaneously plug the etherizer with its cork. The cotton on the cork should feel *moist* with ether but not wet.

6. Observe the flies in the etherizer. Thirty seconds after they have stopped walking around, pour them out onto your counting plate. If the flies have been overetherized (killed), their wings will be extended at right angles to their bodies.

7. Normally, the flies will remain etherized from 5 to 10 minutes. If they begin to awaken on the plate, they can be reetherized as follows: on the inner surface of a petri dish, tape a piece of paper toweling. Moisten the paper toweling with ether, and place the petri dish over the flies to form an etherizing chamber.

D. DISTINGUISHING THE SEX

Examination of the external genitalia under magnification is the best means of distinguishing the sex of flies. Only male flies exhibit darkly colored external genitalia, which are visible on the ventral side of the tip of the abdomen. The following characteristics may also be helpful in distinguishing males from females:

1. *Size:* Females are usually larger than males.

2. *Shape:* The abdomen of the male is round and blunt, whereas that of the female is sharp and protruding. The abdomen of the male is relatively narrow and cylindrical, whereas that of the female is distended and appears spherical or ovate.

3. *Color:* Black pigment is more extensive on the abdomen of the male than on that of the female. On the male, the markings extend completely around the abdomen and meet on the ventral side. On the female, the pigment is present only in the dorsal region.

4. *Sex Combs:* Only males have a small tuft of black bristles called a *sex comb* on the anterior margin at the basal tarsal joint of each front leg. Magnification is necessary to see the sex combs.

E. ISOLATING VIRGINS

Females of *D. melanogaster* can store and use sperm from one insemination for a large part of their reproductive lives. As a result, only virgin females should be used in making crosses. Females of this species can mate 6 hours after they have emerged from the puparium. Therefore, a procedure must be followed that will ensure the collection of females that are no more than 6 hours old. A number of possible procedures are available to you, depending on your preference and the accessibility of the laboratory.

If the laboratory is generally accessible,

1. Shake out and discard into the morgue (a bottle containing ethanol, oil, or detergent) all adult flies (10 or more days after the introduction of the parental flies). This step should be done early in the morning (8–10 A.M.) for most efficient results.

2. Return to the laboratory within 4 to 6 hours, and examine the newly hatched flies. The females in this group may be presumed to be virgin and can be used in experimental matings.

If the laboratory is not generally accessible,

1. Collect darkened pupae from the paper or sides of the bottle with a fine camel-hair brush.

2. Turn each pupa so that the legs are visible through the pupal case. Examine under high power the uppermost or proximal joint of the tarsi of the front legs. The presence of sex combs indicates a male.

3. Collect the pupae that are *without* sex combs, hence female, and place them individually in fresh culture bottles. After the flies have hatched from their pupal cases, examine them to be certain that no male pupae have accidentally been placed in a culture. All the flies that are female can be used in experimental matings.

REFERENCE

Levine, L. and N. M. Schwartz. 1973. *Laboratory Exercises in Genetics.* Mosby.